T0310091

ENERGY PRODUCTION SYSTEMS ENGINEERING

ENERGY PRODUCTION SYSTEMS ENGINEERING

THOMAS H. BLAIR

Tampa Electric Company
and
University of South Florida

IEEE PRESS SERIES ON POWER ENGINEERING

IEEE PRESS

WILEY

Limit of Liability/Disclaimer of Warranty: While the publisher and author have used their best efforts in preparing this book, they make no representations or warranties with respect to the accuracy or completeness of the contents of this book and specifically disclaim any implied warranties of merchantability or fitness for a particular purpose. No warranty may be created or extended by sales representatives or written sales materials. The advice and strategies contained herein may not be suitable for your situation. You should consult with a professional where appropriate. Neither the publisher nor author shall be liable for any loss of profit or any other commercial damages, including but not limited to special, incidental, consequential, or other damages.

For general information on our other products and services or for technical support, please contact our Customer Care Department within the United States at (800) 762-2974, outside the United States at (317) 572-3993 or fax (317) 572-4002.

Wiley also publishes its books in a variety of electronic formats. Some content that appears in print may not be available in electronic formats. For more information about Wiley products, visit our web site at www.wiley.com.

Library of Congress Cataloging-in-Publication Data is available.

ISBN: 978-1-119-23800-3

Printed in the United States of America

10 9 8 7 6 5 4 3 2 1

Dedicated to the memory of
Professor Joseph Peter Skala
Adjunct Professor at USF and St. Petersburg Junior College
for your support and encouragement to your many students.

CONTENTS

LIST OF FIGURES

LIST OF TABLES

LIST OF ANNEX

ACKNOWLEDGMENTS

THE AUTHOR wishes to thank the many people who contributed their time, expertise, and encouragement to the development of the course material for the Energy Production Systems Engineering course at the University of South Florida (USF) Master's Degree Power Engineering Program and, subsequently, this textbook. I especially would like to thank Professor Joe Skala and Dr. Ralph Fehr.

Back in 1980, Professor Joe Skala, while still working as a full-time Professional Engineer at Florida Power Corporation, planted the seed that has since grown into the Power Engineering Program at USF. When Professor Skala started at USF as an Adjunct Professor, there were only two power courses offered at the university as electives. By the time Professor Skala retired in 2000, he had started the Power Engineering Program at USF and developed it into an independent master's degree offering. Under Professor Skala's guidance, the power program grew by about eight courses. Even after retirement in 2000, Professor Skala continued to support education by becoming an Adjunct Professor in the Mathematics Department at St. Petersburg Junior College (SPJC). During his time at SPJC, he, along with Professor Warren DiNapoli, donated his entire salary from teaching mathematics at the Clearwater campus of SPJC to the "DiNapoli and Skala Families Scholarship." This scholarship is awarded to Clearwater campus students who have a demonstrated financial need, a GPA of 3.0 or higher and have completed a minimum of 24 semester hours. He was also an avid supporter to the "Women on the Way" (WOW) program at SPJC, which is a resource and support center developed to help women succeed in college.

Professor Skala touched many people over his lifetime including mine. While he is gone from us now, his influence is within all of his students and will remain with us for a long time to come.

Dr. Ralph Fehr's request to develop a course covering equipment and systems utilized in the electrical power generation industry is the reason there is an Energy Production Systems Engineering course at USF and this textbook. Dr. Fehr was instrumental in the further development of the USF Power Engineering Program after Professor Skala retired. Dr. Fehr joined the power program at USF in 1997. During his tenure at USF, Dr. Fehr has added eight more courses to the power program and, in 2005, Dr. Fehr successfully developed the power program into a PhD offering at USF. As part of this expansion of the power program, Dr. Fehr recommended adding the Energy Production Systems Engineering course to the power engineering curriculum to cover topics associated with the generation side of the utility industry. The intention was that this would be one piece to round out the program to cover all aspects of power engineering: generation, transmission, distribution, and utilization. Dr. Fehr invited me to develop the Energy Production Systems Engineering course material at USF and I was glad to take on the task. Dr. Fehr has been a major contributor of materials for both the university course and the textbook and has spent many hours providing valuable feedback to me. The success of the Energy Production Systems Engineering course at USF is in large part due to the efforts of Dr. Fehr.

Additionally, I would like to thank Joe Simpson with Duke Energy for providing valuable information for Chapter 6. I also would like to thank Bob Buerkel with Parker Pneumatic Division, North America, for his review and suggestions for the valve actuator section. Additionally, Ralph Painter with Tampa Electric has been a great mentor and provided me with technical information in the design, installation, operation, and maintenance of an energy production facility that I have incorporated into the course material. I appreciate the many hours of assistance that Paul Yauilla with Tampa Electric put into editing the images and figures in this textbook. Thank you also to Jane Hutt with National Electric Coil for her efforts on the graphics for the generator section. Divya Narayanan along with all the staff at Wiley-IEEE Publishers spent many hours working with me to develop the final version of this textbook and I greatly appreciate all of their efforts.

I also wish to thank Bill Fowler, Tracy McLellan, John Sheppard, Jack White, Fred Wyly, David Kiepke, Tim Pedro, Tim Parsons, Charles (Terry) Kimbrell, James Cooksey, Michael Burch, Jim Mitchell, Jim Johnson, Dave Ford, Peter Teer, Tim Hart, and all the other many engineers, operators, technicians, electricians, and mechanics that I have worked with over the years and who have freely shared their valuable wisdom and experience. Their many hours of guidance and support have provided me with the background which has allowed me to develop this college course and textbook.

Thomas H. Blair

INTRODUCTION

THOMAS EDISON opened the first commercial electric power generation station in the United States on September 4, 1882, in New York City. This station generated "direct current" electrical energy for distribution locally in Manhattan. Soon after, on November 16, 1896, Nicholas Tesla and George Westinghouse opened a generation station in Niagara, NY, that generated "alternating current" energy. Initially, generation was located near the load center and the various load centers operated independently. Over time, it was determined that, to improve the reliability of the electric supply system and reduce costs, the many load centers and generation stations should be interconnected to a common "transmission" system thus leading to the interconnected systems of generation, transmission, distribution, and utilization that we have today.

Over the past century, power generation has undergone dramatic changes and innovation continues to drive changes and improvements in the electric generation industry. Today, sources of energy to generate electrical energy include coal, oil, natural gas, geothermal, wind, solar, biomass, hydro, tidal, and nuclear power.

Society has become very dependent on the availability of energy and electrical energy has become the primary means of distributing this energy.

The function of the generation station electrical engineer is to ensure a safe and reliable generation facility. The order of these two functions is not arbitrary. Safety is of primary concern for the generation utility engineer. Therefore, safety is the first chapter in this book. If a facility is not a safe facility for employees or the public, then it will not be a reliable facility. Unsafe conditions may not only result in personal injury but often involve equipment failure. An unsafe facility will likely have less reliable equipment and be a less reliable plant. While the primary goal of safety is to ensure the personal health and wellbeing of both the employees and the public, it also must be the primary focus for the utility engineer to ensure both safety and reliability.

This textbook is designed to provide a general introduction to the various facilities, systems, and equipment used in the power generation industry. It provides both theoretical and practical information for various utility systems. This text should provide a solid foundation on which a power generation facility engineer can continue to build.

It is my sincerest hope that this text will be useful in assisting utility electrical engineers to ensure safe and reliable operation of their facilities.

Thomas H. Blair

CHAPTER 1

ELECTRICAL SAFETY

GOALS

- To understand the basic requirements of OSHA 1910.269 and Subpart S
- To apply recommendations of NFPA 70 (National Electrical Code®) and NFPA 70E (Electrical Safety in the Workplace) for compliance with OSHA 1910 Subpart S
- To apply recommendations of IEEE C2 (National Electrical Safety Code) for compliance with OSHA 1910.269
- To be able to determine minimum approach distance (MAD), limited approach boundary, restricted approach boundary, and arc flash boundary for installation
- To be able to determine minimum safety clearance for electric supply station fences
- To be able to determine the minimum illumination requirements for electric supply station locations
- To be able to determine the proper electrical PPE (personal protective equipment) required for various tasks
- To be able to determine the correct classification for areas where hazardous materials may be present

IF ONE were to try to reduce the function of the electrical engineer in the electric power generation industry to one sentence, it would be "*to ensure the design, implementation, and operation of a SAFE and RELIABLE electrical system.*" Electrical safety is of primary importance in the electric utility generation industry. The generation industry is unique from other industrial environments. The available short-circuit fault currents can be very large since the generation source is close and can supply a large amount of fault current. The service voltages at various pieces of equipment can be greater in magnitude for the larger electrical machines utilized in the generation station. Combustible materials may be handled, stored, and utilized in power generation facilities. The above conditions require the power plant electrical engineer

Energy Production Systems Engineering: An Introduction for Electrical Engineers to Electrical Power Generation Facilities, Systems, and Equipment, First Edition. Thomas H. Blair.
© 2017 by The Institute of Electrical and Electronics Engineers, Inc. Published 2017 by John Wiley & Sons, Inc.

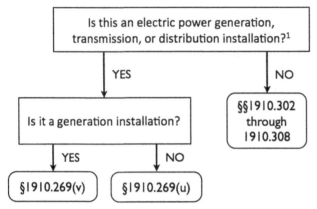

Figure 1.1 Appendix A-1 – Application of 1910.269 and 1910 Subpart S to Electrical Installations. *Source*: Reproduced with permission of U.S. Department of Labor.

to be very familiar with governmental regulations and industry standards regarding safety requirements to ensure the safe operation of the generation facility.

OSHA (Occupational Health and Safety Administration) (osha.gov) issues regulations that cover occupational health and safety. These regulations have the same effect as law. For general industry, which includes utilities, the applicable regulation is OSHA CFR 1910 – General Industry Standards. For general industry, electrical safety is covered under Subpart S. However, under OSHA CFR 1910, there is a separate section for special industries under Subpart R and the electric utility industry is covered under OSHA 1910.269 of Subpart R. This section covers the operation and maintenance of electric power generation systems and equipment and applies to installations utilized for the generation of electrical energy that are accessible only to qualified employees. One might think that all of the requirements for a generation facility fall under OSHA CFR 1910.269 and not OSHA CFR 1910 Subpart S since OSHA CFR 1910.269 regulations were written for electric generation, transmission, and distribution systems, but that is not always the case. So how does a plant engineer know when to apply OSHA 1910 Subpart S (general industry) or OSHA 1910 Subpart R 269 or possibly both regulations? OSHA provides guidance with that question in Appendix A of 1910.269.

To understand how Appendix A addresses this, we need to understand that OSHA segregates its safety requirements into two general categories: electrical safe installation methods and electrical safe work practices.

1910.269 Appendix A-1 as shown in Figure 1.1 answers the question of which regulation (1910.269 or 1910 Subpart S) applies to electrical installation requirements and 1910.269 Appendix A-2 as shown in Figure 1.2 answers the question of which regulation (1910.269 or 1910 Subpart S) applies to electrical safe work practices.

For regulations regarding the safety of the electrical installation, if the facility is directly associated with a generation, transmission, or distribution system, then OSHA CFR 1910.269 is the regulation that describes safe installation requirements. If the facility is not directly associated with generation, transmission, and distribution,

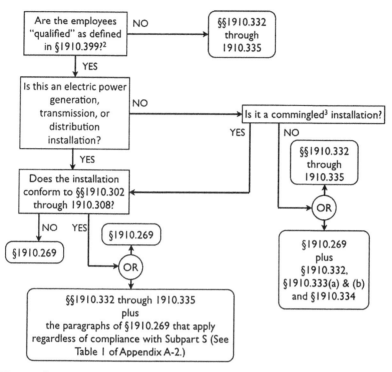

Figure 1.2   Appendix A-2 – Application of 1910.269 and 1910 Subpart S to Electrical Safety-Related Work Practices. *Source*: Reproduced with permission of U.S. Department of Labor.

then 1910 Subpart S applies. For example, if we were to install a motor control center (MCC) in the turbine building of an electric utility generation station, then the safe installation regulations would be defined by 1910.269. However, if we were to install an MCC in a warehouse that is not located on the generation station facility, then the safe installation regulations would be defined by 1910 Subpart S.

The choice of which OSHA standard to use for electrical safe work practices depends both on the location of the work and the qualification of the personnel performing the electrical work. First, we must answer the question, what defines a "qualified person." OSHA defines a *qualified person* as one who has "received training in and has demonstrated skills and knowledge in the construction and operation of electric equipment and installations and the hazards involved" (OSHA 1910.399).

For electrical safe work practice regulations, if the personnel performing the task are not qualified, then the safe work practices of Subpart S are the regulations that govern the work regardless of whether the installation where the work is being performed is on a generation facility or not. For example, if we were to install an air compressor in a generation facility and the equipment supplier is sending a field engineer to assist with startup of the compressor and the field engineer is not trained on the electrical safe work practices of OSHA 1910.269, then the safe work practices for this task must comply with OSHA Subpart S, even though this task is being performed at a generation facility.

However, if the personnel performing the task are qualified AND if the installation is a generation facility, then the safe work practices of 1910.269 apply to the task. So for normal maintenance done at the generation station where the task is performed by station personnel that are trained according to OSHA 1910.269, the tasks are covered under OSHA 1910.269. For commingled installations, the reader is directed to OSHA 1910.269, Appendix A for guidance.

As the reader can see, even though OSHA 1910.269 is written for generation, transmission, and distribution systems, there are instances where OSHA 1910 subpart S applies, so the electric utility engineer must know and understand the installation requirements and safe work practices of both OSHA 1910 Subpart S (general industry) and OSHA 1910.269 (electric utilities).

So where does the electric utility engineer obtain guidance on how to comply with all of these regulations? For the safe installation requirements for general industry described by OSHA 1910 Subpart S (sections 302 to 308), NFPA 70® (National Electric Code® or NEC®) provides guidance on how to comply with these federal regulations. (NFPA 70®, National Electrical Code®, and NEC® are registered trademarks of the National Fire Protection Association, Quincy, MA). For the safe installation requirements for the electric utility industry of OSHA 1910.269, IEEE C2 (National Electrical Safety Code or NESC) provides guidance on how to comply with federal regulations at generation station installations. Specifically, IEEE C2 – Part 1 describes electric supply station installation requirements, IEEE C2 – Part 2 describes overhead line installation requirements, IEEE C2 – Part 3 describes underground line installation requirements.

For the safe work practice requirements for general industry of OSHA 1910 Subpart S (sections 332 to 335), NFPA 70E (Electrical Safety in the Workplace) provides guidance on how to comply with federal regulations. For the safe work practice requirements for the electric utility industry of OSHA 1910.269, IEEE C2 – Part 4 – (NESC) provides guidance on how to comply with federal regulations.

While these codes are not law or regulation, they are referenced by OSHA federal regulation as methods to ensure compliance with the law. Additionally, OSHA regulation updates take many years to occur due to the process of public notification and comment before changing official regulations. However, NEC® is updated every 3 years, NFPA 70E is updated every 4 years, and NESC is updated every 5 years. Therefore, these codes have the most up-to-date information regarding electrical safety. The electric generation utility engineer should have a current copy available of all three of these standards and be familiar with the information contained within. In Chapter 1, we describe just some of the installation and safe work practice information contained in these codes. The reader is reminded to always check the latest version of the codes for updates.

One last note of caution, the OSHA standards and associated codes are NOT design guides, but MINIMUM requirements. When applying these codes, remember that these are the MINIMUM standards to ensure safety; but good engineering practice may suggest additional safety measures in some instances.

INSTALLATION SAFETY REQUIREMENTS—GENERAL INDUSTRY (NEC®)

Access to Working Space <600 V

To ensure the safety of personnel when accessing exposed electrical parts of equipment, minimum equipment clearances are given in codes and standards depending on the nominal voltage of the circuit in question and the equipment. If the structures surrounding the area are neither energized nor grounded (if structures are insulated), then the second column in Table 1.1 applies. If the structure surrounding the area is grounded, then the third column in Table 1.1 applies (note that conductive or partially conductive materials such as concrete are considered grounded). If the structure surrounding the area is energized, then the fourth column in Table 1.1 applies. Additionally, the working space must be of sufficient width, depth, and height to permit all equipment doors to open 90 degrees. The width of the working space must be at least the width of the equipment and cannot be less than 30 in.

Remember that these are minimum requirements. There may be instances where the working space needs to be even larger than the values required by codes. One example where this commonly occurs is in front of switchgear where the breakers can be racked out and removed (this is described in more detail in Chapter 19: Switchgear). The space required for the racking device (including clearance to other energized equipment in the area) commonly is larger than the minimum requirements of code. The design engineer should evaluate both the equipment to be installed and the typical types of maintenance that will be performed and ensure that the working space provided is adequate for personnel safety.

In regard to access to rooms containing electrical equipment, if the service provided is less than 1200 A, then there must be at least one entrance consisting of a minimum of 24-in.-wide by 6-ft-high door provided with panic hardware that allows for egress from the working space by pushing on door from the inside of the building. Where the electrical equipment is wider than 6 ft, there must be two doors.

If the service provided is greater than 1200 A, then a minimum of two, 24-in.-wide by 6-ft-high doors must be provided with panic hardware that allows for egress from either side of working space by pushing on door. This again is an area where

TABLE 1.1   Minimum Aisle Working Space for Nominal Voltage Less than 600 V

Nominal Voltage to Ground (V)	Live to Not Ground or Not Live	Live to Ground	Live to Live
0–150	900 mm	900 mm	900 mm
	3 ft	3 ft	3 ft
151–600	900 mm	1.1 m	1.2 m
	3 ft	3.5 ft	4 ft

Source: Reproduced with permission from NFPA 70®. *National Electrical Code*®, Copyright © 2014, National Fire Protection Association. This is not the complete and official position of the NFPA on the referenced subject, which is represented only by the standard in its entirety. The student may download a free copy of the NFPA 70® standard at: http://www.nfpa.org/codes-and-standards/document-information-pages?mode=code&code=70.

these are minimum requirements. If, during the process of performing maintenance such as racking out equipment, one means of egress may become blocked, the design engineer would be justified in adding additional means of egress for personnel.

Access to Working Space >600 V

In rooms and enclosures where the voltage exceeds 600 V, NEC® requires rooms to be locked and signage to be posted that reads,

DANGER – HIGH VOLTAGE – KEEP OUT

For equipment that nominally operates in excess of 600 V, Table 1.2 describes the minimum working space requirements. The same conditions for surrounding equipment that applied to Table 1.1 apply to Table 1.2. If the structures surrounding the area is neither energized nor grounded (if structures are insulated) then the second column in Table 1.2 applies. If the structure surrounding the area is grounded, then the third column in Table 1.2 applies. If the structure surrounding the area is energized, then the fourth column in Table 1.2 applies.

Example 1.1   You are designing an MCC that is powered from a 480 Vac phase-to-phase, 277 Vac phase-to-ground source. You will place the MCC in an electrical equipment room where the wall across the aisle from the 480 V circuits is grounded. What is the minimum aisle working space for this application?

Solution: Referring to Table 1.1, we find the minimum aisle working space to be 3.5 ft.

For exposed energized conductors located outdoors, NEC® requires that the conductors be enclosed in a fence that is 7 ft minimum in height. Table 1.3 describes the minimum distance from the fence to the live part being protected.

NEC® defines the minimum allowable branch circuit conductor size for a given branch circuit breaker. Following matrix is a summary of branch circuit minimum conductor sizes based on branch circuit protection.

Breaker	15 A	20 A	30 A	40 A	50 A
Conductor	14 AWG	12 AWG	10 AWG	8 AWG	6 AWG

TABLE 1.2   Minimum Aisle Working Space for Nominal Voltage >600 V Around Electrical Equipment

Voltage ph-gnd	Live vs. Insulated	Live vs. Gnd	Live vs. Live
601 V to 2,500 V	900 mm (3 ft)	1.2 m (4 ft)	1.5 m (5 ft)
2,501 V to 9,000 V	1.2 m (4 ft)	1.5 m (5 ft)	1.8 m (6 ft)
9,001 V to 25 kV	1.5 m (5 ft)	1.8 m (6 ft)	2.8 m (9 ft)
25,001 V to 75 kV	1.8 m (6 ft)	2.5 m (8 ft)	3.0 m (10 ft)
Above 75 kV	2.5 m (8 ft)	3 m (10 ft)	3.7 m (12 ft)

Source: Reproduced with permission from NFPA 70®, *National Electrical Code*®, Copyright © 2014, National Fire Protection Association. This is not the complete and official position of the NFPA on the referenced subject, which is represented only by the standard in its entirety. The student may download a free copy of the NFPA 70® standard at: http://www.nfpa.org/codes-and-standards/document-information-pages?mode=code&code=70.

TABLE 1.3   Minimum Distance from Fence to Live Parts

	Minimum Distance from Fence to Live Parts	
Nominal Voltage	m	ft
601–13,799 V	3.05	10
13,800–230,000 V	4.57	15
Over 230,000 V	5.49	18

Source: Reproduced with permission from NFPA 70®, *National Electrical Code*®, Copyright © 2014, National Fire Protection Association. This is not the complete and official position of the NFPA on the referenced subject, which is represented only by the standard in its entirety. The student may download a free copy of the NFPA 70® standard at: http://www.nfpa.org/codes-and-standards/document-information-pages?mode=code&code=70.

Just to reinforce the notion that codes are not design standards, the size of the conductor recommended by NEC® is not the required conductor size, but the minimum allowable conductor size. For long runs where voltage drop might be an issue, the design engineer might choose to oversize the branch circuit conductor which is acceptable. Codes such as NEC® and NESC are not design guides, but only provide the minimum safety requirements for installation.

Grounding and bonding is so important that it has its own chapter and is described in Chapter 21. But for safety, it is worth mentioning here that NEC® requires the electrical distribution system to be grounded and bonded and requires the electrical distribution system to have one main bonding jumper between the main supply neutral bar and ground bar. The functions of the grounding conductor (also known as neutral conductor) and bonding conductor (sometimes referred to as the grounded conductor) are distinctly different. Downstream equipment enclosures are bonded back to the main supply *ground* bar via the *grounded* conductor and downstream equipment neutral terminations are connected back to the main supply *neutral* bar via the *grounding* conductor. The purpose of the *grounding* conductor is to provide a path for current to return to the source and, if there is a phase-to-neutral fault, this current will clear the upstream breaker to remove the fault from the energized electrical system. The purpose of the *grounded* conductor is not to provide a path for fault current, but to maintain electrical continuity between system ground and equipment enclosures in the field. This is done to minimize touch potentials during faults to protect personnel.

INSTALLATION SAFETY REQUIREMENTS—SPECIAL INDUSTRY – UTILITY (NESC)

Access to Exposed Energized Conductors Outdoor

In some areas of a generation facility, there may be exposed overhead conductors. A typical application is where the plant interfaces with the transmission and distribution system. There are minimum distances that need to be maintained from the general public to these exposed circuit conductors. In order to keep non-qualified personnel from accessing the energized circuits, NESC provides guidance on minimum fence

TABLE 1.4 Values of Minimum Distance from Fence to Outdoor Exposed Energized Conductor for Use with NESC Figure 110-1

Nominal Voltage Between Phases (V)	Typical BIL	Dimension "R"	
		m	ft
151–7,200	95	3	10
13,800	110	3.1	10.1
23,000	150	3.1	10.3
34,500	200	3.2	10.6
46,000	250	3.3	10.9
69,000	350	3.5	11.6
115,000	550	4	13
138,000	650	4.2	13.7
161,000	750	4.4	14.3
230,000	825	4.5	14.9
230,000	900	4.7	15.4

Source: Reproduced with permission of IEEE.

space requirements between energized conductors and the fence. This is very similar to the requirements of NEC® mentioned in Table 1.3.

Table 1.4, along with Figure 1.3, shows the minimum radius from a point on the fence 5 ft from grade to the exposed energized conductor to ensure non-qualified people are not exposed to the hazard presented by exposed energized conductors. Note, the fence height must be 7 ft minimum and the fence must be grounded to

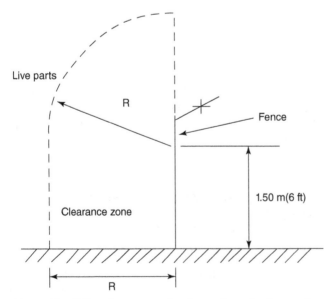

Figure 1.3 Safety clearance to electric supply station fences. *Source*: Reproduced with permission of IEEE.

TABLE 1.5   Minimum Overhead Clearance from Ground to Outdoor Exposed Energized Conductor (for V < 750 Vac)

- 3.6 m (12 ft) – sidewalks and areas accessible to pedestrians only
- 4.9 m (16 ft) – residential driveways, parking lots, and alleys
- 4.9 m (16 ft) – roads, streets, and areas exposed to truck traffic
- 7.3 m (24 ft) – track rails of rail roads

Source: Reproduced with permission of IEEE.

the ground grid to limit touch potential (touch potential is explained in Chapter 21: Ground System).

NESC also provides guidance for exposed overhead lines and minimum clearance from ground depending on the type of traffic that is expected below the exposed lines. This information is found in NESC Part 2, section 232, and summarized in Table 232-1. For example, for exposed conductors energized at less than 750 Vac, the following minimum clearances apply. For areas where only pedestrian traffic is expected, 3.6 m (12 ft) is the minimum clearance from ground to the overhead lines. For areas that transverse residential driveways, parking lots, and alleys, 4.9 m (16 ft) is the minimum clearance from ground to overhead lines. For areas where exposed conductors transverse roads, streets, and areas exposed to truck traffic, 4.9 m (16 ft) is the minimum clearance from ground to overhead lines. For areas where exposed conductors transverse track rails of rail roads, 7.3 m (24 ft) is the minimum clearance from ground to overhead lines. All of these distances are to be maintained during periods of maximum line sag. This data is summarized in Table 1.5.

Example 1.2   You are designing an exposed overhead distribution line that is crossing a sidewalk that will be accessible to pedestrian traffic only. What is the minimum from ground to the overhead line for this application?

Solution: Minimum clearance from ground to overhead line for this application per Table 1.5 is 12 ft.

NESC requires a minimum level of illumination depending on the location or the work place in the generation facility and NESC Table 111-1 provides this guidance. This is to ensure that the area is adequately illuminated for the typical tasks performed in these areas to be performed safely. Table 1.6 is a partial listing of locations in a typical generation facility and the minimum required illumination levels.

In addition to normal lighting requirements listed in Table 1.6, NESC requires emergency lighting to be provided that is energized from an independent source (typically a battery) at locations of egress to provide for safe exit during emergencies such as fire where normal lighting is de-energized. The minimum illumination level at these exits is 11 lux (1 foot-candle) and the independent source must keep the emergency light energized for at least 90 minutes.

Just like NEC®, the NESC requires a minimum amount of working space around electrical equipment. A minimum of 7 ft of head room is required with a

TABLE 1.6 Illumination Levels Required in an Electric Utility Power Generation Station

Location	Lux	Foot-Candles
Generating Station (Interior)		
Highly critical areas occupied most of the time	270	25
Areas occupied most of the time	160	15
Critical areas occupied infrequently	110	10
Areas occupied infrequently	55	5
Generating Station (Exterior)		
Building pedestrian main entrance	110	10
Critical areas occupied infrequently	55	5
Areas occupied occasionally by pedestrians	22	2
Areas occupied occasionally by vehicles	11	1
Areas occupied infrequently	5.5	0.5
Remote areas	2.2	0.2
Substation		
Control building interior	55	5
General exterior horizontal end equipment vertical	22	2
Remote areas	2.2	0.2

Source: Reproduced with permission of IEEE.

minimum working depth in front of equipment of 3 ft. For voltages <600 Vac, the values in NESC are the same as the values in NEC® as shown in Table 125-1 of NESC as described in Table 1.7. For voltages >600 V, the values in NESC listed in Table 124.1 of NESC as described in Table 1.8.

NESC has requirements for the safety of rotating equipment, safety of battery systems, safety of transformers, conductors, circuit breakers, fuses, and switchgear. These safety requirements are discussed in the respective sections of this book to reinforce the importance of safety in the design and operation of this equipment.

Now that we have discussed installation safety requirements, we will discuss electrical safe work practice requirements.

TABLE 1.7 Working Space <600 V

	Clear Distance					
	Condition 1		Condition 2		Condition 3	
Voltage to Ground (V)	mm	ft	mm	ft	mm	ft
0–150	900	3	900	3	900	3
151–600	900	3	1070	3 − 1/2	1200	4

Source: Reproduced with permission of IEEE.
Condition 1: Exposed energized parts on one side and no energized or grounded parts on the other side of the working space.
Condition 2: Exposed energized parts on one side and grounded parts on the other side of the working space.
Condition 3: Exposed energized parts on both sides of the working space.

TABLE 1.8   Working Space >600 V

NESC Table
124-1 – Working
space >600 V

Column 1	Column C	Column 2		Column 3		Column 4	
Max Design Voltage Between Phases	Basic Impulse Insulation Level (BIL)	Vertical Clearance of Unguarded Parts		Horizontal Clearance of Unguarded Parts		Clearance Guard to Live Parts	
kV	kV	ft	in	ft	in	ft	in
0.3	–	Not specified		Not specified		Not specified	
0.6	–	8	8	3	4	0	2
2.4	–	8	9	3	4	0	3
7.2	95	8	10	3	4	0	4
15	95	8	10	3	4	0	4
15	110	9	0	3	6	0	6
25	125	9	1	3	7	0	7
25	150	9	3	3	9	0	9
35	200	9	6	4	0	1	0
48	250	9	10	4	4	1	4
72.5	250	9	10	4	4	1	4
72.5	350	10	5	4	11	1	11
121	350	10	5	4	11	1	11
121	550	11	7	6	1	3	1
145	350	10	5	4	11	1	11
145	550	11	7	6	1	3	1
145	650	12	2	6	8	3	8
169	550	11	7	6	1	3	1
169	650	12	2	6	8	3	8
169	750	12	10	7	4	4	4
242	550	11	7	6	1	3	1
242	650	12	2	6	8	3	8
242	750	12	10	7	4	4	4
242	900	13	9	8	3	5	3
242	1050	14	10	9	4	6	4

Source: Reproduced with permission of IEEE.

SAFE WORK PRACTICE REQUIREMENTS

So what are the potential hazards in a power plant environment in regard to safe work practices? Some of the possibilities are electrical shock, arc flash, arc blast, fall, projectiles, and fire ignition.

Electrical shock occurs when a part of the body comes in contact with an energized circuit. To avoid this hazard, the employee must know the various electrical energy sources, be able to identify the voltage level associated with the hazard, determine what actions are necessary to reduce and/or eliminate the hazard, and be allowed to take those actions to avoid the hazard. The preferred method for prevention of electrical shocks is to ensure that the circuit is de-energized by placing it in an electrically safe work condition prior to performing work on that circuit. If it is determined that the circuit cannot be de-energized, then NFPA 70E and NESC Part 4 provide guidance of procedures and equipment to be utilized when *working on* a system energized. This is known as energized electrical work.

In one recent survey of electrical workers, 97% of workers had experienced a shock at work sometime in the past and 58% were exposed to energized circuits every day. One might think that a majority of the incidents involve younger, less experienced workers. However, one survey shows the following distribution of electrocutions by age.

- <20 = 10%
- 20–24 = 18%
- 25–34 = 34%
- 35–44 = 22%
- 45–54 = 10%
- >55 = 6%

As can be seen from the above data, the most likely age range is the 25–34 year range. Therefore, the concept that experience makes one less likely to be involved with an electrical contact incident is incorrect. In fact, there are more than 30,000 electrical contact events that happen every year in the United States. A majority of these occur on system voltages less than 600 V. One may think that higher voltages may present higher hazards, but from a frequency of incident viewpoint, this is not the case. The hazard exists on any energized circuit.

Identifying circuit parts that are "exposed live parts" may not be as straight forward as one would think. An energized uncovered bus bar in a piece of electrical equipment is an "exposed live part" for certain. However, how about the covered bus shown in the photo in Figure 1.4?

The cable jacket that provides the covering on the cable has a dielectric rating for the voltage involved and is therefore considered insulated. However, the covering on the bus does not have a dielectric rating and is not considered insulated and therefore should be treated as an energized circuit until it is placed in an electrically safe working condition.

In order to avoid electrical shock hazards, a shock hazard analysis should always be performed whenever work is to be performed on or near energized electrical equipment. A shock hazard analysis is a procedure where the voltage level of the circuit is determined; all potential sources supplying the circuit are determined and then all sources are isolated by putting the circuit in an electrically safe working condition. The shock hazard analysis should also determine the necessary PPE required to safely test the circuit for the absence of voltage during the process of placing the circuit in an electrically safe work condition.

Table 1.9 describes the effect that current passing through a body has on the body. The higher value is for larger framed bodies and the lower value is for smaller framed bodies.

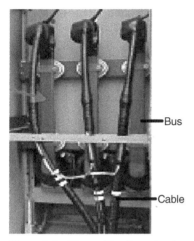

Figure 1.4   Photo of load termination compartment in typical medium voltage switchgear.

To place a circuit in an electrically safe work condition, the employee must perform the following minimum steps.

- Determine all possible sources (include stored or induced energy)
- Interrupt the load, if the isolation device is not rated to interrupt load
- Isolate the conductor or circuit part from the energized electrical system
- Visually verify the isolation
- Lock and/or tag the isolation device(s)
- Test the circuit to ensure the absence of voltage
- If the potential for induced voltages exist, ground the circuit

Control circuits are never to be used as a means of lockout. Only equipment designed for the purpose of lockout should be used as isolation boundaries. For

TABLE 1.9   Effects on the Human Body from Contact with Energized Circuits

	AC Current (A)	
Effect of Current	High	Low
Perception threshold (tingling)	0.001	0.0007
Slight shock – not painful (no loss of ctrl)	0.0018	0.0012
Shock – painful (no loss of ctrl)	0.009	0.006
Shock – severe (muscle ctrl loss, breathing difficulty)	0.023	0.015
Possible ventricular fibrillation (3 sec shock, "let go" threshold)	0.1	0.1
Possible ventricular fibrillation (1 sec shock)	0.2	0.2
Heart muscle activity ceases	0.5	0.5
Tissue and organs burn	1.5	1.5

Source: Reproduced with permission of National Institute for Occupational Safety and Health (NIOSH).

example, a pull cord on a conveyor belt is interlocked with a contactor to de-energize the contactor when the pull cord is activated. While the pull cord is an important safety interlock, it should never be used as a means for isolation for lockout. The reason for this is that the device that isolates power (the contactor) is not locked and could be mechanically pulled in to energize the circuit, thereby bypassing the locking device and the protection of the lock.

Until all of the these steps are completed (this includes testing the circuit and grounding the circuit if required), the circuit must be considered energized and working on the circuit must be considered as energized electrical work. You may realize at this point that the task of testing the circuit to ensure the absence of voltage alone is considered energized electrical work and requires the proper personnel protective equipment (PPE) and safe work practices, since the process of testing a circuit to ensure the absence of voltage may require contact with the circuit before all steps are completed. Electrical safe work practices should be utilized when verifying the absence of voltage. Additionally, when verifying absence of voltage, the meter being utilized should be tested on a known source to verify it is functioning first, then tested on the circuit that is to be put in the electrically safe working condition, and then tested again on a known source to verify if it is still functioning. This is also known as the "test the tester, test the circuit, test the tester" method. This is done to ensure the integrity of the meter used and the validity of the absence of voltage reading. Also ensure that the voltage meter being used is rated for the anticipated voltage level of the system. There have been cases where a worker obtained the incorrect meter and tested a circuit that the meter was not rated for with catastrophic results.

There are two basic methods of lockout/tagout defined by OSHA. The first method known as the simple lockout method which involves only one set of circuit conductors or circuit parts. Each worker is responsible for their own isolation and lock. Under the simple lockout method, a written procedure is not required for each lock. Alternately (and more typical in a power plant environment), the second lockout method is known as the complex lockout/tagout procedure and it involves the use of a lockout procedure. This system is required to be documented. Instead of the individual performing the work being responsible for the isolation, there are three people minimum involved in the isolation procedure. The first is the qualified person in charge of the lockout. The second is the operator that performs the isolation, installs the lock, and verifies that the circuit is de-energized. The third person is the one who is working on the circuit and they are known as the affected person. This system allows for a method to communicate the change of a lockout status to all affected persons.

When it is determined that work has to be performed on equipment that is not put in an electrically safe working condition, then an Electrical Safe Work Permit (ESWP) should be filled out for the task at hand. This is a very useful document, as it guides the affected employee to analyze the circuit and determine the shock and arc flash protection boundaries and the required PPE to work safely. It also forces the persons asking the affected employee to work a system energized to justify why the system cannot be placed in an electrically safe work condition. Specifically, the ESWP has the following items covered and should be filled out and any safety concerns resolved before any energized work is commenced.

- A description of the circuit
- Justification or need to work energized
- A description of the safe work practices
- Results of the shock hazard analysis
- Determination of shock protection boundaries
- Results of the flash hazard analysis
- The *flash protection boundary*
- The necessary personal protective equipment (PPE)
- Means employed to restrict access
- Evidence of job briefing
- Energized work approval

When working on energized circuits, select the proper equipment and PPE required for the voltage level of the system. Do not use a 600 V-rated volt meter to test a 5000 V circuit just because you "know" it is de-energized. If during the test you find that the equipment is not de-energized, severe injury or death can arise for the misuse of the wrong class of meter. Also, the PPE selected should be rated for the voltage level encountered. Details on the correct PPE to select for various types of hazards can be found from the Occupational Safety and Health Administration informational booklet OSHA 3151-12R 2003.

ELECTRICAL PPE

Electrical PPE can be broken down into two basic categories that are defined by the type of hazard that the PPE is intended to protect against. The first category is *shock protection* PPE. The purpose of shock protection PPE is to provide an electrically insulating barrier between the employee and the energized part. Any part of the employee's body must be protected from the hazard of electrical shock by the use of electrical PPE when reaching into the *restricted approach boundary* during electrical energized work. Shock protection PPE includes non-conducting hard hats, non-conducting safety glasses, voltage-rated insulating gloves with leather protectors, and insulating sleeves for higher voltages. In addition to these items, types of supplemental shock protection that may be used are dielectric safety shoes, insulated blankets, and insulating floor mats. Figure 1.5 shows the typical type of personal shock protection PPE utilized in the generation station.

The second category of electrical PPE is *arc flash protection* PPE. The purpose of arc flash protection PPE is to provide a thermally insulating barrier between the employee and the energized part to reduce the thermal energy exposure of the employee's skin in the event of an electrical arcing event. Any part of the employee's body must be protected from the hazard of an electrical arc flash by the use of arc flash PPE when reaching into the arc flash boundary during electrical energized work or when the employee is interacting with the electrical equipment in a fashion that

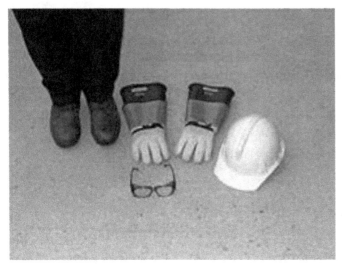

Figure 1.5   Photo of typical shock protection PPE. *Source*: Reproduced with permission of OSHA.

could result in an electrical arc flash. The material that forms this thermal barrier is treated to inhibit the material from igniting when exposed to elevated temperatures up to the energy values that the material is rated for. The energy value that the material has been tested to is known as the *arc rating* (AR) or *arc thermal performance value* (ATPV). ATPV is defined as the arc incident energy required to cause the onset of second-degree burn under the material that is being tested. Arc-rated PPE should not be confused with *flame-resistant* (FR) PPE. FR PPE is treated such that it does not support combustion once the ignition source is removed, but FR PPE is not tested for its response to electrical arcs like AR-rated PPE is. When selecting PPE for protection against the hazards of electrical arcs, ensure that the PPE has an AR rating. Figure 1.6 shows the type of personal arc flash protection PPE utilized in the generation station.

Arc flash protection PPE includes arc-rated (AR) shirt and pants or a full AR-rated coverall suit, AR-rated face shield and balaclava or AR-rated hood, and AR-rated gloves. For tasks that involve both electrical hazards and arc flash hazards, electrically rated gloves with leather protectors can be used instead of AR-rated gloves as the leather protectors provide thermal protection and the electrically rated glove provides the electrical protection.

Just as we stressed previously that the equipment utilized should be rated for the voltage exposure of the task, the shock protection PPE that is utilized must be rated for the voltage exposure of the task. For example, insulating gloves have various ratings or classes depending on their test voltage value and maximum allowable working voltage level. See Table 1.10 for information on proper glove selection based on the system working voltage.

Example 1.3   You have been tasked to test a circuit for potential where the circuit is rated for 4160 V. What is the minimum glove class acceptable for this task?

Figure 1.6   Photo of typical arc flash protection PPE. *Source*: Reproduced with permission of OSHA.

Solution: From Table 1.10, the minimum acceptable class of glove is Class 1 which is rated for use up to 7500 Vac. (Note that while Class 0 glove is tested at 5000 V, it is only allowed to be used on voltages up to 1000 V.)

Any object that enters the restricted approach boundary of an exposed ener-gized conductor must provide protection for the employee using the object. We have discussed electrical shock protection PPE, but if an employee uses a tool inside the restricted approach boundary, that tool must also provide insulation between the

TABLE 1.10   Marking of Rubber-Insulated Glove Classes

Definition of Lineman Insulated Glove Classes

Class Number	Test Voltage	AC Maximum Use Voltage
00	2,500 Vac	500 Vac
0	5,000 Vac	1,000 Vac
1	10,000 Vac	7,500 Vac
2	20,000 Vac	17,000 Vac
3	30,000 Vac	26,500 Vac
4	40,000 Vac	36,000 Vac

Source: Reproduced with permission of OSHA.

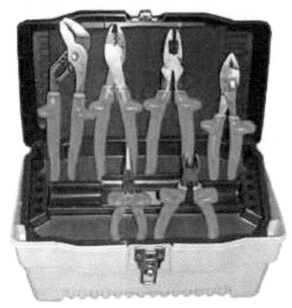

Figure 1.7   Typical insulated tools. *Source*: Reproduced with permission of OSHA.

exposed energized conductor and the employee. Insulated tools are rated to withstand at least 1000 V. Note that a metal screwdriver wrapped in electrician's tape does not constitute an insulated tool. Insulated tools have their dielectric strength tested just like shock protection PPE. When performing energized work, only use rated and tested insulated tools. Most types of tools utilized in electrical work are available in insulated versions as shown in Figure 1.7.

In addition to the hazard from electrical shock, there also exists a hazard from an arcing fault from the heat, pressure wave, sound wave, and projectiles that arise from such a blast. When an arcing fault occurs, the temperature of the arc can exceed 35,000°F, which is about four times hotter than the temperature of the sun, and create plasma. Plasma is an ionized gas consisting of positive ions and free electrons. Arcing faults can and do kill at distances exceeding 10 ft depending on the conditions at the time of the event.

To discuss the arc flash hazard, we have to understand some definitions. The hazard associated with heat is defined by the incident energy exposure level which is expressed in calories per square centimeter (cal/cm^2). To put some understanding of the numbers, 1 cal/cm^2 is equivalent to the exposure on the tip of a finger by a cigarette lighter for approximately one second. Also, when looking at damage to the skin that is exposed to heat energy from an arcing event, an exposure energy of only 1.2 cal/cm^2 may cause a second-degree burn on the human skin.

In addition to the hazard presented by the heat of an arcing fault, there are hazards associated with the blast as well. When copper is exposed to an arcing event and expands, it can expand to 67,000 times its solid volume. This rapid expansion creates a pressure wave and sound wave. This pressure wave can create a fall hazard.

Also the pressure wave can cause internal organ damage such as collapsed lungs. The sound wave can cause the sound levels to exceed 160 dB, rupturing ear drums. Associated with the pressure wave is the shrapnel that is in the gas cloud that can leave the source of the arcing fault at speeds exceeding 700 miles/hr which is fast enough to allow shrapnel to completely penetrate the human body.

Below are some definitions specific to arc fault.

- **ATPV**: Maximum incident energy of a PPE system prior to break open.
- **Arc rating**: The value attributed to materials that describe their ability to withstand exposure to an electrical arc discharge. The arc rating is expressed in cal/cm^2 and is derived from the determined value of the ATPV or energy of break-open threshold (EBT) (should a material system exhibit a break-open response below the ATPV value). Arc rating is reported as either ATPV or EBT, whichever is the lower value.
- **Flash hazard:** Condition caused by the release of energy caused by an electric arc.
- **Flashes protection boundary:** Distance from live parts where second-degree burns would result if an arc flash occurred. By definition, this is the distance at which the incident energy level is at 1.2 cal/cm^2.
- **Limited approach boundary:** Distance from live parts where shock hazard exists.
- **Restricted approach boundary:** Distance beyond which only qualified persons may enter because it requires shock protection equipment or techniques.
- **Exposed live part:** Not suitably guarded, isolated, or insulated.
- **Working near (live parts):** Any activity inside a limited approach boundary.
- **Working on (live parts):** Any activity inside a restricted approach boundary including coming in contact with live parts with body or equipment regardless of PPE.

The arc flash boundary is the distance from the source of an arc where the incident energy level will be 1.2 cal/cm^2, which is enough thermal energy to cause a second-degree burn.

OSHA updated CFR 1910.269 in 2014 (the previous revision was in 1994) and added the following regulation requiring employers to protect their employees from the hazards of arc flash.

1910.269 (I) (8): Protection from flames and electric arcs

(i) The employer shall assess the workplace to identify employees exposed to hazards from flames or from electric arcs.

(ii) For each employee exposed to hazards from electric arcs, the employer shall make a reasonable estimate of the incident heat energy to which the employee would be exposed.

As mentioned previously, an employee must utilize shock protection PPE whenever they are in the restricted approach boundary (for tasks that conform to

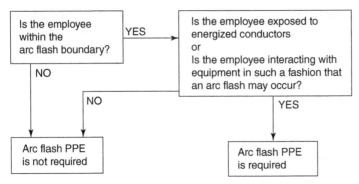

Figure 1.8 Decision matrix if a task requires arc flash PPE.

Subpart S) or whenever they are in the minimum approach distance (for tasks that conform to Part 269). How about arc flash PPE? An employee must utilize arc flash-rated PPE whenever they are exposed to a potential arc flash. So when is that? Whenever the employee is within the arc flash boundary and if the employee is exposed to energized conductors OR if the employee is interacting with the equipment in such a fashion as they could be exposed to an arc flash, then the employee must utilize arc-rated PPE. The flow chart in Figure 1.8 depicts this explanation.

OSHA allows for the use of tables such as those found in NESC or by analysis. Appendix E of OSHA 1910.269, as is shown in Table 1.11, lists which methods of incident energy calculation are applicable for various voltage levels whether the exposure is in open air or in an enclosure.

For voltages less than 15 kV, OSHA recommends the use of NFPA 70E tables or IEEE 1584 equations as methods to determine the incident energy levels. For voltages in excess of 15 kV, OSHA does not recommend the use of NFPA 70E or IEEE1584, but rather recommends ARCPRO for analysis or NESC tables can alternately be utilized. IEEE 1584 and ARCPRO analysis methods tend to be calculation-intensive.

TABLE 1.11 Allowable Methods for Selection of Arc Flash Boundary and PPE

Incident Energy Calculation Method	600 V and Less			601 V to 15 kV			More than 15 kV		
	1P	3PA	3PB	1P	3PA	3PB	1P	3PA	3PB
NFPA 70E-2012 Annex D (Lee equation)	Y-C	Y	N	Y-C	Y-C	N	N	N	N
Doughty, Neal, and Floyd	Y-C	Y	Y	N	N	N	N	N	N
IEEE Std 1584b-2011	Y	Y	Y	Y	Y	Y	N	N	N
ARCPRO	Y	N	N	Y	N	N	Y	Y	Y

Source: Reproduced with permission of OSHA.
1P: Single-phase arc in open air.
3PA: Three-phase arc in open air.
3PB: Three-phase arc in an enclosure (box).
Y: Acceptable – produces a reasonable estimate of incident heat energy from this type of electric arc.
N: Not acceptable – does not produce a reasonable estimate of incident heat energy from this type of electric arc.
Y-C: Produces a reasonable, but conservative, estimate of incident heat energy from this type of electric arc.

ARCPRO and IEEE 1584 analysis software takes into account many variables such as the voltage level of the system, the gap between buses, the clearing time of the device, and other parameters that affect the value of the incident energy level. Due to the complexity of these calculations, the details of these calculations are beyond the scope of this introductory material. Normally, a computer-based software is utilized to obtain results (Note: IEEE provides spreadsheet along with IEEE 1584 to assist in calculations of incident energy levels). Below we will explain the use of the table method in NFPA 70E that OSHA allows for systems less than 15 kV and we will explain the use of the table method in NESC that OSHA allows for systems over 15 kV.

ARC FLASH ANALYSIS UTILIZING NFPA 70E TABLES

The arc flash PPE category and arc flash boundary can be determined using NFPA 70E Table 130.7(C)(15)(A)(b) – Arc-Flash Hazard PPE Categories for Alternating Current (ac) Systems or Table 130.7(C)(15)(B) – Arc-Flash Hazard PPE Categories for Direct Current (DC) Systems.

These tables break the task down to three questions: What is the type of equipment that the task involves? What is the maximum short-circuit current available? And what is the maximum fault clearing time? With those three pieces of information, the tables can be used to determine the arc flash PPE category (based on a working distance given in the table) and the arc flash boundary. Remember that if the task is being done outside the arc flash boundary, then arc flash PPE is not required because the incident energy level is less than 1.2 cal/cm^2. The student can access NFPA 70E® online at http://www.nfpa.org/codes-and-standards/document-information-pages?mode=code&code=70E. In NFPA 70E, Table 130.7(C)(15)(A)(b) (Arc-Flash Hazard PPE Categories for Alternating Current (AC) Systems) provides guidance to the arc flash PPE category and arc flash boundary for AC systems for various types of equipment. Note that the short-circuit current available, the maximum clearing time, and the working distance are all important parameters to the table. To utilize the tables, the actual short-circuit current must be equal to or less than the value given in the table, the actual clearing time must be equal to or less than the value given in the table, and the working distance must be equal to or greater than the value given in the table. Similarly, for DC systems, Table 130.7(C)(15)(B) (Arc-Flash Hazard PPE Categories for Direct Current (DC) Systems) provides guidance to the arc flash PPE category and arc flash boundary for DC systems for various types of equipment. Just as for the AC table, the short-circuit current and clearing time of the actual application must be equal to or less than the value in the table and the working distance must be equal to or greater than the value in the table.

Example 1.4   You have been given a task of pulling a molded case circuit breaker from a 120/240 Vac panel. The fault current at the panel is 10 kA and the panel is protected by a current limiting fuse, so the clearing time is 1 cycle. What is the arc flash boundary and what is the arc flash PPE required to perform the above task?

Solution: To solve, the student will need to view the current NFPA 70E standard at: http://www.nfpa.org/codes-and-standards/document-information-pages?mode=code&code=70E.

From Table 130.7(C)(15)(A)(b) first row, which lists the requirements for panels with voltages at 240 Vac and below and short-circuit currents of 25 kA and clearing time less than 2 cycles, we find that the arc flash PPE required is Category 1 at a distance of 18 in. and the arc flash boundary is 19 in.

So what is arc flash PPE Category 1? NFPA 70E breaks down arc flash categories into four levels based on a range of energy exposure possible. Arc flash PPE Category 1 must protect again an incident energy level of at least 4 cal/cm^2. Arc flash PPE Category 2 must protect again an incident energy level of at least 8 cal/cm^2. Arc flash PPE Category 3 must protect again an incident energy level of at least 25 cal/cm^2. Arc flash PPE Category 4 must protect again an incident energy level of at least 40 cal/cm^2. Notice that there is no category above 40 cal/cm^2. This is because at energy levels that high, heat is not the only hazard to personnel but also hot gases, projectiles, pressure wave, and UV radiation. Due to the magnitude of these other hazards, there is no level of PPE that can protect a person at energy levels in excess of 40 cal/cm^2.

Example 1.5   A task has been assigned to remove a molded case circuit breaker from a 120 Vac panel where there is a maximum bolted fault value of 8 kA and the clearing time is 5 cycles; can you use the tables in NFPA 70E for the determination of the correct arc flash PPE for this task? If you can, what is the arc flash PPE level required for this task?

Solution: To solve, the student will need to view the current NFPA 70E standard at: http://www.nfpa.org/codes-and-standards/document-information-pages?mode=code&code=70E.

You could not use the table to determine the PPE, but have to perform a full arc flash analysis. This is because the table lists only one row for 240 Vac panels and it has the limits of a maximum fault current of up to 25 kA and maximum clearing time of up to 2 cycles. Since clearing time of this application is actually 5 cycles, the tables cannot be used and an analysis must be performed.

Example 1.6   You have been given a task of pulling a molded case circuit breaker from a 125 Vdc panel. The fault current at the panel is 10 kA and the panel is protected by a current limiting fuse, so the clearing time is 1 cycle. What is the arc flash boundary and what is the arc flash PPE required to perform the above task?

Solution: To solve, the student will need to view the current NFPA 70E standard at: http://www.nfpa.org/codes-and-standards/document-information-pages?mode=code&code=70E.

From Table 130.7(C)(15)(B) third row, which lists the requirements for panels with voltages between 100 Vdc and 250 Vdc and short-circuit currents between 7 kA and 15 kA and clearing time less than 2 seconds, we find that the arc flash PPE required is Category 3 at a distance of 18 in. and the arc flash boundary is 6 ft.

The hazard of arc flash can be minimized by proper equipment design and proper operational procedures. PPE should be a last resort to meeting the requirements of arc flash protection. The incident energy of the arc flash hazard is primarily a function of current and time. By specifying low current let through fuses, equipment on the downstream side of fuses can have reduced incident energy levels. Current limiting fuses can provide a dramatic reduction of incident energy levels if the value for arcing current is within the current limiting range of the fuse. When the fuse is applied such that the arcing current is within the current limiting range of the fuse, the fuse will clear this fault current in less than one cycle. One drawback of using the current-limiting fuse is the difficulty in trying to coordinate downstream protection devices with the fuse. Therefore, current-limiting fuses are used only when they are the last circuit-protective device on the radial feeder.

Relay settings are another method of providing faster trip times for electrical equipment. The challenge for the protection engineer is to provide long enough trip settings to allow for both normal system operations such as motor locked rotor currents when starting and to allow for coordination of the protective system such that, when a fault occurs, the relays operate fast enough to provide minimum amount of incident energy levels at the equipment where the fault occurs and still allow for acceptable coordination.

The above discussion addressed arc flash as defined by NFPA 70E which is the standard OSHA recommends for voltages of 15 kV and below. For voltages in excess of 15 kV, OSHA recommends use of NESC. NESC arc flash tables were substantially updated in the 2012 version. Table 410-1, as shown in Table 1.12, addresses arc flash PPE requirements for various types of equipment for voltages of 50 Vac to 1000 Vac. Table 410-2, a portion of which is shown in Table 1.13, addresses arc flash PPE requirements for voltages from 1.1 kV to 46 kV based on the system voltage level, maximum fault current level, and maximum fault clearing time. Table 410-3 addresses arc flash PPE requirements for voltages in excess of 46 kV based on the system voltage level, maximum fault current level, and maximum fault clearing time.

TABLE 1.12   Clothing Systems for Voltages <1 kV

	Nominal Voltage Range and cal/cm^2		
Equipment Type	50–250 V	251–600 V	601–1000 V
Self-contained meters/cabinets	4	20	30
Pad-mounted transformers	4	4	6
CT meters and control wiring	4	4	6
Metal-clad switchgear/MCCs	8	40	60
Pedestals/pull boxes/hand holes	4	8	12
Open air (includes lines)	4	4	6
Network protectors	4	>60	>60
Panel boards – single-phase (all)/three-phase (<100 A)	4	8	12
Panel boards – three-phase (>100 A)	4	>60	>60

Source: Reproduced with permission of IEEE.

TABLE 1.13   Clothing Systems for Voltages >1 kV

Phase-to-Phase Voltage (kV)	Fault Current (kA)	4-Cal System Max Clearing Time (cycles)	8 Cal-System Max Clearing Time (cycles)	12-Cal System Max Clearing Time (cycles)
1 to 15	5	46.5	93	139.5
	10	18	36.1	54.1
	15	10	20.1	30.1
	20	6.5	13	19.5
15.1 to 25	5	27.6	55.2	82.8
	10	11.4	22.7	34.1
	15	6.6	13.2	19.8
	20	4.4	8.8	13.2
25.1 to 36	5	20.9	41.7	62.6
	10	8.8	17.6	26.5
	15	5.2	10.4	15.7
	20	3.5	7.1	10.6

Source: Reproduced with permission of IEEE.

Example 1.7   You have been given the task of performing a phase check between two circuits energized at 22 kV. The maximum fault current level is 10 kA and the maximum clearing time is 5 cycles. What is the minimum arc flash PPE required for this task?

Solution: From NESC Table 410-2 (Table 1.16 in text book) we can see that for 22 kV, a fault current level of 10 kA, and a 5 cycle clearing time, a 4 calorie suit will provide adequate protection.

Approach Boundaries

The minimum distance that a person is permitted to get to an exposed energized conductor is defined by the approach boundary. This boundary is for shock protection, not arc flash protection. As such, this only applies when energized conductors are exposed. Keep in mind that the arc flash boundary may lie within the approach boundaries or may be outside of the approach boundaries. The definitions of the approach boundaries are listed below.

- **Limited approach boundary:** Distance from live parts where shock hazard exists. Persons must either be qualified or be escorted by a qualified person to enter the limited approach boundary. Note, there is a value for a fixed conductor and another value for a movable conductor. Unqualified persons are not permitted to cross the restricted approach boundary even when escorted by a qualified person.

- Restricted approach boundary: A distance beyond which only qualified persons may enter because it requires shock protection equipment or techniques. It is considered the same as making contact with the energized conductor. Inside the restricted approach boundary, the employee must utilize shock protection PPE, insulated tools, and safe energized electrical work practices.

Prior to the 2015 version of NFPA 70E®, there was also a "prohibited approach boundary," but this term was removed as of the 2015 version of NFPA 70E®. The values for boundary distances are given in NFPA 70E, Table 130.4 (D) (a) for AC systems and Table 130.4 (D) (b) for DC systems. The student is encouraged to view the current NFPA 70E standard at: http://www.nfpa.org/codes-and-standards/document-information-pages?mode=code&code=70E.

HAZARDOUS/CLASSIFIED AREAS

In environments where ignitable and/or combustible materials are stored, handled, or utilized, where the materials may be in sufficient quantities to introduce a fire or explosion hazard, the area must be classified as to the type of hazard. Equipment utilized in a classified area must be rated for service in this area. The equipment must be labeled for the area classification, the gas classification group, and the operating temperature of the equipment. For Class I equipment, the operating temperature of the equipment must be less than the autoignition temperature of the material presenting the hazard for that area. Additionally, threaded metallic conduit that presents a conductive path must be wrench tight. Equipment with conductive panels and surfaces must have panels and surfaces solidly secured such that a good path for conduction of fault current is present. Flexible metallic conduit must utilize a grounding bushing on each side of the metallic conduit and have a *bonding jumper* in parallel with the flexible metallic conduit to provide this path for the conduction of fault currents. The goal is to ensure that, should fault current flow in the metallic materials, there are no high impedance connections between metallic parts that can present and arc/spark hazard in the hazardous environment providing an ignition source. NEC® also requires that any equipment utilized in a classified area be selected under the supervision of a registered professional engineer. The process of calculations and the results of the calculations for area classification are required to be documented.

The NESC refers the user back to the NEC® articles 500 to 517 for classification of hazardous areas but NESC provides additional requirements in section 127. For example, NESC127G prohibits gaseous hydrogen storage from being located beneath electric power lines. The National Electrical Code® provides for two types of area classification, the "Class" system and the "Zone" system. To date, a majority of electric utility locations in the United States that require compliance with NEC® utilize the "Class" system.

Because of this, the "Class" system will be introduced first and the "Zone" system second. However, either system, when properly administered, will satisfy the requirements of NEC®, NFPA standards, and OSHA requirements.

CLASSIFIED AREA – "CLASS" SYSTEM

In the "Class" system, there are three classes of materials. The "class" of the system is based on the material that poses the hazard to the area. Under the "class" designation, there is a sub-designation of division. The division of the system is based on whether the material is found in the area under normal operation in quantities and form sufficient to pose a hazard, or whether the material would only be present in an abnormal condition. Additionally, a Division 1 area cannot be adjacent to an unclassified area. There must be a Division 2 area between the unclassified area and the Division 1. The NEC®-defined values for the "Class" system of area classification are listed below.

- Class I location where flammable gases or vapors may be present in air to an explosive level
  - Division 1 – Flammable gases or vapors exist under normal operations or frequently or faulty equipment operation may cause simultaneous failure of electrical equipment
  - Division 2 – Flammable gases or vapors are used, but not normally in explosive concentrations or concentration prevented by positive mechanical ventilation or is adjacent to Class I, Division 1
- Class II location where combustible dust may be present
  - Division 1 – Combustible dusts exist under normal operations or faulty equipment operation may cause simultaneous failure of electrical equipment or where combustible dusts are electrically conductive
  - Division 2 – Combustible dusts not normally in explosive concentrations or concentration may prevent cooling of electrical equipment
- Class III location where easily ignitable fibers may be present
  - Division 1 – Where fibers are produced, handled, or used
  - Division 2 – Areas where stored or handled other than manufacturing process

Example 1.8   The filling station for the hydrogen storage tanks for a power generation facility has been deemed as a location where, under normal operations of filling the storage tank with hydrogen (a flammable gas), the area will frequently contain an explosive mixture of hydrogen when the relief value lifts. What are the correct class and division for this hazardous area classification?

Solution: This would be Class I (flammable gas or vapor), Division 1 (Flammable gases or vapors exist under normal operations or frequently).

For the "Class" system of area classification, the appropriate standards to provide guidance on the classification of the locations are NEC® – National Electrical Code®, NFPA 497 – Classification of Flammable Liquids, Gases, or Vapors and of Hazardous (Classified) Locations for Electrical Installations in Chemical Process Area, and NFPA 499 – Classification of Combustible Dusts and of Hazardous (Classified) Locations for Electrical Installations in Chemical Process Area.

NFPA 497 provides guidance to the classification of flammable liquids, gases, and vapors. Below are some definitions that are useful to understand the application of classified systems.

The autoignition temperature (AIT) of a material is the minimum temperature required to initiate or cause self-sustained combustion of the material independently of the external heating device.

The flash point of a material is the minimum temperature at which a liquid gives off vapor in sufficient quantity to form an ignitable mixture.

A combustible liquid is any liquid that has a flash point at or above 100°F, (37.8°C). Combustible liquid classifications are broken down into three levels:

- Class II liquid – Any liquid that has a flash point at or above 100°F (37.8°C) and below 140°F (60°C)

- Class IIIA liquid – Any liquid that has a flash point at or above 140°F (60°C) and below 200°F (93°C)

- Class IIIB liquid – Any liquid that has a flash point at or above 200°F (93°C)

Combustible liquids will form an ignitable mixture only when heated above its flash point. Flammable liquids will evaporate at a rate that depends on its volatility. The lower the flash point of a material, the greater the volatility, and the faster the evaporation.

A flammable liquid is any liquid with a flash point below 100°F, (37.8°C). Their classification is broken down into four levels:

- Class I liquid – Any liquid that has a flash point below 100°F (37.8°C).

- Class IA liquid – Any liquid that has a flash point below 73°F (22.8°C) and a boiling point below 100°F (37.8°C).

- Class IB liquid – Any liquid that has a flash point below 73°F (22.8°C) and a boiling point at or above 100°F (37.8°C).

- Class IC liquid – Any liquid that has a flash point at or above 73°F (22.8°C), but below 100°F (37.8°C).

The maximum experimental safe gap (MESG) is the maximum clearance between two parallel metal surfaces that has been found to prevent an explosion from being propagated from one chamber to another chamber normalized to methane. The minimum igniting current ratio (MIC ratio) is the ratio of the minimum current required from an inductive spark discharge to ignite the most easily ignitable mixture of a gas or vapor, divided by the minimum current required from an inductive spark discharge to ignite methane under the same conditions. A lower MIC ratio means less current is required to instigate ignition.

Combustible materials in Class I (liquids and vapors) are divided into four groups depending on their MESG or their MIC ratio. These groups are as follows:

- Group A – Acetylene.

- Group B – Flammable gas or vapor that has either an MESG less than or equal to 0.45 mm or an MIC ratio of 0.40 or less. A typical Class I, Group B material is hydrogen.

- Group C – Flammable gas or vapor that has either an MESG greater than 0.45 mm and less than or equal to 0.75 mm or an MIC ratio greater than 0.40 and less than 0.80. A typical Class I, Group C material is ethylene.

- Group D – Flammable gas or vapor that has either an MESG greater than 0.75 mm or an MIC ratio greater than 0.80. A typical Class I, Group D material is Propane.

The minimum ignition energy (MIE) is the minimum energy required from a capacitive spark discharge to ignite the most easily ignitable mixture of gas or vapor.

Air density is an important property as it determines if gases will settle to the ground level, collect and displace air, or if the gases will tend to raise and disperse. Vapor density values less than 1.0 means the gas is lighter than air and will tend to dissipate rapidly in the atmosphere. Vapor density values greater than 1.0 means the gas is heavier than air and will tend to fall to grade elevation and collect in low-lying areas displacing air until they are force-ventilated away from the area.

NFPA 497, as shown in Table 1.14, provides tables with physical properties of flammable liquids, gases, and vapors to help with the classification process. Typical information is shown in Table 1.14 for several gases.

Example 1.9   The filling station for the hydrogen storage tanks has been deemed a classified area. What group letter is this gas?

Solution: This is a Group B gas (see definition of combustible material groups).

TABLE 1.14   NFPA 497 Table 4.4.2 – Selected Chemical Properties

Chemical	CAS No.	Class I Division Group	Flash Point (°C)	AIT (°C)	% LFL	% UFL	Vapor Density (Air = 1)
Hydrogen	1333-74-0	Bd		520	4.0	75.0	0.1
Hydrogen cyanide	74-90-8	Cd	−18	538	5.6	40.0	0.9
Hydrogen selenide	7783-07-5	C					
Hydrogen sulfide	7783-06-4	Cd	−18	260	4.0	44.0	1.2
Isoamyl acetate	123-92-2	D	25	360	1.0	7.5	4.5
Isoamyl alcohol	123-51-3	D	43	350	1.2	9.0	3.0
Isobutane	75-28-5	Dg		460	1.8	8.4	2.0
Isobutyl acetate	110-19-0	Dd	18	421	2.4	10.5	4.0
Isobutyl acrylate	106-63-8	D		427			4.4
Isobutyl alcohol	78-83-1	Dd	−40	416	1.2	10.9	2.5

Source: Reproduced with permission from NFPA 497, *Recommended Practice for the Classification of Flammable Liquids, Gases, or Vapors and of Hazardous (Classified) Locations for Electrical Installations in Chemical Process Areas*, Copyright © 2012, National Fire Protection Association. This is not the complete and official position of the NFPA on the referenced subject, which is represented only by the standard in its entirety. The student may download a free copy of the NFPA 497 standard at: http://www.nfpa.org/codes-and-standards/document-information-pages?mode=code&code=497.

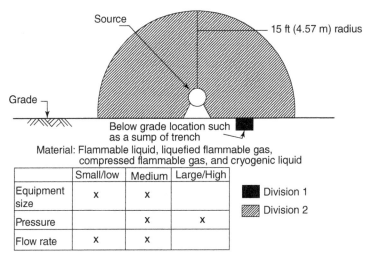

Figure 1.9   NFPA 497 Table 5.9.2 (a) – Leakage located outdoors, at grade. The material being handled could be a flammable liquid, a liquefied or compressed flammable gas, or a flammable cryogenic liquid. *Source*: Reproduced with permission from NFPA 497, *Recommended Practice for the Classification of Flammable Liquids, Gases, or Vapors and of Hazardous (Classified) Locations for Electrical Installations in Chemical Process Areas,* Copyright © 2012, National Fire Protection Association. This is not the complete and official position of the NFPA on the referenced subject, which is represented only by the standard in its entirety. The student may download a free copy of the NFPA 497 standard at: http://www.nfpa.org/codes-and-standards/document-information-pages?mode=code&code=497.

Example 1.10   The filling station for the hydrogen storage tanks has been deemed a classified area. Will the hydrogen gas tend to displace the surrounding air and settle or will the air tend to displace the hydrogen gas and the hydrogen gas will disperse?

Solution: Hydrogen has a vapor air density of 0.1. This means that the density of hydrogen at standard atmospheric temperature and pressure is 10% of the value of air. Because of this, hydrogen would not settle to the ground level, but disperses quickly in the air.

Note that the upper flammability level in air for hydrogen is 75% and the lower flammability level in air for hydrogen is 4% from Table 1.14. This means that for concentrations below 4% hydrogen or above 75% hydrogen, the mixture of hydrogen and air is not flammable. Also, note the AIT of hydrogen is high at 520°C. Hydrogen is a Class I, Group B material.

NFPA 497 also provides guidance depending on the type of material and the physical layout of the area for dimensions for area classification. Figure 1.9 shows an example of Figure 4.9.2(a) out of the standard which applies to gases with vapor densities greater than 1 (i.e., gases tend to settle to the ground) for outdoor installations for a facility at grade. Notice for this application that the below grade location is a Division 1 location, since gases with vapor densities greater than 1 will tend to accumulate in this location.

NFPA 499 provides guidance for the classification of combustible dusts. By definition under the "Class" system, combustible dusts in sufficient quantity to present a hazard are Class II areas. Below are some definitions that are useful to understand the application of classified systems. Combustible dust is any finely divided solid material 420 microns or less in diameter that presents a fire or explosion hazard when dispersed. Class II combustible dusts are divided into three groups as follows.

- Group E – Atmospheres containing combustible metal dusts. Typical dusts in this group are aluminum, magnesium, or other metallic alloys.

- Group F – Atmospheres containing combustible carbonaceous dusts that have more than 8% total trapped volatiles or that have been sensitized by other materials so that they present an explosion hazard. Typical dusts in this group are coal, carbon black, charcoal, and coke dusts.

- Group G – Atmospheres containing combustible dusts that do not fall into the Group E or Group F designation. Typical dusts in this group are flour, grain, wood, plastic, and other similar chemicals.

Dust particles tend to settle out of the atmosphere and form a layer on equipment. This dust layer acts as a blanket and can reduce the heat removal efficiency of equipment causing equipment skin temperature of operating equipment to rise to values greater than design. This should be taken into account when specifying equipment for Class II areas. Additionally, while dusts suspended in atmosphere may present a hazard, once they drop out of the atmosphere and form layers, if left undisturbed and not in direct contact with the ignition source, they will not present a hazard. However, if a small amount of dust is in the air and provides fuel for a small explosion, the pressure wave may cause the layered material to become airborne, providing substantial amount of additional fuel causing a second and oftentimes more damaging explosion. For this reason, *good housekeeping is essential* in areas classified as Class II hazardous locations. Keeping the amount of material that has settled out on equipment to a minimum will help minimize the damage from any secondary explosions that occur due to the pressure wave displacing layered dusts.

In Class II locations, there are three common methods of ignition. The first method of ignition is having combustible dusts suspended in the atmosphere and an ignition source presents itself. The second method of ignition is when combustible dusts have formed a layer on electrical equipment, reducing the heat removal efficiency of the electrical equipment and the skin of the electrical equipment reaches the AIT of the dust. The third method is when a Group E dust (metal dust) is present either layered or suspended in air and a sufficient amount of current flows through the conductive dust to cause ignition of the dust. In determining whether a Class II area is Division 1 or Division 2, the quantity of dust, its physical and chemical properties, its dispersion properties, and the location of walls and surfaces must all be considered. If either under normal or abnormal operation of equipment, a dust cloud may form, the area should be classified as a Division 1 area. Also if a layer forms on equipment that is greater than 1/8th of an inch in thickness, then this area should be classified

TABLE 1.15 Portion of Table 4.5.2, from NFPA 499

Chemical Name	CAS No.	NEC Group	Code	Layer or Cloud Ignition Temperature (°C)
Chlorinated polyether alcohol		G		460
Chloroacetoacetanilide	101-92-8	G	M	640
Chromium (97%) electrolytic, milled	7440-47-3	E		400
Cinnamon		G		230
Citrus peel		G		270
Coal, Kentucky bituminous		F		180
Coal, Pittsburgh experimental		F		170
Coal, Wyoming		F		
Cocoa bean shell		G		370
Cocoa, natural, 19% fat		G		240

Source: Reproduced with permission from NFPA 499, Recommended Practice for the Classification of Combustible Dusts and of Hazardous (Classified) Locations for Electrical Installations in Chemical Process Areas, Copyright © 2013, National Fire Protection Association. This is not the complete and official position of the NFPA on the referenced subject, which is represented only by the standard in its entirety. The student may download a free copy of the NFPA 499 standard at: http://www.nfpa.org/codes-and-standards/document-information-pages?mode=code&code=499.

as a Division 1 area since, in the event of a pressure wave, the layered material may become airborne and present a secondary explosion hazard.

There are also additional hazards due to dusts that do not directly associate with the explosion hazard. Group E dusts, being metallic dusts, can conduct electricity. Therefore, these dusts may provide path for shorts in electrical equipment causing failure of the electrical equipment. The resulting arc from the short may provide the ignition source for explosion. Additionally, certain ferrite-based metals are or may become magnetized by electromagnetic fields and concentrate in areas where EM field exists. Zirconium, thorium, and uranium dusts have extremely low AIT values (some as low as 20°C). This may lead to challenges when handling such dusts.

NFPA 499 provides data on various materials that can form dusts in a facility. It defines the NEC® materials group designation and provides the value of the AIT of the material. It also provides classification diagrams that show recommended dimensions for various area configurations depending on the dust properties and the surrounding area properties. Table 1.15 shows properties for several types of dust from NFPA 499 and Figure 1.10 shows a classification diagram for a Group F or Group G dust in an indoor area from NFPA 499, unrestricted area with operating equipment.

For example, from the data in Table 1.15 for Kentucky bituminous coal, NFPA defines this as a Class F dust with an AIT of 180°C. If a piece of operating equipment were located in an area that handles Kentucky Bituminous Coal and it is determined that this area is to be classified, due to concentration of the dust, the area at the equipment would be a Division 1 area. Then from Figure 1.10, the Division 1 area should extend 20 ft from the equipment. Beyond this, for an additional 10 ft, this would be classified as a Division 2 area. Beyond this, the area is unclassified assuming there are no other sources for hazardous material other than the equipment of this example.

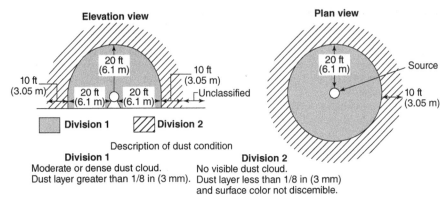

Figure 1.10   Figure 4.8(a) from NFPA 499 – Group F or Group G Dust – Indoor, Unrestricted Area; Open or Semi-Enclosed Operating Equipment. *Source*: Reproduced with permission from NFPA 499, *Recommended Practice for the Classification of Combustible Dusts and of Hazardous (Classified) Locations for Electrical Installations in Chemical Process Areas*, Copyright © 2013, National Fire Protection Association. This is not the complete and official position of the NFPA on the referenced subject, which is represented only by the standard in its entirety. The student may download a free copy of the NFPA 499 standard at: http://www.nfpa.org/codes-and-standards/document-information-pages?mode=code&code=499.

CLASSIFIED AREA – "ZONE" SYSTEM

In the "Zone" system, the hazardous area zone is defined by three items, the type of hazard that is present in the area (defined by the "Class"), the possibility of the hazard being present in sufficient quantities to present ignitable/explosive concentrations (defined by the "Zone"), and the AIT of the hazardous material (defined by the "Group).

In the "Zone" system, the "group" defines the ignitability of the material that presents the hazard to the area. For gases and vapors, the group information provides information as to the energy required to support the combustion process for the material involved. This is denoted by the MIC or the MESG value. NFPA 497 defines three groups.

- Group IIC – Flammable gas or vapor that has either an MESG less than or equal to 0.50 mm or an MIC ratio of 0.45 or less. A typical Class I, zone, Group IIC material is acetylene or hydrogen.

- Group IIB – Flammable gas or vapor that has either an MESG greater than 0.50 mm and less than or equal to 0.90 mm or an MIC ratio greater than 0.45 and less than 0.80. A typical Class I, zone, Group IIB material is acetaldehyde or ethylene.

- Group IIA – Flammable gas or vapor that has either an MESG greater than 0.90 mm or an MIC ratio greater than 0.80. A typical Class I, zone, Group IIA material is acetone, ammonia, ethyl alcohol, gasoline, methane, or propane.

The actual "zone" is a description of the probability of the hazardous material being present in sufficient quantities to present an ignitable/explosive mixture in the environment. There are three zones for gases and vapors.

- A Zone 0 location is where a flammable atmosphere is highly likely to be present and/or may be present for long periods of time or even continuously.

- A Zone 1 location is where a flammable atmosphere is possible but not likely and is most likely not to be present for long periods of time.

- A Zone 2 location is where a flammable atmosphere is unlikely to be present except for very short periods of time. This is also used where under normal equipment operations, the flammable atmosphere is not present, but in the event of equipment malfunction, the flammable atmosphere may become present until equipment is restored.

There are also three zones for dusts and ignitable fibers.

- A Zone 20 location is an area where combustible dust or ignitable fibers and flyings are present continuously for long periods of time and in quantities sufficient to be hazardous.

- A Zone 21 location is an area where combustible dust or ignitable fibers and flyings are likely to exist occasionally under normal operation and in quantities sufficient to be hazardous.

- A Zone 22 location is an area where combustible dust or ignitable fiber and flyings are not likely to occur under normal operation in quantities sufficient to be hazardous.

As a side note, since dusts can settle on equipment and reduce the heat removal efficiency of equipment causing the equipment to run at higher temperature than normal, there may be an equipment derating applied to the AIT of the material for application of equipment.

In addition to classifying the area as to the hazardous material (group) and the likelihood of the material being present in sufficient quantities for ignitable atmosphere (zone) there is an additional classification for equipment temperature classes to ensure the equipment does not approach the AIT of the material presenting the hazard. Equipment rated for operation in classified areas has a rating "T-Class" to define the maximum operating temperature of the equipment. All equipment temperature ratings are based on a temperature rise above a certain maximum ambient temperatures. Unless otherwise listed by the equipment manufacturer, the basis for the T-Class rating is with a maximum ambient temperature of 40°C. Operation of equipment above the maximum allowable rated ambient temperature is not allowed and the equipment must have a T rating based on actual ambient conditions. This is the reason for "derate" when equipment is installed in areas with dusts that can cover equipment and lower the heat removal efficiency of the equipment leading to higher operating temperatures. Below is list of the six basic values for T-Class (there are additional T-Class ratings that some OEMs can provide).

- $T_1 - 450°C$
- $T_2 - 300°C$

- $T_3 - 200°C$
- $T_4 - 135°C$
- $T_5 - 100°C$
- $T_6 - 85°C$

Example 1.11 A classified area will contain Kentucky bituminous coal (see Table 1.15). The lighting fixtures in the room must be rated to operate at a surface temperature below the AIT of the coal. Assuming the maximum ambient temperature of the room is the same as the equipment rating of 40°C, which of the equipment T-Classes will be acceptable?

Solution: For Kentucky Bituminous Coal, the AIT is 180°C (From Table 1.15). For equipment to be utilized in an atmosphere that may contain this material, the equipment should have a T4, T5, or T6 rating. Equipment with a T1, T2, or T3 rating may result in equipment skin temperatures that exceed the AIT of the coal. If the coal is expected to cover the electrical equipment, then a derate may be applied preventing the use of the T4 rating.

BOILER CONTROL AND BURNER MANAGEMENT

Another very important subject in power plant safety is the safe operation of boiler and combustion systems and the unique hazards these systems present. NFPA 85 – Boiler and Combustion System Hazards Code provides guidance to the safe control of these systems to prevent hazards such as implosion or explosion. This standard addresses the minimum requirements for combustion systems and controls. This is a critical system for safety since, under normal operation of a boiler, we intentionally combine air, a combustible material, and an ignition source creating a hazardous atmosphere in the furnace section of the boiler. The hazardous mixture must be safety controlled to ensure that the conditions do not arise that can present an explosive or implosive event.

For most boilers, the positive transient design pressure must be at least +35 inches of water and the negative transient design pressure must be at least −35 inches of water. The positive value is to ensure that the boiler remains intact during a positive pressure excursion and does not explode. The negative value is to ensure that the boiler will remain intact during a negative pressure excursion and does not implode.

NFPA 85 covers areas such as normal startup and shut down procedures as well as procedures used in the event of a boiler trip to ensure that an explosive or implosive condition does not occur. The details for control depend on the type of system involved. Below is an introductory discussion of some of the requirements for NFPA 85. For more detailed information, the reader is advised to refer to the latest copy of NFPA 85.

At minimum, there are several interlocks required to ensure proper boiler controls. When the control system detects a loss of flame in the burner, the system must remove the fuel source to the burner and isolate the ignition source. This is because, if the fuel being supplied to the burner system is not being combusted in the furnace section, then the explosive mixture of air and fuel may remain in the boiler and in the duct system downstream. Any point that presents an ignition source (like the rotating air preheater member causing a momentary spark during rotation) could lead to an explosion of the system.

If an induced draft (ID) fan is tripped, the associated dampers to that fan should be closed unless that is the last ID fan in service. Additionally, if the ID and forced draft (FD) fans are paired, then the associated FD fan should be tripped and its associated damper closed as well. This is done to minimize the pressure transient that may occur if an ID fan tripped but the associated FD fan did not. If the last ID fan in service is tripped, then the dampers should remain full open. This is to allow for the removal of combustibles from the boiler utilizing natural ventilation and/or the ventilation of the associated FD fan. Also on the loss of all ID fans, the master fuel trip should be activated to remove fuel from the burner. Additionally, if furnace pressure exceeds (either positive or negative) pressure values provided by the boiler OEM, the master fuel trip will also be activated to safely shut the boiler down. When starting and stopping ID/FD fans, a specific sequence is required to ensure that the boiler pressure rating is not exceeded. First, an open-air path from suction to discharge of the air system must be established and verified. Since the FD fan is on the inlet side of the boiler and regulates the airflow through the boiler, and the ID fan is on the outlet side of the boiler regulating boiler pressure, the ID fan is started first so that we are able to maintain boiler pressure control. This is to ensure that the boiler does not experience a positive pressure transient during startup of the associated FD fans. Once the ID fan is started, the associated FD fan is started. Dampers are partially closed during startup to minimize load on the motors and minimize the associated time required for acceleration to speed. When shutting down an FD and ID fan train, the FD fan is stopped first, followed by the associated ID fan for the same reasons (ensure we can maintain boiler pressure with ID fan).

The above discussion presented some of the interlocks from the air side of the combustion equation. From the fuel side of the combustion equation, there are also specific responses that NFPA 85 calls out for during certain events. The primary air (PA) fan provides the transport of the fuel from the pulverizer to the combustion chamber. If the PA fan trips, the fuel valve or feeder equipment shall also be tripped that is associated with that PA fan. For a pulverized system, failure of a pulverizer shall trip the associated PA fan. Loss of coal feed to the burner must trip the associated feeder, unless the igniters are Class 1 igniters. The reason for the exception for Class 1 igniter is the amount of energy output from the igniter is sufficient to sustain a flame in the burner area until the coal feed is reestablished. In NFPA 85, igniters are broken down into the following classes.

- Class 1 igniter – An igniter that is applied to ignite the fuel input to a burner and support ignition under any burner conditions. It will provide sufficient ignition energy usually in excess of 10% of full-load burner output at its associated burner to raise any combination of burner fuel–air mixtures above the minimum ignition temperature.

- Class 2 igniter – An igniter that is applied to ignite the fuel input to a burner and support ignition only under certain prescribed light-off conditions. It may also be used to support burner operations during operating conditions. The range of ignition energy is usually between 4% and 10% of full-load burner output at its associated burner.

- Class 3 igniter – An igniter that is applied to ignite the fuel input to a burner and support ignition only under certain prescribed light-off conditions. The range of ignition energy is usually less than 4% of full-load burner output at its associated burner.

- Class 3 special igniter – A special Class 3 high energy electrical igniter that is capable of directly igniting the main fuel burner.

Example 1.12   An igniter for a burner is chosen such that the range of ignition energy is usually between 4% and 10% of full-load burner output at its associated burner. What class of igniter is this?

Solution: This is a Class 2 igniter as defined in NFPA 85.

Once the FD and ID fans are started, the dampers and burner registers are adjusted to the purge position and airflow is adjusted to the "purge airflow" position and the unit purge is started. The purpose of the unit purge is to purge the air inside the boiler sufficiently to ensure that the level of combustibles is reduced to the point where an explosive environment does not exist and the burners can be safely lit off by the igniters. Once the purge is complete, the air register or damper on the burner is adjusted to the "light-off" position, the ignitor valve is opened, and the source of ignition for the igniter will be energized. If a flame is not established within 10 seconds, the igniter valve is closed and "light off" is not reattempted until the cause for the failure to ignite is determined and corrected. After a stable igniter flame is established in the combustion zone, the air register or damper is adjusted to its "operating" position.

The above discussion is a very brief overview of one type of boiler/burner system from NFPA 85. NFPA 85 is broken into several sections and has much more detailed information pertaining to the control of each type of equipment. Table 1.16 lists the various sections of NFPA 85 and to which type of boiler they apply.

The reader is encouraged to review in detail the requirements of the standard for their specific system.

In summary, the primary function of an energy production systems engineer is to ensure both safety and reliability in the systems and procedures utilized in a generation facility. Of these two functions, safety is paramount. This includes both safe installation of equipment and the utilization of safe work practices. The typical electric utility generation station may contain several areas of concern such as electrical hazards, combustible atmospheres and materials, confined space requirements, slip, trips, and falls. As the safety of plant operations is improved, the reliability of the station will benefit also.

TABLE 1.16   NFPA 85 Section Descriptions

Section Number	Description
8501	Single-burner boiler operations
8502	Multiple-burner boiler furnaces
8503	Pulverized fuel systems
8504	Atmospheric fluidized bed boiler operation
8505	Stoker operation
8506	Heat recovery steam generators

Source: Reproduced with permission from NFPA 85, *Boiler and Combustion Systems Hazards Code*, Copyright © 2015, National Fire Protection Association. This is not the complete and official position of the NFPA on the referenced subject, which is represented only by the standard in its entirety. The student may download a free copy of the NFPA 85 standard at: http://www.nfpa.org/codes-and-standards/document-information-pages?mode=code&code=85.

GLOSSARY OF TERMS

- Arc Rating – The value attributed to materials that describe their ability to withstand to an electrical arc discharge. The arc rating is expressed in cal/cm^2 and is derived from the determined value of the arc thermal performance value (ATPV) or energy of break-open threshold (EBT) (should a material system exhibit a break-open response below the ATPV value). Arc rating is reported as either ATPV or EBT, whichever is the lower value.

- Arc Thermal Performance Value (ATPV) – Max incident energy of a PPE system prior to break-open.

- Class 1 Igniter – An igniter that is applied to ignite the fuel input to a burner and support ignition under any burner conditions. It will provide sufficient ignition energy usually in excess of 10% of full-load burner output at its associated burner to raise any combination of burner fuel/air mixtures above the minimum ignition temperature.

- Class 2 Igniter – An igniter that is applied to ignite the fuel input to a burner and support ignition only under certain prescribed light-off conditions. It may also be used to support burner operations during operating conditions. The range of ignition energy is usually between 4% and 10% of full-load burner output at its associated burner.

- Class 3 Igniter – An igniter that is applied to ignite the fuel input to a burner and support ignition only under certain prescribed light-off conditions. The range of ignition energy is usually less than 4% of full-load burner output at its associated burner.

- Class 3 Special Igniter – A special Class 3 high energy electrical igniter that is capable of directly igniting the main fuel burner.

- Class I location where flammable gases or vapors may be present in air to an explosive level

- ○ Division 1 – Exists under normal operations or frequently or faulty equipment operation may cause simultaneous failure of electrical equipment.
- ○ Division 2 – Liquids/gases used, but not normally in explosive concentrations or concentration prevented by positive mechanical ventilation or is adjacent to Class I, Division 1.
- Class I Liquid – Any liquid that has a flash point below 100°F (37.8°C).
- Class IA Liquid – Any liquid that has a flash point below 73°F (22.8°C) and a boiling point below 100°F (37.8°C).
- Class IB Liquid – Any liquid that has a flash point below 73°F (22.8°C) and a boiling point at or above 100°F (37.8°C).
- Class IC Liquid – Any liquid that has a flash point at or above 73°F (22.8°C), but below 100°F (37.8°C).
- Class II location where combustible dust may be present
  - ○ Division 1 – Exists under normal operations or faulty equipment operation may cause simultaneous failure of elect equipment or where combustible dusts are eclectically conductive.
  - ○ Division 2 – Combustible dusts not normally in explosive concentrations or concentration may prevent cooling of electrical equipment.
- Class II Liquid – Any liquid that has a flash point at or above 100°F (37.8°C) and below 140°F (60°C).
- Class IIIA Liquid – Any liquid that has a flash point at or above 140°F (60°C) and below 200°F (93°C).
- Class III location where easily ignitable fibers may be present
  - ○ Division 1 – where fibers are produced handled, or used.
  - ○ Division 2 – areas where stored or handled other than manufacturing process.
- Class IIIB Liquid – Any liquid that has a flash point at or above 200°F (93°C)
- Exposed Live Part – Not suitably guarded, isolated, or insulated.
- Flame-Resistant (FR) – Combustion is prevented, terminated, or inhibited with or without the removal of the ignition source.
- Flash Hazard – Condition caused by the release of energy caused by an electric arc
- Flash Protection Boundary – Distance from live parts where second-degree burns would result if an arc flash occurred. By definition, this is the distance at which the incident energy level is at 1.2 cal/cm^2.
- Group A – Acetylene.
- Group B – Flammable gas or vapor that has either an MESG less than or equal to 0.45 mm or an MIC ratio of 0.40 or less. A typical Class I, Group B material is hydrogen.
- Group C – Flammable gas or vapor that has either an MESG greater than 0.45 mm and less than or equal to 0.75 mm or an MIC ratio greater than 0.40 and less than 0.80. A typical Class I, Group C material is ethylene.

- Group D – Flammable gas or vapor that has either an MESG greater than 0.75 mm or an MIC ratio greater than 0.80. A typical Class I, Group D material is Propane.

- Group E – Atmospheres containing combustible metal dusts. Typical dusts in this group are aluminum, magnesium, or other metallic alloys.

- Group F – Atmospheres containing combustible carbonaceous dusts that have more than 8% total trapped volatiles or that have been sensitized by other materials so that they present an explosion hazard. Typical dusts in this group are coal, carbon black, charcoal, and coke dusts.

- Group G – Atmospheres containing combustible dusts that do not fall into the Group E or Group F designation. Typical dusts in this group are flour, grain, wood, plastic, and other similar chemicals.

- Limited Approach Boundary – Distance from live parts where shock hazard exists.

- Restricted Approach Boundary – Distance beyond which only qualified persons may enter because it requires shock protection equipment or techniques.

- Working Near (Live Parts) – Any activity inside a limited approach boundary.

- Working On (Live Parts) – Coming in contact with live parts with body or equipment regardless of PPE (inside restricted approach boundary).

PROBLEMS

1.1 What are the four minimum steps to establish an electrically safe work condition?

   A. _____

   B. _____

   C. _____

   D. _____

1.2 What activities below constitute energized electrical work?

   A. Taking voltage readings using a contact voltage meter on a panel that has just been isolated from the source but not yet locked out

   B. Racking out circuit breaker from energized bus

   C. Removal of MCC buckets from energized bus

   D. All the above

1.3 A task has been assigned to remove a breaker bucket from a 480 V three-phase MCC where there is a bolted fault current of 35 kA and the clearing time is 5 cycles; can you use the tables in NFPA 70E for determination of the correct arc flash PPE for this task? If you can, what is the arc flash PPE level for this task?

1.4 A task has been assigned to remove a breaker from a 125 Vdc switchboard panel where there is a bolted fault current of 2 kA and the clearing time is 5 cycles; can you use the tables in NFPA 70E for determination of the correct arc flash PPE for this task? If you can, what is the arc flash PPE level for this task?

1.5   A task has been assigned to verify the phasing of a three-phase, 22 kV circuit where the bolted fault current is 15.1 kA and the clearing time is 5 cycles. What is the arc flash PPE level that is required for this task?

1.6   Given the following activity, using table 130.4(C) of NFPA 70E what is the limited approach boundary and the restricted approach boundary in feet and inches? The activity is racking out a three-phase, 480 V line-to-line, power circuit breaker with doors open from an energized switchgear bus; therefore, the exposed bus is fixed and not movable.

Limited approach boundary _____

Restricted approach boundary _____

1.7   Given the following activity, using table 130.4(D) of NFPA 70E what is the limited approach boundary and the restricted approach boundary in feet and inches? The activity is to perform voltage testing of a 250 Vdc panel board circuit breaker with doors open from an energized switchgear bus; therefore, the exposed bus is fixed and not movable.

Limited approach boundary _____

Restricted approach boundary _____

1.8   An area where flammable gases or vapors can exist under normal operating conditions is an example of

   A.   Class I, Division 1 location

   B.   Class I, Division 2 location

   C.   Class II, Division 1 location

   D.   Class II, Division 2 location

   E.   Class III, Division 1 location

   F.   Class III, Division 2 location

1.9   Using Table 1.1 in text for a situation where there is a 480 V MCC assembly across from a concrete wall on opposite side of switchgear, what is the minimum aisle space required between the assembly and the concrete wall?

   A.   3 ft

   B. $3\frac{1}{2}$ ft

   C.   4 ft

1.10   An energized electrical work permit is required to perform energized work. What are some of the items that should be listed in the energized electrical work permit? (Name at least 5)

_____

_____

_____

_____

_____

1.11 The motor controller that drives a coal pulverizer motor for a power generation facility has been deemed as a location where, under normal operations coal dust will not be present suspended in air in sufficient quantities to form an explosive concentrations and due to housekeeping controls, the layer of dust will not be sufficient to prevent cooling of the motor; however, on the failure of one of the multiple seals in the area, coal dust can become suspended in the atmosphere in quantities sufficient to form an explosive concentration should those seals fail. What are the correct class and division for this hazardous area classification?

RECOMMENDED READING

Bronze Book IEEE 739: IEEE Recommended Practice for Energy Management in Industrial and Commercial Facilities, Institute of Electrical and Electronics Engineers, 1996.

Brown Book IEEE 399: IEEE Recommended Practice for Industrial and Commercial Power Systems Analysis, Institute of Electrical and Electronics Engineers, 1998.

Buff Book IEEE 242: IEEE Recommended Practice for Protection and Coordination of Industrial and Commercial Power Systems, Institute of Electrical and Electronics Engineers, 2001.

Department of the Army, TM 5-682, Facilities Engineering, Electrical Facilities Safety, November 8, 1999.

Emerald Book IEEE 1100: IEEE Recommended Practice for Powering and Grounding Electronic Equipment, Institute of Electrical and Electronics Engineers, 2006.

Gold Book IEEE 493: IEEE Recommended Practice for the Design of Reliable Industrial and Commercial Power Systems, Institute of Electrical and Electronics Engineers, 2007.

Green Book IEEE 142: IEEE Recommended Practice for Grounding of Industrial and Commercial Power Systems, Institute of Electrical and Electronics Engineers, 2007.

IEC 61241: Electrical Apparatus for Use in the Presence of Combustible Dust, International Electro-Technical Commission, 1994.

IEC 60079:-Electrical Apparatus for Explosive Gas Atmospheres, International Electro-Technical Commission, 2011.

NFPA 101: Life Safety Code Handbook, National Fire Protection Association, 2012.

NFPA 20: Centrifugal Fire Pumps, National Fire Protection Association, 2013.

NFPA 70®: National Electrical Code®, National Fire Protection Association, 2014.

NFPA 70B: Electrical Equipment Maintenance, National Fire Protection Association, 2016.

NFPA 70E: Electrical Safety Requirements for Employee Workplaces, National Fire Protection Association, 2015.

NFPA 72: National Fire Alarm Code, National Fire Protection Association, 2016.

NFPA 75: Protection of Electronic Computer/Data Processing Equipment, National Fire Protection Association, 2013.

NFPA 85: Boiler and Combustion Systems Hazards Code, National Fire Protection Association, 2015.

NFPA 497: Recommended Practice for the Classification of Flammable Liquids, Gases, or Vapors and of Hazardous (Classified) Locations for Electrical Installations in Chemical Process Areas, National Fire Protection Association, 2008.

NFPA 499: Recommended Practice for the Classification of Combustible Dusts and of Hazardous (Classified) Locations for Electrical Installations in Chemical Process Areas, National Fire Protection Association, 2013.

Orange Book IEEE 446: IEEE Recommended Practice for Emergency and Standby Power Systems for Industrial and Commercial Applications, Institute of Electrical and Electronics Engineers, 1996.

Red Book IEEE 141: IEEE Recommended Practice for Electric Power Distribution for Industrial Plants, Institute of Electrical and Electronics Engineers, 1993.

CHAPTER 2

BASIC THERMAL CYCLES

GOALS

- To understand the basic Brayton and Rankine thermodynamic process
- To understand fundamental units of specific volume, specific heat, specific entropy, and specific enthalpy
- To understand the difference between latent and sensible heat transfer
- To understand the first and second law of thermodynamics
- To perform unit conversions of pressure and temperature for thermodynamic analysis
- To build a T-s diagram for a Rankine cycle
- To build a P-v diagram for a Brayton cycle
- To utilize ASME steam tables to perform thermodynamic analysis
- To calculate the net plant heat rate for a given thermodynamic process

STEAM THERMODYNAMIC ANALYSIS FUNDAMENTALS

A power plant is basically an energy conversion station. The terminology "power generation" when used to describe the function of a power plant is misleading. Power is the rate of change of energy or, described in another way, it is the rate at which energy is transferred. In physics, the law of conservation of energy states that energy is neither created nor destroyed but only altered in form. Therefore, since energy cannot be created or "generated," power cannot be "generated." Most power plants are converting some type of energy in the form of chemical energy stored in a fuel, or thermal energy from sunlight, or potential energy such as hydroelectric plants, or nuclear energy such as nuclear power plants, or kinetic energy contained in the wind or tidal motion, into electrical energy. Therefore, what is termed a "power generation station" is really an "energy conversion station."

Energy Production Systems Engineering: An Introduction for Electrical Engineers to Electrical Power Generation Facilities, Systems, and Equipment, First Edition. Thomas H. Blair.

Most methods of energy conversion involve the conversion of one form of energy to thermal energy, and generation stations use water and steam as a media to transmit this thermal energy from one part of the system to another where it is then converted to mechanical energy. This mechanical energy is used to drive a generator which converts the mechanical energy to electrical energy. Since water and steam are one of the most common methods used for the thermal part of this energy conversion process, a review of steam fundamentals and basic thermodynamic laws along with the typical types of equipment used in this process is essential.

Temperature is the measure of the average molecular kinetic energy of a material. Energy in thermal form is always transmitted from a higher temperature system to a lower temperature system. The transfer of thermal energy is known as heat. The study of this process is called *Thermodynamics*. There are several units of measurement of temperature. The most commonly encountered in daily life are Fahrenheit and Celsius (SI). However, for both of these units, there is an offset. Zero in the Fahrenheit and Celsius scales is not at absolute zero and both of these scales can have negative values. For engineering calculations, it is more common to use units of Rankine and Kelvin (SI) as these have a value of zero at absolute zero and therefore do not have an offset. Not having an offset allows us to use properties of linear equations to solve engineering thermodynamics problems using Rankine and Kelvin scales. Kelvin and Celsius are units in the International System of Units or in the French vernacular Système International d'Unités (SI). Rankine and Fahrenheit are known as United States customary units. In this chapter, we utilize the American Society of Mechanical Engineers (ASME) steam tables which list properties of water and steam in United States customary units; so the discussion of thermodynamics utilizes United States customary units. The units for Fahrenheit and Celsius are given in degrees and the symbols are °F and °C respectively. The units for Kelvin are not given in degrees and the symbol is simply K. Rankine can be expressed either way depending on the author, but in this text, we will express Rankine in degrees Rankine with the symbol °R.

To convert from the Fahrenheit scale to the Rankine scale, use the following formula.

$$T_R = T_F + 460 \tag{2.1}$$

where

T_R = temperature measurement (Rankine)

T_F = temperature measurement (Fahrenheit)

To convert from the Celsius to the Kelvin scale, use the following formula.

$$T_K = T_C + 273 \tag{2.2}$$

where

T_K = temperature measurement (Kelvin)

T_C = temperature measurement (Celsius)

To convert from the Fahrenheit scale to the Celsius scale, use the following formula.

$$T_F = (9/5) \times T_C + 32 \tag{2.3}$$

where

T_C = temperature measurement (Celsius)

T_F = temperature measurement (Fahrenheit)

To convert from the Celsius scale to the Fahrenheit scale, use the following formula.

$$T_C = (5/9) \times (T_F - 32) \tag{2.4}$$

where

T_C = temperature measurement (Celsius)

T_F = temperature measurement (Fahrenheit)

To convert from the Kelvin scale to the Rankine scale, use the following formula

$$T_R = (9/5) \times T_K \tag{2.5}$$

where

T_K = temperature measurement (Kelvin)

T_R = temperature measurement (Rankine)

To convert from the Rankine scale to the Kelvin scale, use the following formula.

$$T_K = (5/9) \times T_R \tag{2.6}$$

where

T_K = temperature measurement (Kelvin)

T_R = temperature measurement (Rankine)

Example 2.1   What is the Rankine equivalent of 80°C?

Solution:

$T_F = (9/5) \times T_C + 32$

$T_F = (9/5)\,(80) + 32$

$T_F = (9/5)\,(80) + 32$

$T_F = 176°F$

$T_R = T_F + 460$

$T_R = 176 + 460$

$T_R = 636°R$

Example 2.2 What is the Kelvin equivalent of 80°F?

Solution:

$$T_C = (5/9) \times (T_F - 32)$$
$$T_C = (5/9)\,(80 - 32)$$
$$T_C = 26.7°C$$
$$T_K = T_C + 273$$
$$T_K = 26.7 + 273$$
$$T_K = 299.7\ K$$

Just like temperature, pressure which is a measure of force per unit area, has various units of measure and not all start at absolute zero. Unit values such as % vacuum, torr, micron, pounds per square inch (psi) (absolute or gauge), inches of mercury (absolute or gauge), and pascal may be used. The SI unit for pressure is pascal (Pa) or micron (1 million microns are equal to 1 pascal). One pascal is equal to one newton of force applied to one square meter of surface area. Units such as torr, micron, psia, inches of mercury absolute, and pascal all have their zero point at absolute vacuum. Units such as psig, % vacuum, and inches of mercury gauge all have their zero point at standard atmospheric pressure (14.7 psia at sea level). The units of PSI are units of pound force over the surface area that the force is distributed over. A torr is equal to the displacement of a millimeter of mercury (mm Hg) in a manometer. One inch of mercury (Hg) is defined as the pressure exerted by a circular column of mercury of 1 inch in height at 32°F (0°C). Percent vacuum is defined as a measurement in percentage of atmospheric pressure where 100% vacuum is absolute zero pressure and 0% vacuum is the standard value of atmospheric pressure under standard temperature and pressure conditions. The bar is a measurement of pressure where zero bar is at 100% vacuum and 1 bar is the standard value of atmospheric pressure under standard temperature and pressure conditions.

In the study of thermodynamics, we select a unit of measurement that has a zero value at absolute zero pressure or 100% vacuum to remove the difficulty of dealing with an offset in our calculations just as we do for units of temperature. For the SI system, the most commonly used units that are zero at absolute zero pressure are pascal for pressures above atmospheric and microns for pressures below atmospheric. The most commonly used United States customary units for pressure that has a zero value at absolute zero pressure are pounds force per square inch absolute (psia) for pressures above atmospheric or in some cases % vacuum for pressures below atmospheric. Units of pounds force should not be confused with pounds mass. Pounds force is a measurement of force whereas pounds mass is a measurement of the mass of material where the mass is measured by the gravitational force that the earth exerts on the mass at sea level. All thermodynamic calculations performed in United States customary units are performed using pounds force (psi).

In this chapter, we utilize the ASME steam tables which list properties of water and steam in United States customary units, so the discussion of thermodynamics in this chapter utilizes the United States customary units of psia.

To convert from vacuum (%) to torr (mm Hg), use the following formula.

$$P \text{ (torr)} = (-760/100) \times P \text{ (vacuum)} + 760 \tag{2.7}$$

To convert from vacuum (%) to micron, use the following formula.

$$P \text{ (micron)} = (-760,000/100) \times P \text{ (vacuum)} + 760,000 \tag{2.8}$$

To convert from vacuum (%) to psia, use the following formula.

$$P \text{ (psia)} = (-14.7/100) \times P \text{ (vacuum)} + 14.7 \tag{2.9}$$

To convert from vacuum (%) to inches mercury absolute, use the following formula.

$$P \text{ (inches mercury absolute)} = (-29.92/100) \times P \text{ (vacuum)} + 29.92 \tag{2.10}$$

To convert from vacuum (%) to inches mercury gauge, use the following formula.

$$P \text{ (inches mercury gauge)} = (29.92/100) \times P \text{ (vacuum)} \tag{2.11}$$

To convert from vacuum (%) to kilopascal, use the following formula.

$$P \text{ (kPa)} = (-101.4/100) \times P \text{ (vacuum)} + 101.4 \tag{2.12}$$

Table 2.1 summarizes these conversions and provides a cross-reference between various systems for pressures between absolute vacuum and atmospheric

TABLE 2.1   Pressure Conversion Chart

% Vacuum	torr (mm Hg)	Micron	psia, (lb/in.^{2})	Inches Mercury Absolute	Inches Mercury Gauge	kPa (absolute)
0	760	760,000	14.7	29.92	0	101.4
1.3	750	750,000	14.5	29.5	0.42	99.9
1.9	735.6	735,600	14.2	28.9	1.02	97.7
7.9	700	700,000	13.5	27.6	2.32	93.5
21	600	600,000	11.6	23.6	6.32	79.9
34	500	500,000	9.7	19.7	10.22	66.7
47	400	400,000	7.7	15.7	14.22	53.2
50	380	380,000	7.3	15	14.92	50.8
61	300	300,000	5.8	11.8	18.12	40
74	200	200,000	3.9	7.85	22.07	26.6
87	100	100,000	1.93	3.94	25.98	13.3
89.5	80	80,000	1.55	3.15	26.77	10.7
93	51.7	51,700	1	2.03	27.89	6.9
96.1	30	30,000	0.58	1.18	28.74	4
97.4	20	20,000	0.39	0.785	29.14	2.7
98.7	10	10,000	0.193	0.394	29.53	1.3
99	7.6	7,600	0.147	0.299	29.62	1
100	0	0	0	0	29.92	0

pressure. Just as in temperature, not having an offset in our pressure measurements allows us to use properties of linear equations to solve engineering thermodynamics problems and the units of psia are utilized for thermodynamic analysis.

For example, from Table 2.1, a pressure of 1 psia equates to a pressure of 2.03 inches of mercury absolute or 93% vacuum.

Another physical property of a material is *density*. Density is a measure of the mass of a material that is in a specific volume of the material. The unit for density is pound mass per unit volume (lbm/ft^3). Density is a function of the properties of the material as well as the pressure and temperature of the material.

Specific volume is the reciprocal of density and is a measure of the volume that a material takes up for a specific amount of mass of the material. The units for specific volume are volume per unit pound mass (ft^3/lbm). Specific volume and density are reciprocals of each other. Just like density, specific volume is a function of the properties of the material as well as the pressure and temperature of the material.

Specific heat is a measure of the amount of thermal energy that is transferred in the form of heat (measured by the change in temperature) per unit mass.

In an electric utility generation station, the goal is to convert energy from one form to another, until we achieve the final energy conversion to electrical energy. When energy is stored in the form thermal energy, materials such as water, oil, and salt solutions are used as a transfer media to transfer this thermal energy from one system to another. By far, water is the most common media utilized for this energy transfer. Thermal energy transfer occurs in steam in two distinctly different heat transfer processes. *Sensible heat transfer* is a process when energy is transferred and the transfer of energy results in a change in the temperature of the material, but the state of the material (i.e., solid, liquid, or gas) remains the same. An example of sensible heat transfer is placing a pot of water that is at room temperature on a stove, placing a thermometer in the water and turning the burner on. Initially, the water remains in the liquid phase and, as energy is added to the water from the burner, the temperature of the water increases as indicated on the thermometer. *Latent heat transfer* is a process when energy is transferred and the transfer of energy results in a change in the state of the material (i.e., change from liquid to gas), but the temperature remains constant. An example of latent heat transfer is seen in the same pot of water, which is now at the boiling temperature of water for the value of atmospheric pressure. As energy is added to the pot of water, the temperature of the water as indicated on the thermometer does not change, but now the water changes from a liquid state to a vapor state. This occurs at the boundary where the water is receiving the energy from the burner (at the bottom of the pot) and the water in vapor form rises in bubbles.

In regard to the path for thermal energy transfer, there are three methods that energy can be transferred: *conduction, convection*, and *radiation*. Conduction is the transfer of energy through matter from particle to particle. It is the transfer and distribution of heat energy from atom to atom within a substance. Conduction is most effective in solids, but it can happen in fluids. Convection is the transfer of heat by the actual movement of the warmed matter. Convection is the transfer of heat energy in a gas or liquid by movement of currents of the transfer media. Radiation is a process where electromagnetic waves directly transport energy through space (no media is needed for heat transfer).

Now that we have some of the fundamental terms of the study of thermodynamics identified, we can start discussing the thermal energy transfer process in the steam cycle and the associated units for this process. The measure of the amount of useful thermal energy in a material is known as *Enthalpy* (*h*). It is a measure of the thermal energy per unit mass and its units are BTU/lbm (kJ/kg). A BTU is the amount of heat required to increase the temperature of 16 ounces of water by 1°F. This includes both flow energy (kinetic energy) and internal energy (potential energy) as shown by equation (2.13).

$$h = (u + p \times v) \tag{2.13}$$

where

h = enthalpy

u = internal energy

p = pressure of the system

v = volume of the system

The value of enthalpy is determined by the physical properties of the material (i.e., pressure, temperature, etc.). In practice, we do not use equation (2.13) to calculate the value of enthalpy, but values for different transfer medias are given in empirically derived tables or databases. The ASME publishes steam tables that can be used to determine the enthalpy of steam for a given set of physical conditions. To determine the total amount of useful energy (enthalpy) in a given amount of material, you multiply the value of the enthalpy by the mass of the material as shown in equation (2.14).

$$E = h \times m \tag{2.14}$$

where

E = energy (BTU)

h = enthalpy (BTU/lbm)

m = mass (lbm)

In the power generation industry, we are more concerned with the power flow than the energy stored. Power is the rate of energy transfer. To determine the rate of energy transfer (power) through a material, you multiply the value of the enthalpy by the mass flow rate of the material as shown in equation (2.15).

$$E/t = P = h \times F \tag{2.15}$$

where

E = energy (BTU)

t = time (hr)

P = power (BTU/hr)

h = enthalpy (BTU/lbm)

F = mass flow rate (lbm/hr)

Example 2.3   If we have 100 lbm of water and its enthalpy value is 68 BTU/lbm, determine the total amount of energy in the material.

Solution: Using equation (2.14),

$$E = h \times m$$

$$E = 68 \text{ BTU/lbm} \times 100 \text{ lbm} = 6800 \text{ BTU}$$

In the study of thermodynamics for power plant *efficiency* and operation, enthalpy (h) is the property used as this is a measure of the amount of useful energy. These calculations are known as *heat rate* calculations. The fact that we define a term for useful energy implies there is also energy in a material that is not useful (i.e., cannot be utilized in the energy conversion process of the plant). This is known as entropy (s). Entropy is a measure of the amount of energy in a material that is not available for work. The classic definition of work is the energy transferred by force acting through a distance. Therefore, entropy (s) is a measure of the amount of energy in a material that is not available to be transferred (i.e., does no work). The units for entropy are (BTU/(lbm × °R)).

The enthalpy values for water for various pressures and temperatures are well known and there are several types of charts and software that allow the user to determine the materials' enthalpy for certain physical properties. For our use, we will use the T-s diagram (a plot of temperature (T) versus entropy (s)). The vertical axis is temperature (in Rankine so that zero on the plot is absolute zero in temperature). The horizontal axis is the value of entropy (s). Figure 2.1 shows the *T-s* diagram for water. There are three distinct phase regions shown in this diagram. The area to the left of the saturated steam region is known as the liquid region or subcooled water region. In this region, water is in the liquid phase. The area to the right of the saturated

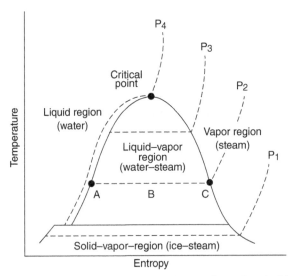

Figure 2.1   T-s diagram for water. *Source*: Reproduced with permission of U.S. Department of Energy.

steam region is known as the vapor region or superheated steam region. In this region, water is entirely in the vapor phase. The area between the subcooled water region and the superheated steam region is known as the liquid-vapor region or saturated steam region. In this region, water is partially in the form of water and partially in the form of steam.

Referencing Figure 2.1, we can see some interesting properties. The lines P_1, P_2, and P_3 are lines of constant pressure. Looking at pressure P_2, at very low temperature, the material is in the subcooled or water region. Any addition of energy will raise the value of enthalpy of the material (measure of energy per unit mass) and the temperature of the material. This process where the addition of energy causes the temperature of the substance to increase but the phase remains constant is known as sensible heat transfer. The addition of energy will raise temperature of the material until the "saturation" temperature is reached. This is the temperature at point A in Figure 2.1. At this point, the material is still 100% liquid. This is also known as 0% steam quality as the percentage of the mass of water and steam mixture that is in the form of steam is 0%. Any additional energy transfer into the water at this point will cause us to move from point A toward point B. Notice that this additional heat energy transfer into the media does not result in a higher temperature of the material, but does move us into the liquid/vapor region where a combination of water and steam exist. As more energy is transferred into the system, more of the water is converted to steam. This is known as latent heat transfer. When enough energy has been added to the system, the material reaches a point where all of the material is in the form of steam. This is point C on Figure 2.1. This is known as 100% steam quality as the percentage of the mass of water and steam that is in the form of steam is 100%.

As we will see later when we look at specific values of enthalpy, the amount of heat added to get the material from point A to point C is substantial. Since our goal is to transfer heat, most thermal processes concentrate their heat transfer using latent heat transfer to maximize the amount of energy transferred.

At point C, all of the water is converted to steam. This is also known as saturated steam since any removal of energy from the steam will not cause a reduction in temperature but will cause some of the steam to condense into water; the magnitude of how much would depend on the mass of steam and the amount of energy removed from the system. From point C, any additional energy transferred into the system will result in an increase in temperature but the phase of the material remains steam. This is known as sensible heat transfer. This will move us to the right of the saturated steam line to the superheated steam region.

In summary, in the subcooled region and the superheat region, the type of heat transfer that occurs is sensible heat transfer where the phase of the media remains constant, and the temperature changes. In the saturated steam region, the type of heat transfer that occurs is latent heat transfer where the temperature of the media remains constant, and the phase of the media changes.

At very low temperatures, the material enters the solid region. In normal power plant processes, we wish to use convection (i.e., material flow) as part of our heat transfer process. Therefore, having the material in solid form is not very useful and we will not spend any time on the properties of the material in a solid phase. Also, take note that at pressure P_4, the transfer from liquid to gas is at the same value of enthalpy. Due to material limitations, most power plant thermal processes do not operate at these high pressures. Therefore, we will not discuss material behavior at or above point P_4, but realize that behavior above this pressure does not contain the saturated region.

While the T-s diagram is a useful tool to understand how the physical properties of water are affected by the addition and/or removal of energy from the media, engineering calculations utilize tables or software databases for determination of steam properties. The ASME publishes steam tables that provide various properties of water and steam based on the physical condition of the material. Below is a discussion on how to use the steam tables to perform some of these basic calculations. Please reference ASME steam tables for the saturated steam–temperature table, the saturated steam–pressure table, and the superheated steam table. Next, we will describe how to utilize the steam tables to perform basic thermodynamics calculations.

Table 1 from the ASME steam tables is a listing of various properties of saturated steam at various temperatures. Remember from the T-s diagram in Figure 2.1 that, for saturated steam, the temperature remains constant from the area where all the steam is condensed to water (0% quality steam) to the area where all the steam is in gaseous form (100% quality steam). In this table, the subscript "l" denotes the property at a condition of 0% quality steam (or 100% liquid) and the subscript "v" denotes the property at a condition of 100% quality steam (or 100% vapor). The following is a description of the various properties listed in Table 2.2.

- Temp = temperature of the saturated steam in °F
- Pressure = absolute pressure of the saturated steam in psia
- v_l = specific volume (ft^3/lbm) when the steam is at 0% quality
- v_v = specific volume (ft^3/lbm) when the steam is at 100% quality

Table 2.3 is a continuation of the same row of information from Table 2.2. The following is a continuation of the description of the various properties listed in Table 2.3.

TABLE 2.2 Saturated Steam: Temperature Table, Part 1

Temp. (°F)	Pressure (psia)	Volume (ft^3/lb$_m$)	
		v_l	v_v
32	0.08865	0.016022	3302.0
35	0.09998	0.016020	2945.5
40	0.12173	0.016020	2443.4
45	0.14757	0.016021	2035.6
50	0.17813	0.016024	1702.9
55	0.21414	0.016029	1430.3
60	0.25639	0.016035	1206.1
65	0.30579	0.016043	1020.8
70	0.36334	0.016052	867.19
75	0.43015	0.016062	739.30

Source: Reproduced with permission of American Society of Mechanical Engineers (ASME)

TABLE 2.3   Saturated Steam: Temperature Table, Part 2

Enthalpy, BTU/lb$_m$		Entropy, BTU/(lb$_m$·°R)		Temp.
h_l	h_v	s_l	s_v	°F
−0.018	1075.2	0.0000	2.1868	32
3.004	1076.5	0.0061	2.1762	35
8.032	1078.7	0.0162	2.1590	40
13.052	1080.9	0.0262	2.1421	45
18.066	1083.1	0.0361	2.1257	50
23.074	1085.3	0.0459	2.1097	55
28.079	1087.4	0.0555	2.0941	60
33.080	1089.6	0.0651	2.0788	65
38.078	1091.8	0.0746	2.0640	70
43.074	1094.0	0.0840	2.0495	75

Source: Reproduced with permission of American Society of Mechanical Engineers (ASME).

- h_1 = enthalpy (BTU/lbm) when the steam is at 0% quality
- h_v = enthalpy (BTU/lbm) when the steam is at 100% quality
- s_1 = entropy (BTU/(lbm × °R)) when the steam is at 0% quality
- s_v = entropy (BTU/(lbm × °R)) when the steam is at 100% quality

As we mentioned earlier, latent heat transfer is a process when energy is transferred and this results in a change in the state of the material and temperature remains constant. Now that we know the state of a material when it is at a steam quality of 0% and 100%, we can find the change in the state of the material as it evaporates (changes steam quality from 0% to 100%) or as it condenses (changes steam quality from 100% to 0%). This is defined in equations (2.16), (2.17), and (2.18) for specific volume, enthalpy, and entropy.

$$v_{lv} = (v_v - v_l) \qquad (2.16)$$

where

v_{lv} = change in specific volume (ft^3/lbm) from 0% to 100% steam quality
v_l = specific volume (ft^3/lbm) when the steam is at 0% quality
v_v = specific volume (ft^3/lbm) when the steam is at 100% quality

$$h_{lv} = (h_v - h_l) \qquad (2.17)$$

where

h_{lv} = change in enthalpy (BTU/lbm) from 0% to 100% steam quality
h_l = enthalpy (BTU/lbm) when the steam is at 0% quality
h_v = enthalpy (BTU/lbm) when the steam is at 100% quality

$$s_{lv} = (s_v - s_l) \qquad (2.18)$$

where

s_{lv} = change in entropy (BTU/(lbm × °R)) from 0% to 100% steam quality

s_l = entropy (BTU/(lbm × °R)) when the steam is at 0% quality

s_v = entropy (BTU/(lbm × °R)) when the steam is at 100% quality

While specific volume does not affect efficiency calculations, it is interesting to note that, as the energy of (or enthalpy of) a saturated liquid/steam mixture is increased, more of the liquid is converted to steam and the specific volume increases substantially. In other words, as energy is added to a system, the density is reduced. This is an important driving force for *natural circulation* which we will address later in the topic of boilers and nuclear reactors. Change in specific volume is also the driving force that causes boiler drum swell and shrink during load transients as will be discussed later in this text book.

In energy production systems engineering, we are most interested in the transfer of useful energy. Therefore, we are not commonly concerned with specific volume or entropy but are most interested in the values of enthalpy (useful energy per unit mass) at various conditions. Therefore, most of our work with the steam tables will involve determining the enthalpy of the steam and water media at various pressures, temperatures, and states.

The values in table 1 of the ASME steam tables directly give us the enthalpy (energy per unit mass) of saturated steam at 0% quality (h_l) and at 100% quality (h_v). However, what if we wanted to know the value of enthalpy of the saturated steam when it is somewhere between 0% and 100% steam quality? Equation (2.19) describes how to calculate the value for enthalpy of the saturated steam when it is somewhere between 0% and 100% steam quality.

$$h = h_l + h_{lv} \times (\%SQ/100\%) \qquad (2.19)$$

where

h = enthalpy of material at existing state (BTU/lbm)

h_l = enthalpy (BTU/lbm) when the steam is at 0% quality

h_{lv} = change in enthalpy (BTU/lbm) from 0% to 100% steam quality

$\%SQ$ = steam quality of the material (%)

Example 2.4   If we have saturated steam water mixture that is at 50% quality and at a temperature of 45°F, what is the enthalpy of the steam at a steam quality of 50%?

Solution: We need to find the values of enthalpy of the steam at 0% quality (h_l) and the value of the change in enthalpy of the steam as it changed from 0% quality to 100% quality (h_{lv}). From Table 2.3, the value for saturated steam at 45°F and 0% quality is 13.052 BTU/lbm and the value for saturated steam at 45°F and 100% quality is 1080.9 BTU/lbm. Next, we need to define the change in enthalpy from 0% quality steam to 100% quality steam. This is achieved using equation (2.17).

$h_{lv} = (h_v - h_l)$

$h_{lv} = (1080.9 - 13.052)$ BTU/lbm

$h_{lv} = 1067.848$ BTU/lbm

TABLE 2.4   Saturated Steam: Temperature Table, Part 3

Temp. (°F)	Pressure (psia)	Enthalpy (BTU/lb$_m$) h_l	h_v
205	12.782	173.13	1147.6
210	14.136	178.17	1149.5
215	15.606	183.20	1151.4
220	17.201	188.25	1153.3
225	18.928	193.30	1155.1
230	20.795	198.35	1157.0
235	22.811	203.41	1158.8
240	24.985	208.47	1160.5
245	27.326	213.54	1162.3
250	29.843	218.62	1164.0

Source: Reproduced with permission of American Society of Mechanical Engineers (ASME).

Now we can use equation (2.19) to define the enthalpy of the steam water mixture at a 50% steam quality.

$$h = h_l + h_{lv} \times (\%SQ/100\%)$$
$$h = 13.052 + 1080.9 \times (50\% / 100\%) \text{ BTU/lbm}$$
$$h = 546.976 \text{ BTU/lbm}$$

Example 2.5   If we have saturated water (implies 0% quality) at a temperature of 210°F), determine the following.

A. What is the enthalpy for a sample with 0% steam quality?
B. What is enthalpy for sample with 100% steam quality?
C. What is the change in enthalpy as the sample changes from 0% to 100% SQ?
D. What is enthalpy for sample with 50% steam quality?
E. What is pressure for above?

Solution:

A. From Table 2.4, $h_l = 178.17$ (BTU/lbm)
B. From Table 2.3, $h_v = 1149.5$ (BTU/lbm)
C. Using equation (2.17),
$$h_{lv} = (h_v - h_l)$$
$$h_{lv} = (1149.5 - 178.170) \text{ BTU/lbm}$$
$$h_{lv} = 971.33 \text{ BTU/lbm}$$
D. Using equation (2.19),
$$h = h_l + h_{lv} \times (\%SQ/100\%)$$
$$h = 178.17 + 971.33 \times (50\% / 100\%) \text{ BTU/lbm}$$
$$h = 663.835 \text{ BTU/lbm}$$
E. From Table 2.3, pressure = 14.136 psia

TABLE 2.5   Saturated Steam: Pressure Table

Pressure (psia)	Temp. (°F)	Enthalpy (BTU/lb$_m$) h_l	h_v
0.1	35.00	3.009	1076.5
0.2	53.13	21.204	1084.4
0.3	64.45	32.532	1089.4
0.5	79.55	47.618	1095.9
0.7	90.05	58.100	1100.4
1.0	101.69	69.728	1105.4
1.5	115.64	83.650	1111.4
2.0	126.03	94.019	1115.8
3.0	141.42	109.39	1122.2
4.0	152.91	120.89	1126.9
6	170.00	137.99	1133.9
8	182.81	150.83	1139.0
10	193.16	161.22	1143.1
12	201.91	170.02	1146.4
14	209.52	177.68	1149.4

Source: Reproduced with permission of American Society of Mechanical Engineers (ASME).

The steam tables also provide the same data but referenced to a given pressure instead of a given temperature. This is shown in table 2 of the ASME steam tables. Table 2.5 is a sample of this data.

Example 2.6   If we have saturated water (implies 0% quality at a pressure of 14 psia), determine the following.

A. What is the enthalpy for a sample with 0% steam quality?
B. What is enthalpy for sample with 100% steam quality?
C. What is the change in enthalpy as the sample changes from 0% to 100% SQ?
D. What is enthalpy for sample with 50% steam quality?
E. What is temperature for above?

Solution:

A. From Table 2.4, $h_l = 177.678$ (BTU/lbm)
B. From Table 2.3, $h_v = 1149.4$ (BTU/lbm)
C. Using equation (2.17),
$$h_{lv} = (h_v - h_l)$$
$$h_{lv} = (1149.4 - 177.678)\ \text{BTU/lbm}$$
$$h_{lv} = 971.722\ \text{BTU/lbm}$$
D. Using equation (2.19),
$$h = h_l + h_{lv} \times (\%SQ/100\%)$$
$$h = 177.678 + 971.722 \times (50\% / 100\%)\ \text{BTU/lbm}$$
$$h = 663.539\ \text{BTU/lbm}$$
E. From Table 2.3, temperature $= 209.52°F$

TABLE 2.6   Superheated Steam, Part 1

Pressure, psia (Sat. T)		Temperature, °F				
		200	250	300	350	400
1	V	392.53	422.42	452.28	482.11	511.93
(101.69)	h	1150.1	1172.8	1195.7	1218.6	1241.8
	s	2.0510	2.0842	2.1152	2.1445	2.1723

Source: Reproduced with permission of American Society of Mechanical Engineers (ASME).

ASME Tables 1 and 2 provide data for the saturated region of the T-s diagram, but how about the subcooled region and the super heater region? The ASME tables do not address the subcooled area. There are various database programs available to calculate these values. The ASME tables cover the superheated region in table 3 of the steam tables. In Table 2.6, we show a small sample of data from table 3 of the ASME steam tables.

To utilize the superheated steam tables, both the pressure and temperature of the superheated steam must be known. Given a certain pressure and temperature, the values of specific volume (v), entropy (s), and enthalpy (h) are provided by the tables. Superheat is defined as the difference between the measured temperature of steam at a given pressure and the saturation temperature of the steam at the same pressure.

Example 2.7 If we measure steam pressure and temperature and find that we have superheated steam at a temperature of 400°F at atmospheric pressure (15 psia), determine the following.

A. What is the enthalpy?

B. What is the amount of "superheat?"

Solution:

A. From Table 2.7, enthalpy (h) is 1239.9 (BTU/lbm)

B. Superheat is the difference between the measured temperature of steam at a given pressure and the saturation temperature of the steam at the same pressure. The measured temperature is given in the problem as 400°F. On the left of the ASME table 3 (shown in Table 2.7) under the pressure of 15 psia, we see the

TABLE 2.7   Superheated Steam, Part 2

Pressure, psia (Sat. T)		Temperature, °F				
	—	200	250	300	350	400
15	V		27.846	29.906	31.943	33.966
(212.99)	h		1168.7	1192.7	1216.3	1239.9
	S		1.7811	1.8137	1.8438	1.8721

Source: Reproduced with permission of American Society of Mechanical Engineers (ASME).

saturation temperature is 212.99°F. The amount of superheat is the difference between these two values, or

$$SH = 400°F - 212.99°F$$
$$SH = 187.1°F$$

Example 2.8 If we measure the temperature of superheated steam and find it is 350°F at atmospheric pressure (15 psia), determine the amount of energy required to raise 2 lbm of steam from 350°F to 400°F?

Solution: Using equation (2.14),

$$E = h \times m$$

From Table 2.7, the value of enthalpy at 400°F and the value of enthalpy at 350°F can be found.

$$E = (1239.9 - 1216.3) \, (BTU/lbm) \times 2(lbm)$$
$$E = 47.2 \, BTU$$

The reader may be wondering at this point why we are spending so much time understanding how to calculate enthalpy. Enthalpy is the basis for thermodynamic energy transfer calculations and, ultimately, for calculations of thermal cycle efficiency as the next example will show.

Example 2.9 If we have superheated steam at a temperature of 400°F and at atmospheric pressure (14.7 psia) that is flowing into a turbine at a rate of 100,000 lbm/hr, determine the ideal rate of energy delivery (power) into the turbine.

Solution: From Table 2.7, the value of enthalpy at 400°F can be found.

$$h = 1239.9 \, BTU/lbm$$

Using equation (2.15),

$$E/t = P = h \times F$$
$$P = (1239.9 \, BTU/lbm) \times (100,000 \, (lbm/hr))$$
$$P = 123.99 \, MBTU/hr$$

where 1 MBTU = 1 million BTU. In some literature, 1 million BTU is also expressed as MMBTU. This is derived from the Roman numeral system where the letter M represents one thousand. Using this notation, 1 MMBTU is 1000 × 1000 × 1 BTU or 1 million BTU.

In power plant operations, we are commonly more interested in the amount of power flow into and out of systems than the amount of energy in a mass of steam. We can calculate the power flow by multiplying the enthalpy of the material by the mass flow rate of the material into or out of a system as shown in equation (2.15).

Knowing the power into a system and the power out of a system allows us to calculate the efficiency of the system using the following formula.

$$\text{Eff} = P_{\text{out}}/P_{\text{in}} \times 100\%, \tag{2.20}$$

where

$\text{Eff} = \text{efficiency} (\%)$

$P_{\text{out}} = \text{power out of a system (BTU/hr)}$

$P_{\text{in}} = \text{power into a system (BTU/hr)}$

This is why enthalpy is such an important part of energy production engineering as it allows us to determine the efficiency of the energy transfer process at various points in the thermodynamic system. This includes both steam-based systems such as boilers and heat exchanges as well as combustion systems where the combustion process adds to the enthalpy of the air which is the media for transportation of thermal energy in combustion processes. The T-s diagram of an air system appears different from the T-s diagram of the steam system, but the same calculations using enthalpy and mass flow rate apply to the determination of power flow and efficiencies.

Example 2.10   For the system described in Example 2.9, if the turbine exhaust steam is at 1 psia, saturated steam conditions, and a steam quality of 90%, what is net power delivered to the turbine by the steam? (Net power of the turbine is power into the turbine from the boiler less the power removed from the turbine by the condensate system.)

Solution: From Example 2.9, enthalpy into turbine was (1239.9) (BTU/lbm) From Table 2.6, the value of the enthalpy of 0% steam quality at 1 psia is found to be 69.728 BTU/lbm and the change in the enthalpy of steam at 1 psia as it changes phase from 0% steam quality to 100% steam quality is (1105.4 BTU/lbm – 69.728 BTU/lbm) or 1035.672 BTU/lbm.

Using equation (2.19), we can find the value of saturated steam at 1 psia and 90% steam quality as,

$h = h_1 + h_{\text{lv}} \times (\%\text{SQ}/100\%)$

$h\,(90\%) = 69.728 \text{ BTU/lbm} + 1035.672 \text{ BTU/lbm} \times (90/100)$

$h\,(90\%) = 69.73 \text{ BTU/lbm} + 932.1048 \text{ BTU/lbm}$

$h\,(90\%) = 1001.833 \text{ (BTU/lbm)} = \text{enthalpy out of turbine}$

Net enthalpy drop is 1239.9 BTU/lbm – 1001.833 BTU/lbm

$= 238.0672 \text{ BTU/lbm}.$

Given the mass flow is 100,000 lbm/hr from Example 2.9, net power delivered to turbine can be calculated using equation (2.15)

$E/t = P = h \times F$

$P_{\text{out}} = 238.0672 \text{ BTU/lbm} \times 100,000 \text{ lbm/hr}$

$P_{\text{out}} = 23.8 \text{ MBTU/hr}$

So from Example 2.9, power into the turbine is 123.99 MBTU/hr and the mechanical power out of the shaft of the turbine is 23.8 MBTU/hr as found in Example 2.10. We can calculate the efficiency of this system using equation (2.20),

$$Eff = P_{out}/P_{in} \times 100\%$$
$$Eff = (23.8 \text{ MBTU/hr} /123.99 \text{ MBTU/hr}) \times 100\%$$
$$Eff = 19.2\% \text{ efficient}$$

A power plant that operates at an efficiency of 19% would not stay in business very long. To increase efficiency of the system, we raise the temperature and pressure of the steam entering the steam turbine. This has the effect of raising the enthalpy of the steam entering the turbine which increases the value of P_{in} for a certain mass flow rate. We also draw a vacuum in the condenser and condense the steam at a lower pressure (and therefore temperature). This has the effect of lowering the enthalpy of the steam leaving the turbine which decreases the value of the power entering the condenser for a certain mass flow rate. By raising the value of Pin and lowering the value of Pout, this has the effect of increasing efficiency of the turbine. This is why we operate boilers at very high pressures and we operate condensers at very low vacuum.

PRESSURE, TEMPERATURE, AND VOLUME RELATIONSHIPS

The ideal gas law or general gas equation states that the pressure and volume product is proportional to the temperature of the ideal gas. While steam, being compressible, is not an ideal gas, the proportionality is still a good approximation for the physical response for steam in a vessel. Under this assumption, the reaction of steam to changes in pressure, temperature, and volume can be described the following equation.

$$PV = nRT \text{ or } PV/T = \text{constant} \tag{2.21}$$

Therefore

$$P_1 \times V_1/T_1 = P_2 \times V_2/T_2 \tag{2.22}$$

The result of the above proportionality is that with a constant volume, with an increase in steam temperature, steam pressure will rise. Also, if temperature is held constant, an increase in steam pressure will decrease steam volume. This relationship is known as *Boyle's law*. If pressure is held constant, an increasing steam temperature will cause an increase in steam volume. This relationship is known as *Charles' law*.

With the discussion behind us of enthalpy and associated calculations of energy flow and component efficiencies, we can now address the topic of system efficiencies.

First, when looking at the overall system, it is helpful to remember two fundamental laws of thermodynamics.

First law of thermodynamics – Energy is neither created nor destroyed. Energy is only altered in form.

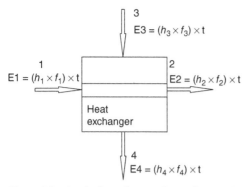

Figure 2.2   Basic flow diagram for equipment thermodynamic performance evaluation.

Second law of thermodynamics – In an ideal, reversible process, the entropy of the system will remain the same. This is known as an *isentropic* process. In a non-ideal, non-reversible process, the entropy of the system always increases. Looking at the units of entropy (BTU/(lbm × °R)), we can see that the non-isentropic case represents non-usable energy or BTU. All thermodynamic processes are irreversible or non-isentropic, but in some cases, the isentropic model simplifies thermodynamic evaluation.

From the first law of thermodynamics, since energy is neither created nor destroyed, all systems can be evaluated using the conservation of energy and conservation of mass principal. A block diagram for a very basic thermodynamic cycle is shown in Figure 2.2.

Assuming that the flow of mass from input 1 is discharged from the equipment at output 2 and assuming that the flow of mass from input 3 is discharged from the equipment at output 4, we calculate the power (rate of change of energy with unit time) into and out of the system by the product of enthalpy (h) (units of energy per unit mass) and flow (f) (units of mass per unit time) over a given unit of time. If there is not an exchange of energy from the system at points 1 and 2 to the system at points 3 and 4, and assuming the system is at steady state, that is, that energy in = energy out for each system, the enthalpy at point 1 would equal the enthalpy at point 2 and the enthalpy at point 3 would equal the enthalpy at point 4.

Of course, this is not the normal operation of a device such as a heat exchanger as its function is to transfer energy from one system to another. In a heat exchanger, some of the energy from one system is transferred to the other system. For the above example, let's assume that energy is transferred from the 3 and 4 ports to the 1 and 2 ports. In this instance, since mass flow at point 1 is the same as point 2, the enthalpy of the system at point 1 would be less than the enthalpy of the system at point 2. Similarly, since mass flow at point 3 is the same as point 4, the enthalpy of the system at point 3 would be greater than the enthalpy of the system at point 4. If this system is in equilibrium, then the energy into the system from ports 1 and 3 is equal to the energy leaving the system at ports 2 and 4.

From a system standpoint, we can look at the basic thermodynamic cycle shown in Figure 2.3 and see the four basic processes of the system.

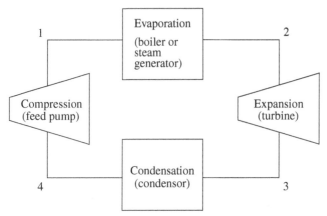

Figure 2.3   Basic thermodynamic cycle.

In the evaporation process (the boiler or steam generator), thermal energy is added to the boiler water and the water is evaporated to generate steam. In the expansion process (turbine), the heated steam is expanded across turbine blades releasing energy to the turbine converting the thermal energy to mechanical energy. The condensation process (condenser) occurs where the exhaust from the turbine section is condensed into liquid form by removal of additional thermal energy from the steam. The compression process (pump) is where mechanical energy is added to the water to increase the pressure of the water. Since mass flow is unchanged at any of the points in the basic thermodynamic cycle, this tells us that as we move from point 1 to point 2 across the evaporation section and add energy to the system, enthalpy increases. As we move from point 2 to point 3 across the expansion section and remove energy from the system, enthalpy decreases. As we move from point 3 to point 4 across the condensation section and remove energy from the system, enthalpy decreases. As we move from point 4 to point 1 across the compression section and add energy to the system, enthalpy increases.

So where do the points 1, 2, 3, and 4 in Figure 2.3 land on our T-s diagram? Of course it depends on where the system pressure and temperature values are at the four points. The most basic representation is where the system all remains in the saturated region. This is known as the Carnot cycle and is represented in Figure 2.4.

This is an ideal cycle in that the compression and expansion cycles happen isentropically (constant entropy). At this point, we should provide some definitions of specific thermodynamic processes.

Isothermal process is a process that occurs at constant temperature

Isobaric process is a process that occurs at constant pressure

Isometric process is a process that occurs at constant volume

Isentropic process is a process that occurs at constant entropy

Adiabatic process is a process that occurs with no heat transfer

Throttling process is a process that occurs at constant enthalpy

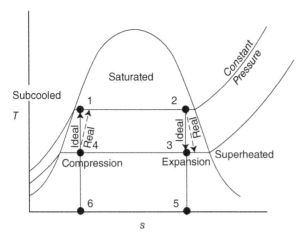

Figure 2.4   Basic Carnot cycle. *Source*: Reproduced with permission of U.S. Department of Energy.

For the Carnot cycle shown in Figure 2.4, the amount of heat into a system (Q_{in}) is the cube 1 2 5 6 and the amount of heat out of a system (Q_{out}) is the cube 4 3 5 6. The net amount of heat transferred into a system Q_{net} is the difference between Q_{in} and Q_{out} which is the cube 1 2 3 4. This is mathematically stated in equation (2.23).

$$Q_{net} = Q_{in} - Q_{out} \qquad (2.23)$$

In the Carnot cycle shown in Figure 2.4, the theoretical maximum efficiency of the steam cycle can be described using temperatures (in units of Rankin) and is given by the equation (2.24). This is because the Carnot cycle has the entire process occurring in the saturated steam region where temperature does not change with the change in enthalpy.

$$\text{Eff}_{thmax} = (1 - T_{out}/T_{in}) \times 100\% \qquad (2.24)$$

where

T_{out} = the absolute temperature for heat rejection (°R)

T_{in} = the absolute temperature for heat addition (°R)

(Fahrenheit temperature is converted to absolute temperature in Rankine by adding 460°.)

As we will learn later in Chapter 9 covering pumps, pumps are designed to pump fluids only and not steam. Pumps designed to pump water can quickly fail due to a phenomenon known as cavitation when steam forms in the pump. This is discussed in more detail in Chapter 9.

Since pumps are not designed to pump steam, the system must be designed such that the compression function is performed in the subcooled region where evaporation in the pump suction cannot occur. Therefore the Carnot cycle is not representative of

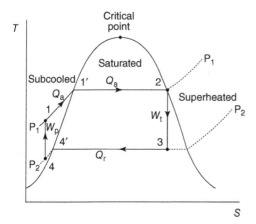

Figure 2.5 Basic Rankine cycle showing operation in subcooled region for pumps. *Source*: Reproduced with permission of U.S. Department of Energy.

the actual thermodynamic cycle of a plant but was a good place to start to understand the basic thermodynamic cycle.

The Rankine cycle, as shown in Figure 2.5, describes the more realistic thermodynamic cycle. There are several differences between the ideal (Carnot) and actual (Rankine) thermodynamic cycles that we will discuss below.

In the basic Rankine cycle, energy is transferred in the form of heat during the evaporation process and the condensation process. Energy is transferred in the form of work during the compression process and the expansion process.

Heat is added (Q_a) to the system between points 1 and 2 during the evaporation process. There are two types of heat transfer during this process. From point 1 to point 1', the heat transfer is in the form of sensible heat transfer where, as energy is added to the system, the temperature of the subcooled water is increased to the point where the water becomes saturated at point 1'. From point 1' to 2, the heat transfer is in the form of latent heat transfer where, as energy is added to the system, the temperature of the material remains constant but the state of the material is changed from liquid state to a gaseous state. Note that this process occurs on a line of constant pressure. Therefore, this is an example of an isobaric process.

Work is done by the system (W_t) between points 2 and 3 during the expansion process. Note that, for this ideal model, this process occurs on a line of constant entropy. Therefore, this is an example of an isentropic process.

Heat is removed (Q_r) from the system between points 3 and 4 during the condensation process. From point 3 to 4', the heat transfer is in the form of latent heat transfer where, as energy is removed from the system, the temperature of the material remains constant but the state of the material is changed from a mostly gaseous phase to a saturated liquid phase. Subcooling occurs between points 4' and 4. Heat transfer is in the form of sensible heat transfer where, as energy is removed from the system, the temperature of the water decreases. This is done to prevent pump cavitation. Note that this process occurs on a line of constant pressure. Therefore, this is an example of an isobaric process.

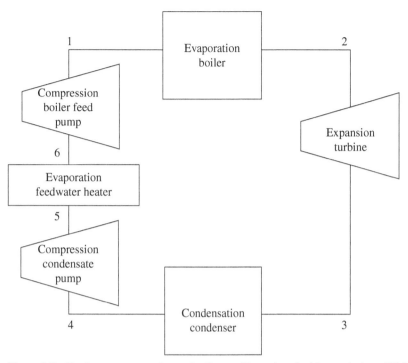

Figure 2.6   Feedwater regenerative cycle. *Source*: Reproduced with permission of U.S. Department of Energy.

Work is done to the system (W_p) between points 4 and 1 during the compression process. Note that, for this ideal model, this process occurs on a line of constant entropy. Therefore, this is an example of an isentropic process.

The thermal efficiency of the system is defined as the amount of net work done by the turbine divided by the amount of heat added to the system and is a measure of the efficiency of energy transfer of the system ($Eff_{th} = W_n/Q_a$). Note that by either maximizing the net amount of work produced by the system or by minimizing the heat into the system, efficiency can be maximized.

In actual application, the actual thermodynamic cycle is not a perfectly isentropic process due to the second law of thermodynamics which states that no process is irreversible. This implies that for any process, the entropy of the system always increases. However, assuming an isentropic process is a reasonable initial approximation to help us initially understand the physics involved in the thermodynamic cycle.

Below are the basic concepts addressed in the discussion above and shown in Figures 2.3 and 2.5.

$$Q_a - \text{heat added to the system} = h_2 - h_1 \qquad (2.25)$$

$$Q_r - \text{heat removed from system} = h_3 - h_4 \qquad (2.26)$$

$$W_p - \text{work done to the system by the pump} = h_1 - h_4 \qquad (2.27)$$

$$W_t - \text{work done by the turbine} = h_2 - h_3 \qquad (2.28)$$
$$W_n - \text{net work of the system} = W_t - W_p \qquad (2.29)$$
$$\text{Eff}_{th} - \text{thermal efficiency of the system} = (W_n / Q_a) \qquad (2.30)$$

As mentioned above, the thermal efficiency of the system can be improved if we increase the net work of the system which is the difference between the work done by the system (W_t) and the work done by the pump (W_p). Therefore, if we reduce the net work done by the pump (W_p), we will increase the overall thermal efficiency of the system. This is the function of the feedwater heater system. The addition of the feedwater heater system is shown in Figure 2.6. As can been seen from Figure 2.7, out of the condenser, the condensate pump increases the energy in the system from point 4 to point 5. Then the feedwater heater increases the energy in the system from point 5 to point 6. Then the boiler feed pump increases the energy of the system from point 6 to point 1. Since the feedwater heater performs some of the work that was previously done by the pump, this has the effect of reducing the work done to the system by the pump (W_p) which has the effect of increasing the net work done by the system (W_n) and increases the overall efficiency of the system. There is an additional benefit in having the work of the condensate pump reduced in that it reduces the amount of pressure reduction in the suction side of the condensate pump. Reducing the pressure drop at the suction of the pump reduces the amount of subcooling required for the condensate pump to ensure we do not start cavitating the pump. If we reduce the subcooling requirement, we increase the efficiency of the thermal process since we are rejecting less heat in the condenser.

Just as pumps are designed to only pump liquids, turbine blades are designed to only have steam pass through the blade stages. If the steam becomes saturated and begins to condense out water in the turbine blade sections, the water droplets will impinge on the blades and cause turbine blade damage. Therefore, it is common practice to superheat the steam entering the expansion stage at the steam turbine

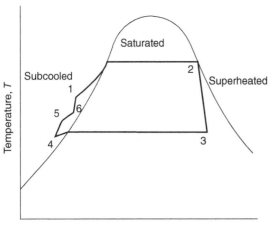

Figure 2.7   Feedwater regenerative T-s diagram. *Source*: Reproduced with permission of U.S. Department of Energy.

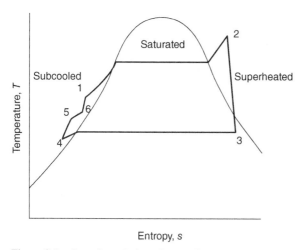

Figure 2.8   Superheat design. *Source*: Reproduced with permission of U.S. Department of Energy.

as is shown in Figure 2.8. Additionally, this increases the amount of heat into the thermodynamic cycle, which increases the thermal efficiency of the process. Now, as the steam is expanded through the turbine, energy is released and the temperature of the steam drops, but the steam remains superheated and does not condense out in the turbine blade sections.

Much like the discussion of using multiple pumps in the compression section to limit the pressure drop across any one pump stage, it is typical to use several turbines and turbine stages in the expansion process. This allows for the removal of more energy from the steam as, in the intermediate section, the steam is reheated at a lower pressure and therefore, once the steam condenses through the second, lower pressure turbine, the saturation temperature of the steam is reduced. This allows us to remove more energy from the steam before the steam enters the saturated region where we no longer want to put the steam through the turbine to prevent water impingement issues. When multiple turbines are used and a reheat is used, the Rankine cycle now appears as shown in Figures 2.9 and 2.10.

The input to the first high pressure turbine is shown at point 8. The discharge from the high pressure turbine is shown at point 1. At this point, the steam is sent back to the boiler in the "cold reheat" lines where energy is added back to the steam to increase the temperature of the steam along the lower intermediate pressure line. The steam from the cold reheat system enters the boiler at point 1 and exits the boiler at point 2. At this point the reheated steam leaves the boiler through the hot reheat line and enters the intermediate or low pressure turbine at point 2. Then the reheated steam is expanded through a second lower pressure stage of the turbine. It leaves the low pressure turbine at point 3 where it enters the condenser. Note that, at this lower pressure, the saturation temperature of the fluid is less than could be achievable at the higher pressure section. Therefore, using reheat design, more energy can be extracted from the steam during the expansion process which increases the thermal efficiency of the system. The generation station equipment that achieves the various functions

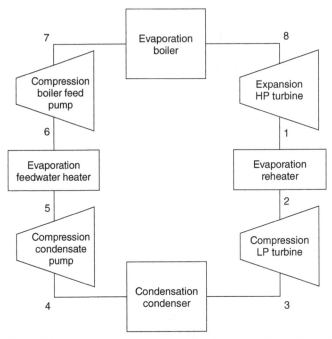

Figure 2.9   Feedwater regenerative cycle with reheat. *Source*: Reproduced with permission of U.S. Department of Energy.

of the feedwater regenerative thermodynamic cycle that is shown in Figure 2.10 can be seen in Figure 2.11.

As mentioned previously, when the condensate enters the condensate pump suction, the pressure drops slightly at the suction side due to losses in the suction piping and the design of the pump impeller. If the liquid is saturated at a specific temperature, reducing the pressure at the suction of the pump will increase the steam quality causing some of the water to evaporate. This rapid evaporation of water to steam in the suction of the pump causes cavitation which occurs when steam bubbles develop in the suction of the pump impeller, and then collapses in the impeller as pressure increases on the surface of the impeller. This would result in destruction of the pump impeller. To prevent this from occurring, the condensate is cooled below the point of saturation. This is known as condensate depression or condensate subcooling. The benefit of this process is to ensure pump protection by preventing cavitation in the pump, but the drawback of this process is that it reduces the thermal efficiency of the system. The act of subcooling the saturated condensate removes more energy from the fluid which then has to be re-added later in the evaporation process and this additional heat does not procure any additional work. In summary,

- Increasing "subcooling" increases the reduction of condensate temperature and reduces system efficiency.
- Decreasing "subcooling" decreases the reduction of condensate temperature and increases system efficiency.

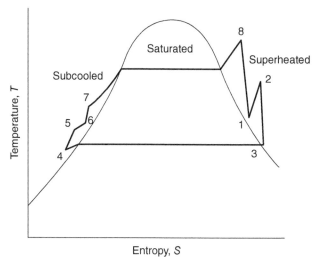

Figure 2.10   Feedwater regenerative cycle with reheat T-s diagram. *Source*: Reproduced with permission of U.S. Department of Energy.

As mentioned previously, the compression and expansion processes are not truly isentropic. During both processes, the entropy of the system is increased. Therefore as the temperature/pressure of the fluid is changed, the entropy of the system increases. Since entropy always increases, this shift is always along the positive direction of entropy.

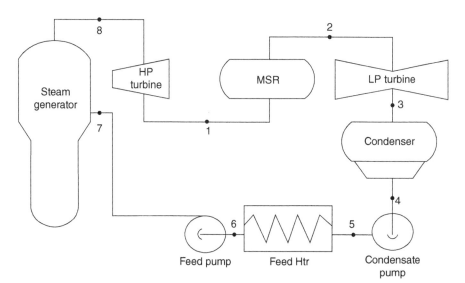

Figure 2.11   Typical steam turbine thermodynamic cycle equipment. *Source*: Reproduced with permission of U.S. Department of Energy.

HEAT RATE

It is desired to operate the power plant as efficiently as possible for two reasons. First, a higher efficiency means, for a certain amount of electrical energy out of the plant (MWh), we minimize the amount chemical energy input needed into the system (BTU). Since the energy into the system is directly related to the amount of fuel needed, a more efficient plant will require less fuel for the same amount of electrical energy delivered. Less fuel means less fuel cost. Therefore, we minimize costs by operating a more efficient plant. In addition, for power plant topologies that relay on fossil fuel combustion, the combustion of the fuel releases various constituents into the environment as a byproduct of combustion. For carbon-based fuels, a primary constituent is the release of carbon dioxide which is a greenhouse gas. Additionally, depending on the type of fuel utilized, other byproducts such as NO and NO_2 (known as NO_x) are released. If we operate a more efficient plant, for a certain amount of electrical energy out of the system, we use less fuel and, therefore, release less byproduct into the environment.

In a power plant, we talk about efficiency in terms of *heat rate*. In general, heat rate is a measure of the energy into the system divided by the energy out of a system. Specifically, the *net plant heat rate* (NPHR) is the amount of heat energy into a boiler divided by the net amount of energy out of the plant and is mathematically described by the following formula.

$$NPHR = Q_B/NPO \qquad (2.31)$$

where

NPHR = net plant heat rate (BTU/kWh)

Q_B = heat into boiler (BTU/hr)

NPO = net plant output (kW)

Efficiency is a measure of the amount of energy out of a system divided by the amount of energy into a system. This is the inverse function of NPHR. Therefore, a plant with a higher efficiency has a lower heat rate.

Just as we can look at the efficiency or heat rate of the plant overall, we can look at the efficiency or heat rate of any system within the plant. For example, the efficiency of the turbine system is described by the *net turbine heat rate* (NTHR) as the ratio of the heat energy into a turbine divided by the net mechanical energy output of the turbine. It is mathematically described by the following formula.

$$NTHR = Q_T/NTO \qquad (2.32)$$

where

NTHR = net turbine heat rate (BTU/kWh)

Q_T = heat into turbine (BTU/hr)

NTO = net turbine output (kW)

The net plant output (NPO) is the difference between the net mechanical energy delivered by the turbine to the generator and the energy utilized by the plant auxiliary systems required to operate the plant. It is mathematically described by the following formula.

$$NPO = NTO - AP \qquad (2.33)$$

where

NPO = net plant output (kW)

NTO = net turbine output (kW)

AP = auxiliary power (kW)

The efficiency of the boiler (η_B) is defined as the ratio of the heat energy into the turbine (which is ideally the heat energy out of the boiler) divided by the heat energy into the boiler. It is mathematically described by the following formula.

$$\eta_B = Q_T/Q_B \qquad (2.34)$$

where

Q_T = heat into turbine (BTU/hr)

Q_B = heat into boiler (BTU/hr)

With the above system efficiencies defined, we can now combine these equations to derive our overall NPHR in terms of the efficiencies of our various systems. This is mathematically derived below.

First, starting with equation (2.31), we find,

$$NPHR = Q_B/NPO$$

Using equation (2.33) that defines our net plant output, we can substitute the values of net turbine output and auxiliary power into equation (2.31) to find,

$$NPHR = Q_B/(NTO - AP)$$

If we multiply both numerator and denominator by (1 / NTO), we find,

$$NPHR = [1/NTO] \times [Q_B/(1 - AP/NTO)]$$

Next, we can rearrange equation (2.32) for net turbine output and find that it is

$$NTHR = Q_T/NTO$$

or

$$NTO = Q_T/NTHR$$

We can substitute this back into the equation for NPHR to find

$$NPHR = [NTHR/Q_T] \times [Q_B/(1 - AP/NTO)]$$

Rearranging the above equation to combine the heat into the turbine and the heat into the boiler, we find,

$$NPHR = [Q_B/Q_T] \times [NTHR/(1 - AP/NTO)]$$

We can now use equation (2.34) and replace the ratio of heat into the boiler and heat into the turbine with boiler efficiency with the following substitution.

$$\eta_B = Q_T/Q_B$$

$$\text{NPHR} = [1/\eta_B] \times [\text{NTHR}/(1 - \text{AP}/\text{NTO})]$$

Rewriting this equation, we can find the NPHR in terms of the efficiencies of our individual systems. We now find that the NPHR is a function of the NTHR, the boiler efficiency, the auxiliary power and the NTHR as shown in equation (2.35).

$$\text{NPHR} = \frac{\text{NTHR}}{\eta_B \times [1 - (\text{AP}/\text{NTO})]}. \qquad (2.35)$$

where

NPHR = net plant heat rate (BTU/kWh)

NTHR = net turbine heat rate (BTU/kWh)

AP = auxiliary power (kW)

NTO = net turbine output (kW)

η_B = efficiency of boiler

To maximize plant efficiency and minimize the NPHR, we want to keep the NTHR as small as possible, keep the plant auxiliary energy utilization (AP) as small as possible, keep the net turbine output (NTO) as large as possible, and keep the boiler efficiency (η_B) as large as possible. Remembering that a smaller NPHR is better performance and a higher NPHR is worse performance, a reduction in the net turbine output or boiler efficiency, an increase in the auxiliary power utilization, or an increase in NTHR will adversely affect our overall NPHR.

Just as steam undergoes the processes of compression, evaporation, expansion, and condensation to make up the Rankine cycle, in a combustion engine, the gas (air typically) used in the combustion process also goes through a cycle of compression, combustion, expansion, and exhaust that make up the Brayton cycle. We will discuss this in more detail in Chapter 4 on combustion.

Some plant topologies incorporate both cycles, where prior to the exhaust process of the Brayton cycle of a combustion turbine, the heated gas or air flows past a "steam generator" where the gas heats up the water and performs the evaporation process of the Rankine cycle. This is known as a combined cycle plant as it combines the functions of both the Brayton and Rankine cycles. This is also known as a topping/bottoming cycle where the Brayton cycle is the topping cycle and the Rankine cycle is the bottoming cycle. The Brayton cycle is discussed in Chapter 12 on combustion turbines and engines.

GAS THERMODYNAMIC ANALYSIS FUNDAMENTALS

The thermodynamic cycle of the gas turbine is known as the Brayton cycle as shown in Figure 2.12.

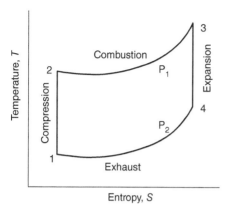

Figure 2.12   T-Diagram for isentropic Brayton cycle. *Source*: Reproduced with permission of NASA.

Enthalpy values for the gas at the 4 points of interest can be derived from the T-s Diagram for the specific gas used for the Brayton cycle. The four basic processes are compression, combustion, expansion, and exhaust.

The compression process increases the pressure of the gas from pressure P_2 to Pressure P_1 at points 1 and 2 where the pressure is increased leading to an increase in the enthalpy value of the gas media. The compressed gas then undergoes the combustion process between points 2 and 3 where the temperature of the gas media is increased leading to a further increase in the enthalpy value of the gas media. The heated compressed gas then is expanded across the gas turbine from points 3 to point 4 where the pressure is reduced from pressure P_1 to pressure P_2 across the turbine blades leading to a reduction of enthalpy across the turbine blades. The product of the mass flow rate of air across the turbine blade and the difference in enthalpy across the turbine blades defines the amount of energy delivered by the gas media to the turbine. The final stage is the exhaust between points 4 and 1 which is simply the difference in the enthalpy of the incoming air and the exhaust air and represents energy lost to the environment during the exhaust process. The compressor that performs the compression is connected to the same shaft as the main turbine and the main turbine provides the energy for compression. This amount of energy is the difference in enthalpy between points 1 and 2 times the mass flow rate of the gas media. The net available power of the turbine to drive the generator is the difference between the gross energy developed during expansion minus the energy used during compression. The amount of work available to turn the generator is the gross work of the turbine minus the amount of work done by the compressor (in applications where the turbine and compressor are on the same shaft) as shown in equation (2.36).

$$W_{net} = \text{Work available to drive generator} = W_{gross} - W_{compressor} \qquad (2.36)$$

where

W_{net} = net work sent to the generator

W_{gross} = gross work developed by turbine

W_{comp} = work absorbed by compressor

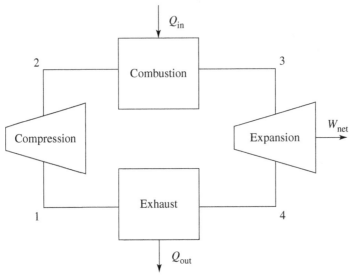

Figure 2.13   Brayton cycle. *Source*: Richard C. Dorf, 1995. Reproduced with permission of Taylor & Francis Group LLC Books.

Typically 50% to 60% of the gross energy developed by the turbine goes to drive the compressor stage. The amount of heat into the system is defined as Q_{in} and the amount of heat rejected from the system is defined as Q_{out} as shown in Figure 2.13.

With these terms defined we can define the thermal efficiency of the combustion turbine as the amount of net work available to drive the generator by the amount of heat added to the system.

$$\eta_{th} = W_{net}/Q_{in} = (Q_{in} - Q_{out})/Q_{in} = 1 - (Q_{out}/Q_{in}) \tag{2.37}$$

where

η_{th} = thermal efficiency of Brayton cycle

W_{net} = net work sent to the generator

Q_{in} = heat into system

Q_{out} = heat rejected from system

If we ignore the compressibility of the gas stream, we can define the ideal thermal efficiency in terms of the temperatures of the gas stream (using absolute values of K or °R).

$$\eta_{th} = 1 - (T_1/T_2) = 1 - (T_4/T_3) \tag{2.38}$$

where

η_{th} = thermal efficiency of Brayton cycle

T_1 = gas temperature into compressor (K or °R)

T_2 = gas temperature out of compressor (K or °R)

T_3 = gas temperature into turbine (K or °R)

T_4 = gas temperature out of turbine (K or °R)

Therefore, if we maximize T_2 and T_3 and/or minimize the values of T_1 and T_4, we maximize the thermal efficiency of the gas thermal cycle. There are thermal material limits that must be maintained to ensure the integrity of the equipment just like a steam turbine so there are constraints that are placed on how high we can make T_2 and T_3.

The gas stream is compressible and, as such, the compressibility of the gas affects the efficiency of the Brayton cycle. The thermal efficiency η_{th} can also be expressed in terms of the compressor ratio.

$$\eta_{\text{th}} = 1 - [1/(P_2/P_1)^{(k-1)/k}] \tag{2.39}$$

where

η_{th} = thermal efficiency of Brayton cycle

P_1 = discharge pressure of air compressor

P_2 = suction pressure of air compressor

k = compressibility of air factor

As we increase the ratio of output compressor pressure P_2 to inlet compressor pressure P_1, the value of P_2/P_1 increases. This decreases the second term of equation (2.39) which increases the thermal efficiency of the gas turbine. It would seem that we could continue to increase the differential pressure across the compressor and continue to make our system more efficient by simply maximizing the compressor ratio P_2/P_1. However, to obtain higher values of differential pressure across the compressor, more work must be done by the turbine to drive the compressor which leaves less work available to drive the generator. At the same time, while increasing P_2/P_1 increases efficiency, the magnitude of the increase is more pronounced at lower ratios than at higher ratios. In other words, doubling the ratio of P_2/P_1 from a value of 2 to 4 has more of an improvement in efficiency than does doubling the ratio of P_2/P_1 from a value of 20 to 40. There is some point where it takes more power to obtain a larger compressor ratio than is gained by the increase in the thermal efficiency due to the compressor pressure differential. A compressor ratio P_1/P_2 of between 10 and 20 results in the most efficient thermal cycle with still adequate power left to drive the generator.

The thermodynamic Brayton cycle is shown in Figure 2.14 with both the ideal isentropic process shown as cycle 1, 2S, 3, 4S and the realistic non-isentropic process shown as cycle 1, 2, 3, 4. Just as in the case of the steam turbine, the Brayton cycle is not ideal (non-isentropic). The enthalpy increases from point 1 to 2 as the compressor does work on the gas stream. In the isentropic case, the compressor would compress the gas from point 1 to point 2S. This change in entropy results in a loss of usable energy in the compressor. The efficiency of the compressor is the ideal change in enthalpy divided by the actual change in enthalpy due to the increase in entropy. The efficiency of the compressor is mathematically modeled by

$$\eta_{\text{comp}} = (h_{2S} - h_1)/(h_2 - h_1) \tag{2.40}$$

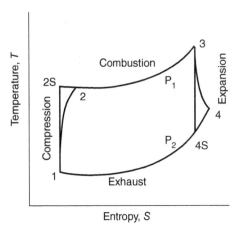

Figure 2.14 T-s Diagram for isentropic and non-isentropic Brayton cycle. *Source*: Reproduced with permission of NASA.

where

η_{comp} = compressor efficiency

h_1 = enthalpy at compressor inlet

h_{2S} = enthalpy at constant entropy at discharge of compressor

h_2 = actual enthalpy at discharge of compressor

Similarly, as the turbine receives work from the gas stream, the enthalpy decreases from point 3 to point 4. In the isentropic case, the turbine would expand the gas from point 3 to point 4S. This change in entropy results in a loss in the turbine. The efficiency of the turbine is the actual change in enthalpy due to the increase in entropy divided by the ideal change in enthalpy. The efficiency of the turbine is mathematically modeled by

$$\eta_{turb} = (h_3 - h_4)/(h_3 - h_{4S}) \tag{2.41}$$

where

η_{turb} = turbine efficiency

h_3 = enthalpy at turbine inlet

h_{4S} = enthalpy at constant entropy at exit pressure

h_4 = actual enthalpy at exit pressure

Example 2.11 Determine the compressor efficiency if

h_1 = 150 BTU/lbm

h_2 = 1000 BTU/lbm

h_{2S} = 850 BTU/lbm

Solution: Using equation (2.40) we find

$$\eta_{comp} = (h_{2S} - h_1) / (h_2 - h_1)$$
$$\eta_{comp} = (850 \text{ BTU/lbm} - 150 \text{ BTU/lbm}) / (1000 \text{ BTU/lbm} - 150 \text{ BTU/lbm})$$
$$\eta_{comp} = 700 \text{ BTU/lbm} / 850 \text{ BTU/lbm}$$
$$\eta_{comp} = 0.82 \text{ or } 82\%$$

Example 2.12   Determine the turbine efficiency if

$$h_3 = 2500 \text{ BTU/lbm}$$
$$h_4 = 1000 \text{ BTU/lbm}$$
$$h_{4S} = 600 \text{ BTU/lbm}$$

Solution: Using equation (2.41) we find

$$\eta_{turb} = (h_3 - h_4) / (h_3 - h_{4S})$$
$$\eta_{turb} = (2500 \text{ BTU/lbm} - 1000 \text{ BTU/lbm}) / (2500 \text{ BTU/lbm} - 600 \text{ BTU/lbm})$$
$$\eta_{turb} = 1500 \text{ BTU/lbm} / 1900 \text{ BTU/lbm}$$
$$\eta_{turb} = 0.78 \text{ or } 78\%$$

GLOSSARY OF TERMS

- Adiabatic process is a process that occurs with no heat transfer.
- Conduction is the transfer of energy through matter from particle to particle.
- Convection is the transfer of heat by the actual movement of the warmed matter.
- Density – The measure of the mass of a material that is in a specific volume of material.
- Efficiency is a measure of the amount of energy out of a system divided by the amount of energy into a system. This is the inverse function of NPHR.
- Heat rate is a measure of the energy into the system divided by the energy out of a system.
- Isentropic process is a process that occurs at constant entropy.
- Isobaric process is a process that occurs at constant pressure.
- Isometric process is a process that occurs at constant volume.
- Isothermal process is a process that occurs at constant temperature.
- Latent Heat Transfer – A process when energy is transferred and results in a change in the state of the material (i.e., change from liquid to gas), but the temperature remains constant.
- Natural Circulation – The circulation of water in a boiler caused by differences in density.

- Net plant heat rate (NPHR) is the amount of heat energy into a boiler divided by the net amount of energy out of the plant.
- Net turbine heat rate (NTHR) is the ratio of the heat energy into a turbine divided by the net mechanical energy output of the turbine.
- Pressure – The measure of force per unit area.
- Radiation is a process where electromagnetic waves directly transport energy through space.
- Sensible Heat Transfer – A process when energy is transferred and the transfer of energy results in a change in the temperature of the material, but the state of the material (i.e., solid, liquid, or gas) remains the same.
- Specific Volume – The measure of the volume that a material takes up for a specific amount of mass of the material.
- Specific Heat – The measure of the amount of energy that is transferred in the form of heat per unit mass.
- Temperature – The measure of the average molecular kinetic energy of a material.
- Throttling process is a process that occurs at constant enthalpy.

PROBLEMS

2.1 A combined cycle power plant utilizes two thermal cycles to drive the two generators. One generator is driven by a combustion turbine and the other generator is driven by a steam turbine. What are the names of the two thermal cycles that are referenced by the term "combined cycle?"

    A. Rankine – Carnot

    B. Rankine – Brayton

    C. Carnot – Brayton

2.2 Complete this sentence. Overall power plant thermal efficiency will decrease if…

    A. the steam temperature entering the turbine is increased

    B. the temperature of the feedwater entering the steam generator is increased

    C. the amount of condensate depression (subcooling) in the main condenser is decreased

    D. the temperature of the steam at the turbine exhaust is increased

2.3 The theoretical maximum efficiency of a steam cycle is given by equation (2.24). A power plant is operating with a stable steam generator pressure of 900 psia at saturated steam conditions with a saturation temperature of 532.02°F. What is the approximate theoretical maximum steam cycle efficiency this plant can achieve by establishing its main condenser vacuum at 1.0 psia at saturated steam conditions with a saturation temperature of 101.69°F? (*Hint*: Assume both the steam generator and condenser are at saturated steam conditions. Use ASME steam tables and assume saturated steam conditions in the steam generator and condenser to determine the temperature of the working fluid in the two states.)

A. 35%

B. 43%

C. 57%

D. 65%

2.4 Main condenser pressure is 1.0 psia. During the cooling process in the condenser, the temperature of the low pressure turbine exhaust decreases to 100°F, at which time it is a…

A. saturated liquid

B. saturated vapor

C. subcooled liquid

D. superheated vapor

2.5 A liquid is saturated with 0% quality. Assuming pressure remains constant, the addition of a small amount of heat will…

A. raise the liquid temperature above the boiling point

B. result in a subcooled liquid

C. result in vaporization of the liquid

D. result in a superheated liquid

2.6 Which one of the following is the approximate steam quality of a steam-water mixture at 250°F with an enthalpy of 1000 BTU/lbm?

A. 25%

B. 27%

C. 83%

D. 92%

2.7 If a saturated vapor is at 205°F and has a steam quality of 90%, its specific enthalpy is approximately…

A. 173 BTU/lbm

B. 271 BTU/lbm

C. 1050 BTU/lbm

D. 1147 BTU/lbm

2.8 If the steam's pressure is 230 psia and is at a temperature of 900°F, what is the approximate amount of superheat? (Given the saturation temperature of saturated steam at 230 psia is 393.71°F.)

A. 368°F

B. 393°F

C. 506°F

D. 535°F

2.9 Which one of the following is the approximate amount of thermal energy required to convert 2 lbm of water at 100°F and 100 psia to a saturated vapor at 100 psia? (Given the value of enthalpy of subcooled water at 100°F and 100 psia is 68.3 BTU/lbm and the value of enthalpy of saturated vapor at 100% steam quality, and 100 psia is 1187.5 BTU/lbm.)

A. 560 BTU

B. 1,120 BTU

C. 2,238 BTU

D. 3,356 BTU

2.10 In the basic heat cycle there are four processes, compression, expansion, evaporation, and condensation. Referring to Figure 2.4, these four processes are represented by four lines. Which line presents the compression process of the basic steam cycle?

A. Line 1–2

B. Line 2–3

C. Line 3–4

D. Line 4–1

2.11 In the basic heat cycle shown in Figure 2.4, the heat into the system (Q_{in}) is defined as which of these blocks?

A. Box 1, 2, 3, 4, 1

B. Box 1, 2, 5, 6, 1

C. Box 4, 3, 5, 6, 4

2.12 Given a temperature of 0°F, what is this temperature in the Rankine scale?

2.13 Condensate depression is the process of…

A. removing condensate from turbine exhaust steam

B. spraying condensate into turbine exhaust steam

C. heating turbine exhaust steam above its saturation temperature

D. cooling turbine exhaust steam below its saturation temperature

2.14 The law that states that any thermodynamic process is irreversible as the net entropy of a system and its surroundings always increases is which law?

A. The first law of thermodynamics

B. The second law of thermodynamics

C. Ohms Law

D. Stuarts Law

2.15 Given the following values for compressor inlet and outlet isentropic and non-isentropic enthalpies, determine the compressor efficiency.
Determine the compressor efficiency if

$h_1 = 250$ BTU/lbm

$h_2 = 1250$ BTU/lbm

$h_{2S} = 1000$ BTU/lbm

2.16 Given the following values for turbine inlet and outlet isentropic and non-isentropic enthalpies, determine the turbine efficiency.

$h_3 = 3000$ BTU/lbm

$h_4 = 1000$ BTU/lbm

$h_{4S} = 700$ BTU/lbm

RECOMMENDED READING

ASME Steam Tables-Compact Edition, American Society of Mechanical Engineers, www.asme.org, 2006.

DOE Fundamentals Handbook, Department of Energy, 1999, http://energy.gov/

Electrical Machines, Drives and Power Systems, 6th edition, Theodore Wildi, Prentice Hall, 2006.

The Engineering Handbook, 2nd edition, Richard C. Dorf (editor in chief), CRC Press, 1995.

Machinist's Mate 3 & 2 (Surface), NAVEDTRA 14151, Naval Education and Training Professional Development and Technology Center, 2003.

Power Plant Engineering, Black & Veatch, edited by Larry Drbal, Kayla Westra, and Pat Boston, Chapman & Hall/Springer, 1996.

Standard Handbook of Powerplant Engineering, 2nd edition, Thomas C. Elliott, McGraw-Hill, 1998.

CHAPTER 3

BOILERS AND STEAM GENERATORS

GOALS

- To identify the equipment used for in the Rankine thermodynamic process
- To understand the basics of feedwater heater level control
- To identify the basic difference between a pressurized boiler design and balanced draft boiler design
- To develop a temperature profile for a heat recovery steam generator and identify the location of the superheater approach temperature and the economizer approach temperature
- To identify the difference between a watertube and firetube boiler design
- Identify auxiliary equipment utilized in boiler control systems
- Utilizing a psychometric chart, perform temperature conversions from dry bulb to wet bulb temperatures

NOW THAT we have a fundamental understanding of the steam cycle we can begin to discuss some of the equipment that is used in the thermodynamic system in a power generation facility. The steam generator or *boiler* is a device used in the evaporation process to add energy to a water system to increase the enthalpy of the system, causing the water to evaporate to steam (and superheat the steam) to prepare the steam for delivery to the turbine section. We saw in Chapter 2 on the Rankine cycle that we will preheat the feedwater to the boiler to improve overall plant efficiency. This is accomplished with a *feedwater heater*.

Figure 3.1 shows a typical feedwater heater arrangement. The feedwater system enters the "water box" on the right side and enters the tubes through the tube sheet. The tubes and tube sheet form the barrier between the feedwater side and steam side. Extraction steam is fed to the shell side and it enters a de-superheating zone. In this area, the extraction steam temperature is reduced to the saturation temperature for the operational pressure of the steam side of the feedwater heater. Shrouds and baffles are

---

Energy Production Systems Engineering: An Introduction for Electrical Engineers to Electrical Power Generation Facilities, Systems, and Equipment, First Edition. Thomas H. Blair.

Typical feedwater heater internal arrangements
Desuperheating, condensing, and subcooling zones

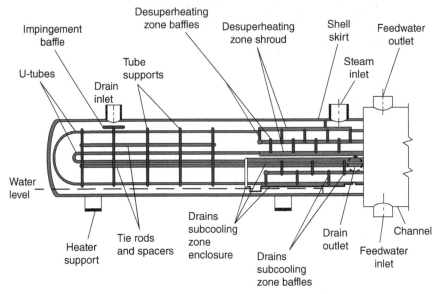

Figure 3.1   Typical feedwater heater arrangement. *Source*: Reproduced with permission of Tampa Electric Company. Reproduction is forbidden without the express consent of Tampa Electric Company.

used to allow for sufficient time for the extraction steam to be exposed to the tubes containing the feedwater to allow for condensation. After condensation, the liquid enters the subcooling section and then leaves the heater via the drains.

There are two main parameters to control to ensure the performance of the feedwater heater. The first parameter to monitor is the heat transfer capability of the heat exchanger and the second parameter to monitor is the feedwater heater level to ensure the shell side level remains at the normal level.

Energy in the feedwater heater is being transferred from the extraction steam system to the feedwater system. On the extraction steam side, heat transfer occurs as latent heat transfer (i.e., it occurs at the saturation temperature for the steam pressure on the shell side of the feedwater heater). Therefore, except for any subcooling zone activity, the shell side temperature does not change from the extraction steam that enters the shell side to the area where condensate is collected on the shell side. As thermal energy is transferred in the form of heat from the steam side to the feedwater side, the steam side condenses but remains at saturation temperature, but since the feedwater side heat transfer is sensible heat transfer, the feedwater side temperature increases. A measure of the efficiency of this heat transfer is the terminal temperature difference (TTD). TTD is defined as the saturation temperature of the extraction steam minus the feedwater outlet temperature. In a feedwater heater with good heat transfer characteristics, the value of TTD will be very small. Should the heat transfer characteristics deteriorate in a feedwater heater (for instance, if the feedwater heat tubes start to build up a layer of calcium increasing the thermal resistivity of the tube to heat transfer), then the value of TTD will increase. Typical ranges for TTD on a high pressure heater is 5° F to 10° F depending on the design of the feedwater heater. Low pressure feedwater heaters have a value of TTD of around the 5° F point.

Level control on the steam (shell) side is critical for safe and efficient feedwater heater operation. The condensate on the shell side cannot be allowed to reach the steam inlet or the condensate may feed back to the turbine drains causing impingement damage to the turbine blades. At the same time, the level of the shell side of the feedwater heater should be maintained at a low level. This allows for maximum tube surface area exposure to the extraction steam. Since latent heat transfer is capable of greater amounts of energy transfer for the same mass flow rate than sensible heat transfer, we try to maximize the number of tubes that are exposed to saturated steam and transfer heat to the feedwater via latent heat transfer instead of allowing the steam to condense completely. The level of the condensate in the shell side must never reach the drain outlet. Should this happen, steam could be introduced into the condenser and a loss of vacuum may occur in the condenser causing reduced efficiencies and heating concerns with the low pressure turbine blade stage.

The second method of monitoring the performance of a feedwater heater is by monitoring the level of condensed extraction steam in the shell side of the feedwater heater. Some feedwater heaters contain level sensors to monitor this and more can be learned about level control in Chapter 24 on instrumentation. An alternate means of determining the level in the feedwater heater is by measuring the difference between the drain cooler outlet and the feedwater inlet. We can infer condensate level in the drain cooler section of the feedwater heater by the temperature difference between the temperature of the condensate in the drain cooler temperature and the feedwater inlet temperature. The drain cooler approach temperature (DCA) is the temperature of the steam drain cooler section less the feedwater inlet temperature. An increasing DCA temperature difference indicates the level in the shell side of the feedwater heater is decreasing and a decreasing DCA temperature indicates the level in the shell side of the feedwater heater is rising as is shown in the Figure 3.2.

Figure 3.2 shows the relationship of shell side condensate level versus the DCA. The level of the feedwater heater must be maintained at a minimum level to provide a minimum amount of subcooling. This ensures that the condensate exiting the steam drains does not flash into steam as it passes the drain valve. The measure of DCA measures the difference between the feedwater and steam drain system. If there is no direct indication of feedwater heater shell side liquid level, the DCA can be utilized to estimate the level of liquid in the feedwater heater shell side utilizing Figure 3.2. The value of DCA is maintained at a low level ensuring a minimum shell liquid level in the feedwater heater. As the level on the shell side reduces, the DCA increases and can be used as indication that condensate level in the shell side is too low.

The furnace section of a boiler is where the fuel and air is mixed and combusted. The water and steam is contained inside the boiler tubes (this is known as a *watertube design*) and the hot gas from the combustion process passes by the outside of the tubes. The boiler has a "back pass" region that contains the superheater section and economizer section. All of the *water wall* tubes combine at a *header* at the top of the

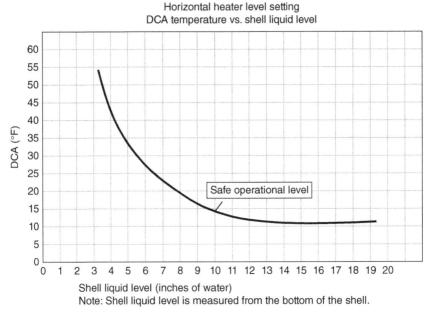

Figure 3.2   DCA temperature vs. shell liquid level. *Source*: Reproduced with permission of Tampa Electric Company. Reproduction is forbidden without the express consent of Tampa Electric Company.

boiler called the upper wall header and feed the "steam drum" at the top or penthouse of the boiler. Similarly, all the water wall tubes combine at a common header at the bottom of the boiler called a lower wall header that connects to a water drum or "mud drum" (if provided). There are several large diameter pipes connecting the bottom of the steam drum to the water or mud drum called "downcomers". If the boiler is a forced circulation design, then there will be "boiler water recirculation" pumps located in the downcomer region. The equipment used in the basic steam flow cycle is shown in Figure 3.3.

Steam *attemperators* are located in the penthouse section of the boiler (space above the top of the boiler gas path) for the main steam and hot reheat steam lines returning to the turbine to provide spray for control of steam temperature leaving these sections and to be delivered to the turbine section. Turbine blade materials have operational thermal limitations and the steam attemperators are used to ensure that maximum allowable steam temperatures are not exceeded.

The condensate pumps and boiler feed pump supply feedwater into the boiler to the main steam drum. Exhaust gases from the boiler in the back pass section preheat this water in the economizer section prior to the feedwater entering the steam drum. The water in the steam drum is in saturation conditions for the operating pressure and temperatures of the boiler. Therefore, there is a steam and water mixture in the steam drum. Typical values for pressure in the main steam system are 2500–3500 psig.

The cooler water in the steam drum is transported via forced or natural circulation (depending on the design of the downcomer section of the boiler) to the water

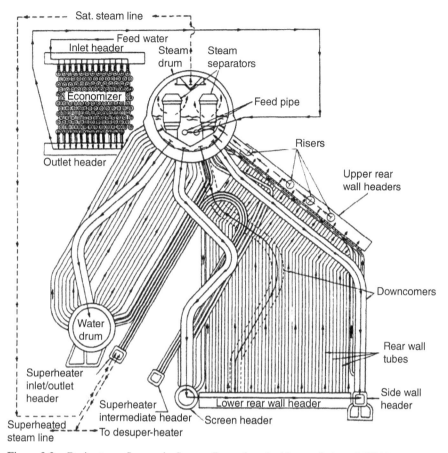

Figure 3.3 Basic steam flow path. *Source*: Reproduced with permission of US Navy Training Manual.

or mud drum at the bottom of the boiler where the water is distributed to the water wall tubes. The water wall tubes pick up energy from the hot gases in the furnace and raise the temperature of the water in the water wall tubes. The water wall tubes also act as cooling mechanism for the walls of the boiler. The water in the water wall region rises back to the steam drum at the top of the boiler. During the heating process, some of the saturated water changes phase to steam. This water–steam mixture reenters the steam drum. The water in the downcomer section is denser than the water in the water wall section. This sets up a natural circulation of the water between the steam drum and the water or mud drum. The more dense water in the downcomer section is forced down to the water or mud drum and the less dense water in the water wall section is forced up to the steam drum due to the density difference between these two locations.

The *steam drum* performs two functions. First, it is the water reservoir for this circulating system between the steam drum, the water wall tubes, and the water or mud drum. Second, it contains the moisture separators which separate the steam from the water in the saturated steam region of the steam drum. The water is returned to the steam drum for further circulation through the water wall tubes and the steam is allowed to continue on to the primary superheat section that is located in the back pass of the boiler. The steam picks up energy in the primary superheat section and enters the superheat region of the steam tables. This superheated steam leaves the primary superheat section in the back pass and enters the secondary superheat section in the convection pass at the top of the boiler. The gases passing the secondary superheat section are closer physically to the furnace section than the primary super-heat section and therefore, more energy can be added to the superheat section in the secondary superheat tubes. When the steam transfers from the primary superheater to the secondary superheater, it passes the attemperator section where condensate is injected into the superheated steam as necessary to maintain steam temperature at the desired value. This is done to ensure that the temperature of the steam, as it enters the turbine, is maintained within the physical constraints of the metallurgy of the turbine blades. Once the steam is past the attemperator, the steam enters the secondary super-heat section and picks up even more energy from the hot gases in the convection pass of the boiler and leaves as superheated steam. Steam from this point leaves the boiler and is transported to the turbine for the expansion process in steam lines known as main steam lines.

Some designs incorporate a *boiler water recirculation pump* in the downcomers of the boiler. This is a forced circulation design. Other designs depend on the density difference of the water in the downcomers and the water in the water wall tubes to provide the driving force for fluid flow between the steam drum and the water or mud drum.

In regard to the air system that feeds the boiler, there are various designs. For very small boilers, most boiler designs are *natural draft*. Air heated in the boiler provides the driving force for airflow through the boiler. Air is naturally sucked into the boiler inlet and is heated in the furnace section. As the air temperature increases, the air becomes less dense and this causes the air to rise and leave the boiler through the exhaust duct. For larger-rated boilers, this natural circulation is not adequate to ensure sufficient airflow for the rate of combustion of the boiler. These are known as *forced draft* systems. There are two basic types of forced draft systems.

The first system is a *pressurized boiler* system that has one or more forced draft (FD) fan(s) on the air inlet side of the furnace and no induced draft (ID) fan on the discharge side. Since the FD fans are on the intake side of the boiler, the operation of these FD fans raises the air pressure from atmospheric pressure on the suction side of the FD fans to the pressure delivered to the furnace. Therefore the boiler furnace section operates at a positive pressure with respect to atmospheric pressure. These types of systems are called pressurized boilers. The air is then heated in the furnace section, passes the various boiler tubes for heat exchange, and is exhausted through the duct work back to the atmosphere.

A second type of system is called a *balanced draft system*. In this system, there are one or more FD fans on the intake side of the furnace and there is one or more ID fans on the discharge side of the furnace somewhere between the economizer section and the stack. As will be discussed in Chapter 23 on controls, the FD fan regulates the amount of airflow required for the amount of boiler combustion and the ID fan regulates the pressure inside the furnace to the amount required for boiler operation. The ID fans will be operated to maintain the pressure in the boiler furnace section at a slightly negative pressure.

In addition, some systems utilize a *gas recirculation fan* that takes its suction from the boiler discharge ductwork and reinjects this heated air back to the furnace section. This helps improve boiler efficiency as the reinjected air is at an elevated temperature and reduces the amount of fuel needed to heat the air from the FD fans to operational temperatures.

Over the last several decades, additional equipment has been added to the gas stream at the exit of the boiler to remove unwanted materials from the exhaust gases. More will be discussed about these requirements in Chapter 26 on environmental control. However, most of the systems provide additional pressure drop across the exhaust duct work and associated equipment. To compensate for this additional pressure drop, booster fans are installed downstream of the ID fans and, as their name implies, their purpose is to boost the pressure in the discharge duct work to compensate for the additional pressure drop incurred across pollution control equipment in the discharge duct system.

Much like the purpose of the gas recirculation fan, most systems utilize an *air preheater* that is basically a basket of metal that rotates between the boiler inlet duct system from the FD fan and the boiler outlet duct system going to the ID fan. Part of the time the metal spends in the discharge duct and is exposed to the higher furnace air discharge temperatures and this raises the metal temperature of the air preheater baskets. Then the metal passes to the suction ductwork with the cooler supply air for the furnace and heats up the incoming air to the furnace. Just like the gas recirculation fan, the air preheater increases the temperature of the air entering the furnace. This will reduce the fuel requirement from the burner section needed to operate the system at the desired operating point. Since the air is preheated, less fuel is needed to raise the air temperature in the boiler to the necessary value for correct operation. Less fuel required leads to improved overall system efficiency and reduced emissions.

As the fuel–air mixture is combusted in the furnace, noncombustible ash material is generated and carried by the air across the various sections of boiler tubes. Some of this ash material settles out on the boiler tubes. If the temperature of the boiler tube exceeds the ash melting temperature, the ash material melts and forms soot. Most boilers utilize various types of mechanical systems to remove the soot from the tubes. Most commonly, *soot blowers* are used to inject steam through a soot blower lance in the furnace toward the heat transfer surfaces of the tubes to remove the soot and maintain the heat transfer capability of the tubes. If this soot is not removed, it can have two adverse effects. First, it will reduce the heat transfer capability of the tube reducing the amount of heat transferred to the water. Second, as the thermal resistivity of the tube wall increases, this will lead to elevated tube temperatures and eventual tube failure.

In a combined cycle plant, the *heat recovery steam generator* (HRSG) functions much like the boiler in a fossil fuel plant, but the combustion occurs in a gas turbine, not a furnace. The hot combustion gases exit the turbine section of the gas turbine engine at elevated temperatures. These hot gases then pass through the steam generator that contains boiler tubes. Since the gas temperature at the inlet of the steam

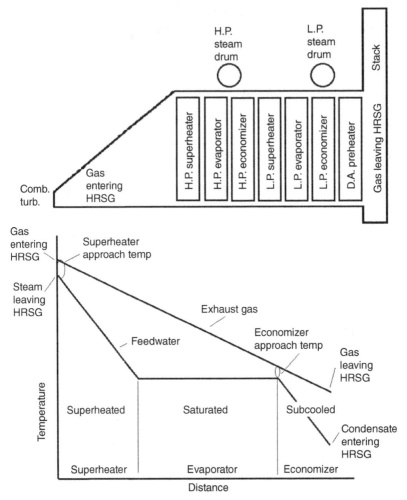

Figure 3.4 Typical temperature profile for a heat recovery steam generator (HRSG).

generator is greater than the gas temperature at the outlet of the steam generator, the high pressure steam supply tubes in the superheater section are oriented toward the air inlet of the steam generator, the intermediate reheat pressure tubes in the evaporator section are located in the mid-section of the HRSG and the low pressure economizer tubes in the economizer section are located at the air output of the steam generator.

Figure 3.4 shows the typical temperature profile of the HRSG. Energy transfer in the form of heat is always from the hotter gas surrounding the steam generator tube to the steam in the steam generator tube. Since, on the gas inlet side (left side of Figure 3.4), the gas temperature is greater than the gas outlet side (right side of Figure 3.4), the hotter steam tubes are on the gas inlet side (the left side of Figure 3.4) and the cooler feedwater tubes are on the gas outlet side (the right side of Figure 3.4). Turbine exhaust gas enters the HRSG at the superheat section at about 1000°F and leaves the superheater section at a temperature of about 900°F. The steam enters the steam generator tubes of the superheater section in a saturated condition at 100% *steam quality* at a temperature of about 500°F and leaves the superheater section as super-heated steam at a temperature of about 900°F. The difference between the turbine exhaust gas temperature entering the superheater section and the superheated steam temperature leaving the superheater section defines *the superheater approach tem-perature*. For the above values, this temperature difference is about 100°F (1000°F − 900°F). In the evaporator section, the inlet gas is about 900°F (the outlet gas temper-ature of the superheater section) and leaves the evaporator section at a temperature of about 600°F. The feedwater enters the evaporator section, from the economizer section in a saturated condition at 0% steam quality of 500°F and leaves the evapora-tor section in a saturated condition at 100% steam quality at the same temperature of 500°F. Note that the temperature of the feedwater does not change in the evaporation section since the heat transfer process is latent heat transfer where the temperature is held constant, but the phase of the water changes to steam. Also, energy is always transferred in the form of heat from a point of higher temperature to a point of lower temperature. Therefore, throughout the gas path in the HRSG, the gas temperature has to be greater than the steam/water temperature at all locations. The two points where these two temperatures are closest between the gas stream and the water/steam stream occur at the feedwater outlet of the economizer section and at the steam outlet of the superheater section. This defines the approach points shown in Figure 3.4. The gas enters the economizer section at a temperature of about 600°F (from the outlet of the evaporation section) and leaves the economizer section (and the HRSG) to the exhaust ductwork at a temperature of about 500°F. The feedwater enters the econo-mizer section as a subcooled liquid at a temperature of about 250°F and leaves at a saturation temperature of just below 500°F.

Boiler Auxiliaries

Burners – The function of a burner is to combine the fuel and the air at the proper mixture and to provide the source of ignition to ensure complete combustion of the fuel, and minimize the formation of thermal NO_x emissions (emissions are addressed in Chapter 26 on pollution controls).

Burners control the admittance of oxygen to the combustion chamber. The cor-rect amount of oxygen needs to be controlled for proper combustion operation. Too little oxygen (a rich mixture) will lead to incomplete combustion of the fuel and the release of carbon monoxide (CO). CO is a combustible molecule and still contains a substantial amount of available chemical energy. Therefore, providing too little oxy-gen results in some of the combustible material not being utilized, but being carried away with the flue gas. Too much oxygen allows for the complete combustion of the fuel, but may increase the amount of NO_x generated in the burner. Additionally, excess air not used in the combustion process only absorbs heat from the furnace but does not contribute to the combustion process. The additional amount of heated air is exhausted to the atmosphere and this loss of thermal energy represents a reduction of efficiency of the boiler. Figure 3.5 shows the relationship for a typical boiler between excess air and the reduction of boiler efficiency.

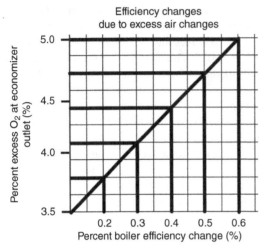

Figure 3.5 Reduction of boiler efficiency due to excess air. *Source*: Reproduced with permission of Tampa Electric Company. Copyright 2016 Tampa Electric Company. Reproduction is forbidden without the express consent of Tampa Electric Company.

Therefore, the burner control system is always trying to balance just the correct amount of air necessary to allow for complete combustion, while maintaining excess air to a minimum.

Burners are also designed to control the temperature at which complete combustion takes place to minimize the formation of NO_x molecules.

Burners operating properly provide the following functions to the boiler.

Oxygen control – stoichiometric quantity (not too much and not too little)

Minimize NO_x and SO_x formation

Provide for uniform combustion

Provide a wide and stable firing range

Provide a fast response

Provide high availability and low maintenance

Provide ignition source for startup and shutdown

There are two main types of boiler design, *watertube* and *firetube*. In smaller boilers (nonutility), it is common for the hot gases from the flame to be contained within the tubes and the water and steam mixture to surround the tubes. For larger-sized boilers, it is more common for the water and steam mixture to be in the boiler tube and for the hot gases from the furnace flame to surround the outside of the tubes. Since the watertube design is most common in utility applications, this is what will be presented in this text book.

In the wall-fired boiler furnace design. The burners are horizontally oriented and direct their flame perpendicularly into the boiler (in actuality, there is a slight tilt in the downward direction to direct the flame slightly lower in the furnace). There are

Figure 3.6   Photo of burner front area with boiler wall removed and water wall tubes exposed.

multiple burners on any one level and there may be several levels of burners depending on the furnace section design.

Figure 3.6 shows a boiler with a *horizontally fired* or *wall-fired* burner design under construction. The boiler wall insulation system is not yet installed and the burner auxiliaries and secondary air *dampers* are not yet installed. The water wall tubes that surround the furnace section are shown from both the burner front area as well as from the inside of the boiler. You can see the tubes are modified to make room for the burner components. The burner directs the flame perpendicular to the boiler wall where the burners are installed. If the wall was a perfectly horizontal wall, this would imply the flame is aimed horizontally directly into the furnace section. However, as can be seen from Figure 3.6, there is a slight turn in the downward direction to compensate for the flame rising in the furnace section.

In the typical horizontally fired boiler burner, the coal is transported to the burner via air called "*primary air*" system. In the burner, the coal is agitated in a vortex by the distribution vanes of the burner and directed to the combustion zone. Air from the "secondary air" system is added to and mixed with this coal at the combustion zone and the igniter provides the ignition source. The amount of coal admitted to the burner is controlled by varying the speed of the conveyors feeding coal to the coal mills. The amount of secondary air admitted to the burner is controlled by the secondary air dampers. The function of the secondary air system is to add sufficient air at the combustion zone to ensure complete combustion of the fuel. Figure 3.7 shows a typical secondary air damper.

Figure 3.7   Secondary air damper before installation.

Some burner designs also include a SOFA system or *separated over fired air* system to lengthen the flame and ensure that complete combustion occurs in an area of lower flame temperature to minimize the generation of NO_x.

The function of the igniter at the burner is to provide the source of energy for ignition of the fuel. As discussed in Chapter 1 on safety, there are several classes of igniter based on the energy level the igniter provides.

Figure 3.8 shows a different type of furnace design called a *tangentially fired boiler* furnace. Instead of the burners being oriented along the furnace walls as in the case of the horizontally fired boiler furnace, the burners are arranged along the four corners of the furnace, with several levels of burners. The burner vertical angle can be

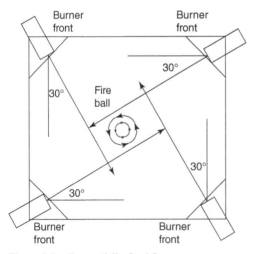

Figure 3.8   Tangentially fired furnace.

controlled to raise or lower the flame in the furnace. The burners at the four corners form a fire ball in the center of the boiler.

The burners can be raised or lowered to move the fireball in the center of the boiler vertically which can be used to control the gas temperatures around the top of the boiler area at the convection pass. This feature can be used to control the heat transfer to the secondary superheater section of the boiler. This will control the temperature of the main steam that is delivered to the high pressure turbine.

For boilers that burn fuels with less energy content, a longer flame is required to completely combust the fuel. In these applications, either a *stoker boiler* or a *vertical furnace* firing pattern is utilized. In the furnace with vertical burners, the flame is directed in a downward direction and allowed to rise back into the furnace section of the boiler. This provides for an overall longer flame and more time for complete combustion of the fuel to occur. In the stoker style boiler, the fuel is placed on a traveling grate and the fuel is transported into the furnace section where it is exposed to the high temperatures of the furnace for a long period of time. Both of these methods provide the fuel with a long residence time in the furnace to allow for complete combustion of the fuels. This is the typical type of boiler employed in applications where the fuel is combustible refuse.

Pulverized coal has a smaller final particle size than crushed coal. For coal plants that utilize dry crushed coal instead of pulverized coal, cyclone burners are used to combust the coal and air mixture. The finer the coal granules, the quicker the coal is completely combusted. Therefore, pulverized coal burners such as horizontally fired or tangentially fired burners do not require the residence time that crush coal plants needed to ensure complete combustion of the coal. To get the longer residence time, the crushed coal is injected to the combustion chamber of the cyclone burner and primary air is injected to push the coal around the burner area where it is combusted. This allows for longer exposure times for the coal to completely combust. For more details on the process of combustion, see the information in Chapter 4 on combustion.

Next, we will discuss some of the auxiliary loads that utilize energy in the plant to support plant operations.

AIR PREHEATER

The energy contained in hot exhaust gases leaving the furnace represent losses to the overall plant efficiency. If we had a method of using that hot exhaust gas to preheat the air entering the furnace, then we would be able to recover some of this energy and redirect it back to the furnace. Therefore, less energy would be required to get furnace gas temperature from atmospheric temperature up to required furnace gas temperatures. The air preheater is such a heat exchanger. This is a rotating device with baskets. This device has several sections. A two-section air preheater is known as a bisector and a three-section air preheater is known as a trisector.

One section of the air preheater has hot exhaust gases passing through it from the economizer section of the boiler which heats up the metal baskets in the air preheater. This collection of baskets is then slowly rotated so that the hot baskets move to the other section of the air preheater where cold incoming air to the boiler furnace

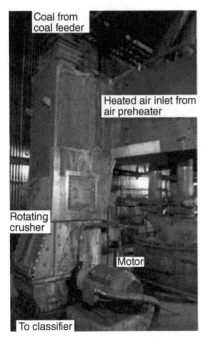

Figure 3.9   Coal dryer/crusher.

is passed through the heated metal baskets. In this section, the metal baskets give off their heat to the air that is then sent to the furnace. By preheating the air that will be entering the furnace section with energy from the exhaust gases, we improve the efficiency of the operation of the boiler by requiring less energy in the form of fuel combustion needed to allow gases to reach required temperature. We also reduce the environmental impact the exhaust gases may have to the environment by lowering the temperature that the gases are discharged at.

In addition to adding energy in the form of heat to the air entering the furnace area, preheated air can also be used for drying functions. Figure 3.9 shows a coal crusher that has a heated air duct supplied to the crusher to perform a drying function while the coal is being crushed.

There are several types of coal *pulverizers* in service and which one is used depends on the design of the furnace and the particle size of coal needed in the burner section. One common type is the ball mill pulverizer as shown in Figure 3.10.

In the *ball mill coal pulverizer*, raw coal from the feeders is fed to the pulverizer at both ends of the pulverizer and it enters the mill area. Ball mills are used to grind raw coal material from a size of 1/4 inch (the discharge of the drier crusher), down to the particle size of 20–75 microns. The mill is turned at a slow speed. The balls that perform the coal grinding are inserted in the mill at the ball-charging hopper. These balls are stored in 55 gal drums and added one drum at a time. As the mill rotates, the balls rise along the side of the mill and fall back onto the coal grinding the coal to finer sizes. Industrial ball mills are typically 15 ft in diameter but can be as large as 25 ft in diameter. The material of the balls depends on the material being ground and, for applications of coal milling, the balls are made from stainless steel.

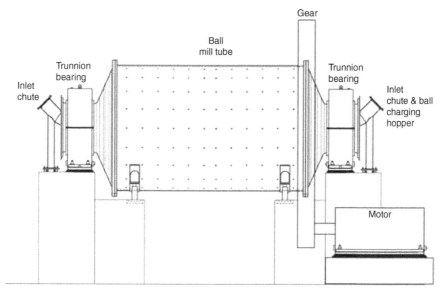

Figure 3.10   Ball mill coal pulverizer.

The ball material must be denser than the material being ground so the balls do not float on top of the material being ground and the ball material must be harder than the material being ground to ensure long life of the balls. The size of the balls must be substantially larger than the size of the particles being ground and the typical size for the balls in a ball mill is 1 inch to 3 inches in diameter depending on the size of the coal entering the ball mill. Ball mills normally operate with an approximate ball charge of 30%. The primary physical factor that drives the power drawn by the ball mill is not the amount of coal in the mill but the weight of the balls inside the mill. Over time, the balls erode and lose both size and weight and become less efficient at grinding the coal. Motor power is monitored to detect when the total weight of the balls in the mill is reduced to the point where grinding efficiency is reduced. This is also known as mill ball charge. As the mill ball charge is reduced over time, the motor power drawn is reduced indicating that the mill ball charge is in need of replacement. More balls can be added to increase the mill ball charge at the ball-charging hopper. The typical motor size that drives the ball mill is 1000–5000 HP.

Primary air is injected in the mill at both ends. This primary air is heated to provide a drying function. The primary air is the transportation means to lift the pulverized coal from the pulverizer outlet ductwork to the classifiers. A photo of a classifier is shown in the Figure 3.11. The function of the *classifier* is to allow the finer particles of coal to leave the classifier via the coal pipes and to be transported to the furnace burners. The larger particles of coal that require further processing are separated in the classifier and returned to the pulverizer for further processing. Steam lines are supplied to the classifier and mills for inerting functions. This is intended to minimize the risk of a coal fire in the mill or classifier.

Another type of coal pulverizer is the *roll wheel pulverizer* that is shown in Figure 3.12. In this equipment, the coal enters the mill from the feeders and falls onto

Figure 3.11   Coal classifier.

the yoke or grinding table. The grinding table is rotated in the mill and the coal is pulverized between the yoke and the roll assembly. The spring loading system on the roll assembly maintains the force on the roll assembly to pulverize the coal. The pulverized coal is then transported from the grinding area to the classifier on top of the mill via primary air that enters through the windbox.

The function of the classifier is to allow the finer particles of coal to leave the classifier in the coal pipes and to be transported to the furnace burners. The larger particles of coal that require further processing are separated in the classifier and returned to the mill for further processing. A typical spindle from a mill grinding roll assembly is shown in the Figure 3.13.

COOLING TOWERS

A *closed loop cooling s*ystem is necessary to remove the waste heat that is generated in the equipment in the electric utility generation station. Examples of equipment that may require cooling water are excitation rectifiers, lube oil coolers, and any large TEWAC (totally enclosed water to air cooled) motors. Additionally, some condensate cooling systems may utilize "helper" cooling towers to minimize the temperature of the condenser cooling water before it is returned to the environment. The heat energy that is picked up by the cooling water system from this plant equipment has to be released at some point in the cooling system. This is the function of cooling towers. Cooling towers are basically water-to-air heat exchange systems. There are many designs in use and some of the more common systems are described below.

In a typical *counter-flow (forced circulation) cooling tower* arrangement, heated cooling system water enters the tower and is discharged vertically down through spray headers in the tower and falls into a basin. The spray headers

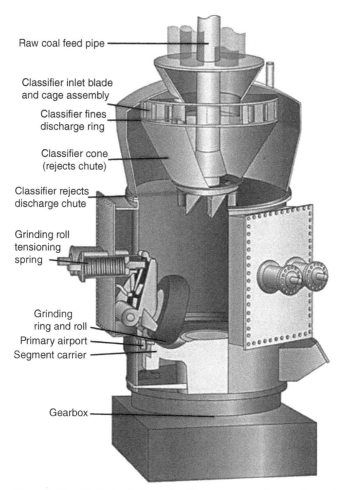

Raw coal feed pipe

Classifier inlet blade
and cage assembly

Classifier fines
discharge ring

Classifier cone
(rejects chute)

Classifier rejects
discharge chute

Grinding roll
tensioning
spring

Grinding
ring and roll

Primary airport

Segment carrier

Gearbox

Figure 3.12   Vertical spindle coal mill arrangement. *Source*: Reproduced with permission of
SAS Global Corporation.

effectively increase the surface area of the water allowing for enhanced heat transfer
to the air stream. Air enters the bottom of the tower and flows vertically up through
the spray system, thus the name counter flow. The fan motor is located at the top of
the tower to force the airflow through the water stream.

A more common arrangement is the *double cross flow cooling tower* arrangement. In a double cross flow (forced circulation) cooling tower arrangement, heated
cooling system water enters the tower and is discharged vertically down across a
"splash fill" or louvers. The louvers increase the surface area of the water, allowing
for enhanced heat transfer to the air stream. The louvers also increase the amount of
time the water is exposed to the air, further enhancing the heat transfer capability of
the system. Air enters the side of the tower and flows horizontally across the louvers.
This is where the "cross flow" name comes from. The fan motor is located at the top
of the tower to force the airflow through the water stream.

Figure 3.13   Vertical spindle coal mill roll assembly.

Both the counter flow and cross flow designs are forced air-cooled. Forced air-cooled means there is a motor-driven fan to drive the air through the cooling tower. Alternately, there is a natural draft cooling tower. The way the *natural draft tower* works is that cooler air is allowed to enter the bottom of the tower. The hot water is sprayed in a counter flow fashion across fill that allows for long residual time for heat exchange between cool air and hot water. Energy is transferred from water to air (equation 3.1). The heated air is now less dense than cooler air at the bottom of the tower and on the outside of the tower. As such, the less dense heated air inside the tower is forced to rise in the tower from the more dense cooler air at the bottom of the tower. This is the force causing natural circulation of air in the cooling tower. Additionally, the natural draft cooling tower has a hyperboloid or inwardly curved shape where the diameter reduces as the warmed air approaches the top of the tower. The hyperboloid shape provides structural strength and reduces the cross-sectional air path as the elevation of the tower increases. Assuming a constant mass flow rate, as the diameter reduces at the upper level of the tower, the speed of the warmed air begins to increase. Therefore, the speed of the air at the top of the tower is greater than the speed of the air just above the water sprays helping to assist with the natural circulation process.

Just as in the study of thermodynamics, we can look at the amount of mass and energy entering the cooling tower and this must equal the amount of mass and energy leaving the cooling tower due to the laws of conservation of energy and mass. The energy entering the cooling tower consists of two streams. The two streams consist of the cool air entering at a lower temperature and the cooling water entering the tower at a warmer temperature. The energy leaving the cooling tower consists of three streams. The three streams consist of the heated air leaving the cooling tower, the cooler cooling water leaving the tower, and the steam (from the water side) that was generated during latent heat transfer and leaves the cooling tower in the air stream. The energy (or heat) balance equation is defined by equation (3.1).

Energy gained by air = Energy lost by water + Energy lost by evaporation.

$$G \, \Delta h = L \, \Delta t + G \, \Delta H \, (t_2 - 32) \tag{3.1}$$

where

G = mass flow air

Δh = change in enthalpy

L = mass flow water

t = temperature

ΔH = change in humidity ratio

We control the airflow through the cooling tower and the feedback parameter we use for the control loop is the water out temperature (cold water from the tower) as this is cooling water to equipment and we want to maintain a steady cooling water temperature to this equipment. The ability of a cooling tower to remove the heat energy from the water is based on equation (3.1). This equation uses the concept of energy conservation. The left side of the equation is the energy (change in enthalpy × mass flow rate) gained by the air in the exchange process. The right side of the equation is the energy given up by the water side and this has two parts. The first part is the sensible heat transfer (heat transfer that results in a change in temperature) as the water temperature drops. The second part is due to the latent heat transfer (heat transfer that results in a change of phase) as the water evaporates. The evaporation process is dependent on the "wet bulb" temperature, not "dry bulb" temperature. This is the reason for the humidity ratio (H) term in equation (3.1) to convert the dry bulb temperature measured and used in the equation to the wet bulb temperature.

We control the amount of heat exchange between the water stream and the air stream by controlling the airflow stream. We could use flow control of the water stream, but that could possibly result in restricting the water flow. The concern about using the water flow for temperature control of the cold water temperature out of the cooling tower is that we do not want to starve equipment in the facility for cooling water. By controlling airflow, we do not put restrictions on the water flow through the plant equipment. The feedback signal for control is the water system output (cold side) temperature.

The "wet bulb" temperature is simply a means of defining the amount of humidity in the air at a certain temperature. The wet bulb temperature is an experimentally determined value. It is determined by taking a wet cloth and wrapping it around a mercury temperature bulb, and then air is forced across it with some amount of humidity in the air. Some of the water in the cloth will evaporate, and with this evaporation, some heat transfer occurs. This heat transfer results in a lower temperature in the thermometer after the water evaporates due to latent heat transfer which is a measure of the amount of energy removed by the evaporation process.

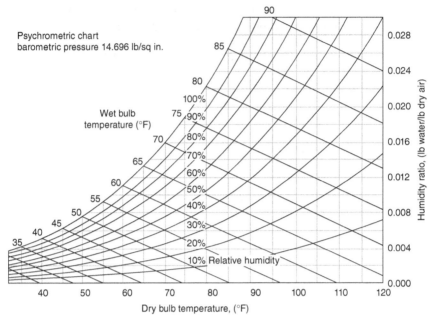

Figure 3.14   Chart of dry bulb temperature to wet bulb temperature Reprinted by permission of Tampa Electric Company. Copyright 2016 Tampa Electric Company. Reproduction is forbidden without the express consent of Tampa Electric Company.

At 100% humidity, evaporation cannot occur since the air is already in saturation conditions. Since evaporation does not occur, at the 100% humidity point, the dry bulb and wet bulb temperatures are the same. At 0% humidity, evaporation easily occurs and the maximum amount of heat transfer occurs resulting in the most reduction from dry bulb temperature to wet bulb temperature. Therefore, the wet bulb temperature is lower than the dry bulb temperature by the largest amount at 0% humidity. A chart comparing "wet bulb" temperature to "dry bulb" temperature is shown in the simplified psychometric chart shown in Figure 3.14.

Example 3.1   Given a dry bulb temperature of 70°F and a relative humidity ratio of 40%, determine the wet bulb temperature.

Solution: Using Figure 3.14, find the value of 70°F dry bulb temperature on the horizontal axis. Then find the humidity curve of 40%. Find the intersection for these two points and the diagonal wet bulb temperature that is associated with these two points is found to be about 56°F.

GLOSSARY OF TERMS

- Air Preheater – Heat-transfer apparatus through which air is passed and heated by a media of higher temperature, such as the products of combustion or steam.

- Attemperator – Apparatus for reducing and controlling the temperature of a superheated vapor or of a fluid.
- Balanced Draft – The maintenance of a fixed value of draft in a furnace at all combustion rate by control of incoming air and outgoing products of combustion.
- Boiler – A closed vessel in which water is heated, steam is generated, steam is superheated, or any combination thereof, under pressure or vacuum by the application of heat from combustible fuels, electricity, or nuclear energy.
- Damper – A device for introducing a variable resistance of regulating the volumetric flow of gas or air.
- Deaerating Heater – A type of feedwater heater operating with water and steam in direct contact. It's designed to heat the water and to drive off oxygen.
- Downcomer – A tube or pipe in a boiler or water wall circulation system through which fluid flows downward between headers.
- Forced Draft (FD) Fan – A fan supplying air under pressure to the fuel burning equipment.
- Gas Recirculation (GR) – The reintroduction of part of the combustion gas at a point upstream of the removal point, in the lower furnace for the purpose of controlling steam temperature.
- Header – A distribution pipe supplying a number of smaller lines tapped off of it. A main receiving pipe supplying one or more main pipe lines and receiving a number of supply lines tapped into it.
- Mud or Lower Drum – A pressure chamber of a drum or header type located at the lower extremity of a watertube boiler convection bank which is normally provided with a blow-off valve for periodic blowing off of sediment collecting in the bottom of the drum.
- Primary Air (PA) – Air introduced with the fuel at the burners.
- Pulverizer – A machine which reduces a solid fuel to a fineness suitable for burning in suspension.
- Steam Drum – A pressure chamber located at the upper extremity of a boiler circulatory system in which the steam generated in the boiler is separated from the water and from which steam is discharged at a position above a water level maintained there.
- Steam Quality – The percentage by weight of vapor in a steam and water mixture.
- Tangential Firing – A method of firing by which a number of fuel nozzles are located in the furnace walls so that the centerlines of the nozzles are tangential to a horizontal circle. Corner firing is usually included in this type.
- Tertiary Air – Air for combustion supplied to the furnace to supplement the primary and secondary air.
- Water Tube Boiler – A boiler in which the water or other fluid flows through the tubes and the products of combustion surround the tubes. This kind of boiler is

mainly used for high pressure steam (utility application) but also can be used to produce low pressure steam (industrial application).

• Water Wall – A row of watertubes lining a furnace or combustion chamber, exposed to the radiant heat of the fire, used to protect refractory and to increase capacity of the boiler.

PROBLEMS

3.1 In a balanced draft boiler, the _____ fan is used to control boiler pressure (vacuum).

    A. forced draft

    B. induced draft

    C. primary air

    D. gas recirculation

3.2 In a balanced draft boiler, the _____ is used to control the fuel flow.

    A. primary air fan

    B. forced draft fan

    C. induced draft fan

    D. pulverizer

3.3 Given a wet bulb temperature of 55°F and a relative humidity ratio of 0%, determine the dry bulb temperature.

3.4 The furnace of the boiler is where _____.

    A. coal is ground or pulverized

    B. water is converted to steam

    C. fuel and air is mixed and combusted

3.5 Complete this sentence:
In an HRSG, the superheater approach temperature is the difference in temperature between the gas _____ the _____ section and the steam _____ the _____ section of the steam generator.

    A. entering, superheater, leaving, superheater

    B. leaving, superheater, entering, superheater

    C. entering, economizer, leaving, economizer

    D. leaving, economizer, entering, economizer

    E. entering, evaporator, leaving, evaporator

    F. leaving, evaporator, entering, evaporator

3.6 Complete this sentence:
In an HRSG, the economizer approach temperature is the difference in temperature between the gas _____ the _____ section and the steam _____ the _____ section of the steam generator.

    A. entering, superheater, leaving, superheater

    B. leaving, superheater, entering, superheater

    C. entering, economizer, leaving, economizer

    D. leaving, economizer, entering, economizer

    E. entering, evaporator, leaving, evaporator

    F. leaving, evaporator, entering, evaporator

3.7 What is the function of a classifier?

3.8 Complete this sentence:
A downcomer is a tube or pipe in a boiler through which _____ flows downward.

    A. primary air

    B. coal

    C. fluid or water

    D. secondary air

RECOMMENDED READING

Electric Power Plant Design, Technical Manual TM 5-811-6, Department of the Army, USA, 1984.

Electrical Machines, *Drives and Power Systems*, 6th edition, Theodore Wildi, Prentice Hall, 2006.

The Engineering Handbook, 3rd edition, Richard C. Dorf (editor in chief), CRC Press, 2006.

Foster Wheeler, www.fwc.com

Power Plant Engineering, Black & Veatch, edited by Larry Drbal, Kayla Westra, and Pat Boston, Chapman & Hall / Springer, 1996.

Riley Power, www.rileypower.com

Standard Handbook of Powerplant Engineering, 2nd edition, Thomas C. Elliott, McGraw Hill, 1998.

Plant Performance Data and Calculations, *Training Handbook*, Tampa Electric, Tampa, FL.

US Navy Training Manual Machinist Mate 2nd Class, Naval Education and Training Professional Development and Technology Center, 2003, http://navybmr.com/studymaterial%204/NAVEDTRA%2014151.pdf

CHAPTER 4

FOSSIL FUELS AND THE BASIC COMBUSTION PROCESS

GOALS

- To identify the three elements of the combustion process
- To identify the various combustion reaction equations and the amount of heat of combustion released for each reaction
- To calculate the molecular weights of various elements
- To calculate the stoichiometric quantity of air necessary for complete combustion
- To understand the basic integrated gasification combined cycle (IGCC) process
- To be familiar with the various types of fuels utilized for chemical combustion

---

THE **PROCESS** of combustion requires three items: (1) a combustible fuel containing a certain amount of available energy that can be released during the *combustion process*, (2) oxygen in sufficient quantity to ensure *complete combustion* (and the volume residence time required for complete combustion), and (3) a heat source. This is commonly called the fire triangle as shown in Figure 4.1.

Below we will discuss the combustible fuel and the oxidizing agent (air) as they pertain to the typical fossil fuel steam generator. The heat source was discussed in Chapter 1: Electrical Safety, under section Boiler Control and Burner Management, in the subsection on ignitors.

COMBUSTIBLE FUEL

The two most common types of fossil fuels utilized for electricity generation are natural gas and coal as shown in Figure 4.2.

We will focus our attention on the most common fuel, coal, but the science is similar for the various fuels as they all are combusting one or more elements in a

---

Energy Production Systems Engineering: An Introduction for Electrical Engineers to Electrical Power Generation Facilities, Systems, and Equipment, First Edition. Thomas H. Blair.

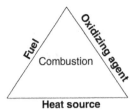

Heat source

Figure 4.1   Fire triangle. *Source*: Reproduced with permission of Department of Energy (DOE) Fundamentals Handbook.

redox (oxidation-reduction) reaction to release the chemically stored energy in the fuel. For coal, the main components are *carbon*, sulfur, and hydrogen. When carbon is mixed with a sufficient amount of air containing oxygen for a long enough period of time and is completely combusted, it provides 14,100 BTU/lbm of carbon combusted and it releases carbon dioxide CO_2 gas as shown in Table 4.1 in the first row.

Carbon dioxide is an inert gas and cannot be combusted to release any further chemical energy. When carbon is combusted, if the amount of oxygen is insufficient for complete combustion, then carbon monoxide (CO) gas is released as is shown in the reaction column of Table 4.1 in the second row. Unlike carbon dioxide, carbon monoxide can undergo further combustion reaction with oxygen to release more energy and release carbon dioxide (CO_2) gas as is shown in the reaction column of Table 4.1 in the third row.

Table 4.1 shows the combustion equations for various types of elements contained in fossil fuels. It also shows the theoretical maximum amount of energy released for one pound mass of element combusted.

Example 4.1   What is the amount of energy released when 10 lb of carbon is completely combusted releasing CO_2?

Solution: From Table 4.1, the energy released per pound mass for the complete combustion of 1 lbm of carbon is 14,100 BTU/lbm. The total energy released is the product of the energy per pound mass multiplied by the mass of material combusted.

$$E = 14,100 \text{ BTU/lbm} \times 10 \text{ lbm}$$

$$E = 141,000 \text{ BTU}$$

OXYGEN

As shown in the fire triangle, combustion requires oxygen. The most abundant source of oxygen is air, but air is a combination of different elements. The composition of air by percent weight and percent mass and the molecular weights of these constituents are shown in Table 4.2.

From Table 4.2, percent volume is a measure of the percentage of volume of air that a particular element makes up. Molecular weight is a measure of the mass of the

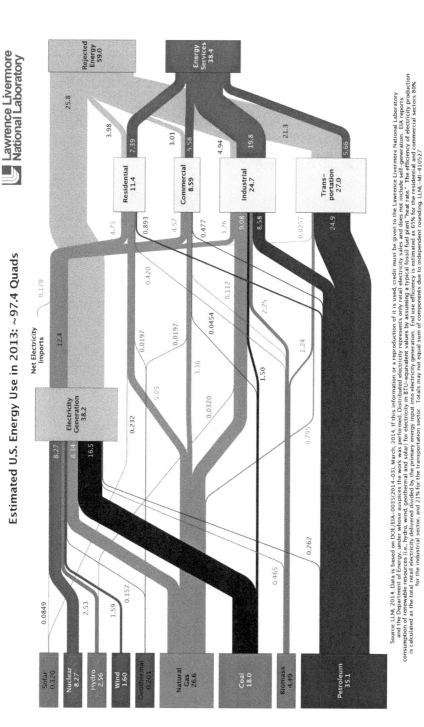

Estimated U.S. Energy Use in 2013: ~97.4 Quads

Source: LLNL 2014. Data is based on DOE/EIA-0035(2014-03), March, 2014. If this information or a reproduction of it is used, credit must be given to the Lawrence Livermore National Laboratory and the Department of Energy, under whose auspices the work was performed. Distributed electricity represents only retail electricity sales and does not include self-generation. EIA reports consumption of renewable resources (i.e., hydro, wind, geothermal and solar) for electricity in BTU-equivalent values by assuming a typical fossil fuel plant "heat rate." The efficiency of electricity production is calculated as the total retail electricity delivered divided by the primary energy input into electricity generation. End use efficiency is estimated as 65% for the residential and commercial sectors 80% for the industrial sector, and 21% for the transportation sector. Totals may not equal sum of components due to independent rounding LLNL-MI-410527

Figure 4.2   Estimated US energy usage for 2013. *Source:* Reproduced with permission of Lawrence Livermore National Laboratory.

TABLE 4.1 Selected Values for Heat of Combustion

Combustible → Resultant	Molecular Weight	Reaction	Heat Release (BTU/lbm)
Carbon → carbon dioxide	12	$C + O_2 \rightarrow CO_2$	14,093
Carbon → carbon monoxide	12	$2C + O_2 \rightarrow 2CO$	8694
Carbon monoxide → carbon dioxide	28	$2CO + O_2 \rightarrow 2CO_2$	19,506
Hydrogen → Water	2	$2H_2 + O_2 \rightarrow 2H_2O$	61,000
Sulfur → sulfur dioxide	32	$S + O_2 \rightarrow SO_2$	3983
Hydrogen sulfide → sulfur	34	$2H_2S + 3O_2 \rightarrow 2S_2 + 2H_2O$	7100
Methane → carbon dioxide	16	$CH_4 + 2O_2 \rightarrow CO_2 + 2H_2O$	23,900
Ethane → carbon dioxide	30	$2C_2H_6 + 7O_2 \rightarrow 4CO_2 + 6H_2O$	22,300
Propane → carbon dioxide	44	$C_3H_8 + 5O_2 \rightarrow 3CO_2 + 4H_2O$	21,500
Butane → carbon dioxide	58	$2C_4H_{10} + 13O_2 \rightarrow 8CO_2 + 10H_2O$	21,300
Pentane → carbon dioxide	72	$C_5H_{12} + 8O_2 \rightarrow 5CO_2 + 6H_2O$	22,000

Source: Black & Veatch, 1996. Reproduced with permission of Springer.

element contained in air. Since there are a certain number of moles of oxygen needed to completely combust a certain number of moles of a particular type of molecule, molecular weight is utilized in combustion calculations and not the volume. Molecular weight is the mass of one molecule normalized to 1/12 of the mass of carbon which has by definition a molecular weight of 12.011 (reference Table 4.3 for the periodic table of elements).

Example 4.2   What is the molecular weight of methane (CH_4) in moles?

Solution: One molecule of methane (CH_4) consists of one carbon (C) and four hydrogen (H_4) atoms. One carbon (C) atom has a molecular weight of 12. Four hydrogen atoms (H_4) have a total molecular weight of $4 \times 1 = 4$. The total molecular weight for a methane molecule is the sum of the individual components or $12 + 4 = 16$ moles.

From the fundamental combustion equations provided in Table 4.1, for any particular element or molecule used for a fuel source, the reaction process takes a certain number of fuel molecules to combine with a certain number of oxygen molecules to completely combust and release the energy available along with the products of combustion. We can multiply the number of molecules by the molecular weight of the material to determine the mass of the material used for fuel. Using this, we can

TABLE 4.2 Fundamental Composition of Air

Component	Volume (%)	Molecular Weight	Weight (%)
Nitrogen	78.09	28.016	19.5
Oxygen	20.95	32	22.2
Argon	0.93	39.944	27.7
Carbon dioxide	0.03	44.01	30.6

Source: Black & Veatch, 1996. Reproduced with permission of Springer.

TABLE 4.3 Periodic Table of the Elements

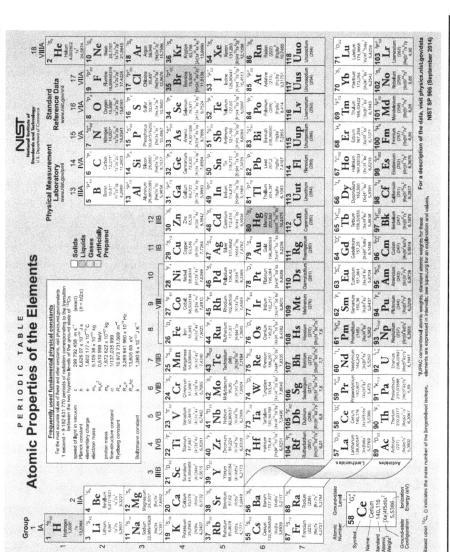

Source: Reproduced with permission of National Institute of Standards and Technology, U.S. Department of Commerce.

determine the mass of oxygen, and therefore, the mass of air needed for complete combustion.

For example, to completely combust carbon and release carbon dioxide, from Table 4.1, the chemical reaction equation is

$$C + O_2 \rightarrow CO_2$$

It takes one molecule of carbon (C) and one molecule of oxygen (O_2) to completely combust the carbon. From Table 4.1, one molecule of carbon (C) has a molecular weight of 12. This can also be found in the periodic table of elements in Table 4.3. In the periodic table, each element displays the molecular weight of the element just under the name of the element.

For example, from Table 4.3 a molecule of oxygen (O_2) has a molecular weight of 32 as each oxygen atom (O) has a molecular weight of 16 and there are two oxygen atoms (O) that make up an oxygen molecule (O_2). The mass ratio is the same regardless of whether we are talking about moles or pounds. From the reaction equation above, we see we need 12 moles of carbon (C) for 32 moles of oxygen (O_2). We can restate this to say that, to completely combust 12 lb of carbon (C), it requires 32 lb of oxygen. Normalizing this to 1 lb of carbon (C), we can state that to combust one pound of carbon (C), it requires 32/12 or 2.667 lb of oxygen (O_2).

As stated above, since oxygen constitutes about 23% of air, by mass, if we divide the mass of oxygen required by 23%, we find the mass of air needed to supply sufficient oxygen needed for complete combustion. In this example, we needed 2.66 lbm of oxygen (O_2). Therefore, we will need 11.6 lbm of air (2.667 lbm O_2 / 23% = 11.6 lbm air) to provide the minimum amount of oxygen required to completely combust 1 lbm of carbon. Therefore, it takes 11.6 lbm of air to completely combust 1 lbm of carbon and release carbon dioxide. This is an example of the method to calculate the "stoichiometric" quantity of air or *theoretical air* for complete combustion of fuel. Table 4.4 summarizes the "stoichiometric" quantity of air necessary for various elements as determined by the above process.

Example 4.3   To completely combust 100 lbm of Pentane (C_5H_{12}) gas requires how much minimum oxygen in pounds?

Solution: From Table 4.4, the ratio of pounds of pentane (C_5H_{12}) to pounds of oxygen is 4.27/1.

Multiplying the mass of fuel by the ratio of oxygen to fuel give us

$$\text{Pounds oxygen} = 3.55 \text{ pounds of oxygen/1 pound of pentane}$$
$$\times 100 \text{ pounds of pentane}$$

$$\text{Pounds oxygen} = 355 \text{ pounds of oxygen}$$

Therefore, to combust 100 lb of Pentane (C_5H_{12}) gas requires, at minimum, 355 lb of oxygen.

Example 4.4   To completely combust 100 lb of Pentane (C_5H_{12}) gas requires how much minimum air in pounds?

TABLE 4.4 Stoichiometric Amounts of Air Required for Various Fundamental Molecules

Combustible	Symbol	Mole of Combustable	Reaction Equation	Mole of Oxygen	lbm O_2/lbm Combustable	lbm Air/lbm Combustable
Carbon → CO_2	C	12.011	$C + O_2 \rightarrow CO_2$	31.998	2.66	11.58
Carbon → CO	C	24.022	$2C + O_2 \rightarrow 2CO$	31.998	1.33	5.79
Hydrogen	H_2	4.032	$2H_2 + O_2 \rightarrow 2H_2O$	31.998	7.94	34.50
Sulfur	S	32.06	$S + O_2 \rightarrow SO_2$	31.998	1.00	4.34
Hydrogen sulfide	H_2S	68.152	$2H_2S + 3O_2 \rightarrow 2SO_2 + 2H_2O$	95.994	1.41	6.12
Methane	CH_4	16.043	$CH_4 + 2O_2 \rightarrow CO_2 + 2H_2O$	63.996	3.99	17.34
Ethane	C_2H_6	60.14	$2C_2H_6 + 7O_2 \rightarrow 4CO_2 + 6H_2O$	223.986	3.72	16.19
Propane	C_3H_8	44.097	$C_3H_8 + 5O_2 \rightarrow 3CO_2 + 4H_2O$	159.99	3.63	15.77
Butane	C_4H_{10}	116.248	$2C_4H_{10} + 13O_2 \rightarrow 8CO_2 + 10H_2O$	415.974	3.58	15.56
Pentane	C_5H_{12}	72.151	$C_5H_{12} + 8O_2 \rightarrow 5CO_2 + 6H_2O$	255.984	3.55	15.43
CO → CO_2	CO	56.02	$2CO + O_2 \rightarrow 2CO_2$	31.998	0.57	2.48

Source: Black & Veatch, 1996. Reproduced with permission of Springer.

Solution: From Table 4.4, the ratio of pounds of Pentane (C_5H_{12}) to pounds of air is 15.43/1.

Multiplying the mass of fuel by the ratio of air to fuel give us

Pounds oxygen = 15.43 pounds of air/1 pound of Pentane $\times$ 100 pounds of pentane

Pounds oxygen = 1543 pounds of air

Therefore, to combust 100 lb of pentane (C_5H_{12}) gas requires, at minimum, 1543 lb of air.

Example 4.5 Given that your fuel is pure carbon. Further assume that you completely combust the carbon via the following reaction and the only byproduct of the combustion process is carbon dioxide.

$$C + O_2 \rightarrow CO_2$$

How much CO_2 is released for 1 kWh of electrical energy delivered? Assume a net plant heat rate of 10,000 BTU/kWh for your calculation. Further, assume that the only chemical process is the complete combustion of carbon and the emission is pure CO_2.

Solution: Starting with what is given, we are generating 1 kWh of energy and given the heat rate of the plant, this means we need to burn enough carbon to obtain 10,000 BTU of energy.

Chemical energy = Electrical energy $\times$ heat rate

Chemical energy = 1 kWh $\times$ 10,000 BTU/kWh = 10,000 BTU

Now we need to find out how much carbon it takes to obtain 10,000 BTU energy input. From Table 4.1, if we burn one pound of carbon, we release 14,093 BTU of energy (Table 4.1). Therefore, the amount of carbon we need to combust for a release of 10,000 BTU is

Carbon = 10,000 BTU/(14,093 BTU/lbm) = 0.7095 lbm carbon

From Table 4.4, for every molecule of carbon combusted, we release one molecule of CO_2. From the periodic Table, the molecular weight of carbon is 12.011 moles and this releases one mole of carbon dioxide with a molecular weight of 1 $\times$ 12.011 + 2 $\times$ 15.999 = 44.009 moles of carbon dioxide. The molecular ratio of carbon dioxide to carbon is 44.009/12.011 = 3.664. We can use this ratio to determine the amount of CO_2 released when we combust 0.7092 lbm of carbon.

CO_2 = 0.7095 lbm carbon/(1 lbm C/3.664 lbm CO_2) = 2.6 lbm CO_2

Therefore, to produce 1 kWh of energy, we release 2.6 lbm of carbon dioxide when combusting pure carbon.

It is worth mentioning from the discussion above that, if the heat rate of the plant is improved (i.e., we have a lower BTU/kWh value), then it takes less BTU to generate 1 kWh of energy. Reducing the amount of BTU required means we burn less carbon, which in turn means we produce less CO_2 for the same kWh generated. This leads to a more environmentally friendly plant (as well as reduced fuel cost per kWh).

The above example is an ideal case. It made two assumptions to simplify the calculation. First assumption was that the fuel was completely carbon. This is never the case as fossil fuels are a combination of elements. For example, there are many combustibles in coal, but the three highest in quantity are carbon, hydrogen, and sulfur (in that order). Also, there are many non-combustibles that add to the weight of the coal, but not the BTU content. This adversely affects the BTU/lbm heating value of the coal. Additionally, there is the moisture content that adds to the weight and actually takes away from the BTU content. Different types of coal that are obtained from different regions have different mixtures of these materials, so the BTU per lbm (heating value) of coal varies greatly. This is why sampling of the BTU value of coal is important to determining its heating value. The second assumption was that we completely combusted all the carbon and the only byproduct was carbon dioxide.

Lastly, in practical application, the "stoichiometric" quantity of air for complete combustion of fuel is controlled by monitoring for one of two gases in the exhaust gas of the *boiler*. The first method is to monitor for the amount of oxygen or O_2. If the amount of secondary air is less than necessary for complete combustion, then the value of O_2 will be low and the amount of CO will be high. If the amount of secondary air is more than necessary for complete combustion, then the value of O_2 will increase. This method works well in pressurized boilers but may suffer in balanced draft boilers due to air-in leakage at the boiler. Balanced draft boilers operate at a slightly negative pressure. Any openings in the boiler between the furnace and the point where O_2 probes are installed would cause the reading of the O_2 probes to read a higher value of O_2 than what actually exists as a byproduct of combustion. Another method for monitoring for the "stoichiometric" quantity of air for complete combustion of fuel is by monitoring the amount of CO. This method does not suffer from errors due to air in leakage in balance draft boiler designs. The combustion air is controlled to maintain the amount of CO at a small amount dependent upon the design of the combustion system.

FOSSIL FUELS

Referring to Figure 4.2, the two most used common fossil fuels in a power generation facility are coal and natural gas. Fossil fuels are formed by the decomposition of organics (materials that contain carbon) over millions of years. The main combustible material in fossil fuel that is used to obtain chemical energy is carbon, although other molecules such as hydrogen and sulfur may provide some amount of energy during the combustion process. As of 2013 in the United States, about 46% of fossil fuel used in power generation facilities is coal, 20% is natural gas, and about 1% is fuel oil. Fossil fuels are considered as *nonrenewable resources* due to the fact that their formation takes millions of years. *Renewable resources* are defined as resources that can be formed in a short period of time.

There are environmental concerns with the combustion of fossil fuels. Fossil fuels are carbon-based and, as such, the combustion of fossil fuels releases carbon dioxide (CO_2). In Chapter 4 on combustion, we covered the combustion equations that show that the complete combustion of carbon-containing fuels releases carbon dioxide (CO_2) into the environment. Carbon dioxide (CO_2) has been identified as a greenhouse gas and a significant change in the amount of carbon dioxide (CO_2) in the atmosphere can cause adverse effects. (Reference the section on gasification where the process of carbon sequestration is discussed.) There are fuels that do not contain carbon and, as a result, they do not emit carbon dioxide (CO_2). Hydrogen is one of these elements where the combustion of hydrogen results in water vapor (H_2O).

Also, coal has the disadvantage of requiring a great deal of handling and processing before it can be combusted in the furnace of a boiler. However, it has the advantage of being able to be stored in large quantities locally at the power generation facility. A typical power generation facility coal field will be able to store 3 to 4 months' worth of coal onsite. This enhances the reliability of the power generation facility operations as it makes the power generation facility less dependent on the availability of fuel transportation. Should the supply of coal be interrupted, the plant can continue to operate for several months before the need for more shipments of coal would interfere with plant operations. Oil, similarly, can be stored on site, but in smaller quantities. Instead of months of fuel being able to be stored in the case of coal, for fuel oil, the quantity is several days, up to several weeks where multiple storage tanks are maintained on the power generation facility. Natural gas is utilized as it is delivered. Therefore, an interruption to the natural gas delivery system forces a loss of the generation capability that uses natural gas for its energy source. However, from an environmental standpoint, the combustion of natural gas results in reduced emissions when compared with the emissions of fuel oil and coal.

Coal Gasification

While the typical coal plant pulverizes the coal and then delivers it to the furnace for combustion of the carbon content, another method for combustion of coal is coal gasification. This is known as a "clean coal" technology as it has the ability to strip the carbon and other materials from coal and deliver pure hydrogen to the combustion turbine for the combustion process. The byproduct of combustion of hydrogen is pure water. A typical Integrated Gasification Combined Cycle (IGCC) process is shown in Figure 4.3.

In this process, the coal is milled and then mixed with water and delivered to a preliminary combustion chamber called a "gasifier." Pure oxygen is admitted to the gasifier and the coal slurry and oxygen (O_2) are partially combusted to release the two main combustible elements that are of interest, carbon monoxide (CO) and hydrogen (H_2). The chemical reaction that generates the carbon monoxide (CO) and hydrogen (H_2) molecules is shown below. First, the carbon and oxygen are combusted to generate carbon dioxide as shown by equation (4.1).

$$C + O_2 \rightarrow CO_2 + \text{Heat} \tag{4.1}$$

Integrated gasification combined cycle process

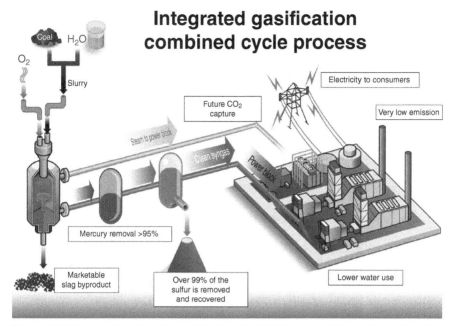

Figure 4.3   Integrated gasification combined cycle (IGCC) process. *Source*: Reproduced with permission of Tampa Electric Company.

The thermal energy or heat generated in the reaction described by equation (4.1) is then used to convert carbon dioxide into carbon monoxide as shown equation (4.2).

$$C + CO_2 + Heat \rightarrow 2\,CO \qquad (4.2)$$

Additionally, the thermal energy or heat generated in the reaction described by equation (4.1) is also used to convert carbon in the coal and water in the slurry mixture into hydrogen and carbon monoxide as shown in equation (4.3).

$$C + H_2O + Heat \rightarrow H_2 + CO \qquad (4.3)$$

The partial combustion of carbon to carbon monoxide releases some energy and this energy is recovered using heat exchangers to generate steam. This steam cycle is known as the bottoming cycle or Rankine cycle. This steam is then sent to a steam turbine to convert this thermal energy into mechanical energy in the turbine and, finally, into electrical energy in the generator.

This mixture of carbon monoxide (CO) and hydrogen (H_2) is called synthetic gas or simply "syn-gas." This combustible gas is then delivered to a combustion turbine to transfer the chemical potential energy of the carbon monoxide (CO) and hydrogen (H_2) into the mechanical energy needed to drive the generator connected to the combustion turbine. This combustion cycle is known as the topping cycle or Brayton cycle.

However, the syn-gas can be further refined to "capture" the carbon and generate pure hydrogen. Referencing Table 4.1, you will see that both carbon monoxide (CO) and hydrogen (H_2) have a substantial amount of BTU content available, but the energy released for one pound of hydrogen is substantially greater than the energy released for one pound of carbon monoxide. Referencing Table 4.1, one pound of carbon (C) releases 14,100 BTU of energy, but one pound of hydrogen (H_2) releases 61,000 BTU of energy. Additionally, capture of the carbon before combustion prevents carbon dioxide from being released during the combustion process.

Further processing of the carbon monoxide (CO) with water (H_2O) can convert these two molecules into carbon dioxide and hydrogen, resulting in the final combustible element of hydrogen (H_2). The chemical equation that achieves this conversion is known as a water gas shift reaction and is shown equation (4.4).

$$CO + H_2O \rightarrow CO_2 + H_2 \qquad (4.4)$$

Now, only the hydrogen is sent to the combustion turbine and, as a result of the combustion of hydrogen molecules, the emissions of the combustion turbine are only water (H_2O). The combustion turbine spins another generator to convert this mechanical energy into electrical energy. The resultant carbon dioxide from the reaction described by equation (4.4) may be stored (called carbon sequestration). The water gas shift reaction process makes gasification a technology of interest in the ability to strip or clean the carbon from the coal and only deliver hydrogen compounds to the combustion turbine, thereby eliminating the generation of carbon dioxide (CO_2) in the emission of the combustion turbine and allowing for the removal and long-term storage of the carbon dioxide. This is commonly known as "clean coal technology."

Additionally, during the gasification process, the other elements such as sulfur and mercury are stripped from the coal. The sulfur can be processed to create a marketable side product of sulfuric acid which is utilized in many industries.

NATURAL GAS

Natural gas is composed mostly of methane (CH_4) and ethane (C_2H_6) and a few other hydrocarbons. One of the greatest advantages of natural gas over fuel oil and coal as an energy source for power generation is that the emissions produced by natural gas have lower values of regulated sources than fuel oil or coal. Oil and coal both contain some amount of sulfur. Combustion of oil or coal results in the formation of sulfur dioxide (SO_2). Since the main constituents of natural gas do not contain sulfur (S), the combustion of natural gas does not produce sulfur dioxide as a byproduct of combustion. The largest disadvantage to natural gas for an energy resource is the inability to store large quantities of natural gas onsite at the power generation facility. In instances where onsite storage is not possible, the generation is dependent on the immediate delivery of the natural gas at the time of consumption.

Natural gas is found in deep underground rock formations or associated with other hydrocarbon reservoirs in coal beds. It is found in the same vicinity as oil deposits. Natural gas pulled from the ground contains many impurities and is processed to remove all the other materials except for the hydrocarbons such as propane, butane, pentane, sulfur, and carbon dioxide. Natural gas is a colorless and odorless media. Additives are utilized to give the gas the "rotten egg" odor to help with detection of gas leaks for safety reasons.

Natural gas is the second most common fossil fuel utilized in electric power generation. The most common application is to drive combustion turbines utilizing the Brayton thermodynamic cycle, but alternately, some boilers utilize natural gas in their furnace sections to generate the heat needed to drive the thermal Rankine cycle. Utilization of natural gas to drive a combustion turbine especially when utilizing a heat recovery steam generator is more efficient than combustion of natural gas in a boiler. The higher efficiency of natural gas combined cycle combustion turbine plants results in less carbon combusted for a given value of electrical energy (kWh) produced, which leads to a reduction of carbon dioxide released by combined cycle combustion turbine plants than coal-fired boiler plants.

For reduced emissions, natural gas is preferred over fuel oil and coal technologies. It produces less carbon dioxide (CO_2) than other fuels and has very low sulfur content, so the amount of sulfur dioxide (SO_2) is also reduced. Another emission from the combustion of fossil fuels with air is the formation of nitrogen monoxide (NO) and nitrogen dioxide (NO_2), which is more commonly known as NO_x *formation*. The amount of NO_x is controlled using both pre- and post-combustion processes. For more information on emission controls, please reference Chapter 26 on emission controls.

FUEL OIL

The popularity of fuel oil for a fuel for electric generation has decreased over the years, but it is still popular as an ignition fuel when starting and shutting down a boiler. This is mostly due to the fact that oil can be stored for long periods onsite until needed for startup and shutdown. Also, since oil is in liquid form, the equipment necessary to handle the oil is simply a forwarding pump, piping, and associated control valves, so handling of the oil to transport it to the furnace section is simple.

Fuel oil is a distillate of petroleum oil. When petroleum oil is received, it is processed into various distillates, gasoline, diesel oil, fuel oil, and others. The American Society for Testing and Materials (ASTM) has six classifications of fuel oil quality utilizing parameters such as *viscosity*, ignitability, and *vaporization temperature* numbered 1 through 6. The vaporization temperature (the temperature at which the oil will begin to evaporate at atmospheric pressure) increases with increasing numbers. The viscosity also decreases with increasing temperatures. When fuel oil is injected in the combustion chamber, it is atomized to increase the surface area of the fuel oil and ensure complete combustion during the combustion process. An atomizing air compressor is used to provide the air pressure required to inject the fuel oil into the combustion chamber.

Fuel oils with higher viscosity atomize less efficiently. This leads to less surface area exposed in the combustion section. To ensure adequate atomization, the fuel oils that have a higher viscosity are heated to reduce the viscosity prior to the fuel oil entering the combustion chamber. Additionally, oils with higher viscosity will be injected into the combustion chamber with high pressure air that atomizes the oil to increase the surface area of the oil and ensure complete combustion of the oil. The following is a description of the ASTM numbering system.

ASTM # 1 fuel oil has a low vaporization temperature and low viscosity. It is intended for stove and range type burners.

ASTM # 2 fuel oil has a slightly higher vaporization temperature. It is most commonly used for home heating oil applications.

ASTM # 3 fuel oil is no longer manufactured, but was used where low viscosity fuel oil was required. This grade of oil has been merged with ASTM #2.

ASTM # 4 fuel oil viscosity is greater than ASTM # 1 through #3, but it can still flow without the assistance of heating. Applications for ASTM # 4 included industrial and commercial heating applications where the system requires the oil to flow at low temperatures.

ASTM # 5 fuel oil's (also known as Bunker B oil) viscosity is greater than ASTM #4 and, as such, it requires preheating to allow for proper atomization of the oil in the combustion chamber. The amount of preheating is between 160°F and 220°F. Due to the requirement for preheating, this oil is normally used by large industrial facilities and power generation facilities.

ASTM # 6 fuel oil (also known as residual fuel oil (RFO) or Bunker C) has the highest of the six classifications. The amount of preheating to allow for proper atomization of the oil in the combustion chamber is between 220°F and 260°F. Due to the requirement for preheating, this oil is normally used by large industrial facilities and power generation facilities.

Example 4.6   Which ASTM number oil is also known as Bunker B oil?

Solution: ASTM #5 fuel oil.

Example 4.7   What is the temperature range required to preheat ASTM #6 fuel oil at to ensure adequate viscosity?

Solution: Typical preheat is between 220°F and 260°F

GLOSSARY OF TERMS

- ASTM # 1 fuel oil has a low vaporization temperature and low viscosity.
- ASTM # 2 fuel oil has a slightly higher vaporization temperature.
- ASTM # 3 fuel oil is no longer manufactured.

- ASTM # 4 fuel oil has a viscosity greater than ASTM # 1 through #3, but it can still flow without the assistance of heating.

- ASTM # 5 fuel oil has a viscosity greater than ASTM #4 and requires preheating between 160°F and 220°F. This is also known as Bunker B oil.

- ASTM # 6 fuel oil has the highest viscosity of the six classifications and requires preheating between 220°F and 260°F. It is also known as residual fuel oil (RFO) or Bunker C oil.

- Ash – The incombustible inorganic matter in the fuel.

- Bituminous coal is soft coal that may vary from low to high volatile content, with calorific values ranging from 10,500 to 14,000 BTU/lb on a moist, mineral-matter-free basis. Bituminous coal is prevalent in the eastern United States.

- Boiler – A closed vessel in which water is heated, steam is generated, steam is superheated, or any combination thereof, under pressure or vacuum by the application of thermal energy.

- British Thermal Unit (BTU) – A BTU is essentially 252 calories or 1055 joules. The BTU is the amount of energy needed to raise the temperature of one pound of water by one degree Fahrenheit.

- Carbon – An element. The principal combustible constituent of most fuels.

- Coal Gasification – The process of partially combusting coal and processing the resultant elements into a synthetic gas which is normally utilized in a combustion turbine.

- Combustion process requires three items, a combustible fuel, oxygen, and an ignition source.

- Complete Combustion – The complete oxidation of all the combustible constituents of a fuel.

- Incomplete Combustion – The partial oxidation of the combustible constituents of a fuel.

- Lignite, or brown coal, is the lowest-rank solid coal. Lignite has high moisture content and calorific values of less than 8300 BTU/lb on a moist, mineral-matter-free basis. US lignites are found in North Dakota, Montana, Texas, and other Gulf Coast states.

- Nonrenewable Resources – Resources that are considered as nonrenewable since it takes millions of years to produce the resource. Examples of nonrenewable resources are coal, oil, and gas.

- Sub-bituminous coal is a black coal with calorific values ranging from 8300 to 10,500 BTU/lb on a moist, mineral-matter-free basis. Sub-bituminous coals are found in the western United States, primarily in Montana, Wyoming, and Alaska, with significant additional deposits in New Mexico and Colorado.

- Renewable Resources – Resources that are considered as renewable since it takes a very short amount of time to produce the resource. Examples of renewable resources are bio fuels, solar, and wind energy.

- Theoretical Air – The quantity of air required for complete combustion with no excess oxygen.
- Vaporization Temperature – the temperature at which oil will begin to evaporate at atmospheric pressure.
- Viscosity – The state of being thick, sticky, and semifluid in consistency, due to internal friction.
- Waste Heat – Sensible heat in non-combustible gases discharged to the environment.

PROBLEMS

4.1 What is the amount of heat energy released in BTU when 10 lb of hydrogen is completely combusted releasing water?

4.2 What is the molecular weight of pentane (C_5H_{12})?

4.3 To completely combust 100 lb of hydrogen sulfide (H_2S) gas requires how much minimum oxygen in pounds?

4.4 To completely combust 100 lb of hydrogen sulfide (H_2S) gas requires how much minimum air in pounds?

4.5 Which ASTM number oil is also known as Bunker C oil?

4.6 What is the temperature range required to preheat ASTM #5 fuel oil at to ensure adequate viscosity?

4.7 What thermodynamic cycle is utilized in an integrated gasification combined cycle plant?
- **A.** Brayton
- **B.** Rankine
- **C.** Carnot
- **D.** Both Brayton and Rankine

RECOMMENDED READING

ASTM D396: Standard Specification for Fuel Oils, American Society for Testing and Materials, 2015.
Combustion Fossil Power: A Reference Book on Fuel Burning and Steam Generation, 4th edition, Joseph G. Singer, Combustion Engineering, 1993.
Electric Power Plant Design, Technical Manual TM 5-811-6, Department of the Army, USA, 1984.
The Engineering Handbook, 3rd edition, Richard C. Dorf (editor in chief), CRC Press, 2006.
Power Plant Engineering, Black & Veatch, edited by Larry Drbal, Kayla Westra, and Pat Boston, Chapman & Hall/Springer, 1996.
Standard Handbook of Powerplant Engineering, 2nd edition, Thomas C. Elliott, McGraw-Hill, 1998.

CHAPTER 5

HYDRAULIC TURBINES

GOALS

- To understand the three basic types of hydro-turbine plants: impoundment, diversion, and pumped storage
- To calculate the amount of power, head, or flow developed by a hydraulic turbine
- To understand the three basic types of hydro-turbine unit blade designs: reaction, impulse, and kinetic energy units

---

ONE OF The oldest methods of driving a prime mover to drive a generator is the use of the water flowing through a hydraulic turbine. The hydraulic turbine converts the potential energy associated with a *head* of water to kinetic energy in the *flow* of water and then converts the kinetic energy of the water flow into mechanical energy. This mechanical energy is transmitted to a generator via the turbine shaft. The electric generator converts this mechanical energy to electrical energy. Hydraulic turbines rotate at a much slower speed than other turbines using other technologies such as steam. Therefore, the generators connected to hydraulic turbines are constructed with a greater number of poles to achieve the same system frequency as other prime movers based on other technologies. The larger number of poles effects the construction of the generators such that they tend to be larger in diameter but smaller in length for the same MW rating when compared with higher speed generators. Design of generators will be covered in more detail in Chapter 16.

Hydraulic turbines convert the potential energy contained in a head of water to mechanical energy in the rotor of the turbine. The amount of power transferred is proportional to the amount of head across the turbine blades and the flow through the turbine blades. Therefore, if we increase the differential pressure across the turbine blades (or net head) at the same flow rate, or if we increase flow of water through the turbine at the same net head, then the amount of power transfer will increase. The amount of mechanical power available at the turbine shaft is

$$P = (H \times Q \times \eta)/8.81 \tag{5.1}$$

Energy Production Systems Engineering: An Introduction for Electrical Engineers to Electrical Power Generation Facilities, Systems, and Equipment, First Edition. Thomas H. Blair.
© 2017 by The Institute of Electrical and Electronics Engineers, Inc. Published 2017 by John Wiley & Sons, Inc.

where

P = turbine output (HP)

H = net head (ft)

Q = turbine discharge (ft³/sec)

η = turbine efficiency

Example 5.1   What is the power developed by a hydraulic turbine if the net head of water is 100 ft, turbine discharge flow is 1000 ft³/sec, and turbine efficiency is 0.85?

Solution:

$$P = (H \times Q \times \eta)/8.81$$

$$P = (100 \text{ ft}) \times (1000 \text{ ft}^3 / \text{ sec}) \times (0.85)/8.81$$

$$P = 9648 \text{ HP}$$

The proportionality laws for hydraulic turbines define the change in power, speed, and flow in a hydraulic turbine as we change the net head and/or *runner* diameter. These results are shown in equations (5.2) to (5.4).

$$P \propto D^2 H^{1.5} \tag{5.2}$$

$$n \propto H^{0.5}/D \tag{5.3}$$

$$Q \propto D^2 H^{1.5} \tag{5.4}$$

where

P = turbine output (HP)

D = runner discharge diameter (ft)

n = turbine rotating speed, (rpm)

Q = turbine discharge flow (ft³/sec)

H = net head (ft)

By combining equations (5.2) and (5.4), we find that power is proportional to flow as shown in equation (5.2a).

$$P \propto Q \tag{5.2a}$$

where

P = turbine output (HP)

Q = turbine discharge flow (ft³/sec)

Therefore, in the application of hydraulic turbines, control of the flow of water through the turbine will control the amount of mechanical power developed by the turbine shaft which will, in turn control the amount of electrical power delivered by the hydraulic turbine generator. Fundamentally, flow control is achieved by the use of either *wicket gates* that control the flow to the turbine blades and/or by the use of variable pitch blades on the turbine rotor.

Example 5.2   If the power developed by a turbine with a head of 100 ft is 10,000 HP and the head is reduced to 80 ft, but the runner discharge diameter remains constant, what is the new HP developed by the hydraulic turbine?

Solution: Using equation (5.2) we find the relationship

$$P \propto D^2 H^{1.5}$$

In the problem statement, head is changed, but runner diameter is held constant. Therefore, D does not change and equation (5.2) simplifies to

$$P \propto H^{1.5}$$

Putting this in equation form, we find

$$(P_1/P_2) = (H_1/H_2)^{1.5}$$

Solving for P_2

$$P_2 = P_1 \times (H_1/H_2)^{-3/2}$$

$$P_2 = 10,000\,\text{HP} \times (100\,\text{ft}/80\,\text{ft})^{-3/2}$$

$$P_2 = 7155\,\text{HP}$$

When performing problems of this nature, always make sure that the change is in the right direction. For example, from Example 5.2, we were given that net head decreased. Since power is the product of net head and flow, if net head is reduced, then the power level must be reduced. If, after performing the calculation, the result does not change in the correct direction, double check your calculation.

Example 5.3   If the speed developed by a turbine with a head of 100 ft is 180 rpm and the head is reduced to 80 ft, but the runner discharge diameter remains constant, what is the new speed developed by the hydraulic turbine?

Solution: Using equation (5.3), we find the relationship

$$n \propto H^{0.5}/D$$

In the problem statement, head is changed, but runner diameter is held constant. Therefore, D does not change and equation (5.3) simplifies to

$$n \propto H^{0.5}$$

Putting this in equation form, we find

$$(n_1/n_2) = (H_1/H_2)^{1/2}$$

Solving for n_2,

$$n_2 = n_1 \times (H_1/H_2)^{-1/2}$$

$$n_2 = 180\,\text{rpm} \times (100\,\text{ft}/80\,\text{ft})^{-1/2}$$

$$n_2 = 161\,\text{rpm}$$

There are three types of hydropower facilities: impoundment, diversion, and pumped storage. Of the three types of facilities, the impoundment is the most common type of facility for an electric generation utility. The impoundment facility utilizes a large dam to create a water reservoir. This storage of water provides a source of potential energy. When water is released through the dam, this potential energy stored by the water in the reservoir is converted to kinetic energy associated with the flow of the water from a higher potential to a lower potential. This kinetic energy is imparted on the turbine blades converting the kinetic energy into mechanical energy in the turbine.

The second type of facility is known as a diversion or a run-of-river facility. This is where only a portion of the natural river waters is diverted through a canal or *penstock* that is in parallel with the natural river run. This diverted water is utilized to drive the hydraulic turbine. Typically, the net available heads for diversion facilities are less than those for impoundment facilities and therefore smaller turbine generator sets are utilized in these applications.

The third type of facility is known as a pumped storage facility. This type of facility can act as an energy storage facility. During periods of a net surplus of electrical energy in the electrical transmission and distribution system, the pumped storage facility will draw electrical energy from the electrical grid and pump water from a reservoir at a lower elevation to another reservoir at a higher elevation. In effect, the pumping station is taking electrical energy from the grid and storing it in the form of potential energy in the water. Periods of net surplus of energy on the electrical grid occur during periods of low load, or when generation is high, such as during the day time when solar plants are providing energy to the electrical grid and during windy periods when wind power plants are providing energy to the electrical grid. During periods of time when there is a net deficit of energy available on the electrical grid, the pumping station can be used to deliver electrical energy to the electrical grid. During these periods, the water in the higher elevation reservoir is released through a hydraulic turbine to the lower reservoir and the flow of water drives a hydraulic turbine which, in turn, drives a generator to convert this potential energy, to kinetic energy, to mechanical energy, and finally to electrical energy for distribution.

There are two main types of rotating turbine blade designs: reaction and impulse. In a hydraulic turbine that uses a reaction turbine blade, in the stationary portion of the turbine, the nozzle cross sectional area is reduced as the water flows through the stationary nozzle. This reduction of cross sectional area converts the potential energy of the water (pressure) into kinetic energy (velocity). As such, in the nozzle section of the hydraulic turbine, the pressure drops and the velocity increases. Then the water imparts its energy onto the rotating blade and the rotating blade converts the kinetic energy of the water into mechanical energy on the shaft of the turbine. The runner is placed directly in the water stream flowing over the blades rather than striking each individually. Reaction turbines are generally used for sites with lower head and higher flows than the impulse turbines.

The other type of hydraulic turbine is the impulse turbine. The impulse turbine uses a turbine blade design where the pressure and volume are constant in the rotating buckets and velocity decreases, but the water impinges on the rotating bucket blades and the kinetic energy of the water is converted to mechanical energy in the turbine shaft. The water flows out of the bottom of the turbine housing after hitting the runner. An impulse turbine is generally suitable for high head, low flow applications. Most water turbines that are utilized in lower head applications (<100 ft of head) are reaction turbines. Impulse turbines are often used in very high (>1000 ft) head applications.

HYDRAULIC REACTION TURBINES

In a reaction turbine, a pressure drop occurs in both fixed and moving blades. There are four major designs of hydraulic reaction turbines in service. The Francis turbine (developed by James Francis) is the most common water turbine in use today. A Francis turbine has a runner with fixed buckets (vanes). This type of hydraulic turbine uses wicket gates to control the flow of water to the turbine to control the amount of power that a hydraulic turbine generator delivers to the electrical system. Water is introduced just above the runner and all around it and then falls through, causing the turbine rotor to spin. The Francis turbine has one of the widest ranges of applications for flow and net head. It is utilized for net heads of 40–500 m and for flow rates of 1–1000 m³/sec. This ability to use a wide range of flow and head make the Francis turbine one of the most utilized designs today in hydraulic reaction turbines. The Francis turbine has a range of power ratings from 100 kW up to 1000 MW.

The Kaplan turbine (developed by Viktor Kaplan) uses a propeller-type turbine blade where the pitch of the turbine blade can be adjusted to control the amount of force developed by the turbine blade. Additionally, it is provided with adjustable wicket gates allowing for a wider range of operation The Kaplan turbine is designed for lower values of net head. It is utilized for net heads of 5–70 m and for flow rates of 1–1000 m³/sec. The Kaplan turbine has a range of power ratings from 100 kW up to 100 MW.

The bulb turbine propeller is similar to the Kaplan with pivoting propeller type turbine blades, but the bulb unit (including the generator) is submersed entirely in the flow of water that drives the turbine. The bulb turbine gets its name from the shape of the watertight casing which contains both the generator and associated transmission to the propeller. The bulb unit turbine is also known as an axial flow turbine since the flow of water is parallel with the shaft of the turbine. The bulb unit turbine is designed for very low values of net head. It is utilized for net heads of 5–20 m and for flow rates of 1–500 m³/sec. The bulb unit turbine has a range of power ratings from 100 kW up to 50 MW.

The Deriaz turbine (developed by Paul Deriaz) is similar to the Kaplan, but the propeller blades are inclined to allow for operation at higher heads. The Deriaz pump/turbine unit turbine is designed and most efficient when utilized in applications with moderate values of net head. It is utilized for net heads of 20–140 m and for flow rates of 1–1000 m³/sec. The Deriaz pump/turbine unit has a range of power ratings from 200 kW up to 200 MW.

HYDRAULIC IMPULSE TURBINES

Impulse turbines change the velocity of the water in the turbine blade just like a garden hose nozzle. The water jet pushes on the turbine's curved blades which changes the direction of the flow. The resulting change in momentum (impulse) causes a force on the turbine blades. Since the turbine is spinning, the force acts through a distance (work) and the diverted water flow is left with diminished kinetic energy. An impulse turbine is one where the pressure of the fluid flowing over the rotor blades is constant and all the work output is due to the change in the velocity (kinetic energy) of the fluid. Prior to hitting the turbine blades, the water's pressure (potential energy) is converted to kinetic energy by a nozzle. No pressure change occurs at the turbine blades, and the turbine does not require an enclosure for operation.

There are three major categories of hydraulic impulse turbines in service. These are the Pelton turbine, the cross flow turbine, and the Turgo turbine. Each uses a different design of turbine blade section that is designed to optimize the efficiency of the turbine for certain values of flow and head.

A Pelton wheel has one or more free jets discharging water and impinging on the buckets of a runner. The Pelton turbine is designed for high range values of net head. It is utilized for net heads of 140–2000 m and for flow rates of 0.5–50 m^3/sec. The Pelton turbine has a range of power ratings from 100 kW up to 200 MW.

The Ossberger cross flow turbine uses a drum-shaped rotor that resembles a squirrel cage. A rectangular nozzle section directs the water flow against the curved vanes on the cylindrically shaped runner. A guide vane at the entrance to the turbine controls the amount of water flow through the turbine. The cross flow turbine is designed for lower range values of net head. It is utilized for net heads of 5–100 m and for flow rates of 0.5–5 m^3/sec. These are smaller hydraulic turbines with a range of power ratings from 10 kW up to 500 kW.

The Turgo turbine is also designed for low range values of net head. The Turgo runner is a cast wheel whose shape generally resembles a fan blade that is closed on the outer edges. The water stream is applied on one side, goes across the blades and exits on the other side. It is utilized for net heads of 5–100 m and for flow rates of 0.5–5 m^3/sec. These are smaller hydraulic turbines with a range of power ratings from 10 kW up to 500 kW

KINETIC ENERGY HYDRAULIC TURBINES

Kinetic energy hydraulic turbines are a developing technology. Kinetic energy turbines, which are also known as free-flowing turbines, do not require the impoundment of a potential energy source from a reservoir of water. Similar to how wind turbines capture the kinetic energy of atmospheric currents, kinetic hydroelectric turbines capture the kinetic energy contained in the flow of natural waters. Recent developments in the field of wind energy have provided insight into advances for similar types of technology for hydraulic kinetic energy turbines. Hydraulic kinetic energy turbines are driven by the kinetic energy of free-flowing water rather than form a pressure or

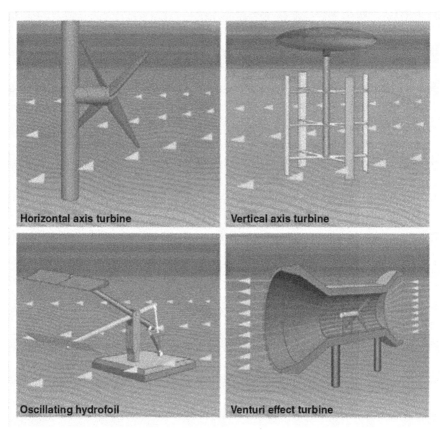

Figure 5.1   Typical kinetic energy conversion devices *Source*: Reproduced with permission of National Renewable Energy Laboratory (NREL).

head from an elevated reservoir of water. These types of systems are installed in rivers, tidal water, and ocean streams and convert the kinetic movement of these waters to electrical energy. They tend to be smaller in size but do not require the capital investment that an impoundment arrangement might cost. Some recent designs are shown in Figure 5.1.

GLOSSARY OF TERMS

- Hydraulic Reaction Turbine – A turbine whose rotating blades utilize a reaction design where the pressure drops in the rotating section.
- Hydraulic Impulse Turbine – A turbine whose rating blades utilize an impulse design where the pressure is constant in the rotating buckets.
- Efficiency – A percentage obtained by dividing the actual power or energy by the theoretical power or energy. It represents how well the hydropower plant converts the potential energy of water into electrical energy.

- Head – Vertical change in elevation, expressed in feet or meters, between the head (reservoir) water level and the tail water (downstream) level.
- Flow – Volume of water, expressed as cubic feet or cubic meters per second, passing a point in a given amount of time.
- Headwater – The water level above the powerhouse or at the upstream face of a dam.
- Low Head – Head of 66 ft or less.
- Penstock – A closed conduit or pipe for conducting water to the powerhouse.
- Runner – The rotating part of the turbine that converts the energy of falling water into mechanical energy.
- Tailrace – The channel that carries water away from a dam.
- Tail Water – The water downstream of the powerhouse or dam.
- Ultra-low Head – Head of 10 ft or less.
- Wicket Gates – Adjustable elements that control the flow of water to the turbine passage.

PROBLEMS

5.1 Using equation (5.1), calculate the power developed (in HP) by a hydraulic turbine if the net head of water is 250 ft, turbine discharge flow is 1000 ft^3/sec, and turbine efficiency is 0.8?

5.2 If the power developed by a turbine with a head of 200 ft is 20,000 HP and the head is reduced to 100 ft, but the runner discharge diameter remains constant, what is the new HP developed by the hydraulic turbine?

5.3 If the speed developed by a turbine with a head of 200 ft is 360 rpm and the head is reduced to 100 ft, but the runner discharge diameter remains constant, what is the new speed developed by the hydraulic turbine?

5.4 If the flow through a turbine with a head of 200 ft is 1000 ft^3/sec and the head is reduced to 100 ft, but the runner discharge diameter remains constant, what is the new flow through the hydraulic turbine?

RECOMMENDED READING

Types of Hydropower Turbines, DOE (Department of Energy), http://energy.gov/eere/water/types-hydropower-turbines

Electrical Machines, Drives and Power Systems, 6th edition, Theodore Wildi, Prentice Hall, 2006.

The Engineering Handbook, 3rd edition, Richard C. Dorf (editor in chief), CRC Press, 2006.

NREL (National Renewable Energy Laboratory), http://www.nrel.gov/emails/wind/utws_newsletter_2012-03.html

Power Plant Engineering, Black & Veatch, edited by Larry Drbal, Kayla Westra, and Pat Boston, Chapman & Hall/Springer, 1996.

Standard Handbook of Powerplant Engineering, 2nd edition, Thomas C. Elliott, McGraw-Hill, 1998.

CHAPTER 6

NUCLEAR POWER

GOALS

- To understand the fundamentals of nuclear reactions

- To understand the differences between alpha, beta, gamma, and neutron byproducts of nuclear reactions

- To be able to calculate the decay constant or half-life of a radio nuclide and determine the amount of material remaining

- To understand the difference and purpose of the reactor coolant and reactor moderator

- To understand the difference between fissile and fertile material

- To be able to calculate the neutron multiplication factor or reactivity of the nuclear fuel and determine whether the reactor is subcritical, critical, or supercritical

- To understand the six basic types of nuclear power plant designs

- To identify several types of accidents or events and systems utilize to mitigate the effects of these modes of failure

- To develop an understanding on the difference between units of radioactivity and units of exposure

- To understand the maximum permissible occupational radiation exposure for adults and children

---

NUCLEAR POWER plants utilize the energy released in a nuclear reaction as the source of thermal energy to heat steam which in turn drives a steam turbine that is connected to a generator to generate electricity. The most common fuel used as a source of this nuclear reaction is uranium (U^{235}). Ninety-nine percent of natural uranium mined from the earth contains the isotope uranium (U^{238}). To obtain the isotope uranium (U^{235}), the uranium is processed. Uranium-235 is an unstable isotope. Thermally moderated neutrons collide with the U^{235} atom and cause the atom to decay into other atoms. One type of decay of the U^{235} atom is to decay into bromine (Br^{90}) and xenon (Xe^{143}) and other minor particles. This decay of U^{235} into Br^{90} and

---

Energy Production Systems Engineering: An Introduction for Electrical Engineers to Electrical Power Generation Facilities, Systems, and Equipment, First Edition. Thomas H. Blair.

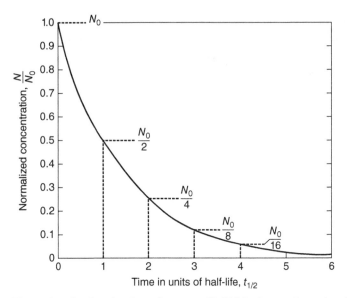

Figure 6.1   Radioactive decay in terms of half-life. *Source*: Reproduced with permission of Department of Energy (DOE) Fundamentals Handbook.

Xe^{143} releases neutrons N, gamma (γ) particles, and beta (β)⁻ particles and energy in the form of heat. These particles continue to decay into other elements and each decay releases *decay heat*. Alpha decay is the emission of alpha particles which are helium nuclei with an atomic number of 2 and an atomic mass of 4 without the two electrons of helium. Beta (β) decay is the emission of electrons of nuclear rather than orbital origin. These particles are electrons that have been expelled from the nucleus. Gamma (γ) radiation is a high energy electromagnetic radiation that originates in the nucleus. It is emitted in the form of photons which are discrete bundles of energy that have both wave properties and particle properties.

The basic features of decay of a radionuclide sample are shown by the graph in Figure 6.1.

The *half-life* ($t_{1/2}$) of an unstable element is the average time for one half of a population of that element to decay to the next level of elements. The decay constant is defined by the following formula.

$$\lambda = \frac{0.693}{t_{1/2}} \tag{6.1}$$

where

λ = decay constant

$t_{1/2}$ = half-life

The formula to calculate the amount of radioactive material left after a certain amount of time, given the half-life of the material is

$$N = N_0 \times e^{(-\lambda \times t)} \tag{6.2}$$

where

N = the final amount of radioactive material

N_0 = the initial amount of radioactive material

λ = decay constant (time^{-1})

t = time between initial population and final population (time)

Notice that the units of time and of the decay constant can utilize any unit of time, but they must utilize the same units of time. If the decay constant is given in units of seconds^{-1} then time must be in units of seconds. If the decay constant is given in units of years^{-1} then time must be in units of years.

The decay heat is transferred to water or other fluid system via a steam generator that then generates steam to drive a steam turbine(s) to drive an electrical generator.

Example 6.1   A sample of material contains 20 micrograms of Bromine-77 (Br^{77}). Bromine-77 has a half-life of 57 hours. Calculate the following.

a. Calculate the decay constant (λ) of Bromine-77

b. Calculate the amount of Bromine-77 that will remain after 1 week

Solution:

a. Using equation (6.1) we find that the decay constant is

$$\lambda = \frac{0.693}{t_{1/2}}$$

$$\lambda = \frac{0.693}{57 \text{ hr}}$$

$$\lambda = 0.012158 \text{ hr}^{-1}$$

b. Using equation (6.2), we find that the final amount of Bromine-77 is;

$$N = N_0 \times e^{(-\lambda \times t)}$$

In the exponent, we need to use the same units for both the decay constant (λ) and time (t). We found the decay constant in part (a) to be 0.012158 hours^{-1} and the problem gives us the time as 1 week. We can convert either 1 week to hours, or 0.012158 hours^{-1} to weeks^{-1}. Converting 1 week to hours, we find the time passed is

$$1 \text{ week} \times (7 \text{ days}/1 \text{ week}) \times (24 \text{ hours}/1 \text{ day}) = 168 \text{ hours}$$

Now we can use equation (6.2) to find out the amount of Bromine-77 after 1 week.

$$N = 20 \text{ micrograms} \times e^{(-0.012158 \text{ hours}^{-1} \times 168 \text{ hours})}$$

$$N = 20 \text{ micrograms} \times e^{(-2.042526316)}$$

$$N = 20 \text{ micrograms} \times 0.1297 = 2.594 \text{ micrograms}$$

There are two basic types of nuclear reactions, fission and fusion. Nuclear *fusion* is a nuclear reaction in which two or more atoms collide and form a new element. Due primarily to the high density and high temperature environments required to support a fusion reaction, this process is not currently utilized in commercial nuclear electrical generation stations and will not be covered in any detail in this text. Nuclear *fission* is a reaction where the incident neutron collides with an unstable element. This unstable elements then "splits" into two or more smaller elements and energy and other particles are released in the reaction. The fusion reaction has several benefits over the fission reaction. One benefit is that the energy released by fusion is three to four times greater than the energy released by fission. Another benefit is that, for the same energy released, fusion produces substantially fewer radioactive particles than fission. Due to the substantial benefits of fusion over fission, fusion well may be the future of nuclear energy once the challenges of the higher temperature and density requirements of fusion technology have been resolved.

A reactor is a vessel that contains the nuclear fuel, *control rods*, and other systems to support the nuclear reaction. A *moderator* is used to reduce the neutron energy. Moderated neutrons (lower energy neutrons) have a higher probability of resulting in a nuclear reaction with the nuclear fuel than high energy neutrons (for most types of reactor plants). Some materials (water being the most common) are used to remove some of the energy associated with the neutron population contained in the reactor fuel area to increase the ability of neutrons to cause a nuclear reaction with the fuel. A *coolant* in the *reactor vessel* is a material that absorbs the thermal energy released in the nuclear reaction and transfers that energy to a secondary location for further delivery to the steam turbine. Some systems transfer this steam directly to the turbine for use, while other systems utilize a heat exchanger called a steam generator to transfer heat to a secondary loop before it is delivered to the turbine.

There are two types of nuclear fuel materials based on the type of reaction it undergoes during decay and the material that results from that decay. *Fissile material* generates more neutrons and also non-fissile materials. *Fertile materials* generate neutrons and other fissile materials that can then undergo further decay. This is a slightly different concept than the concept of a breeder reactor or converter reactor. A *breeder reactor* is a reactor that generates more fuel in the decay process than it consumes in the nuclear reaction. A *converter reactor* is a reactor that generates less fuel in the decay process than it consumes in the nuclear reaction.

The process of the nuclear reactions inside the reactor vessel results in a certain population of neutrons in this area. If the population of neutrons is increased, then there are more neutrons available to cause a reaction with the fuel and the *reactivity* of the reactor increases. If the population of neutrons is decreased, then there are fewer neutrons available to cause a reaction with the fuel and reactivity of the reactor decreases. The quantity of neutrons that are produced in the reactor vessel that are generated from the nuclear reaction is known as *rate of production*. The neutrons that are lost from the reactor vessel and not available for further reactions are lost to two destinations. One area of loss is when the neutrons involved in a nuclear reaction are absorbed by other elements. This is known as *rate of absorption*. The other area of loss is when neutrons leave the reactor vessel without causing a reaction and this is known as the *rate of leakage*. The neutron multiplication factor (k) is defined by equation (6.3).

$$k = (ROP)/(ROA + ROL) \qquad (6.3)$$

where

> k = neutron multiplication factor
>
> ROP = rate of production
>
> ROA = rate of absorption
>
> ROL = rate of leakage

If the rate of neutron production equals the rate of neutron loss (due to absorption and leakage) then the value of the neutron population in the reactor is constant and the neutron multiplication factor (k) is 1. Since the overall population of neutrons in the reactor is not changing at this point, this state is known as *critical*. If the rate of neutron production exceeds the rate of neutron loss (due to absorption and leakage) then the value of the neutron population in the reactor is increasing and the neutron multiplication factor (k) is greater than 1. Since the overall population of neutrons in the reactor is increasing at this point, this state is known as *supercritical*. If the rate of neutron production is less than the rate of neutron loss (due to absorption and leakage), then the value of the neutron population in the reactor is decreasing and the neutron multiplication factor (k) is less than 1. Since the overall population of neutrons in the reactor is decreasing at this point, this state is known as *subcritical*.

A term related to the neutron multiplication factor (k) is the reactivity (ρ) of the reactor. The reactivity is defined in equation (6.4).

$$\rho = (k - 1)/k \qquad (6.4)$$

With some algebraic manipulation of equation (6.3) and (6.4), we find that the reactivity of the reactor is also defined as shown in equation (6.5).

$$\rho = [\text{ROP} - (\text{ROA} + \text{ROL})]/\text{ROP} \qquad (6.5)$$

where

> ρ = reactivity
>
> ROP = rate of production
>
> ROA = rate of absorption
>
> ROL = rate of leakage

Example 6.2   Given a neutron multiplication factor (k) of 1, what is the reactivity (ρ)?

Solution: Using equation (6.4), we find reactivity as follows.

$$\rho = (k - 1)/k$$

$$\rho = (1 - 1)/1$$

$$\rho = 0$$

Equation (6.5) mathematically defines the term reactivity. The reactivity (ρ) of the reactor is the change in the neutron population normalized to the rate of neutron production in the reactor. Therefore, if the rate of neutron production equals the rate of neutron loss (due to absorption and leakage) then the value of the reactivity (ρ) is 0. Since the overall population of neutrons in the reactor is not changing at this point, this state is known as *critical*. If the rate of neutron production exceeds the rate of neutron loss (due to absorption and leakage) then the value of reactivity (ρ) is greater than 0. Since the overall population of neutrons in the reactor is increasing at this point, this state is known as *supercritical*. If the rate of neutron production is less than the rate of neutron loss (due to absorption and leakage) then the value of the neutron population in the reactor is decreasing and the reactivity (ρ) is less than 0. Since the overall population of neutrons in the reactor is decreasing at this point, this state is known as *subcritical*.

This is summarized in the equations (6.6), (6.7), and (6.8);

$$k = 1, \ \rho = 0; \ \text{system is critical} \tag{6.6}$$

$$k < 1, \ \rho < 0; \ \text{system is subcritical} \tag{6.7}$$

$$k > 1, \ \rho > 0; \ \text{system is supercritical} \tag{6.8}$$

Example 6.3   For a reactor operating in a "subcritical" state, the reactivity (ρ) is _____ and the neutron multiplication factor (k) is _____

A. $\rho > 1, k > 0$
B. $\rho < 0, k < 1$
C. $\rho = 0, k = 1$
D. $\rho = 1, k = 0$

Solution: From equation (6.7) the answer can be found to be

B. $\rho < 0, k < 1$

As can be seen from discussion above, the main method of controlling the reaction rate in the *nuclear reactor* is by controlling the population of thermal neutrons available for reactions in the reactor vessel. This can be done by adjusting the rate of production, rate of leakage, or rate of absorption. The primary method of controlling these is by the use of control rods which controls the rate of absorption. Control rods have neutron absorbing materials (poisons) that absorb available neutrons. The control rods are inserted into the reactor fuel region or removed as needed to regulate the quantity of neutrons absorbed. Under normal operation, these control rods are slowly driven in or out of the reactor fuel region to slowly adjust the rate of neutron absorption. In the event that the reactor needs to be quickly shut down, the controls rods are quickly inserted into the fuel region to reduce the thermal neutron population and the number of reactions occurring in the fuel region. This is known as a reactor *SCRAM*. The acronym SCRAM stands for "safety control rod axe man". Back in the early period of nuclear development, the control rods were located above the *reactor vessel head* (the top of the reactor vessel). In the event that the reactor had to be shut down quickly for safety, a person could use an axe to cut the supports holding the control rods, thereby allowing gravity to quickly inserting the control rods into the reactor vessel and forcing the reactor into a subcritical state.

In addition to the control rods, a soluble *neutron poison* (typically boron) can be introduced into the fuel region to absorb some of the neutron population and control the reactivity of the fuel region. Also fixed *burnable fission poisons* are permanently distributed throughout the fuel region. These fixed burnable poisons help ensure that the power density is evenly distributed throughout the fuel region to minimize the occurrence of thermal hot spots in the reactor core.

Most reactor designs have an inherent negative temperature feedback design. As the temperature of the moderator increases, the density of the moderator decreases. The reduced density in the moderator allows more neutrons to leave the fuel area without becoming thermally moderated. Therefore, fewer thermal neutrons are available for reaction. This results in reduced number of reactions and reduced thermal output of the reactor. The temperature of the moderator can change by either an initial increase in reactor power level with no change in steam flow to the turbine by withdrawing reactor control rods or, more commonly with no change in control rod position, if the steam flow out of the steam generator is reduced (say due to reduced mechanical load on the steam turbine), then the temperature of the moderator increases which inherently reduces the reactivity of the fuel area due to increased moderator temperature and reduced moderator density.

There are six basic types of nuclear power plant designs that are listed below.

BWR = Boiling water reactor

PWR = Pressurized water reactor

PHWR = Pressurized heavy water reactor

PTGR = Pressurized tube graphite reactor

HTGR = High temperature gas-cooled reactor

LMFBR = Liquid metal fast breeder reactor

BOILING WATER REACTOR

The boiling water reactor BWR design utilizes light water (H_2O) for both coolant and moderator. The BWR only has one loop and the reactor is the "boiler" in the thermodynamic cycle. Water enters the reactor vessel and is heated in the reactor core at location 1 of Figure 6.2. At the top of the reactor vessel at location 3 of Figure 6.2 is the steam generator section where the water undergoes change of state from water to steam. There is a steam separator and a super heat section at the top of this area to remove any moisture from the steam and add thermal energy to superheat the steam. Since the steam separator is located at the top of the reactor vessel, there is no room on top of the reactor for the control rods and the control rods for the BWR are driven in from the bottom of the reactor. Water that does not form steam in the steam separator is recirculated back to the reactor core using recirculation pumps. These pumps help to maintain cooling to the reactor core. The steam that leaves the steam separator leaves the containment vessel and is delivered to the steam turbine. In the pressurized water reactor (PWR) design, the steam that drives the turbine never passes through the reactor vessel and therefore is never in close proximity to the radiation source. The same is not true for the BWR design. The water that flows

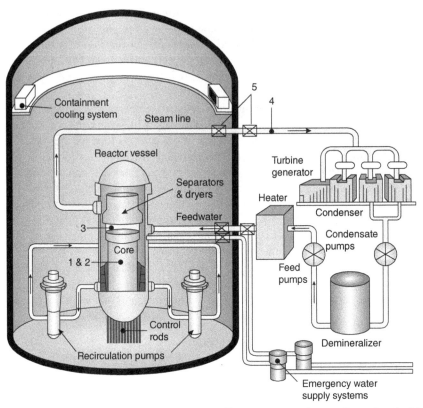

Figure 6.2 Typical boiling water reactor (BWR) arrangement. *Source*: Reproduced with permission of U.S. Nuclear Regulatory Commission.

through the reactor core becomes slightly radioactive due to the collisions between the radiation (mostly gamma particles) and the elements contained in the water. In the PWR design, this water remains in the containment vessel and circulates between the reactor and steam generator, but in the BWR design, this same water is converted to steam and is used to drive the steam turbine. Therefore, in the BWR design, the steam is slightly radioactive. Additional shielding is utilized on steam and water piping and the steam turbine in BWR designs that is not needed in the PWR design. Boiling water reactors contain between 370 and 800 fuel assemblies depending on the design of the reactor vessel. Figure 6.2 shows the typical arrangement of the BWR. The steam then leaves the containment vessel and, at this location (location 5 of Figure 6.2), there are isolation values in the steam piping to maintain containment across the containment vessel. The steam is then sent to the steam turbine in the power plant for use and the condensate from the turbine generator in the power plant is fed back to reactor vessel though the containment vessel where, again, there are valves for isolation at the containment vessel.

The BWR was developed by the Idaho National Laboratory and General Electric (GE) in the mid-1950s but the majority of units in service are of the early 1980's vintage. Since the first design of the BWR, there have been six changes to the basic design to enhance system safety. The latest design, known as the advanced boiling water reactor design (ABWR), utilizes recirculation pumps that are physically inside the reactor vessel instead of the recirculation pumps that are inside the containment vessel but external of the reactor vessel shown in Figure 6.2. Having these recirculation pumps internal to the reactor vessel eliminates several penetrations in the reactor vessel that the old system required for the external recirculation jet pumps. The ABWR design also allows for finer control rod adjustments leading to finer adjustment of the reactivity of the fuel section. There have also been changes made to increase the required response time to a *loss-of-coolant accident* (LOCA). In some designs of the ABWR, the system can provide cooling water for up to 3 days. These include a design that uses natural circulation to allow coolant to circulate in the reactor vessel without the need for electrically powered external recirculation pumps. Another safety improvement is a modified condenser which allows for the use of the condenser which normally takes the exhaust steam from the steam turbine. The modification allows for steam to be sent directly from the reactor vessel to the condenser for emergency cooling. Also, some designs have a large pool of borated water (boron is a natural neutron-absorbing material or poison) and this borated water can be used to flood the reactor vessel in a LOCA to reduce the reactivity of the fuel region thereby reducing the power density of the reactor.

PRESSURIZED WATER REACTOR

The pressurized water reactor (PWR) is similar to the BWR except that this design contains two loops for coolant. The primary loop circulates completely inside the containment vessel between the reactor vessel and the steam generator. The steam generator forms the boundary between the primary and secondary coolant loops. The steam generator is the heat sink to the primary loop system and the heat source for the secondary loop system. The installation of a primary loop that is completely contained in the reactor vessel and the secondary loop provides isolation between the two systems and thereby provides one more containment boundary between the fuel and the environment. The core inside the reactor vessel at location 1 of Figure 6.3 is the heat source for the primary loop. The steam generator at location 3 of Figure 6.3 transfers the thermal energy from the primary loop to the secondary loop. The reactor core fuel assemblies are cooled by the circulation of the primary coolant. The reactor pumps provide the pressure or head required for forced circulation of the primary coolant loop. Additionally, the arrangement of the reactor vessel and steam generator is such that a natural circulation exists between these two pieces of equipment which is discussed later. Pressurized water reactors contain between 150 and 200 fuel assemblies.

Unlike the BWR design, the steam lines that exit the containment vessel and the condensate lines that enter the containment vessel do not enter the reactor vessel but enter the steam generator on the secondary side. As such, the steam generator forms one of the containment boundaries. Since the secondary loop is isolated from

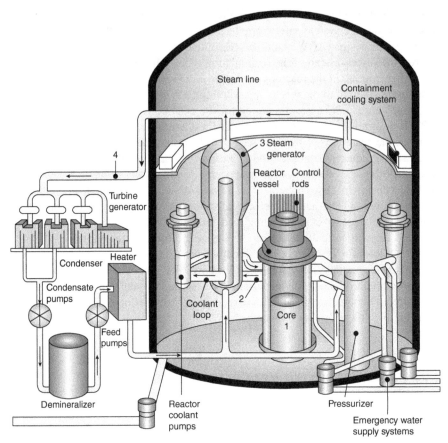

Figure 6.3   Typical pressurized water reactor (PWR) arrangement. *Source*: Reproduced with permission of U.S. Nuclear Regulatory Commission.

the primary loop, there is no need for the isolation values in the steam and condensate lines as they enter and exit the containment vessel that exist in the BWR design. In the PWR design, the coolant in the primary loop remains in the liquid state except for the pressurizer where a saturated water and steam mixture exists. The purpose of the pressurizer is to provide temperature and pressure control for the primary loop. In the pressurizer, the primary coolant is in a saturated condition. In the pressurizer, heaters are available to convert some of the saturated steam water mixture in the pressurizer to steam thereby increasing the pressure of the primary loop system. Also in the pressurizer, nozzles that provide water sprays are also available to condense some of the steam to water in the pressurizer which has the effect of lowering the pressure of the primary loop system. The heaters and water sprays in the pressurizer therefore provide pressure control and therefore temperature control of the primary loop. Referencing the T-s diagram of Figure 2.1, if we change the line of pressure we are operating at, then we change the saturation temperature of the loop.

Just like the BWR design, the PWR utilizes light water (H_2O) for both moderator and coolant. Since the PWR does not have the steam separator unit on top of the reactor vessel, the control rods are located on top of the reactor vessel in this design. This has the benefit of letting gravity assist the motion of control rod insertion during a reactor SCRAM. In the BWR design, the reactor control rods are inserted from the bottom of the reactor vessel and are driven by gas pressure. In the BWR design, the control rods must overcome gravity to be driven into the reactor against the force of gravity.

In the PWR design, there are three types of steam generators used. There is a once through vertical design, the horizontal design, and the U-tube design. The steam generator has both an operational function and a safety function. The operational function is the transfer of heat from primary loop to secondary loop to create steam to drive the steam turbine. The safety function is the cool-down of the reactor coolant system to the point where the decay heat removal system (DHRS) can be placed in service after a reactor SCRAM to remove the decay heat produced in the reactor. Another safety function of the steam generator is that it provides a containment barrier between the primary and secondary loops to prevent fission products that may be contained in the primary loop system from release outside the containment vessel. Loss of tube integrity or tube sheet integrity in the steam generator results in loss of two out of the three fission product barriers of the system. As such, the integrity of the tubes and the tube sheet in the steam generator is critical and is monitored frequently. A tube tester using eddy current loss info is used to determine the thickness of the tubes during outages to ensure the thickness of this barrier and the integrity of the tubes in the steam generator.

In the U-tube design of steam generator, the primary coolant enters the steam generator at the bottom of the steam generator and leaves at the bottom of the steam generator. The primary coolant flows through the inside of the tubes. The secondary loop enters the steam generator at the top of the tube bundle, is forced down past the tube wrapper, and then rises through the tube bundle. The process of allowing the secondary loop fluid to rise as it absorbs the thermal energy from the tube bundle assists in the natural circulation of the secondary loop fluid. As the secondary loop rises, it evaporates and becomes less dense. The more dense fluid entering the steam generator on the secondary loop side tends to displace the heated steam water mixture of less density. At the top of the steam generator is the steam separator section which is used to remove any moisture content from the steam before it is sent to the steam turbine. The secondary steam leaves the top of the steam generator.

In a PWR, the reactor vessel is physically at the bottom of the *containment system* or structure and the steam generator is at the top of the containment structure. This arrangement allows for natural circulation of coolant in the primary loop through the reactor vessel and steam generator. The coolant in the reactor is heated in the reactor vessel. This heated water becomes less dense than water outside the reactor vessel and this heated water is forced out of the reactor vessel up to the steam generator. In the steam generator, heat from the primary loop water is removed which increases the density of the primary loop coolant. The cooler, denser primary loop water now falls down back to the reactor vessel. This is the process of natural circulation and the reason for the physical location of the reactor vessel and the steam generator.

However, this natural circulating force is not adequate for full power operations of the reactor and typically four or more reactor circulating pumps are installed in this loop to provide the additional pressure needed to drive the flow of fluid in the primary loop.

PRESSURIZED HEAVY WATER REACTOR

The pressurized heavy water reactor (PHWR) uses heavy water as the moderator. *Heavy water* is a molecule that contains deuterium and oxygen (D_2O) as compared with *light water* which is a molecule that contains hydrogen and oxygen (H_2O). Deuterium (D or ^2H) has one neutron and one proton whereas the far more common hydrogen isotope (also known as protium), has one proton and no neutron in the nucleus. In its pure form, heavy water has a density about 11% greater than light water (H_2O). The additional neutrons in heavy water make it less likely to absorb free neutrons which, in turn, will leave more thermal neutrons available for reactions with other nuclear fuel material in the reactor core. Light water moderator absorbs more neutrons which decreases the population of thermal neutrons in light water reactors as compared with heavy water reactors. This feature makes heavy water a more efficient means of thermally moderating high energy neutrons into low energy thermal neutrons in a reactor. The moderator and coolant form two loops in the PHWR.

The PHWR design uses tubes in the reactor vessel for flow and, therefore, the fuel cells in at PHWR can be replaced online. This allows for less downtime and more frequent replacement of spent fuel cells which also allows for better management of the thermal density distribution of the reactor fuel area. When refueling online, the reactivity control and thermal density of the reactor vessel can be more tightly controlled than the previous designs discussed that can only have their fuel cells replaced during an outage. Reactivity in the PHWR design can also be achieved by control rods as discussed previously or by dumping of the moderator to reduce the population of thermal neutrons.

Below we discuss the three remaining types of reactors, but since they constitute only a few of the active designs in service, the discussion is brief.

PRESSURE TUBE GRAPHITE REACTOR

The pressure tube graphite reactor (PTGR) has two separate materials for the moderator and coolant. The PTGR utilizes light water (H_2O) for its cooling media and graphite for its moderator. The cooling system forms two loops much like the PWR design and, as such, the PTGR utilizes a steam generator in the containment vessel for removal of heat energy from the reactor vessel. Again, since the cooling media and moderator media form different loops and are installed in tubes in the reactor vessel, online fuel replacement is possible with the PTGR just like the PHWR. Control of the reactivity of this design of reactor is via control rods mounted on the reactor vessel.

HIGH TEMPERATURE GAS-COOLED REACTOR

The high temperature *gas-cooled reactor* (HTGCR) uses graphite as the neutron moderator. Instead of using water for its cooling media, the HTGCR uses air (or other gases) as the cooling media for heat removal from the reactor vessel and delivery to a steam generator that is a gas-to-water heat exchanger. The gas is on the primary loop side of the steam generator. The water is on the secondary loop side of the steam generator. The water is heated by the gas of the primary loop cooling system and forms steam to drive the steam turbine. Since the primary loop is a gas and not a liquid, this design does not require a pressurizer, like the PWR design. Like most designs, this design uses control rods for reactivity control.

LIQUID METAL FAST BREEDER REACTOR

The liquid metal fast breeder reactor (LMFBR) uses graphite as its neutron moderator. All the previous designs are fission reactors that produce non-fissionable materials in the reaction. The LMFBR is an example of a breeder reactor. The most common material used for fuel in the LMFBR is uranium-235 (U^{235}) with uranium-238 (U^{238}). Uranium-238 (U^{238}) is the most common naturally occurring isotope of uranium. In fission reactors, uranium-238 does not react with thermal neutrons like uranium-235. Therefore, in spent fuel from fission reactors, while the concentration of uranium-235 is depleted, the concentration of uranium-238 is still significant. One proposal for the recycling of depleted fuel from a fission reactor is reprocessing to obtain the uranium-238 and utilization in a breeder reactor. When uranium-238 reacts with a fast neutron, it decays into plutonium 239 (PU^{239}), which is a fissile material and can be used as nuclear fuel. Breeder reactors carry out such a process of transmutation to convert the fertile isotope ^{238}U into fissile Pu-239. The LMFBR uses both fissile Plutonium and fertile uranium for the fuel. Both of these elements react with fast neutrons to produce further nuclear reactions and, therefore, a moderator is not required with the LMFBR design. For coolant, this design uses liquid sodium (a metal on the periodic table, table 4.3). By definition, since this is a breeder reactor, unlike the previous designs discussed, the LMFBR produces fissionable material that can further generate nuclear reactions in the fuel assembly. Currently, there are four breeder reactors in operation.

NUCLEAR POWER SAFETY

In nuclear power, safety is of primary concern and one of the primary topics of safety is containment of the radioactive materials. There are multiple barriers between the nuclear fuel and the outside environment that form the containment system. The reactive fuel is contained inside pellets which are then assembled into the fuel rod cladding. The first boundary is the fuel pellets and the second boundary is the cladding. These fuel rods are then assembled into the reactor core vessel. The reactor vessel has several penetrations needed to supply reactor cooling water, connections for recirculation pumps, steam generator vessel, pressurizer, and other equipment. The reactor vessel and all the associated systems that connect to the reactor vessel form the third boundary between the fuel and the outside environment. Lastly, these systems are housed inside a reinforced containment structure and this containment structure provides the fourth boundary between the fuel and the outside environment. All materials have certain mechanical (temperature and pressure) limits which must be maintained to ensure the integrity of the material. Operation of the reactor inside the thermal limits of these materials ensures the integrity of the *fuel cladding*. To help minimize the operational temperature of the fuel pellets and cladding, the thermal resistance between the fuel cells and the fuel rod is minimized by the use of helium (He) gas that resides inside the fuel rod. The helium gas uses convection to enhance the heat transfer capability between the fuel pellet and the fuel rod cladding.

There are several types of accidents or events that could occur in a nuclear reactor, but the reactor is controlled to ensure the safe and reliable operation of the nuclear power plant. Reactors operate at a steady-state (critical) power level reasonably well. However, during transients or changes to reactor power levels, the reactor system response is not a linear function. During the transient, production and decay of the byproducts of nuclear fission occur. Some of the fission byproducts produced by the subsequent decays are neutron poisons. Since the production of these poisons is a subsequent decay of the primary fission reaction, there is a time delay between the rate of primary fission reactions and the production of poisons. If the reactor is driven subcritical, initially the primary fission reactions are reduced, but the production of poisons is still based on the previous power level and initially does not change. Therefore, the number of neutrons absorbed by control rods and the produced poisons increases initially. After some time, the production of neutron poisons begins to reduce due to the fewer decays coming from the reduced fission reaction rate. This reduction in the population of poisons results in fewer thermal neutrons absorbed and, since there are more neutrons available for fission, there is a slight increase in the fission reactions in a reactor. This causes the power level of the reactor to slightly increase. The more rapid the change in power level, the more pronounced this response to poison level changes can be. Due to this, there are limits to the rate of change of reactor power levels during normal operation to tightly control the reactive transients due to neutron poison population changes.

Loss of cooling water flow is another concern in the safe operation of reactors. The nuclear reaction generates a large amount of thermal energy and the removal of this energy from the reactor vessel is critical to ensuring that the materials in the reactor vessel that form the boundaries between the fuel and the outside environment maintain their integrity. The normal method of thermal energy removal is via the coolant system. The heat is removed from the reactor vessel by the coolant system and eventually delivered to the steam turbine. In the event of a loss of the steam turbine (unit trip), this heat needs to be removed and various cooling systems exist to remove this decay heat from the reactor vessel. Some systems provide piping to circulate the coolant so the decay heat can be directly removed and sent to the condenser that normally takes exhaust steam from the turbine. This will have a heat exchanger in the containment compartment to provide a boundary between the reactor coolant loop and the secondary loop that enters the condenser for removal of heat. There is a backup system that can be used to fill the reactor vessel with borated water (boron is a neutron poison) to provide both cooling as well as absorption of neutrons in the reactor vessel.

It may not initially be obvious, but overcooling is also an event that can damage the reactor system and controls must be in place to prevent overcooling the moderator. As the moderator is cooled, it becomes denser. A more dense moderation will react with a larger population of high energy neutrons and produce a higher population of thermal neutrons which will increase the reaction rate in the reactor. Therefore, overcooling of the moderator can cause the reactor reaction rate to increase causing the reactor to become supercritical. Fortunately, since the reactor design has a negative temperature relationship, as the power level increases, with no change in flow or pressure of the coolant system, the temperature of the moderator should increase, decreasing moderator density resulting in a reduction of the production of thermal neutrons leading to a reduction of reactivity.

Integrity of the containment boundaries is critical to ensure the reactive material is contained within the boundaries. The steam generator presents the largest amount of surface area that forms one of these boundaries. As such, the integrity of the tube sheets and tubes in the steam generator is of primary concern. The integrity of the steam generator materials is monitored closely to ensure its adequacy for service.

During outages, spent fuel rods are removed from the reactor vessel and transported to a holding pond with borated water to provide both cooling and absorption of neutrons. The spent fuel rods still contain fissionable material and still generate neutrons resulting in nuclear reactions and the associated heat even after removal. Therefore, the storage of these fuel rods must account for the removal of this decay heat and control of the neutrons produced during the decay process.

Additionally, external events pose potential concerns for safe reactor operation. The reactor is designed in a manner to minimize the effects an external event might have on the safe and reliable operation of the reactor. Emergency generators are provided to provide backup power in the event of a loss of system power. Additionally, critical systems are normally fed from battery and inverter systems to ensure that during a complete loss of power, the battery is available to power up these critical systems.

The systems in the reactor that are designed to provide the ability to control the above events are many. The control rods are primary method of controlling the reactivity inside the reactor vessel. During a trip event, the control rods can be driven into the reactor vessel to reduce reactivity. After this initial event, systems such as the *emergency core cooling system* (ECC) shown in Figure 6.4 can be used to remove the decay heat from the reactor to maintain material temperatures inside the reactor vessel within their safe limits should there be a loss of primary cooling media event.

For longer-term cooling, there is a backup system known as the post-accident heat removal system (PAHR) shown in Figure 6.5 to remove decay heat from the reactor vessel. This system utilizes a residual heat removal heat exchanger to remove heat generated in the reactor core from decay products. The reactor coolant is circulated between the reactor vessel and the residual heat exchanger to remove decay heat from the reactor. In the residual heat exchanger, this heat energy is exchanged

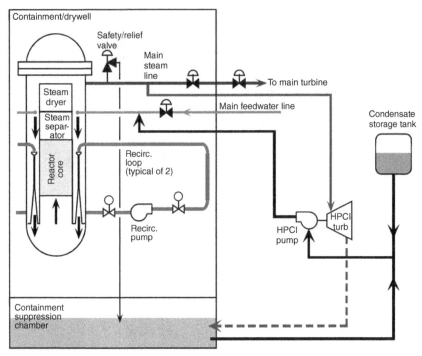

Figure 6.4   Reactor emergency core cooling system. *Source*: Reproduced with permission of U.S. Nuclear Regulatory Commission.

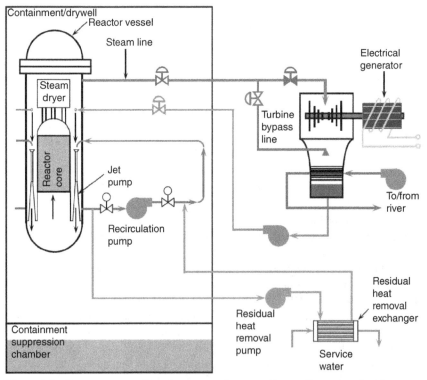

Figure 6.5   Reactor post-accident heat removal system. *Source*: Reproduced with permission of U.S. Nuclear Regulatory Commission.

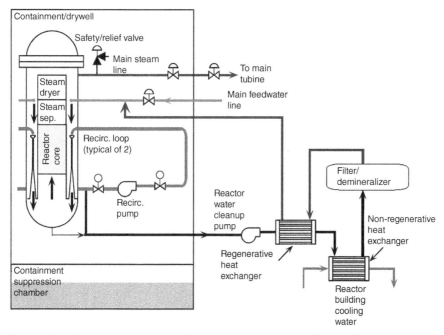

Figure 6.6   Reactor post-accident radioactivity removal system. *Source*: Reproduced with permission of U.S. Nuclear Regulatory Commission.

with some secondary coolant system such as the service water system as shown in Figure 6.5.

Should the reactor vessel boundary fail, heat from the reactor vessel can increase the temperature and pressure of the atmosphere in the containment vessel. Additionally, if the fuel cladding and fuel rod boundaries have been compromised, a failure of the reactor vessel can result in release of decay products into the containment chamber. To prevent over pressurization of the containment vessel, the *post-accident radioactivity removal* system (PARR) shown in Figure 6.6 provides sprays and filters inside the containment vessel that will collect radioactive particles in the atmosphere of the containment vessel for containment and reduce the change of release to the outside atmosphere. The spray also allows the steam inside the containment vessel to be condensed to liquid, thereby reducing the pressure inside the containment vessel. Lastly, the integrity of the containment building is maintained to ensure that reactive material is not released to the outside environment.

Nuclear reactions produce various types of radiation. There are two systems of measurement depending on if we are measuring the amount or activity of radioactive decay or if we are measuring the equivalent exposure to human tissue.

UNITS OF ACTIVITY

The amount of radiation (disintegrations or decays per second) is measured by two common units: curie (Ci) and becquerel (Bq). In the United States, the units used are the curie or mill curie (mCi, 0.001 Ci) or micro curie (μCi, 0.000001 Ci). The Curie is an amount of radioactive material emitting 2.22×10^{12} decays (particles or photons) per minute (dpm) or 3.7×10^{10} decays per second (dps). Becquerel (SI units) is an amount of radioactive material emitting 60 dpm or 1 dps.

$$1 \text{ curie (Ci)} = 3.7 \times 10^{10} \text{ decays per second}$$

$$1 \text{ becquerel (Bq)} = 1 \text{ decay per second}$$

The conversion between units is

$$1 \text{ curie (Ci)} = 3.7 \times 10^{10} \text{ becquerel (Bq)} \tag{6.9}$$

The nuclear fuel source presents a source of radioactive particles and the quantity of particles is measured in the decays per minute (DPM). Meters such as the Geiger-Mueller meter measure the quantity of radioactivity in the area of the sample and provides a measure in counts per minute (CPM). The meter does not detect every decay, but has some value of efficiency that is the ratio of the counts per minute (CPM) the instrument detects compared to the number of decays per minute (DPM) that produce the measured number of counts. Or

$$\eta = \text{CPM/DPM} \tag{6.10}$$

where

η = efficiency

CPM = counts per minute.

DPM = decays per minute.

The unit of measure for the nuclear particle quantity is the micro curie (μCi). One micro curie is defined as;

$$\mu\text{Ci} = \text{DPM}/2.22 \times 10^{6} \tag{6.11}$$

It is required by NRC law that all records relevant to NRC licensed activities must be maintained in units of DPM or micro curies. Another unit used for ionizing X-ray and gamma radiation is the roentgen.

UNITS OF EXPOSURE

The different types of particles produced in a fission reaction have different reactions on the human body. Even though different particles react differently with the human body, each particle is a count. Therefore the measure of decays per minute or counts per minute is not a good indicator of the damage to human tissue. The collision of a neutron with human tissue is more damaging than the collision of a gamma particle. This is due to the fact that the neutron particle is denser than the gamma particle and, therefore causes more damage to human tissue. Therefore, the quantity of particles (CPM) or Curie is not a useful measure of the amount of exposure to human tissue.

Units used to describe the quantity or amount of radiation exposure to the human body are the radiation absorbed dose (RAD). The RAD refers to the energy deposition by any type of radiation in any type of material. (The international unit for absorbed dose is gray and it is defined as being equal to 100 rads.) Units used to measure the quality or effect of the radiation type on the human body and there-fore, the exposure level is the radiation equivalent man (REM) dose. The REM is a *dose equivalent* (DE) measurement. (The international or SI unit for human exposure is sievert, which is defined as equal to 100 rem.) The radiation damage to tissue is different for different types of particles. The relationship between RAD and REM is defined by equation (6.12).

$$REM = RAD \times QF \times \text{modifying factors} \tag{6.12}$$

The qualifying factor (QF) is a measure of the damage one type or particle has on the human body and allows for measure of the quality of radiation exposure (REM) instead of just the quantity of radiation exposure (RAD). The amount of damage a particle can cause to the human body is a function of both the mass or size of the particle as well as the energy level of the particle. A low mass particle causes less damage to human tissue than a high mass particle of equal energy. A low energy particle causes less damage to human tissue than a high energy particle of equal mass. Table 6.1 shows the value of the qualifying factor (QF) for various types of particles.

Alpha particles consist of two protons and two neutrons. By looking on the periodic table (Table 4.3) you will see that an alpha particle is basically a helium (He) nucleus without the two outer shell electrons of a helium atom and has a molecular weight of approximately 4. The molecular weight is approximately the number of neutrons plus the number of protons in an atom. Therefore, compared with other types of decay particles, the alpha particle is very large. Since it is large, it is blocked by the first layer of skin and does not penetrate deep to damage critical organs. The most energetic alpha particles are stopped by a few centimeters of air or a sheet of paper. However, due to the large mass of an alpha particle, the damage it can do when colliding with human tissue is large, therefore the quality factor is 20 for this type of particle.

Beta particles consist of high energy electrons (e−) that have been ejected at a high velocity from an unstable nucleus. The electron has a much smaller mass than

TABLE 6.1   Types of Radiation and Their Respective Quality Factors

Type of Radiation	Quality Factor (QF)
X-ray, gamma, or beta radiation	1
Alpha particles, multiple-charged particles, fission fragments and heavy particles of unknown charge	20
Neutrons of unknown energy	10
High energy protons	10

Source: Reproduced with permission of Nuclear Regulatory Commission (NRC).

the neutron or proton. As such, the beta particle has an overall substantially lower mass than the alpha particle. Therefore, the quality factor (QF) for the beta particle is 1, substantially less than the alpha particle. Due to the size and nature of the beta particle, it can be shielded or blocked by clothing.

Gamma particles are electromagnetic radiation and are very high energy and very low mass particles. Due to the fact that they are very low mass, their potential damage to human tissue is minimal. Therefore, the quality factor (QF) of a gamma particle is 1 (similar to the beta particle which also has very low mass). Since the gamma particle is so small, neither skin nor clothing will shield this particle. The main shielding used is lead due to the molecular density of the lead element. Very few other materials will shield gamma particles well.

Neutron particles are high energy and high mass particles. Due to the fact that they are both high energy and high mass, neutron particles have a great probability of causing damage to human tissue during the collision process. The mass is not as high as an alpha particle, but larger than beta or gamma particle. Due to this, the quality factor (QF) of neutron particles is 10 (more than beta or gamma particle, but less than the massive alpha particle). Water provides an effective shield to neutron particles.

The SI unit for the RAD value is gray and the SI unit for the REM value is sievert.

1 gray = 100 rad

1 sievert = 100 rem

Radiation is received by humans in everyday life outside of the nuclear power industry. For a point of reference, the typical annual exposure levels in mrem (0.001 rem) are provided in Table 6.2 for various situations.

So how much radiation does it take to affect the human body in an adverse way? The quality of the radiation as measured in rem and the effects that the radiation has on the human body are listed in Table 6.3.

Note these values are in rem not mrem as in the previous table.

TABLE 6.2   Typical Annual Exposure Levels

milli-rem	Source
5	Statutory limit on radiation from operating a nuclear power plant
25	Internal exposure from radioactive material ingested into the body
45	Cosmic rays
75	Diagnostic medical exposure (X-rays)
60	External radiation from radioactive ores, etc.
120	Natural radiation sources (combined)
200	Average total exposure in the United States
500	Average occupational dose for radiologists
500	Maximum permissible occupational exposure for children < 18 years old
1250	Natural exposure in mountainous regions of Brazil
5000	Maximum permissible occupational exposure for adults (5 rem)

Source: Reproduced with permission of Nuclear Regulatory Commission (NRC).

TABLE 6.3   Effects of Radiation on the Human Body

rem	Effect
0–25	No observable effect
25–100	Slight blood changes
100–200	Significant temporary reduction in blood platelets and white blood cells
200–500	Severe blood damage, nausea, hair loss, hemorrhage, death in many cases
>600	Death in less than two months for over 80% of people

Source: Reproduced with permission of Nuclear Regulatory Commission (NRC).

Example 6.4   What is the range of REM exposure where slight blood changes are noticeable?

   A. 0–25 rem

   B. 25–100 rem

   C. 100–200 rem

   D. 200–500 rem

Solution: From Table 6.3, answer is B. 25 – 100 rem

So, if we use the Geiger–Mueller to measure the quantity of particles, what instrument do we use to measure the quality or effect of the particles? One of the most common devices is the *thermoluminescent dosimeter* (TLD) badge. TLDs can be designed to measure the quality or effect of the particles the device is exposed to. The TLD dosimeter can measure skin dose (the effect the particles have to the layer of human skin), the eye dose (the effect the particles have to the layer of human eye tissue), and deep dose (the effect the particles have to major organ tissue deep in the body). Thermo luminescence (TL) is the ability of a material to convert the energy received from a collision with a radioactive particle to radiation of a different wavelength that can be measured. The collision with the particle forces a valence electron into a higher excited orbit. When the TLD is to be read, it is heated and this heating returns the valence electron to is original orbit and in this process, the TLD releases energy in the form of light. The wavelength of this light release is measured and converted to the REM dose. After the readout, the TLD is annealed back to a zero state and ready for reuse.

The advantages of the TLD device over other methods of monitoring REM are that it has a greater range of doses (types of tissue exposure) than other devices and the readout is easily obtained by the readout process. The TLD does not have to be sent off site to read the dose value. Also, after readout is annealed back to is zero state, the TLD is available for reuse. The main disadvantage of the TLD is that, once read, the device is zeroed out, so there is no history of exposure past the last readout of the device.

GLOSSARY OF TERMS

   • Auxiliary Building – A building at a nuclear power plant, adjacent to the reactor containment structure, that houses the reactor auxiliary and safety systems, such as the radioactive waste system, the chemical and volume control systems, and the emergency cooling system.

- Boiling Water Reactor (BWR) – A reactor design in which the coolant water flows to the core where it is heated and is allowed to boil. Steam that is produced is passed through a heat exchanger where it is condenses to water and returned to the reactor.

- Breeder Reactor (BR) – A nuclear reactor designed to produce more fuel (fissile material) than it consumes. Such reactors surround the fuel with a fertile (non-fissile) material that, when irradiated, produces fissile material. It is also known as a converter reactor because it converts fertile material into fissile material. A reactor that generates more fuel in the decay process than it consumes in the nuclear reaction.

- Burnable Fission Poison – A strong neutron absorber added to reactor fuel or coolant water to control reactivity. The poison initially reduces the reactivity of new fuel. Neutron production by the fuel decreases over time, occurring as the concentration of the poison decreases due to neutron capture. This burning of the poison compensates for the loss of reactivity of the fuel, helping maintain the overall reactivity of the reactor.

- Cleanup System – A continuous water filtration and demineralization system for reactor coolant systems. It serves to reduce contamination levels in the water and to reduce corrosion.

- Containment System – Those systems, including ventilation, that act as barriers between areas containing radioactive substances and the environment.

- Control Rod – A component of a reactor that contains strong neutron absorbers and which is used to control reactor reactivity.
  - General control rods are used to control and regulate the reactor power.
  - Regulator rods are used for fine adjustments.
  - Shim rods are used for large changes to the reactivity.
  - Safety rods are used to rapidly decrease the reactivity in the event of accidents.

- Converter Reactor – A reactor that converts fertile material into fissile material. The term converter often is used to classify reactors that produce fissile material that is different than the reactor fuel. An example is a reactor fueled with ^{235}U that converts ^{238}U into ^{239}Pu (neutron capture followed by beta decay). When the reactor produces a fissile material that is used as the fuel, it is known as a breeder reactor.

- Coolant – A liquid or gas circulated through a reactor to remove heat. Coolants can also act as neutron moderators. The material in a reactor vessel that absorbs the thermal energy released in the nuclear reaction and transfers that energy to a secondary location for removal of energy in the form of heat.

- Critical – The state of having just enough neutrons to sustain a fission chain reaction at current levels. A system whose effective multiplication factor is equal to 1.0. Rate of production of neutrons in a reactor is equal to the rate of loss of neutrons in a reactor. In this condition, the population of neutrons in the reactor is constant.

- Decay Heat – Heat produced by the decay of radioactive materials in a reactor that has been shut down.
- Dose Equivalent (DE) – The product of the absorbed dose in tissue, quality factor, and all other necessary modifying factors at the location of interest. The units of dose equivalent are the rem and Sievert.
- Emergency Core Cooling System (ECC) – System used to remove the decay heat from the reactor to safely maintain material temperatures inside the reactor vessel.
- Fertile Material – Material that generates neutrons and fissile materials that can then undergo further decay.
- Fissile Material – Material that generates more neutrons and also non-fissile materials.
- Fission Poison – Materials with ultra-high neutron capture cross sections that significantly reduce neutron fields in a reactor, reducing reactivity, thereby creating a barrier to reactor operations. The nuclides ^{149}Sm and ^{135}Xe are two examples.
- Fuel Cladding – Material used to construct reactor components and designed to maintain a separation between their contents and the coolant. Zirconium and zirconium alloys (e.g., zircaloy) are common cladding materials.
- Gas-Cooled Reactor (GCR) – A nuclear reactor design which uses gas as a coolant.
- Half-life – The average time for one half of a population of an element to decay to the next level of elements.
- Heavy Water Reactor – A reactor that uses heavy water (D_2O) as its moderator.
- LOCA – An acronym for loss-of-coolant accident. Inadvertent escape of water from the primary coolant system.
- Moderator – Used to reduce the neutron energy to convert high energy neutrons into thermal neutrons.
- Neutron Leakage – Neutrons that escape the moderator of a reactor or a neutron source without undergoing capture by the fissile material or other material in the system.
- Neutron Moderator – A substance used to reduce the energy of neutrons through inelastic scattering. Good moderators have a large scattering cross section and a small absorption cross section. Graphite, water, and heavy water are examples of good neutron moderators.
- Neutron Poison – A synonym for fission poison.
- Nuclear Reactor – A device in which a sustained fission reaction can be maintained. The core is made of a fissile material such as uranium enriched in the

isotope ^{235}U. It is usually surrounded by water which moderates neutrons and removes heat from the core.

- Post-accident Heat Removal System (PAHR) – System used to remove decay heat from the reactor after event.

- Post-accident Radioactivity Removal System (PARR) – Systems that will collect radioactive particles in the atmosphere of the containment vessel for containment.

- Pressurized Water Reactor (PWR) – A reactor design in which water flows through the core at very high pressures. The water is not allowed to boil in the core but flows to a steam generator.

- Reactivity – A measure of the departure of a reactor from criticality.

- Reactor Vessel – Vessel that contains the nuclear fuel, control rods, and other systems to support the nuclear reaction.

- Reactor Vessel Head – The top section of a reactor pressure vessel. It is bolted in place during reactor operation and is removed to provide access to the core during maintenance and refueling.

- SCRAM (Safety Control Rod Axe Man) – The sudden shutting down of a reactor, by the rapid insertion of control rods. It may occur either automatically or manually by the reactor operator. It originally stood for safety control rod axe man, a title given to personnel who were assigned to insert the emergency rod at the original Chicago pile.

- Subcritical – The state of having insufficient neutrons to sustain a fission chain reaction. A system whose effective multiplication factor is less than 1.0. Rate of production of neutrons in a reactor is less than the rate of loss of neutrons in a reactor. In this condition, the population of neutrons in the reactor is decreasing.

- Supercritical – The state of a system whose effective multiplication factor is greater than 1.0. Rate of production of neutrons in a reactor is greater than the rate of loss of neutrons in a reactor. In this condition, the population of neutrons in the reactor is increasing.

- Thermoluminescent Dosimeter (TLD) – A type of radiation dosimeter. A TLD measures ionizing radiation exposure by measuring the intensity of visible light emitted from a crystal in the detector when the crystal is heated.

PROBLEMS

6.1 You have four radioactive cookies – one an alpha emitter cookie, one a beta emitter cookie, one a gamma emitter cookie, and one a neutron emitter cookie. You must eat one, hold one in your hand, put one in your pocket, and give the last one you throw away. Which cookie do you eat, which cookie do you hold in your hand, which cookie do you put in your pocket and which cookie do you throw away to minimize your radiation exposure?

6.2 In a fission reactor fuel cell, the cladding contains the fuel pellets and what type of material to improve the thermal conductivity between the fuel pellets and cladding?

A. Hydrogen

B. Helium

C. Oxygen

D. Nitrogen

6.3 A saturated steam–water mixture with an inlet steam quality of 60% is flowing through a moisture separator. The moisture separator is 100% efficient for removing moisture. How much moisture will be removed by the moisture separator from 50 lbm of the steam–water mixture?

A. 10 lbm

B. 20 lbm

C. 30 lbm

D. 40 lbm

6.4 A reactor coolant system is being maintained at 1000 psia. A pressurizer safety relief valve is slowly discharging to a collection tank, which is maintained at 5 psig. Use the ideal assumption that, since the flow is so low, the process is a throttling process. Assuming 100% quality steam in the pressurizer vapor space, what is the approximate enthalpy of the fluid entering the tank?

A. 1210 BTU/lbm

B. 1193 BTU/lbm

C. 1178 BTU/lbm

D. 1156 BTU/lbm

6.5 The thermodynamic cycle efficiency of a nuclear power plant can be increased by…

A. decreasing power from 100% to 25%

B. removing a high pressure feed water heater from service

C. lowering condenser vacuum from 29 in. to 25 in.

D. decreasing the amount of condensate depression (sub cooling)

6.6 To achieve maximum overall nuclear power plant thermal efficiency, feed water should enter the steam generator (S/G) _____ and the pressure difference between the S/G and the condenser should be as _____ as possible.

A. as subcooled as practical; great

B. as subcooled as practical; small

C. close to saturation; great

D. close to saturation; small

6.7 A nuclear power plant is operating at 85% reactor power when the extraction steam to a high pressure feedwater heater is isolated. After the transient, the operator returns reactor power to 85% and stabilizes the plant. Compared to conditions just prior to the transient, current main turbine generator output (kW) is…

    A. higher because increased steam flow is causing the turbine to operate at a higher speed

    B. lower because decreased steam flow is causing the turbine to operate at a lower speed

    C. higher because plant thermal efficiency has increased

    D. lower because plant thermal efficiency has decreased

6.8 A pressurizer is operating in a saturated condition at 636°F. If a sudden pressurizer level decrease of 10% occurs, pressurizer pressure will _____ and pressurizer temperature will _____.

    A. remain the same; decrease

    B. remain the same; remain the same

    C. decrease; decrease

    D. decrease; remain the same

6.9 Nuclear reactor fuel rods are normally charged with _____ gas to improve the heat transferred by _____ from the fuel pellets to the cladding.

    A. helium; convection

    B. helium; conduction

    C. nitrogen; convection

    D. nitrogen; conduction

6.10 If a nuclear reactor is operated within core thermal limits, then...

    A. plant thermal efficiency is optimized

    B. fuel cladding integrity is ensured

    C. pressurized thermal shock will be prevented

    D. reactor vessel thermal stresses will be minimized

6.11 Establishing natural circulation requires that a heat sink be _____ in elevation than a heat source and that a _____ difference exist between the heat sink and heat source.

    A. lower; pressure

    B. lower; temperature

    C. higher; pressure

    D. higher; temperature

6.12 For a reactor operating in a "critical" state, the reactivity (ρ) is _____ and the neutron multiplication factor (k) is _____

    A. $\rho > 1, k > 0$

    B. $\rho < 0, k < 1$

    C. $\rho = 0, k = 1$

    D. $\rho = 1, k = 0$

6.13 The primary purpose of the "moderator" is
   A. primary heat extraction
   B. lubrication
   C. slow neutrons produced by fission
   D. shielding

6.14 What two types of reactor types utilize "light water" as both coolant and moderator?
   A. BWR and PWR
   B. PHWR and PTGR
   C. GCR and LMFBR

6.15 The purpose of a "pressurizer" in a PWR design is to control
   A. primary loop temperature and pressure
   B. secondary loop temperature and pressure
   C. primary loop flow and pressure
   D. secondary loop flow and pressure

6.16 The reactor type whose design provides for online refueling is
   A. PWR
   B. BWR
   C. PHWR
   D. HTGR

6.17 What is the annual maximum permissible occupational radiation exposure for a worker in rem?
   A. 1 rem
   B. 5 rem
   C. 10 rem
   D. 15 rem

RECOMMENDED READING

Electric Power Plant Design, Technical Manual TM 5-811-6, Department of the Army, USA, 1984.

DOE Fundamentals Handbook – Nuclear Physics and Reactor Theory, 1993, http://energy.gov/ehss/downloads/doe-hdbk-10191-93

Electrical Machines, Drives and Power Systems, 6th edition, Theodore Wildi, Prentice Hall, 2006.

Nuclear Glossary, 2006, Scientific Digital Visions, Inc., http://www.nuclearglossary.com/nuclearglossary_reactors.html

Power Plant Engineering, Black & Veatch, edited by Larry Drbal, Kayla Westra, and Pat Boston, Chapman & Hall/Springer, 1996.

Standard Handbook of Powerplant Engineering, 2nd edition, Thomas C. Elliott, McGraw-Hill, 1998.

U.S. Nuclear Regulatory Commission (NRC), http://www.nrc.gov/reading-rm/basic-ref/teachers/03.pdf

CHAPTER 7

CONVEYORS

GOALS

- To understand the basic purpose and types of conveyors in a power generation facility
- To develop awareness of safety hazards associated with conveyors systems and the equipment utilized to mitigate the dangers
- To perform system calculations for belt length, tension, belt speed, and material flow for a belt conveyor
- To develop a general understanding of pneumatic, rotary screw, and vibratory conveyor systems

---

CONVEYORS **ARE** used in power generation facilities to efficiently convey or move large quantities of materials from one point to another. Typical applications are the movement of coal from the delivery system (rail, barge, or truck) to the long-term storage facility as well as movement of the coal from the long-term storage facility to the plant. One typical application is for the controlled feeding of coal to the coal pulverizers. The speed of the conveyor feeding coal to the pulverizers and, eventually to the furnace, dictates the rate of fuel injected into the furnace section and, therefore, the firing rate of the boiler. In Chapter 4 on combustion, we discussed how the amount of chemical energy converted to thermal energy in the furnace of the boiler was based on the type of combustible element along with the mass of the amount of material. Coal feeding conveyors are designed to measure both the weight of the coal on the conveyor *belt* (for normal operation) and the volume of coal on the belt (for abnormal operation). For normal operation of the coal feeding conveyor, the control system operates in a gravimetric mode. The gravimetric control mode is where the conveyor speed is controlled to regulate the amount of mass of coal delivered to the furnace. Should the gravimetric system fail, the conveyor can be operated in an abnormal state called a volumetric state. The volumetric control mode is where the conveyor speed is controlled to regulate the amount of volume of coal delivered to the furnace. The normal gravimetric mode of operation is much more accurate in

---

Energy Production Systems Engineering: An Introduction for Electrical Engineers to Electrical Power Generation Facilities, Systems, and Equipment, First Edition. Thomas H. Blair.
© 2017 by The Institute of Electrical and Electronics Engineers, Inc. Published 2017 by John Wiley & Sons, Inc.

the control of the mass of combustible material delivered to the furnace section than the volumetric control mode.

Belt conveyors, while being one of the most common type of conveyor system in the power generation facility, are not the only type of conveyor system in the power generation facility. Fly ash is collected in the hoppers at the bottom of the bag house or electrostatic precipitator and is conveyed to the fly ash storage facility using a *pneumatic conveyor* transportation system where air provides the means of transport. Similarly, primary air is used to transport pulverized coal from the coal mill to the furnace for combustion. In a stoker boiler, a *chain conveyor* is used to transport the fuel into the furnace section. These various types of conveyor systems are described below and some of the applications found in the power generation facility are described.

First, we must discuss the safety concerns with conveyors. Below are some minimum safety requirements. Conveyors present safety hazards that must be addressed. OSHA CFR 1926.555 requires that the employee be protected against entrainment by moving rotating equipment and NESC section 122 requires that moving parts that are likely to injure a person must be guarded or isolated. Conveyors must be provided with equipment guards to prevent personnel from becoming entangled in the conveyor system. Methods for stopping the conveyor must be available locally at the conveyor and control station and the method of stopping must latch in the off state until manually reset. Devices such as *emergency pull chords* and *emergency stop push buttons or switches* are distributed along the conveyor length. These devices are latching in that, once activated to stop the conveyor, they remain in their off state until intentionally reset. These devices directly disable the conveyor motor controls when activated for this safety function. Another safety concern is the use of *counterweights*. Long belt conveyors will utilize counterweights to maintain tension on the belt during periods of transient torque such as startup. This counterweight may present a crushing hazard should an employee find themselves in the vicinity of the counterweight when the conveyor starts. The area around the counterweight must be guarded to prevent employees from accessing the area around the counterweight to prevent such an injury from occurring.

When performing maintenance on conveyors (or any mechanical equipment that can present a physical hazard such as impacting, crushing, abrading, or shearing), conveyors shall be locked out or otherwise rendered inoperable, and locked out and tagged with a "Do Not Operate" tag during repairs and when operation is hazardous to employees performing maintenance work. *Simply depressing an emergency stop or pull chord does not place the conveyor in a safe work condition for maintenance.* The conveyor equipment must be locked out to allow for maintenance activities (see Chapter 1 on safety for requirements for lockout). Lastly, before a conveyor system starts, it is required to activate an alarm indicating the conveyor system will begin moving, before the system actually starts. This alarm system must be audible but may include both audible and visual indication of pending conveyor system startup. The duration of the audible warning shall be long enough to allow anyone who is endangered by an activated conveyor system to move to safety. Further guidance on methods for safeguarding moving mechanical equipment can be found in ANSI/ASME B15.1.

BELT CONVEYOR

This is the most common type of conveyor in the power generation station. The conveyor belt consists of two or more pulleys. One pulley, known as the *head pulley*, is connected to the motor and provides the energy to move the belt. The second pulley on the other end of the belt conveyor is known as the *tail pulley*. Another pulley in the belt system is known as the *take-up* pulley. The take-up pulley's function is to maintain constant belt tension and is connected via a spring or weight to the rest of the conveyor. The purpose of the take-up pulley is to provide constant tension on the belt and take up any slack that occurs in the belt. This is very important on long conveyors that are started across the line, as the accelerating torque of the motor may tend to stretch the belt and the take-up pulley moves to maintain tension on the belt during this transient. Along the belt length, there are multiple idler wheels that support the belt and the material on the belt. These must be maintained and lubricated as any additional friction in the idler wheels will add to the torque required to run the belt. Additionally, where one belt conveyor feeds another belt conveyor, logic is developed to monitor the state of the downstream conveyor. The downstream conveyor must be verified to be energized and operating before the upstream conveyor is permitted to start and, should the downstream conveyor stop, the upstream conveyor(s) is(are) stopped. This is to prevent material accumulation at a conveyor that has stopped from an upstream conveyor that is still energized and operating.

To determine the required length of the belt for the belt conveyor, we can evaluate the surface length of the belt.

For the belt conveyor shown in Figure 7.1, the conveyor belt connects the outer surface of the head pulley and the outer surface of the tail pulley. The belt covers one half of the circumference of the head pulley. If we know the diameter of the head pulley, then we can calculate the circumference of the head pulley using equation (7.1)

$$C_h = \pi \times D_h \tag{7.1}$$

where

C_h = circumference of the head pulley

D_h = diameter of the head pulley

π = approximately 3.141592654

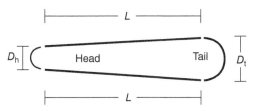

Figure 7.1   Simplified belt conveyor arrangement with different diameters.

Since the belt only connects to one half of the circumference of the head pulley, then the length of the belt along the head pulley is found utilizing equation (7.2)

$$\tfrac{1}{2} \times C_h = \tfrac{1}{2} \times \pi \times D_h \qquad (7.2)$$

Similarly, if we know the diameter of the tail pulley, then we can calculate the circumference of the tail pulley using equation (7.3)

$$C_t = \pi \times D_t \qquad (7.3)$$

where

C_t = circumference of the tail pulley (in.)

D_t = diameter of the tail pulley (in.)

π = approximately 3.141592654

Since the belt only connects to one half of the circumference of the tail pulley, then the length of the belt along the tail pulley is

$$\tfrac{1}{2} \times C_t = \tfrac{1}{2} \times \pi \times D_t \qquad (7.4)$$

where

C_t = circumference of the tail pulley (in.)

D_t = diameter of the tail pulley (in.)

π = approximately 3.141592654

The final remaining amount of conveyor belt needed is twice the distance between the centerlines of the head and tail pulleys as can be seen in Figure 7.1. This is defined as

$$2 \times L \qquad (7.5)$$

where

L = distance between centerlines of head and tail pulleys (ft)

The reader may realize at this point that the actual length of the conveyor belt between the center line of the head and tail pulleys is slightly longer than the value of the distance between the centerlines when the diameters of the heat and tail pulleys are not the same. However, in cases where the difference in diameter between the head and tail pulleys is much, much smaller than the centerline to centerline distance between the pulleys, the centerline to centerline distance is a reasonable estimate of actual belt length.

The total length of conveyor belt needed is the summation of the amount on the head pulley plus the amount on the tail pulley, plus the amount between the centerline of the head and tail pulleys. By combining equations (7.2), (7.3), (7.4) and (7.5), we find the total length necessary to be

$$\text{Belt length} = (\tfrac{1}{2} \times \pi \times D_h)/12 + (\tfrac{1}{2} \times \pi \times D_t)/12 + 2 \times L$$

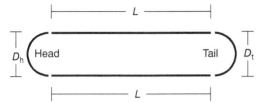

Figure 7.2   Simplified belt conveyor arrangement with same diameters.

Using distribution we can pull out the constants and simplify the above equation to

$$\text{Belt length} = \{1/2 \times \pi \times (D_h + D_t)\}/12 + 2 \times L \tag{7.6}$$

where

  belt length = in units of feet
  D_h = diameter of the head pulley (in.)
  D_t = diameter of the tail pulley (in.)
  π = approximately 3.141592654
  L = distance between centerlines of head and tail pulleys (ft)

Equation (7.6) can be used to determine the minimum belt length necessary for a conveyor system shown in Figure 7.1. For the special case where the head and tail pulleys have the same diameter as is shown in Figure 7.2, then equation (7.6) can be simplified further as shown in equation (7.7).

$$\text{Belt length} = (\pi \times D_h)/12 + 2 \times L \tag{7.7}$$

where

  D_h = diameter of the head pulley (in.)
  π = approximately 3.141592654
  L = distance between centerlines of head and tail pulleys (ft)

Example 7.1   Given a belt conveyor where the diameter of the head and tail pulleys will be 6 in. and the distance between the head and tail pulleys will be 10 ft, determine the minimum length of conveyor belt needed.

Solution: Utilizing equation (7.7), we find the minimum belt length to be

  belt length = $(\pi \times D_h)/12 + 2 \times L$
  belt length = $\pi \times 6$ in. (1 ft / 12 in.) + 2×10 ft
  belt length = $\pi \times 0.5$ ft + 2×10 ft
  belt length = 1.6 ft + 20 ft
  belt length = 21.6 ft

The linear speed of the belt is a function of the rotational speed of the head pulley. To convert from rotation speed of the head pulley in revolutions per minute (rpm) to the linear speed of the belt in feet per minute (FPM), we can use equation (7.1) to determine the linear speed of the belt and, since diameter is given in inches and *belt speed* in feet, we use unit conversion to cancel these units.

$$V = \text{RPM} \times (\pi \times D_h / \ 1 \ \text{rev}) \times (1 \ \text{ft}/12 \ \text{in.})$$

$$V = \text{RPM} \times \pi \times D_h / 12$$

Or by combining the constants, we find the equation to determine belt speed (FPM) for a given head pulley speed (RPM) to be

$$V = 0.2618 \times \text{RPM} \times D_h \tag{7.8}$$

where

V = linear speed of conveyor belt (FPM)

RPM = rotational speed of pulley (rpm)

D_h = diameter of the head pulley (in.)

In practical application, we will be given the needed belt speed and we need to determine the head pulley speed. To find head pulley speed, we can simply rearrange equation (7.8) for head pulley speed as shown in equation (7.9)

$$\text{RPM} = V/(0.2618 \times D_h)$$

or

$$\text{RPM} = 3.819719 \times V/D_h \tag{7.9}$$

where

V = linear speed of conveyor belt (FPM)

RPM = rotational speed of pulley (rpm)

D_h = diameter of the head pulley (in.)

Example 7.2   Given a belt conveyor where the diameter of the head and tail pulleys will be 6 in., determine the rotational speed of the head pulley necessary to drive the belt at a linear speed of 94 (FPM).

Solution: Utilizing equation (7.9), we find that the necessary rotation speed of the head pulley is

RPM = $3.819719 \times V / D_h$

RPM = 3.819719×94 FPM / 6 in.

RPM = 60 rpm

So to drive the belt at a linear speed of 94 FPM given we have pulleys with a diameter of 6 in., we need to drive the head pulley at a speed of 60 rpm.

Example 7.3   Given another belt conveyor where the diameter of the head and tail pulleys will be twice the value given in Example 7.2 or 12 in., determine the rotational speed of the head pulley necessary to drive the belt at a linear speed of 94 FPM.

Solution: Utilizing equation (7.9), we find that the necessary rotation speed of the head pulley is

$$RPM = 3.819719 \times V / D_h$$
$$RPM = 3.819719 \times 94 \text{ FPM} / 12 \text{ in.}$$
$$RPM = 30 \text{ rpm}$$

So to drive the belt at a linear speed of 94 FPM, given we have pulleys with a diameter of 12 in., we need to drive the head pulley at a speed of 30 rpm.

Examples 7.2 and 7.3 show an important relationship between pulley diameter and head pulley speed for a given belt speed. We found that to keep belt speed constant, as we increase the diameter of the head pulley, the rotational speed of the head pulley decreases proportionally. Now let us revisit Figure 7.1 where the two pulleys are of different diameter. If our belt speed in Figure 7.1 is 94 FPM and the head pulley is 6 in. in diameter and the tail pulley is 12 in. in diameter, what is the rotational speed of the head pulley and tail pulley? From Example 7.2, we see that the head pulley with a diameter of 6 in. is rotating at a speed of 60 rpm and from Example 7.3, we see that the tail pulley with a diameter of 12 in. is rotating at a speed of 30 rpm. This shows us the relationship for a given belt speed that as we increase the diameter of a pulley, the rotational speed of the belt decreases or

$$\text{(for constant belt velocity) } RPM_1 / RPM_2 = D_2 / D_1 \qquad (7.10)$$

RPM = rotational speed of pulley (rpm)

D = diameter of the pulley

Similarly, from the fact that belt velocity is proportional to speed and pulley diameter, we find that, for a given pulley speed, the velocity of the belt is directly proportional to the diameter of the pulley.

$$\text{(for constant pulley speed) } V_1 / V_2 = D_1 / D_2 \qquad (7.11)$$

V = linear speed of conveyor belt (FPM)

D_h = diameter of the pulley

Similarly, from the fact that belt velocity is proportional to speed and pulley diameter, we find that, for a given pulley diameter, the velocity of the belt is directly proportional to the speed of the pulley.

$$\text{(for constant pulley diameter) } V_1 / V_2 = RPM_1 / RPM_2 \qquad (7.12)$$

V = linear speed of conveyor belt (FPM)

RPM = rotational speed of pulley (rpm)

The flow rate of the material on the belt conveyor is proportional to the speed of the belt, assuming the amount of material per unit length does not change as we change the belt speed. Belt conveyors ideally are constant-torque loads. Constant torque means that as the belt is increased or decreased in speed, the amount of torque required to drive the material is constant. Since the flow rate of material increases or decreases in direct proportion to the speed of the machine, we can mathematically express this as

$$Q_1/Q_2 = V_1 V_2 \tag{7.13}$$

where

Q_1 = flow through conveyor at speed V_1

Q_2 = flow through conveyor at speed V_2

V_1 or V_2 = linear speed of conveyor (FPM or linear feet per minute)

As mentioned previously, the amount of torque demanded by the conveyor is independent of the speed of the conveyor. Since power is the product of torque and speed, the power requirement for a conveyor is increased or decreased in direct proportion to the change of speed. This is mathematically expressed as

$$P_1/P_2 = V_1/V_2 \tag{7.14}$$

where

P_1 = power drawn by the conveyor at speed V_1

P_2 = power drawn by the conveyor at speed V_2

V_1 or V_2 = linear speed of conveyor (FPM)

Example 7.4 Given a belt conveyor where the initial feed rate is 100 tons per hour (tph) at a belt speed of 50 LFPM, that is pulling 100 kW of power, if we change the motor speed such that the new belt speed is 100 FPM, calculate the following.

a. The new value for feed rate

b. The new value for power

Solution:

a. Utilizing equation (7.13);

$$Q_1 / Q_2 = V_1 / V_2$$

Rearranging

$$Q_2 = Q_1 (V_2 / V_1)$$
$$Q_2 = (100 \text{ tph}) (100 \text{ FPM}) / (50 \text{ FPM})$$
$$Q_2 = 200 \text{ tph}$$

b. Utilizing equation (7.14);

$$P_1 / P_2 = V_1 / V_2$$

Rearranging

$$P_2 = P1 \; V_2 / V_1$$
$$P_2 = (100 \text{ kW}) (100 \text{ LFPM}) / (50 \text{ LFPM})$$
$$P_2 = 200 \text{ kW}$$

The amount of energy required to move an object over a distance is the product of the applied force on the object and the distance that the object is moved. Since power is the rate at which energy is used or work is done, the amount of power required to move an object over a distance is the product of the applied force on the object and the rate of change of distance that object is moving (or speed of the object). The equation that relates the weight of the material on the conveyor belt and the velocity that the belt is moving is

$$P = (F \times V)/33,000 \qquad (7.15)$$

where

$P =$ the power required to drive the conveyor (HP).

$F =$ effective tension of belt to drive system (lbm)

$V =$ velocity or speed of the conveyor (FPM)

The value of 33,000 is needed for unit conversion. We have units of HP on the left side of equation (7.15) and we have units of ft lbm/min on the right side of the equation. By definition, one HP is defined as 550 foot-pounds per second (ft lb/sec) or 33000 foot-pounds per minute (ft lb/min), so the factor of 33,000 takes care of our unit conversion.

Example 7.5   Given a belt conveyor with a linear velocity of 5000 FPM and a belt tension of 400 lb, calculate the power required to run this conveyor at steady state speed.

Solution: Using equation (7.15), we find the power to be as follows.

$$P = (F \times V) / 33,000$$
$$P = 400 \text{ lb } 5000 \text{ FPM } / 33,000$$
$$P = 60 \text{ HP}$$

The amount of power required to drive the load can also be derived from the torque and rotational speed of the *drive pulley*. The amount of power to rotate an object is the product of the torque applied to the pulley and the rotational speed of the pulley. The equation that relates the torque applied to the pulley and the rotational speed of the pulley is

$$P = (T \times \text{RPM})/63,025 \qquad (7.16)$$

where

$P =$ power required to drive the conveyor (HP)

$T =$ Torque applied to the pulley (in. lb)

$\text{RPM} =$ rotational speed of the pulley (rpm)

The value of 63,025 is needed for unit conversion. The value of 63,025 comes from the fact that we have units of HP on the left side of equation (7.16) and we have units of in. lbm rev/min on the right side of the equation. By definition, one HP is defined as 550 foot-pounds per second (ft lb/sec) or 33,000 foot-pounds per minute (ft lb/min) or 396,000 inch-pounds per minute (in. lb/min). Therefore, if we divide the right side of equation (7.16) we will have units of HP rev. Now we just need to take care of the units of revolutions. We can use the identity that one revolution is equal to $2 \times \pi$ radians and we can achieve this by multiplying the right-hand side by the identity that (2π radians / 1 rev). This will now provide us with units of HP on both sides. Summarizing, if we multiply by $2 \times \pi$ radians / rev and divide by 396,000 inch-pounds per minute (in. lb/min), we wind up with our desired units of HP. We can simplify 396,000 / (2π) as a single number of 63,025 which is shown in equation (7.16).

If we assume an ideal system with no losses between the power used to drive the head pulley and the power required to move the material along the conveyor, we can combine equations (7.15) and (7.16) to find the relationship between conveyor linear power (force and velocity) and conveyor rotational power (torque and rotational speed) as follows.

$$T \times \text{RPM} \propto F \times V \tag{7.17}$$

T = torque applied to the pulley (in. lb)

RPM = rotational speed of the pulley (rpm)

F = effective tension of belt to drive system (lbm)

V = velocity or speed of the conveyor (FPM)

The effective tension is the summation of all the individual physical components that present tension or resistance to the belt movement. Some of these factors that create forces on the conveyor belt are the weight of the load on the conveyor, the friction of the components that support and drive the conveyor such as *rollers* and gear reducers, the frictional resistance of the material being conveyed, and the force required to move a material continuously as it is fed onto and off of the conveyor.

Equation (7.17) provides the necessary power to drive the conveyor at steady state but does not address the amount of energy and, therefore, power to accelerate the conveyor from a stopped condition to a full-speed condition and the design of the conveyor should account for the required acceleration energy or torque for the conveyor system. To ensure adequate starting torque from the motor, the motor-locked rotor torque value should be designed to exceed the sum of the required torque to lift the material and twice the torque required to overcome total conveyor friction. As mentioned in Chapter 17 on motors, the torque of the motor on the motor data sheet is defined at nominal voltage and, as the motor voltage is reduced, the torque at any given slip speed reduces by the square of the reduction of voltage. Therefore, this analysis should be done looking at the worst-case voltage drop to the motor during the starting process and, thereby, the worst-case available motor torque value.

Also, as mentioned in Chapter 17 and shown in Figure 17.17, the motor torque-speed curve should not drop below the load-torque curve line, or the motor will not be able to successfully start the conveyor under those conditions.

The tension on the belt should be maintained at a value to limit the sag on the belt to prevent material spillage. Historical data indicates that, for belt sags greater than 3% of the span between the belt idlers, the load spillage becomes significant. Equations (7.18), (7.19), and (7.20), define the minimum tensions for a 3% sag, 2% sag, and 1.5% sag given a certain idler spacing and belt and material weight. (for specification applications, refer to vendor provided methods for calculation of belt sag).

$$T(3\% \text{ sag}) = 4.2 \, S_i \, W \tag{7.18}$$

$$T(2\% \text{ sag}) = 6.25 \, S_i W \tag{7.19}$$

$$T(1.5\% \text{ sag}) = 8.4 \, S_i W \tag{7.20}$$

where

W = weight of belt and material (lb/ft)

S_i = idler spacing (ft)

T = tension in belt (lb)

$T(3\% \text{ sag})$ = tension in belt for 3% sag in belt between idlers (lb)

$T(2\% \text{ sag})$ = tension in belt for 2% sag in belt between idlers (lb)

$T(1.5\% \text{ sag})$ = tension in belt for 1.5% sag in belt between idlers (lb)

Example 7.6   For a belt with an idler spacing of 2 ft and a combined belt and material weight of 100 lb/ft, calculate the tension in the belt for maximum sag of 3%.

Solution: Using equation (7.18), we find the following.

$T(3\% \text{ sag}) = 4.2 \, S_i \, W$
$T(3\% \text{ sag}) = 4.2 \, (2 \text{ ft}) \, (100 \text{ lb/ft})$
$T(3\% \text{ sag}) = 840 \text{ lb}$

Example 7.7   For a belt with an idler spacing of 2 ft and a combined belt and material weight of 100 lb/ft, calculate the tension in the belt for maximum sag of 2%.

Solution: Using equation (7.19), we find the following.

$T(2\% \text{ sag}) = 6.25 \, S_i \, W$
$T(2\% \text{ sag}) = 6.25 \, (2 \text{ ft}) \, (100 \text{ lb/ft})$
$T(2\% \text{ sag}) = 1250 \text{ lb}$

Example 7.8   For a belt with an idler spacing of 2 ft and a combined belt and material weight of 100 lb/ft, calculate the tension in the belt for maximum sag of 1.5%.

Solution: Using equation (7.20), we find the following.

$$T(1.5\% \text{ sag}) = 8.4 \, S_i \, W$$
$$T(1.5\% \text{ sag}) = 8.4 \, (2 \text{ ft}) \, (100 \text{ lb/ft})$$
$$T(1.5\% \text{ sag}) = 1680 \text{ lb}$$

The idler spacing should be designed to maintain a maximum of 3% sag when the belt is operating under normal load and maintain a maximum of 4.5% sag when the belt is at standstill. Additional to the requirements for idler spacing for correct belt tension and sag, the idler spacing should never exceed the idler load ratings provided by the manufacturer.

A special application of the belt conveyor is the typical coal feeder belt (called a gravimetric feeder) in the power generation station that uses coal for its fuel source. These conveyors are designed with scales integrated into the control of the conveyor. As the material passes over the scale on the belt, the material is weighed so that the mass of coal that is being fed to the pulverizer is monitored and measured. Since this represents the mass (lbm) of fuel being provided to the pulverizer and burner, if we know the heating value of the coal (BTU/lbm), then we accurately can calculate the amount of energy (BTU) we are providing to the mills and thus into the furnace.

In the power generation facility, trippers are commonly used to discharge the material on belt conveyors into their final destination which is a hopper or allow material to bypass the hopper if material is to be fed to another location. As shown in Figure 7.3, a diverter gate is utilized to direct material flow either through the bypass chute or through the discharge chute depending on if the tripper is in discharge or bypass mode. The tripper is provided on rails so that it can move from hopper to hopper and discharge the material into the hopper. The position of the tripper is indicated by limit switches along the conveyor or by the use of a radar or laser distance

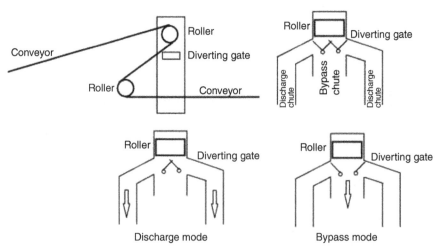

Figure 7.3   Typical tripper arrangement.

sensor. The tripper elevates the conveyor and discharges the material into the hoppers below and the diverter gate determines if the material is discharged into the bin under the tripper or if the bin below is bypassed by dumping the material back onto the conveyor belt.

Starting a Belt Conveyor

While the starting torque for the conveyor needs to be adequate to successfully start the conveyor, sudden changes in torque on the conveyor belt can lead to reduced equipment life and additional material spillage issues. There are several methods available to smoothly control the change in belt torque during the acceleration process. Please refer to Chapter 17 on motors for information on methods of reducing motor starting torque.

PNEUMATIC CONVEYOR SYSTEMS

Pneumatic conveyors make use of compressed air to transport a material to a desired location through a system of pipes. A common utility generation station application is the conveyance system that supplies the furnace with pulverized coal. The pulverizer reduces the particle size of the coal and the primary air fan provides the air required to transport the pulverized coal from the mills to the burners in the furnace section via coal pipes. In pneumatic conveyors the material being transported is usually a very light material that is easily entrained in the air stream.

ROTARY SCREW CONVEYOR SYSTEM

A *rotary screw conveyor* uses a rotating helical screw or auger assembly to transport material from one side of the conveyor to the other side. The auger assembly is contained inside a pipe and material is collected in the blades of the screw. As the shaft turns the rotary screw, the material is captured by the rotary screw and material is forced from one end of the pipe to the other end of the pipe. The speed of rotation of the shaft of the helical screw defines the flow rate through this conveyor. Screw conveyors are used with a variable speed motor or variable frequency drive when control of the volume of material is desired. One of the greatest advantages of this conveyor over the belt conveyor is that the rotary screw conveyor can be used to transport materials both vertically as well as horizontally.

VIBRATING CONVEYOR SYSTEM

Vibrating conveyors are commonly used when, in addition to transporting materials from one system to another, the material must be sorted or classified. One example of this is slag treatment. The slag material is ground down to a small size such that the material can be sold and used in processes such as sand blasting operations. A vibrating screen (combination conveyor and classification screen) is used to perform this type of operation. The raw slag is placed on the vibrating screen. As the slag byproduct is transported between locations, the wire mesh of the vibrating screen allows the finer particles to fall under the vibrating screen and these are transported to their storage facility. The particles that are too large to fit through the wire mesh will be delivered to a grinder where the particle size is reduced and then the grounded slag is resent to the source side of the vibrating conveyor screen for reprocessing.

There are numerous other conveyor systems in use, but the belt conveyor system, pneumatic conveyor system, vibrating screen conveyor system, and the rotary screw conveyor system are the most common in the power generation facility.

CONVEYOR SAFETY

To ensure accident-free conveyor operation, use the following procedures.

1. Inspect all interlocks and safety devices prior to starting the conveyor.
2. Verify that all warning plates are in place.
3. Establish continuous communications between remote locations and the central control location.
4. For safety, it is recommended that at least two people be at any one location. This is to ensure that, if one person becomes incapacitated due to injury, the second person can act to ensure the safety of the first individual.

GLOSSARY OF TERMS

- Axle – A non-rotating shaft on which wheels or rollers are mounted.
- Bearing – A machine part in or on which a shaft, axle, pin, or other part rotates.
- Belt – A flexible band placed around two or more pulleys for the purpose of transmitting motion, power or materials from one point to another.
- Belt Conveyor – A conveyor belt consists of two or more pulleys, with a continuous loop of material – the conveyor belt – that rotates about them.
- Belt Speed – The length of belt, which passes a fixed point within a given time.
- Chain Conveyor – Any type of conveyor in which one or more chains act as the conveying element.
- Conveyor Drive – An assembly of the necessary structural, mechanical, and electrical parts which provide the motive power for a conveyor.
- Drive Pulley – A pulley mounted on the drive shaft that transmits power to the belt with which it is in contact.

- Emergency Pull Cord – Cord that runs along the side of the conveyor that can be pulled at any time to stop the conveyor.
- Emergency Stop Switch – Electrical device used to stop the conveyor in an emergency.
- Pneumatic Conveyor Systems – A conveyor which transports dry, free-flowing, granular material in suspension.
- Roller – A round part free to revolve about its outer surface. The face may be straight, tapered or crowned. Rollers provide the rolling support for the load being conveyed.
- Rotary Screw Conveyor System – A duct along which material is conveyed by the rotational action of a spiral vane which lies along the length of the duct.
- Vibrating Conveyor System – A feeder for bulk materials (pulverized or granulated solids), which are moved by the vibration of a slightly slanted, flat vibrating surface.

PROBLEMS

7.1 Given a belt conveyor where the diameter of the head and tail pulleys will be 18 in., determine the rotational speed of the head pulley necessary to drive the belt at a linear speed of 94 FPM.

7.2 Given a belt conveyor where the initial feed rate is 10 tons per hour (tph) at a belt speed of 20 LFPM that is pulling 10 kW of power, if we change the motor speed such that the new belt speed is 40 LFPM, the new feed rate.

7.3 Given a belt conveyor where the initial feed rate is 10 tons per hour (tph) at a belt speed of 20 LFPM that is pulling 10 kW of power, if we change the motor speed such that the new belt speed is 40 LFPM, the new power drawn at the new speed.

7.4 Given a belt conveyor where the initial feed rate is 10 tons per hour (tph) at a belt speed of 20 LFPM, that is pulling 10 ft lbm of torque, if we change the motor speed such that the new belt speed is 40 LFPM, calculate the new value of toque delivered by the motor.

7.5 Given a belt conveyor with a linear velocity of 10,000 FPM and a belt tension of 500 lb, calculate the power required to run this conveyor at steady state speed.

7.6 For a belt with an idler spacing of 3 ft and a combined belt and material weight of 500 lb/ft, calculate the tension in the belt for a maximum sag of 3%.

7.7 For a belt with an idler spacing of 3 ft and a combined belt and material weight of 500 lb/ft, calculate the tension in the belt for a maximum sag of 2%.

7.8 For a belt with an idler spacing of 3 ft and a combined belt and material weight of 500 lb/ft, calculate the tension in the belt for a maximum sag of 1.5%.

RECOMMENDED READING

CEMA Belt Book, 7th edition, 2016, Conveyor Equipment Manufacturers Association, http://www.cemanet.org

Electric Power Plant Design, Technical Manual TM 5-811-6, Department of the Army, USA, 1984.

Electrical Machines, Drives and Power Systems, 6th edition, Theodore Wildi, Prentice Hall, 2006.

The Engineering Handbook, 3rd edition, Richard C. Dorf (editor in chief), CRC Press, 2006.

Power Plant Engineering, Black & Veatch, edited by Larry Drbal, Kayla Westra, and Pat Boston, Chapman & Hall / Springer, 1996.

Standard Handbook of Powerplant Engineering, 2nd edition, Thomas C. Elliott, McGraw Hill, 1998.

CHAPTER 8

FANS

GOALS

- To understand the basic design of axial and radial fans
- To calculate the amount of air power associated with a given value of volume and differential pressure
- To build a fan system curve and identify the operating point
- To utilize the centrifugal fan laws to calculate change in fan flow, differential pressure, and/or power

F**ANS ARE** widely utilized in power generation stations to move compressible gasses such as atmospheric air to various systems in the plant. Some of the larger fan applications are the *forced draft* (FD), *induced draft* (ID), primary air (PA), booster and gas recirculation fans.

The power (rate of energy utilization) of a fan is a product of the flow through the fan and the change in pressure (developed head) across the fan. Since air is a compressible media, there is also a factor that takes into account the compressibility of air. These three factors are described mathematically by the following formula.

$$\text{Ideal Air HP} = (k \times V \times H)/6356$$

where

k = compressibility factor

V = volume or flow (CFM)

H = head (in. H_2O)

The value of 6356 is necessary for the conversion of units from units of (CFM) $\times$ (in. H_2O) to units of HP, where CFM is cubic feet per minute. One HP is equivalent to 746.7 W, so to convert from HP to watts, we multiply the HP value by a value of 746.7. One watt is equivalent to (1/47.82) (CFM $\times$ atm), so 1 HP is equivalent to (746.7/47.82) (CFM $\times$ atm) or 1 HP is equivalent to (15.6) (CFM $\times$ atm), where atm is 1 atmosphere of pressure. Notice that the units are now in the

Energy Production Systems Engineering: An Introduction for Electrical Engineers to Electrical Power Generation Facilities, Systems, and Equipment, First Edition. Thomas H. Blair.

form of the product of units of (CFM) which is a unit of volumetric flow and units of (atm) which is a unit of differential pressure. This is as expected since we know that mechanical power is the product of flow through a piece of equipment and the differential pressure across the piece of equipment and we have an equation with units of power on one side and the product of units of differential pressure and flow on the other side. Lastly, we need to convert the units of differential pressure from units of (atm) to units of (in. H_2O). 1 (atm) is equivalent to 407.189 (in. H_2O), so 1 HP is equivalent to ($\{407.189 \times 746.7\}/47.82$) (CFM × in. H_2O) or 1 HP is equivalent to (6356) (CFM × in. H_2O).

The equation for ideal air HP is the ideal amount of mechanical power, but there are inefficiencies to a fan *system* due to various losses such as friction. These losses are accounted for by the addition of an efficiency term. This gives us the final equation for mechanical fan power as shown in equation (8.1).

$$\text{Air HP} = (k \times V \times H)/(6356 \times \text{eff}) \tag{8.1}$$

where

k = compressibility factor

V = volume or flow (CFM)

H = head (in. H_2O)

eff = mechanical efficiency

The compressibility factor (k) is an experimentally defined term and is a function of the differential pressure across the fan and the design of the fan. The compressibility factor for a fan application changes as the change in pressure across the fan is changed. Since air is a compressible media, as the difference in pressure increases across the fan, the air is compressed and this affects the amount of air power (HP) mechanically that is provided by the fan as shown in equation (8.1).

CENTRIFUGAL FAN (RADIAL AIRFLOW)

There are several types of centrifugal fan blade designs as demonstrated in Figure 8.1 and different designs have different benefits and drawbacks as described by Table 8.1 Centrifugal fans direct the airflow radially away from the center of the fan along the radius of the fan utilizing *radial blades*, thus the term *radial airflow*. There are several designs for the blades each with certain efficiency and erosion susceptibility.

The design of the blade affects both the efficiency of the fan and the tolerance for erosion from particulate matter in the air stream. Table 8.1 describes both the efficiency of the design as well as the erosive tolerance of the blade design.

For an air stream that has little particulate, such as a booster fan downstream of a bag house or precipitator, the airfoil design might be better choice since it has higher efficiency providing reduced power requirements for a certain airflow and differential pressure. At the same time the airfoil design has the low tolerance for erosion but that is not as much of an issue with the low amount of particulate matter in

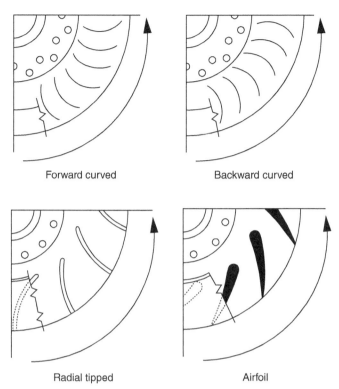

Figure 8.1   Types of centrifugal fan blade designs. *Source*: Reproduced with permission of Tampa Electric Company. Reproduction is forbidden without the express consent of Tampa Electric Company.

the air stream downstream of the ash collection system (bag house or precipitator). However, if the fan is to be used as a gas recirculation fan that is taking exhaust air directly from the economizer section (with a high ash content) and re-injecting it into the FD air stream, this air stream may contain higher amounts of particulate matter and, as such, the fan design should have a very high tolerance for erosion. In this application, a radial tipped fan blade design would be a better choice. Even though the

TABLE 8.1   Comparison of Centrifugal Fan Tip Design Efficiencies and Erosion Tolerances

Blade Type	Efficiency (%)	Erosive Tolerance
Radial tipped	60–70	High
Forward curved	45–60	Medium
Backward curved	75–85	Medium
Airfoil	80–90	Low

Source: Reproduced with permission of Tampa Electric Company. Reproduction is forbidden without the express consent of Tampa Electric Company.

efficiency is lower for the radial tipped fan blade design, the higher erosive tolerance may lead to reduced maintenance costs and higher equipment availability over the life of the fan.

In the typical arrangement of a centrifugal fan, air is admitted on the side of the fan housing and enters the center of the fan blade at the inside diameter of the fan blades. The blades of the fan add kinetic energy to the air in the *impeller* section of the fan increasing the speed of the air and directing the air in a radial direction away from the fan *rotor*. The air mass is directed toward the discharge duct work where the kinetic energy (velocity) is converted to potential energy (pressure). This movement of air from the inside diameter to the outside diameter of the fan creates a low pressure area at the inside diameter of the fan blade drawing in more air to the suction side of the fan. The amount of flow is controlled either by an inlet guide vane assembly, by varying the speed of the shaft which can be accomplished via a variable speed drive controlling the speed of the motor, a variable slip clutch drive between motor and shaft, or a special design motor such as wound rotor motor. Motor speed controls are covered in detail in Chapter 17 covering motors. In this section, we will discuss airflow control utilizing discharge *damper* control.

With discharge damper control, flow is controlled by changing the system resistance. Flow is reduced by closing the discharge damper and flow is increased by opening the discharge dampers. Adjustment of the discharge damper also changes the differential pressure across the fan as well as flow through the fan. Since power is the product of flow and differential pressure, this affects the mechanical power drawn by the fan and is defined by the intersection of the fan performance curve and the system curve and this point is known as the *point of operation*. Initially the increase in pressure has greater effect than reduction in flow and this has the effect of increasing HP requirement to a certain point in the *fan curve*. Below this point, the reduction in flow exceeds the increase in pressure and the HP requirement actually drops.

Looking at Figure 8.2, we can discuss the difference between discharge damper flow control and variable speed flow control and how these two control methods affect the power requirements of the fan. Assuming an initial flow of about 3× ACFM (actual cubic feet per minute) on the curve in Figure 8.2, we see that the static pressure rise across the fan is about 6.5 in. H_2O. The unit of inches of water (in. H_2O) is defined as the differential pressure across a column of water that is one inch high, for water density at standard temperature and pressure (density of about 1.00).

This is shown as point A in Figure 8.2. If we want to reduce flow, we can close the discharge damper in series with the fan to increase system resistance to airflow. This changes the operating point along the *system operating curve* to a new value shown as Point B Figure 8.2. Now the fan static pressure has increased to about 7.5 in. H_2O, but the flow has been reduced to about 2.5 ACFM. If we want to increase flow, we can open the discharge damper in series with the fan to decrease the system resistance to airflow. This new operating point is shown as point C in Figure 8.2. Now the fan static pressure has decreased to about 5 in. H_2O, and the flow has increased to about 3.5 ACFM.

Going back to equation (8.1), we see power is proportional to both flow and head. Restating this, the power required is the area under the operating curve as it is

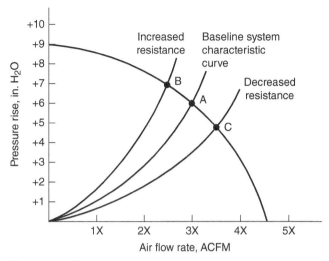

Figure 8.2   Chart showing fan system curve for three system resistance values that depend on damper position with constant fan speed. *Source*: Richard C. Dorf, 1995. Reproduced with permission of Taylor & Francis Group LLC Books.

the product of flow and head. The value of mechanical power for each of the three operating flow points is shown in Figure 8.3.

Alternatively, if we keep inlet guide vane position or damper position at the same position to keep the system resistance line the same, but we reduce or increase the speed of the fan, this will change the fan differential pressure. Since the system resistance curve does not change, but the fan differential pressure changes, the system operating point will lie in the intersection of the system resistance curve and the fan differential pressure curve for the speed at which the fan is operating. This is shown in Figure 8.4. Now the fan static pressure response is different and, as such, the power drawn by the fan when using variable speed control is different than when using damper or inlet guide vane control. Looking at Figure 8.4, we can assume an initial operating point at point A. Assuming an initial flow of about 3 ACFM on the curve in Figure 8.4, we see that the static pressure is about 6.5 in. H_2O. This is shown as point A on Figure 8.4 which is at the intersection of the system resistance line and the fan curve at a speed of Y rpm. If we want to reduce flow down, we can reduce the speed of the fan by 100 rpm. This new operating point (Y – 100 rpm) is shown as point E in Figure 8.4. Now the fan static pressure has reduced to about 5.5 in. H_2O, and flow has reduced to about 2.8 ACFM. If we want to increase flow, we can increase the speed of the fan to a value of Y + 100 rpm. This new operating point is shown as point D in Figure 8.4. Now the fan static pressure has increased to about 7.5 in. H_2O, and the flow as increased to about 3.2 ACFM.

Again, going back to equation (8.1), we see power is proportional to both flow and head. Restating this, the power required is the area under the operating curve as it is the product of flow and head. With the flow control by speed, when flow is reduced,

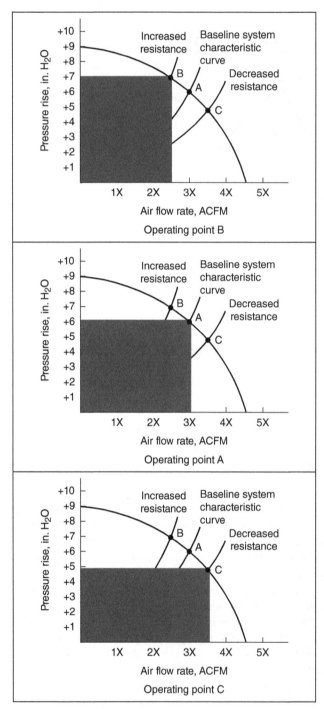

Figure 8.3   Power for operating points A, B, and C. *Source*: Richard C. Dorf, 1995. Reproduced with permission of Taylor & Francis Group LLC Books.

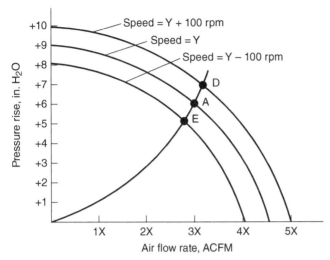

Figure 8.4   Chart showing fan system curve for three fan speeds with constant system resistance utilizing fan speed for control. *Source*: Richard C. Dorf, 1995. Reproduced with permission of Taylor & Francis Group LLC Books.

static head is reduced. Since power is proportional to both flow and head, power is reduced substantially. With damper or inlet guide vane control, with a reduction in flow, power is reduced. However, since the reduction of flow occurs with an increase in static head, the reduction of power is less dramatic.

The value of mechanical power for each of the three operating flow points is shown in Figure 8.5.

In the above discussion, we discussed only outlet damper control of flow and not inlet guide vane control. In reality, there is a substantial difference in the system resistance curve when using outlet damper versus inlet guide vane control. Outlet damper control always has higher power requirement at any reduction of flow than other two methods of flow control. When normal operational speed is between 80% and 100% of full speed, the variable speed control and variable inlet vane control both have about the same power requirements. Since inlet guide vane control is simpler to implement and less costly, this is the method of flow control used when the flow does not normally drop below 80%. In applications that may spend many operational hours below the 80% threshold, variable speed control is chosen due to the substantial reduction in power requirements at these low flow conditions.

In addition to the above discussion, speed control for the motor has the additional benefits of reduction of erosion on fan blades when operating at lower speeds. Also, if an electronic variable frequency drive is used, it will tend to reduce the mechanical shock of starting the fan and, depending on drive design, may remove the fan motor short circuit contribution from the upstream bus during a fault condition. Additionally, damper controls tend to be more noisy than variable speed controls. The air restriction of a damper tends to enhance the noise of airflow.

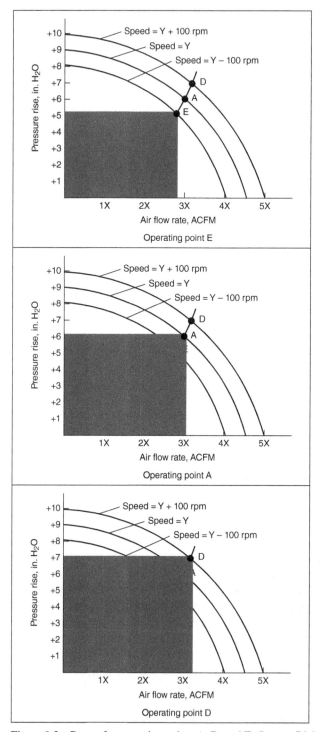

Figure 8.5 Power for operating points A, D, and E. *Source*: Richard C. Dorf, 1995. Reproduced with permission of Taylor & Francis Group LLC Books.

In addition to the option of electronic variable frequency drive control of the motor for speed control as discussed above, the options of using a fluid drive, multiple speed motor, or steam turbine also allows for variable speed of the fan and control of airflow.

AXIAL FAN (AXIAL AIRFLOW)

Another type of fan technology is the axial fan. While in the centrifugal fan design, the airflow is perpendicular or radial to the shaft of the fan, in the *axial flow fan*, the airflow is parallel to the shaft of the fan. Airflow is directed to the inlet of the fan housing and flow follows the shaft of the fan. Unlike the centrifugal fan, the blades of the axial flow fan are airfoil-type blades and the construction of the assembly is such that the pitch of the airfoil blades is controlled. By increasing the pitch of the airfoil blades, the airflow can be increased and by reducing the pitch of the airfoil blades, the airflow can be reduced for a specific motor speed. Some of the drawbacks to the axial flow fan as compared with the centrifugal fan designed for radial airflow are that this design has the additional mechanical linkage to be able to adjust the pitch of the blades on the rotating shaft of the fan and requires more maintenance than the centrifugal fan with inlet guide vane control. The airfoil blades are mounted on the rotor (rotating) of the fan assembly, but the controls are mounted on the stationary part of the fan and linkages must allow for connection between these two assemblies. With the inlet guide vane control of the centrifugal fan assembly, both the inlet guide vanes and controller are mounted stationary which simplifies interconnection. Also the axial fan does not lend itself well to damper control as the additional system impedance of a damper would provide increased fan power requirements. Lastly, the axial flow fan has low tolerance for erosion and must be used with relatively clean air due to the use of the airfoil fan blades. See Table 8.1 for tolerance of airfoil blades to erosion.

One of the benefits of the axial flow fan technology is that, since axial fans use the airfoil blades, they tend to be much more efficient than the centrifugal fan. A greater efficiency means that, for a given required amount of mechanical power out of the fan, less power is needed to be supplied to the fan motor for the same flow and differential pressure requirements. Additionally, the function of blade pitch to airflow on an axial flow fan is a linear function. By comparison, the function of inlet guide vane position to flow on a centrifugal fan is a nonlinear function.

CENTRIFUGAL FAN FUNDAMENTAL LAWS

As we discussed previously, the power required by a *centrifugal fan* is proportional to the volume of air through the fan, the pressure across the fan and the compressibility factor which takes into account the change in density of the air as the pressure changes through the fan (reference equation 8.1). For a centrifugal fan design with radial airflow, the *centrifugal fan laws* define for us the change in flow, pressure, and power as we change other fan parameters. They define that, for a given size fan, system resistance, and air density, if we change speed or pressure, the other parameters change as shown below.

Fan Laws

1. For given fan size, system resistance, and air density
- When speed varies
  - Flow varies directly with speed
  - Pressure varies to square of speed
  - Power varies to cube of speed
- When pressure varies
  - Flow varies as the square root with pressure
  - Speed varies as the square root with pressure
  - Power varies by a factor of 1.5

Remember when pressure is increased, density increases and when temperature increases, density decreases. Therefore, when a fan undergoes air density changes (due to either pressure or temperature change) the following relationships apply.

Fan Laws – Density Changes

2. For constant pressure
- Speed, flow, and power vary as the square root of density (directly for pressure, inverse for temperature)

3. For constant flow and speed
- Power and pressure vary directly to density (directly for pressure, inverse for temperature)

4. For constant flow
- Capacity, speed, and pressure vary inversely to density
- Power varies inversely as square of density

Defined mathematically, we find the fan laws to be:
When speed is varied

$$F_1/F_2 \propto N_1/N_2 \tag{8.2}$$
$$P_1/P_2 \propto (N_1/N_2)^2 \tag{8.3}$$
$$HP_1/HP_2 \propto (N_1/N_2)^3 \tag{8.4}$$

When pressure is varied

$$F_1/F_2 \propto (P_1/P_2)^{1/2} \tag{8.5}$$
$$N_1/N_2 \propto (P_1/P_2)^{1/2} \tag{8.6}$$
$$HP_1/HP_2 \propto (P_1/P_2)^{3/2} \tag{8.7}$$

When density is varied and pressure constant

$$N_1/N_2 \propto (D_1/D_2)^{1/2} \tag{8.8}$$
$$F_1/F_2 \propto (D_1/D_2)^{1/2} \tag{8.9}$$
$$HP_1/HP_2 \propto (D_1/D_2)^{1/2} \tag{8.10}$$

When density is varied and flow and speed are constant

$$P_1/P_2 \propto D_1/D_2 \tag{8.11}$$
$$HP_1/HP_2 \propto D_1/D_2 \tag{8.12}$$

When density is varied and flow is constant

$$C_1/C_2 \propto D_2/D_1 \tag{8.13}$$
$$N_1/N_2 \propto D_2/D_1 \tag{8.14}$$
$$P_1/P_2 \propto D_2/D_1 \tag{8.15}$$
$$HP_1/HP_2 \propto (D_2/D_1)^2 \tag{8.16}$$

where

F = flow (CFM)

P = pressure (in. H_2O)

HP = power (HP)

N = speed (rpm)

D = density or mass per unit volume (lb/ft^3)

C = capacity (CFM)

GLOSSARY OF TERMS

- Air Velocity – Rate of speed of an airstream, expressed in feet per minute or FPM.
- Axial Flow – In-line air movement parallel to the fan or motor shaft.
- Centrifugal Fan – A fan design in which air is discharged perpendicular to the wheels rotational axis.
- Curve, Fan Performance – A graphic representation of static or total pressure and fan BHP requirements over an airflow volume range at a stated inlet density and fan speed.
- Curve, System – A graphic representation of the pressure versus flow characteristics of a given system and density.
- Damper – An accessory to be installed at the fan inlet or outlet for air-volume modulation.
- Fan Laws – Theoretical constant relationships among flow, speed, pressure, and power for a given fan used in a given system.

- Forced Draft – A fan supplying air under pressure to the fuel burning equipment. Typically on the air intake side of the boiler.
- Impeller – Another term for fan "wheel." The rotating portion of the fan designed to increase the energy level of the gas stream.
- Impingement – Striking or impacting, such as material impingement on a fan wheel.
- Induced Draft – A fan removing air from a boiler. Typically on the air discharge side of the boiler.
- Point of Operation – The intersection of a fans static pressure curve and the system curve to which the fan is being applied.
- Radial Blade – Fan wheel design with blades positioned in straight radial direction from the hub.
- Rotor – The rotating part of most motors and fans.
- System, Gas flow – A series of ducts, conduits, elbows, filters, diffusers, etc., designed to guide the flow of air, gas, or vapor to and from one or more locations. A fan provides the energy necessary to overcome the system's resistance to flow and causes air or gas to flow through the system.

PROBLEMS

8.1 Using equation (8.1), given that the needed volume of a fan application is 100,000 CFM and the pressure difference across the fan is 20 in. H_2O, the ideal air HP of the fan is most closely what value? For this problem, assume a compressibility factor of 1 and assume the fan is 100% efficient.

    A. 210 Air HP

    B. 315 Air HP

    C. 385 Air HP

    D. 410 Air HP

8.2 Given the centrifugal fan laws in the chapter, for a given fan size, system resistance, and air density, as fan speed is reduced by a factor of 2, fan power is reduced by a factor of

    A. 2

    B. 4

    C. 8

    D. 16

    E. 1.414 (square root of 2)

8.3 Given the centrifugal fan laws in the chapter, for a given fan size, system resistance, and air density, as fan speed is reduced by a factor of 2, fan pressure is reduced by a factor of

    A. 2

    B. 4

C. 8

D. 16

E. 1.414 (square root of 2)

8.4 Given the centrifugal fan laws in the chapter, for a given fan size, system resistance and air density, as fan speed is reduced by a factor of 2, fan flow is reduced by a factor of

   A. 2

   B. 4

   C. 8

   D. 16

   E. 1.414 (square root of 2)

8.5 Which fan design has the highest efficiency?

   A. Radial tipped

   B. Forward curved

   C. Backward curved

   D. Airfoil

8.6 Which fan design has the highest tolerance for erosion?

   A. Radial tipped

   B. Forward curved

   C. Backward curved

   D. Airfoil

8.7 For an axial flow fan, what is the most common method of controlling airflow?

   A. Discharge damper

   B. Inlet damper

   C. Inlet guide vanes

   D. Controlling blade pitch

RECOMMENDED READING

Electric Power Plant Design, Technical Manual TM 5-811-6, Department of the Army, USA, 1984.

Electrical Machines, Drives and Power Systems, 6th edition, Theodore Wildi, Prentice Hall, 2006.

The Engineering Handbook, 2nd edition, Richard C. Dorf (editor in chief), CRC Press, 1995.

Power Plant Engineering, Black & Veatch, edited by Larry Drbal, Kayla Westra, and Pat Boston, Chapman & Hall/Springer, 1996.

Quick Reference Guide, 2012, New York Blower, http://www.nyb.com/UWM/QuickRefGlossary.pdf

Standard Handbook of Powerplant Engineering, 2nd edition, Thomas C. Elliott, McGraw-Hill, 1998.

CHAPTER 9

PUMPS

GOALS

- To understand the basic design of centrifugal, axial, and positive displacement pumps
- To build a pump system curve and identify the operating point
- To calculate pump mechanical power, given the flow, suction, and discharge pressure
- To calculate pump efficiency
- To calculate net positive suction head required to prevent cavitation

---

PUMPS ARE used to deliver energy to a fluid, which increases the discharge pressure or head of the pump. This pressure is then used to move the fluid from one point to another.

There are three main types of pumps used in the power generation facility. These are the *centrifugal pump*, the *axial flow pump*, and the *positive displacement pump*. By far, the most common of these is the centrifugal pump and this will be discussed in detail. The axial flow pump and positive displacement pumps will also be briefly addressed in this chapter.

SYSTEM RESISTANCE CURVES

The amount of pressure or head that a pump must develop to deliver a certain amount of flow is determined by the system resistance to flow that the pump is connected to. To produce flow, the pump must first develop sufficient pressure to overcome the static pressure and frictional losses at the discharge of the pump that the system imposes on the pump. This is defined by the system resistance curve. The *system resistance curve* is plotted on a chart where the flow is on the horizontal axis and the head or pressure on the vertical axis.

The system described by Figure 9.1 depicts a very basic pump application. It shows two tanks at the same elevation with the same fluid level in both tanks. Since the

Energy Production Systems Engineering: An Introduction for Electrical Engineers to Electrical Power Generation Facilities, Systems, and Equipment, First Edition. Thomas H. Blair.

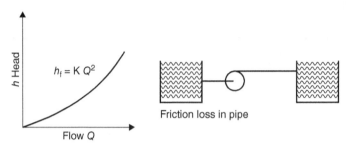

Figure 9.1 Pump system resistance with zero differential head and no throttling. *Source*: Reproduced with permission of Tampa Electric Company. Reproduction is forbidden without the express consent of Tampa Electric Company.

suction and discharge side head or pressure is the same, the pump must only overcome the frictional losses in the pipe, valves, and fittings to achieve flow. Those losses vary with the square of flow as shown on the flow versus head chart in Figure 9.1.

When a valve is placed in the discharge path of the pump and is used to control flow, this affects the system resistance curve. When the position of the valve in the flow path changes, the resistance of the system to flow also changes. Reducing the opening of the valve increases system resistance and increasing the opening of the valve decreases system resistance. The losses of the system still vary with the square of flow, but now the additional resistance of the discharge valve changes the slope of the curve as shown in Figure 9.2. With the value fully open (100%), the discharge head is low and the flow is high. With the value partially closed (50%), the discharge head is increased and the flow is reduced.

Now that we have added flow control to the standard pump configuration where there is no difference between suction and discharge head and analyzed the effect the discharge value has on the system resistance curve, let us discuss a more practical application where the suction and discharge heads are different.

A typical pumping application will have the pump suction connected to a source of fluid from some lower head or pressure value and discharging the fluid into a system

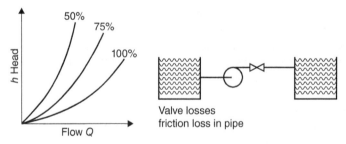

Figure 9.2 Pump system resistance with zero differential head and throttling. *Source*: Reproduced with permission of Tampa Electric Company. Reproduction is forbidden without the express consent of Tampa Electric Company.

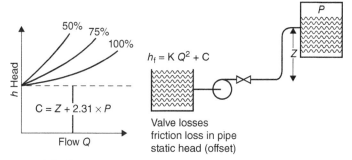

Figure 9.3   Pump system resistance with non-zero differential head and throttling. *Source*: Reproduced with permission of Tampa Electric Company. Reproduction is forbidden without the express consent of Tampa Electric Company.

with a higher head or pressure value. Additionally, the tank that is being discharged to may also have some positive pressure associated with the tank. This adds to the static head of the pump. This arrangement is depicted in Figure 9.3. The pump must overcome the static head (difference between suction side pressure and discharge side pressure) when pumping into a system with a higher positive discharge pressure.

The combination of the pump and connected system will operate at a point defined by the head and flow that satisfies both the pump curve and system curve simultaneously (at the intersection of the two curves). By adjusting a flow control valve in series with the pump, flow will be controlled and pump discharge pressure will change. By throttling down (closing) on the flow control valve, flow will decrease with a side effect of a higher discharge pressure at the pump for the reduced flow conditions through the system. By throttling up (opening) the flow control valve, flow will increase with a side effect of a lower discharge pressure at the pump for the increased flow conditions through the system. Figure 9.4 depicts this application.

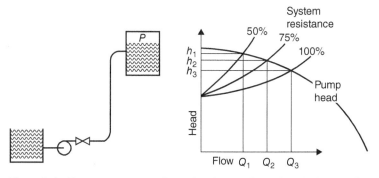

Figure 9.4   Pump system operating point. *Source*: Reproduced with permission of Tampa Electric Company. Reproduction is forbidden without the express consent of Tampa Electric Company.

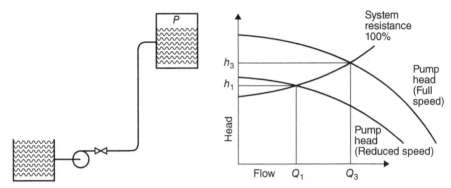

Figure 9.5   Pump flow control utilizing pump speed. *Source*: Reproduced with permission of Tampa Electric Company. Reproduction is forbidden without the express consent of Tampa Electric Company.

Recalling that power is the product of flow and differential pressure, we can see that the act of throttling down or closing a valve increases the resistance across the valve, increasing the overall system resistance to flow. This increases the discharge pressure of the pump while reducing flow. Even though the flow is reduced through the valve, due to the increased differential pressure across a valve, the throttling process introduces losses into the overall system and reduces the efficiency of the pumping system when reducing flow.

An alternate method of flow control is to control the speed of the pump. By adjusting the speed of the pump, the pump discharge head is reduced, but the system resistance curve remains the same. This is shown in Figure 9.5.

Notice when using pump speed for flow control that as the pump speed is reduced, the system resistance curve does not change. What changes is the discharge head of the pump, which is reduced. So how can we determine the difference in net energy required to maintain a certain flow rate utilizing throttling control versus pump speed control? We know that the mechanical power utilized by the system is the product of head and flow. Since power is the product of flow through the pump and head developed across the pump, this can be visualized by the area under the pump curve when drawn at the pump operating point. Utilizing throttling flow control, we see that this area for the flow point Q_1 is shown in Figure 9.6.

Now if we take the same flow rate but achieve this by controlling the speed of the pump to reduce the discharge head of the pump, we can obtain the same flow point Q_1, but since the discharge pressure of the pump is reduced, the product of head and flow using pump speed for control is substantially reduced. This is shown in Figure 9.7 utilizing speed control when comparing the area shown for the same flow point in Figure 9.6 utilizing throttling.

The energy saved by using pump speed to obtain flow rate Q_1 as compared with throttling to obtain the flow rate Q_1 is the difference between these two areas defined by Figures 9.5 and 9.6 and is visualized as shown in Figure 9.8.

In order to know the operating point of a pump and fluid system and develop the pump curve described above, we need to monitor the pump suction pressure,

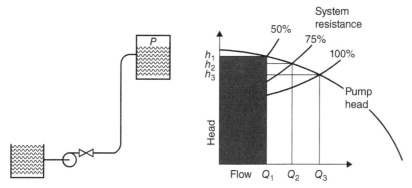

Figure 9.6   Pump flow control utilizing throttling. *Source*: Reproduced with permission of Tampa Electric Company. Reproduction is forbidden without the express consent of Tampa Electric Company.

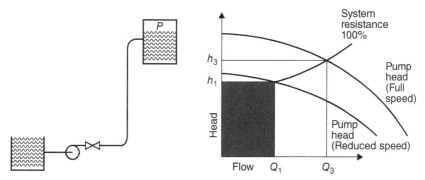

Figure 9.7   Pump flow control utilizing pump speed. *Source*: Reproduced with permission of Tampa Electric Company. Reproduction is forbidden without the express consent of Tampa Electric Company.

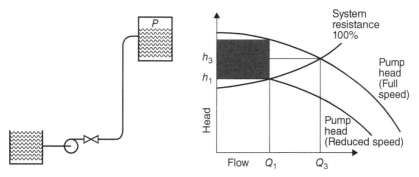

Figure 9.8   Pump flow control energy savings. *Source*: Reproduced with permission of Tampa Electric Company. Reproduction is forbidden without the express consent of Tampa Electric Company.

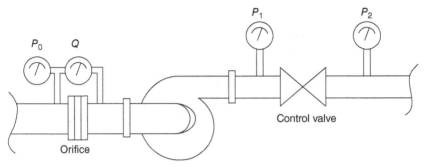

Figure 9.9   Pump monitoring configuration. *Source*: Reproduced with permission of Tampa Electric Company. Reproduction is forbidden without the express consent of Tampa Electric Company.

the pump discharge pressure, the system pressure, and flow through the pump. A simplified pump monitoring system is shown in Figure 9.9.

The flow Q is obtained by monitoring the differential pressure across an orifice (see Chapter 24 on field instrumentation for more detailed description of pressure and flow instruments). The suction pressure P_0, the pump discharge pressure P_1, and the throttling valve discharge pressure P_2 are monitored. The pump discharge head can be calculated utilizing equation (9.1).

$$H_1 = 2.31 \times (P_1 - P_0) \qquad (9.1)$$

where

H_1 = head (ft of H_2O)
P_1, P_0 = pressure (psi)
2.31 is a factor to convert units from psi to feet of H_2O @ 40°F

The system head can be calculated utilizing equation (9.2).

$$H_2 = 2.31 \times (P_2 - P_1) \qquad (9.2)$$

where

H_2 = head (feet of H_2O)
P_2, P_1 = pressure (psi)
2.31 is a factor to convert units from psi to feet of H_2O @ 40°F

By varying the control valve and plotting pump head (H_1) and system head (H_2) for various values of flow (Q), we can plot both the pump capability curve and the system resistance curve (not including valve losses) as shown in Figure 9.10.

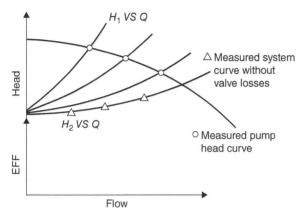

Figure 9.10   Pump monitoring configuration. *Source*: Reproduced with permission of Tampa Electric Company. Reproduction is forbidden without the express consent of Tampa Electric Company.

CENTRIFUGAL PUMP

A centrifugal pump uses a rotating impeller to create flow by adding kinetic energy to the fluid. The fluid enters the pump in the vortex or center of the pump impeller (similar to the centrifugal fan impeller, but water is treated as an incompressible fluid while air is considered compressible). As the pump impeller spins, any fluid contained in the impeller has a force applied to it from the motor and causes the fluid to spin in a circle or centrifugally around the pump casing. This action increases the fluid speed by adding kinetic energy to the fluid. This increase of fluid speed through the pump impeller section causes a lower pressure at the suction of the pump (center of the impeller). New fluid enters to replace the fluid forced from the suction side of the pump to the discharge side of the pump on the outside diameter of the pump casing.

As the fluid reaches the discharge section of the centrifugal pump, the velocity decreases. Since energy is neither created nor destroyed, but only altered in form, this reduction in velocity represents a reduction in kinetic energy of the fluid. This is because the kinetic energy of the fluid is converted to potential energy at the pump discharge, causing an increase in pressure at the discharge of the pump.

Centrifugal pumps can provide a large amount of water, just as centrifugal fans are used when a large volume of air must be moved. However, the pressure head developed by the centrifugal pump is smaller than other pump designs. Additionally, to increase the amount of discharge pressure of the centrifugal pump, we must increase the rate of increase in speed of the fluid at the pump impeller (add more kinetic energy faster) which will result in a decreased suction pressure at the inlet of the pump. At some point, the fluid in the center of the impeller (as shown in Figure 9.11), under the correct circumstances, can undergo latent heat transfer and form small gas bubbles. Then when the fluid moves to the discharge side of the pump where pressure increases, these bubbles will collapse. This creates large localized forces that can wear the impeller prematurely. This process is commonly known as cavitation.

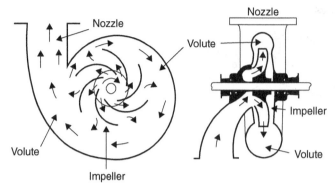

Figure 9.11   Typical centrifugal pump. *Source*: Reproduced with permission of U.S. Navy.

This limits the maximum allowable differential pressure across a centrifugal pump and limits the application of this pump in high discharge pressure applications.

In applications where the pump must develop large amounts of pressure on the discharge, a series multistage centrifugal pump will be used. *A tandem multistage centrifugal pump* consists of one motor and one shaft that have multiple centrifugal pumps connected to the common shaft. The benefit of the tandem design is that only one motor is required to drive all of the pumps. A *cross compound multistage centrifugal pump* has multiple motors and shafts to drive the multiple pumps. The benefit of the cross compound design is that, since each pump has a different motor and shaft driving the pump, the speed of each pump can be a different value and can be ideally matched to the head and differential pressure demands of the pump. In both designs, the pumps are connected in series with each other such that the overall pump head is the summation of each pump stage head. If you equate pump head to an electrical DC circuit voltage and pump flow to electrical DC circuit current, this is similar to placing multiple battery cells in series to obtain larger voltages with the same flow rate or ampacity.

In applications where larger amounts of flow are needed, the pumps are connected in parallel (with a check valve in the discharge of each pump to prevent reverse rotation of the pump). If you equate pump head to an electrical DC circuit voltage and pump flow to electrical DC circuit current, this is similar to placing multiple battery cells in parallel to obtain larger ampacity rating (more flow) with the same voltage (or differential pressure).

The amount of power used by the pump is the product of the fluid density, the head developed across the pump, and the flow rate through the pump. The following discussion utilizes the SI standard units of measurement as this reduces the need for constants for converting units. Equation (9.3) describes the relationship between these parameters.

$$P_m = \rho g \, HQ/\eta_p \qquad (9.3)$$

where

P_m = input power required by the mechanical pump (W)

ρ = fluid density (kg/m^3)

g = standard acceleration of gravity (9.80665 m/sec^2)

H = energy head added to the flow (m)

Q = flow rate (m^3/sec)

η_p = efficiency of the mechanical pump in per unit

The value P_m is also known as the *brake horsepower* (bhp) of the pump. This is not the same as the water horsepower of the pump. The water horsepower of the pump is the amount of power that the mechanical pump delivers to the fluid being pumped and is the product of (ρ g H Q). The mechanical power of the pump (bhp) is the horsepower (water horsepower) out of the pump divided by the efficiency of the pump (η_p).

The amount of electrical power required by the pump motor is the mechanical power demanded by the pump divided by the efficiency of the motor.

$$P_e = P_m/\eta_m \qquad (9.4)$$

where

P_e = input power required by the electrical pump motor (W)

P_m = input power required by the mechanical pump (W)

η_m = efficiency of the pump motor in per unit

Combining the two equations (9.3) and (9.4), we can find the relationship between the electrical power required by the pump motor and the fluid density, the head developed across the pump and the flow through the pump.

$$P_e = \rho g \, H \, Q/(\eta_m\eta_p) \qquad (9.5)$$

where

P_e = input power required by the electrical pump motor (W)

ρ = fluid density (kg/m^3)

g = standard acceleration of gravity (9.80665 m/sec^2)

H = energy head added to the flow (m)

Q = flow rate (m^3/sec)

η_m = efficiency of the pump motor in per unit

η_p = efficiency of the mechanical pump in per unit

Cavitation and Net Positive Suction Head

Another important issue with centrifugal pumps is the avoidance of cavitation in the impeller. Pump *cavitation* occurs when the pressure in the suction of the pump falls below the saturation *vapor pressure* of the fluid for the fluid type, temperature, and suction pressure. Should the pressure drop below the saturation pressure, the fluid partially vaporizes forming local vapor bubbles. Then as the fluid moves to the higher pressure on the discharge side of the pump, the fluid is no longer in the saturated state and the bubbles violently collapse causing mechanical damage to the impeller.

Therefore, it is important to ensure that the pressure at the suction of the pump never drops below the saturation pressure of the fluid.

The amount of pressure in the suction of a pump is the summation of the static pressure (potential energy) of the fluid at the suction of the pump (P_s) and the velocity pressure (kinetic energy) at the suction of the pump (P_v). The *net positive suction head available* (NPSHA) is defined as the difference between the suction pressure of the pump ($P_s + P_v$) and the saturation or vapor pressure of the fluid at the temperature of the fluid (P_{vp}) at the suction of the pump. This is mathematically defined in equation (9.6).

$$NPSHA = P_s + P_v - P_{vp} \qquad (9.6)$$

where

NPSHA = net positive suction head available

P_s = pressure associated with the static pressure at the suction side of the pump

P_v = pressure associated with the velocity pressure at the suction of the pump

P_{vp} = fluid vapor pressure for the specific temperature of the fluid

The pump manufacturer will provide a minimum value for the *net positive suction head required* (NPSHR) at the suction of the pump. To prevent cavitation, the NPSHA value must always be greater than then manufacturer-provided NPSHR value. Should the NPSHA value drop below the manufacturer-provided value for NPSHR, then cavitation will occur in the pump reducing pump performance and shortening pump impeller life.

Looking at equation (9.6), we can see that we can increase the NPSHA by several means listed below.

a. We can increase the suction pipe diameter to reduce the flow rate through the suction pipe for a given value of mass flow rate. We can also reduce the number of bends in the valves and suction piping. This will have the effect of reducing the losses in the piping and valves increasing suction head pressure.

b. If we raise the height of the fluid container, we will raise the positive head pressure from the physical height of the source fluid, thereby increasing the value of the positive suction head which also increases suction head pressure.

c. We can increase the surface pressure on the source side fluid, thereby increasing the value of the pressure at the suction of the pump. Thereby also increasing the value of the suction head pressure.

Alternately, if we are unable to achieve a higher NPSHA, then we must select a pump design that has a lower value of NPSHR. A typical example of this is the condensate pump. The condenser is under very low vacuum. Therefore, the value of NPSHA is very low. This limits the maximum allowable differential pressure for a one-stage pump to be used in this application. Condensate pumps are of a multistage design which allows for a reduced NPSHR value from the OEM for any single stage to meet the NPSHA of the application.

We can further break up equation (9.6), for NPSHA, into its components. The value for the static pressure at the suction of the pump is given in equation (9.7a).

$$P_s = p/SW \qquad (9.7a)$$

where

p = impeller inlet pressure (Pa)

SW = specific weight of fluid or liquid (N/m^3)

The value for the velocity pressure at the suction side of the pump is given by equation (9.7b).

$$P_v = V^2/2g \qquad (9.7b)$$

where

V = velocity of fluid or liquid (m/sec)

g = acceleration of gravity (9.80665 m/sec^2)

The value for the vapor pressure of the fluid at the suction of the pump at the given temperature of the fluid can be calculated by equation (9.7c).

$$P_{vp} = p_v/SW \qquad (9.7c)$$

where

p_v = vapor pressure of fluid or liquid (Pa)

SW = specific weight of fluid or liquid (N/m^3)

Now we can combine equations (9.7a–9.7c) into equation (9.6) and find a final equation for NPSHA as shown in equation (9.8).

$$\text{NPSHA} = (V^2/2g) + (p/SW) - (p_v/SW) \qquad (9.8)$$

where

NPSHA = net positive suction head available (m)

V = velocity of fluid or liquid (m/sec)

p = impeller inlet pressure (Pa)

p_v = vapor pressure of fluid or liquid (Pa)

SW = specific weight of fluid or liquid (N/m^3)

g = acceleration of gravity (9.80665 m/sec^2)

Note that the net positive suction pressure is given in units of meters. To convert to a more common unit of feet, use the conversion one meter is approximately 3.28 feet. Once we have units of feet, how do we convert a column of water height into more common units of pressure such as pounds per square inch? The height of a fluid in a column and the pressure it exerts on the bottom of the column is a function of the *specific gravity* (or density) of the fluid. For the conversion of units of psia to feet, the conversion is shown in equation (9.9).

$$\text{Feet} = \text{psia} \times 2.31/\text{sg} \tag{9.9}$$

where

Feet = pressure in feet of water

psia = pressure in pounds per square inch absolute

sg = specific gravity of fluid at given pressure and temperature

2.31 is a factor to convert units from psi to feet of H_2O @ 40°F

Example 9.1   We are given an application where we need to pump water at a velocity of 30 m/sec. The temperature of the water is at 20°C and the suction pressure is 10 Pa. When water is at 20°C (68°F), it has a specific weight of 998 N/m³ (62.303 lb/ft³) and a saturation vapor pressure of 2333 Pa (0.339 psi). What is the value of the NPSHA? Also, if the pump manufacturer provides a required net positive suction head value of 20 m, will this pump experience cavitation under the given circumstances?

Solution: Using equation (9.8), we calculate the available net positive suction head as follows.

$$\text{NPSHA} = (V^2/2g) + (p/\text{SW}) - (p_v/\text{SW})$$

$$\text{NPSHA} = ((30)^2/2(9.80665)) + (10/998) - (2333/998)$$

$$\text{NPSHA} = 43.6 \text{ m}$$

The required net positive suction head from the pump manufacturer is 20 m. Since NPSHA > NPSHR, this pump will not experience cavitation.

Example 9.2   We are given an application where we need to pump water at a velocity of 10 m/sec. The temperature of the water is at 20°C and the suction pressure is 10 Pa. When water is at 20°C (68°F), it has a specific weight of 998 N/m³ (62.303 lb/ft³) and a saturation vapor pressure of 2333 Pa (0.339 psi). Also, if the pump manufacturer provides a required net positive suction head value of 20 m, will this pump experience cavitation under the given circumstances?

Solution: Using equation (9.8), we calculate the available net positive suction head as follows.

$$\text{NPSHA} = (V^2/2g) + (p/\text{SW}) - (p_v/\text{SW})$$

$$\text{NPSHA} = ((10 \text{ m/s})^2/2(9.80665 \text{ m/s}^2)) + (10 \text{ pa}/998 \text{ kg/m}^3) - (2.339/998 \text{ kg/m}^3)$$

$$\text{NPSHA} = 2.77 \text{ m}$$

The required net positive suction head from the pump manufacturer is 20 m. Since NPSHA < NPSHR, this pump will experience cavitation.

Pump Affinity Laws

Centrifugal pumps follow proportionalities known as the centrifugal pump affinity laws. These laws define the properties of operation of a centrifugal pump as we change various parameters.

The volume or flow of a centrifugal pump is directly proportional to the change in speed of the pump impeller and the change in pump diameter. This is mathematically expressed as

$$Q_1/Q_2 \propto (n_1/n_2)\&(d_1/d_2) \tag{9.10}$$

where

Q_1 = flow at impeller speed n_1 and impeller diameter d_1
Q_2 = flow at impeller speed n_2 and impeller diameter d_2
n_1 or n_2 = wheel velocity
d_1 or d_2 = wheel diameter

The head or differential pressure of a centrifugal pump is directly proportional to the square of the change in speed of the pump impeller and the square of the change in pump diameter. This is mathematically expressed as

$$H_1/H_2 \propto (n_1/n_2)^2\&(d_1/d_2)^2 \tag{9.11}$$

where

H_1 = flow at impeller speed n_1 and impeller diameter d_1
H_2 = flow at impeller speed n_2 and impeller diameter d_2
n_1 or n_2 = wheel velocity
d_1 or d_2 = wheel diameter

The power consumed by a centrifugal pump is directly proportional to the cube of the change in speed of the pump impeller and the cube of the change in pump diameter. This is mathematically expressed as

$$P_1/P_2 \propto (n_1/n_2)^3\&(d_1/d_2)^3 \tag{9.12}$$

where

P_1 = flow at impeller speed n_1 and impeller diameter d_1
P_2 = flow at impeller speed n_2 and impeller diameter d_2
n_1 or n_2 = wheel velocity
d_1 or d_2 = wheel diameter

Example 9.3   Given a centrifugal pump where the initial flow is 100 gpm, the initial head is 100 ft, the initial power is 5 bhp and the initial speed is 1800 rpm; if we change the pump speed to a speed of 3600 rpm, based on the assumption that impeller diameter does not change, calculate the following.

   a. The new value for flow
   b. The new value for head
   c. The new value for power

Solution:

   a. Using equation (9.10),

$$Q_1/Q_2 \propto (n_1/n_2) \& (d_1/d_2)$$

   Rearranging for Q_2

$$Q_2 = Q_1[(n_2/n_1)(d_2/d_1)]$$

   Since diameter does not change in this problem, this simplifies to

$$Q_2 = Q_2(n_2/n_1)$$
$$Q_2 = (100 \text{ gpm})(3600 \text{ rpm})/(1800 \text{ rpm})$$
$$= 200 \text{ gpm}$$

   b. Using equation (9.11),

$$H_1/H_2 \propto (n_1/n_2)^2 \& (d_1/d_2)^2$$

   Rearranging,

$$H_2 = H_1[(n_2/n_1)^2(d_2/d_1)^2]$$

   Since diameter does not change in this problem, this simplifies to

$$H_2 = H_1(n_2/n_1)^2$$
$$= (100)(3600/1800)^2$$
$$= 400 \text{ ft}$$

   c. Using equation (9.12),

$$P_1/P_2 \propto (n_1/n_2)^3 \& (d_1/d_2)^3$$

   Rearranging,

$$P_2 = P_1[(n_2/n_1)^3(d_2/d_1)^3]$$

   Since diameter does not change in this problem, this simplifies to

$$P_2 = P_1(n_2/n_1)^3$$
$$= 5(3600/1800)^3$$
$$= 40 \text{ bph}$$

Example 9.4   Given a centrifugal pump where the initial flow is 100 gpm, the initial head is 100 ft, the initial power is 5 bhp and the initial impeller diameter is 8 in.; if we change the pump impeller diameter to 6 in., based on the assumption that shaft speed does not change, calculate the following:

   a. The new value for flow

   b. The new value for head

   c. The new value for power

Solution:

   a. Using equation (9.10),

$$Q_1/Q_2 \propto (n_1/n_2)\&(d_1/d_2)$$

Rearranging,

$$Q_2 = Q_1[(n_2/n_1)(d_2/d_1)]$$

Since shaft speed does not change in this problem, this simplifies to

$$Q_2 = Q_1(d_2/d_1)$$
$$= (100)(6/8)$$
$$= 75$$

   b. Using equation (9.11),

$$H_1/H_2 \propto (n_1/n_2)^2 \&(d_1/d_2)^2$$

Rearranging,

$$H_2 = H_1[(n_2/n_1)^2(d_2/d_1)^2]$$

Since shaft speed does not change in this problem, this simplifies to

$$H_2 = H_2(d_2/d_1)^2$$
$$= (100)(6/8)^2$$
$$= 56.3 \text{ ft}$$

   c. Using equation (9.12),

$$P_1/P_2 \propto (n_1/n_2)^3 \&(d_1/d_2)^3$$

Rearranging,

$$P_2 = P_1[(n_2/n_1)^3(d_2/d_1)^3$$

Since shaft speed does not change in this problem, this simplifies to

$$P_2 = P_2(d_2/d_1)^3$$
$$= 5(6/8)^3$$
$$= 2.1 \text{ bhp}$$

TABLE 9.1 Comparison of Pump Design Efficiencies for Typical Generation Station

Pump Type	Efficiency (%)
Boiler feed pump	74–83
Condensate pump	72–80
Circulating water pump	70–90

Source: Reproduced with permission of Tampa Electric Company. Reproduction is forbidden without the express consent of Tampa Electric Company.

Flow control for a centrifugal pump is achieved by throttling a discharge value (note a bypass line may be required to maintain the pump minimum flow if this method is used for flow control), or a variable speed motor control system.

Centrifugal pumps are the most common pumps found in the power generating station. Applications include cooling water pumps, condensate pumps, feedwater pumps, boiler feed pumps, and sluice pumps. All these applications requires high flow rate which the centrifugal pump is ideally suited to meet. The design of the pump also affects the efficiency of the pump and typical efficiency values are listed in Table 9.1 for several types of pumps. Notice that the circulating water pump efficiency is slightly higher than the others listed in the table. That is because this pump is an axial flow design rather than a radial flow pump design.

AXIAL FLOW PUMP

Axial flow pumps are much rarer in the power generation facility. There is one application in a generation station where the pump is almost universally an axial flow pump and that is the circulating water pump application. (For more detail on the purpose of circulating water pumps, please see Chapter 10 on condenser cooling). These pumps must provide a very large quantity of water (high flow rate) at a very low head (low discharge pressure) to the tubes inserted in the condenser to condense the exhaust steam from the low pressure turbine into a subcooled liquid. The axial flow pump is ideally suited for this purpose as the pump can operate at very high efficiencies with high flow and low head.

The difference between axial flow pump and the radial flow centrifugal pump is the direction of water flow from the suction to the discharge of the pump. In the centrifugal pump, water enters the center of the impeller and travels radially away from the shaft to the outside diameter of the impeller. In the *axial flow pump*, the fluid travels axially along the shaft of the pump. Typical circulating water pumps are oriented vertically. In combination with the weight of the pump mechanical components due to installation being vertical, additional axial thrust is imposed by the pump impeller on the shaft of the motor pump assembly as it applies axial force to the fluid for pumping. These forces cause a net axial thrust on the motor that does not occur in most radial flow pumps. The motor is designed with an internal thrust bearing to be able to tolerate the axial thrust of this application.

Flow control for an axial flow pump, when the application requires flow control, is achieved by variable vane pitch or a variable speed motor. Unlike the radial flow centrifugal pump that can use a discharge valve for flow control, the axial flow pump does not use this method for flow control as the pump discharge pressure increases substantially when system resistance is increased by throttling, leading to very poor efficiencies. The axial flow pump is most efficient in applications with low system resistance to flow.

POSITIVE DISPLACEMENT PUMP

The third type of pump used in power generation facilities is the positive displacement pump as is shown in Figure 9.12. The *positive displacement pump* operates by trapping fluid inside a chamber at the suction of the pump and then moves this fixed volume of fluid from the suction side to the discharge side. As the fluid is transported from the suction to the discharge side, the volume of the chamber is reduced. Since pressure and volume are inversely proportional, the reduction of chamber volume results in a pressure increase in the fluid at the discharge side of the positive displacement pump. Small clearances in positive displacement pumps are critical to ensure an adequate seal of the chamber that is used to increase the pressure of the positive displacement pumps. There are two main types of positive displacement pumps: a rotary style and a reciprocating style. The rotary style has screws that form a compression chamber between the screw element and the pump case to compress the fluid. This is similar in mechanical motion as the rotary screw compressor, but with much smaller clearances to prevent compression chamber leakage.

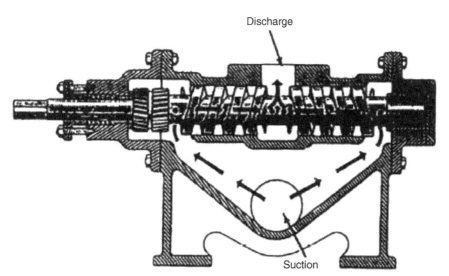

Figure 9.12   Typical rotary screw positive displacement pump. *Source*: Reproduced with permission of U.S. Navy Training Manual.

The reciprocating pump consists of a cylinder in a chamber. The cylinder compresses the fluid in the chamber and moves the fluid from suction to discharge.

Flow control for a positive displacement pump is almost always via variable speed control. The design of the pump is such that it will maintain a certain flow for a certain speed regardless of the head developed. If the discharge side of the positive displacement pump were to become suddenly blocked, the discharge pressure of the positive displacement pump could reach destructively high pressures quickly. For safety, in a positive displacement pump application, one or more pressure relief valves are installed on the discharge header of a positive displacement pump to protect the piping system from over-pressurization in the event the discharge header becomes blocked.

Pump Safety

Never operate a positive-displacement rotary pump with the discharge valve closed. This may lead to a rapid over-pressurization of the discharge header. Ensure that all relief valves are not isolated and functional before placing a positive pressure pump into service.

GLOSSARY OF TERMS

- Axial Flow Pump – Uses foil blade impellers to force the fluid axially along the shaft of the pump.
- Centrifugal Pump – A pump that uses an impeller to move water or other fluids. The impeller imparts the kinetic energy of the rotor into the liquid media that is being pumped.
- Cavitation – Cavitation is the formation of cavities or bubbles in the suction side of a pump and the subsequent collapse of these cavities or bubbles in the pump impeller.
- Net Positive Suction Head (NPSH) – The measurement of liquid pressure at the pump end of the suction system, including the design of the pump.
- Net Positive Suction Head Available (NPSHA) – The difference between the suction pressure of the pump ($P_s + P_v$) and the saturation or vapor pressure of the fluid at the temperature of the fluid (P_{vp}) at the suction of the pump. The result must be equal to or greater than NPSHR.
- Net Positive Suction Head Required (NPSHR) – This is the amount of atmospheric pressure required to move liquid through the suction side of the pump. NPSHR is directly related to pump design.
- Safety Factor – This value is used in the NPSH calculations to take in to account for fluctuations in atmospheric pressure.
- Specific Gravity (SG) – The weight of any liquid relative to that of water.

- Standard Atmospheric Pressure – The weight of atmosphere at sea level under normal atmospheric conditions (14.7 psi)
- Vapor Pressure (VP) – The pressure at which a liquid will vaporize. This pressure is relative to the liquid's temperature. It is also known as the saturation pressure for a given liquids temperature.
- Positive Displacement Pump – Operates by trapping fluid inside a chamber at the suction of the pump and then moves this fixed volume of fluid from the suction to the discharge.

PROBLEMS

9.1   For a positive displacement pump, if the speed of the pump is doubled, which doubles the flow through the pump, what is the effect on pump power assuming the head across the pump is constant?

9.2   Given a centrifugal pump where the initial flow is 100 gpm and the initial speed is 1800 rpm, if we change the pump speed to a speed of 900 rpm, based on the assumption that impeller diameter does not change, calculate the value of the new flow from the pump.

9.3   Given a centrifugal pump where the head is 100 ft and the initial speed is 1800 rpm, if we change the pump speed to a speed of 900 rpm, based on the assumption that impeller diameter does not change, calculate the value of the new head developed by the pump.

9.4   Given a centrifugal pump where the initial power is 500 bhp and the initial speed is 1800 rpm, if we change the pump speed to a speed of 900 rpm, based on the assumption that impeller diameter does not change, calculate the value of the new power drawn from the pump.

9.5   We are given an application where we need to pump water at a velocity of 100 m/sec. The temperature of the water is at 20°C. When water is at 20°C (68°F), it has a specific weight of 998 N/m^3 (62.303 lb/ft^3) and a saturation vapor pressure of 2333 Pa (0.339 psi).

RECOMMENDED READING

Electrical Machines, Drives and Power Systems, 6th edition, Theodore Wildi, Prentice Hall, 2006.
The Engineering Handbook, 3rd edition, Richard C. Dorf, (editor in chief), CRC Press, 2006.
Electric Power Plant Design, Technical Manual TM 5-811-6, Department of the Army, USA, 1984.
Power Plant Engineering, Black & Veatch, edited by Larry Drbal, Kayla Westra, and Pat Boston, Chapman & Hall/Springer, 1996.
Standard Handbook of Powerplant Engineering, 2nd edition, Thomas C. Elliott, McGraw-Hill, 1998.
US Navy Training Manual, Machinist's Mate 2nd class, Naval Education and Training Professional Development and Technology Center, 2003.

CHAPTER 10

CONDENSER COOLING SYSTEM

GOALS

- To understand the basic design and components of the condenser in a steam cycle
- To understand the purpose of the vacuum pump or air ejector in a condensate system
- To be aware of safety hazards associated with condensate system equipment
- To understand the importance of the gland seal system in maintaining condenser vacuum
- To calculate the minimum condensate temperature necessary for a given value of condenser vacuum to prevent condensate pump cavitation

---

CONDENSER COOLING

To maximize the thermal efficiency of our thermal cycle, we want to maintain the exhaust of the steam turbine at as low a pressure and temperature as possible (see Chapter 2 on thermodynamics for discussion of why condenser pressure and temperature is important to thermal cycle efficiency). To maintain the exhaust of the turbine at low temperature and low pressure, we have to remove the heat from the exhaust steam from the steam turbine. The *condenser* is the *heat exchanger* that provides this heat transfer path. A typical main condenser assembly is shown in Figure 10.1. There are several designs in use.

Just as in the case of the feedwater heat exchanger, the most efficient heat transfer method is via *latent heat* transfer on the steam side. This latent heat transfer occurs at the surface of the tubes, where the steam is condensed. To maximize the tube surface area exposed to the exhaust steam, the condensate level is maintained below the tubes. However, as we discussed in Chapter 9 on pumps, we need to maintain the *net positive suction head available* (NPSHA) at a greater value than the *net positive suction head required* (NPSHR) for the condensate pumps. This is achieved by both

---

Energy Production Systems Engineering: An Introduction for Electrical Engineers to Electrical Power Generation Facilities, Systems, and Equipment, First Edition. Thomas H. Blair.
© 2017 by The Institute of Electrical and Electronics Engineers, Inc. Published 2017 by John Wiley & Sons, Inc.

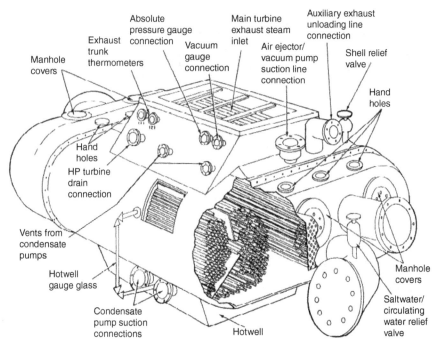

Figure 10.1   Typical condenser arrangement. *Source*: Reproduced with permission of U.S. Navy Training Manual.

subcooling the condensate in the hotwell as well as maintaining the level of the condensate at the upper end of the hotwell. Also, this is controlled by design by locating the condensate pumps as low as physically possible and utilizing large diameter piping between the hotwell and the suction of the condensate pump to maximize the suction pressure or head to the suction of the condensate pumps. As shown in Figure 10.1, the tap in the hotwell for the suction of the condensate pumps is at the lowest point possible in the hotwell.

The function of the condenser cooling system is to remove the thermal energy from the exhaust steam and transfer this energy to some heat sink outside of the facility. One of the most common methods of removing the thermal energy in the form of heat from the condenser is to provide water flow from an outside water reservoir such as a river, bay, or estuary. This water flows through the tubes in the condenser and removes heat from the exhaust steam. Then this heated water is discharged back to the river, bay, or estuary downstream of the intake to ensure the intake does not receive heated water from the discharge. A modification of this system is sometimes used when it is desired to maintain discharge water temperatures within certain environmentally friendly values. A "helper" cooling tower may be added to the discharge between the condenser and water reservoir to lower the temperature of the water prior to discharge back to the river, bay, or estuary. A typical equipment arrangement for a condenser is shown in Figure 10.2.

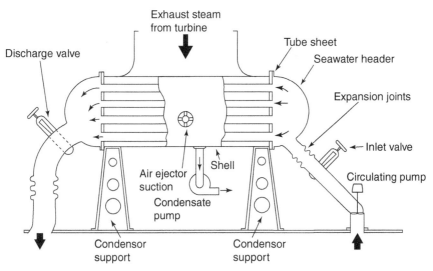

Figure 10.2   Typical condenser arrangement. *Source*: Reproduced with permission of U.S. Navy Training Manual.

Sometimes either due to restrictions of the mass of cooling water available or due to environmental impacts to the water resource, a closed loop system is used where the water is heated in the condenser when it removes the thermal energy from the exhaust steam of the turbine and then the water is cooled in a cooling tower. Since a majority of cooling takes place in the cooling tower utilizing latent heat transfer, evaporation occurs in the cooling tower and *makeup* water to the tower must be supplied to keep the mass of water in the closed loop cooling system steady. Also, as water is evaporated and makeup is supplied, the impurities in the water tend to increase in concentration. These impurities will adversely affect the condenser tubes and must be removed from the boiler water. A blowdown line is provided to allow for the release of these impurities to maintain chemistry of the cooling water within the constraints of the tube technology employed in the condenser.

Condensers operate in very low pressures (typically 1 psia) to maximize the efficiency of the thermal process. Since the steam on the shell side of the condenser is in a saturated state, maintaining condensate temperatures low will keep the condenser shell-side pressure low as well. However, the circulating water system alone is not sufficient to maintain the condenser vacuum at desired values and a secondary air removal system is employed. Either *air ejection pumps* or *air removal vacuum pumps* are utilized to maintain the condenser shell side under a strong vacuum.

The *vacuum pump* uses a water seal and pump assembly to pump air out of the condenser shell side while the air ejector uses steam flow and a venturi element to draw a vacuum on the shell of the condenser. A two-stage air ejector system is shown in Figure 10.3.

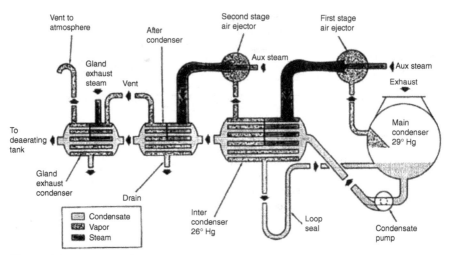

Figure 10.3   Two-stage air ejector system. *Source*: Reproduced with permission of U.S. Navy Training Manual.

Air Ejector Safety

The following safety precautions should be used with air ejectors:

1. When starting an air ejector to prevent overpressurization, always open the discharge valves before admitting steam to the nozzles. When securing an air ejector, always close tightly the steam supply valves to the nozzles before closing the discharge valves.

2. Before starting an air ejector, always drain the steam supply line, and open the drain valves in the inter-condenser and after-condenser drain lines. This is to prevent an explosion that might occur should steam be allowed to enter these chambers with condensate in the chamber. Should that occur, the condensate will flash into steam and create a rapid pressure rise.

Even with very small clearances between rotating and stationary sections of the turbine, without some type of seal between rotating and stationary sections, air may leak into the steam side and find its way to the condenser, thereby increasing the pressure in the condenser shell side. In order to maintain the barrier between the steam side and air side of these boundaries, a *gland steam* seal system is utilized. The gland steam seal system provides low pressure sealing steam to these locations and utilizes a multitude of labyrinth rings. Under normal operations, the steam in the turbine provides adequate pressure to prevent air intrusion into the steam side.

However, for periods of low load operation and startup/shutdown periods, the gland steam system injects the low pressure steam (about 2 psi) into this labyrinth system. The labyrinth rings provide a high flow impedance path and, since the pressure of the steam seal system is slightly higher than atmospheric conditions, the seal steam is forced into the seals displacing any air. A figure showing the arrangement of the gland steam seal system is shown in Figure 10.4.

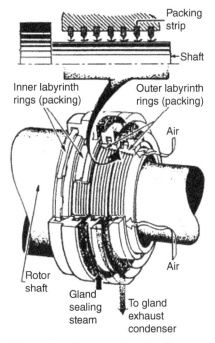

Figure 10.4   Steam turbine gland steam seal system. *Source*: Reproduced with permission of U.S. Navy Training Manual.

CONDENSER OPERATION

The two primary operational rules that apply to the operation of the condenser are as follows:

1. The discharge circulating water temperature rise has environmental limits that must not be exceeded (typical value is 10°F rise from suction to discharge).
2. The condensate temperature leaving the hotwell should be within 2°F of the condensing temperature corresponding to the vacuum in the condenser. This is an adequate amount of subcooling necessary to ensure the condensate pump does not cavitate without excessive subcooling. If we increase the condensate temperature depression excessively below the saturation temperature, we will reduce the efficiency of the thermal cycle (see Chapter 2 on the thermodynamic cycle for details of condensate depression and efficiency effects). Table 10.1 lists the value of saturation temperature for various values of condenser vacuum.

Example 10.1   The condenser vacuum is measured at 29 inches of mercury. What is the desired temperature of the condensate in the hotwell to ensure we do not create cavitation in the condensate pump, while at the same time, we do not adversely affect the thermal efficiency of the steam cycle?

TABLE 10.1   Vacuums and Corresponding Saturation Temperatures

Condenser Vacuum	Corresponding Saturation	
	Temperature	
Inches of Mercury	(°F)	(°C)
29.6	53	11.67
29.4	64	17.78
29.2	72	22.22
29	79	26.11
28.8	85	29.44
28.6	90	32.22
28.4	94	34.44
28.2	98	36.67
28	101	38.33
27.8	104	40.00
27.6	107	41.67
26.5	120	48.89

Source: Reproduced with permission of U.S. Navy Training Manual.

Solution: Ideally, the amount of subcooling should be about 2°F. This value will ensure we do not create cavitation while at the same time, we will not adversely affect the thermal efficiency of the steam cycle. From Table 10.1, at a condenser vacuum of 29 inches of mercury, the corresponding saturation temperature is 79°F. Adding 2°F subcooling to the saturation temperature results in an ideal hotwell temperature of 77°F.

If the condenser vacuum is not as high as it should be in relation to the condenser load, this would indicate that something in the system is not operating normally. One possibility may be that the air ejectors may not be properly removing air from the condenser resulting in a reduced condenser vacuum. Alternately, if the level of the hotwell is not maintained below the tubes, this could cause a reduction in heat transfer surface where latent heat transfer occurs resulting in a reduction in heat removal efficiency of the condenser. Another possibility is a leak in the condenser shell or boundary between steam side and air side. Such a leak will allow air from the air side of the shell to leak into the steam side of the condenser. This is also known as air in leakage to the condenser. If the gland seal system is not operating properly, then this may lead to increased condenser shell-side pressure and reduced thermodynamic efficiency.

CONDENSER SAFETY PRECAUTIONS

As with all equipment in the utility generation station, there are safety hazards associated with the condenser system. The following safety precautions should be followed for the safe operation of the condenser.

1. Always be on the alert to detect and eliminate air leaks in the vacuum system.
2. If, at any time, a loss of vacuum is accompanied by a hot or flooded condenser, shut down the steam turbine. The concern of a flooded condenser is condensate

carryover into the low pressure steam turbine blades. This can cause catastrophic failure of the low pressure turbine blade section. The concern of a hot condenser is thermal expansion of the low pressure turbine section. Elevated temperature in the low pressure turbine stage may allow the turbine blades to grow and make contact with the low pressure turbine shell.

3. During shutdowns, lift relief valves or vent the water chest to prevent over pressurization.

4. Keep the saltwater side of operating condensers free of air. If the water chest is allowed to collect air, then some of the condenser tubes may become air locked. This will result in a reduction of cooling capability of the condenser.

5. In systems where the cooling water is salt water, test the condensate for any increases in salinity. A tube or tube sheet leak would allow for water contamination of the condensate which is the makeup water to the boiler. The cooling water through the tube is not treated to remove contaminants and, in the case where the cooling water is salt water, a failure of a tube or tube sheet will increase the salinity of the boiler water and adversely affect boiler water chemistry. Early detection of cooling water leaks in the condenser is critical to reliable boiler operation.

6. NEVER bring an open flame or anything that will cause a spark close to a freshly opened saltwater side of a condenser until it has been thoroughly blown out with steam or air. Hydrogen and/or methane (sewer) gases may be present.

7. The condenser represents a confined space and OSHA CFR 1910.146 requirements for confined space entry must be followed whenever performing maintenance in a condenser.

In some locations, a body of water may not be available to provide cooling for the condenser. In these applications, condensers can be cooled using air. Since the heat transfer using air is less efficient that water, large volumes of air and a larger condenser surface area are required to achieve the required amount of heat removal.

Lastly, some steam plants utilize the exhaust steam from the low pressure turbine in other processes in the plant. These systems may not utilize a condenser at the exhaust of the steam turbine.

GLOSSARY OF TERMS

- Air Ejector – A device that removes air and other gases from steam condensers through the suction action of a stream jet.

- Blow Down – Removal of a small quantity of water. The purpose of this system is to prevent the water concentration of dissolved solids from increasing.

- Condenser – The commonly used term for a heat exchanger installed on the exhaust steam from a steam turbine in thermal cycles.

- Heat Exchanger – A device for transferring heat from one media to another.

- Latent Heat – energy released or absorbed, by a body or a thermodynamic system, during a constant-temperature process. (Refer to chapter 2 – thermodynamics for more detail).

- Makeup – Water that is supplied to a system to replenish water that is lost from the system due to either evaporation, blowdown, leakage or other losses to the system.

- Net Positive Suction Head Required (NPSHR) – The minimum pressure required at the suction port of the pump to keep the pump from cavitating. NPSHR is a function of the pump and must be provided by the pump manufacturer.

- Net Positive Suction Head Available (NPSHA) – The amount of pressure at the suction port of a pump. This is a function of the system and must be calculated. (Refer to chapter 9 – pumps for more detail).

- Vacuum Pump – A pump used for creating a vacuum.

PROBLEMS

10.1 What is the term for water that is removed from a condenser that uses evaporation in cooling towers as part of the cooling process to prevent the buildup of suspended solids in the water stream?

10.2 The condenser vacuum is measured at 29.2 inches of mercury. What is the desired temperature of the condensate in the hotwell to ensure we do not create cavitation in the condensate pump, while at the same time, we do not adversely affect the thermal efficiency of the steam cycle?

10.3 What is the term for the minimum pressure required at the suction port of the pump to keep the pump from cavitating?

10.4 List at least three safety procedures when operating a condenser.

RECOMMENDED READING

Electric Power Plant Design, Technical Manual TM 5-811-6, Department of the Army, USA, 1984.

Electrical Machines, Drives and Power Systems, 6th edition, Theodore Wildi, Prentice Hall, 2006.

The Engineering Handbook, 3rd edition, Richard C. Dorf, (editor in chief), CRC Press, 2006.

Power Plant Engineering, Black & Veatch, edited by Larry Drbal, Kayla Westra, and Pat Boston, Chapman & Hall/Springer, 1996.

Standard Handbook of Powerplant Engineering, 2nd edition, Thomas C. Elliott, McGraw-Hill, 1998.

US Navy Training Manual, Machinist's Mate 2nd class, Naval Education and Training Professional Development and Technology Center, 2003.

CHAPTER 11

STEAM TURBINES

GOALS

- To understand the basic design and components of a high pressure, intermediate pressure, and low pressure steam turbine

- To be able to generate a simplified single-line diagram of a steam plant to include the turbine, condenser, feedwater heaters, deaeration tank, condensate pump, and the boiler feed pump

- To identify the difference between an impulse turbine blade and a reaction turbine blade design

- To be able to describe the difference between partial arc and full arc steam admission

- To identify the four types of steam flow control methods

- To be aware of safety hazards associated with turbine system equipment

---

TURBINES PROVIDE the primary means of converting the thermal energy contained in the steam system to the mechanical energy in the turbine shaft. This is used to drive the main generator in a power plant as well as various auxiliary pieces of equipment such as *boiler feed pumps*. The amount of energy exchanged between the steam system and the turbine system during this thermal to mechanical energy conversion process is a function of both the velocity of the steam flow as well as the differential pressure drop across the steam blades (energy is proportional to flow and pressure drop). As we expand the steam (expansion implies a pressure drop across the turbine blades), energy is transferred from the steam system to the turbine system.

Some of the typical equipment used to support the turbine is shown in Figure 11.1. Superheated steam from the *boiler* or *steam generator* enters the high pressure (HP) section of the turbine where the steam is expanded across the high pressure turbine blades. The steam then leaves the high pressure turbine section and returns to the reheater section of the boiler in the cold reheat lines. Back at the boiler, additionally, energy is added to this steam and the reheated steam is sent back to the turbine and enters the intermediate pressure (IP) section of the turbine. Once expanded

Energy Production Systems Engineering: An Introduction for Electrical Engineers to Electrical Power Generation Facilities, Systems, and Equipment, First Edition. Thomas H. Blair.
© 2017 by The Institute of Electrical and Electronics Engineers, Inc. Published 2017 by John Wiley & Sons, Inc.

Typical turbine water and steam cycle

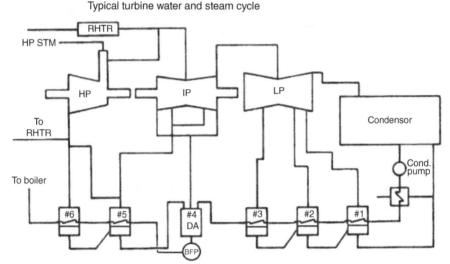

RHTR = Reheater; STM = Steam; HP = High pressure turbine; IP = Intermediate pressure turbine; LP = Low pressure turbine; DA = Deaerating tank.

Figure 11.1   Typical steam turbine equipment. *Source*: Reproduced with permission of Tampa Electric Company.

through the IP section the steam leaves the IP section and enters the low pressure (LP) section of the turbine via the crossover steam piping. The exhaust for the LP section is then sent to the condenser. The *condensate* pump takes its suction from the hotwell of the condenser and feeds the first set of feedwater heaters (known as the low pressure feedwater heaters since these are on the suction side of the boiler feed pump). In Figure 11.1, there are three feedwater heaters. After the feedwater heaters, condensate is sent to the *deaeration tank* (DA). The DA tank performs two functions. First, it removes dissolved gases from the condensate and second, it is mounted at an elevated position to provide net positive suction head for the boiler feed pump. The boiler feed pump (BFP) takes condensate from the DA tank and sends this water to the high pressure feedwater heaters. In Figure 11.1, these heaters are labeled heaters 5 and 6. After the high pressure feedwater heaters, the feedwater is sent to the economizer section of the boiler.

In the turbine, there are both stationary blade sections (mounted to the casing of the turbine) and rotating blade sections (mounted to the rotor of the turbine). The stationary portion consists of nozzle vanes where the cross-sectional area of the steam flow path is reduced. Much like a garden hose at home, this has the effect of increasing the velocity of the steam flow through the nozzle and this effect results in a subsequent reduction of steam pressure across the stationary nozzle (remember that energy is proportional to the product of flow and pressure). Since no energy is removed in the fixed blade section, for an increase in velocity, energy conservation says that there must be an associated decrease in steam pressure across the nozzle. This stationary section is followed by one or more rotating sections. There are two basic types of

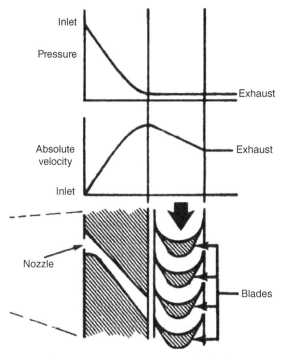

Figure 11.2   Impulse stage pressure and velocity response. *Source*: Reproduced with permission of U.S. Navy Training Manual.

rotating turbine blade stages depending on the design of the turbine blades and the mechanical pressure drop and velocity changes that occur across the turbine blade.

The *impulse stage* shown in Figure 11.2 is a turbine blade where the pressure and volume are constant in the rotating buckets. Steam speed is reduced in the rotating bucket section as the steam impinges on the rotating blades and, as the steam changes direction, kinetic energy is transferred from the steam to the rotating turbine blade causing a reduction in steam velocity. In practicality, typical design of a turbine may have one stationary nozzle section, followed by several rotating impulse blade and stationary blade sections. This is known as a *velocity compound stage*. This is shown in Figure 11.3. The name velocity compound stage is due to the fact that while pressure and volume remain constant across the multiple rotating sections of turbine blades, the velocity drops across each section of rotating turbine blade. *The fundamental thing to remember is that for a rotating impulse stage, the pressure out of the rotating turbine blade section is equal to the pressure into the rotating turbine blade section.*

Another type of combination of stages is the *pressure compound stage* where a stationary nozzle section is followed by one rotating impulse section, followed by another set of stationary nozzles and rotating impulse blade sections. This is shown in Figure 11.4. The term pressure compound stage comes from the fact that in each stationary nozzle section, the pressure drops and the velocity increases, while in each impulse rotating section, the pressure remains constant while the velocity drops.

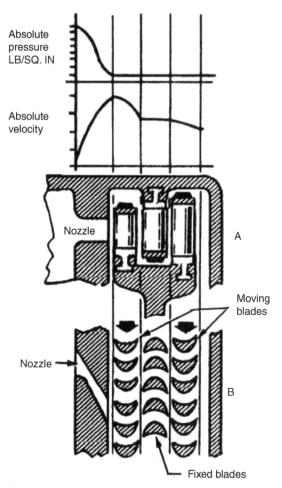

Figure 11.3 Velocity compounded impulse stage. *Source*: Reproduced with permission of U.S. Navy Training Manual.

The above discussion addressed the rotating impulse type of turbine blade. There is another type of rotating turbine blade design called a reaction blade. In a *reaction turbine blade* design, the pressure drops and the volume increases in the rotating section due to the reduction in cross-sectional area across the rotating turbine blade (much like the stationary nozzle section). Both the reaction and impulse stages use nozzle vanes in their stationary section to increase the steam speed through the stationary nozzle section. Figure 11.5 shows a cross section of a typical simple reaction stage. *The fundamental thing to remember is that for a rotating reaction stage, the pressure out of the rotating turbine blade section is less than the pressure into the rotating turbine blade section.*

Referring back to the basic thermodynamic cycle, the expansion through the turbine section is represented by the change in lines of pressure on the T-s diagram

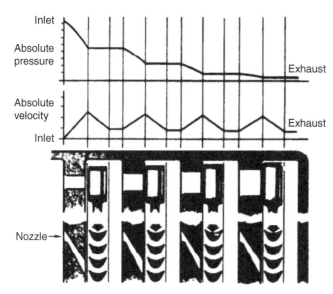

Figure 11.4   Pressure compounded stage. *Source*: Reproduced with permission of U.S. Navy Training Manual.

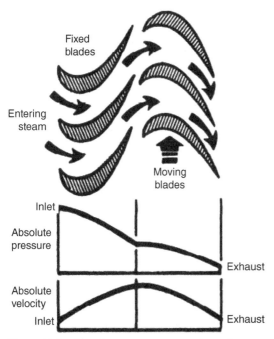

Figure 11.5   Reaction stage pressure and speed response. *Source*: Reproduced with permission of U.S. Navy Training Manual.

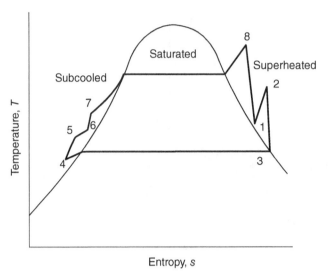

Figure 11.6   Rankine thermodynamic cycle. *Source*: Reproduced with permission of Department of Energy (DOE) Fundamentals Handbook.

from the pressure at the inlet or governor valves to the outlet of the turbine or inlet to the condenser. In Figure 11.6, this is the line from 8 to 1 for the high pressure turbine section, and also the line from 2 to 3 for the combination of the intermediate and low pressure turbine sections. Heat is added at the reheater along the line from 1 to 2. Notice as mentioned in the discussion of the thermodynamic process that this process is not isentropic (constant entropy) but the entropy increases from the inlet of the turbine (points 8 and 2) to the outlet of the turbine (points 1 and 3).

The following describes the various processes that are represented in Figure 11.6.

Pump work added: 4–5 and 6–7

Heat added: 5–6, 7–8, and 1–2

Work produced: 8–1 and 2–3

Heat rejected: 3–4

From Figure 11.6, the rate at which energy is delivered to the turbine is simply the change in enthalpy between points 3 and 2 (for the LP turbine) and the change in enthalpy between points 1 and 8 (for the high pressure (HP) turbine) multiplied by the mass flow rate of steam through the turbine. This is the power into the turbine that is converted to mechanical power in the rotating section of the turbine. This is the ideal case, but in reality, some amount of energy loss, both mechanical and thermal, occurs in the turbine section which represents a small portion of this energy.

There are various types of *steam turbine* arrangements depending on the overall system requirements. The simplest type is the straight flow condensing steam turbine. In this design, the steam enters the turbine at higher pressure. As the steam expands through each turbine stage, the steam pressure drops. To get energy out of the next section at lower pressure, the diameter of the next section of turbine blades is increased for each stage, increasing the surface area of the blades, thereby compensating for the lower pressure available at the next stage.

More common is the *single extraction condensing steam turbine*. This is similar in design to the straight flow condensing steam turbine except at several turbine sections, some of the steam is extracted from the steam turbine for the purpose of supplying the feedwater heaters and auxiliary steam systems in the plant. The extraction steam system is depicted in Figure 11.1.

Due to the difference in construction between the intermediate pressure turbine section and the low pressure turbine section, it is common practice to have these two turbines constructed separately and coupled together at the shaft. The steam that exhausts the intermediate pressure section has to get to the input of the low pressure section and this steam piping is known as the crossover section.

To improve efficiency, most turbine designs will send the exhaust steam of the high pressure turbine back to a set of tubes in the boiler call the reheater tubes where the steam picks up more thermal energy in the boiler and this increases the enthalpy or energy contained in the steam. The lines from the discharge of the high pressure steam turbine to the reheat section of the boiler are called the cold reheat lines. Once thermal energy is added to the steam in the reheat section of the boiler or steam generator, this reheated steam is sent back to the intermediate pressure turbine through piping known as the hot reheat lines. In this design, in the low pressure turbine the steam enters the center of the low pressure turbine and expands in both directions across the LP turbine blades. This is done to minimize the axial thrust on the turbine shaft thereby reducing the requirements of the thrust bearing on the turbine. Similarly, the direction of steam flow in the high pressure (HP) turbine is opposite from the flow of steam in the IP turbine again to minimize the net axial thrust generated on the shaft of the steam turbine.

For larger rated machines, one reheat line back through the boiler does not provide adequate improvement in the efficiency of the machine and a second reheater is added along with a reheat turbine section.

As can be seen from the above discussion, it is very common to utilize multiple turbines in a power plant. These turbines can be connected in either a tandem compounded unit or a cross-compounded unit. The *tandem compounded unit* has only one generator and the generator is coupled to all the turbines. All the turbines and the generator run at the same shaft speed since they are coupled together to form one common shaft.

Sometimes to improve efficiencies of the low pressure turbine, the system may be designed as a cross-compounded system. In a *cross-compounded unit*, the low pressure turbine drives one generator and the high and intermediate turbines drive a different generator. In the cross-compounded unit, low pressure turbine can operate at a different speed (lower) than the speed of the high and intermediate turbines which can allow for a more efficient low pressure turbine design. The lower speed generator that is driven by the low pressure turbine will have more poles to maintain the generated frequency of both generators at the same value.

To control the power transmitted to the turbine rotor, we control the flow of steam through the turbine. Remember that the rate at which energy is transferred to the turbine from the steam system was the change in enthalpy of the steam from inlet of the turbine to the outlet of the turbine times the mass flow rate. Therefore, by changing the flow rate, even with the same enthalpy values at the inlet and outlet of the turbine, we can control the rate of energy transfer from the steam system to the mechanical turbine system. Alternately, we can maintain flow rate the same, but if we adjust the header pressure of the boiler, this changes the enthalpy entering the steam turbine (by changing the constant pressure line on the Ts diagram that we are operating on) and by adjusting header pressure we are able to control the rate of energy transferred to the turbine from the steam system.

Most steam turbines have two or more sets of valves that admit steam to the high pressure turbine. Depending on the manufacturer, the names of the valves are different but their function is the same. Manufacturer "G" calls their two valves the *main steam stop valve* followed by the *main steam control valves*. Manufacturer "W" calls their two valves the *throttle valves* followed by the *governor valves*.

The *main steam stop* or *throttle valve* is the backup to the *main steam control* or *governor valve* for trip functions. During a turbine trip, should one of the main steam control valves fail to close, steam would continue to flow through the turbine. Once the load was removed from the generator then the power into the turbine via the leaking main steam control valve would exceed the power output of the generator (which after tripping the generator breaker, the generator power is effectively zero) and the machine speed would increase to the point of catastrophic failure. To protect against this, the main steam stop valve also closes on a trip providing two isolation points between the boiler or steam generator and the input to the high pressure turbine. Depending on turbine design there are anywhere from one to four main steam stop valves on the turbine. The main steam stop valve or throttle valve also may have a function in the startup mode that we will discuss in next paragraph. On a turbine trip, the fuel to the furnace section of the boiler is immediately tripped or shut down. However, there is still thermal energy being added to the steam in the boiler due to residual heat in the boiler and the steam and water in the boiler tubes. Steam is vented to atmosphere either by control valves or relief valves to dissipate this thermal energy until a steady state thermal condition is reached in the boiler or steam generator.

The main steam control or governor valves are the primary means of controlling the steam flow to the high pressure turbine. There are always as many main steam control valves or governor valves on the machine as there are *arc sections* in the nozzle box (casing that contains the first stage of stationary nozzles). Each governor valve on the machine provides steam to one of these arc sections. See Figure 11.7 for arrangement of valves and the nozzle box arrangement for partial arc admission.

There are two methods of control of the main steam control or governor valves, partial arc admission or full arc admission. See Figure 11.7 for arrangement of steam valves for this discussion. During full arc admission, all the control valves open or close by equal amounts to admit the same flow of steam to all arc sections. This is done during startup to ensure that all sections of the first row of turbine blades see about the same steam flow and this ensures even heating of the first section of turbine blades during the startup process. Since, during the startup process the steam flow required is small, the main steam stop valve or throttling valve is closed and a main steam stop valve bypass line allows the flow of steam to the main steam control

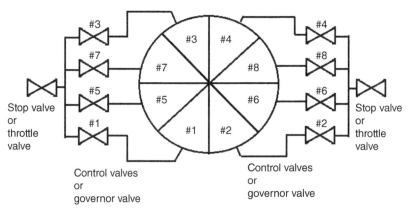

Figure 11.7   Location of stop (or Throttle) valves and control (or governor) valves and concept of partial arc admission.

valves. Alternately, if the steam flow requirement exceeds the ability of the stop valve bypass line, the main steam stop or throttle valve can be controlled to control the flow to the main steam control valves. The drawback of throttling the main steam control valve in this process is that the valve losses are higher. Energy losses occur anytime you have a differential pressure across a valve with some amount of flow through the valve. During full arc admission, when all main steam control or governor valves are partially open with flow through all the valves, this presents a pressure drop and associated loss at each valve. As we will see in the next paragraph, during partial arc admission, some of the control valves are closed and, with no flow through these valves, there are no throttling losses across these valves. Therefore, valve losses during full arc admission exceed the value of valve losses during partial arc admission.

During normal steady state operation of the turbine, to improve the turbine efficiency, the operation of the turbine valves is changed over to partial arc admission. During partial arc admission the control valves are controlled independently. Some valves are fully closed while others are opened from the 0% open state to 100% open state. Once one governor valve hits the fully open state, then the next governor valves begins to open. The advantage of this system is that it reduces the valve losses across the governor valves. A fully closed valve will have zero valve losses because, even though there is a pressure drop across the valve, the flow through the valve is zero. Since the throttling loss is proportional to the product of the differential pressure across the valve and the flow through the valve, the loss is zero since the flow is zero. A fully open valve will have near-zero valve losses because, even though there is flow through the valve, the pressure drop across the valve is almost zero. Since the pressure drop energy loss is proportional to the differential pressure across the valve and the flow through the valve, the loss is zero since the differential pressure across the valve is zero. The degree of arc admission is defined in equation (11.1).

$$\text{Degree of arc admission} = 360° / \# \text{ of valves} \tag{11.1}$$

Similarly in the inlet to the reheat or intermediate pressure steam turbine, there are two valves to provide isolation to the IP and LP turbines. The first valve is the

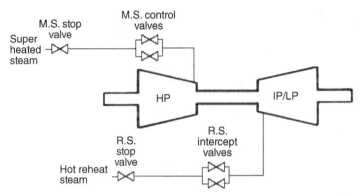

Figure 11.8   Simplified steam turbine valve arrangement.

reheat steam stop valve. This valve serves as a backup to the reheat steam intercept valve for trip to isolate the steam flow to the IP and LP turbines during a trip. The reheat steam stop valve feeds steam to the reheat steam intercept valve. The *reheat steam intercept valve* is the primary valve that shuts during a trip to remove steam flow from the IP and LP turbine sections. It also serves as steam control to these sections during load swings to maintain turbine mechanical power and torque within the needs of the generator that is driven by the turbine shaft. Figure 11.8 shows a simplified model of the arrangement of the steam control valves.

In addition to valves that isolate and control the steam admitted to the steam generator, there are valves designed to vent off steam from the turbine stages. These are known as ventilator valves. These valves will bleed trapped steam off from the turbine after a trip event. This is done to ensure that trapped steam in the turbine will not continue to drive the shaft after a trip thereby resulting in an overspeed condition.

There are basically four types of steam flow control methods for controlling the flow of steam through the steam turbine.

1. Throttling control

2. Governing control

3. Variable pressure control

4. Hybrid variable pressure and governing control

Throttling control is where the main steam control valves are controlled in the full arc admission mode. The main steam line pressure is held constant. In this mode, all the control valves are at the same position and the same amount of steam flows into each arc section in the nozzle box. This is the least efficient of the four control types listed, but does have the benefit of even heating of the first stage turbine blades and is used during startup to ensure even thermal expansion of the turbine blades and casing.

Governing control is where the main steam control valves are controlled in the partial arc admission mode. The main steam line pressure is held constant. In this mode, the control valves are individually controlled to minimize the pressure drop loss across the valve assembly. This mode of control is more efficient than the throt-tling control mode. This is the control mode most utilized under full-load conditions. These are the two most utilized modes of control. In addition to these methods, there are two more modes of operation.

The *variable pressure control* is where the boiler or steam generator main steam pressure is varied and the control valves are left in a fixed open position. Instead of maintaining the pressure constant and controlling the flow of steam through the turbine in the throttle control or governing control modes, the variable pressure control mode controls the inlet pressure to the first stage and does not control flow. Remember back to thermodynamics, the rate of energy transferred to the turbine is the difference in enthalpy between the steam inlet of the turbine and the steam outlet of the turbine times the mass flow rate. In the governing control and throttling control mode, we are maintaining the pressure difference and, therefore the enthalpy differ-ence across the turbine blades, but we control flow to control the energy transferred. In the variable pressure control mode, we maintain flow relatively constant, but by a reduction of the main steam side pressure, we reduce the enthalpy difference between the inlet and outlet of the steam turbine thereby controlling the transfer of energy from the steam system to the turbine system. The benefit of this system is that by leaving the main steam control valves in the open position, we avoid the losses asso-ciated with valve loss. The main drawback of this system is the limited load response. Using governing or throttling control modes, the valve position can quickly change to follow load changes. With the variable pressure control mode, to change the pressure in the main steam line, the firing rate of the boiler must first be changed to change the heat transfer to the superheat plenum section. This change of firing rate is much slower than simply changing valve position. Therefore, the variable pressure mode of control is inherently significantly slower than either governing or throttling control.

Since variable pressure control has the best efficiency, but governing control has the quicker response, a possible fourth mode of operation is the *hybrid variable pressure and governing control* system. In this mode, under steady or slow transition operation, the steam generator or boiler pressure is varied and the main steam control valves are fully open for full arc admission. During transients, the control valves are allowed to move to allow for quick response to system load changes.

Figure 11.9 shows the typical physical arrangement of the turbine section from a side view.

The shaft of the high pressure turbine is connected to a shaft with instrumenta-tion and equipment to support the operation of the turbine. This section of the turbine is called the *front standard* as is shown in Figure 11.10. The front standard contains the turbine lube oil pump, mechanical tachometer feedback for the overspeed protec-tion, the thrust bearing and the #1 radial bearing. The bearings are numbered from the front standard to the generator end with the front standard bearing being the #1 radial bearing to the other end of the shaft where the generator exciter exists that has the highest numbered bearing. Figure 11.11 shows the typical bearing numbering arrangement for a typical tandem compounded turbine generator arrangement.

To protect the low pressure turbine casing from an overpressure condition, there are *overpressure diaphragms* or *rupture discs* on top of the turbine shell. These are designed as weak points in the low pressure turbine casing and, in the event of an

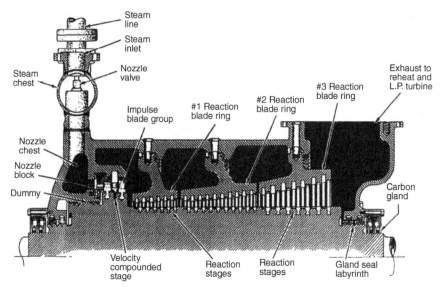

Figure 11.9   High pressure (HP) steam turbine side view. *Source*: Reproduced with permission of U.S. Navy Training Manual.

overpressurization of the low pressure steam line feeding the low pressure turbine, these are designed to blow out, thereby protecting the low pressure turbine casing from damage due to the overpressure condition.

The turbine is cooled by the flow of steam through the turbine. This is important to the last stage of the low pressure turbine blades. During low steam flow conditions, another system must be used to remove heat from the last stage of turbine blades. There is an exhaust hood system in the last stage used to provide a cooling spray for the last stage under these low flow conditions.

After shutdown of the turbine while the turbine metal is still hot, the top of the rotor is heated by the turbine metal, while the bottom of the rotor is cooled from

Figure 11.10   Turbine front standard.

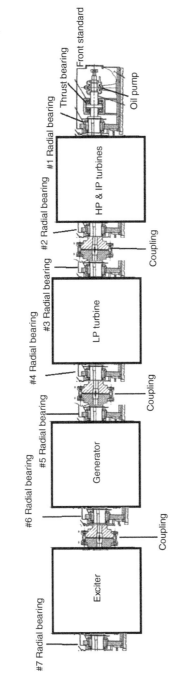

Figure 11.11   Radial bearing numbering arrangement for turbine generator.

the condenser. If the turbine shaft is stopped and not rotated, the top of the turbine rotor will tend to expand more than the bottom of the rotor due to this temperature difference. If left unaccounted for, this will cause the turbine rotor to bow upwards which can lead to rotor deformation. To prevent this, after the turbine is shut down, the turbine's rotor is slowly turned by a turning gear to ensure even rotor heating until the turbine temperatures are reduced to safe values where this phenomena does not occur. Once the turbine metal temperatures reduce to below 200–300°F (depending on turbine manufacturer recommendations), the turning gear can be turned off. Normally an electric motor drives the turning gear to rotate the shaft. However, in the event that the turbine trips and stops during an event where all electrical power is lost, there is either a DC motor that can drive the turning gear or there may be an air drive that can drive the turning gear in the event the AC electric motor is unavailable.

Various systems are needed to support the operation of the steam turbine. The space between the turbine casing and the rotor provides a possible location for steam to leak out of the turbine casing or air to leak into the turbine casing causing a loss of vacuum in the condenser. To prevent this, radial shaft seals are used to provide a barrier between the atmosphere in the turbine casing and the outside air. The stationary portion of these radial shaft seals are shown in Figure 11.12.

Induction of water into a hot spinning turbine is an event that can cause severe catastrophic damage to the turbine blades and casing. Water in the turbine can impinge on the spinning blades causing them to break off while spinning at or near full speed and damage both rotating and stationary members. To prevent water induction, the main steam lines and reheat lines have low point drains or traps. These drains function to collect water and moisture from the lines. When the level in the drain trap reaches the high point, the drain valve opens automatically until the water in the trap reaches the low point where the drain valve closes. This prevents water from being

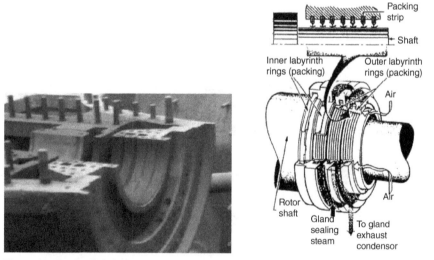

Figure 11.12   Radial shaft seals. *Source*: Reproduced with permission of U.S. Navy Training Manual.

entrained in the steam flow, while at the same time providing a seal to prevent steam from leaking to the atmosphere through the low point drain trap.

Maintaining metal temperatures of the steam plant equipment within design limitations is critical to reliable turbine operations. The temperature of the main steam entering the high pressure turbine must be closely controlled. De-superheat spray or attemperation lines are used to spray condensate into the main steam lines when main steam temperatures exceed rated values to reduce those temperatures back under operational limits.

TURBINE SAFETY

In electrical systems, a protective relay does not provide any control functionality, but continuously monitors the operation of the electrical system for any abnormal or unsafe condition. In the event the relay detects an abnormal condition, the protective relay provides some mechanism to remove that unsafe condition. Similarly, turbines have safety devices. These safety devices do not provide any normal control functions but continuously monitor the operation of the turbine for any unsafe condition. When the safety device detects an unsafe condition, the safety device provides some mechanism to remove that unsafe condition. All turbines have at least four safety devices:

- An overspeed trip
- A back-pressure trip
- A low-oil pressure trip or alarm
- An emergency hand trip

The *overspeed trip* shuts off the steam supply to the turbine after a predetermined speed has been reached, thus stopping the unit. Overspeed trips are set to trip out at approximately 110% of normal operating speed. This is to prevent catastrophic failure of the turbine due to excessive vibration and centrifugal forces that may occur in an overspeed condition. The most likely cause for an overspeed condition is a full-load rejection that would occur if the main generator breaker was to open while the unit was operating at full-load condition.

The *back-pressure trip* closes the throttle automatically whenever the pressure at the exhaust of the low pressure turbine reaches a set pressure. The most likely cause for a back-pressure trip would be loss of condenser vacuum due to either loss of vacuum pumps or air ejectors or possibly loss of the circulating water system.

The *low-oil pressure trip* closes the throttle if the lubricating oil pressure drops below a specified pressure. Some turbines only have low-oil pressure alarms. The typical lubricating oil system of a steam turbine generator has three sources for required pump head to provide lubricating oil the system. The primary pump is directly connected to the turbine shaft. As shown in Figure 11.10, the shaft driven oil pump for bearing lubrication is located in the front standard of the turbine. This is the normal source of lubricating oil pressure when the turbine is operating at rated speed. A secondary pump may be provided that is driven by an electric motor fed from the station AC electric system. This pump provides lubricating oil during periods of startup and shutdown, when the shaft-driven lubricating oil pump is not able to provide this pressure. A third pump may be provided that is driven by an electric motor from the station DC system and it is intended to be the emergency backup source of lubricating oil pressure should the AC system fail.

The *emergency hand trip* provides for a method to manually close the throttle quickly in case of damage to either the turbines or generators.

The purpose of the *exhaust hood spray* as mentioned before is to maintain the last stage turbine blade temperature during periods of low flow. If this stage exceeds rated temperatures it can thermally expand and make contact with the seal in the casing surface causing damage to the rotating and stationary sections. If the exhaust temperature out of the turbine section exceeds manufacturer recommended values, the turbine should be tripped. Additionally, the low pressure turbine exhausts into the condenser where a low vacuum is maintained to maintain the heat rate of the steam cycle. Should vacuum be lost in the condenser, the heat rate will suffer greatly and, in addition, damage can occur to the last stage turbine blade section. To prevent this, on a loss of condenser vacuum, the turbine should be tripped.

For proper operation, various system parameters must be monitored to ensure reliable turbine operation. The shaft speed must be monitored to ensure speed is maintained within safe operational limits. The position of the control valves must be monitored to ensure adequate steam flow. Additionally on a normal unit shutdown the power of the generator is monitored along with the position of the control and stop valves. Once the stop and control valves are completely shut and once we sense a power reversal at the generator, the generator breaker is tripped. This ensures that, when the generator breaker opens and the power demand by the generator from the turbine is removed, that the steam turbine will not overspeed. As mentioned previously, we must monitor the vibration at all the bearings of the turbine generator. We also monitor the expansion of the turbine shell and the bearing metal temperatures. We monitor and trend at the steam pressure and temperature both at the input to the turbine as well as at the exhaust to the condenser as this provides information on the enthalpy of the steam entering and leaving the turbine.

Bearings are the devices that support the weight and restrain the operating forces of the turbine. They are very basic in their design but very critical to safe operation of the turbine. The following safety precautions will help prevent most casualties.

- Never use a piece of machinery if the bearings are known to be in poor condition.
- Where possible, be sure that bearings have the proper quality and quantity of lube oil before starting the machinery.
- Rapid heating of the bearing is a danger sign. A bearing that feels hot to the hand after an hour's operation may be all right, but the same temperature reached in 10 or 15 min indicates trouble. Monitor bearing temperatures and vibration for early indication of impending bearing failure.
- When disassembling a bearing, take care to document the order of disassembly and document the location and orientation as each part is removed to ensure it is installed correctly. Also insure all parts are reassembled especially any parts that may be providing electrical insulation between the bearing shoes and ground potential.

The turbine has very close clearances internally in the machine between the stationary and rotating sections. The turbine grows due to thermal expansion from a cold state to an in service state. Sometimes this expansion can be up to 1 in. or more. If the expansion of the turbine casing is not even, then the clearances between stationary and rotating members of the machine can be reduced to the point where contact is made between the stationary and rotating sections. In the event the turbine expansion is not within manufacturer tolerances, the turbine should be tripped to protect the machine.

TURBINE VIBRATION

Steam turbines have a very large amount of inertia in their rotating mass and vibration should be closely monitored. Issues with either the balance of the rotating section, the generator field, the oil lubricating system, or the bearing surfaces can cause excessive vibration which can damage the machine quickly. Depending on the technology utilized, vibration can be monitored utilizing proximity probes (providing vibration in units of displacement peak to peak), velocity sensors (providing vibration in units of in./sec) or accelerometers (providing vibration in units of in./sec^2). All three types of vibration monitoring sensors provide valuable data regarding machinery vibration and, given that the movement of vibration is sinusoidal, the three units of measurement are related to each other. The relationship between displacement and velocity is similar to the relationship between the diameter of a circle (displacement peak to peak) and the circumference of the circle. Knowing that the relationship between the circumference of a circle to the diameter of the circle is the value π, we can convert from units of displacement to units of velocity utilizing equation (11.2):

$$V = \pi \times F \times D \tag{11.2}$$

where

$V =$ velocity (in./sec)

$F =$ frequency (Hz) or (cycles/sec)

$D =$ displacement (in. peak to peak)

Example 11.1   What is the vibration in units of in./sec of a piece of equipment rotating at 1200 rpm given that the displacement is measured as total displacement of 10 mils peak to peak?

Solution: First, we must convert our units of measurement of displacement from mils peak to peak to inches peak to peak. Knowing that 1000 mils = 1 in., we find the displacement to be

$$D = 10 \text{ mils peak to peak} \times (1 \text{ in.}/1000 \text{ mil})$$
$$D = 0.01 \text{ in. peak to peak}$$

Now, we utilize equation (11.2) to find the velocity of vibration to be

$$V = \pi \times F \times D$$
$$= \pi \times 1200 \times 0.01$$
$$= 37.68 \text{ in.}/\text{min}$$

However, the standard units for velocity are in inches per second, so converting to standard units, we find

$$V = (37.68 \text{ in.}/\text{min}) \times (1 \text{ min}/60 \text{ sec})$$
$$V = 0.628 \text{ in.}/\text{sec}$$

Example 11.2   What is the vibration in units of inches per second of a piece of equipment rotating at 3600 rpm given that the displacement is measured as total displacement of 10 mils peak to peak?

Solution: Just as before, we must convert or units of measurement of displacement from mils peak to peak to inches peak to peak. Knowing that 1000 mils = 1 in., we find the displacement to be

$$D = 10 \text{ mils peak to peak} \times (1 \text{ in.}/1000 \text{ mil})$$
$$D = 0.01 \text{ in. peak to peak}$$

Now we utilize equation (11.2) to find the velocity of vibration to be

$$V = \pi \times F \times D$$
$$V = \pi \times 3600 \text{ cycles}/ \text{min} \times 0.01 \text{ in. peak to peak}$$
$$V = 113 \text{ in.}/\text{min}$$

However, the standard units for velocity are in inches per second, so converting to standard units we find

$$V = (113 \text{ in.}/\text{min}) \times (1 \text{ min}/60 \text{ sec})$$
$$V = 1.884 \text{ in.}/\text{sec}$$

Notice from Examples 11.1 and 11.2 that, with the same value of vibration when given in terms of displacement, these two examples result in different units of vibration when given in terms of velocity. Typically, mechanical vibration limits for machinery are based on units of velocity. For the same vibration limit in units of velocity, a slower machine will be able to tolerate a larger value of vibration when measured utilizing a displacement probe.

If vibration levels exceed those provided by the turbine manufacturer, then the machine should be tripped off-line quickly. Typical values of vibration for turbine trip are around 1–2 in./sec. For a 3600 rpm turbine, using equation (11.2), we find this to be equivalent to displacement values of around 5–10 mils of displacement (1 mil is 1/1000th of an inch). Note, just as we discussed above, the value of limits of vibration based on units of displacement may be larger for turbines that rotate at slower speeds than higher speed machines even if the limits of vibration in units of velocity are similar as Example 11.3 below will show.

Example 11.3   What is the vibration limits in units of displacement (mils peak to peak) of a hydro turbine that rotates at a speed of 360 rpm if the manufacturer has specified that the hydro turbine be tripped if vibration exceeds 0.5 in./sec?

Solution: In order to solve this, we must rearrange equation (11.2) in terms of displacement. Performing this algebraic exercise, we find

$$V = \pi \times F \times D$$
$$D = V/(\pi \times F)$$
$$D = (0.5 \text{ in./sec})/\{\pi \times (360 \text{ cycles/min}) \times (1 \text{ min}/60 \text{ sec})\}$$
$$D = 0.027 \text{ in. peak to peak}$$

Normally, displacement is given in units of mils displacement peak. Utilizing the conversion that 1 mil is 1/1000th of an inch, we find the trip point in units of displacement to be

$$D = 0.027 \text{ in. peak to peak}/(1 \text{ in.}/1000 \text{ mil})$$
$$D = 27 \text{ mils peak to peak}$$

For general rotating mechanical machinery, Table 11.1 displays guidelines for interpreting the condition of the equipment based on vibration in units of velocity. Note that Table 11-1 and Table 11.2 are general industry consensus values. Always consult the original equipment manufacturer operation and maintenance manual and drawings for specific limitations to vibration.

For a machine in the "smooth" to "fair" range, vibration is low and system is working well. For a machine in the "rough" range, an evaluation should be performed as to the cause of vibration and units that are "very rough" should be considered in need of immediate shutdown to evaluate and repair the cause of vibration. However, in evaluating machine condition utilizing vibration, it is not just the magnitude, frequency, or phase relationship of vibration, but just as importantly, the change in vibration over time. A large turbine generator with a large number of couplings and bearings may normally operate in the "fair" to "slightly rough" range, while a smaller turbine such as a steam turbine driven boiler feed pump with only two support bearings may normally run in the "smooth" to "good" range. It is important to continuously monitor vibration and not only alarm on magnitude but also on a sudden change of vibration.

TABLE 11.1   General Machinery Vibration (velocity) Severity Table

Condition	Vibration (units of velocity)(in./sec)
Extremely smooth	< 0.0049
Very smooth	0.0049–0.0098
Smooth	0.0098–0.0196
Very good	0.0196–0.0392
Good	0.0392–0.0784
Fair	0.0784–0.1568
Slightly rough	0.1568–0.3136
Rough	0.3136–0.6272
Very rough	> 0.6272

If we are using proximity probes to monitor and determine the condition of the machine, the categories of machine condition are now based on displacement values which are in units of mils peak to peak. As we have learned, for a certain limit of vibration given in units of velocity, this will lead to a limit of displacement that is dependent on machine speed and, therefore, frequency of vibration. Table 11.2 converts values of vibration in units of velocity to values of vibration in units of displacement.

Notice that a turbine operating at 360 rpm with an indicated vibration displacement of 4 mils peak to peak is considered as being in a "good" condition. However

TABLE 11.2 General Machinery Vibration (Displacement) Severity Table

Velocity (in./sec)	0.0049	0.0098	0.0196	0.0392	0.0784	0.1568	0.3136	0.6272
Speed (rpm)	Displacement (mils peak to peak)							
100	0.9358	1.8717	3.7433	7.4866	14.9733	29.9466	59.8932	119.7864
200	0.4679	0.9358	1.8717	3.7433	7.4866	14.9733	29.9466	59.8932
300	0.3119	0.6239	1.2478	2.4955	4.9911	9.9822	19.9644	39.9288
360	0.2600	0.5199	1.0398	2.0796	4.1592	8.3185	16.6370	33.2740
400	0.2340	0.4679	0.9358	1.8717	3.7433	7.4866	14.9733	29.9466
450	0.2080	0.4159	0.8318	1.6637	3.3274	6.6548	13.3096	26.6192
500	0.1872	0.3743	0.7487	1.4973	2.9947	5.9893	11.9786	23.9573
514	0.1821	0.3641	0.7283	1.4565	2.9131	5.8262	11.6524	23.3047
600	0.1560	0.3119	0.6239	1.2478	2.4955	4.9911	9.9822	19.9644
700	0.1337	0.2674	0.5348	1.0695	2.1390	4.2781	8.5562	17.1123
720	0.1300	0.2600	0.5199	1.0398	2.0796	4.1592	8.3185	16.6370
800	0.1170	0.2340	0.4679	0.9358	1.8717	3.7433	7.4866	14.9733
900	0.1040	0.2080	0.4159	0.8318	1.6637	3.3274	6.6548	13.3096
1000	0.0936	0.1872	0.3743	0.7487	1.4973	2.9947	5.9893	11.9786
1200	0.0780	0.1560	0.3119	0.6239	1.2478	2.4955	4.9911	9.9822
1800	0.0520	0.1040	0.2080	0.4159	0.8318	1.6637	3.3274	6.6548
2000	0.0468	0.0936	0.1872	0.3743	0.7487	1.4973	2.9947	5.9893
3000	0.0312	0.0624	0.1248	0.2496	0.4991	0.9982	1.9964	3.9929
3600	0.0260	0.0520	0.1040	0.2080	0.4159	0.8318	1.6637	3.3274
4000	0.0234	0.0468	0.0936	0.1872	0.3743	0.7487	1.4973	2.9947
5000	0.0187	0.0374	0.0749	0.1497	0.2995	0.5989	1.1979	2.3957
6000	0.0156	0.0312	0.0624	0.1248	0.2496	0.4991	0.9982	1.9964
7000	0.0134	0.0267	0.0535	0.1070	0.2139	0.4278	0.8556	1.7112
8000	0.0117	0.0234	0.0468	0.0936	0.1872	0.3743	0.7487	1.4973
9000	0.0104	0.0208	0.0416	0.0832	0.1664	0.3327	0.6655	1.3310
10000	0.0094	0.0187	0.0374	0.0749	0.1497	0.2995	0.5989	1.1979
20000	0.0047	0.0094	0.0187	0.0374	0.0749	0.1497	0.2995	0.5989
30000	0.0031	0.0062	0.0125	0.0250	0.0499	0.0998	0.1996	0.3993
40000	0.0023	0.0047	0.0094	0.0187	0.0374	0.0749	0.1497	0.2995
50000	0.0019	0.0037	0.0075	0.0150	0.0299	0.0599	0.1198	0.2396
60000	0.0016	0.0031	0.0062	0.0125	0.0250	0.0499	0.0998	0.1996
70000	0.0013	0.0027	0.0053	0.0107	0.0214	0.0428	0.0856	0.1711
80000	0.0012	0.0023	0.0047	0.0094	0.0187	0.0374	0.0749	0.1497
90000	0.0010	0.0021	0.0042	0.0083	0.0166	0.0333	0.0665	0.1331
100000	0.0009	0.0019	0.0037	0.0075	0.0150	0.0299	0.0599	0.1198

Diagonal category labels (by column): Extremely smooth, Very smooth, Smooth, Very good, Good, Fair, Slightly rough, Rough, Very rough.

a turbine operating at 3600 rpm with the same indicated vibration displacement of 4 mils peak to peak is considered as being in a "very rough" condition.

In addition to measuring vibration utilizing displacement and velocity, we can also measure vibration utilizing acceleration. The relationship between vibration in units of acceleration (g) and vibration in units of velocity (in./sec) is described in equation (11.3).

$$V = g \times A/(2 \times \pi \times F) \tag{11.3}$$

where

V = velocity (in./sec)

A = acceleration (multiples of g)

F = frequency (Hz) or (cycles/sec)

g = gravitational acceleration (386.0886 in./sec^2)

Now that we have a relationship between displacement and velocity and also between acceleration and velocity, we can derive the relationship between displacement and acceleration. Given that $V = \pi \times F \times D$ and that $V = g \times A/(2 \times \pi \times F)$, we can state the following:

$$V = \pi \times F \times D = g \times A/(2 \times \pi \times F)$$

or

$$\pi \times F \times D = g \times A/(2\pi \times F)$$

or

$$2 \times \pi^2 \times F^2 \times D = g \times A$$

Lastly, rearranging for displacement we find the following equation.

$$D = g \times A/(2 \times \pi^2 \times F^2) \tag{11.4a}$$

Alternately we can rearrange for acceleration and find the following equation.

$$A = (2 \times \pi^2 \times F^2 \times D)/g \tag{11.4b}$$

where

A = acceleration (multiples of g)

F = frequency (Hz) or (cycles/sec)

D = displacement (inches peak to peak)

g = gravitational acceleration (386.0886 in./sec^2)

Example 11.4   What is the vibration in units of velocity (in./sec) of a hydro-turbine that operates at 360 rpm and that has a measured value of vibration of 0.2 g utilizing an accelerometer?

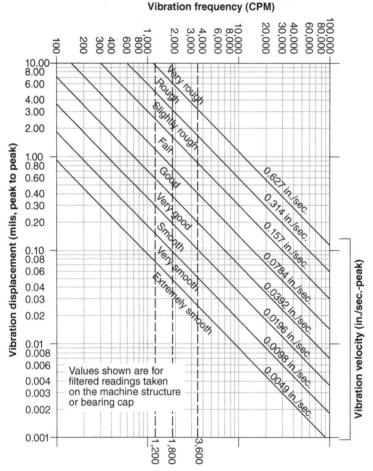

For use as guide in judging vibration as a warning of impending trouble
IRD mechanalysis, Inc. 6150 Huntley road, Columbus, Ohio 43229
Copyright 1984, international research and development corp.

Figure 11.13   General machinery vibration severity chart. *Source*: Reproduced with permission of IRD Mechanalysis Ltd. 1/5 Marol Co-op Industrial Estate Ltd.

Solution: We can utilize equation (11.3) directly to calculate velocity from acceleration as shown below.

$$V = g \times A/(2\pi \times F)$$
$$V = (386.0886 \text{ in./sec}^2) \times (0.2 \text{ g})/\{2\pi \times 360 \text{ rpm} \times 1 \text{ min}/60 \text{ sec})$$
$$V = 2.05 \text{ in./sec}$$

Figure 11.13 shows graphically the same information displayed in Table 11.2 in tabular format. It depicts acceptable limits of vibration for generator mechanical equipment as a function of both units of displacement and units of velocity.

So how do we select whether to utilize proximity sensors, velocity sensor, or accelerometers for our vibration sensing probes? One rule of thumb is based on the anticipated frequency of vibration that we are interested in monitoring. For frequencies between 0 and 10 Hz, typically, the displacement probe is utilized as it measures the actual distance from the probe to the rotating element being monitored. For frequencies between 10 Hz and 1 kHz, the velocity sensor is normally utilized as the speed of movement limits the effectiveness of the proximity sensor. For frequencies above 1 kHz, the accelerometer is normally utilized as the force associated with the vibration limits the effectiveness of the velocity sensor in high frequency applications (force = mass × acceleration).

GLOSSARY OF TERMS

- Back-Pressure Trip – Closes the throttle automatically whenever the exhaust pressure at the exhaust of the low pressure turbine reaches a set pressure.
- Boiler – A fuel-burning apparatus or container for heating water to generate steam.
- Boiler Feed Pump (BFP) – A boiler feedwater pump is a specific type of pump used to pump feedwater into a steam boiler. The water may be freshly supplied or returning condensate produced as a result of condensation of the steam produced by the boiler.
- Condensate – A liquid obtained by condensation of a gas or vapor.
- Condensate Pump – A condensate pump is a specific type of pump used to pump the condensate (water) produced in a condensing boiler furnace or steam system.
- Cross-Compounded Turbine Generator Unit – Utilizes multiple generators and turbines where, one set of turbines run at a high speed and drive one generator and another set of turbines run at a lower speed and drive another generator with more poles (for the same output frequency). In this system, the low pressure turbine drives one generator and the high and intermediate turbines drive a different generator.
- Deaeration Tank – A tank that provides the suction for the boiler feed pump and also provides systems to remove dissolved gasses from the condensate. There are two basic types of deaerators, the tray-type and the spray-type.
- Emergency Hand Trip – Provides for a method to manually close the throttle quickly in case of damage to either the turbines or generators.
- Exhaust Hood Spray – Condensate spray between last stage of low pressure turbine blade section and condenser section. Exhaust hood spray is used to maintain the last stage turbine blade temperature during periods of low flow.
- Feedwater Heater – A feedwater heater is a power plant component used to pre-heat water delivered to a steam generating boiler. Preheating the feedwater reduces the irreversibilities involved in steam generation and therefore improves the thermodynamic efficiency of the system.

- Governing Control – Governing control is where the main steam control valves are controlled in the partial arc admission mode.
- Hotwell – A reservoir in a turbine installation for receiving the warm condensed steam drawn from the condenser.
- Impulse Turbine – Impulse turbines have their rotating turbine blades shaped in the form of "buckets". Impulse turbines change the direction of flow of a high velocity fluid or gas jet. The resulting impulse converts some kinetic energy of the steam to mechanical energy in the turbine rotor. This spins the turbine and leaves the steam flow with diminished kinetic energy.
- Journal Bearing – Journal bearings that consist of a shaft or journal which rotates freely in a supporting metal sleeve or shell. There are no rolling elements in these bearings.
- Low-Pressure Trip – Closes the throttle if the lubricating oil pressure drops below a specified pressure.
- Main Steam Control Valve – A valve or group of valves designed to regulate the supply of steam to a high pressure turbine during normal running conditions using partial arc admission.
- Main Steam Stop Valve – A valve or group of valves designed to isolate supply of steam to a high pressure turbine and also to provide some control during transients or startup.
- Nozzle – A cylindrical or round spout at the end of a pipe, hose, or tube, used to control a jet of gas or liquid.
- Overspeed Trip – Shuts off the steam supply to the turbine after a predetermined speed has been reached, thus stopping the unit.
- Pressure Compounded Stage – A turbine that has a stationary blade section followed by one rotating section, followed by another set of stationary and rotating sections is known as a pressure compounded stage. This is due to the fact that in each stationary section, the pressure drops and the velocity increases, while in each rotating section, the pressure remains constant while the velocity drops.
- Reaction Turbine – Reaction turbines have their rotating turbine blades shaped in the form of "nozzles." Reaction turbine nozzles decrease the cross-sectional area. This transfers some of the steam potential energy (pressure) into kinetic energy (velocity). The turbine blades are reacting with the gas or fluid's pressure or mass. The pressure of the gas or fluid changes as it passes through the turbine rotor blades.
- Reheat Steam Control Valve – A valve or group of valves designed to regulate the supply of steam to the intermediate and low pressure turbine during normal running conditions using partial arc admission.
- Reheat Steam Stop Valve – A valve or group of valves designed to isolate supply of steam to the intermediate and low pressure turbine and also to provide some control during transients or startup.
- Steam Generator – A large apparatus for converting hot water into steam at high pressure and often with supplementary sections to superheat the steam.

- Steam Turbine – A device that extracts thermal energy from pressurized steam and uses it to do mechanical work on a rotating output shaft.
- Tandem Compounded Turbine Generator Unit – Turbine generator assembly that has only one generator and it is coupled to all the turbines. All the turbines and the generator run at the same shaft speed since they are coupled together to form one common shaft.
- Throttling Control – Throttling control is where the main steam control valves are controlled in the full arc admission mode.
- Turbine Front Standard – The front standard contains the turbine lube oil pump, mechanical tachometer feedback for the overspeed protection, the thrust bearing, and the #1 radial bearing.
- Velocity Compounded Stage – Turbine blade section where the stationary blade section is followed by multiple impulse stages is known as a velocity compounded stage. This is due to the fact that while pressure and volume remain constant across the multiple rotating sections of turbine blades, the velocity drops across each section of rotating turbine blade.
- Variable Pressure Control – Variable pressure control is where the boiler or steam generator main steam pressure is varied and the control valves are left in a fixed open position.

PROBLEMS

11.1 A steam turbine stage, the design where pressure drops only across the fixed nozzle and not across the moving buckets is

    A. reaction turbine

    B. impulse turbine

11.2 A turbine installation where there are two LP turbines connected on a common shaft driving one generator is known as a

    A. tandem compound turbine

    B. cross-compound turbine

11.3 The valve that is used to control steam flow to the high pressure (HP) section turbine for control during full-load operation is known as the

    A. governor valve

    B. intercept valve

    C. stop valve

    D. throttle valve

11.4 The purpose of the turbine turning gear is to rotate the turbine during shutdown and startup to

    A. prevent bowing of turbine rotor due to weight of rotor

    B. build up oil film in journal bearings

    C. prevent bowing of turbine rotor due to uneven heating

    D. check for rotor rotation direction

11.5 The type of turbine control that refers to the sequential opening of the steam control (governor) valve (partial arc admission) is

    A. throttle control

    B. governing control

    C. variable pressure control

11.6 The pump that takes suction from the deaeration tank and provides water to the boiler is the

    A. condensate pump

    B. boiler recirculation pump

    C. boiler feed pump

    D. cooling water pump

11.7 The device that converts the thermal energy of steam into mechanical energy on a shaft is

    A. boiler

    B. steam generator

    C. steam turbine

    D. electric generator

11.8 The type of control method where the boiler or steam generator main steam pressure is varied and the control valves are left in a fixed open position is known as

    A. throttling control

    B. governing control

    C. variable pressure control

11.9 Using Figure 11.13 for typical vibration analysis, on a turbine running at 3600 rpm, we indicate a displacement of 0.10 mils. What is the most likely condition of the bearings?

    A. Extremely smooth

    B. Very smooth

    C. Smooth

    D. Very good

    E. Good

    F. Fair

    G. Slightly rough

    H. Rough

    I. Very rough

11.10 Using Figure 11.13 for typical vibration analysis, on a turbine running at 3600 rpm, we indicate a displacement of 3.0 mils. What is the most likely condition of the bearings?

    A. Extremely smooth

    B. Very smooth

C. Smooth

D. Very good

E. Good

F. Fair

G. Slightly rough

H. Rough

I. Very rough

11.11 Using Figure 11.13 or Table 11.2 for typical vibration analysis, on a turbine running at 1800 rpm, we indicate a displacement of 0.4 mils. What is the most likely condition of the bearings?

A. Extremely smooth

B. Very smooth

C. Smooth

D. Very good

E. Good

F. Fair

G. Slightly rough

H. Rough

I. Very rough

11.12 What is the vibration in units of inches per second of a piece of equipment rotating at 720 rpm given that the displacement is measured as total displacement of 2 mils peak to peak?

A. 0.7536 in./sec

B. 0.04578 in./sec

C. 0.07536 in./sec

D. 0.4578 in./sec

RECOMMENDED READING

Electric Power Plant Design, Technical Manual TM 5-811-6, Department of the Army, USA, 1984.

Electrical Machines, Drives and Power Systems, 6th edition, Theodore Wildi, Prentice-Hall, 2006.

The Engineering Handbook, 3rd edition, Richard C. Dorf (editor in chief), CRC Press, 2006.

Power Plant Engineering, Black & Veatch, edited by Larry Drbal, Kayla Westra, and Pat Boston, Chapman & Hall / Springer, 1996.

Standard Handbook of Powerplant Engineering, 2nd edition, Thomas C. Elliott, McGraw-Hill, 1998.

US Navy Training Manual, Machinist's Mate 2nd class, Naval Education and Training Professional Development and Technology Center, 2003.

CHAPTER 12

GAS TURBINES

GOALS

- To understand the basic design and components of a combustion turbine
- To identify the six major sections of a combustion turbine
- To describe the typical flow diagram for a simple cycle and combined cycle combustion turbine
- To calculate the Brayton cycle heat rate of a combustion turbine
- To describe the two methods of NO_x control for a combustion turbine

---

GAS TURBINES operate in a fashion similar to steam turbines. Both the steam turbine and gas turbine convert the thermal energy contained in the media that is supplied to the turbine into mechanical energy. Instead of using steam as the media for the transfer of energy, gas turbines use air or gas. The term "gas turbine" does not refer to the fuel used for the *combustion* process but to the media used to transport the energy to the *turbine section*. In most instances, this "gas" is air. Gas turbines can operate as peaking or base load units. A peaking unit is defined as one that operates less than 2000 hr/year whereas a base loaded unit will normally be online for more than 5000 hr/year. The conversion between years and hours is

$$1 \text{ year} = 8760 \text{ hr}$$

Please reference Chapter 2 on basic thermodynamic cycles for an explanation of the Brayton cycle. Recalling from thermodynamics, there are four stages to the Brayton thermodynamic cycle: compression, combustion, expansion, and exhaust. Below, we will describe the equipment utilized and a typical combustion turbine to achieve this process.

In the typical combustion turbine, there are at least six fundamental components to the combustion turbine required to achieve the conversion of energy to mechanical energy. These are

1. Air inlet
2. Compressor

Energy Production Systems Engineering: An Introduction for Electrical Engineers to Electrical Power Generation Facilities, Systems, and Equipment, First Edition. Thomas H. Blair.

3. Combustion system

4. Turbine

5. Exhaust

6. Support systems

The function of the *air inlet assembly* is to condition the air being supplied to the compressor stage of the system to ensure that the quality and cleanliness of the inlet air is adequate for the compression stage. The typical inlet assembly contains a filter media such as a screen to mechanically filter out debris from the incoming air. This assembly will also include a means to clean these screens of built-up debris. Additionally, the inlet assembly may contain an inlet duct heating system. Heating of the incoming air adds energy to the air mass which improves the compressor and overall turbine efficiency. In some instances, air coolers may be installed in the air inlet system. The purpose of air coolers is to increase the maximum available power from the turbine, since with elevated ambient air temperatures, the incoming air to the compressor is less dense. If we cool the air stream entering the compressor stage, this will increase the density of the incoming air. As the density of the incoming air stream increases to the compressor stage, more mechanical power can be extracted by the combustion turbine.

The function of the *compressor section* is to raise the pressure of the incoming air stream to achieve the compression process of the Brayton cycle. Looking at the Brayton cycle as shown in Figure 2.14, the compression section increases the enthalpy of the air stream by increasing the pressure of the gas stream as it flows through the compressor section. Most compressors on utility-sized combustion turbines are of the axial flow design. However, unlike axial flow fans that utilize variable pitch blades for the control of airflow, the typical compressor section of the combustion turbine utilizes fixed pitch air foil blades in the compressor section. During periods of startup when the shaft begins to roll, but we do not yet want to compress the air stream, an inlet guide vane is used to isolate the suction side of the compressor stage. Due to the large differential pressure that the compressor section must achieve, there are many stages of rotor blades in the compressor section. This reduces the increase in pressure to significantly smaller and achievable values for each stage of the compressor. Figure 12.1 shows the multiple compressor sections on the shaft of the compressor.

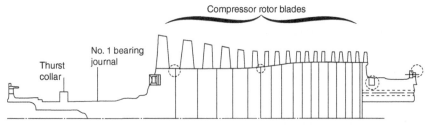

Figure 12.1   Combustion turbine: compressor rotor assembly. *Source*: Reproduced with permission of General Electric Company.

In the figure, we can see a *thrust collar*. Both the compression section and turbine section are mounted on a common shaft assembly. The design of the compressor and turbine is such that the axial force on the shaft from the compressor section opposes the axial force on the shaft from the turbine section. However, these two forces are not completely equal in magnitude and do not completely cancel out. There will be some small amount of next axial force on the shaft of the rotor. To maintain the combustion turbine *rotor* in the correct axial position, the rotor has a thrust collar machined into it and the casing will have a *thrust bearing* installed to allow for control of the axial position of the machine.

Between each row of rotating air foil blades is a row of stationary nozzles in a section connected to the turbine casing. These stationary nozzles are mounted on the casing of the compressor section. The function of the rotating blades is to increase the pressure across the rotating blade section and the purpose of the stationary nozzle section is to redirect the airflow so that it enters the next rotating section at the correct angle. Extraction air may be utilized from the compressor section for various functions such as cooling of combustion section components.

The function of the *combustion* section is to raise the temperature of the incoming air stream supplied from the compressor section to achieve the combustion process of the Brayton cycle. Looking at the Brayton cycle as shown in Figure 2.14, the *combustion section* increases the enthalpy of the air stream by increasing the temperature of the gas stream as it flows through the combustion section.

To achieve combustion, we need the three elements of the combustion triangle; an oxygen source, a fuel source, and a heat source (reference Chapter 4 on combustion for more detail on this process). The *combustor* is a complex piece of equipment where these three systems meet to perform multiple functions. The combustion section is where the three elements of the combustion triangle are combined in the correct combination to achieve proper combustion. Each combustion chamber receives the high pressure air from the compression section and disperses the correct amount of fuel in the combustion zone. For certain fuels such as oil, in addition to receiving the oil, the combustor also atomizes the oil with compressed air (called *atomizing air*) to maximize the exposed surface area of the oil fuel. This is done to ensure complete combustion of the fuel in the combustion section. Additionally, the combustor may have some type of dilatant such as steam or nitrogen to reduce the temperature of the mixture where complete combustion takes place to minimize the formation of NO_x. Figure 12.2 shows a typical arrangement for a combustor.

In addition to the necessary amount of air and fuel for combustion, the combustor contains the ignition source in the form of a spark plug. These plugs receive their spark from a high energy power supply. Several of these combustors are arranged around the periphery of the turbine shaft. Some combustors may be interconnected with a crossfire tube that connects several combustor sections together. The function of the crossfire tube is to transmit the ignited gas to other chambers to support combustion in those chambers. Lastly, the combustor contains a flame detector. If during the combustion process the flame is lost, this could lead to a buildup of combustible materials in the turbine section and exhaust section. NFPA 85 requires that the ignition source and fuel source be removed should a loss of flame be detected and the turbine and exhaust sections purged with air prior to the reapplication of ignition

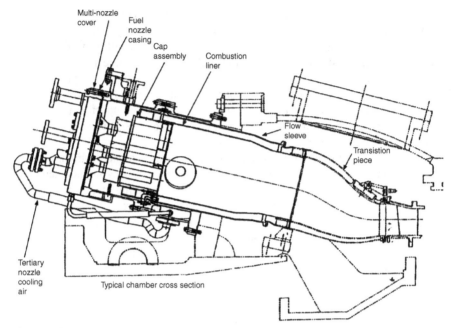

Figure 12.2   Combustion turbine: typical combustor assembly. *Source*: Reproduced with permission of General Electric Company.

energy to prevent the existence of an explosive atmosphere in the turbine and exhaust section. Figure 12.3 shows a typical arrangement the combustion chambers around the periphery of the turbine shaft.

The function of the turbine section is to expand the gas stream and convert the energy contained in the gas stream to mechanical energy delivered to the shaft of the turbine. During this process, the turbine removes energy from the gas stream. This is described in the expansion process of the Brayton cycle. Looking at the Brayton cycle as shown in Figure 2.14, the turbine section decreases the enthalpy of the gas stream by decreasing the pressure of the gas as it flows through the turbine sections. Just an in the compressor section, there are multiple sections to the turbine section. This allows for a smaller pressure drop across each turbine section. Figure 12.4 shows the typical turbine blade arrangement on the rotor.

Between each rotating turbine blade section is a *stationary nozzle* section. The stationary turbine nozzles are mounted on the casing of the turbine section. The function of the *rotating blades* is to convert the energy of the air stream to mechanical energy on the rotating shaft and the function of the stationary nozzle section is to redirect the airflow so it enters the next rotating section at the correct angle. Cooling air from the compressor section is utilized to maintain the temperatures of the turbine casing, nozzles, and turbine blades to ensure material thermal limits are not exceeded. Figure 12.5 shows the typical cooling air path through the turbine section.

The *exhaust system* is the section of the turbine just after the turbine blade section and before the exhaust stack. This section may or may not contain a *heat recovery*

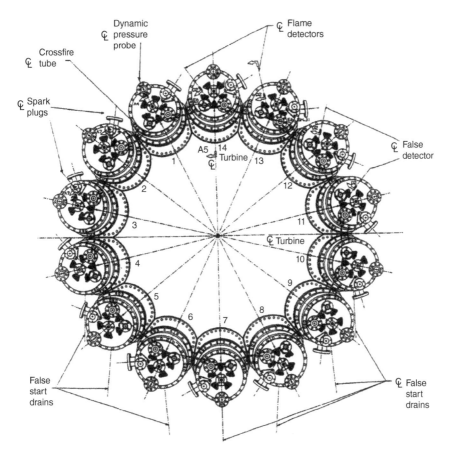

Figure 12.3 Combustion turbine: typical combustor arrangement. *Source*: Reproduced with permission of General Electric Company.

steam generator (HRSG) depending on if the machine is a *simple cycle* or *combined cycle* machine. The function of the exhaust system is to cool the exhaust gasses from the turbine section and provide noise attenuation in the exhaust gases prior to those gasses being released into the environment. If a HRSG is included, then an additional function of the HRSG in the exhaust system is to convert the thermal energy contained in the exhaust gas stream to thermal energy in a steam cycle. This thermal energy is then utilized to drive a separate steam/turbine *generator*, thus combining the Brayton and Rankine cycles in a "combined-cycle" design. During this process, the HRSG removes energy from the gas stream reducing the enthalpy of the gas stream. The exhaust system is exposed to elevated air temperatures and, as such, has one or more expansion joints in the ductwork to accommodate thermal expansion and contraction of the duct. Additionally, the exhaust duct may contain additional sound barriers to reduce the acoustic emissions from the combustion turbine exhaust duct. Lastly, at the top of the exhaust stack, the instruments for the *continuous emissions monitoring system* (CEMS) are installed.

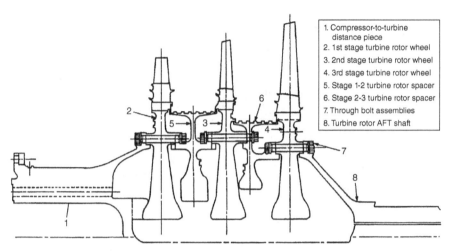

Figure 12.4 Combustion turbine: turbine rotor assembly. *Source*: Reproduced with permission of General Electric Company.

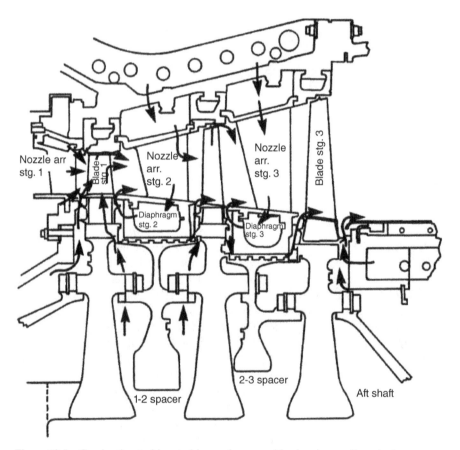

Figure 12.5 Combustion turbine: turbine casing assembly showing cooling air. *Source*: Reproduced with permission of General Electric Company.

There are various support systems necessary to support the operation of the combustion turbine. The thrust bearing provides axial support of the position of the turbine rotor. Additionally, several bearings are provided to support the weight of the turbine and maintain the radial position of the turbine rotor. These bearings are journal bearings which consist of a machined journal on the shaft of the rotor that faces a soft babbitt material inside a stationary bearing housing. The function of the bearing assembly is to provide the necessary lubrication required to the journal and babbitt surface to prevent metal to metal contact. The babbitt material is a softer metal such as lead, tin, or copper to prevent damage to the journal section on the rotor should metal to metal contact occur. The bearing assembly will have lubricating oil provided to the bearing housing under pressure provided by a pump. In addition to providing the oil necessary to the bearings for support of the turbine mechanical forces, the oil system may provide lubrication to other components such as generator bearings. Additionally, for generators that contain hydrogen as the atmosphere in the generator casing, the area of penetration by the generator rotor through the end bells of the generator casing are sealed by a labyrinth seal system. Oil is provided to this labyrinth seal system under a pressure that is greater than the pressure of the hydrogen in the generator casing. Having the oil pressure supplied to the generator hydrogen seals greater than the hydrogen pressure in the generator ensures that the oil will retain the hydrogen inside the generator casing.

Various types of instrumentation are necessary to monitor the status of the operating combustion turbine and provide the control system with the necessary feedback. Temperatures of bearings, gas streams, lubricating oil, pressures of oil pumps and hydrogen, levels in the oil reservoir, hydrogen purity in the generator casing, generator excitation current, stator current, and stator voltage are just a few of the many signals the control system needs to provide safe and reliable control of the turbine generator. The user is directed to Chapter 24 and Chapter 23 of this book for more information on the topics of instrumentation and control, respectively. Figure 12.6 shows a typical complete assembly of a combustion turbine showing the compression section, the combustion section, the turbine expansion section, and the exhaust section.

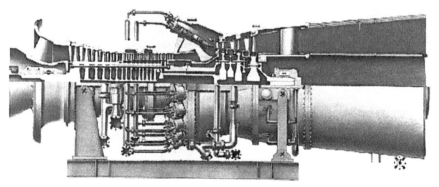

Figure 12.6   Combustion turbine: complete turbine assembly. *Source*: Reproduced with permission of General Electric Company.

There are three basic types of gas turbine arrangements:

1. Simple cycle
2. Combined cycle
3. Cogeneration

As mentioned previously, gas turbines can be operated as peaking units (<2000 hr of operation annually) or base load units (>5000 hr of operation annually). Due to the annual cost for operating a unit, simple cycle machines are operated as peaking units as their annual costs are least when operated less than 2000 hr/year. The combined cycle machine has the lowest annual costs when operated between 2000 and 5000 hr/year. For base loaded units operating above 5000 annual hours per year, the lowest cost units are typically steam turbines.

A simple cycle machine sends the exhaust gases from the discharge of the compressor stage directly to the exhaust stack. A simple cycle process flow diagram is shown for the Brayton cycle diagram shown in Figure 12.7.

A *combined cycle* system uses a heat recovery steam generator (HRSG) between the exhaust of the gas turbine and the exhaust stack to remove some of the remaining thermal energy from the gas stream (Brayton cycle) and transfer this energy to a steam cycle (Rankine cycle) for use in driving a steam turbine. The fact that this design uses both the Brayton cycle (also known as the upper or topping cycle) and the Rankine cycle (also known as the lower or bottoming cycle) leads to the term "combined cycle" power plant. Figure 12.7 shows the typical flows of the combined cycle power plant.

In the design of a combined cycle power plant, one of the primary concerns that affects the design of the combustion turbine is whether the gas turbine will only be used in the combined cycle mode or if it will be desired to operate the combustion turbine in simple cycle mode without the use of the HRSG and the steam turbine. If it is desired to be able to operate the combustion turbine in simple cycle mode, then a bypass stack is usually provided to bypass the hot exhaust gasses around the heat recovery steam generator (HRSG). The tubes in the HRSG depend on the flow of water from the steam cycle to remove heat from the tubes. If the steam process is not

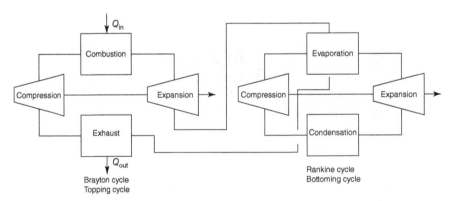

Figure 12.7   Typical flow diagram for combined cycle power plant.

being utilized (such as when the plant is not operating the steam turbine), then either the exhaust gases from the combustion turbine must be bypassed around the tube section of the HRSG or water will be continued to be supplied to the HRSG for the purpose of cooling the tubes. However, when this water picks up heat and becomes steam, the steam is vented to atmosphere to remove the heat from the HRSG tube section which represents a large loss of high purity boiler water.

In addition to a simple cycle arrangement and a combined cycle arrangement for a gas turbine, the third possible configuration is a *cogeneration* plant. This is very similar to the combined cycle plant, but instead of the steam from the HRSG feeding a steam turbine to drive an electric generator, the steam from the HRSG goes to feed other plant steam loads.

Comparing the gas turbine/generator plant to a standard coal-fired steam turbine/generator plant, the gas turbine has the advantage of a shorter design and construction schedule. Also, when looking at a combined cycle gas turbine, the efficiency of the combined cycle gas turbine exceeds the efficiency of the fossil fuel plant. The gas turbine can be started and brought on line much faster than a fossil fuel plant and the gas turbine can be cycled more times with less maintenance than the typical fossil fuel plant. Additionally, the emissions from the gas turbine that utilizes natural gas for the fuel require less post-treatment systems to meet environmental regulations than the standard fossil fuel plant utilizing coal for fuel which is the result of the fuel used in the process. The fossil fuel plant requires substantial real estate and equipment to handle the fossil fuel as well as the post-combustion gas treatment systems. The main attraction for fossil fuel plants is for base-loaded units with the cost per kWh delivered is less for base loaded units (>5000 annual operating hours). Another attraction for coal plants is the ability to store months of fuel on site, versus the natural gas plant that requires delivery of the fuel as it is utilized.

Gas turbine performance is ambient temperature dependent. In months of warmer ambient temperatures, the outside air is less dense. Without compensation, this results in reduced mass flow rate through the combustion turbine during warmer months. Since the rate of energy transfer to the combustion turbine is the product of the change in enthalpy across the turbine section times the mass flow rate, as the mass flow rate is reduced, the rate of energy transfer by the turbine becomes limited. One method to compensate for this in warmer environments involves the use of evaporative coolers on the suction side of the compressor. These have the effect of reducing the temperature of the air prior to the air entering the compression stage. The reduction in air temperature results in increased air density entering the compressor. This compensates for the effect of warmer air temperatures and minimizing the reduction of turbine output.

Also, to enhance the efficiency of the Brayton cycle, some combustion turbines use a *recuperator* to recover some of the energy in the exhaust gas stream and return it to just before the combustor section. Much like the gas recirculation fan in the combustion boiler, this recuperator has the effect of improving the efficiency of the thermal cycle. A recuperative heat exchanger is provided and some of the hot exhaust gas passes one side of the recuperative heat exchanger, while the cooler-compressed gas that leaves the compressor passes on the other side of the recuperative heat exchanger. The hot turbine exhaust gas gives up some of its thermal energy to the cooler-compressed gas side, thus improving the overall efficiency of the combustion turbine.

To maintain the efficiency of the compressor and prevent accelerated wear on the turbine blades, the air that flows through the compressor must be kept free of contaminants. Most gas turbines contain some type of filter media to remove dust and other contaminants from the air stream.

The overall heat rate of the gas turbine, Brayton cycle is defined as the product of the fuel lower heating value and the fuel consumption rate normalized to the generator output as shown in equation (12.1).

$$HR = (FCR \times FHV)/GO \qquad (12.1)$$

where

HR = heat rate (BTU/kWh)

FCR = fuel consumption rate (lbm/hr)

FHV = fuel heating value (BTU/lbm)

GO = generator output (kW)

Just as in the Rankine cycle, the rate of energy transfer (or power) in the Brayton cycle can be determined using the temperatures and pressures at key points in the cycle. The enthalpy of the gas media can be determined from these two parameters. Once the enthalpy at each point is determined, we can multiply the enthalpy by the mass flow rate to determine the rate of energy transfer through the system.

The most common necessary emission control technology required for combustion turbines is NO_x control. NO_x is a term for nitrogen oxides. The two most common nitrogen oxides formed from fossil fuel combustion are nitric oxide (NO) and nitrogen dioxide (NO_2). Increased release of nitrogen oxides has been identified as a contributor to smog and acid rain. Please see Chapter 26 for more information on emissions control systems and technologies.

The first method of controlling the NO_x in the exhaust gas is known a precombustion control. The amount of NO_x is related to the temperature of the flame where complete combustion takes process. At higher temperatures, a greater amount of NO_x is generated. At lower temperatures, a lower amount of NO_x is generated. One method of reducing the flame temperature where complete combustion occurs is by the injection of water or steam into the combustion chamber. Alternately, some systems inject nitrogen (N_2). Initially it may not make sense that, to reduce the formation of NO_x, we inject N_2. However, the injection of N_2 reduces the temperature of the flame and this reduction in temperature results in a lower amount of NO_x produced. The second method of controlling the NO_x in the exhaust gas is known as post-combustion control. A selective catalytic reduction system (SCR) (see Chapter 26 on environmental controls for more on SCRs) can be used to reduce the NO_x released in the exhaust stream to the environment. The SCR functions to convert the NO_x molecules into inert N_2 gas such that N_2 gas is released instead of NO_x.

During operation of the combustion turbine, combustion particles tend to deposit onto the turbine blades as the heated gas expands across these devices. These deposits tend to reduce the efficiency of the turbine section and they need to be occasionally removed from the surface of the turbine blades. Most combustion turbines provide a system for the cleaning of the turbine blades. Some systems can only be used while the turbine is off-line while other systems can be used during operation of the turbine. These cleaning systems remove deposits from the turbine blade section restoring the efficiency of the turbine section.

GLOSSARY OF TERMS

- Air Inlet Assembly – The function of the air inlet assembly is to condition the air being supplied to the compressor stage of the system to ensure the quality and cleanliness of the inlet air is adequate for the compression stage.

- Atomizing Air – High pressure air utilized to convert to minute particles or to a fine spray the fuel oil supplied to a combustor.

- Cogeneration Gas Turbine – Type of gas turbine that has heat recovery steam generator to recover energy from waste heat stream and use energy for plant thermal loads other than to drive steam generator.

- Combined Cycle Plant – An electric generating station that produces electricity from a combustion turbine generator (Brayton cycle) and a steam turbine generator (Rankine cycle). The steam is generated using waste heat from the combustion turbine to produce steam in a heat recovery steam generator (HRSG).

- Combustion – The burning or rapid oxidation of fuel resulting in the release of energy in the form of heat and light.

- Combustion Section – The burning or rapid oxidation of fuel resulting in the release of energy in the form of heat and light. The function of the combustion section is to raise the temperature of the incoming air stream supplied from the compressor section to achieve the combustion process of the Brayton cycle.

- Combustor – A combustor is a component or area of a gas turbine where combustion takes place. It is also known as a burner, combustion chamber or flame holder. The combustor or combustion chamber is fed high pressure air by the compression system.

- Compressor Section – The function of the compressor section is to raise the pressure of the incoming air stream to achieve the compression process of the Brayton cycle.

- Continuous Emissions Monitoring System (CEMS) – A continuous emission monitoring system (CEMS) is the total equipment necessary for the determination of a gas or particulate matter concentration or emission rate using pollutant analyzer measurements and a conversion equation, graph, or computer program to produce results in units of the applicable emission limitation or standard.

- Exhaust System – The function of the exhaust system is to cool the exhaust gasses from the turbine section and provide noise attenuation in the exhaust gases prior to those gasses being released into the environment.

- Gas Turbine – Gas turbines are internal combustion engines that employ a continuous combustion process with hot gases sent to turbine section that converts the kinetic energy of these hot compresses gases into mechanical energy contained in the rotor of the turbine.

- Generator – A rotating electric machine which transfers mechanical torque energy into electrical power/energy.

- Heat Recovery Steam Generator (HRSG) – A device used in cogeneration gas turbines and combined cycle gas turbines to recover some of the energy in the exhaust stream. This device is a gas to water heat exchanger.

- Recuperator – A recuperator captures waste heat in the turbine exhaust system to preheat the compressor discharge air before it enters the combustion chamber.

- Rotating Turbine Blades – The function of the rotating blades is to convert the energy of the air stream to mechanical energy on the rotating shaft.

- Rotor – The rotor of a turbine is the rotating shaft that is connected to the rotating turbine blades. The rotor of a generator is the rotating shaft that contains the electric field circuit.

- Simple Cycle Gas Turbine – Type of gas turbine that has no provision for waste heat recovery.

- Stationary Turbine Nozzles – The function of the stationary nozzle section is to re-direct the airflow so it enters the next rotating section at the correct angle.

- Turbine Section – The function of the turbine section is to expand the gas stream and convert the energy contained in the gas stream to mechanical energy delivered to the shaft of the turbine.

- Thrust Bearing – Thrust bearings keep the rotor in its correct axial position.

PROBLEMS

12.1 What are the four parts of the Brayton combustion cycle?

_____

_____

_____

_____

12.2 What are the six major components of a combustion turbine?

_____

_____

_____

_____

_____

_____

12.3   Given that the plant generator output is 400 MW (or 400,000 kW), the fuel used has a heating value of 10,000 BTU/kWh, and the fuel is consumed by the plant at a rate of 200,000 lbm/hr, determine the plant heat rate.

12.4   What is the function of the air inlet assembly?

12.5   What is the function of the combustion section?

12.6   What is the function of the turbine section?

12.7   What is the function of the recuperator?

12.8   What is pre-combustion method to minimize NO_x formation?

12.9   What is post-combustion method to minimize NO_x formation?

RECOMMENDED READING

Electric Power Plant Design, Technical Manual TM 5-811-6, Department of the Army, USA, 1984.

Electrical Machines, Drives and Power Systems, 6th edition, Theodore Wildi, Prentice Hall, 2006.

The Engineering Handbook, 3rd edition, Richard C. Dorf, (editor in chief), CRC Press, 2006.

General Electric Company – Gas Turbine-Generator Familiarization Training Manual for Tampa Electric Company, Polk County, FL, 2000.

Power Plant Engineering, Black & Veatch, edited by Larry Drbal, Kayla Westra, and Pat Boston, Chapman & Hall/Springer, 1996.

Standard Handbook of Powerplant Engineering, 2nd edition, Thomas C. Elliott, McGraw-Hill, 1998.

CHAPTER 13

RECIPROCATING ENGINES

GOALS

- To understand the basic ideal Otto thermodynamic cycle
- To understand the basic operation of two cycle and four cycle reciprocating engines
- To be able to calculate the ideal power available from a reciprocating engine
- To be able to calculate the efficiency of a reciprocating engine
- To understand the basic ideal Diesel thermodynamic cycle

SMALLER **GENERATORS** may use diesel oil or gas-fired *reciprocating engines* for their prime movers. Reciprocating engines are smaller in power ratings than their turbine counterpart prime movers, so the connected generator will have a smaller kVA rating. Additionally, to maximize the efficiency of the engine in large electrical generation applications, reciprocating engines operate at speeds slower than their turbine counterpart prime movers. Typical speeds for steam or combustion turbines are 3600 rpm or 1800 rpm. However, the typical speed for a reciprocating engine is on the order of 1200 rpm down to 120 rpm. Looking at Chapter 16 on generators, for the generator to produce the same operational frequency, as the speed of the shaft of the generator is decreased, the number of poles of the generator must increase to provide a given frequency. Therefore, the generators connected to turbine prime movers are cylindrical rotor machines rated for the 1800 rpm or 3600 rpm speed while the generators connected to diesel engines are salient pole rotor machines rated for the lower speeds of the rotor. As such, the generators connected to reciprocating engines tend to be shorter in length but wider in diameter.

The *thermodynamic process* for the ignition combustion engine is defined by the *Otto cycle* as shown in Figure 13.1, which is similar to the Brayton combustion cycle that describes the combustion process for gas turbines where there are four processes in the combustion cycle: *compression*, *combustion*, *expansion*, and *exhaust*. The primary difference between the gas turbine (Brayton cycle) and the reciprocating engine (Otto cycle) is the device used for the expansion process. In the gas turbine, the hot gas is expanded across a row of turbine blades (or multiple rows of turbine

Energy Production Systems Engineering: An Introduction for Electrical Engineers to Electrical Power Generation Facilities, Systems, and Equipment, First Edition. Thomas H. Blair.
© 2017 by The Institute of Electrical and Electronics Engineers, Inc. Published 2017 by John Wiley & Sons, Inc.

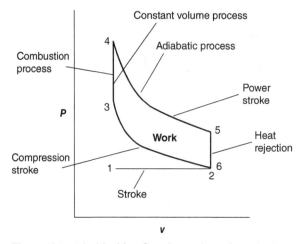

Figure 13.1 Ideal ignition Otto thermodynamic cycle. *Source*: Reproduced with permission of NASA.

blades) to drive the rotor. In the reciprocating engine, the hot gas combusts in an enclosed *cylinder* which increases the pressure of the gas and this forces a *piston* that is connected to *crankshaft* or camshaft by a connecting rod to be driven down to the bottom of travel thereby turning the camshaft or rotor. Since the reciprocating engine defines a specific volume of gas as the position of the piston changes, it is more convenient for us to evaluate the thermodynamic Otto cycle on a P-v diagram or *Pressure volume (P-v) curve* instead of the T-s diagram that we used for the Rankine or Brayton cycle.

On the horizontal axis, the volume of the cylinder is shown. On the vertical axis, the pressure of the gas is shown. As the piston moves from bottom center position (point 2) to top center position (point 1), the change in volume is the difference between the volume at point 2 and the volume at point 1 which is known as the *stroke* of the piston. The thermodynamic cycle is described by points 3, 4, 5, and 6. As the air in the cylinder is compressed from point 6 to point 3, the enthalpy of the gas is increased, reducing the volume that the gas occupies which results in an increase in gas pressure (and thereby temperature) at point 3. The work done on the gas to compress the gas from point 6 to point 3 represents work done by the shaft of the engine through the connecting rod and piston on the gas in the cylinder and is a net negative amount of work done by the engine. The compression ratio of an engine is the ratio of the volume of the gas in the cylinder at point 6 with respect to the volume of the gas in the cylinder at point 3. The *compression ratio* is mathematically defined as

$$CR = V_6/V_3 \qquad (13.1)$$

where

CR = compression ratio

V_6 = volume of the cylinder at the bottom of the cylinder travel

V_3 = volume of the cylinder at the top of the cylinder travel

The cylinder displacement is defined as the difference between volume of the cylinder at the bottom of the cylinder travel and the volume of the cylinder at the top of the cylinder travel. This is mathematically stated as

$$\text{disp/cylinder} = V_6 - V_3 \qquad (13.2)$$

where

disp/cylinder = displacement of one piston from top to bottom

V_6 = volume of the cylinder at the bottom of the cylinder travel

V_3 = volume of the cylinder at the top of the cylinder travel

We can increase the efficiency of the reciprocating engine by increasing the compression ratio of the engine. However, there is a limit to how high we can design the compression ratio. As the compression ratio is increased, the peak pressure (and thereby peak temperature) in the cylinder when the piston is at the top position is increased. If this pressure (and therefore temperature) is allowed to encroach on the auto ignition temperature of the fuel, then the fuel may begin combustion before our intended point of ignition. This causes knocking or pre-ignition of the fuel in the combustion chamber and can result in serious damage to the engine. In internal combustion engines that utilize spark rods or *spark plugs*, the typical maximum value for the compression ratio is 12.

There are two main types of reciprocating engines: two-stroke and four-stroke engines.

The *two-stroke engine*, which is described in Figure 13.2, completes one thermodynamic cycle in two strokes of the piston or one shaft revolution. At the beginning of the first stroke, the air intake port is opened and air enters the combustion chamber.

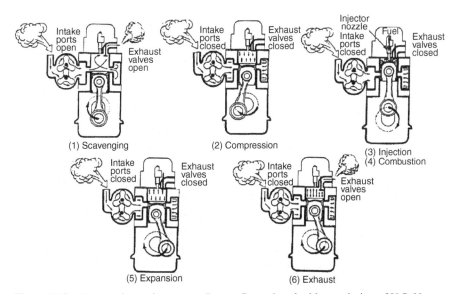

Figure 13.2 Two-stroke engine stages. *Source*: Reproduced with permission of U.S. Navy.

As the piston is driven up, the air intake port is closed by the piston. The cam drives the piston up to compress the air thus completing the compression process. Near the top of travel of the piston, the fuel is injected into the cylinder. In the ideal combustion engine, thermodynamic cycle (called the Otto cycle), when the piston is near the top of its stroke, the spark is initiated and combustion takes place instantaneously. In the actual thermodynamic cycle, combustion does not take place instantaneously, but takes a few milliseconds for complete combustion of the air and fuel mixture to occur. Since we want this combustion to occur on average around *top dead center* (TDC), the spark actually occurs a few degrees before TDC and the combustion is complete a few degrees after TDC. The *ignition advance* is the number of degrees before TDC when the spark is activated.

After the gas is combusted, it drives the piston down the cylinder expanding the hot gas (this is the expansion process). When the piston is near the end of travel to the bottom of the cylinder but before the intake ports are opened, the exhaust valves are opened to allow the combusted gas to exhaust to the atmosphere (this is the exhaust process). As can be seen from the above description, in a two-stroke engine, the piston undergoes two strokes (one up and one down) for one cycle and the shaft rotates one time for one thermodynamic cycle.

The *four-stroke engine* completes one thermodynamic cycle in four strokes of the piston or two shaft revolutions as shown in Figure 13.3. During the first stroke, the air intake valves are opened and the piston is driven down admitting outside air into the combustion chamber. Once the piston reached the bottom of the cylinder, the intake air valve closes. Then the piston is driven up in the cylinder by the connected camshaft compressing the air (this is the compression process). Fuel injection occurs near the end of the compression stage. When the fuel and gas mixture is compressed, the mixture is ignited (this is the combustion process). After the gas is combusted, it drives the piston down the cylinder expanding the hot gas (this is the expansion process). When the piston is at the end of travel to the bottom of the cylinder, the exhaust valves open. Then the piston is driven up in the cylinder by the camshaft

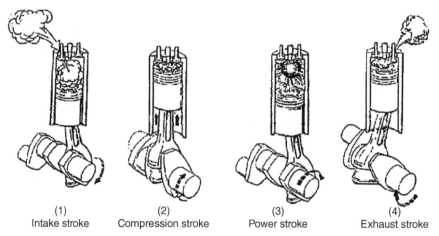

| (1) | (2) | (3) | (4) |
| Intake stroke | Compression stroke | Power stroke | Exhaust stroke |

Figure 13.3   Four-stroke engine stages. *Source*: Reproduced with permission of U.S. Navy.

to allow for the combusted gases to exhaust to the atmosphere (this is the exhaust process). When the piston is at the top of its stroke, the exhaust valves are closed and the intake valves are opened and we are ready for the next cycle. As can be seen from the above description, in a four-stroke engine, the piston undergoes four strokes (one up, one down, a second one up, and a second one down) for one cycle and the shaft rotates two times for one thermodynamic cycle.

Theoretically, in the process of the ideal Otto cycle, the entire combustion process takes place instantaneously at top dead center (TDC). Since the piston does not change its position during this process, we represent this combustion process in the Otto cycle as occurring at a constant volume (isometric process). The combustion process is represented as the difference between the enthalpy of the gas stream between point 3 and point 4 in Figure 13.1. During the combustion process, since volume does not change and enthalpy increases from combustion, the pressure of the gas increases at one value for volume as shown in the P-v diagram. The pressure inside the cylinder places a force on the top of the piston driving the piston toward the bottom of the cylinder. This represents the expansion process or the power stroke of the piston. The power stroke is where the thermal energy contained in the combusted gas is transferred to the mechanical shaft in the piston which is transferred to the crankshaft by the connecting rod. The work done on the gas to drive the piston from top center to bottom center position from point 4 to point 5 in Figure 13.1 represents the work done by the combusted gas on the shaft of the engine through the connecting rod and piston. This represents a net positive amount of work done by the engine. The total amount of work done by the reciprocating engine is the difference between the work done by the engine on the shaft of the engine represented by the area under the curve defined by point 4 and point 5 and the work done by the shaft on the engine represented by the area under the curve defined by point 6 and point 3.

We can also evaluate the ideal amount of work done by the combusted gas on the shaft of the engine by looking at the force on the piston and the displacement of the piston. Remembering back to basic physics, work is defined as force acting through a distance. The ideal amount of work that a single cylinder can provide is the product of the force on the cylinder and the distance between the position of the piston at the top of the cylinder and the position of the piston at the bottom of the cylinder. Looking at the P-v diagram, we see that, as the piston moves from the top of the cylinder to the bottom of the cylinder, the pressure in the cylinder changes. If we define an average amount of pressure in the cylinder, then we can state that the amount of work done by the expansion of the gas in the cylinder is the product of the average pressure in the cylinder and the displacement of the piston as shown in equation (13.3).

$$W_{1\text{cylinder-1cycle}} = P_{\text{avg}} \times \text{disp/cylinder} \qquad (13.3)$$

where

$W_{1\text{cylinder-1cycle}}$ = work done by one thermodynamic cycle of one cylinder

P_{avg} = average pressure that the piston sees during the expansion process (lbf/in.^2)

disp/cylinder = displacement of one piston from top to bottom

A reciprocating engine will have multiple cylinders, so the ideal amount of work that all of the cylinders perform on the shaft of the engine is the product of the work done by one cylinder and the number of cylinders (n) in the engine.

$$W_{\text{engine-1cycle}} = n \times W_{\text{1cylinder-1cycle}} \tag{13.4}$$

where

$W_{\text{engine-1cycle}}$ = work done by the engine for one thermodynamic cycle

n = number of cylinders in engine

$W_{\text{1cylinder-1cycle}}$ = work done by one thermodynamic cycle of one cylinder

Power is the amount of work done by the engine per unit time. As we mentioned previously, for a two-stroke engine, one revolution of the shaft represents one thermodynamic cycle, but for the four-stroke engine, one thermodynamic cycle is completed in two shaft rotations. Therefore, for a four-stroke engine with the same number of cylinders and same displacement and average pressure, the ideal power delivered by the combusted gas of the two-cycle engine is double that of the four-cycle engine. To account for this, we define a constant K which is the number of mechanical rotations of the engine shaft for one thermodynamic cycle. Therefore, we can represent the ideal amount of power delivered by an engine as

$$P_{\text{in}} = (V \times \text{disp/cylinder} \times N \times P_{\text{avg}})/(K \times 396{,}000) \tag{13.5}$$

where

P_{in} = ideal power available from the combusted gas (HP)

V = rotational speed of shaft of engine (rpm)

disp/cylinder = displacement of one cylinder (in.^{3})

N = number of cylinders in engine

P_{avg} = average pressure that the combusted gas places on the piston during the expansion process (lbf/in.^{2})

$K = 1$ for two-cycle engine and $K = 2$ for four-cycle engine.

The value of 396,000 is necessary for unit conversion from units of (in. × lbf/min) on the right side of the equation to units of HP on the left side of the equation. By definition, one HP is defined as 550 foot-pounds per second (ft-lb/sec) or 33,000 foot-pounds per minute (ft-lb/min) or 396,000 inch-pounds per minute (in.-lb/min).

Engine displacement is not given in the displacement for one cylinder, but the displacement for all cylinders. The total displacement of the engine is the product of the displacement of one cylinder and the number of cylinders in the engine and is defined by

$$D = \text{disp/cylinder} \times N \tag{13.6}$$

where

D = displacement of the engine (in.^{3})

disp/cylinder = displacement of one cylinder (in.^{3})

N = number of cylinders in engine

Substituting equation (13.6) into equation (13.5), we can represent the ideal amount of power delivered by an engine as

$$P_{in} = (V \times D \times P_{avg})/(K \times 396{,}000) \tag{13.7}$$

where

P_{in} = ideal power available from the combusted gas (HP)

K = 1 for two-cycle engine and K = 2 for four-cycle engine.

V = rotational speed of shaft of engine (rpm)

D = displacement of the engine (in.^{3})

P_{avg} = average pressure that the combusted gas places on the piston during the expansion process (lbf/in.^{2})

Example 13.1   What is the total displacement of a 16-cylinder engine where each cylinder displaces a volume of 10 in.^{3}?

Solution: Using equation (13.6), we find the total engine displacement to be

D = disp/cylinder $\times N$

D = 10 (in.^{3}) $\times$ 16

D = 160 (in.^{3})

Example 13.2   What is the ideal power available from the combusted gas for a four cycle, 16-cylinder engine with a total engine displacement of 160 in.^{3} that operates at 1200 rpm and has an average pressure that the combusted gas places on the piston during the expansion process of 165 lbf/in.^{2}?

Solution: Given the following values from the problem statement,

K = 2 for a four-cycle engine

V = 1200 rpm

D = 160 (in.^{3})

P_{avg} = 165 (lbf/in.^{2})

Using equation (13.7), we find the ideal power available from the combusted gas to be

$P_{in} = (V \times D \times P_{avg}) / (K \times 396{,}000)$

$P_{in} = (1200 \text{ rpm}) \times (160 \text{ in.}^3) \times (165 \text{ lbf} / \text{in.}^2) / (2 \times 396{,}000)$

$P_{in} = 40 \text{ HP}$

Equation (13.7) provides us with the amount of power available from the combustion of the gas. However, the actual power available at the shaft of the engine to perform useful work is less than this. There are two general categories where power is lost. First, looking back at the Otto thermodynamic cycle described in Figure 13.3, we see that the work done on the gas to compress the gas from point 6 to point 3 comes from the shaft of the engine. This work over a unit time is the power required to achieve this compression function. This leaves less power available on the shaft of the engine to perform work by the engine. Additionally, there are losses in the form of heat in the operation of the engine that also detract from the amount of available power at the shaft of the engine. This leads us to the definition of mechanical efficiency. Mechanical efficiency is simply the ratio of the power available at the shaft of the engine (P_{out}) divided by the ideal power available from the combusted gas (P_{in}), or mathematically stated:

$$\eta\text{-engine} = P_{out}/P_{in} \qquad (13.8)$$

η-engine = efficiency of the reciprocating engine

P_{in} = ideal power available from the combusted gas (HP)

P_{out} = power available at the shaft of the reciprocating engine (HP)

Example 13.3 What is the efficiency of a reciprocating engine that has a value for ideal power available from the combusted gas of 40 HP and a measured available shaft power of 20 HP?

Solution: Using equation (13.8), we find the efficiency of this engine to be

η-engine = P_{out} / P_{in}

η-engine = 20 HP / 40 HP

η-engine = 0.5 or 50%

For some types of fuel, the combustion is initiated with a spark plug while for other types, the compression alone generates enough heat to cause combustion. In the above discussion for the reciprocating engine that incorporates an ignition system, we assumed that the combustion takes place instantaneously when the piston is at the top of the cylinder and the spark plug or spark rod ignites the mixture. Therefore, we assumed that combustion occurs at a constant volume as we can see in the Otto thermodynamic cycle. In the diesel engine, combustion does not occur instantaneously. The diesel engine does not utilize a spark plug or spark rod to ignite the combustible gas, but rather during the compression process, the combustible gas is heated up to its auto ignition temperature. Once the pressure and thereby the temperature required to instigate ignition of the combustible gas is reached, the gas ignites automatically. The fuel is injected during the process of ignition to control the rate of ignition. Due to this need to control the injection of the fuel during the combustion process, diesel engines do not utilize carburetors to mix the fuel and air, but utilize fuel injection to control the process of injection of the fuel into the cylinder. During this combustion, the piston begins to travel to the bottom of the cylinder. The pressure in the cylinder is maintained at approximately the same value until the gas is completely combusted. Therefore, the diesel engine combustion process takes place under constant pressure (isobaric process), not constant volume (isometric process). Once the gas is completely combusted, the piston continues to travel toward the bottom of the combustion process and the pressure begins to decrease with this expansion. This forces the piston to begin to travel from the top of the cylinder toward the bottom of

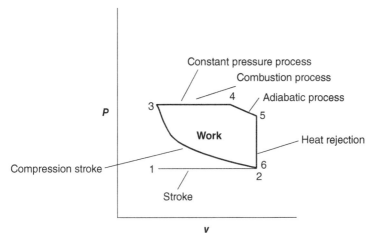

Figure 13.4   Ideal Diesel thermodynamic cycle. *Source*: Reproduced with permission of NASA.

the cylinder during the combustion process. After combustion is complete, the expansion continues to drive the piston to the bottom of the cylinder during the expansion process. The *Diesel thermodynamic cycle* is described in Figure 13.4.

Defining the efficiency of the engine as the amount of heat into the engine by the work delivered by the engine, we can compare the efficiencies of the Otto and Diesel-based engines by comparing the amount of work done by both the Otto and Diesel thermodynamic cycle. If we assume the same compression ratios for the Otto and Diesel thermodynamic cycle engines and overlay Figure 13.3 and Figure 13.4, it would appear that the Otto thermodynamic cycle engine is more efficient than the Diesel since for the same compression ratio, we obtain less work from the Diesel thermodynamic cycle than we do for the Otto thermodynamic cycle. This is shown in Figure 13.5.

In Figure 13.5, the amount of net work for the Otto thermodynamic cycle is defined by the figure of 3–4–5–6–3. The net work for the Diesel thermodynamic cycle is defined by the figure of 3–4′–5′–6–3. We can visually see that the Otto thermodynamic cycle produces more work. However, unlike the engine based on the Otto thermodynamic cycle, the diesel engine has larger compression ratio values. We stated previously that as we increase the compression ratio of the engine, we increase the efficiency of the machine. A typical value of compression ratio for a diesel engine can reach values of 20 or higher (almost double the values of the standard Otto thermodynamic based machine). If we now overlay the two thermodynamic cycles and assume a larger value of compression ratio for the diesel engine, we should see that we can obtain a greater amount of work out of the diesel engine and this is in fact true as Figure 13.6 shows.

In Figure 13.6, the amount of net work for the Otto thermodynamic cycle is defined by the figure of 3–4–5–6–3. The net work for the Diesel thermodynamic cycle

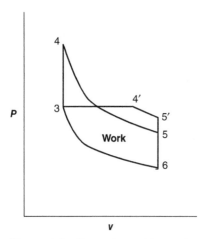

Figure 13.5 Comparison of Otto and Diesel thermodynamic cycle with same compression ratio. *Source*: Reproduced with permission of NASA.

is defined by the figure of $3'–4'–5'–6–3'$. We can visually see that with more realistic values of compression ratios, the Diesel thermodynamic cycle produces more work.

Lastly, we will discuss the use of *superchargers or turbochargers*. The previous discussions concerning the Otto and Diesel thermodynamic cycles all had the same starting point for the beginning of the compression process. All of the above machines started with the incoming gas at atmospheric pressures as was shown as point 6. What would happen if we were to start the thermodynamic compression process with the incoming air already pressurized above atmospheric pressure? This would shift the compression portion of the P-v curve up, but the expansion side of the

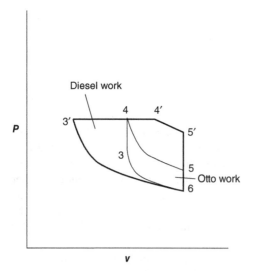

Figure 13.6 Comparison of Otto and Diesel thermodynamic cycle with different compression ratios. *Source*: Reproduced with permission of NASA.

thermodynamic cycle would still exhaust to atmospheric pressures. This would have the effect of increasing the amount of work one thermodynamic cycle could produce and, thereby, this would increase the amount of power that an engine could produce. One typical device utilized to increase the pressure of the air supply to the engine is the turbocharger. The turbocharger is constructed as a turbine and compressor on one common shaft. The turbine is placed in series with the exhaust gas of the reciprocating engine. The turbine drives the compressor and the compressor is placed in series with the air inlet to the engine combustion chamber. This compressor allows for some of the energy in the exhaust gas stream that would otherwise be discharged to the outside air and be a loss to the efficiency of the system and redirect a portion of this energy back to the air intake section of the engine. Turbochargers are very common in diesel engines that drive generators as they can increase the available power from the engine by values of 30% or higher, depending on the design of the engine and turbocharger. A supercharger performs the same function as a turbocharger, but it is connected to the crankshaft of the engine and derives its power from this physical connection. The main difference between a supercharger and a turbocharger is the source of power. The supercharger will take its power from the crankshaft whereas the turbocharger will draw power from exhaust gases that result from combustion.

One method of starting a reciprocating engine is to use compressed air to drive the pistons until the engine reaches a speed adequate for starting, then the fuel is injected into the combustion cylinders and the normal operation of the reciprocating engine begins. Another method is to use a DC motor driven by a set of batteries to initially turn the engine until speed is sufficient for the combustion to commence. One major benefit to reciprocating engines is their ability to be used for black start units. Since they can be started by compressed air or a DC motor fed from a battery and only minor auxiliary loads (lube oil, turbine controls, etc.) are required for operation, in the event of a plant power outage, reciprocating engines can be designed to start only with DC battery power and compressed air. This makes reciprocating engines an ideal technology for a prime mover when the generator must be able to start with no external source of auxiliary power.

Emission controls for reciprocating engines are very similar to the combustion controls for gas turbines. The precombustion controls are by controlling the amount of air and fuel in the combustion chamber to minimize the excess O_2 available. By minimizing excess O_2, we can minimize NO_x formation. However, if we error too far and have a fuel-rich combustion environment, then we run the risk of incomplete combustion where we will be exhausting CO which means we are letting some of the heating value of the fuel out with the gas stream without utilizing this possible energy source. Post-treatment is similar to the gas turbine where we would use an SCR system or catalytic converter to remove the NO_x. Please see the emissions section of this text for more information on emissions control.

GLOSSARY OF TERMS

- Camshaft – Shaft that is moved by the crankshaft with two or more offset cams (lobes) that operate the valves.

- Combustion – Process in which a substance reacts with oxygen to give heat and light (burning).
- Compression – The action of compressing or being compressed. The reduction in volume (causing an increase in pressure) of the fuel mixture in an internal combustion engine before ignition.
- Compression Ratio – Ratio of the volume of the cylinder at the bottom of the cylinder travel normalized to the volume of the cylinder at the top of the cylinder travel.
- Crank/crankshaft – Rod that spins and drives a piston movement in the cylinder.
- Combustion – rapid chemical combination of a substance with oxygen, involving the production of heat and light.
- Cylinder – Central working part of a reciprocating (or piston) engine, the space through which a piston travels.
- Diesel Cycle – The Diesel cycle is a combustion process of a reciprocating internal combustion engine. In it, fuel is ignited by heat generated during the compression of air in the combustion chamber, into which fuel is then injected.
- Exhaust – Gases ejected from an engine as waste products.
- Expansion – The action of becoming larger or more extensive.
- Four-Stroke Engine – A four-stroke engine (also known as four-cycle) is an internal combustion engine in which the piston completes four separate strokes which constitute a single thermodynamic cycle. A stroke refers to the full travel of the piston along the cylinder, in either direction.
- Ignition Advance – The ignition advance is the number of degrees before top dead center (TDC) when the spark is activated.
- NOx – A chemical term for nitrogen oxides produced during combustion and a contributor to smog and acid rain.
- Otto Cycle – A four-stroke cycle for internal combustion engines that incorporates the four processes of compression, combustion, expansion, and exhaust.
- Piston – Rod inside a cylinder that is moved by pneumatic pressure.
- Pressure-Volume (P-v) curve – A graphical representation of the changing relationship between pressure and volume in the cylinder during engine cycles.
- Reciprocating Engine – An engine in which one or more pistons move up and down in cylinders; a piston engine.
- Spark Plug or Ignitor Gun – Electrical device that fits into the cylinder head of some internal combustion engines and ignites compressed aerosol gasoline by means of an electric spark.
- Supercharger or Turbocharger – Air compressor used for forced induction of an internal combustion engine. The greater mass flow rate provides more oxygen to support combustion than would be available in a naturally aspirated engine, which allows more fuel to be provided and more work to be done per cycle, increasing power output. The main difference between a supercharger and a

turbocharger is the source of power. The supercharger will take its power from the crankshaft whereas the turbocharger will draw power from exhaust gases that result from combustion.

- Thermal Efficiency – Ratio of engine work to heat energy of consumed fuel.
- Thermodynamics – The study of relationship between thermal energy (heat) and all other forms of energy (i.e., mechanical, electrical).
- Top Dead Center (TDC) – The position of a piston that is farthest from the crankshaft (at its highest point).
- Two-Stroke Engine – A two-stroke, or two-cycle, engine is a type of internal combustion engine which completes a power cycle with two strokes (up and down movements) of the piston during only one crankshaft revolution.

PROBLEMS

13.1 For a two-stroke gas or oil reciprocating engine, it takes how many shaft revolutions to complete one full cycle?

    A. One

    B. Two

    C. Three

    D. Four

13.2 For a four-stroke gas or oil reciprocating engine, it takes how many shaft revolutions to complete one full cycle?

    A. One

    B. Two

    C. Three

    D. Four

13.3 What is the total displacement of an eight-cylinder engine where each cylinder displaces a volume of 15 in.^{3}?

13.4 What is the ideal power available from the combusted gas for a four-cycle, eight-cylinder engine with a total engine displacement of 120 in.^{3} that operates at 2000 rpm and has an average pressure that the combusted gas places on the piston during the expansion process of 165 lbf/in.^{2}?

13.5 What is the efficiency of a reciprocating engine that has a value for ideal power available from the combusted gas of 50 HP and a measured available shaft power of 30 HP?

13.6 What is the volume of the cylinder at the bottom of the cylinder travel if the volume of the cylinder at the top of the cylinder travel is 2 in.^{3} and the compression ratio (CR) is 10?

13.7 What is the displacement of the cylinder at the bottom of the cylinder travel if the volume of the cylinder at the top of the cylinder travel is 2 in.^{3} and the compression ratio (CR) is 10?

RECOMMENDED READING

Electrical Machines, Drives and Power Systems, 6th edition, Theodore Wildi, Prentice Hall, 2006.

The Engineering Handbook, 3rd edition, Richard C. Dorf (editor in chief), CRC Press, 2006.

National Aeronautics and Space Administration (NASA) web page, http://www.grc.nasa.gov/WWW/K-12/airplane/otto.html

Power Plant Engineering, Black & Veatch, edited by Larry Drbal, Kayla Westra, and Pat Boston, Chapman & Hall/Springer, 1996.

Standard Handbook of Powerplant Engineering, 2nd edition, Thomas C. Elliott, McGraw-Hill, 1998.

US Navy Training Manual, Machinist's Mate 3rd class, Naval Education and Training Professional Development and Technology Center, 2003.

CHAPTER 14

ELECTRICAL SYSTEM

GOALS

- To understand the basic design and components of an electrical distribution system

- To understand some of the safety requirements of NESC in regard to the electrical distribution system

- To calculate the daily and annual load factors of a distribution system

- To describe the basic arrangement of a radial distribution system, primary selective system, primary loop system, secondary selective system, and sparing transformer distribution system

- To identify the various types of NEMA enclosure ratings

- To describe the difference between non-segregated bus duct, segregated bus duct, and isolated phase bus duct

- To calculate the ampacity of a conductor

- To identify proper wiring methods for hazardous classified locations

- To calculate cable pull tensions based on IEEE 1185 – Recommended Practices for Cable Installation at Generating Stations

- To understand the basic concepts and types of cathodic protection

IN THE design and operation of the electrical system of a power plant, the reliability of the electrical system is critical to the safe and reliable operation of the power plant. Benjamin Franklin once said, "An ounce of planning is worth a pound of cure." This certainly applies to the design of the electrical system for a power generation station.

If our goal is to expect a safe and reliable electrical distribution system in the electric generation station, then adequate analysis, planning, and design must be done to achieve our goal of a safe and reliable electric distribution system.

The first priority is always safety. With that in mind, in addition to all the installation and safe work practices listed in the first chapter of this book, below are some additional minimum safe installation work requirements.

Energy Production Systems Engineering: An Introduction for Electrical Engineers to Electrical Power Generation Facilities, Systems, and Equipment, First Edition. Thomas H. Blair.
© 2017 by The Institute of Electrical and Electronics Engineers, Inc. Published 2017 by John Wiley & Sons, Inc.

NESC section 15 requires current transformer secondaries to be effectively grounded. Additionally, when secondary wiring is connected to a terminal block, NESC requires the terminal block to be a shorting type. This is to prevent the development of dangerous voltage levels should the secondary wire be opened while the CT is under load.

NESC also requires the cases of all power transformers to be effectively grounded or guarded to prevent inadvertent electrical contact. When the power transformer is a liquid-filled transformer, either less flammable liquids, space separation, fire-resistant barriers, automatic extinguishing systems, absorption bends, and/or enclosures shall be used to minimize the degree of the fire hazard. Also, power transformers shall be provided with a means to disconnect automatically the source of the supply of current for high magnitude short circuit faults within the transformer.

NESC requires that all circuit conductors have protection systems to protect the conductors against excessive heating by overcurrent, alarm, indication, or trip devices with the exception of conductors that are normally grounded. Normally, grounded conductors shall NOT have any overcurrent protection or other devices that might interrupt the continuity to ground. Where the installed conductor may be subject to mechanical damage, casing, armor, or other means shall be employed to limit the likelihood of damage or disturbance. All non-shielded conductors of greater than 2500 V to ground and all bare conductors of greater than 150 V to ground must be isolated by elevation or guarding. Ends and joints of insulated conductors shall have insulating covering equivalent to that of other portions of the conductor.

NESC requires that all circuit breakers, circuit switchers, reclosers, and fuses shall be rated for the momentary and interrupting ratings that the electrical distribution system would impose on these devices. Circuit breakers and switches shall be arranged so that they can be locked in the open condition. Fuses shall be arranged so that, before fuse removal, the fuse can be disconnected from all sources of electric energy or the fuses can be removed by means of insulating handles.

The design of the electrical system does not start at the beginning of the project, but more commonly follows behind the design and specification of the loads that the system must support. When looking at the requirements for the electrical system one of the first things that must be done is to develop a load schedule. This should list the types of equipment that are required to be supported by the electrical system and also the modes of operation of the equipment. Not all equipment is energized at the same time and, not all equipment is fully loaded at the same time. For example, for a combustion turbine that can run on natural gas or oil, there is an atomizing air compressor that is used to adequately atomize the oil fuel as it enters the combustion chamber. The atomizing air compressor would only be required when the unit is running on the oil fuel. The atomizing air compressor is not required when the unit is running on natural gas as the fuel source.

In addition to the system modes of operation, the load development schedule should include items such as

1. Peak load requirements

2. Temporary power requirements

3. Installation timing requirements

Additionally, most distribution systems tend to grow over the life of the power generation facility. For example, as new environmental control technologies become available, there might be a future requirement to install these technologies to reduce the emissions from the plant. Much of this equipment may require electrical power sources to support the operation of the equipment. It is much faster, easier, and less expensive to have spare electrical distribution system capability installed when the plant is first designed and constructed than to have to modify the existing electrical system to allow for the increased load requirements at later dates.

Also the type of load is very important. Whether the load is a continuous load or an intermittent or cyclic load can determine the rating of the distribution equipment such as transformers and switchgear feeding the system. If we define the *daily load factor* (DLF) as the percentage of the daily average energy usage (DAEU) normalized to the daily peak energy usage (DPEU), the daily load factor can be calculated by the following formula:

$$DLF = DAEU/DPEU \tag{14.1}$$

where

DLF = daily (kWh for 24 hr period / peak kW during the 24 hr period)

DAEU = daily average energy usage for a 24-hr period

DPEU = daily peak energy usage for a 24-hr period

Similarly, there may be variations to the load over a yearly period. If we define the *annual load factor* (ALF) as the percentage of the average annual energy usage (AAEU) normalized to the annual peak energy usage (APEU), the annual load factor can be calculated by the following formula:

$$ALF = AAEU/APEU \tag{14.2}$$

where

ALF = annual (kWh for a 8670 hr period / peak kW during the 8670 hr period)

AAEU = annual average energy usage for a 8670-hr period

APEU = annual peak energy usage for a 8670-hr period

Another consideration for adequate electrical distribution system planning is loads that surge such as motors. Motors have locked rotor currents in the order of six times their full-load current values. For an across-the-line start of a motor, the MVA drawn in locked rotor conditions is the same proportion as locked rotor to full-load current or six times the running MVA of the motor. The rating of the transformer and system feeding motors and the conductors that feed motors should be analyzed to ensure that the system voltage does not dip low enough where it effects either the successful starting of the motor or causes other loads on the same bus to drop out on the undervoltage condition that may occur when starting large motors. (For more information on motor design criteria for reliable and safe control and operating of motors in a power generation facility, refer to Chapter 17 on motors).

Also, there are some special loads that deserve care in their design. Any load that is nonlinear should be carefully designed to ensure that the conductors and the transformer feeding the load are rated for the non-sinusoidal current waveform. Triplen harmonics in particular are of concern as they tend to travel back on the neutral conductor and, if the neutral conductor is not adequately sized, the neutral conductor can overheat.

As mentioned in previous chapters, there are certain loads that are critical to the safe and reliable operation of the power plant such as backup lube oil pumps, hydrogen seal oil pumps, control system computers, and burner management system controls. The electrical distribution to these systems should be from a very reliable source. Possible sources with enhanced reliability may be the station battery or an inverter that is fed from the station battery. Additionally, there may be backup sources should the normal source fail such as an emergency generator that is used to service such loads.

Once the electrical load list has been developed, a determination of the appropriate voltage classes of the distribution system can be made. Manufacturers follow IEEE standards for the requirements of manufacturing and testing of electrical equipment. According to IEEE 141, IEEE Recommended Practice for Electric Power Distribution for Industrial Plants, there are three basic voltage classes of equipment: low voltage, medium voltage, and high voltage. *Low voltage* systems are systems with voltages that are less than 1000 V. *Medium voltage* systems are system with voltages that are greater than or equal to 1000 V but less than 100,000 V. *High voltage* systems are systems that are greater than or equal to 100,000 V up to and including 230,000 V. Systems in excess of 230,000 V up to 1,000,000 V are known as *extra high voltage systems*. Voltage systems in excess of 1,000,000 V are known as *ultra-high voltage systems*. Be aware that the terms low voltage, medium voltage, and high voltage have different meanings to different electrical professionals depending on the topic being discussed. For example, when discussing electrical safe installation and the requirements of the National Electrical Code® (NEC®), anything exceeding 600 V is considered high voltage.

Larger loads tend to be served from higher voltages to reduce the current demand of the load which, in turn reduces the I^2R losses and voltage drop on the feeder to the load. Smaller loads tend to be serviced by lower voltages as this reduces the expense of the equipment and installation. Low voltages are used to supply utilization equipment. Medium voltage class voltages are used for distribution to transformers that step this voltage down to low voltage equipment. Medium voltages also directly supply larger loads such as FD and ID fan motors. High voltage class voltages are used to transmit large amounts of electric power between transmission substations. In the generation station facility, high voltage systems are limited to the connections between the transmission system and the generator step-up (GSU) transformer and the interconnect between the generation station reserve electrical distribution system and the transmission system.

This brings up the difference between nominal system voltage and nominal utilization voltage. *Nominal system voltage* is the voltage by which a portion of the system is designated and to which certain operating characteristics of the system are related. *Nominal utilization voltage* is the voltage rating of certain utilization equipment used in the system. For example, for a three-wire 120 Vac/240 Vac distribution panel, the nominal system voltage is 120 Vac/240 Vac, but the nominal utilization voltage is 115 Vac/230 Vac. While the system is rated for 120 Vac/240 Vac, the equipment is actually utilized at 115 Vac/230 Vac due to voltage drop in the conductors feeding the distribution panel. Typical values for low voltage nominal system voltages are 120 Vac, 240 Vac, 480 Vac, and 600 Vac while typical values for low voltage nominal utilization voltages are 115 Vac, 230 Vac, 460 Vac, and 575 Vac. For medium voltages, the typical values for nominal system voltage are 2400 Vac, 4160 Vac, 13800 Vac, and 69,000 Vac. For high voltage the typical values for nominal system voltages are 115 kVac, 138 kVac, and 230 kVac.

Once the load schedule has been developed and the necessary nominal system and equipment utilization voltages have been determined, we must start to design the distribution system that will feed the plant loads. There are a number of distribution systems available, each with their own advantages and disadvantages. The various configurations available for a distribution system are the radial system, primary-selective system, and the secondary selective system.

DISTRIBUTION SYSTEM CONFIGURATION

The first type of electrical distribution system is known as the simple *radial distribution system*. Figure 14.1 shows the typical configuration of a simple radial electrical distribution system. In this system, the path for energy flow from source to load is radial from the source and only has one path to each load. Some of the advantages of this system are the simplicity of operation and protection design and the low initial capital cost. Since there is only one source, the effort required to ensure coordination between fault interrupting devices is minimal. Additionally, the operation of the system is simpler. To isolate a piece of equipment and perform lockout/tag out, there is

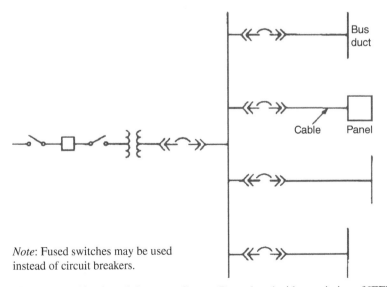

Note: Fused switches may be used instead of circuit breakers.

Figure 14.1   Simple radial system. *Source*: Reproduced with permission of IEEE.

only one source to consider and this system is less prone to the possibility of not identifying a source for the lockout/tag out. The major disadvantage of this system is the lack of an alternate source. On a failure of a piece of equipment in the source such as the transformer or main breaker, all of the load downstream of the failed device will be de-energized. If the main breaker, main transformer, or high side transformer breaker fails in service, the entire low voltage bus will be de-energized. Another drawback is the lack of the ability to perform maintenance on plant equipment and still keep equipment energized. To maintain the source, main breaker or transformer requires the low voltage bus to be de-energized. Typical utility generation station distribution design requires that the electrical distribution system be maintained while the unit is providing electricity to the electrical grid. Therefore, the radial system is not found in equipment critical to unit operation, but may be utilized in non-critical equipment and systems.

The second type of electrical distribution system is known as the expanded radial system. Figure 14.2 shows the typical configuration of an *expanded radial electrical distribution system*. This system is similar to the simple radial system, except the load is now distributed between several buses. For each bus, the path for energy from source to load is still radial and only has one path. However, now since not all equipment is on one bus, the failure of one of the source transformers or the maintenance of one of the source transformers means that only one low voltage bus is de-energized and the other low voltage buses remain energized. Just as in the simple radial system, the main advantage of this system is the simplicity of operation. The cost for this system is higher than the simple radial system as the number of main breakers and source transformers has increased. Also, to make this system

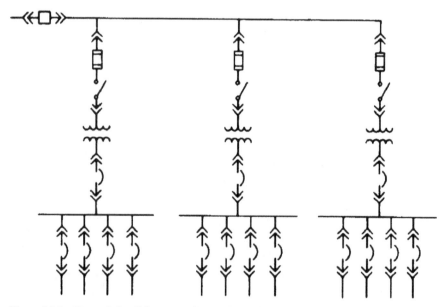

Figure 14.2   Expanded radial system. *Source*: Reproduced with permission of IEEE.

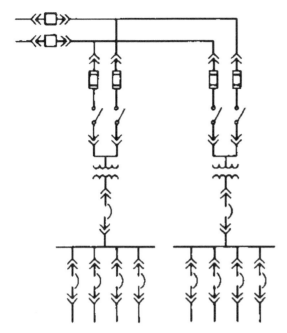

Figure 14.3   Primary selective system. *Source*: Reproduced with permission of IEEE.

functionally redundant in a power plant, typically the loads will be designed with an $N + 1$ safety factor. For example, if there are two blowers needed for full-load operation of the power plant ($N = 2$), this system would install three blowers ($N + 1 = 2 + 1 = 3$) and feed one blower from each bus. Designed in this way, on the de-energization of one bus and the loss of one blower, the remaining two buses and blowers will be able to support plant operations without a derate. This additional load equipment adds initial costs to the project as well.

The third type of electrical distribution system is known as the primary selective system. Figure 14.3 shows the typical configuration of the primary selective system. The *primary selective system* is similar to the expanded radial system, except the bus transformer has multiple sources to select from. This allows for a bus transformer to be fed from more than one source. This increases reliability because on a loss of one source, the bus can be maintained energized by switching to the alternate source. However, as with the simple and expanded radial systems, a failure in the transformer and main secondary breaker will force the load bus to be de-energized. Also, now since there are two sources with two isolation devices, the effort to design the protective system becomes more time-consuming as we need to look at the fault current available from both sources and the coordination of other protective devices for a feed from either source. Additionally, normally there are interlocks between the source isolating devices to prevent closure of both sources to the same transformer. This would parallel the sources and increase the available short circuit currents at the primary isolation devices location, possibly exceeding the interrupt rating of these devices.

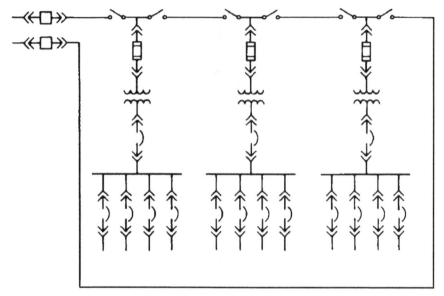

Figure 14.4   Primary loop system. *Source*: Reproduced with permission of IEEE.

The fourth type of electrical distribution system is known as the primary loop system. Figure 14.4 shows the typical configuration of the primary loop system. The *primary loop system* is similar to the primary selective system, except the two sources to the bus transformers are no longer radial feeds to each transformer but form a loop or ring bus. This system has slightly lower costs than the primary selective system, since two radial feeds are not required for each transformer. If the ring bus is to be maintained as a closed loop, then the momentary and interrupting devices on the source side of all the transformers must be designed and constructed to withstand the short circuit available from both sources at the same time. Normally, one of the switches in the loop is designed to be a "normally open" switch. This allows for a lower momentary and interrupt rating for the transformer isolating devices.

The fifth type of electrical distribution system is known as the secondary selective system. Figure 14.5 shows the typical configuration of the secondary selective system. The *secondary selective system* is similar to the primary selective system except each secondary bus has two possible sources. There are two subtypes of secondary selective system. Figure 14.5 shows a secondary selective system with a tiebreaker between buses. On a loss of one transformer, the downstream breaker from the failed transformer is opened and the tiebreaker is closed to maintain power to the bus.

The other type of secondary selective system shown in Figure 14.6 is designed with two secondary circuits in each transformer. Each secondary circuit feeds a different bus via an isolating secondary transformer. The benefit of this system over the secondary selective system is that it does not depend on one breaker to tie the two buses together. In the tie arrangement, a failure of the tiebreaker has the possibility of affecting both secondary buses. The disadvantage of this arrangement over the tie

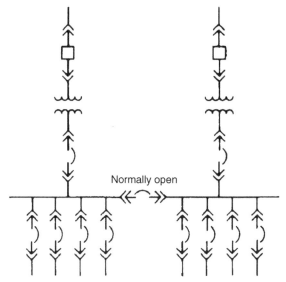

Figure 14.5   Typical secondary selective system. *Source*: Reproduced with permission of IEEE.

arrangement for secondary selective systems is that this arrangement requires the use of one more breakers than the secondary system utilizing the tie breaker design.

The secondary selective system now allows for a transformer or a bus main breaker to fail or be de-energized for maintenance and an alternate source is available to maintain the low voltage bus energized. This system has enhanced reliability since more of the equipment can be maintained while the rest of the system remains energized and, failure of the source or transformer does not affect the energization of the load equipment since an alternate source is available. Now that we have multiple sources for the secondary equipment buses, the design of the protection system takes more time and effort to ensure proper coordination as we need to evaluate the short circuit currents for various sources and load configurations and we need to evaluate the protection system setting coordination for the various source system configurations. In this system, when one transformer is isolated and the other transformer provides the power for both low voltage buses, the transformers must be sized for the load of both buses. This may result in a transformer that is up to twice the size (MVA rating) of the transformer used in the primary selective system. This adds increased initial cost to this system due to the higher required transformer rating when one transformer is feeding two buses.

The sixth type of electrical distribution system is known as the sparing transformer system. Figure 14.7 shows the typical configuration of the sparing transformer system. The *sparing transformer system* is similar to the secondary selective system, except now the alternate source for all low voltage buses is the same spare transformer. This saves on the cost of doubling the size of every transformer that has to be done on the secondary selective system, since, in the sparing transformer design, each transformer only has to provide the power for one bus. However, this system requires

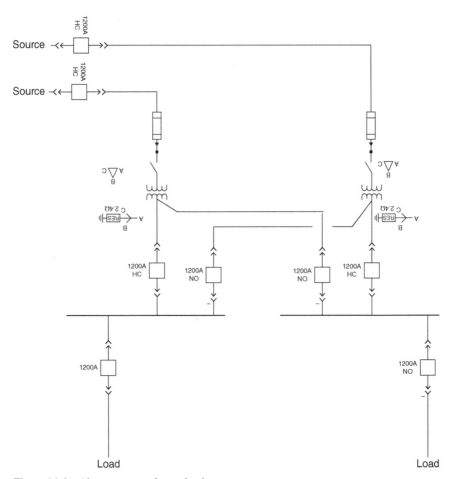

Figure 14.6 Alternate secondary selective system.

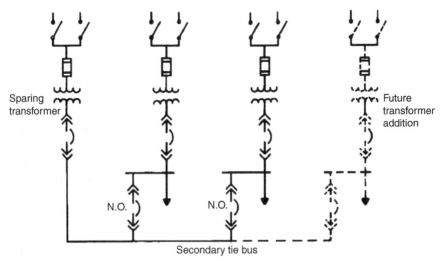

Figure 14.7 Sparing transformer. *Source*: Reproduced with permission of IEEE.

one more transformer than the secondary selective system. Therefore, if doubling the size of the number of transformers is less initial cost, the secondary selective system is chosen. However, if adding one spare transformer of lower size is less cost, then the sparing transformer scheme is chosen. The sparing transformer system is also a possible retrofit to add to an existing simple radial system to enhance the reliability of the system.

ENCLOSURES

Electrical equipment enclosures are designed and built for various types of operating environments. Table 14.1 compares various NEMA enclosure ratings and their features.

Selection of the correct enclosure for electrical equipment for the environment that it will be installed in is critical to the safe and reliable operation of the equipment. For enclosures installed in hazardous classified areas, refer to Chapter 1 regarding proper bonding and grounding of the enclosure. In hazardous locations, when completely and properly installed and maintained, Type 7 and 10 enclosures are designed to contain an internal explosion without causing an external hazard. Type 8

TABLE 14.1   Comparison of Specific Applications of Enclosures for Indoor Nonhazardous Locations

Provides a Degree of Protection Against the Following Conditions	Type of Enclosure									
	1^a	2^a	4	4X	5	6	6P	12	12K	13
Access to hazardous parts	X	X	X	X	X	X	X	X	X	X
Ingress of solid foreign objects (falling dirt)	X	X	X	X	X	X	X	X	X	X
Ingress of water (dripping and light splashing)	...	X	X	X	X	X	X	X	X	X
Ingress of solid foreign objects (circulating dust, lint, fibers, and flyings^b)	...	...	X	X	...	X	X	X	X	X
Ingress of solid foreign objects (settling airborne dust, lint, fibers, and flyings^b)	...	...	X	X	X	X	X	X	X	X
Ingress of water (hose down and splashing water)	...	...	X	X	...	X	X	...	...	...
Oil and coolant seepage	...	...	...	..	...	...	...	X	X	X
Oil or coolant spraying and splashing	...	...	...	...	...	...	...	...	...	X
Corrosive agents	...	...	...	X	...	...	X	...	...	...
Ingress of water (occasional temporary submersion)	...	...	...	...	...	X	X	...	...	...
Ingress of water (occasional prolonged submersion)	...	...	...	...	...	...	X	...	...	...

Source: Reproduced with permission of National Electrical Manufacturers Association.

^{a}These enclosures may be ventilated.

^{b}These fibers and flyings are nonhazardous materials and are not considered Class III type ignitable fibers or combustible flyings.

TABLE 14.2 Comparison of Specific Applications of Enclosures for Indoor Hazardous Locations

Provides a Degree of Protection Against the Following Conditions	Type of Enclosure									
	3	3X	3R[a]	3RX[a]	3S	3SX	4	4X	6	6P
Access to hazardous parts	X	X	X	X	X	X	X	X	X	X
Ingress of water (rain, snow, and sleet[b])	X	X	X	X	X	X	X	X	X	X
Sleet[c]	...	...	...	...	X	X	...	...	...	...
Ingress of solid foreign objects (windblown dust, lint, fibers, and flyings)	X	X	...	...	X	X	X	X	X	X
Ingress of water (hosedown)	...	...	...	...	...	...	X	X	X	X
Corrosive agents	...	X	...	X	...	X	...	X	...	X
Ingress of water (occasional temporary submersion)	...	...	...	...	...	...	...	...	X	X
Ingress of water (occasional prolonged submersion)	...	...	...	...	...	...	...	...	...	X

Source: Reproduced with permission of National Electrical Manufacturers Association.
[a]These enclosures may be ventilated.
[b]External operating mechanisms are not required to be operable when the enclosure is ice covered.
[c]External operating mechanisms are operable when the enclosure is ice covered.

enclosures are designed to prevent combustion through the use of oil-immersed equipment. Type 9 enclosures are designed to prevent the ignition of combustible dust. Table 14.2 compares various NEMA enclosure ratings and their features. See Annex A for more detailed specifics on NEMA enclosure ratings.

Example 14.1 You are specifying a control cabinet that is going to be installed at the seawall of a generation facility. The atmosphere is considered corrosive due to the high salinity content of the seawater in the area. What type(s) of enclosure would be adequate for this installation?

Solution: From Table 14.1, either type NEMA 4X or NEMA 6P would provide anti-corrosion characteristics needed.

BUSWAY APPLICATIONS

Busway is a common name applied to equipment utilized for the transfer of large amounts of electrical energy in a reliable method between two locations. Typical busway uses either aluminum or copper conductors and these bus systems are enclosed in a metal housing. The bus may or may not be insulated (insulated bus is more common with newer busway systems than with older busway systems). There are two main types of busway. Feeder bus is used to transmit large amounts of power between a source and a destination. Available current ratings range from 600 A to 5000 A and by paralleling bus higher currents can be achieved. There are not often many connections to the feeder busway between the source and load. In a generation plant, these are usually the connection methods utilized between the step-down transformer and the main low voltage bus. The plug-in busway system consists of changeable load connections where loads can be easily added or removed as the system and plant configurations require. These are more commonly applied at industrial facilities where manufacturing occurs and the layout of the processing equipment changes frequently to allow the electrical distribution system to accommodate the frequent load change configuration. Plug-in busway systems are not normally utilized in a power generation facility and are not covered in this chapter. This chapter covers busway applications designed for the transmission of large magnitudes of power between two points.

There are three basic types of bus ducts utilized in the utility electric generation station. Which type of bus system used is a function of the reliability needed by the application and the availability of finances. The first type of bus duct is *nonsegregated phase bus duct*. This consists of bus bar supported by non-conducting insulators in an overall environmental enclosure. The insulators must be spaced close enough to provide the support for the bus work to be able to withstand the momentary fault current of the application. Figure 14.8a and Figure 14.8b show a section of non-segregated bus duct. The advantage of this system over the other two is that this tends to be the least costly of the three. However, the disadvantage of this system is that there is the potential for a phase-to-phase fault in the bus duct as there are no barriers between phases.

The second type of bus duct is *segregated phase bus duct*. This is similar in construction to the non-segregated phase bus duct, except that there is a grounded metal barrier between each phase. The addition of the grounded metal barrier between phases helps reduce the chance of a phase-to-phase fault; however, it does not eliminate the possibility. Most faults in ducts start as a phase-to-ground, but once the fault occurs and develops a plasma ball, the copper expands to about 67,000 times its original density. This causes a large pressure wave and, if this pressure wave causes the metal segregation to fail, this could lead to a phase-to-phase fault. The segregated phase bus is more costly than non-segregated phase but less costly than isolated phase bus duct. Figure 14.9a and Figure 14.9b show a section of segregated phase bus duct.

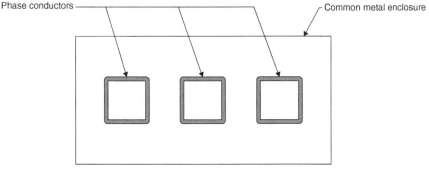

Figure 14.8a   Non-segregated phase bus duct. *Source*: Reproduced with permission of IEEE.

Figure 14.8b   Non-segregated phase bus duct. *Source*: Reproduced with permission of Crown Electric Engineering and Manufacturing LLC.

The third type of phase bus duct is the *isolated phase bus duct*. In this system, each phase has its own enclosure. This eliminates the possibility of a phase-to-phase failure (except for the event where one phase shorts to ground and another phase at the same time shorts to ground). This system provides the greatest protection against a phase-to-phase fault but is the most costly. Because of the enhanced protection against a phase-to-phase fault, segregated phase bus duct is the most common type of system utilized for the connection between a large generator and the generator step-up (GSU) transformer. Most large utility generators are wye-connected with an impedance in the neutral designed to keep a phase-to-ground fault current at only a few amps (the minimum requirement is to keep the zero sequence resistance equal to or less than the capacitive reactance of the machine to minimize voltage transients developed in the machine during a fault). However, the generator fault current on

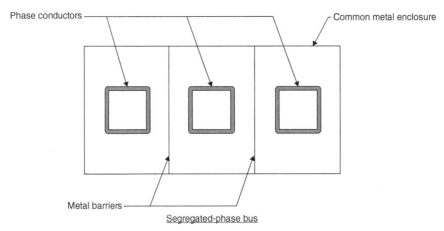

Figure 14.9a   Segregated phase bus duct. *Source*: Reproduced with permission of IEEE.

Figure 14.9b   Segregated phase bus duct. *Source*: Reproduced with permission of Crown Electric Engineering and Manufacturing LLC.

a phase-to-phase fault is only limited by the generator reactance. The subtransient reactance (X_s'') of the typical generator may be in the order of 15 to 20% leading to a first few cycles of current being about 500–600% of the rated current of the machine. The magnetic forces associated with this fault current are the square of the current. These large currents can quickly damage a machine, especially in the end turn region of the generator. That is the reason that most utility generators use isolated phase bus duct (sometimes called isophase) for protection against phase-to-phase faults. Figure 14.10a and Figure 14.10b show a section of isolated phase bus duct.

For points that connect up to enclosures that contain hazardous gases such as generator lead boxes where the generator casing contains hydrogen, seals are provided and the bus duct is purged with air to ensure that an explosive amount of gas does not collect in the phase bus duct. For bus duct that passes through walls with fire ratings (common occurrence for the GSU where there is a firewall between the GSU

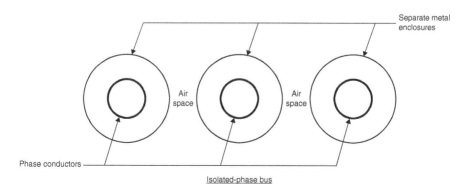

Figure 14.10a   Isolated phase bus duct. *Source*: Reproduced with permission of IEEE.

Figure 14.10b Isolated phase bus duct. *Source*: Reproduced with permission of Crown Electric Engineering and Manufacturing LLC.

and the generator) the barrier of the bus duct must have the same fire rating as the wall. NFPA 850: Recommended Practice for Fire Protection for Electric Generating Plants and High Voltage Direct Current Converter Stations recommends a 2 hour fire rating for this barrier.

IEEE/ANSI C37.23 standard covers the design, manufacture, testing, and rating of the all-metal enclosed bus duct, including cable duct systems. This standard also contains information on the associated equipment such as inter connections, enclosures, switches, supporting structures, and disconnecting links.

Typical values for thermal currents and insulation tests for bus duct systems are listed below in the Table 14.3.

In addition to the above bus systems, there is one system used for the transfer of large amounts of power using cable systems, but installed in an enclosure with adequate bracing to support the cable system during a fault. This is known as a cable

TABLE 14.3 Standard Bus Duct Ratings for Non-segregated Phase Bus Duct

Standard Bus Duct Ratings				
Rated Voltage (kV)	BIL (kV)	60 Hz Hipot (kV)	DC Hipot (kV)	Thermal Current (A)
0.635	N/A	2.2	3.1	1200, 1600, 3000, 4000
4.76	60	19	27	1200, 1600, 3000
15	95	36	50	1200, 1600, 3000
25.8	125	60	85	1200, 2000
38	150	80	N/A	1200, 2000

Source: Reproduced with permission of IEEE.

bus duct system. This is similar in design to the non-segregated phase bus duct, but instead of bus work providing the path for conduction, cables are installed in the enclosure to provide the path for current flow.

CABLES

Cable selection is an important part of ensuring a safe and reliable electrical distribution system installation. There are many factors that are part of specifying the correct type of cable for any application. The one factor most engineers think about is the requirement to ensure that the cable insulation is thermally rated for the magnitude of the current flowing in the conductor. Another factor that should be considered is the type of load (whether non-motor or motor load). NEC® requires that cables for motor feeders must be rated at a minimum of 125% of full-load current of the motor due to the ability of the motor to draw greater than full-load current for short periods of time under normal operating conditions. Another consideration to selecting cable parameters would be to allow for any future growth on the feeder. For instance, consider a feeder that feeds a motor control center. A motor control center (MCC) is an assembly of one or more enclosed sections having a common power bus which contains multiple motor control units and some non-motor control units. Typical practice is to have some installed spare capability for future loads in an MCC. Additionally, motor control centers are designed to allow for the addition of vertical sections for future growth. Therefore, the feeder should not be sized just for the load connected to the MCC today, but the maximum capability of the bus of the MCC. This might be done to ensure that, during future electrical distribution system growth, the size of the feeder cable does not restrict the use of the MCC and we can obtain the full-load capability of the MCC.

Another important item to consider is the fault clearing time that may be imposed on a cable. If a fault occurs on the load side of a power cable, we want to ensure that we have adequately sized the cable to be able to withstand the fault current magnitude for the maximum time that the relaying and breaker or fuse protection needs to clear the fault without damage to the cable. If the cable becomes damaged during a fault downstream, it can have an adverse effect on plant operations in both dollars and down time of the equipment to replace the cable. Voltage drop in the cable is an important issue especially in motor-type loads. For a non-motor load, as voltage is reduced, the power drawn (and the current drawn) is reduced. Most non-motor loads appear as constant impedance loads such as resistive heating elements. Other non-motor loads draw a constant current regardless of the variation of voltage such as battery chargers. By comparison, motor loads appear as constant power loads while running. Remember in an application where a motor is connected to a load, the load mechanical power is proportional to load torque and load rotational speed. If the torque remains constant and speed remains constant, then the load is trying to pull a fixed amount of power from the motor. If we reduce the voltage to a motor under load, the power requirement of the load does not change, so the current drawn by the motor will increase, making the voltage drop condition worse.

Another issue that arises with voltage drop is motor starting. When a motor is running near full load, the current drawn by the motor is at or below the motor full-load current and the power factor of the motor is close to 1.0 (depending on application and motor load). However, when a motor starts, the current drawn by the motor can be six to eight times the value of motor full-load current. Additionally, until the motor begins to rotate and provide power to the connected load, the amount of real power drawn by the motor is small. The combination of high current and low real power causes the initial power factor of a motor while starting to be very low (power factor of 0.1 or lower are not uncommon). Most engineers are aware of the concept that the larger magnitude of current drawn in the line during motor starting will result in a larger voltage drop at the motor ($V = I \times R$). Not so obvious, however, is the fact that the low power factor of a starting motor will also cause a significant additional increase on the voltage drop at the motor in addition to the amount due to current magnitude. Let's explore this concept with an example.

First let us model an energized motor running near full load. For our example, we will model a 480 V, three-phase, 7.5 HP motor with a full-load current of 10 A and a locked rotor of 60 A. To develop our single-line equivalent circuit, we will divide our three-phase voltage by 1.732 to find the single-line equivalent circuit voltage of 480 V/1.732 = 277 V. Additionally, since we defined our motor full-load current as 10 A, the motor running impedance can be estimated by using an infinite source voltage (277 V) divided by the motor full-load current (10 A), or our running internal impedance is 277 V/10 A = 27.7 ohms. Since we know the motor operates near unity power factor when fully loaded, we will model this impedance as a pure resistance. The single-line equivalent circuit is shown in Figure 14.11.

For this energized condition, we can calculate the magnitude and phase angle of the source current I_S as

$$I_S = E_S/(R_M + jX_L) = 277\angle 0°/(27.7 + j\,2.77) = 277\angle 0°/27.8\angle 4.68°$$
$$I_S = 9.96\ A\angle - 4.68°$$

The equivalent voltage at the motor is the voltage drop across the system inductance due to line current subtracted from the source voltage.

$$V_M = E_S - I_S \times jX_L = 277\angle 0° - 9.96\angle - 4.68° \times 2.77\angle 90° = 277\angle 0° - 27.62\angle 85.3°$$
$$V_M = (277 - 27.56\cos(85.3)) - j27.56\sin(85.3) = 275.15 - j22.55 =$$
$$V_M = 276\angle - 4.67°$$

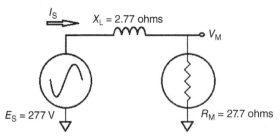

Figure 14.11   Single-line equivalent circuit for an energized 480 V, three-phase, 7.5 HP motor operating near full speed.

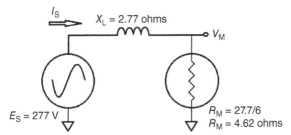

Figure 14.12   Single-line equivalent circuit for starting 480 V, three-phase, 7.5 HP motor assuming a high power factor.

So the magnitude of the single-line equivalent voltage at the motor while the motor is energized drawing close to full-load current is 276 Vac.

The percentage voltage drop from the source to the motor for the running condition is found to be

$$V_D = 100\% \times (V_S - V_M)/V_S = 100\% \times (277 - 276)/277 = 0.33\%$$

Notice there is very little voltage drop across the system when this motor is operating near full load. Next, we will calculate the locked rotor condition. However, even though we know the motor starts with a low power factor, to evaluate the effect power factor has, let us initially calculate system parameters with locked rotor current, but the same high power factor that we see when the motor is running. Since we are assuming a high power factor, we can still represent the motor impedance as a resistance. The single-line equivalent circuit is shown in Figure 14.12.

For this starting condition, we first find the equivalent motor impedance by either dividing the infinite source voltage by the locked rotor current (277 V/60 A = 4.62 ohms) or we could simply divide the running impedance by a factor of 6 (27.7 ohms/6 = 4.62 ohms). In this example, since we are evaluating the system given a motor with a high power factor, we can calculate the magnitude and phase angle of the source current I_S as

$$I_S = E_S/(R_M + jX_L) = 277\angle 0°/(4.62 + j2.77) = 277\angle 0°/5.14\angle 26.2°$$
$$I_S = 53.84\ A\angle - 26.2°$$

The equivalent voltage at the motor is the voltage drop across the system inductance due to line current subtracted from the source voltage.

$$V_M = E_S - I_S \times jX_L = 277\angle 0° - 53.84\angle - 26.2° \times 2.77\angle 90° = 277\angle 0° - 122.2\angle 63.8°$$
$$V_M = (277 - 122.2\cos(63.8)) - j122.2\sin(63.8) = 223.1 - j109.7 =$$
$$V_M = 248\angle - 23.8°$$

So the magnitude of the single-line equivalent voltage at the motor while the motor is starting (assuming a starting power factor close to unity) is 248 Vac.

The percentage voltage drop from the source to the motor for the starting condition assuming a high power factor during start is found to be

$$V_D = 100\% \times (V_S - V_M)/V_S = 100\% \times (277 - 248)/277 = 10.3\%$$

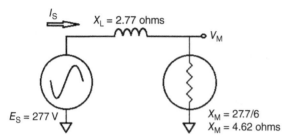

Figure 14.13   Single-line equivalent circuit for starting 480 V, three-phase, 7.5 HP motor assuming a low power factor.

We can see from the above calculation that larger magnitudes of currents seen when starting a motor, even with a constant power factor, will cause a significant voltage drop at the motor. However, in actuality we know that, until the motor begins to rotate and deliver power to the load, the starting power factor of the motor is very small (typically below 0.1). Therefore, the motor impedance during starting appears reactive, not resistive. Let us assume the same locked rotor impedance value of the motor, but now treat it as inductance and once again model the system to find out what effect a low power factor has on voltage drop. The single-line equivalent circuit is shown in Figure 14.13.

For this starting condition, we will use the same impedance we found for the locked rotor condition previously, but this time, we will treat this impedance as inductive due to the low power factor the motor actually presents to the electrical system while in a locked rotor condition. We can calculate the magnitude and phase angle of the source current I_S as

$$I_S = E_S/(jX_M + jX_L) = 277\angle 0°/(j4.62 + j2.77) = 277\angle 0°/6.89\angle 90°$$
$$I_S = 40.22 \text{ A}\angle - 90°$$

Notice that with the same value of locked rotor impedance for the motor (4.62 ohms), but now treating the impedance as reactive instead of resistive, we have a lower value of system current. With the motor drawing mostly reactive power, the line current is 40.22 A. When we calculated the same motor impedance, but assumed it was drawing mostly real power, the line current was 53.8 A. (This is the same principle why current limiting reactors are used instead of current limiting resistors to limit fault currents in some electrical system applications.) Before we continue to calculate the voltage drop, one might initially think that the lower current drawn from the system might lead to a lower voltage drop across the system and subsequently, the motor voltage may be higher for the case where the motor is modeled as a low power factor. Let us continue our modeling effort and find out of if this is true or false.

The equivalent voltage at the motor is the voltage drop across the system inductance due to line current subtracted from the source voltage.

$$V_M = E_S - I_S \times jX_L = 277\angle 0° - 40.22\angle - 90° \times 2.77\angle 90° = 277\angle 0° - 94.3\angle 0°$$
$$V_M = 185.7\angle 0°$$

So the magnitude of the single-line equivalent voltage at the motor while the motor is starting (assuming a low value of starting power factor) is 185.7 Vac.

The percentage voltage drop from the source to the motor for the starting condition assuming a low power factor during start is found to be

$$V_D = 100\% \times (V_S - V_M)/V_S = 100\% \times (277 - 185.7)/277 = 33\%$$

The magnitude of the single-line equivalent voltage at the motor while the motor is starting (assuming that the starting power factor is low) is less (185 V) than when we assumed a high power factor (248 Vac). Even though our magnitude of current drawn from the system was less than before, the voltage drop was greater than previously calculated. Does that make sense? The reason this occurs is due to the relationship between the system impedance and the load impedance. When the motor appears inductive as the system appears, this causes a greater voltage drop at the motor even in the case where the magnitude of current is the same or slightly less.

As a result, both the magnitude of the system current as well as the phase angle of the current (i.e., the power factor of the load) effect the magnitude of the voltage drop at the motor. For the same magnitude of current, when starting a motor, the voltage drop will be larger than if the power factor were high during starting since the reactance of the motor is in phase with the inductance of the system. When a motor is at full speed, the impedance of the motor appears resistive so the voltage drop is less severe.

The routing of the cable makes a difference in the ampacity of the cable as well. Cables in free air versus cable in *conduit* or covered tray or direct buried cables will all have different maximum safe current values due to the effect of cooling on the cable. Also, cables ampacity table ratings are based on a nominal temperature of 30°C. For an ambient temperature other than 30°C, there is a correction factor that should be used to obtain the true ampacity of the cable system. Table 14.4 shows one of several ampacity tables from the National Electrical Code and Table 14.5 shows the derating factor necessary when cables are installed in environments that exceed 30°C. Lastly, if more than three current carrying conductors are installed in a conduit or raceway, there is an additional ampacity derate required by the National Electrical Code due to the combined heating effects of the multiple conductors. For up to 3 current carrying conductors installed, there is no ampacity derating. For raceways with 4 to 6 current carrying conductors installed, the derate is 80%. For raceways with 7 to 9 current carrying conductors installed, the derate is 70%. For higher number of current carrying conductors in a raceway, the reader is encouraged to refer to the latest version of the National Electrical Code.

Alternatively, IEEE 835 (IEEE Standard Power Cable Ampacity Tables) contains tables that specify cable ampacity for various cable sizes, ambient temperatures, and wiring methods used and can be utilized to size conductors.

The two primary conductors used in industry are copper and aluminum and ampacity tables are available for both types of conductors. Aluminum has a higher resistivity than copper and for the same cross-sectional area of metal. Therefore, for the same value of current, aluminum generates more heat than copper. Therefore, for the same insulation system, the ampacity of an aluminum conductor will be less than for the same physical size of copper conductor. Aluminum also has a lower melting

TABLE 14.4 Ampacity of Conductors in Raceway

	Copper Conductors			Aluminum Conductors			
	Temperature Rating of Conductor			Temperature Rating of Conductor			
Size	60°C	75°C	90°C	60°C	75°C	90°C	Size
	Types	Types	Types	Types	Types	Types	
AWG OR kcmil	TW UF	RHW THW THWN	RHH RHW-2 XHHW XHHW-2 XHH / THHW THWN-2 THW-2 "THHN USE-2	TW UF	RHW THW THWN	RHH RHW-2 XHHW XHHW-2 XHH / THHW THWN-2 THW-2 THHN USE-2	AWG OR kcmil
14[b]	20	20	25	—	—	—	—
12[b]	25	25	30	20	20	25	12[b]
10[b]	30	35	40	25	30	35	10[b]
8	40	50	55	30	40	45	8
6	55	65	75	40	50	60	6
4	70	85[a]	95[a]	55	65	75	4
3	85	100[a]	110[a]	65	75	85	3
2	95	115[a]	130[a]	75	90[a]	100[a]	2
1	110	130[a]	150[a]	85	100[a]	115[a]	1
1/0	125	150[a]	170[a]	100	120[a]	135[a]	1/0
2/0	145	175[a]	195[a]	115	135[a]	150[a]	2/0
3/0	165	200[a]	225[a]	130	155[a]	175[a]	3/0
4/0	195	230[a]	260[a]	150	180[a]	205[a]	4/0
250	215	255[a]	290[a]	170	205[a]	230[a]	250
300	240	285	320	190	230[a]	255[a]	300
350	260	310[a]	350[a]	210	250[a]	280[a]	350
400	280	335[a]	380[a]	225	270	305	400
500	320	380	430	260	310[a]	350[a]	500

Source: Reproduced with permission from NFPA70®, *National Electrical Code®*, Copyright © 2014, National Fire Protection Association. This is not the complete and official position of the NFPA on the referenced subject, which is represented only by the standard in its entirety. The student may download a free copy of the NFPA70® standard at: http://www.nfpa.org/codes-and-standards/document-information-pages?mode=code&code=70.

[a]Refer to NEC 310.15(B)(2), which is shown as Table 14.5 in this textbook, for the ampacity correction factors where the ambient temperature is other than 30°C (86°F).

[b]Refer to NEC 240.4(D) for conductor overcurrent protection limitations. For example, 14 AWG copper conductor must be protected at 15 A and 12 AWG copper conductor must be protected at

TABLE 14.5   Ambient Correction Factors for NEC® Ampacity Tables

Ambient Temperature °C	Conductor Temperature			Ambient Temperature of
	60°C	75°C	90°C	
21–25	1.08	1.05	1.04	70–77
26–30	1.00	1.00	1.00	78–86
31–35	0.91	0.94	0.96	87–95
36–40	0.82	0.88	0.91	96–104
41–45	0.71	0.82	0.87	105–113
46–50	0.58	0.75	0.82	114–122
51–55	0.41	0.67	0.76	123–131
56–60		0.58	0.71	132–140
61–70		0.33	0.58	141–158
71–80			0.41	159–176

Source: Reproduced with permission from NFPA70®, *National Electrical Code*®, Copyright © 2014, National Fire Protection Association. This is not the complete and official position of the NFPA on the referenced subject, which is represented only by the standard in its entirety. The student may download a free copy of the NFPA70® standard at: http://www.nfpa.org/codes-and-standards/document-information-pages?mode=code&code=70.

temperature than copper and, as such, when applying aluminum conductors, evaluation of the thermal damage curve for fault currents should be carefully evaluated. So why would anyone use aluminum? The advantages of aluminum over copper are twofold. First, copper has a higher cost than equivalently sized aluminum conductor. This is a benefit to the initial capital cost of the project. Also there is the advantage in that the use of aluminum also reduces the chance of conductors being stolen for their salvage value. This is mentioned remembering that much of the equipment in a power generation facility is located in remote locations and may be rarely visited by operations personnel. Using a material that has a lower risk of theft may lead to a more reliable installation in the long run. The second benefit of aluminum is its reduced weight per unit volume. One application of this may be the cable bus duct discussed previously. Cable bus typically utilizes aluminum bus. Since the aluminum weight is less, this will reduce the overall weight of the enclosed bus system. This may result in reduced bus support requirements.

Another disadvantage of aluminum over copper is the coefficient of thermal expansion. The coefficient of thermal expansion of aluminum is 36% higher than copper. Simply put, for the same temperature rise, the aluminum conductor will expand a larger amount. Because of this the terminations for the aluminum conductor must be rated for use with aluminum. Another disadvantage of aluminum is that aluminum oxidizes into aluminum oxide which is a non-conductive material. Copper, when it oxidizes, results in copper oxide which is still a conductive material. Therefore, when making terminations with aluminum conductors, an anti-oxidizing compound will be used to prevent the formation of non-conductive aluminum oxides.

In regard to conductor insulation systems, there are two basic types of insulation systems, thermosetting compounds and thermoplastic compounds. *Thermoplastic compounds* are materials that will deform when heated and re-harden when cooled. During this process, their shape can change. These materials tend to be more flexible than thermosetting compounds and, from a constructability standpoint, are preferred for ease of installation. Thermosetting compound is a polymer or plastic material (also known as a thermoset) that, once cured, does not change its shape as long as the material temperature remains below a value that will permanently damage the material.

The most commonly used thermoplastic insulating compound is *Polyvinyl Chloride* (PVC). This is a very inexpensive insulation system. One of the main disadvantages of this system is that the PVC, when exposed to heat, in addition to deformation, produces a thick toxic smoke that is hazardous to personnel. It also tends to continue to burn once ignited. In the power generation facility environment, the use of PVCs is prohibited in any indoor area where there may be exposure to personnel and, due to the PVC materials ability to support ignition, PVC insulation systems are not used anywhere in a power plant facility.

Polyethylene is part of a class of insulating materials called polyolefins. Polyethylene is sensitive to moisture when exposed to high voltage stresses and, as such is also not normally used in a power generation facility. This phenomenon is commonly called "water treeing" where moisture under a strong electric field causes the material to form very fine cracks in the insulation system.

Thermosetting compounds are polymer resins that will not soften as the material is heated up and, as such, retains its shape. Insulation systems utilizing thermosetting compounds are more commonly found in utility environments due to the insulation system being more tolerant of higher temperatures.

Cross-linked polyethylene (XLPE) has cross-linking of polyethylene chains which makes the insulation system stronger and very useful in higher temperature applications. The insulation system is ridged and therefore more difficult to install. However, it is less susceptible to water treeing and has very good aging characteristics.

Ethylene propylene rubber (EPR) is a combination of both ethylene and propylene. This construction makes EPR insulation systems more flexible than XLPE insulation systems and easier for installation while still providing the benefits of a thermosetting compound. Table 14.6 shows a comparison of the various types of insulation systems available for cables.

The National Electrical Code® provides classifications of insulation systems that provide guidance with the allowable application of certain insulation systems for both maximum temperature range allowable and whether the cable can be used in wet and/or dry locations.

For EPR or XLPE insulation systems, with or without a jacket, NEC® provides three classifications of use.

Type RHW can be used for up to 75°C maximum conductor temperatures and can be used in either wet or dry locations.

Type RHH can be used for up to 90°C maximum conductor temperatures and can be used in only dry locations.

Type RHW-2 can be used for up to 90°C maximum conductor temperatures and can be used in either wet or dry locations.

TABLE 14.6   Comparison of Various Insulation Systems

Common Name	Chemical Composition	Properties of Insulation	
		Electrical	Physical
Thermosetting			
cross-linked polyethylene	Polyethylene	Excellent	Excellent
EPR	Ethylene propylene rubber (copolymer and terpolymer)	Excellent	Excellent
Butyl	Isobutylene isoprene	Excellent	Good
SBR	Styrene butadiene rubber	Excellent	Good
Oil base	Complex rubber-like compound	Excellent	Good
Silicone	Methyl chlorosilane	Good	Good
TFE[a]	Tetrafluoroethvlene	Excellent	Good
ETFE[b]	Ethylene tetrafluoroethylene	Excellent	Excellent
Neoprene	Chloroprene	Fair	Good
Class CP rubber[c]	Chlorosulfonated polyethylene	Good	Good
Thermoplastic			
Polyethylene	Polyethylene	Excellent	Good
Polyvinyl chloride	Polyvinyl chloride	Good	Good
Nylon	Polyamide	Fair	Excellent

Source: Reproduced with permission of IEEE.
[a]For example, Teflon or Halon.
[b]For example, Tefzel.
[c]For example, Hypalon.

For EPR or XLPE insulation systems, without a jacket, NEC® provides two classifications of use. (These ratings are restricted to installations in conduit.)

- Type XHHW can be used for up to 75°C maximum conductor temperatures in wet locations and can be used for up to 90°C maximum conductor temperatures in only dry locations.
- Type XHHW-2 can be used for up to 90°C maximum conductor temperatures in wet and dry locations.

For PVC insulation systems, with a nylon jacket, NEC® provides two classifications of use.

- Type THWN can be used for up to 75°C maximum conductor temperatures and can be used in either wet or dry locations.
- Type THHN can be used for up to 90°C maximum conductor temperatures and can be used only in dry locations.

For PVC insulation systems, without a nylon jacket, NEC® provides one classification of use. (These ratings are restricted to installations in conduit.)

- Type THW can be used for up to 75°C maximum conductor temperatures and can be used in either wet or dry locations.

TABLE 14.7   Properties of Cable Jacket Materials

Material	Abrasion Resistance	Flexibility	Low Temperature	Heat Resistance	Fire Resistance
Neoprene	Good	Good	Good	Good	Good
Class CP rubber[a]	Good	Good	Fair	Excellent	Good
Cross-linked polyethylene	Good	Poor	Poor	Excellent	Poor
Polyvinyl chloride	Fair	Good	Fair	Good	Fair
Polyurethane	Excellent	Good	Good	Good	Poor
Glass braid	Fair	Good	Good	Excellent	Excellent
Nylon	Excellent	Fair	Good	Good	Fair
ETFE	Excellent	Poor	Excellent	Good	Fair

Source: Reproduced with permission of IEEE.
Note: Chemical resistance and barrier properties depend on the particular chemicals involved, and the question should be referred to the cable manufacturer.
[a] For example, Hypalon.

For metal-clad cable (Type MC) NEC® provides one classification of use. This cable uses a metallic sheath or armor tape for protection of the cable and is usually classified as type XHHW, XHHW-2, RHH/RHW, or RHW-2.

- Type MC cable may be installed in any raceway, in *cable tray*, as open runs of cable, direct buried, or as aerial cable on a messenger.

For power and control tray cable (Type TC) NEC® provides one classification of use. This cable is provided with a flame-retardant non-metallic jacket. Due to its flame-retardant property, this cable is used in power generation facilities in enclosed areas such as cable tray rooms to help control the spread of any cable tray fire and minimize damage to the cable.

- Type TC may be installed in cable trays, raceways, or where supported in outdoor locations by a messenger wire.

As mentioned above in the NEC® classifications, the insulation system is only one of two concerns for a cable system. The other concern is the protection of this insulation system by either a jacket material or by the installation method such as installed in tray or conduit. The typical jacket materials for a jacketed cable are PVC, polychloroprene (neoprene), or chlorosulfonated polyethylene (CPE). When neoprene is used, the location of the cable should avoid areas exposed to petroleum-based products as petroleum-based products have the tendency to soften the neoprene jacket and reduce its effectiveness in protecting the cable insulation system. Table 14.7 summarizes the properties of jacket materials.

There are various wiring methods that can be used to protect cable installations in a power generation facility. One of the more common systems is the use of conduit. *Conduit* or *Rigid Steel Conduit* is a metallic pipe used to protect cabling from physical damage in the existing installation. In addition to rigid steel conduit, there are two basic types of plastic conduit available depending on wall thickness size. Plastic or PVC conduit is available in either schedule 40 (thinner wall thickness) or schedule 80 (thicker wall thickness). As an example, a one-inch schedule 40 PVC conduit has a 0.133-inch minimum wall thickness and 450 psi rating, while the same diameter schedule 80 PVC conduit has a 0.179-inch minimum wall thickness and 630 psi rating.

There is a risk associated with pulling in cables into conduit of cable damage and calculations are done to ensure that the cable maximum pulling tension is not exceeded. Compounds are used to reduce friction between the cable and conduit during the pull to reduce the tension on the cable. Additionally, after the conduit is installed and before the cable is pulled through the conduit, the conduit is deburred or cleared of any abrasions from the installation of the conduit that may present a risk of damage to the cable system to be installed in the conduit. NEC® defines calculations for minimum conduit size requirements for various size and quantity of conductors to be pulled into the conduit system. IEEE 1185 describes two methods for calculating the cable pulling tensions when pulling cable into conduit. We will review both methods later in this chapter.

NEC® requires metallic conduit systems to be bonded with the ground of the source panel. NEC® does allow the metal-to-metal connections to count for this bonding. However, in a power generation facility, a grounding wire is installed along with the cables and bonded at all locations such that the continuity of the mechanical connection of the conduit to equipment is maintained. This is done to ensure that we do not depend on the mechanical connections between the conduit and electrical equipment to pass fault current. This will minimize the possibility of a shock touch potential during a phase-to-ground fault in the field.

Another method for protecting wiring is the use of *Intermediate Metal Conduit* (IMT). This is a special case of conduit where the wall thickness of the conduit is about 50% of the thickness of standard conduit. Since there is less material used for IMT, it is easier to form and install and the initial material costs for IMT is less than conduit. However, it does not provide the same level of physical protection for the cable system as conduit and therefore IMC is not used on a power generation facility.

Electric metallic tubing (EMT) is similar to IMC but with even thinner walls than IMT. Again, it does not provide the level of protection that conduit does. Its main advantage is the reduced cost of material and installation. Since the level of protection from EMT is inferior to that of conduit, EMT is not used in a power generation facility.

Another method of wiring is the use of cable tray. *Cable tray* is a cable support system that provides mechanical protection for the cable. There are several basic types of cable tray systems available today depending on the industry and application. A solid bottom tray (also known as a cable channel or cable trough) provides enhanced physical protection, but may result in reduced air flow and cooling of the cables in the tray. For power applications where the temperature of the cable may be a concern, a vented tray system or ladder tray may be utilized. A ladder tray is constructed in a similar fashion as a ladder with perpendicular bars supporting the cable weight. The tray may or may not be provided with a cover to prevent collection of materials on the cables in the tray.

Cables generate heat while current is flowing through the cable and the cable system depends on the surrounding air to dissipate the heat generated in the cable. NEC® provides direction to the maximum fill allowed in cable tray to ensure adequate cooling of conductors. NEC® also provides ampacity derating values for cable installed in cable trays that are provided with covers to protect the cable from objects above the cable tray as the cover tends to reduce the amount of cooling for the cable

TABLE 14.8   Wiring Methods for Hazardous Classified Locations

Wiring Method	Class I Division		Class II Division		Class III Division
	1	2	1	2	1 or 2
Threaded rigid metal conduit	X	X	X	X	X
Threaded steel intermediate metal conduit	X	X	X	X	X
Rigid metal conduit				X	X
Intermediate metal conduit				X	X
Electrical metallic tubing				X	X
Rigid nonmetallic conduit					X
Type MI mineral insulated cable	X	X	X	X	X
Type MC metal-clad cable		X		X	X
Type SNM shielded nonmetallic cable		X		X	X
Type MV medium-voltage cable		X			
Type TC power and control tray cable		X			
Type PLTC power-limited tray cable		X			
Enclosed gasketed busways or wireways		X			
Dust-tight wireways				X	X

Source: Reproduced with permission of IEEE.

systems in the tray. Supports for cable trays are located every 10–20 ft and are sized for a static load of at least 200 lb.

Cables are clamped or tied down to the cable tray rungs to provide support to the cables and prevent movement of the cable system during fault conditions. This is done to protect the cable from mechanical damage that may occur during movement due to forces incurred during an electrical fault. Just like metallic conduit, metal cable tray is required to be bonded to the ground system. While some metallic cable tray with the correct installation is listed to be used as a bonding connection to ground, in the power generation facility, a ground conductor is run in the tray with the power cables and bonded to each section of the cable tray system to ensure bonding of the tray back to earth and minimize the shock potential of the tray system during a fault.

As mentioned in Chapter 1, some locations in utility electric generation stations are considered hazardous classified areas. NEC® only allows certain wiring methods to be utilized in classified hazardous areas. Table 14.8 lists the NEC®-approved wiring methods for various hazardous classified locations.

Cables installed as direct buried cable underground should be rated for operation under the ampacity of the 69°C rating of the cable regardless of the thermal rating of the insulation system. This is due to the phenomena of thermal runaway. Cables directly buried must dissipate their heat into the surrounding soil. This heat tends to evaporate any moisture in the ground surrounding the cable which reduces the thermal resistivity of the soil surrounding the cable. As such, any cable installed in the ground will find itself in dry soil over time with reduced capability of removing the heat from the conductor. To ensure that the thermal rating of the cable insulation system and jacket material is not exceeded, the cable should be sized to ensure that

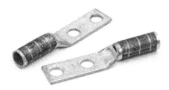

Figure 14.14   Compression lug. *Source*: Reproduced with permission of Eaton/Cooper Industries.

the operating temperature of the cable does not exceed 69°C to minimize the risk of evaporation of the moisture in the soil around the cable.

For reliable cable system installations, proper cable terminations are critical. Terminations of low voltage cables (600 V and below) are with either compression lug or a mechanical lug. A *compression lug* is compressed onto the cable end and mechanically bonds with the cable conductors. Figure 14.14 shows typical compression lugs.

Mechanical lugs on the other hand depend on a set screw being torqued to maintain electrical connection with the conductor. Figure 14.15 shows a mechanical lug.

The compression lug requires the use of a special crimping tool to make the proper connection to the cable system but makes for a longer lasting and lower resistance connection. The mechanical lug is prone to long-term loosening from thermal expansion and contraction as the circuit current is cycled and should be checked for tightness over time. Since the maintenance of the compression lug is less, and the compression lug potentially has greater reliability than the mechanical lug, the compression lug is the preferred method of termination of cables in the power generation facility.

For medium voltage cables, the termination process is complicated by the desire to maintain voltage stresses from the electrostatic field around the cable terminations to a minimum. A stress cone is provided for medium voltage terminations and the end

Figure 14.15   Mechanical lug.

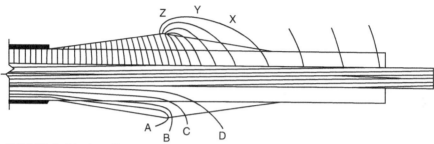

X, Y, Z Dielectric stress lines
A, B, C, D Equipotential lines

Figure 14.16 Stress cone. *Source*: Reproduced with permission of IEEE.

of the cable is carefully prepared to ensure that there are no sharp edges in either the conductor or the semiconducting layer or insulation of the cable and termination kit. As shown in Figure 14.16, the purpose of the stress cone is to control the electrostatic lines of force to ensure that we do not exceed the rated dielectric (*dv/dt*) strength of the insulation. Lines A, B, C, and D display the equipotential lines. By controlling the distance between these lines, we can control the voltage gradient across the insulation.

After termination kits are applied and before medium voltage cables are actually terminated to their end equipment, the cable assembly with terminations undergoes a high voltage test as described in Table 14.9 to ensure that there are no creepage or clearance issues with the cable insulation system including the termination kits applied to the cable ends.

Also, the cable shield may be terminated at both ends or may be terminated at only one end depending on the engineering design. If the power cable shield is terminated at both ends, then there is a derating of the cable ampacity due to circulating currents that may flow on the cable shield. If the cable shield is only terminated on one end, then there is no derate of cable ampacity for the current flowing on the shield but the non-terminated side of the cable must have some type of voltage surge protection as this end will develop voltages on the shield depending on circuit conditions. This can pose a safety hazard at the non-terminated end so in

TABLE 14.9   DC Cable Insulation Testing Levels

L-L System Voltage (kV)	BIL (kV)	Test Voltage (kV)	
		100% Insulation Level	133% Insulation Level
2.5	60	40	50
5	75	50	65
8.7	95	65	85
15	110	75	100
23	150	105	140
28	170	120	—
34.5	200	140	—

Source: Reproduced with permission of IEEE.
Note: These test voltages should not be used without the cable manufacturers' concurrence as the cable warranty will be voided.

power generation facilities, the power cable shield is grounded at both ends and the derating for this wiring method is considered in the sizing of the cable system.

There are three basic types of medium voltage termination kits. A class 1 kit provides a seal for the factory cable from pressure, moisture, and contamination and also provides voltage gradient stress control. A class 2 kit provides a seal for the factory cable from moisture and contamination and also provides voltage gradient stress control. It does not provide a seal for the cable insulation system for pressure. The Class 3 kit provides only stress control and there is no seal provided for pressure, moisture, and contamination. Therefore, class 3 termination kits are only used for clean, dry locations where only voltage stress control is of concern. In a utility generation station, class 3 kits are not utilized.

Cable Installation Engineering Calculations

The first rule is to ensure that we do not exceed the maximum allowable conduit fill allowed by the NEC®. As defined in NEC®, the maximum amount of cross-sectional area in a raceway that a single conductor may occupy is 53% of the raceway area. The maximum amount of cross-sectional area in a raceway that two conductors may occupy is 31% of the raceway area. The maximum amount of cross-sectional area in a raceway that three or more conductors may occupy is 40% of the raceway area. Remember that these are maximum values and other design requirements may reduce the maximum allowable raceway fill area. In addition to NEC® requirements, IEEE provides guidance with recommended installation practices for cables.

IEEE 1185 – Recommended Practices for Cable Installation at Generating Stations provides guidance specific for utility generation sites and is widely used. There are two methods described in IEEE 1185 that can be used to determine the maximum allowable pulling length (and therefore tension) for a cable pull. In addition to maintaining the pulling tension within allowable limits, the sidewall pressure when pulling a cable through a bend must be maintained below values that would potentially damage the cable. Table 14.10 describes the minimum bend radius for non-shielded cable and Table 14.11 provides the minimum bend radius for shielded cable; however, the minimum bend radius may be increased due to the requirement to maintain sidewall pressure below maximum allowable values for a specific cable pull.

The maximum allowable pull tension is provided by the cable manufacturer. If data from the manufacturer are not available, the maximum allowable cable pulling

TABLE 14.10   Minimum Trained Bend Radius for Non-shielded or Non-armored Cable

Non-shielded, Non-armored Cable	Minimum Cable Bend Radius = Multiplier X OD of Cable in Inches (mm)		
Thickness of insulation mils (mm)	<1 in. (<25.4 mm)	1–2 in. (25.5–51 mm)	>2 in. (>51 mm)
155 mils (3.9 mm) or smaller	4	5	6
57 mils (4.0 mm) to 310 mils (12.2 mm)	5	6	7
Over 310 mils (12.2 mm)	N/A	7	8

Source: Reproduced with permission of IEEE.

TABLE 14.11 Minimum Trained Bend Radius for Shielded or Armored Cable

Shielded, Armored Cable	Minimum Cable Bend Radius = Multiplier X OD of Cable
Tape shield	12
Wire shield	8
Interlocked armor	7
Corrugated, welded armor	12
Smooth welded armor	10–15
Extruded aluminum armor	10–15

Source: Reproduced with permission of IEEE.

tension can be calculated as the product of the number of conductors, the conductor area, and a coefficient that depends on the material of the cable (copper or aluminum) and is calculated as shown in equation (14.3).

$$T_{max} = K\, n\, A_c \tag{14.3}$$

where

T_{max} = maximum pulling tension in lbf (N)

A_c = conductor area in circular mils (mm^2)

n = number of conductors

K = 0.008 lbf/cmil (70.27 N/mm^2) for copper

K = 0.006 lbf/cmil (52.71 N/mm^2) for aluminum

When pulling a cable through a conduit system, the pull cord should be attached to both the conductor and the jacket and not just the jacket. Attaching the pull cord to only the jacket will damage the insulation of the cable. The pull should be from the longest, straightest section of conduit first. Pulling the cable from the longest, straightest section will result in the lowest amount of pull tension at the final end of the conduit system.

The pulling tension on the cable for a pull through a straight section of conduit is the product of the weight of the cable and the coefficient of friction between the cable and the conduit and is calculated as shown in equation (14.4).

$$T = W \times 0.5 \tag{14.4}$$

where

T = tension required for straight length (lb)

W = weight of cable total (lb)

0.5 = estimate of friction coefficient (with lubrication)

To calculate the total weight (W) of the cable, multiply the length of the cable by the weight of the cable per unit length that is provided by the cable manufacturer.

Example 14.2 For a straight pull of 100′, pulling a cable weighing 2 lb/ft, what is the pulling tension?

Solution: Using equation (14.4), we find

$T = W \times 0.5$

$T = (2 \text{ lb/ft}) \times (100 \text{ ft}) \times 0.5$

$T = 100 \text{ lb}$

The pulling tension on the cable for a pull through a bend in the conduit is the product of the tension of the cable entering the bend and the exponent of the product of the bend angle in radians and the coefficient of friction and is calculated as shown in equation (14.5).

$$T_c = T_1 \times e^{(0.5 \times a)} \tag{14.5}$$

where

T_c = tension out of a bend in a conduit system, lb

T_1 = tension entering a bend in a conduit system, lb

a = angle of bend, rad (degree / 57.3)

0.5 = estimate of friction coefficient (with lube)

e = exponential function

Example 14.3   What is the pull tension for a cable pull through a 90° bend where the tension at the bend inlet is 100 lb?

Solution: Using equation (14.5), we find

$T_c = T_1 \times e^{(0.5 \times a)}$

$T_c = 100 \text{ lb} \times e^{(0.5 \times (90/57.3))}$

$T_c = 219 \text{ lb}$

When pulling multiple cables through a conduit system, care should be utilized to ensure that the cable will not tend to jam in the conduit. NEC® provides for a calculation of the jam ratio which is simply the ratio of the inside diameter (ID) of the conduit to the outside diameter (OD) of the cable. For jam ratios of 2.8 to 3.0, the cable tends to jam in the conduit. NEC® recommends a jam ratio value of greater than 3 to ensure that cables will not become jammed in the conduit system and the cable pull tension become exceeded.

The above formulas for calculation of the pull tensions for a certain cable pull are well known and are most commonly used. These calculations do not take into account the direction of the cable pull. For example, if we pull a cable in a downward direction (along with the direction of the force of gravity) or if we pull in an upward direction (along with the direction of the force of gravity) the above calculations provide the same result. However, when pulling a cable in a downward direction, the weight of the cable actually provides a driving force and the overall pulling tension is less. Therefore, the calculation method above provides a conservative value. IEEE 1185 provides an alternate method for calculation to ensure we do not exceed the maximum pulling tension of a cable pull into a conduit system while at the same time taking into account the direction of the pull for a more realistic result. In this system, instead of calculating the pull tension value and comparing this to the maximum allowable pull tension, the IEEE 1185 method calculates the "effective conduit length" of a certain direction of pull into a conduit system and then provides the "maximum conduit length" that a cable can be pulled into given a certain side wall bearing pressure (SWBP), coefficient of friction, conduit size and total degrees of bend in the pull. Since this method takes into account the direction of the pull, this method provides a more realistic (less conservative) result than the first procedure.

To use the IEEE 1185 second method, first we must calculate the slope adjustment factor (SAF). This is a factor used to calculate the adjustment to the conduit "effective" length taking into account the direction of the pull. The SAF is calculated as shown in equation (14.6).

$$SAF = [\sin \Theta + K' \cos \Theta]/K' \tag{14.6}$$

where

SAF = slope adjustment factor (see Table 14.12)

Θ = angle of the slope from horizontal (degrees)

K' = effective coefficient of friction

Table 14.12 shows the results of this calculation for slope angle of 15, 30, 25, 60, and 90 degrees.

Now, we can use the SAF to convert the conduit actual length to a conduit effective length depending on the direction of the cable pull through the conduit. If the conduit pull is horizontal ($\Theta = 0$), then the SAF = 1 and the effective conduit length is the same as the actual conduit length. If the conduit pull is vertical in the upward direction (against gravity), then $\Theta = 90$, the SAF = 2 for a coefficient of friction of 0.5, and the SAF = 3.33 for a coefficient of friction of 0.3. In this case, the effective conduit length is either two times the actual length for the case where the coefficient of friction is 0.5 or the effective conduit length is 3.33 times the actual length for the case where the coefficient of friction is 0.3. If the cable pull through the conduit system is in the upward direction at some angle other than 90 degrees, then use either the formula 14.6 to calculate the SAF or Table 14.12 to determine the

TABLE 14.12 Table of Slope Adjustment Factor (SAF)

Slope Angle	Effective Coefficient of Friction (K')		
Degrees	Radians	0.5	0.3
15	0.26	1.48	1.83
30	0.52	1.87	2.53
25	0.44	1.75	2.32
60	1.05	2.23	3.39
90	1.57	2	3.33

Source: Reproduced with permission of IEEE.

TABLE 14.13   Development of Effective Conduit Length Comparison

Type of Conduit Section	"Effective" Conduit Length
Horizontal Conduit Sweep	As measured
Conduit bend	Not included (till determination of maximum allowable effective conduit length)
Vertical conduit up	Multiply by SAF (2 for $K' = 0.5$, 3.33 for $K' = 0.3$)
Vertical conduit down	Not included
Slope up	Multiply by SAF
Slope down	As measured

Source: Reproduced with permission of IEEE.

SAF and multiply the actual cable length by the SAF to determine the effective cable length.

If the conduit pull is vertical in the downward direction, note that gravity is effectively pushing the cable through the conduit. Therefore, the pull tension added for this section does not add to the overall pull tension and for a vertical pull in the downward direction, the effective conduit length is zero. If the cable pull through the conduit system is in the downward direction at some angle other than 90 degrees (case described above), then use the actual conduit length as the effective conduit length. Table 14.13 summarizes the conversion of the actual conduit length to the effective conduit length depending on the direction of the conduit section and type (straight or bend).

Once the effective conduit length has been calculated, then the tables in IEEE 1185 provide information the maximum allowable effective conduit length for a cable pull for a given value of side wall bearing pressure (SWBP), coefficient of friction (K'), conduit trade size, and the total degrees of conduit bends in the overall pull. Notice that in the calculation of effective conduit length, we never performed a calculation for pull tension through conduit bends. This is because in the results where we determine if the effect length of the conduit system is less than the cable maximum allowable effective length, this table provides the correction to the maximum effective conduit length depending on the total number of degrees of conduit bends in a conduit system. Table 14.14 shows one of these tables from IEEE 1185 and will be used in this text book as an example. The reader is encouraged to use IEEE 1185 for other values of SWBP and coefficient of friction values to determine the maximum effective conduit length.

Example 14.4   A control cable is being pulled into $3''$ trade size conduit. See Figure 14.17 for arrangement and cable pull direction for this example. SWBP of cable is 500 lbf/ft and coefficient of friction is not known but assumed to be no greater than 0.5. Determine the following.

a. The total effective conduit length

b. The total degrees of bend

c. The maximum allowable effective conduit length for cable specified

d. The maximum pulling tension for application

TABLE 14.14 Conduit-Cable Pulling Chart for Control Cable SWBP = 500 lbf/ft and K′ = 0.5

Conduit Trade Size	Maximum Effective Conduit Length (ft)						Maximum Allowable Pulling Tension (lbf)
	Total Degrees of Conduit Bend						
	45°	90°	180°	270°	315°	360°	
³⁄₄	935	631	288	131	89	60	66
1	754	509	232	106	71	48	132
1 ¹⁄₂	483	326	149	68	46	31	310
2	327	221	101	46	31	21	353
2 ¹⁄₂	251	169	77	35	24	16	386
3	200	135	62	28	19	13	478
3¹⁄₂	173	117	53	24	16	11	551
4	142	96	44	20	13	9	583
5	138	93	43	19	13	9	895
6	120	81	37	17	11	8	1124

Source: Reproduced with permission of IEEE.
Minimum one single conductor 14 AWG or one multiple conductor 14 AWG conductor size.

Solution: To determine items (a) and (b), we first need to set up a table to list each pull, the actual length and the direction so we can convert the actual conduit length to the effective conduit length. The first pull is horizontal, so the actual conduit length (10 ft) and effective length are the same. The second pull is also horizontal so the actual conduit length (5 ft) and effective length are the same. The third pull is vertically down, so the effective length is zero. The fourth pull is horizontal so the actual conduit length (10 ft) and effective length are the same. Then we sum up the

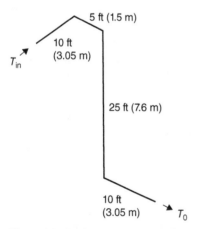

Figure 14.17 Conduit system arrangement for Example 14.4. *Source*: Reproduced with permission of IEEE.

TABLE 14.15 Summary of Results for Example 14.4

Section Type	Angle (degree)	Actual Length (ft)	Effective Length (ft)
Straight horizontal		10	10
Horizontal bend	90		
Straight horizontal		5	5
Bend down	90		
Vertical down		25	0
Bend down	90		
Straight horizontal		10	10
End of pull (totals)	270		25

Source: Reproduced with permission of IEEE.

total degrees of bends (to be used when we look up the maximum allowable effective length in the IEEE 1185 tables) and the total effective length of the conduit. The total effective length is found to be 25 ft. See Table 14.15 for summary. The total number of bends is 270 degrees from Table 14.15.

From the problem description, we were given that we are using a three-inch conduit trade size and pulling a cable with a maximum SWBP of 500 lbf/ft. With this information, we can find the maximum effective conduit length for a pull with 270 degrees total bends in table A1a of IEEE 1185 (see Table 14.15) and find the maximum allowable effective conduit length to be 28 ft (c) and the maximum pulling tension to be 478 lbf (d). Since the total effective length was 25 ft and the maximum allowable effective length is 28 ft, this pull is allowable and will not damage the cable. Also, since the maximum pulling tension was 478 lbf and the maximum pulling tension allowed by the cable manufacturer was 500 lbf/ft, this pull is allowable and will not damage the cable.

So does the direction of the pull matter? Let's use the above example, but now we will pull from the other direction to see if our results are the same or different.

Example 14.5   Same application as Example 14.4, but we now pull the cable from the other direction. A control cable is being pulled into 3″ trade size conduit but starting at the bottom. See Figure 14.18 for arrangement and cable pull direction for this example. SWBP of cable is 500 lbf/ft and coefficient of friction is not known but assumed to be no greater than 0.5. What is

a. the total effective conduit length?

b. the total degrees of bend?

c. the maximum allowable effective conduit length for cable specified?

d. Does pull end selection make difference?

Solution: To determine items (a) and (b), we first need to set up a table to list each pull, the actual length, and the direction so we can convert the actual conduit length to the effective conduit length. The first pull is horizontal, so the actual conduit length (10 ft) and effective length are the same. The second pull is vertically up with an actual

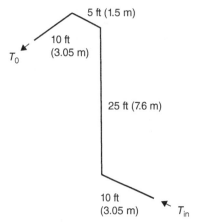

Figure 14.18 Conduit system arrangement for Example 14.5. *Source*: Reproduced with permission of IEEE.

conduit length of 25 ft. With the coefficient of friction of 0.5, the SAF is calculated (or determined from Table 14.12) to be 2, so the effective conduit length is calculated to be 50 ft. The third pull is horizontal, so the actual conduit length (5 ft) and effective length are the same. The fourth pull is horizontal so the actual conduit length (10 ft) and effective length are the same. Then we sum up the total degrees of bends (to be used when we look up the maximum allowable effective length in the IEEE 1185 tables) and the total effective length of the conduit. The total effective length is found to be 75 ft. See Table 14.16 for summary. The total number of bends is 270 degrees from Table 14.16.

From the problem description, we were given that we are using a 3-inch conduit trade size and pulling a cable with an SWBP of 500 lbf/ft. With this information, we can find the maximum effective conduit length for a pull with 270 degrees total bends in table A1a of IEEE 1185 (see Table 14.16) and find the maximum allowable effective conduit length to be 28 ft (c) and the maximum pulling tension to be 478 lbf

TABLE 14.16 Summary of Example 14.5

Section Type	Angle (degree)	Actual Length (ft)	Effective Length (ft)
Straight horizontal		10	10
Bend up	90		
Vertical up		25	50
Bend down	90		
Straight horizontal		5	5
Horizontal bend	90		
Straight horizontal		10	10
End of pull (totals)	270		75

Source: Reproduced with permission of MIL-HDBK-1004/10

(d). Since the total effective length was 75 ft for pulling the cable in the other direction and the maximum allowable effective length is 28 ft, this pull is not allowable and will damage the cable.

Notice that the direction of a cable pull can make a large difference in the maximum pull tension and side wall pressure of the cable, especially when the pull is not completely horizontal. The direction of the pull should be considered when designing a cable pull into a conduit.

Lastly we will review some laws of statics and how they pertain to cable pulls. To pull cable into conduit and tray, pulleys and rollers are used to minimize the tension on the cable as it is pulled into the wiring system. The supports for these devices must be designed to handle the maximum forces that they will see during the cable pull. When an item is in a state of rest, the sum of the forces must sum to zero. Force is defined as the product of mass and acceleration as shown in equation (14.7).

$$F = M a \tag{14.7}$$

where

F = force

M = object mass

a = acceleration of the object

For the object to be at rest, then the acceleration must be zero. If the acceleration is zero, the net force from equation (14.7) states that the net force on the object is zero. Therefore the sum of the forces in the X and Y plane must equal zero.

Example 14.6   We have a cable running through a pulley making a 90 degree bend. We calculate that there will be a maximum of 10 lbf on the cable as shown in Figure 14.19. The pulley we have provided has two support members, one in the X direction and one in the Y direction. Calculate the force that the pulley supports experience both the support in the X direction and the support in the Y direction.

Solution: Looking at Figure 14.19, we have 10 lbf in the X direction and 10 lbf in the Y direction due to cable tension. Therefore, the pulley support member must provide 10 lbf in the opposite X direction and 10 lbf in the opposite Y direction. Therefore, the X support will experience 10 lbf and the Y support will experience 10 lbf.

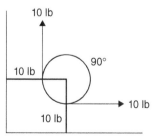

Figure 14.19   Forces on pulley with cable at 90 degrees with two supports.

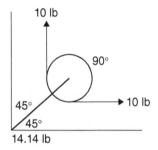

Figure 14.20   Forces on pulley with cable at 90 degrees with one support.

Example 14.7   We have a cable running through a pulley making a 90 degree bend. We calculate that there will be a maximum of 10 lbf on the cable as shown in Figure 14.20. The pulley we have provided has one support member that is 45 degrees from the X direction and 45 degrees from the Y direction. Calculate the force that the pulley support will experience from the cable tension.

Solution: Looking at Figure 14.20, we have 10 lbf in the X direction and 10 lbf in the Y direction due to cable tension. Therefore the pulley support member must provide 10 lbf in the opposite X direction and 10 lbf in the opposite Y direction. Using Pythagoreans theorem, the total force is

$$F = \sqrt{(X^2 + Y^2)} \text{ pounds force}$$
$$F = \sqrt{(10^2 + 10^2)} \text{ lbf}$$
$$F = 14.14 \text{ lbf}$$

The support will experience a net force of 14.14 lbf.

Example 14.8   We have a cable running through a pulley making a 180 degree bend. We calculate that there will be a maximum of 10 lbf on the cable as shown in Figure 14.21. The pulley we have provided has one support member that is parallel with the X direction. Calculate the force that the pulley support will experience from the cable tension.

Solution: Looking at Figure 14.21, we have two sources of 10 lbf in the X direction so the total force of the cable on the pulley in the X direction is 20 lbf. Therefore, the support will experience a net force of 20 lb in the X direction.

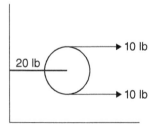

Figure 14.21   Forces on pulley with cable at 180 degrees with one support.

CABLE TESTING

To ensure proper equipment and cable installation, the cable system should be tested. The following section applies to the insulation resistance testing of rotating equipment (AC and DC), power cables and buses, circuit breakers and high voltage switchgear. It does not apply to testing of electrical equipment such as transformers, solid state electronic equipment, and capacitors. Always verify with the original equipment manufacturer (OEM) any insulation testing requirements before commencement of testing to ensure that the equipment warranty is not voided and tests are meaningful.

MEGGER TESTING

Before applying a high potential test set to equipment, a Megger test is applied to detect any defects in the insulation system. The *Megger* is a current-limited test instrument and, as such, if there are any defects in the insulation system, the Megger will detect it without damaging the insulation system. The same may not be true with a high voltage test set. The following is a general procedure for performing insulation testing with a Megger.

- Barricade the limited approach boundary for the equipment under test.
- Select the tester for the proper DC output voltage. Record the test voltage on the insulation resistance (Megger) data record.
- Solidly ground the enclosure of the frame to the station ground.
- Remove all external equipment that may affect insulation resistance readings (e.g., capacitors, lightning arresters, brush rigging, etc.).
- Discharge the specimen to the ground.
- Remove all visible surface dust and moisture from the test specimen when practical.
- Discharge the insulation resistance tester.
- Connect the tester leads.
- Select proper test voltage on the insulation resistance (Megger) tester in accordance with the equipment vendor's instructions. If no specific instructions are available from the vendor, the following test voltages should be selected:

Rated Voltage – AC/DC Volts	Test Voltage - DC Volts
120–250	500
440–600	1000
2.5–4.16 kV	2500
4.16 kV and above	5000

- Announce to all affected personnel that the equipment under test is being tested and to remain outside of the limited approach boundary.
- Verify all personnel are clear of the limited approach boundary.
- Perform the test in accordance with the vendor's instructions.
- Record data.
- Discharge the insulation resistance tester (Megger),
- Disconnect the tester leads, using voltage rated gloves.
- Test equipment under test to verify charge is dissipated before removing limited approach boundary.

There are two acceptance criteria for a successful Megger test. First, the insulation resistance should be a minimum value as shown in equation (14.8) at one minute into the test.

$$R_m = 2 \times kV + 1 \qquad (14.8)$$

where

R_m = recommended minimum insulation resistance in megohms of the entire machine winding (in units of megohms)

kV = rated machine potential, in kilovolts

Second, the polarization index should be greater than 1.5 for class A insulation systems and greater than 2 for class B, F, and H insulation systems. The polarization index (PI) is defined as

$$PI = \text{resistance at 10 minutes/resistance at 1 minute.} \qquad (14.9)$$

A low PI reading may indicate moisture contamination of the insulation system. *Note*: some insulation systems have such a high value that the 1 min reading may already be in the giga-ohm range. In these cases, the PI ratio does not apply.

HIGH POTENTIAL TESTING

Once a successful Megger test is completed, the next step for new installations of wire and cable is to perform a high potential test on the cable insulation system. This paragraph applies to both AC and DC high-potential testing. In the United States, DC high-potential testing has been the preferred method for testing on a periodic basis since it is more than a "go" or "no-go" test, as is the AC test. The AC high potential test is more the accepted standard in countries such as Canada. For more detailed information on high potential testing, the reader is directed to IEEE 62 – IEEE Recommended Guide for Making Dielectric Measurements in the Field. The following is a procedure to perform a high potential test on cable insulation systems.

- Due to high voltage being used, adequate safety precautions should be taken to avoid injury to personnel and damage to property.
- Always use approved and tested safety rubber gloves when connecting and disconnecting test leads to and from the apparatus, and while operating the test equipment.
- Test equipment operation shall be in accordance with the manufacturer's instructions.
- In the case of unshielded cable, the non-uniform electric field between conductor and the metal enclosure (ground) may introduce excessive stresses in the insulation. In the case of unshielded cable, it is therefore recommended that a DC test only be carried out.
- Shielded power cables require electrical stress control when they are terminated. Cables shall be made up with either stress cones or with heat-shrinkable stress control material.
- The cable route or the equipment should be walked down to establish safety boundaries as necessary for personnel and equipment protection.
- Ensure that all voltage transformers are disconnected for the duration of the test and that all current transformers are short-circuited.
- Continuity and phasing of all cables and switchgear bus work should be checked, where practical.
- Provision should be made to ground equipment through a resistance by means of a switch stick after Megger or high potential testing.
- Ensure that all cable shields and switchgear grounds have been made per the installation instructions.
- Conductors of each like phase may be tested together, provided the source has adequate capacity to maintain the specified test voltage for the required period of time. Conductors not being tested shall be tied together, using a soft wire, and grounded at the test end.
- Phases should be properly identified and labeled.
- This test should be performed only by personnel who are properly trained on the test equipment, safety rules, testing procedures, and interpretation of the results.
- Determine the test voltage, kVA, and times and voltage steps to be used (refer to IEEE 62 – IEEE Recommended Guide for Making Dielectric Measurements in the Field).
- De-energize, discharge, and safety tag the cable/equipment to be tested, as necessary.

- Disconnect the cable/equipment from the equipment/cables at each end and position it with maximum possible clearance from objects. Ensure that open cable ends are insulated.

- Ground all equipment, cables, and conductors not under test that are in contact with the cable under test or that can become energized by a flashover.

- Position the test instrument inside the test area, keeping the high voltage lead between the test instrument and the cable/equipment being tested.

- Connect the test set in accordance with the manufacturer's instructions.

- Set the ammeter to its highest range.

- Set the voltage to its lowest range.

- Barricade the limited approach boundary for the equipment under test.

- Announce to all affected personnel that the equipment under test is being tested and to remain outside of the limited approach boundary.

- Verify all personnel are clear of the limited approach boundary.

- Slowly increase the test voltage until the desired test level is reached. During the voltage buildup, the rate of increase of the applied voltage shall be approximately uniform so that the required time shall be no less than 10 sec and no more than 60 sec for DC high potential testing and shall not exceed 10 sec for AC high potential testing.

- *Note*: If cable/equipment insulation failure should occur, as indicated by the high ammeter reading, secure the test voltage immediately.

- Measure and record leakage current at the end of each step. Monitor for any step increases in leakage current and, if leakage current suddenly increases, stop the test.

- Once maximum test voltage is achieved, maintain maximum test voltage as indicated in IEEE 62 – IEEE Recommended Guide for Making Dielectric Measurements in the Field.

- Upon completion of the test, reduce test voltage continuously, but not abruptly.

- Turn off the test instrument and allow the voltage to decay for the minimum time in accordance with the test instrument instructions.

- *Note*: Both the cable conductor and the shield may store a charge after high potential testing. Treat both the cable and the shield as energized until they have been grounded for a long enough period to ensure any stored charge has been dissipated.

- Ground the cable/equipment (use a discharge rod) and allow the ground to remain in place for a period of at least 3 to 5 min.

- Remove the test connections.

- Test equipment under test to verify charge is dissipated before removing limited approach boundary.

- Restore the cable to its as-found condition.

- Remove all grounds installed previously.

ACCEPTANCE

The cable insulation system tested shall withstand the imposed potentials for the prescribed periods of time as indicated in the IEEE standards without distress or failure.

Note: If the leakage current versus voltage are plotted as a curve, as long as this plotted curve is linear for each step, the insulation system is in good condition. At some value of step voltage, if the leakage current begins to increase noticeably, an increase in the slope of the curve will be noticed. If the test is continued beyond this test voltage, the leakage current will increase even more rapidly and immediate breakdown may occur in the cable insulation. Any substantial increase in leakage current that is associated with an increase in applied voltage should be an indication of potential failure of cable and the test should be stopped.

CATHODIC PROTECTION

All metals undergo a process of corrosion which is a result of an electrochemical reaction. When two dissimilar metals or materials meet and there is a conducting path between the metals, there is the potential for one metal to give up electrons (the anode) and for the other metal to accept these electrons (the cathode). This results in corrosion of the anode material that gives up its electron in the chemical reaction.

$$M_o \rightarrow M^+ + e^- \tag{14.10}$$

where

M_o = metal atom

M^+ = metal ion

e^- = electron

At the cathode, the available electron reacts with an ion in solution to form a reduced ion.

$$R^+ + e^- \rightarrow R_o \tag{14.11}$$

where

R^+ = positive ion in solution

e^- = electron

R_o = reduced atom

The goal of a *cathodic protection* system is to more easily provide this electron than the natural metallic anode. This can be done via a passive method or an active method.

The passive cathodic protection system consists of a *sacrificial anode* that more easily gives up its free electrons than the metal being protected. In this method, the sacrificial anode is corroded away instead of the metal being protected. The benefit of the passive cathodic protection system is that there is no external power supply

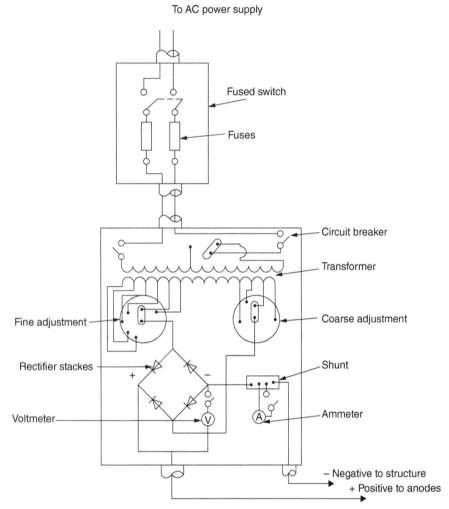

Figure 14.22 Typical active cathodic protection circuit. *Source*: Reproduced with permission of Naval Facilities Engineering Command.

required to maintain protection, but the drawback of this system is that, occasionally, maintenance is required to replace the sacrificial anode to maintain protection.

The active cathodic protection system (or impressed current cathodic protection) will consist of a step-down transformer and a rectifier. An example of one such circuit is shown in Figure 14.22 The negative terminal (with spare available electrons) is connected to the metal that is being protected and the positive terminal is connected to a separate impressed current anode. When the active cathodic protection is energized, free electrons are supplied from the DC power supply and collect on the metal being protected. These electrons are more easily given up than the electrons associated with the metal and this protects the metal from the effects of corrosion. The benefit of the active cathodic protection system is that there is no need for the occasional sacrificial anode replacement that exists with the passive system. The drawback of the active cathodic protection system is that it requires the external power supply to be maintained and the condition of the impressed current anode can affect the performance of this system. The voltage between the protected metal and the impressed current anode is measured and, as long as there is a minimum of approximately 1 Vdc, the system protection is active. Should this voltage potential decrease (either due to a loss of electric supply or due to a lack of conductivity to the impressed current anode), the controller will alarm indicating a loss of protection for the system.

The decision between using a passive or active cathodic protection system comes down to a decision of feasibility and cost. Systems that require a smaller amount of cathodic current (0.5 A or less per 100 linear feet of structure) will utilize a passive system while systems that require a larger amount of cathodic current (1 A or more per 100 linear feet of structure) will utilize an active system. Additionally, if the required current demand will change substantially over time, then an active system will be used.

Typical electric generation station utility structures that are protected by cathodic protection are any system that comes into contact with non-deionized water such as seawater or any system associated with maintaining a boundary for a hazardous material such as natural gas lines or hydrogen lines, especially where these lines make contact with other material or are directly buried in the ground.

GLOSSARY OF TERMS

- Annual Load Factor – The ratio of the average annual energy usage divided by the peak annual energy usage.
- Busway – A prefabricated assembly of standard lengths of bus bars rigidly supported by solid insulation and enclosed in a sheet-metal housing.
- Cable – An insulated wire or wires having a protective casing and used for transmitting electricity.
- Cable Tray – Cable tray is a rung and ladder system that supports the cable along the run of the cable.
- Cathodic Protection (CP) – Cathodic protection is a technique used to control the corrosion of a metal surface by making it the cathode of an electrochemical cell.
- Compression Lug – A compression lug is compressed onto the cable end and mechanically bonds with the cable conductors.
- Conduit – Conduit is a metallic or plastic pipe used to protect cabling from physical damage in the existing installation.
- Cross-Linked Polyethylene (XLPE) – XLPE has cross-linking of polyethylene chains which makes the insulation system stronger and very useful in higher temperature applications.

- Daily Load Factor – The ratio of the average daily energy usage divided by the peak daily energy usage.

- Electric Metallic Tubing (EMT) – EMT is similar to Intermediate Metal Tubing (IMT) but with even thinner walls than IMT.

- Ethylene Propylene Rubber (EPR) – EPR is a combination of both ethylene and propylene. EPR is more flexible than XLPE insulation systems but provides the benefits of a thermosetting compound.

- Expanded Radial Distribution System – Similar to the simple radial system in that the path for energy flow from source to load is radially from the source and there is only one path to each load, except the load is now distributed between several buses.

- Extra High Voltage – A class of nominal system voltages equal from 230,000 V to 1,000,000 V.

- High Voltage – A class of nominal system voltages equal from 100,000 V to 230,000 V.

- Intermediate Metal Tubing (IMT) – IMT is a special case of conduit where the wall thickness of the conduit is about 50% of the thickness of standard conduit.

- Isolated phase bus duct. Bus duct where each phase has its own enclosure and the enclosure is metallic and grounded.

- Low Voltage – A class of nominal system voltages less than 1000 V.

- Mechanical Lug – A mechanical lug depends on a set screw being torqued to maintain electrical connection with the conductor.

- Medium Voltage – A class of nominal system voltages equal to or greater than 1000 V and less than 100,000 V.

- Megger Testing – Megger is current-limited insulation system test instrument and, as such, if there are any defects in the insulation system, the Megger will detect the existence of a defect without damaging the insulation system.

- Nominal System Voltage – The voltage by which a portion of the system is designated and to which certain operating characteristics of the system are related. Each nominal system voltage pertains to a portion of the system that is bounded by transformers or utilization equipment.

- Nominal Utilization Voltage – The voltage rating of certain utilization equipment specifying the acceptable voltage range at the connected load.

- Non-Segregated Phase Bus Duct – Bus bar supported by non-conducting insulators in an overall environmental enclosure.

- Sacrificial Anode – Sacrificial anodes are highly active metals that are used to prevent a less active material surface from corroding. Sacrificial anodes are created from a metal alloy with a more negative electrochemical potential than the other metal it will be used to protect.

- Segregated Phase Bus Duct – Bus that consists of bus bar supported by non-conducting insulators in an overall environmental enclosure and containing phase separation by a grounded metal barrier between each phase.

- Ultra-High Voltage – A class of nominal system voltages in excess of 1,000,000 V.

- Utilization Voltage – The voltage at the line terminals of utilization equipment.

PROBLEMS

14.1 The system configuration shown below is an example of what type of plant distribution system?

    A.   Primary selective system

    B.   Secondary selective system

    C.   Loop system

    D.   Radial system

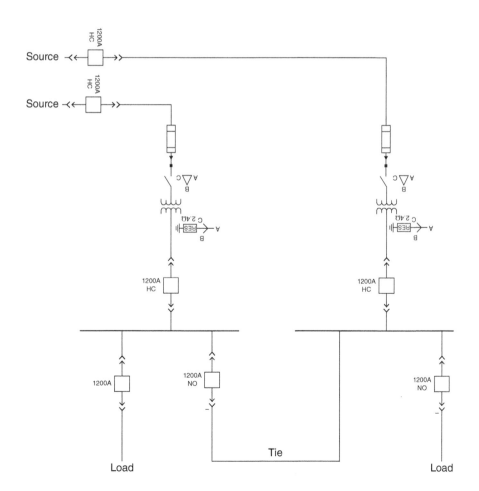

14.2 In the figure below, which figure depicts "segregated phase bus duct"?

    A. Arrangement A

    B. Arrangement B

    C. Arrangement C

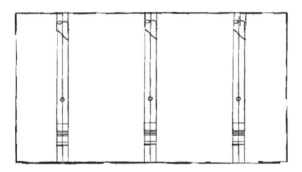

Arrangement A

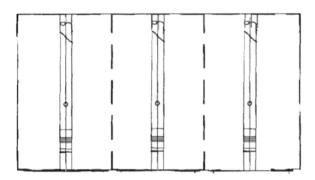

Arrangement B

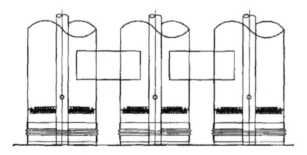

Arrangement C

14.3 For a straight pull of 200′, pulling a cable weighing 1 lb/ft, what is the pulling tension assuming coefficient of friction to be 0.5?

    A. 50 lb

    B. 100 lb

    C. 150 lb

    D. 200 lb

14.4 What is the pull tension for a cable pull through a 90° bend where the tension at the bend inlet is 100 lb assuming coefficient of friction to be 0.5?

    A. 59 lb

    B. 159 lb

    C. 219 lb

    D. 319 lb

14.5 A control cable is being pulled into 30 ft of 3″ trade size conduit. See figure below for physical layout. We have a choice for the direction of the cable pull as shown by part (A) or part (B) of the figure below. SWBP of cable is 500 lbf/ft and coefficient of friction is not known but assumed to be no greater than 0.5. For this problem, use Table 14.14 for cable data. For both the cable pull direction shown in part (A) and the cable pull direction shown in part (B), determine the following.

    a. The total effective conduit length

    b. The total degrees of bend

    c. The maximum allowable effective conduit length for cable specified

    d. The maximum pulling tension for application

    e. Is this cable pull allowable from the direction indicated?

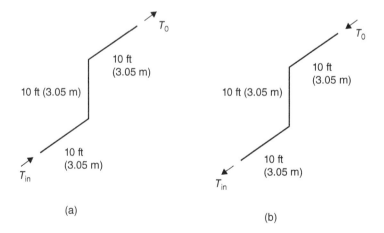

(a)

(b)

14.6 We have a cable that we are pulling with a tension of 20 lb running through a pulley making a 180 degree bend. This is shown in the figure below on the right. That pulley is then connected to another pulley which is shown on the figure below on the left. This second pulley is attached to the wall. Calculate the force that the pulley on the left will impose on the wall from the cable tension of 20 lbs.

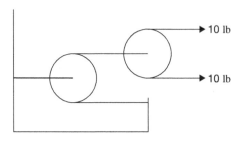

RECOMMENDED READING

Electric Power Plant Design, Technical Manual TM 5-811-6, Department of the Army, USA, 1984.

Electrical Machines, Drives and Power Systems, 6th edition, Theodore Wildi, Prentice Hall, 2006.

The Engineering Handbook, 3rd edition, Richard C. Dorf (editor in chief), CRC Press, 2006.

IEEE 141: IEEE Recommended Practice for Electric Power Distribution for Industrial Plants (IEEE Red Book), 1993.

IEEE 576-2000: IEEE Recommended Practice for Installation, Termination, and Testing of Insulated Power Cable as Used in the Petroleum and Chemical Industry.

IEEE C37.23-2003: IEEE Standard for Metal-Enclosed Bus and Calculating Losses in Isolated-Phase Bus.

IEEE C37.23-2003: IEEE Standard for Metal-Enclosed Bus.

IEEE 1185-2010: IEEE Recommended Practice for Cable Installation in Generating Stations and Industrial Facilities.

Industrial Power Distribution, 2nd edition, Ralph E. Fehr III, Wiley-IEEE Press, 2016. ISBN: 978-1-119-06334-6.

Military Handbook, Electrical Engineering, Cathodic Protection, MIL-HDBK-1004/10, 1994.

NEC® – National Electrical Code®, 2014.

Power Plant Engineering, Black & Veatch, edited by Larry Drbal, Kayla Westra, and Pat Boston, Chapman & Hall / Springer, 1996.

Southwire Power Cable Manual, 2nd edition, 1997.

Standard Handbook of Powerplant Engineering, 2nd edition, Thomas C. Elliott, McGraw-Hill, 1998.

TRANSFORMERS AND REACTORS

$\mathbf{T}$**HE MODERN** power plant requires an internal electrical distribution system to operate all the auxiliary functions needed for reliable operation of the power plant. A typical plant will have a non-critical low voltage (120 Vac) system, a critical (inverter-supplied) low voltage (120 Vac) system, a low voltage three-phase system for smaller motors and loads (480 Vac), a medium voltage three-phase system for the large motors and loads (4, 6, or 13 kVac), and a DC system (48, 125, or 250 Vdc) for both critical control systems as well as critical plant equipment and motors. With the exception of the DC system, the conversion between these different levels of voltage is done with a transformer. The transformer is a very simple machine that has two or more windings mounted on a common core. Before we delve into the details of

Energy Production Systems Engineering: An Introduction for Electrical Engineers to Electrical Power Generation Facilities, Systems, and Equipment, First Edition. Thomas H. Blair.
© 2017 by The Institute of Electrical and Electronics Engineers, Inc. Published 2017 by John Wiley & Sons, Inc.

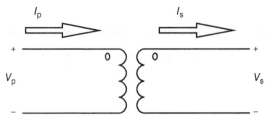

Figure 15.1   Ideal single-phase transformer model.

transformer applications, let us review the fundamentals of transformer theory. The ideal single-phase transformer is represented in Figure 15.1.

In a transformer, the magnitude of voltage developed on a winding is a function of the number of turns the winding makes around the core of the transformer and the rate of change of flux in the core. This is mathematically described by

$$V = N \times (d\varphi/dt)$$

Rearranging for the rate of change of flux, we find

$$(d\varphi/dt) = V/N$$

where

V = voltage

N = number of turns the winding makes around the core

φ = flux

t = time

For the primary side winding, the magnitude of voltage developed is defined by

$$(d\varphi/dt)_P = V_P/N_P$$

For the secondary side winding, the magnitude of voltage developed is defined by

$$(d\varphi/dt)_S = V_S/N_S$$

Since the primary winding and secondary winding are wrapped around the same core, they both see the same rate of change of flux per unit time.

$$(d\varphi/dt)_P = (d\varphi/dt)_S$$

Therefore, we can combine the above two equations to find the relationship between the turns around a core and the voltage developed on the winding as

$$V_P/N_P = V_S/N_S$$

Rearranging, we find that the ratio of primary winding turns to secondary winding turns also defines the ratio of primary voltage to secondary voltage as

$$V_P/V_S = N_P/N_S$$

We define a variable (n) as the turns ratio where the turns ratio of a transformer is simply the number of primary turns to the number of secondary turns or

$$n = N_P/N_S = V_P/V_S \tag{15.1}$$

In the ideal transformer, there are no losses, so the summation of power into the transformer from both primary and secondary windings must be equal to zero (conservation of energy). Mathematically, this is shown as

$$S_P + S_S = 0$$

We define positive voltage as having the positive voltage applied to the polarity terminal as shown in Figure 15.1. We also define positive current as the current flowing into the polarity terminal. Power into the transformer is the product of the voltage and the current. On the primary side, the voltage is positive and the current is positive since it is flowing into the polarity terminal. Therefore, the power into the transformer from the primary side is

$$S_P = V_P \times I_P$$

On the secondary side, the voltage is positive but now the current is negative since it is flowing out of the polarity terminal. Therefore, the power into the transformer from the secondary side is

$$S_S = V_S \times (-I_S)$$

Substituting equations (15.2) and (15.3) into equation (15.1), we find the relationship between primary and secondary voltage and currents.

$$(V_P \times I_P) + (V_S \times (-I_S)) = 0 \quad \text{OR} \quad V_P \times I_P = V_S \times I_S$$

or

$$V_P/V_S = I_S/I_P \tag{15.2}$$

From equation (15.1), we know the relationship between voltage ratio and turns ratio to be

$$V_P/V_S = n \tag{15.3}$$

Substituting equation (15.1) into equation (15.2), we also find the relationship between current ratio and turns ratio to be

$$I_S/I_P = n \tag{15.4}$$

Lastly, if we look at impedance as simply the voltage divided by the current, then the impedance that the transformer sees on the primary side is

$$Z_P = V_P/I_P$$

The impedance that the transformer sees on the secondary side is

$$Z_S = V_S/I_S$$

If we define the impedance ratio as the impedance seen by the transformer primary by the impedance seen by the transformer secondary, then we can mathematically state this as

$$Z_P/Z_S = (V_P/I_P)/(V_S/I_S) = (V_P/I_P) \times (I_S/V_S)$$

Now, let us rearrange this slightly into terms of the ratio of currents and the ratio of voltages.

$$Z_P/Z_S = (V_P/V_S) \times (I_S/I_P)$$

If we substitute the value of the turns ratio we found in equations (15.7) and (15.8) into equation (15.11), we can finally find the relationship between the impedance reflected through a transformer and the turns ratio of the transformer.

$$Z_P/Z_S = (n) \times (n)$$

or

$$Z_P/Z_S = n^2 \tag{15.5}$$

The basic relationships between the transformer's turns ratio and the voltage, current, and impedance of a transformer are summarized in equations (15.3), (15.4), and (15.5).

$$V_P/V_S = n$$
$$I_S/I_P = n$$
$$Z_P/Z_S = n^2$$

where

V_P = voltage applied to the primary winding with positive connected to the polarity terminal

V_S = voltage applied to the secondary winding with positive connected to the polarity terminal

I_P = current flowing into the polarity terminal side of the primary winding

I_S = current flowing out from the polarity terminal side of the primary winding

Z_P = impedance that is seen by the primary side of the transformer and defined as the primary voltage divided by the primary current

Z_S = Impedance that is seen by the secondary side of the transformer and defined as the secondary voltage divided by the secondary current

n = turns ratio of the transformer and is defined as the number of winding turns on the primary side divided by the number of turns on the secondary side of the transformer

Example 15.1   For a 42,000 VA, single-phase transformer that is rated 4200 V on the primary and 120 V on the secondary where the secondary winding is connected to a 1-ohm load, determine the following.

A. Turns ratio (n)

B. Rated primary current

C. Rated secondary current

D. The impedance seen by the circuit connected to the primary of the transformer

Solution:

A. Turns ratio (n): Using equation (15.3),

$$V_P / V_S = n$$
$$n = 4200 \text{ V} / 120 \text{ V}$$
$$n = 35$$

B. Rated primary current:

$$S_P = V_P \times I_P$$
$$I_P = S_P / V_P$$
$$I_P = 42{,}000 \text{ VA} / 4200 \text{ V}$$
$$I_P = 10 \text{ A}$$

C. Rated secondary current:

$$S_S = V_S \times I_S$$
$$I_S = S_S / V_S$$
$$I_S = 42{,}000 \text{ VA} / 120 \text{ V}$$
$$I_S = 350 \text{ A}$$

D. The impedance seen by the circuit connected to the primary of the transformer due to the 1-ohm resistor connected to the secondary circuit can be found utilizing equation (15.5).

$$Z_P / Z_S = n^2$$
$$Z_P = n^2 \times Z_S$$
$$Z_P = (35)^2 \times 1 \text{ ohm}$$
$$Z_P = 1225 \text{ ohm}$$

For more information on applications of transformers in potential and current metering circuits, the reader is directed to Chapter 24 on instrumentation.

With the fundamentals behind us, we turn our attention to some of the practical aspects of transformer applications. The voltage induced on the secondary is a function of the number of turns of conductor around the core and the rate of change of flux in the core. There are two main types of transformers in the industry based on their construction. The first type is the oil-filled transformer. The second type is known as a dry type (there are also other designs such as water-cooled transformers, but their use in the power environment is limited and as such will not be addressed in this text).

The oil-filled transformer for the same kVA rating is a physically smaller device. The oil-filled transformer utilizes oil in its main tank to remove heat from the windings and core of the transformer and also to provide electrical insulation around the windings. The dry-type transformer does not rely on the circulation of oil to remove heat and provide insulation, but instead uses airflow around the insulation of the transformer windings to remove heat from the windings. Since oil is a much better thermal conductor of heat energy than air, for the same kVA rated transformer, the oil-filled transformer is smaller in size than the dry-type transformer. There are various types of oil-filled transformer cooling systems as defined by IEEE. In 2000,

TABLE 15.1 Classes of Oil-Filled Transformer Cooling System Ratings

Class C57.12.00 Before 2000	Class C57.12.00 After 2000	Method of Cooling
OA	ONAN	Liquid immersed, self-cooled
FA	ONAF	Liquid immersed, forced air-cooled
OA/FA	ONAN/ONAF	Liquid immersed, self-cooled/forced air-cooled
OA/FA/FA	ONAN/ONAF/ONAF	Liquid immersed, self-cooled/forced air-cooled/forced air-cooled
OA/FA/FOA	ONAN/ONAF/OFAF	Liquid immersed, self-cooled/forced air-cooled/forced liquid cooled
OA/FOA/FOA	ONAN/ODAF/ODAF	Liquid immersed, self-cooled/forced air, forced liquid cooled/forced air, forced liquid cooled
FOA	OFAF	Liquid immersed, forced liquid cooled with forced air-cooled
FOW	OFWF	Liquid immersed, forced liquid cooled with forced water-cooled
OW	ONWF	Liquid immersed, water-cooled

Source: Reproduced with permission of IEEE

IEEE C57.12.00, which covers the general requirements for liquid-immersed distribution, power, and regulating transformers, was changed to reflect a new method of defining cooling systems of liquid-type transformers. Since large oil-filled transformers have a life well in excess of 30 years, engineers must be knowledgeable of both cooling system classification methods. Table 15.1 shows both the pre-2000 system and the post-2000 system for identifying transformer cooling systems.

The oil-filled transformer brings with it design concerns regarding the flammability of the oil and the potential environmental concern of an oil spill that the dry-type transformer does not bring.

Under the dry-type transformer category, there are two main types of insulation systems. The first is a *vacuum pressure impregnated* transformer winding (VPI). All transformers undergo this process whereby the transformer winding is inserted in a tank. Initially, a vacuum is drawn to pull out the moisture in the winding insulation system. After a certain time, pressure is then applied with an epoxy insulating material to force the insulation material into the voids of the transformer insulation system.

An alternate system is achieved by encasing the entire transformer winding with an epoxy resin. The *epoxy resin transformer* is slightly higher in cost, but has the benefit of a more robust insulation system and reinforcement system. During through faults, the epoxy resin transformer provides more support to the transformer windings to prevent movement and possible failure of the transformer windings during a through fault.

Dry-type transformers use air circulation for heat removal. IEEE C57.12.01 which covers the general requirements for dry-type distribution and power transformers including those with solid cast and/or resin encapsulated windings has also designated several different categories of cooling systems for dry-type transformers. These are listed in Table 15.2.

TABLE 15.2   Classes of Dry-Type Transformer Cooling System Ratings

Class	Method of Cooling
AA	Dry-type, ventilated self-cooled
AFA	Dry-type, ventilated forced air-cooled
AA/FA	Dry-type, ventilated self-cooled / forced air-cooled
ANV	Dry-type, non-ventilated, self-cooled
GA	Dry-type, sealed self-cooled

Source: Reproduced with permission of IEEE.

Transformers are designed to operate satisfactorily when supplying sinusoidal currents. For nonlinear loads such as arc welding, rectifier loads, etc., the current waveform is non-sinusoidal. The non-sinusoidal currents contain harmonic currents that can cause additional heating in the transformer. Therefore, if the load that the transformer will be feeding is a nonlinear load, then the transformer must be specified to supply the nonlinear load.

UL 1561 and UL 1562 defines a K-factor rating to transformers that serve non-linear loads and provides guidance to manufacturers and users as to the quantity of the harmonics involved in the load the transformer will be feeder and, therefore, the construction of the transformer required to ensure reliable operation. For transformers with higher K-rating, the core material quality and core area are designed to operate farther away from the B-H curve saturation knee to ensure core losses are not excessive for the quantity of harmonics involved.

The K-factor is defined in UL 1561 and UL 1562 as shown in equation (15.6).

$$K\text{-factor} = \sum_{h=1}^{\infty} \left[\frac{I_h}{I_R}\right]^2 h^2 = \frac{1}{I_R^2} \sum_{h=1}^{\infty} I_h^2 h^2 \qquad (15.6)$$

where

$K = K$-factor of transformer

$h =$ harmonic number

$I_h =$ value of current at harmonic h

$I_R =$ rated rms load current of transformer

A K-factor of 4 is recommended for lighting, UPS systems, welders, induction heating, SCR controllers, and PLCs. For loads that may present a greater magnitude of harmonic currents to the transformer or require more tolerance to harmonics such as telecommunication equipment, UPS systems, and health care facilities, a K-factor of 13 is recommended. A K-factor of 20 is recommended for variable speed drive applications or critical locations such as critical care areas of hospitals.

Notice from equation (15.6) that the UL method of de-rating transformers based on the rms rating of the transformer I_R. IEEE C57.110 (IEEE Recommended Practice for Establishing Transformer Capability When Supplying Non-sinusoidal Load Currents) utilizes a slightly different method for de-rating a transformer when it is subject to non-sinusoidal currents. C57.110 uses a value called harmonic loss factor to define the de-rate required for various applications. The harmonic loss factor is defined by

$$
F_{HL} = \frac{\displaystyle\sum_{h=1}^{h=h_{max}} \left[\frac{I_h}{I_1}\right]^2 h^2}{\displaystyle\sum_{h=1}^{h=h_{max}} \left[\frac{I_h}{I_1}\right]^2}
\tag{15.7}
$$

where

F_{HL} = harmonic factor of transformer

h = harmonic number

I_h = value of current at harmonic h

I_1 = rated fundamental load current of transformer

I_R = rated rms load current of transformer

For a new transformer with harmonic currents specified as per unit of the rated transformer secondary current, the K-factor and harmonic loss factor have the same numerical values. The numerical value of the K-factor equals the numerical value of the harmonic loss factor only when the square root of the sum of the harmonic currents squared equals the square of the rated secondary current of the transformer as shown by equation (15.8) that shows the relationship between K-factor and harmonic factor.

$$
K\text{-factor} = \left[\frac{\displaystyle\sum_{h=1}^{h=h_{max}} I_h^2}{I_R^2}\right] F_{HL}
\tag{15.8}
$$

where

K = K-factor of transformer

F_{HL} = harmonic factor of transformer

h = harmonic number

I_h = value of current at harmonic h

Typically, transformers in a power plant distribution system are connected in a delta on the primary side and connected in a wye on the secondary side. A delta connection on the primary side prevents any triplen harmonic current (3rd, 6th, 9th, etc.) from flowing in the primary circuit. Any triplen harmonic currents that occur on the secondary side (only can happen with a grounded wye secondary) will be trapped and circulate in the delta-connected primary windings. From a relaying standpoint, it needs to be recognized that a phase-to-ground fault on the secondary side (if secondary side is a grounded, solidly or through impedance, wye connection) would appear on the primary side (delta side of transformer) as a phase-to-phase fault, not

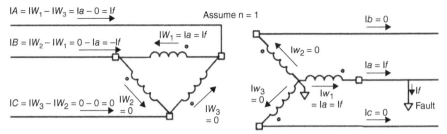

Figure 15.2   Response of three-phase transformer to line-to-ground fault.

a phase-to-ground fault. We can understand this phenomenon by evaluating the currents in a delta-wye transformer during a line-to-ground fault on the wye side of the transformer as is done in Figure 15.2.

In this example, we experience a phase A line-to-ground fault. This causes fault current (I_f) to flow in the secondary A phase conductor. During the fault, the B phase line current and C phase line current which are not part of the fault and are very small and treated as 0 A. Therefore,

$$I_a = I_f$$
$$I_b = I_c = 0 \text{ A}$$

The A phase current (I_f) is supplied by the transformer secondary winding one (w_1) of the transformer while the transformer secondary winding two (w_2) and transformer secondary winding three (w_3) currents are 0 A.

$$I_{w1} = I_a = I_f$$
$$I_{w2} = I_b = 0 \text{ A}$$
$$I_{w3} = I_c = 0 \text{ A}$$

To simplify this discussion, we now will assume the turns ratio is 1, but realize that the actual current is reflected back to the primary by the factor of $1/n$. The transformer primary winding one current is the transformer secondary winding one current divided by the turns ratio which, in this case is 1. Therefore, the fault current (I_f) is supplied totally by the transformer primary winding one (w_1) current. Since the transformer secondary winding two and three currents are zero, the transformer primary winding two and three currents must also be zero since they are the secondary values divided by the turns ratio.

$$I_{W1} = (1/n) \times I_{w1} = (1) \times I_{w1} = I_{w1} = I_a = I_f$$
$$I_{W2} = (1/n) \times I_{w2} = (1) \times I_{w2} = I_{w2} = 0 \text{ A}$$
$$I_{W3} = (1/n) \times I_{w3} = (1) \times I_{w3} = I_{w3} = 0 \text{ A}$$

Now that we have identified the transformer primary winding currents, we can evaluate the line currents feeding the transformer primary. By summing the currents at the A phase terminal of the transformer we find

$$I_A = I_{W1} - I_{W3} = I_f - 0 \text{ A} = I_f$$
$$I_B = I_{W2} - I_{W1} = 0 \text{ A} - I_f = -I_f$$
$$I_C = I_{W3} - I_{W2} = 0 \text{ A} - 0 \text{ A} = 0 \text{ A}$$

Summarizing, the secondary lines see fault current in only one of the three-phase conductors causing the fault to appear as a single line-to-ground fault when viewed from the secondary conductors.

$$I_a = I_f$$
$$I_b = 0 \text{ A}$$
$$I_c = 0 \text{ A}$$

However, the primary line conductors see fault current in two of the three-phase conductors causing the fault to appear as a double line-to-line fault when viewed from the primary conductors.

$$I_A = I_f$$
$$I_B = -I_f$$
$$I_C = 0 \text{ A}$$

Ferroresonance of transformer connections may be an issue in certain applications. For transformers that are shell formed or transformer banks that are made up from three single-phase transformers, there is no magnetic coupling between phases and ferroresonance is not an issue. However, for three-phase three-leg core or five-leg core transformers, especially those connected in delta, where the feeder cable is totally or partially underground, the capacitance of the cable and/or the system and the inductance presented by the core of the transformer can present a ferroresonant circuit should only one of the three transformer leads open. While the system does not have to be ungrounded for a ferroresonance condition to exist, ungrounded systems are more susceptible to this condition. Applications such as aerial cable or cable in trays and/or conduits that have significant distributed system capacitance can also present a ferroresonant condition. To prevent the condition of ferroresonance, transformers in power generation facilities that are connected on the primary side in a delta are only energized by three-phase switchgear and not by individually controlled phase protection such as fuse links. If a transformer is protected by fuse links and only one or two of the three-phase fuses opens, one or two legs of the transformer remains energized. This can lead to high voltages on the open legs of the transformers, leading to saturation of the transformer core and excessive heating and eventual failure of the transformer. To prevent this, circuit protection for transformer windings are almost always three-phase breakers so that either all three phases are open or closed.

There are several transformers installed in a generation station that should be designed to withstand the through fault currents and clearing times that a generator can provide. These are all the transformers directly connected to the generator leads. IEEE C57.116 – IEEE Guide for Transformers Directly Connected to Generators provides guidance to the design requirements for these transformers. These include the following transformers.

The generator step-up (GSU) transformer takes the output voltage of the generator (13–24 kV) and steps the voltage up to transmission levels (138 kV, 230 kV, or 500 kV) for transmission of power to the electrical transmission system. The unit auxiliary transformer (UAT), which is also called the station service transformer (SST), is connected directly to the generator leads and steps the voltage of the generator down to the local distribution levels for support of the unit electrical distribution equipment.

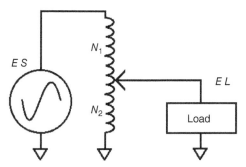

Figure 15.3   Autotransformer single-line equivalent circuit.

Depending on the design of the generator, the decay time of the fault current from a generator can be on the order of 5–30 sec and any transformer directly connected to the generator should be designed to withstand the fault current levels for the decay time values that a generator can provide without damage.

Autotransformers are a special case of transformer where there is only one winding (per phase) on a transformer but there are multiple taps midpoint in the auto-transformer winding. A single line equivalent circuit for an autotransformer is shown in Figure 15.3. The ratio of turns the primary circuit sees compared to the number of turns the secondary circuit sees defines the turn ratio of the transformer and, therefore the voltage ratio across the transformer. However, since the primary and secondary circuits are electrically connected to the same winding, there is no electrical isolation provided by the autotransformer.

Referring to Figure 15.3, the applied voltage (E_S) is equally divided across the total turns ($N_1 + N_2$) of the autotransformer. Therefore, the voltage across each turn of the winding of the transformer is

$$\text{Volts per turn} = E_S/(N_1 + N_2)$$

The load voltage (E_L) is simply the voltage across each turn of the winding of the transformer multiplied by the number of turns between the tap connection on the transformer and the end of the transformer winding that is common to both the load and source. Therefore, the voltage across the load is

$$E_L = N_2 \times \text{volts per turn} = N_2 \times E_S/(N_1 + N_2) = E_S \times \{N_2/(N_1 + N_2)\}$$

If we take a typical power transformer winding and core and eliminate the secondary winding, we are left with only the primary winding wrapped around a core. This is the basic construction of the *current limiting reactor* (CLR). The symbolic representation for a current-limiting reactor is shown in Figure 15.4 This device is constructed as one winding (per phase) around a common core. This winding presents inductive impedance in series with a load. Under normal circumstances, the load current is mostly resistive with a high power factor. Since the impedance of the reactor is mostly reactive with a very low power factor and the magnitude of current drawn by the load is within nominal values, the voltage drop of the reactor is very small. In the event of a fault at the load, the fault current increases in magnitude to values several

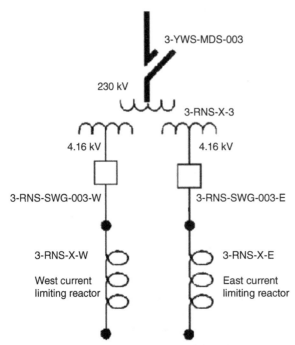

3-YWS-MDS-003

230 kV

3-RNS-X-3

4.16 kV   4.16 kV

3-RNS-SWG-003-W   3-RNS-SWG-003-E

3-RNS-X-W   3-RNS-X-E

West current
limiting reactor   East current
limiting reactor

Figure 15.4   Current limiting reactor. *Source*: Reproduced with permission of Tampa Electric Company. Reproduction is forbidden without the express consent of Tampa Electric Company.

time nominal. Additionally fault currents are very inductive with a low power factor. Therefore, the reactor impedance presents a substantial portion of the overall system impedance and the value of the fault current is reduced. A current limiting reactor is shown in Figure 15.4.

Transformers will have insulation rating levels assigned to them depending on the nature of the surge testing the transformer undergoes during manufacturing. The transformer undergoes a one-minute, power frequency, AC high potential test. This test verifies the insulation system of the transformer under normal system frequencies. The transformer will also undergo tests to verify the insulation system integrity during surges.

Another concern with transformers is the protection of the transformer due to surges induced during transients. Transformers can present large reactive impedances to the electrical system. For any inductor, the voltage developed across the inductor is a function of the inductance and the rate of change of current in the inductor as defined in equation (15.9).

$$V_1 = L(di/dt) \tag{15.9}$$

where

V_1 = voltage induced in inductor winding (V)

L = inductance of winding (H)

di/dt = rate of change of current in inductor winding (A/s)

Example 15.2   A transformer with a primary winding inductance of 5 mH is connected to an electrical system with a nominal voltage of 4.16 kV. A vacuum breaker is used to isolate the transformer in the event of an electrical fault. During an electrical fault, the breaker interrupted 20,000 A in a time of 10 ms. What is the potential voltage developed across the transformer winding in this instance? Utilizing equation (15.9), we find

$$V_1 = L \, (di \, / \, dt)$$
$$V_1 = (0.005 \text{ H}) \times (\, 20,000 \text{ A}) \, / \, (0.010 \text{ sec})$$
$$V_1 = 10,000 \text{ V}$$

Some applications of transformers are very susceptible to very high values of rate of change of current (di/dt). Transformers that are either fed from or feed vacuum circuit breakers or transformers that have the source- or load-side conductors exposed to potential lightning may experience very high values of (di/dt) in their application. In these applications, surge arrestors are installed 5 ft or less from the transformer to limit the peak voltage experienced by the transformer during periods of high (di/dt). This prevents the surge inducted in the inductance of the transformer from damaging insulation systems and connected electrical distribution switchgear. The need for the minimum of 5 ft is to provide some cable impedance between the source of the voltage (the transformer winding) and the surge arrestor.

There are two tests to verify the insulation integrity of a transformer for voltage transients. The first *Basic Impulse Level* (BIL) test is a 1.2/50 full-wave voltage impulse test. During this test, the voltage on the transformer increases in 1.2 µs from a value that is 10% of the peak or crests value to a value that is 90% of the peak or crest value. Then the voltage on the transformer is reduced in a period of 50 µs from the peak or crest value to a value that is 50% of the peak or crest. This test verifies the insulation system integrity for high frequency surges that the transformer may experience during its service life.

The second test is known as a chopped-wave voltage impulse test. During this test, the voltage on the transformer is increases in 1.2 µs from a value that is 10% of the peak or crest value of the full-wave test times 115% to a value that is 90% of the peak or crest value of the full-wave test times 115%. Then, the voltage on the transformer is reduced. At the point 3 µs into the test (1.8 µs after the peak), the voltage is clamped and brought to zero. This test also verifies the insulation system integrity for high frequency surges that the transformer may experience during its service life. Figure 15.5 depicts the full-wave and chopped-wave test parameters.

The standard values of insulation levels for both low frequency testing and surge testing for oil-filled power transformers are listed in Table 15.3. The standard values of insulation levels for dry-type power transformers are listed in Table 15.4.

Since the operation of transformer affects the reliable operation of the power system in the energy production facility and it is expected that these transformers have a life of 30–40 years, it is typical to specify a BIL level that is one or two levels above the minimum ratings recommended by the standards. Table 15.5 shows the standard BIL ratings of insulation systems for dry-type transformers depending on the winding nominal voltage (indicated by an "S") and also what higher levels of BIL are available from manufacturers (indicated by a "1").

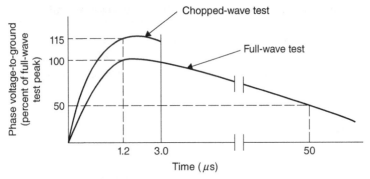

Figure 15.5   Standard impulse test waves. *Source*: Reproduced with permission of IEEE.

The insulation systems for transformers also have a thermal limit depending on the material of construction. All transformers manufactured to IEEE standards have a temperature rise rating based on an ambient of 30°C. Note that this nominal ambient temperature value of 30°C is unique to transformers and different from the maximum ambient ratings of other classes of equipment that have their temperature rise based

TABLE 15.3   Standard Insulation System Ratings for Oil-Filled Transformers

Insulation Class and Nominal Bushing Rating	Windings					Bushing Withstand Voltages		BIL Impulse Full Wave (1.2/50)
	Hi-Pot Tests	Chopped Wave – Minimum Time to Flashover		BIL Full Wave (1.2/50)	Switching Surge Level	60-Cycle 1 min Dry	60-Cycle 10 sec Wet	
kV (rms)	kV (rms)	kV (crest)	μs	kV (crest)	kV (crest)	kV (rms)	kV (rms)	kV (crest)
1.2	10	54 (36)	1.5 (1)	45 (30)	20	15 (10)	13 (6)	45 (30)
2.5	15	69 (54)	1.5 (1.25)	60 (45)	35	21 (15)	20 (13)	60 (45)
5.0	19	88 (69)	1.6 (1.5)	75 (60)	38	27 (21)	24 (20)	75 (60)
8.7	26	110 (88)	1.8 (1.6)	95 (75)	55	35 (27)	30 (24)	95 (75)
15.0	34	130 (110)	2.0 (1.8)	110 (95)	75	50 (35)	45 (30)	110 (95)
25.0	50	175	3.0	150	100	70	70 (60)	150
34.5	70	230	3.0	200	140	95	95	200
46.0	95	290	3.0	250	190	120	120	250
69.0	140	400	3.0	350	280	175	175	350
92.0	185	520	3.0	450	375	225	190	450
115.0	230	630	3.0	550	460	280	230	550
138.0	275	750	3.0	650	540	335	275	650
161.0	325	865	3.0	750	620	385	315	750

Source: Reproduced with permission of IEEE.

Note: Values in parentheses are for distribution transformers, instrument transformers, constant-current transformers, step- and induction-voltage regulators, and cable potheads for distribution cables. The switching surge levels shown are applicable only to power transformers (not distribution transformers). Test voltages are defined in IEEE Standard C57.12.00-1980.

TABLE 15.4   Standard Insulation System Ratings for Dry-Type Transformers

Impulse Test Levels for Dry-Type Transformers

Nominal Winding Voltage (V)		High Potential Test	Standard BIL (1.2/50)
Delta or Ungrounded Wye	Grounded Wye	kV (rms)	kV (crest)
120–1200	1200 Y/693	4	10
		4	10
2520	4360 Y/2520	10	20
		10	20
4160–7200	8720 Y/5040	12	30
		10	30
8320		19	45
12,000–13,800	13,800 Y/7970	31	60
		10	60
18,000	22,860 Y/13,200	34	95
		10	95
23,000	24,940 Y/14,400	37	110
		10	110
27,600	34,500 Y/19,920	40	125
		10	125
34,500		50	150

Source: Reproduced with permission of IEEE.
Note: Nominal voltages shown are exactly as tabulated in IEEE Standard C57.12.01-1979 and are not, in all cases, in accordance with the classifications commonly encountered on industrial and commercial systems.

on an ambient of 40°C. Table 15.6 shows both the allowable average winding temperature rise (calculated by resistance) above an ambient temperature of 30°C and the allowable maximum winding temperature rise above an ambient temperature of 30°C for certain classes of transformer insulation systems.

TABLE 15.5   Standard BIL Levels and Optional BIL Levels for Dry-Type Transformers

Nominal System Voltage (kV)	Basic Lightning Impulse Insulation Levels (BILs) in Common Use (kV crest)									
	10	20	30	45	60	95	110	125	150	200
1.2	S	1	1							
2.5		S	1	1						
5.0			S	1	1					
8.7				S	1	1				
15.0					S	1	1			
25.0						2	S	1	1	
34.5								2	S	1

Source: Reproduced with permission of IEEE.S = Standard value.
1 = Optional higher levels where exposure to overvoltage occurs and improved protective margins are required.
2 = Lower levels where protective characteristics of applied surge arresters have been evaluated and found to provide appropriate surge protection.

TABLE 15.6   Insulation Classes for Dry-Type Transformers[a]

Insulation System Temperature Class (°C)	Winding Hottest-Spot Temperature Rise (°C)	Average Winding-Temperature Rise by Resistance (°C)[b]
130	90	75
150	110	90
180	140	115
200	160	130
220	180	150

Source: Reproduced with permission of IEEE

[a]Based on an average daily ambient temperature of 30°C, with a maximum ambient temperature of 40°C. Insulation system temperature class (°C) Winding hottest-spot temperature rise (°C) Average winding temperature rise by resistance (°C).

[b]Higher average winding temperature rises by resistance may apply if the manufacturer provides thermal-design test data substantiating that temperature limit of the insulation are not exceeded.

Example 15.3   What is the standard BIL level for a transformer winding that will be connected to an electrical system with a nominal voltage of 5 kV?

Solution: From Table 15.5, for a nominal system voltage of 5 kV, the standard BIL is 30 kV

Example 15.4   What value should be the winding hot spot alarm be set for on a dry-type station service transformer that was constructed with a class 130 insulation system and measures the average winding temperature using an RTD?

Solution: From Table 15.6, for a class 130 insulation system, for a transformer that has an average temperature rise monitored by resistance, the alarm should be set at or below 75°C above the 30°C ambient.

Example 15.5   What should be the winding hot spot alarm on a dry-type station service transformer that was constructed with a class 130 insulation system and measures the actual hot spot temperature rise?

Solution: From Table 15.6, for a class 130 insulation system, for a transformer that monitors the actual hot spot temperature rise, the alarm should be set at or below 90°C above the 30°C ambient.

As transformer MVA ratings become larger, the "per unit" impedance of transformers tends to rise. As a rule of thumb, standard transformers rated less than 500 kVA have impedance values in the range of 3–5%. Transformers rated more than 500 kVA have impedance values of 5.75% or larger. In addition to their base rating, dry-type transformers may have second power rating base on forced airflow (AA/FA rated). In this arrangement, the additional power rating at the FA rating is 133% of the AA rating. For example, if a dry-type transformer is rated 1000 kVA on its AA rating then it is expected that the FA rating would be 1333 kVA.

In addition to insulation system testing, the transformer equivalent circuit parameters are identified during testing of the transformer. There are two basic tests to identify the various parameters of a transformer as shown in Figure 15.6.

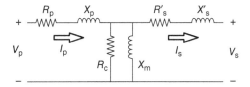

Figure 15.6   Transformer equivalent circuit.

The first test is called a *no-load test* or open circuit test. During this test, the secondary leads are left open-circuited and the rated voltage is applied to the transformer primary. Under these conditions, the current in the winding is very small and the losses in the primary and secondary windings can be neglected. The only significant losses are those associated with the core resistance and core inductance parameters R_c and X_m. We then measure the amount of real power (P_{nl}) and primary voltage (V_p) and primary current (I_p). Knowing the following relationships, we can then calculate the values for the core resistance (R_c) and core reactance (X_m).

$$P_{nl} = V_p^2/R_c$$

or

$$R_c = V_p^2/P_{nl} \tag{15.10}$$

where

R_c = effective resistance of core to eddy current losses and hysteresis losses (ohms)

V_p = no-load voltage applied to primary winding (V)

P_{nl} = no-load real power loss (W)

$$S_{nl}^2 = (V_p \times I_p)^2 = P_{nl}^2 + Q_{nl}^{2'}$$

Rearranging for the reactive power component (Q_{nl}), we find

$$Q_{nl}^2 = (V_p \times I_p)^2 - P_{nl}^2$$

But, we also know that the reactive power is the applied voltage squared divided by the reactance that we are trying to determine or

$$Q_{nl} = V_p^2/X_m$$

Combining these two equations, we find that

$$(V_p^2/X_m)^2 = (V_p \times I_p)^2 - P_{nl}^2$$

Rearranging for the variable of interest, we find our resultant equation to be

$$X_m = V_p^2/ \left[(V_p \times I_p)^2 - P_{nl}^2\right]^{1/2} \tag{15.11}$$

where

X_m = effective core reactance due to energy stored in magnetic core (ohms)

V_p = no-load voltage applied to primary winding (V)

I_p = no-load current measured in the primary winding (A)

P_{nl} = no-load real power drawn in core (W)

The second test is called a *full-load test* or short circuit test. During this test, the secondary leads are shorted and a reduced voltage is applied to the transformer primary sufficient to generate rated current in the primary leads of the transformer. Under these conditions, the voltage of the primary circuit is reduced and the core losses are reduced by the square of the applied voltage. Therefore, during this test, we can assume the core losses (R_c and X_m) are minimal and, therefore, the core losses windings can be neglected. We then measure the amount of real power (P_{fl}) and reactive power (Q_{fl}). Knowing the following relationships, we can then calculate the values for the series combination of the primary and secondary winding resistance ($R_p + R_s$) and for the series combination of the primary and secondary winding reactance ($X_p + X_s$).

$$P_{fl} = I_p^2 \times \left(R_p + R_s'\right)$$

or

$$\left(R_p + R_s'\right) = P_{fl}/I_p^2 \tag{15.12}$$

where

$(R_p + R_s')$ = effective resistance for primary winding and secondary winding reflected through the turns ratio (ohms)

I_p = full-load current through primary winding (A)

P_{fl} = full-load real power loss in primary and secondary windings (W)

$$S_{fl}^2 = (V_p \times I_p)^2 = P_{fl}^2 + Q_{fl}^{2'}$$

Rearranging for the reactive power component (Q_{fl}), we find

$$Q_{fl}^2 = (V_p \times I_p)^2 - P_{fl}^2$$

But we also know that the reactive power is the applied voltage squared divided by the reactance that we are trying to determine or

$$Q_{fl} = V_p^2 / \left(X_p + X_s'\right)$$

Combining these two equations, we find that

$$\left(V_p^2 / \left(X_p + X_s'\right)\right)^2 = (V_p \times I_p)^2 - P_{fl}^2$$

Rearranging for the variable of interest, we find our resultant equation to be

$$\left(X_p + X_s'\right) = V_p^2 / \left[(V_p \times I_p)^2 - P_{fl}^2\right]^{1/2} \tag{15.13}$$

where

$(X_p + X_s')$ = effective reactance for primary winding and secondary winding reflected through the turns ratio (ohms)

V_p = full load, reduced voltage applied to primary winding (V)

I_p = full-load current measured in the primary winding (A)

P_{fl} = full-load real power drawn in core (W)

One will notice that, during the full-load test, the resistance and reactance of the windings cannot be easily separated, but there are two methods to estimate how much is associated with the primary winding and how much is associated with the secondary winding. The first method is to measure directly with a DLRO (digital low resistance ohmmeter), the resistance of the primary and secondary windings and then assume the ratio of reactances are similar. Alternately, the typical transformer primary-to-secondary resistance and reactance can be assumed to follow the square of the turns ratio of the transformer. This is an approximation based on the idea that the turns ratio provides approximate ratio of the length of the conductor combined with the idea the higher turns ratio side has a smaller current rating, and therefore the size of the conductor is smaller and thereby has a large value of resistance per unit length. These parameters are not given in units of ohms but in the per unit system. Conversion of resistance in units of ohms to the per unit system is explained in detail in Chapter 22 and the reader is encouraged to refer to Chapter 22 for this final step.

Example 15.6   A generator step-up transformer consists of three single-phase transformers that are under test to determine their single-line equivalent circuit values of core resistance and core reactance. A no-load test is performed on the transformer and, when rated primary potential (V_p) of 24,000 V is applied to the primary of the GSU, the measured value for real power loss (P_{oc}) was 5760 W and the measured value for primary current (I_p) was 2.412 A. Determine the effective values for transformer core resistance (R_c) and core reactance (X_m).

Solution: Utilizing equation (15.10),

$$R_c = V_p^2/P_{nl}$$
$$R_c = (24,000)^2/5760 \text{ ohms}$$
$$R_c = 100,000 \text{ ohms}$$

Utilizing equation (15.11),

$$X_m = V_p^2 / [(V_p \times I_p)^2 - P_{nl}^2]^{1/2}$$
$$X_m = 24,000^2 / [(24,000 \times 2.412)^2 - 5760^2]^{1/2}$$
$$X_m = 10,000 \text{ ohms}$$

Example 15.7   A generator step-up transformer consists of three single-phase transformers that are under test to determine their single-line equivalent circuit values of core resistance and core reactance. A full-load test is performed on the transformer and, when rated primary current is drawn with a shorted secondary on the transformer, the measured primary potential was found to be 1200 V, the measured value for real power loss was 57,600 W and the measured value for primary current was 200 A. Determine the effective values for transformer primary and secondary winding resistance ($R_p + R_s'$) and primary and secondary winding reactance ($X_p + X_s'$).

Solution: Utilizing equation (15.12),

$$(R_p + R_s') = P_{fl} / I_p^2$$
$$(R_p + R_s') = 57,600 / (1200)^2$$
$$(R_p + R_s') = 1.44 \text{ ohms}$$

Utilizing equation (15.13),

$$(X_p + X'_s) = V_p^2 / [(V_p \times I_p)^2 - P_{fl}^2]^{1/2}$$
$$(X_p + X'_s) = 1200^2 / [(1200 \times 200)^2 - 57,600^2]^{1/2}$$
$$(X_p + X'_s) = 6.18 \text{ ohms}$$

One last item to mention is phase relationships of delta-wye transformers. In a power generation facility that utilizes a secondary or primary selective system where sources can be paralleled, it is critical to get the phase relationships correct between electrical buses. IEEE C57.12.00 standard angular displacement is such that the low voltage side of three-phase delta-wye transformers always lags the high voltage winding by 30 degrees. For wye-wye and delta-delta transformers, the phase shift from primary to secondary is 0 degrees. For wye-delta transformers, the phase shift from primary to secondary is 30 degrees leading or lagging. IEEE C57.12.00 defines the standard angular displacement of wye-delta transformers to be such that the low voltage side lags the high voltage side by 30 degrees. There is one transformer in the electric generation facility that is almost always a non-standard IEEE C57.12.00 angular displacement. Let us see this by evaluating the simplified one-line diagram of a unit generator, generator step-up transformer (GSU), unit auxiliary transformer (UAT) or (SST), and a reserve service transformer (RST).

The RST is a wye-wye connected transformer and, as such does not have any phase shift from primary to secondary. In Figure 15.7, the phase A to neutral orientation is 210 degrees. To limit phase-to-ground fault currents, any transformer directly connected to the generator is connected in a delta. Therefore, both the GSU and the SST are delta-wye connected transformers. Looking at the GSU, the high

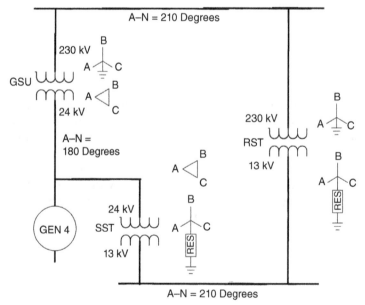

Figure 15.7   Non-standard transformer phase shift application.

voltage side has a phase relationship on the 230 kV system where the angle of phase A voltage to neutral is 210 degrees. The GSU low voltage side lags behind the high voltage side and has a phase relationship where the angle of phase A voltage to neutral is 180 degrees. This agrees with IEEE C57.12.00 standard angular relationship since the low voltage side lags the high voltage side by 30 degrees. But what about the SST? The high voltage side of the SST has a phase relationship on the 24 kV side of 180 degrees phase to neutral. The SST low voltage side leads AHEAD of the high voltage side and has a phase relationship of 210 degrees phase A to neutral. This is a non standard configuration as for the SST, the high voltage side leads the low voltage side by 30 degrees. Most generation station distributions systems that follow this design have an angular relationship that does not conform to IEEE C57.12.00 in regards to angular relationships for the station service transformer. This is addressed in more detail in IEEE C57.116-1989: IEEE Guide for Transformers Directly Connected to Generators.

GLOSSARY OF TERMS

- Autotransformer – An autotransformer is an electrical transformer with only one winding. The "auto" prefix refers to the single coil rather than any automatic mechanism. In an autotransformer, portions of the same winding act as both the primary and secondary winding.

- Basic Impulse Level (BIL) – A measure of the ability of a transformer's insulation system to withstand very high voltage, short-time surges.

- Current Limiting Reactor (CLR) – A reactor connected in series with a circuit or piece of equipment, used to limit the current that can flow under short-circuit (fault) conditions or other switching conditions.

- Epoxy Resin Transformer – The winding of an epoxy resin transformer is first treated just like a VPI transformer, but after the completion of the VPI process, the winding is incased in an epoxy resin shell. This provides a more robust insulation system and reinforcement system.

- Ferroresonance – Ferroresonance or nonlinear resonance is a type of resonance in electric circuits which occurs when a circuit containing a nonlinear inductance (such as the inductance a transformer core can present when undergoing saturation) is fed from a source that has series capacitance.

- Full-Load Test – A test on a transformer that, during this test, the secondary leads are shorted and a reduced voltage is applied to the transformer primary to cause rated secondary current to flow. This test is used to determine primary and secondary winding losses.

- No-Load Test – A test on a transformer that, during this test, the secondary leads are left open-circuited and the rated voltage is applied to the transformer primary. This test is used to determine core losses.

- Transformer – An apparatus for reducing or increasing the voltage of an alternating current.

- Vacuum Pressure Impregnated (VPI) – Vacuum pressure impregnated transformer windings undergo a process whereby the transformer winding is inserted in a tank. Initially, a vacuum is drawn to pull out the moisture in the winding insulation system. After a certain time, pressure is then applied with an epoxy insulating material to force the insulation material into the voids of the transformer insulation system.

PROBLEMS

15.1 For a 28,800 VA, single-phase transformer that is rated 14,400 V on the primary and 120 V on the secondary where the secondary winding is connected to a 1-ohm load, determine the following:

    A. Turns ratio (n)

    B. Rated primary current

    C. Rated secondary current

    D. The impedance seen by the circuit connected to the primary of the transformer.

15.2 A generator step-up transformer consists of three single-phase transformers that are under test to determine their single-line equivalent circuit values of core resistance and core reactance. A no-load test is performed on the transformer and, when rated primary potential of 36,000 V is applied to the primary of the GSU, the measured value for real power loss was 6000 W and the measured value for primary current was 3 A. Determine the effective values for transformer core resistance (R_c) and core reactance (X_m).

15.3 A generator step-up transformer consists of three single-phase transformers that are under test to determine their single-line equivalent circuit values of core resistance and core reactance. A full-load test is performed on the transformer and, when rated secondary current is drawn on the shorted secondary of the transformer, the measured primary potential was found to be 2000 V, the measured value for real power loss was 90,000 W and the measured value for the primary current was found to be 300 A. Determine the effective values for transformer primary and secondary winding resistance ($R_p + R_s'$) and primary and secondary winding reactance ($X_p + X_s'$).

15.4 What should be the winding hot spot alarm on a dry-type station service transformer be that was constructed with a class 180 insulation system and measures the average winding temperature using an RTD?

15.5 What is should be the winding hot spot alarm on a dry-type station service transformer be that was constructed with a class 180 insulation system and measures the actual hot spot temperature rise?

15.6 What is the standard BIL level for a transformer winding that will be connected to an electrical system with a nominal voltage of 15 kV?

15.7 A transformer with a primary winding inductance of 10 mH is connected to an electrical system with a nominal voltage of 4.16 kV. A vacuum breaker is used to isolate the transformer in the event of an electrical fault. During an electrical fault, the breaker interrupted with 30,000 A in a time of 10 ms. What is the potential peak voltage developed across the transformer winding in this instance?

RECOMMENDED READING

Electric Power Plant Design, Technical Manual TM 5-811-6, Department of the Army, USA, 1984.

IEEE 141: IEEE Recommended Practice for Electric Power Distribution for Industrial Plants (IEEE Red Book), 1993.

IEEE C57.116: IEEE Guide for Transformers Directly Connected to Generators.

Electrical Machines, Drives and Power Systems, 6th edition, Theodore Wildi, Prentice Hall, 2006.

The Engineering Handbook, 3rd edition, Richard C. Dorf (editor in chief), CRC Press, 2006.

Industrial Power Distribution, 2nd edition, Ralph E. Fehr III, Wiley-IEEE Press, 2016. ISBN: 978-1-119-06334-6.

Power Plant Engineering, Black & Veatch, edited by Larry Drbal, Kayla Westra, and Pat Boston, Chapman & Hall / Springer, 1996.

Standard Handbook of Powerplant Engineering, 2nd edition, Thomas C. Elliott, McGraw-Hill, 1998.

CHAPTER 16

GENERATORS

> **GOALS**
>
> - To understand the basic design and construction of generators
> - To calculate the synchronous frequency of a generator with a DC field
> - To calculate the sub-synchronous or super-synchronous frequency of a generator with an AC field
> - To describe the difference between a cylindrical and salient pole rotor
> - To calculate the amount of real and reactive power transferred from a generator delivered to or drawn from the connected system
> - To draw the generator capability curve
> - To calculate the required field current for a given operating point given a generator "V-curve"
> - To understand the difference between leading and lagging operation of synchronous generator as compared with a synchronous motor
> - To calculate field temperature based on calculated resistance of the field
> - To understand the various types of excitation systems utilized in synchronous generators
> - To understand the process of synchronizing a generator to an energized bus
> - To properly design and apply generator protection systems

GENERATORS CONVERT the mechanical energy delivered by the prime mover into electrical energy to be transmitted to the electrical system. Most generators are synchronous machines that have an external field applied to the machine and drive voltage and current into the electrical system at synchronous frequency. The synchronous frequency of a generator is a product of the number of poles in a generator and the speed of the generator and defined by

$$f = n \times p / 120 \tag{16.1}$$

Energy Production Systems Engineering: An Introduction for Electrical Engineers to Electrical Power Generation Facilities, Systems, and Equipment, First Edition. Thomas H. Blair.

where

f = frequency of the voltage supplied by the generator (Hz)

n = speed of the machine (rpm)

p = number of poles

Example 16.1   A generator is designed as a two-pole machine and operates at 3600 rpm. What is the frequency of voltage provided by this generator to the electrical system?

Solution: Using equation (16.1), we find

$f = n \times p / 120$

$f = 3600 \text{ rpm} \times 2 \text{ pole} / 120$

$f = 60 \text{ Hz}$

In a limited number of cases, an induction machine can be used as a generator to deliver voltage and current to an electrical system. The benefit of an induction machine is that there is no requirement for a separately powered field. However, while the induction machine can either absorb real power from the electrical system or deliver real power to the electrical system, the *induction generator* always absorbs vars from the system to operate due to the fact that it does not have its own externally powered field. The induction generator typically will utilize an external device such as a capacitor bank or inverter to prevent drawing reactive power from the system. As such, the application of these machines is limited to special applications such as wind power.

The *synchronous generator* can be used to absorb or deliver reactive power or vars to the electrical system by control of the field current to the generator. Increasing the field current to the generator increases the internal voltage developed by the generator, thereby increasing the vars delivered by the generator to the electrical system and decreasing field current to the generator decreases the internal voltage developed by the generator, thereby decreasing the vars delivered by the generator to the electrical system.

One of the more popular applications of an induction machine to be used as a generator is in the field of wind power. A typical arrangement for the use of an induction generator in the field of wind power is to connect the wind turbine to the shaft of an induction generator. The *stator* of the induction generator is then connected to the electrical system. Since the frequency of the system is fixed, and since induction machines need to slip to develop torque (and therefore deliver power), the induction machine is connected to the grid via an AC-to-DC-to-AC converter that is used to compensate the frequency generated by the induction generator to the system frequency. The use of an AC-to-DC-to-AC converter also allows for some var control to or from the system.

One method of connection of an induction generator to an electrical system in a wind turbine application is to connect the wind turbine to the shaft of a wound rotor motor (an induction machine). The stator of the wound rotor motor is directly coupled to the electrical system. The slip rings of the rotor are connected to the electrical system through the use of frequency converter. The frequency of the rotor circuit is modified based on the speed of the machine to maintain a constant output frequency in synchronism with the system frequency.

In this special case, the wound rotor motor has a separately powered field circuit on the rotor and in reality is operating as a synchronous machine and not an induction machine. This is a great example of showing that it is not the construction of the machine that defines if it is a synchronous machine or induction machine but whether or not the machine has an externally applied current source applied to the field of the machine. In this special case, the frequency output of the stator is defined by the speed of the machine, the number of poles of the machine, and also the frequency of the rotor as well as the direction of rotor field rotation.

If the rotor field rotation is in the same order (A B C) as the stator magnetic field rotation (A B C), then the machine is said to be operating in *subsynchronous operational mode*. In this mode of operation, the physical rotor shaft is turning less than the synchronous speed of the machine if the machine had DC current on the rotor. In this mode, the frequency required to be placed on the rotor for a given shaft speed and motor number of poles is given by equation (16.2)

$$f_r = f_s - (n \times p)/120 \qquad (16.2)$$

where

f_r = rotor frequency (Hz)

f_s = stator frequency (Hz)

n = speed of machine (rpm)

p = number of poles in machine

Notice that, if the frequency applied to the rotor is DC (i.e., $f_r = 0$), then the above equation simplifies to our equation for the frequency of a synchronous generator as defined by equation (16.1).

$$f_r = f_s - (n \times p) / 120$$
$$0 = f_s - (n \times p) / 120$$
$$f = n \times p / 120$$

Equation (16.1) defines the frequency of a synchronous machine with DC applied to the field.

If the rotor field rotation is in the opposite order (C B A) as the stator magnetic field rotation (A B C), then the machine is said to be operating in *supersynchronous operational mode*. In this mode of operation, the physical rotor shaft is turning faster than the synchronous speed of the machine if the machine had DC current on the rotor. In this mode, the frequency required to be placed on the rotor for a given shaft speed and motor number of poles is given by the equation

$$f_r = (n \times p)/120 - f_s \qquad (16.3)$$

where

f_r = rotor frequency (Hz)

f_s = stator frequency (Hz)

n = speed of machine (rpm)

p = number of poles in machine

Again, notice that, if the frequency applied to the rotor is DC (i.e., $f_r = 0$), then equation (16.3) simplifies to equation (16.1).

Example 16.2   A wind turbine that utilizes a four-pole, wound rotor motor for its generator is rotating at a speed of 1000 rpm. The power system that the stator of the wound rotor generator is connected to has a system frequency of 60 Hz that is A B C rotation. What mode (subsynchronous or supersynchronous) is the generator operating at, what is the required frequency of the rotor circuit to ensure the stator of the wound rotor motor at system frequency, and what is the phase rotation of the rotor current?

Solution: First, we need to determine the wound rotor generator's synchronous speed and compare it with the actual speed to see if it is in subsynchronous mode or supersynchronous mode. The synchronous speed of the wound rotor motor is found utilizing equation (16.1).

$n = 120 f / p$

$n = 120\ 60\ \text{Hz} / 4\ \text{pole}$

$n = 1800\ \text{rpm}$

Since the machine is turning at 1000 rpm, it is operating in subsynchronous mode. To find the required frequency of the rotor circuit to ensure the stator of the wound rotor motor at system frequency, we calculate using equation (16.2)

$f_r = f_s - (n \times p) / 120$

$f_r = 60\ \text{Hz} - (1000\ \text{rpm} \times 4\ \text{pole})/120$

$f_r = 60\ \text{Hz} - 33.3\ \text{Hz}$

$f_r = 26.7\ \text{Hz}$

Since the machine is operating is subsynchronous mode, the phase rotation of the rotor current is the same as the stator,

$$\text{Phase rotation} = (\text{A B C})$$

Example 16.3   A wind turbine that utilizes a four-pole, wound rotor motor for its generator is rotating at a speed of 2000 rpm. The power system that the stator of the wound rotor generator is connected to has a system frequency of 60 Hz that is A B C rotation. What mode (subsynchronous or supersynchronous) is the generator operating at, what is the required frequency of the rotor circuit to ensure the stator of the wound rotor motor at system frequency, and what is the phase rotation of the rotor current?

Solution: First, we need to determine the wound rotor generator's synchronous speed and compare it with the actual speed to see if it is in subsynchronous mode or supersynchronous mode. The synchronous speed of the wound rotor motor is

$n = 120 f / p$

$n = 120 \times 60\ \text{Hz} / 4\ \text{pole}$

$n = 1800\ \text{rpm}$

Since the machine is turning at 2000 rpm, it is operating in supersynchronous mode. To find the required frequency of the rotor circuit to ensure the stator of the wound rotor motor at system frequency, we calculate using equation (16.3)

$$f_r = (np)/120 - f_s$$
$$f_r = (2000 \text{ rpm} \times 4 \text{ pole })/120 - 60 \text{ Hz}$$
$$f_r = 66.7 \text{ Hz} - 60 \text{ Hz}$$
$$f_r = 6.7 \text{ Hz}$$

Since the machine is operating in supersynchronous mode, the phase rotation of the rotor current is the opposite of the stator,

Phase rotation = (C B A)

With the exception of wind power, the majority of large utility generators that utilize rotating machines for the final energy conversion process utilize synchronous machines, so we will turn our attention back to synchronous machines. Figure 16.1 shows the typical component arrangement of a synchronous generator.

The synchronous generator is made up of the stationary components and rotating components. The stator coils are imbedded in slots of the stator core. The current flowing in the coils of the stator generates a substantial amount of heat due to the I^2R losses in the copper. There are three basic types of stator coil construction in regard to the removal of heat generated in the stator coil. *Indirectly cooled coils* or conventionally cooled coils are solid in construction and have ground wall insulation encapsulating the conductors. Heat is removed by the flow of gas inside the casing of the generator around the outside of the ground wall insulation. Figure 16.2a shows two configurations of indirectly cooled coils. The coil on the left is a nine-turn coil for a hydro-generator. The coil on the right is a two-turn Roebel bar commonly used in two-pole generators.

Gas-intercooled coils have channels constructed inside the coils. The gas inside the casing of the generator is forced through these channels. This is a more efficient means of removing the heat generated in the coil conductor area since the heat does not have to flow through the ground wall insulation to be removed. Figure 16.2b shows

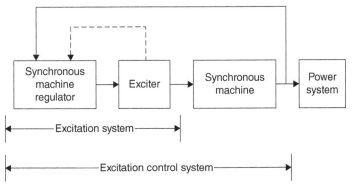

Figure 16.1   Typical synchronous machine arrangement. *Source*: IEEE 421, Reproduced with permission of IEEE.

(a)
Indirectly cooled

(c)
Water intercooled

(b)
Gas intercooled

Figure 16.2 Generator winding configurations. *Source*: Reproduced with permission of National Electric Coil.

three configurations of gas-intercooled coils. The coil on the left is a hydrogen-inner-cooled Roebel bar design known as the double tube stack and gets its name from the number of tube columns in the winding. The coil in the middle is a hydrogen-inner-cooled Roebel bar design known as the single tube stack and also gets its name from the number of tube columns in the winding. The coil on the right is an air-inner-cooled Roebel bar design. Note that the design that uses air as the cooling medium has a larger cross-sectional area in the tube than the design using hydrogen. Since air is less efficient at removing heat than hydrogen, this is compensated for by increasing the cross-sectional area of the tube thereby allowing a greater volume of air to flow through the tube to obtain adequate cooling of the coil.

The third type of coil construction is the *water-intercooled coil* as shown in Figure 16.2c. This design also has channels constructed inside the coils but instead of the gas in the generator being used as the media for removal of heat, water is circulated through these channels. Since the thermal conductivity of the water is much better than that of gas, the channels can be smaller in cross-sectional size and still achieve adequate removal of heat from the coil section. Note that while most commonly water is used for the cooling media, some older designs utilized oil as the cooling media and, as such, this design is sometimes called a liquid-cooled machine. Many of the inner oil-cooled windings have been converted to water-cooled systems, because water is a more efficient cooling media. The picture shown is the inner water-cooled type. Note that in all of the coils shown in Figure 16.2, the asymmetry seen in the strand stacks is the key indicator that the cross-section has a Roebel transposition of the conductors.

In addition to the above discussion on cooling, when one examines Figure 16.2, one will notice that the conductor in the coil is not one solid piece of copper but strands of copper with thin layers of insulation between the strands. This is done to ensure even distribution of current through the coil cross-section area. The magnetic field at the top of the slot area (toward the centerline of the generator) has a different magnitude than the magnetic field at the bottom of the slot. The top of the slot, being closer to the main field and associated air gap, has a lower flux density than the bottom of the slot that is surrounded by the magnetic core. The amount of electromagnetic forced generated in the coil is a function of the speed of the machine and the amount of flux the conductor cuts as defined in equation (16.4).

$$E = Z n \varphi / 60 \qquad (16.4)$$

where

E = voltage

Z = length of conductors perpendicular to magnetic field

n = speed that field cuts the conductor

φ = flux

Since the bottom of the coil slot area is exposed to a larger flux density than the top of the coil slot area, the majority of the current would tend to flow in the top section of the coil if the coil were one solid piece of copper, (area with lower flux density). This would lead to higher I^2R losses in the area of the coil with higher current density. This is shown in Figure 16.3.

To ensure even current distribution, the coil is constructed of many strands with insulation between each strand. Then, along the length of the stator core, the strands are rolled to ensure that, along the length of the slot, each strand occupies every position in the coil slot. This is known as a Roebel Transposition of the strands. This is shown in Figure 16.4. This is the method used to ensure even current density throughout the cross-section of the coil. Up until the mid-1970s, only the area in the slots was transposed as this area is exposed to higher flux density levels and even exposure to the gap flux is most important in this area. Designs for Roebel bars with 360 degree and 540 degree transpositions (360 degree is one revolution of strand position through the length of the core slot area and a 540 degree is a one-and-a-half revolution of

Rotor

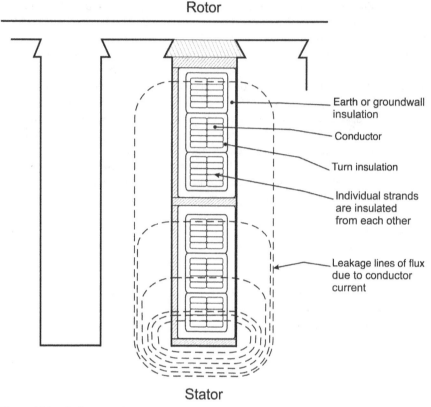

Earth or groundwall insulation

Conductor

Turn insulation

Individual strands are insulated from each other

Leakage lines of flux due to conductor current

Stator

Figure 16.3   Coil assembly in slot area. *Source*: Reproduced with permission of National Electric Coil.

strand position through the length of the core slot area) have been around for a long time. Since the mid-1970s a few manufacturers now also transpose the coil in the end-turn region although most still only transpose the area in the core slot section.

These coils are placed in the slots of a stationary core. Figure 16.5 shows the typical installation of a top and bottom $\frac{1}{2}$ coil in a stator core slot. Between the top and bottom coils is a separator where the resistance temperature detector (RTD) is located for monitoring temperature of the winding. Between the ground wall insulation and the sides of the core slots, there may be side fillers to ensure a tight fit of the coil in the slot. Additionally, fillers or springs may be used between the top of the top coil and the slot wedge for the same purpose. These fillers prevent rubbing of the ground wall insulation on the grounded core. At the top of the slot, a wedge is installed to keep the coils in place and a ripple spring is installed between the top coil and the slot wedge to ensure the coil is firmly held in place and does not vibrate. Vibration can cause premature failure of the ground wall insulation system due to insulation fretting.

The core supports an alternating flux. If this core were one solid piece of iron, then circulating currents would be induced in the core material. These circulating currents are known as *eddy currents*. Eddy currents flow in closed loops within

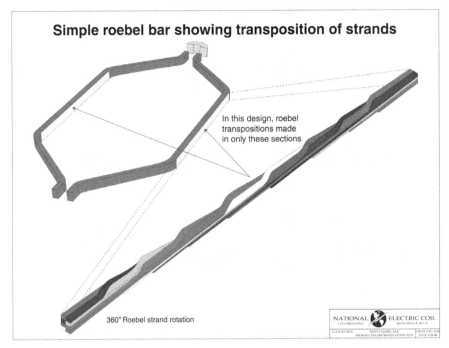

Simple roebel bar showing transposition of strands

In this design, roebel
transpositions made
in only these sections

360° Roebel strand rotation

NATIONAL ELECTRIC COIL

Figure 16.4   Roebel transposition. *Source*: Reproduced with permission of National Electric
Coil.

conductors, in planes perpendicular to the magnetic field. These currents do no useful
work and have associated with them I^2R losses due to the resistance of the iron that
makes up the stationary core. In order to minimize these losses, the core is not one
solid piece of iron, but is stacked up in thin laminations that are perpendicular to the
magnetic field to reduce the cross-sectional area where these circulating current may
flow. Additionally, selection of a core material with a lower resistivity will result in
reduced eddy current losses.

The core is constructed of laminated sections. Each lamination is typically
about 15 mils (15/1000th of an inch) thick. There are several sections or segments
needed to complete one 360 degree section of lamination of a certain thickness (see
Figure 16.6 for an example of one lamination section). Then these segments are
stacked on top of each other to stack the core. The coils are constructed so that half
of the coil lies on the top of one slot and the other half of the coil lies in the bottom
of another slot defined by the slot pitch.

The *coil pitch* is the distance between the top half coil and the bottom half coil
of an individual coil of an AC armature. This is not always the same value as the pole
pitch. The *pole pitch* is the distance measured on the circumference of the armature
from the center of one pole to the center of the next pole. When the angular distance
between the sides of a coil (coil pitch) is exactly equal to the angular distance between
the centers of adjacent field poles (pole pitch), the coil is termed to be a *full-pitch coil*.
To determine the coil pitch in number of slots, take the slot number of the top half
coil and subtract the slot number of the bottom half coil. For example, if the top half

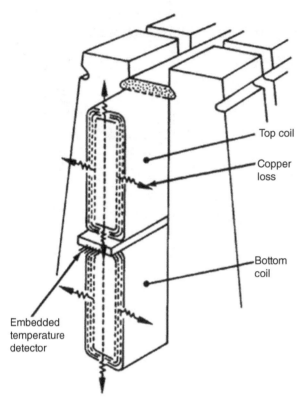

Figure 16.5   Coil assembly in slot area. *Source*: IEEE 67, Reproduced with permission of IEEE.

of coil #1 lies on the top of slot 1 and the bottom half of coil #1 lies in slot 18, then the coil pitch for this generator would be

$$\text{Coil pitch} = \text{Slot } 18 - \text{Slot } 1 = 17 \text{ slots}.$$

By comparison, if the above machine core was a three-phase, two-pole machine and contained 54 slots for the coils, then the pole pitch would be

$$\text{Pole pitch} = 54 \text{ slots} / 2 \text{ poles} = 27 \text{slots} / \text{pole}.$$

The number of slots in the machine equals the number of coils in the machine. Each coil is constructed in two parts or half coils to allow for assembly into the slot section. Since the number of slots equals the number of coils, then the number of half coils equals two times the number of slots.

A coil group is the windings that are associated with a certain pole and certain phase and certain winding of an electrical machine. Mathematically, the number of coil groups is defined as

$$\text{Number of coil groups} = (\text{Number of coils})/(\text{Number of phases}$$
$$\times \text{Number of windings} \times \text{Number of poles})$$

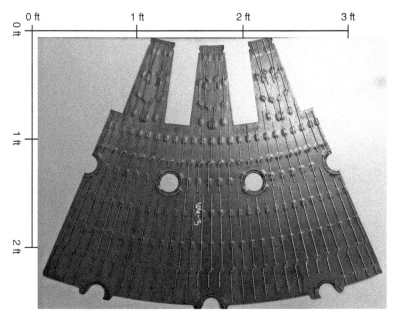

Figure 16.6   Typical lamination section with ventilation channels.

Example 16.4   A three-phase, one-winding, four-pole synchronous generator contains 48 coils. Determine the number of coil groups in the machine.

Solution:

Coil groups = (Number of coils) / (Number of phases × Number of windings × Number of poles)

Coil groups = (48 coils) / (3 phases × 1 winding × 4 poles)

Coil groups = 4 coils per group

So how many lamination pieces might be used to build one machine?

Example 16.5   A generator is to be constructed using nine laminations per disc (one 360 degree section) and each lamination is 15 mil thick. If the length of the core (excluding ventilation fingers) is 225 in., approximately how many pieces of lamination will be needed to construct this core completely?

Solution: It takes nine laminations of 15 mil (0.015 inch) thick to build one circle that is 15-mil thick. To get 225 in., this takes

225 in. / 0.015 in. = 15,000 circles.

Since there are nine laminations to make one circle, the total number of laminations is

15,000 circles × 9 laminations / circle = 135,000 laminations

Figure 16.7   Generator stator with rotor removed.

In the core, in addition to the eddy current losses that are controlled by using thin laminations, hysteresis losses are another power loss in the machine. *Hysteresis losses* occur due to the energy required to align up the magnetic field in the iron each time the flux alternates from a positive orientation to a negative orientation and again from a negative orientation back to a positive orientation. For a 60 Hz system (60 cycles per second), this loss occurs once each half cycle or 120 times per second.

Figure 16.7 shows the stator slot area with the stator coils assembled inside the stator core with the rotor removed from the generator.

The rotor circuit has multiple turns of conductor that form the main field. There are layers of insulating material between these conductors to ensure the insulation integrity between turns. The magnetic field strength (ampere × turns) of the main field is proportional to the number of turns (turns) in the main field and the current flowing (ampere) in the main field. Should this insulation fail, this can cause some of the turns to short together. This reduces the effective number of turns in the main field and, for the same current, the magnetic field strength (ampere × turns) is reduced. This can lead to a reduction of reactive power transfer capability. Additionally, since the opposite pole most likely does not have the same number of turns shorted, this can result in an uneven magnetic field distribution in the machine. This can lead to unbalance on the machine. One of the key indicators of a shorted field is a sudden increase in vibration, especially if the change in vibration is related to the amount of field current applied to the machine or MVA delivered by the machine.

One method of determining if the main field has a turn-to-turn short is by the use of a *flux probe*, as is shown in Figure 16.8. This probe sits in the air gap between the stator and the rotor. It senses the magnetic field strength of each slot in the rotor as it passes the probe.

In Figure 16.9, the dashed line shows the typical signal picked up by the probe as the seven slots in the rotor Pole B pass. This is the normal waveform seen and you

Figure 16.8   Flux probe mounted in generator stator.

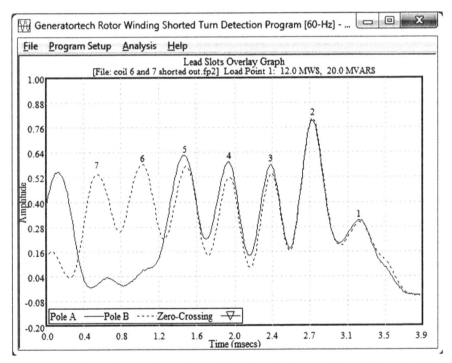

Figure 16.9   Flux probe waveform. *Source*: Reproduced with permission of Generatortech, Inc.

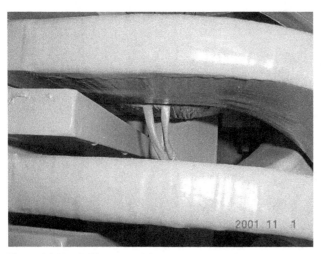

Figure 16.10   RTD mounted in generator stator between top and bottom coils.

can see a peak flux for each slot that passes the probe. Since magnetic flux is function of main field current and the number of turns in a slot, the amplitude of each waveform is also a function of both the main field current and the number of effective turns. Should some turns become shorted, then for the same main field current magnitude, the amplitude displayed on the oscillograph is reduced. In Figure 16.9, for Pole A shown in the solid line, you can see, the amplitude and therefore, the magnetic field strength does not change as slots 6 and 7 pass. This is an indication that all 18 turns in both Coil 6-Pole A and Coil 7-Pole B are shorted in the main field.

In addition to monitoring the flux in the machine we monitor the coil temperatures by the use of an RTD installed between the top and bottom half coils, as is shown in Figure 16.10. At minimum, there are two RTDs per phase in the machine.

Generators use either air or some type of gas to internally remove heat from components. A common gas used inside generators is hydrogen (H_2). The thermal conductivity of H_2 gas is very low allowing for good heat transfer from the heat sources in the generator to the H_2 gas. Also, the specific density of H_2 is about 1/10 of that of air. As the rotating mass of the machine spins in the gas, losses due to friction between the moving members of the rotor and the air inside the machine (known as *windage losses*) reduce the efficiency of the machine and generate heat. By using H_2 gas instead of air, these losses are substantially reduced. Of course, using hydrogen gas has special handing requirements. Hydrogen is explosive in concentrations between 5% and 95%. Therefore, the hydrogen purity in the machine is continuously monitored to ensure the concentration remains at 99% or higher. This prevents the generation of an explosive atmosphere in the machine. Additionally, during outages that require access to the inside of the generator, hydrogen is purged with an inert gas (CO_2). Then this inert gas is purged and outside air is allowed to enter. Once work is complete, this process is reversed. The gas in the machine is forced circulated utilizing blowers that are mounted on the generator shaft. Once the hydrogen gas picks up the thermal energy in the form of heat from internal components, it is directed

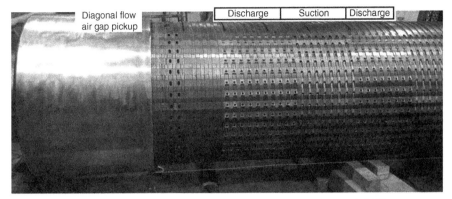

Figure 16.11   Cylindrical rotor (GE) with diagonal flow air gap pickup (DFAGPU) hydrogen cooling system.

to coolers in the generator where the heat is removed from the gas and the thermal energy is transported away with the cooling system.

There are two main types of rotor construction, cylindrical (non-salient pole) and salient pole. *Cylindrical rotors* are used in higher speed machines. Figure 16.11 depicts a typical cylindrical rotor. These rotors have a smaller diameter to limit the centrifugal forces in the rotor. These machines are two-pole and four-pole machines. For 60 Hz generators, these machines spin at 3600 rpm and 1800 rpm respectively. These machines have slots machined in the rotor surface and the main field conductor is placed in these slots. Due to construction, the flux density around a cylindrical machine is more uniform than with a salient pole design. This results in a power angle in the cylindrical machine that is purely sinusoidal. The power angle is the angle between the machines internally generated voltage (EMF) and the terminal voltage of the generator. Figure 16.12 defines the term power angle. Notice the ventilation slots in the rotor coil wedges and the aerodynamic design. This design is known as *diagonal flow air gap pickup* (DFAGPU). The name comes from the fact that the wedges form a scoop design that picks up the cooling gas in the generator gap area and the flow through the rotor body is in a diagonal direction.

From a derivation of the simplified model, it can be seen that the power generated by the cylindrical machine is a sine function of the power angle (angle between

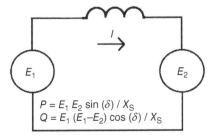

Figure 16.12   Simplified one-line diagram of power flow.

the machine-generated EMF and the terminal voltage of the machine) as defined in equation (16.5). Note that equation (16.5) only applies to cylindrical machines and not non-salient pole machines due to the non-uniform distribution of flux density in the salient pole design.

$$P = E_1 \times E_2 \sin(\delta)/X_s \tag{16.5}$$

where

P = Power transferred from the machine to the system

E_1 = internal EMF of the machine

E_2 = terminal voltage of the machine

δ = power angle

X_s = synchronous reactance of the machine

Looking at equation (16.5), the ideal peak power transfer occurs when the power angle (δ) = 90 degrees.

The *salient pole rotors* are used in lower speed machines and have a larger diameter. These machines are six-pole or more machines. For 60 Hz generators these machines spin at 1200 rpm or less. These machines have poles with field windings wound on the poles and the poles are mounted to the rotor forging using a support system known as a "spider." Due to the construction of the salient poles, the flux density in the air gap is less uniform. This results in a distortion of the power calculation shown for the salient pole machine such that the ideal peak power transfer in a salient pole machine occurs at an angle (δ) = 70 degrees. Equations (16.6), (16.6a), and (16.6b) approximate the power transfer function of a salient pole machine in terms of direct and quadrature axis currents and the peak production of power occurs where the summation of the direct and quadrature summation of these components are at a maximum value. The exact value of this peak power transfer power angle is dependent on the actual rotor construction.

$$P = \{I_d \sin(\delta) + I_q \cos(\delta)\} \times E_2 \tag{16.6}$$

where

$$I_d = \{E_1 - E_2 \times \cos(\delta)\}/X_d \tag{16.6a}$$
$$I_q = \{E_2 \times \sin(\delta)\}/X_q \tag{16.6b}$$

And where

P = power transferred from the machine to the system

E_1 = internal EMF of the machine

E_2 = terminal voltage of the machine

δ = power angle

X_d = direct axis synchronous reactance of the machine

X_q = quadrature axis synchronous reactance of the machine

I_d = direct axis current due to X_d

I_q = quadrature axis current due to X_q

Generators have various ratings associated with them. The rated kVA (thousand or kilo volt amps) of the generator is the rated current and voltage that the machine can deliver safely. A generator may have several kVA ratings that vary depending on the gas pressure inside the machine and/or the temperature of the cold gas leaving the cooler and entering the generator for cooling. The higher the gas pressure is inside the machine, the more heat that can be removed from the internal components of the machine. Therefore, with higher gas pressures, a generator will have higher kVA ratings. Additionally, the lower the cold gas temperature, the more heat that can be removed from the internal components of the machine without exceeding the thermal rating of the components of the machine. Therefore, with lower cold gas temperatures, a generator will have higher kVA ratings. The generator capability curve breaks down the safe kVA rating further into both kW (thousand watts or kilowatt) and kvar (thousand volt amps reactive or kilo volt amps reactive).

Figure 16.13 is a *generator capability curve* from IEEE C67, Guide for the Operation and Maintenance of Turbine Generators. The horizontal axis of the capability curves show real power transmitted by the generator in per unit (pu) kW. For this machine, this ranges from 0 to a positive value of 1.0 pu at a 1.0 power factor and a hydrogen pressure of 75 psi (517 kPa).

One might ask why there is no area on the capability curve for a negative real power (kW). If the machine starts to absorb real power (i.e., negative kW), then the generator will begin to absorb real power from the electrical system and, by definition of electrical machine function, the synchronous generator actually becomes a synchronous motor. In this area of operation, the generator begins to push power into the prime mover. While the generator may not have an issue with being driven by the electrical system as a motor, this is an issue for the prime mover that drives the generator. In the instance of a steam turbine, the amount of power delivered to the generator is a function of the steam flow through the steam turbine. If the generator begins to motor the turbine, this implies that there is little to no steam flow through the steam turbine. When the generator acts as a motor, it drives the turbine as the load. This causes windage losses in the turbine blade section. Without steam flow through the turbine to cool the turbine blades, this heating due to the windage losses will cause the rotating blades to thermally grow. If this condition is not quickly removed, the rotating blades can grow and eventually bind into the turbine casing causing severe damage to the generator. In the case of the diesel engine, driving the engine can quickly cause ignition in the combustion cylinder at a point where it is not designed to combust causing failure of the engine. In applications where the electric machine (generator) is only intended to deliver real power to the system and not absorb real power from the system, the capability curve only shows values for positive kWs to the system.

The vertical axis of the capability curve shows reactive power either transmitted or absorbed by the generator. For positive values of kvars (lagging or overexcited region), the power is delivered from the generator to the transmission system. For negative values of kvars (leading or underexcited region), the power is absorbed by the generator from the electrical system.

Inspecting Figure 16.13, we notice that between a power factor of 0.90 lagging (overexcited) and 0.95 leading (underexcited) (between points B and D), the kVA (current multiplied by voltage) ability of the machine is constant. However, at a power

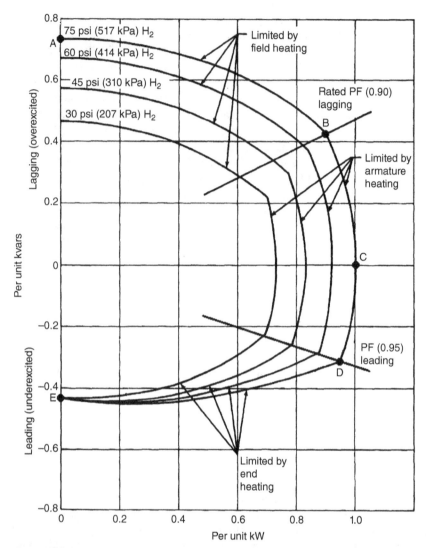

Figure 16.13 Typical capability curve for a hydrogen-intercooled machine. *Source*: IEEE C67, Reproduced with permission of IEEE.

factor less than 0.90 lagging (overexcited) (between points A and B) or 0.95 leading (underexcited) (between points D and E), the kVA rating of the machine is reduced. This forms three distinct sections of generator limitations. For the section between 0 and 0.9 lagging (overexcited) power factor or between points A and B, the main limitation is heating on the generator main field. To get higher values of kvars delivered to system, the current on the main field is increased. Once the power factor exceeds 0.90 lagging, the I^2R losses in the main field are the limiting component to generator operation. For the section between 0.9 lagging (overexcited) power factor and 0.95

leading (underexcited) or between points B and D, the main limitation is heating on the generator *stator winding*. Since the kVA rating of the machine is defined as the product of the generator terminal voltage and the generator current, since the limit in this region is the I^2R losses in the generator stator winding, this implies that the limit is the current in the stator winding and, if we have the same rated stator current from 0.9 lagging (overexcited) power factor and 0.95 leading (underexcited), then it makes sense that between these two power factors, the value of the rated generator kVA is the same. For the section between 0 and 0.95 leading (underexcited) power factor or between points D and E, the main limitation is end-turn heating due to stray flux concentrations in the end-turn area during periods of underexcited operation. At this point, the generator is absorbing vars from the system and this increases the leakage flux at the ends of the machine. This increase in leakage flux leads to increased losses in the end-turn area. It is not typical for a synchronous machine to operate at a leading power factor (i.e., absorbing vars from the electrical system) except in those instances where transmission voltage may run higher than nominal during light loads. In these instances, the generator may be operated in the leading region to absorb vars from the system and reduce the magnitude of the system voltage. Normally synchronous machines are called to deliver vars to the electrical system as voltage support for the electrical system. Therefore, the leading region of operation is avoided in most applications.

Also notice that there are four distinct curves given in Figure 16.13 for the generator depending on the hydrogen pressure in the machine. With hydrogen pressure of 75 psi (517 kPa), the synchronous generator is rated for 1.0 pu at a 1.0 power factor. A second curve shows the generator capability at a lower hydrogen pressure of 60 psi (414 kPa) to be approximately 0.92 pu at a 1.0 power factor. A third curve shows the generator capability at an even lower hydrogen pressure of 45 psi (310 kPa) to be approximately 0.84 pu at a 1.0 power factor. A fourth curve shows the generator capability at an even lower hydrogen pressure of 30 psi (207 kPa) to be approximately 0.76 pu at a 1.0 power factor. This reinforces the concept that, with increased hydrogen pressure in the generator, the ability to remove heat from the generator components is improved.

One point should be made here for those that are familiar with synchronous motor terminology but not synchronous generator terminology.

For a synchronous motor, the condition of being underexcited is called lagging because current lags voltage in this condition. When the synchronous motor is underexcited or has a lagging power factor, the motor is absorbing reactive power (var) from the electrical system. For the synchronous motor, the condition of being overexcited is called leading because current leads voltage in this condition. When the synchronous motor is overexcited or has a leading power factor, the motor is delivering reactive power (var) to the electrical system. For a synchronous generator, because the direction of power flow is in the opposite direction from a motor, the condition of being underexcited is called leading because current leads voltage in this condition. When the synchronous generator is underexcited or has a leading power factor, the generator is absorbing reactive power (var) from the electrical system. The condition of being overexcited is called lagging because current lags voltage in this condition. When the synchronous generator is overexcited or has a lagging power

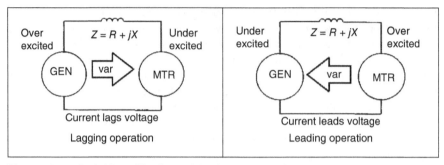

Figure 16.14   Definition of leading and lagging operation for motors and generators and direction of reactive power flow.

factor, the generator is absorbing reactive power (var) from the electrical system. In summary, in a motor, a lagging current condition means reactive power is absorbed by the motor while in a generator, a lagging current condition means reactive power is delivered by the generator. Alternately, in a motor, a leading current condition means reactive power is delivered by the motor while in a generator, a leading current condition means reactive power is absorbed by the generator. Figure 16.14 shows leading and lagging operation for motors and generators.

Knowing that

$$kVA = kW/PF \qquad (16.7)$$

If you specify a lower power factor for a certain kW rated machine, you will get a larger kVA rated machine.

The short circuit ratio (SCR) of a synchronous generator is the ratio of rotor current at full voltage no load (FVNL) normalized to the rotor current at full load (FL) for a three-phase fault or

$$\text{SCR (short circuit ratio)}$$
$$= \text{I rotor @ FVNL/I rotor @ FL for three-phase fault} \qquad (16.8)$$

Synchronous generators with higher short circuit ratios provide increased stability to system transients. They also have reduced efficiencies and higher fault currents than machines of the same kVA rating with a lower short circuit ratio. The synchronous reactance of the synchronous generator affects the transient stability of the machine. The lower the synchronous reactance of the machine, the more stable the machine operation will be to system transients. Additionally, a lower synchronous reactance will result in higher available fault currents from the machine.

Another useful generator chart to assist in the understanding of the capabilities of the generator is the generator *V-curve*. Figure 16.15 describes the V-curve of the same generator that was described by the capability curves in Figure 16.13. In Figure 16.15, field current (and thereby the strength of the main field in the generator) is represented on the horizontal axis of the chart in per unit and the kVA is displayed on the vertical axis in per unit. Lines of constant power factor are drawn on the chart.

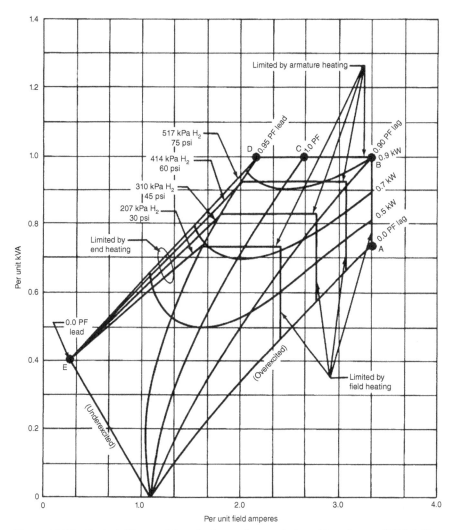

Figure 16.15   Typical "V-curve" for a hydrogen-intercooled machine. *Source*: IEEE C67, Reproduced with permission of IEEE.

Three curves of constant kW (0.9 kW pu, 0.7 kW pu, and 0.5 kW pu) are also displayed in Figure 16.15. The purpose of this chart is to define the required field current for a given generator operating point given in kVA, kW, and power factor. Also, if the real power (kW) is changed by the system, this chart can provide guidance as to the change in the power factor and MVA of the machine during this transient given the value of field current after the change. Also notice on this chart for a given kVA, as you increase field current above unity power factor the machine is operating in the lagging region and delivering vars to the electrical system. As you decrease field current below unity power factor the machine is operating in the leading region and absorbing vars from the electrical system.

Lastly, the generator capability curve (Figure 16.13) and the generator V-curve (Figure 16.15) are not independent, but are related. Let us look at the five points shown in both Figures 16.13 and 16.15. Point A of Figure 16.13 lies on the capability curve for a hydrogen pressure of 75 psi (517 kPa) and shows the generator operating at 0 pu MW, 0.0 power factor lagging and about 0.74 pu kvar lagging or overexcited. This same operating point (point A) is shown in Figure 16.13 to determine the machine field current pu kVA for this point. From Figure 16.15, the required field current to operate at point A is about 3.3 pu current. Point B of Figure 16.13 shows the generator operating at 0.9 pu kW, 0.9 power factor lagging and about 0.43 pu kvar overexcited. This same operating point (point B) is shown in Figure 16.15 to determine the machine field current and per unit kVA for this point. As can be seen from Figure 16.15, the magnitude of field current for point A and point B is the same at about 3.3 pu amperes. This reinforces the concept that the capability of the generator between points A and B is limited by field current. Point C of Figure 16.13 shows the generator operating at 1.0 pu kW, 1.0 power factor lagging and 0 kvar. This same operating point (point C) is shown in Figure 16.15 to determine the machine field current and per unit kVA for this point and can be found to be about 2.67 pu amperes. Point D of Figure 16.13 shows the generator operating at 0.95 pu kW, 0.95 power factor leading or underexcited and about 0.3 pu kvar underexcited. This same operating point (point D) is shown in Figure 16.15 to determine the machine field current and per unit kVA for this point and the field current is found to be about 2.15 pu amperes. Point E of Figure 16.13 shows the generator operating at 0 MW out, 0.0 power factor leading and about 0.4 pu kvar underexcited. This same operating point (point E) is shown in Figure 16.15 to determine the machine field current and per unit kVA for this point and the field current is found to be about 0.3 pu amperes. By inspection of Figure 16.15 we can easily see that the limitation of operation of the generator from operating point A to operating point B is field current which was stated previously when discussing the generator capability curve shown in Figure 16.13.

Example 16.6 Given the V-curve of Figure 16.15, given that the machine is operating at a 1.0 power factor at rated kVA for a 75 psi (517 kPa) hydrogen pressure (1.0 pu kVA), what is the field current required to support this point?

Solution: As shown in Figure 16.15a, Drawing a line from the intersection of 1.0 pu kVA and 1.0 power factor down to the horizontal axis, we see about 2.67 pu amperes are required to support this point of operation.

Example 16.7 Given the V-curve of Figure 16.15, given that the machine is operating at a 1.0 power factor at rated kVA (1.0 pu kVA) and the system suddenly sheds load such that the new power delivered by the machine is 0.7 pu kW and field current has not changed, determine

a. The new power factor of the machine.

b. If the power factor is lagging or leading.

c. Determine if vars are transmitted to system or absorbed.

d. The new per unit kVA operational point of the machine.

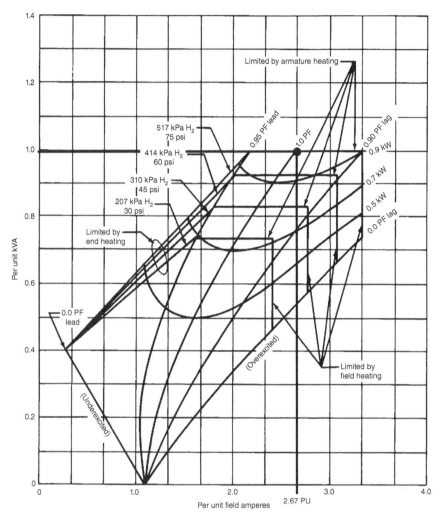

Figure 16.15a   "V-curve" solution for Example 16.6. *Source*: IEEE C67, Reproduced with permission of IEEE.

Solution: As shown in Figure 16.15b, Since field current did not change, we can draw a line from the intersection of 1.0 pu kVA and 1.0 power factor (our original operating point) down to the power line for 0.7 pu kW, this is our new operating point.

   a. The new power factor is approximately 0.90 leading.

   b. The power factor is now on the right of the 1.0 power factor line so this power factor is lagging (overexcited)

   c. Lagging power factor in a generator are vars transmitted from the generator to the system.

   d. Drawing a line horizontally from the new operating point we find the new per unit VA to be approximately 0.76 pu kVA.

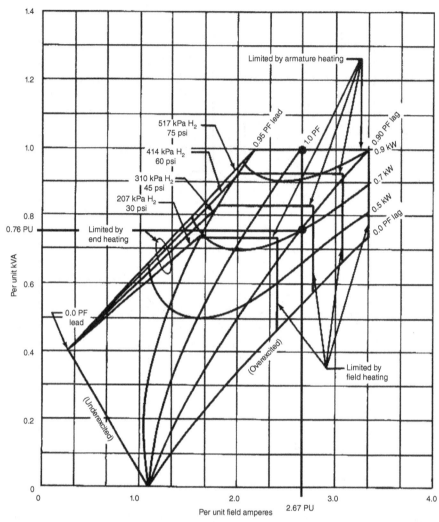

Figure 16.15b "V-curve" solution for Example 16.7. *Source*: IEEE C67, Reproduced with permission of IEEE.

Since the main field in the generator is rotating, direct indication of the main field winding temperature is not measured. The temperature is calculated from determination of the field resistance. As a side note, very recently, isolated thermal detectors that can be mounted in the rotating field winding area have been developed. They are battery-operated and can transmit the thermal value of field temperature via telemetry to a stationary receiver. This is a new development and not many devices are currently in use at this time. The generator manufacturer will provide the value of the field resistance at two different reference temperatures. The change in resistance in the field winding is a straight linear function based on the temperature of the winding.

The following formula can be used to determine the linear slope of change of resistance due to field temperature changes.

$$M = (R_h - R_c)/(T_h - T_c) \qquad (16.9)$$

where

M = slope

T_c = temperature of the colder resistance value given by the manufacturer.

R_c = resistance of the field winding at temperature T_c

T_h = temperature of the hotter resistance value given by the manufacturer.

R_h = resistance of the field winding at temperature T_h

Then using the value of slope from equation (16.9), the following formula can be used to interpolate the value of temperature of the field winding.

$$T_2 = [(R_{T_2} - R_{T_1})/M] + T_1 \qquad (16.10)$$

where

T_1 = temperature of the colder resistance value given by the manufacturer.

R_{T_1} = resistance of the field winding at temperature T_1

T_2 = calculated temperature of the field resistance in service

R_{T_2} = resistance of the field winding at temperature T_2

During operation, the magnitude of field current and brush voltage is measured. A small adjustment is made to remove the effect of brush resistance from the data. Then the resistance is determined using Ohm's law

$$R_{T_2} = Vdc/Idc \qquad (16.11)$$

where

R_{T_2} = resistance of main field (Ω)

Vdc = applied DC voltage to main field (V)

Idc = DC current flowing in main field (A)

Example 16.8   A generator manufacturer provides the following data in respect to the resistance of the generator main field at two values of temperature.

Field resistance = 0.1516 ohms @ 125°C

Field resistance = 0.1094 ohms @ 25°C

Determine the temperature of the main field winding if the measured field voltage (without brush drop) is determined to be 250 Vdc and the measured field current is 2040 amps.

Solution: First, we need to calculate the slope from the manufacturer-provided data using equation (16.9)

$M = (R_h - R_c) / (T_h - T_c)$

$M = (0.1516 \text{ ohm} - 0.1094 \text{ ohm}) / (125°C - 25°C)$

$M = 0.0422 \text{ ohm} / 100°C$

$M = 0.000422 \text{ ohm/°C}$

Now we need to calculate the field resistance from rotor winding voltage and current data using equation (16.11).

$$R_{T_2} = Vdc / Idc$$
$$R_{T_2} = 250\ Vdc / 2040\ Adc$$
$$R_{T_2} = 0.1225\ \Omega$$

Now that we have calculated the operating resistance of the main field, we can calculate the temperature of the main field. We can use equation (16.10).

$$T_2 = [(R_{T_2} - R_{T_1}) / M] + T_1$$
$$T_2 = [(0.1225\ \Omega - 0.1094\ \Omega)/(\ 0.000422\ \text{ohm/}°\text{C})] + 25°\text{C}$$
$$T_2 = [0.131\ \Omega /(\ 0.000422\ \text{ohm/}°\text{C})] + 25°\text{C}$$
$$T_2 = 31°\text{C} + 25°\text{C}$$
$$T_2 = 56°\text{C}$$

The impedance of the generator defines the response of the machine during a fault. The subtransient value of synchronous reactance X_s'' defines the impedance the machine presents to the system during the first few cycles of the fault ($0 < t < 0.2$ sec). This is shown in Figure 16.16 between points 1 and 2. Typical values in per unit for the subtransient value of synchronous reactance X_s'' are 0.1 to 0.4. This means that, for the first few cycles of a fault, the generator can deliver 2.5 to 10 times nameplate current into a fault. The transient value of synchronous reactance X_s' defines the impedance of the machine 12 cycles to between 30 and 120 cycles ($0.2 < t < (0.5$ sec to 2.0 sec)) into the fault. Typical values in per unit for the transient value of synchronous reactance are 0.2 to 0.6. This means that, for the next few seconds of a fault, the generator can deliver 1.5 to 5 times nameplate current into a fault. The steady-state reactance of the machine X_s defines the steady-state impedance the machine presents to the system for times between 30 to 120 cycles and greater. The steady-state synchronous reactance is 0.5 to 2.0 pu.

Since the initial fault current of the synchronous generator can be magnitudes of order larger than nameplate, this fault current can lead to very large magnetic forces in the machine causing windings to move and thereby destroying the machine. Because of this, the generators are connected to transformers via isolated phase bus duct or isophase bus duct to prevent the possibility of a phase-to-phase or three-phase short on the generator leads. The generator windings are connected in a wye where the neutral point of the generator is connected to ground via a grounding resistor or a grounding transformer with a resistor on the secondary. The purpose of this resistor is to limit phase-to-ground fault current of the generator. In the event the generator fails phase-to-ground, the fault current is limited to just a few amps and the subsequent potential damage to the generator windings is limited.

Synchronous generators (and synchronous motors) require a separately powered field to produce the flux required for operation. The magnitude of flux inside a synchronous machine defines the amount of reactive power the machine delivers to or receives from the electrical system by controlling the internal EMF of the machine. For the cylindrical rotor synchronous generator, the magnitude and direction

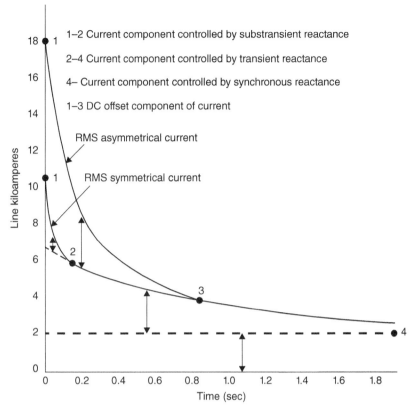

Figure 16.16 Three-phase short circuit decrement curve. *Source*: IEEE 242, Reproduced with permission of IEEE.

of reactor power flow is a product of the internal EMF of the machine, the difference between the internal EMF of the machine and the terminal voltage of the machine, and the cosine of the power angle. Note, for the salient pole synchronous generator, the equation for both real and reactive power is a function of both direct and quadrature synchronous reactance and, as such equation (16.12) only represents reactive power flow for cylindrical rotor synchronous machines. It is inversely proportional to the synchronous reactance of the machine. The reactive power flow for the cylindrical synchronous machine is defined as

$$Q = E_1 \times (E_1 - E_2) \cos \delta / X_s \qquad (16.12)$$

where

Q = reactive power delivered by the machine (Mvar)

E_1 = internal EMF of the machine (V)

E_2 = terminal voltage of the machine (V)

δ = power angle (degrees)

X_s = synchronous reactance of the machine

The internal EMF is a function of the amount of flux in the machine. The magnitude of EMF is the product of the speed of the machine and the flux as we presented earlier in equation (16.4).

$$E = Z n \varphi / 60$$

where

E = voltage

Z = length of conductors perpendicular to magnetic field

n = speed that field cuts the conductor

φ = flux

If speed is constant, then internal EMF is proportional to flux and flux is proportional to the magnitude of the current injected in the synchronous machines main field. Therefore, to increase the amount of reactive power generated by the synchronous generator, we increase the field current to the machine.

There are various types of exciters utilized in the industry. The first and oldest type is the *DC generator commutator exciter*. This is a DC generator directly coupled to the synchronous generator. As the shaft of the machine is rotating, an external field is applied to the dc generator and the DC generator produces DC current. This DC current is transferred to the main field directly. This system tends to be maintenance-intensive especially in the DC commutating section and DC brushes of the DC generator. Gradually, this system is being replaced with more modern systems. This system requires an external DC source for the field of the DC generator when initially energizing or "flashing" the field. Figure 16.17 shows a typical arrangement for a DC generator commutating exciter system.

The second type of system is the *AC generator commutating exciter*. This is very similar to the DC generator commutating exciter, except that this system uses an alternator that is mechanically coupled directly to the shaft that generates AC. The AC current is transmitted to external stationary rectifiers and then rectified. The rectified DC is sent to the main field brush assembly to inject the DC current into the main field. This is very similar to the DC generator system, except that, instead of using the mechanical rectification of the commutating section of the DC generator, this system uses a solid-state rectifier on the stationary part of the exciter to convert the AC of the generator to DC needed for the field. The output DC current is connected to the slip rings on the shaft of the machine. This system requires an external DC source for the field of the DC generator.

The third main type of excitation system is the *alternator rectifier brushless exciter*. This system has a permanent magnet generator (PMG) directly connected to the shaft of the machine where magnets are connected to the shaft of the machine and the armature of the PM generator rotate around these magnets. As the shaft of the main generator rotates at a rated speed, the permanent magnet generator develops an AC current on the stationary armature. This AC current is then sent to an external stationary rectifier that uses silicone-controlled rectifier (SCR) phase angle controller. This rectifier converts the AC output of the PM generator to a variable DC current. This current is then sent to an AC alternator that has its field poles stationary and

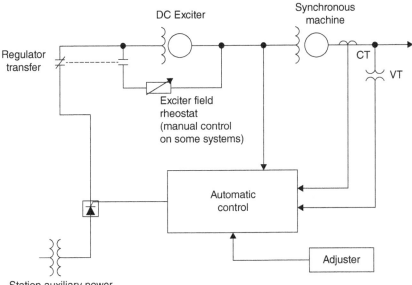

Figure 16.17 DC generator commutator exciter *Source*: IEEE 421, Reproduced with permission of IEEE.

mounted around the alternator armature that is mounted on the shaft. As DC current is applied to the stationary field poles, AC current is induced in the rotating armature. The amount of current induced in the rotating armature is proportional the amount of current applied to the stationary DC field poles. The output of the rotating AC armature is then fed to a group of rectifiers that are mounted on the shaft. This converts the AC current output from the rotating armature winding to DC current. This DC current is fed directly into the synchronous generator main field. Figure 16.18 describes the arrangement of an alternator rectifier brushless exciter.

A fourth type of excitation system is the *potential source static exciter*, which is shown in Figure 16.19. In this type of excitation system, and external AC voltage source is rectified using phase angle-controlled current regulating SCR controls. The output DC current is connected to the slip rings on the shaft of the machine. To prevent loss of excitation on a system short circuit near the machine when the machine voltage is depressed and excitation is most needed, a set of current transformers also feed the current regulating SCR controls. In the event of a low voltage condition in the system, the CTs act to boost power to the SCR controls to ensure adequate field current to the machine during the transient.

In all synchronous machines, whether the machine uses a direct connection, slip rings or brushes to apply the main DC current to the main field at the point on the shaft that enters the generator casing, there are radial studs and a hollow bore in the center of the shaft. The DC current bus bars are connected to the radial studs and connect to bus in the bore of the machine. A second set of radial studs exists inside the machine and connects the bus of the inner boar back to the surface of the shaft

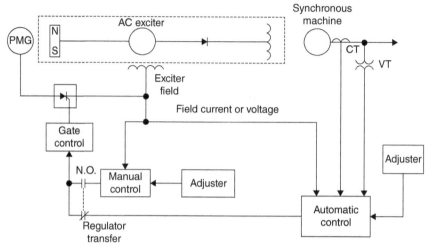

Figure 16.18 Alternator rectifier: brushless exciter. *Source*: IEEE 421, Reproduced with permission of IEEE.

inside the machine where the bus connections to the main rotor are made. This allows for the connection from the outside source for main field current around the bearing and seals at the generator enclosure to the main field inside the generator enclosure. This is shown in Figure 16.20.

The exciter functions to control the internal EMF generated. The terminal voltage of the machine is a function of both the internal EMF as well as the magnitude

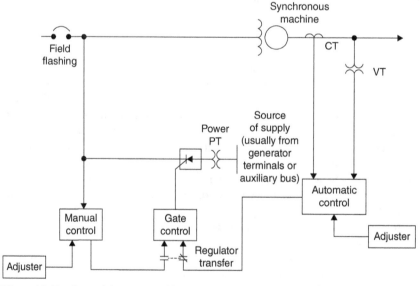

Figure 16.19 Potential source rectifier exciter. *Source*: IEEE 421, Reproduced with permission of IEEE.

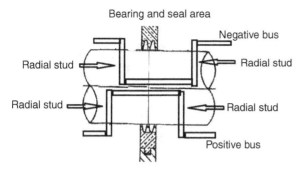

Figure 16.20 Radial stud configuration.

and phase angle of the current flowing through the machine armature windings. The control equipment which controls the terminal voltage of the machine in the exciter is called a voltage regulator. The *voltage regulator* is the device that defines the amount of field current required for the desired amount of output voltage.

Voltage regulators have two modes of control. The first mode of control is the manual or DC mode. In the manual or DC mode, the set-point is the DC current magnitude to the field and there is no feedback signal to the control system for set-point control. The second mode of control is the automatic or AC mode. In the automatic (AC) mode there is a set-point which may be either the generator terminal voltage or the machine delivered reactive power (vars). The control system compares the set-point value with the feedback signal (either voltage or vars depending on design) and the current to the field of the synchronous generator is automatically adjusted to regulate the terminal voltage or system vars at the set-point selected. The feedback signal that the machine is regulating on may also be power factor in some applications, but typical utility generation stations desire to be able to select the magnitude of terminal voltage or system vars delivered to the transmission system. One benefit of using generator terminal voltage is that maintaining generator voltage constant will provide automatic adjustment of vars needed for system voltage regulation. For example, if system voltage reduces, but generator voltage is constant, the reactive power flow (equation 16.12) shows us that reactive power flow into the system will increase since the difference between generator voltage and system voltage increases. To reinforce this concept, from the example that dealt with the generator V-curve you can see from Figure 16.15 that, when the generator power is changed, if the field current is not changed, this makes the machine power factor more leading and the machine tends to deliver more reactive power to the system. In automatic control, as the machine power delivered is reduced, the field current would be automatically controlled to maintain the set point for vars delivered.

The excitation control system also has some protective functions associated with it that are important in nature and will be reviewed. The *underexcitation limiter* (UEL) compares the real and reactive power of the generator to the capability curve while the synchronous generator is operating in the underexcited region and ensures that operation of the generator remains in the safe operating range defined by the generator capability curve and the end-turn region thermal limitations. Similarly the *overexcitation limiter* (OEL) compares the real and reactive power of the generator to the generator capability curve while the synchronous generator is operating in the overexcited region and ensures that the operation of the generator remains in the safe operating range defined by the generator capability curve and field winding heating limits. Additionally, the excitation system can provide control and protection of the generator core. The amount of flux contained inside the core is proportional to the ratio of the magnitude of generator voltage and the frequency of generator voltage. This is known as the volts-to-hertz ratio. Exceeding the designed volts to hertz radio on a synchronous machine can place the core operation in the saturated region of the B-H curve. This can lead to damage of the core of the machine. Since the excitation system defines the voltage induced on the generator armature leads, the excitation system senses the frequency of the generator voltage (defined by system load and turbine steam valve position) and regulates the excitation to the machine to maintain the ratio of generator voltage magnitude to generator voltage frequency within safe limits. Also, during the process of bringing a generator up to voltage and frequency in anticipation of bringing the unit online, the initial field current and therefore, generator voltage is applied at a minimum value to prevent an overvoltage condition on the machine when the field breaker is initially closed. When the field breaker is open, the excitation system offline limiter acts at full-voltage, no-load condition (VFNL) to limit excitation current and limit generator terminal voltage magnitude to no more than 105% of generator nominal voltage.

At the start of the process to bring a generator online, the generator field breaker and generator main breaker are both open. A synchronizing relay (25) monitors the magnitude, phase angle, and frequency of both transmission system voltages and generator system voltages through potential transformers that step the system voltage down to metering levels. It provides control to the generator excitation and turbine control systems to control generator voltage magnitude and frequency during the process of synchronization. Figure 16.21 shows a schematic representation of a synchronizing relay (25) monitoring the voltage from two sides of a circuit breaker (52).

After the synchronous generator is driven up to full speed by the prime mover, the field excitation breaker is closed and the initial main field current magnitude is applied to the synchronous machine. The application of main field current builds voltage on the generator armature terminals. Note the excitation control system is either in DC manual mode or AC automatic. When the excitation control system is in AC automatic mode, it is only regulating generator terminal voltage, not generator reactive power (vars). Reactive power control (when available) requires the generator

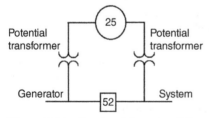

Figure 16.21   Synchronizing relay (25).

breaker to be closed and the generator must be delivering current to the system. The main field current of the generator is gradually increased until the magnitude of the generator terminal voltage on one side of the generator breaker slightly exceeds the system voltage on the other side of the generator breaker. Initially, one would think we want to exactly match the voltage magnitudes on both sides of the generator main breaker. The reason that we excite the machine such that the generator side voltage magnitude is slightly higher than the voltage magnitude on the system side of the generator breaker is the initial direction of reactive power flow the instant after the generator breaker is closed. Remember that reactive power flow direction is primarily defined by the difference between two voltages (and which voltage is more positive in magnitude) as was shown in equation (16.12).

$$Q = E_1(E_1 - E_2)\cos\delta/X_s$$

where

Q = reactive power delivered by the machine (Mvar)

E_1 = internal EMF of the machine (V)

E_2 = terminal voltage of the machine (V)

δ = power angle (degrees)

X_s = synchronous reactance of the machine.

From equation (16.12) you can see if the voltage magnitude on the generator side of the generator main breaker is slightly higher than the voltage magnitude on the system side of the generator main breaker, then the initial value of reactive power flow (Q) is positive which means that the generator will be delivering vars to the system when the generator breaker is closed. Remembering the discussion about the generator capability curve, the overexcited region is the normal area of operation.

The frequency of the generator voltage is determined by the speed of the machine as was defined in equation (16.1).

$$f = n \times p/120$$

where

f = frequency of the voltage supplied by the generator (Hz)

n = speed of the machine (rpm)

p = number of poles.

Before the generator is synchronized, the speed of the machine is determined by the speed control system of the prime mover. Once the excitation system applies main field current and the generator develops voltage, the control system monitors the speed of the machine, and thereby the frequency on the generator side of the generator main breaker and compares this with the frequency on the system side of the generator main breaker.

The position of the speed control system (governor) is gradually modified until the magnitude of the generator terminal frequency on one side of the generator breaker slightly exceeds the system frequency on the other side of the generator

Figure 16.22   Synchronizing meter.

breaker. Looking at the typical synch scope, the scope is rotating slowing in the "fast" (clockwise) direction as shown in Figure 16.22.

Initially, one would think we want to exactly match the frequencies on both sides of the generator main breaker. The reason that we have the generator frequency slightly higher than system frequency is the initial direction of real power flow the instant after the generator breaker is closed. Remember that the direction of real power flow is primarily defined by the power angle and whether the sine function is positive or negative as was defined in equation (16.5).

$$P = E_1 E_2 \sin(\delta)/X_s$$

where

E_1 = internal EMF of the machine

E_2 = terminal voltage of the machine

δ = power angle

X_s = synchronous reactance of the machine

By having the synchronous generator frequency slightly higher than the system frequency, we know when we initially close the generator main breaker, the value of the power angle (δ) will be positive and real power will flow from the synchronous generator to the system instead of the generator initially absorbing real power.

In addition to the voltage magnitude and voltage frequency, during generator synchronization, we monitor the phase rotation across the main generator breaker to ensure the same rotation on both sides. Most of these functions are done with a "synch check" relay.

We also look at the phase angle of the voltage across the breaker. Using the breaker closing time (3 to 5 cycles) and relaying delay time ($^1/_2$ to 1 cycle), and knowing the slip speed which is the speed at which the phase angle from generator side voltage compared with the system side voltage is changing, the control system anticipates when the next zero cross will occur across the generator breaker. The close command to the generator breaker is given slightly before this point where the breaker voltage phase angles are matched. When the generator breaker is finally closed, the system and generator voltages are matched.

GENERATOR PROTECTION

In an energy production facility, the electrical generator is both an expensive and a critical piece of equipment. In the typical generation facility, for other plant equipment, there are multiple pieces of equipment to provide redundancy. For example, there may be two forced draft (FD) fans, two coal pulverizers, two primary air (PA) fans, etc. This is done for reliability since with the loss of one piece of equipment, the unit can continue to produce electricity (but perhaps at a reduced maximum load). However, the synchronous generator is not redundant and is critical to the production of electricity. Therefore, the protection of the generator and turbine is critical in the design and operation of the power plant facility.

The generator will have current transformers on both the high side terminations as well as the neutral side terminations. These current transformers feed a differential relay (87) as shown in Figure 16.23. These current transformers sense the current into and out of each winding. As long as the current into one side of a winding matches the current flowing out of the other side of the winding, all the current flows through the restraint coils and no current flows through the operate coil and the differential relay does not operate. Note the polarity of the current transformers and connections to the relay as this is critical for the correct operation of a differential relay. Remember that current flowing into the polarity mark of the current transformer on the primary side is the current flowing out of the polarity mark on the secondary side. However, if an internal fault occurs in the generator where the currents on the two ends of the windings do not match, then current flows through the operate coil. When the ratio of operate coil current to restraint coil current exceeds a set value (defined by the slope setting of the relay), the differential relay activates and causes the generator to trip, protecting the machine.

The thermal protection for the stator winding is done by ensuring that the generator is operating within the specific capability curve for the gas pressure inside the machine (recall that the excitation system controller contains the UEL and OEL protection functions). While generators have RTDs to monitor the temperature of the windings of the generator, it is not common practice to trip on these values but to alarm to allow for operations to correct the cause of elevated temperatures.

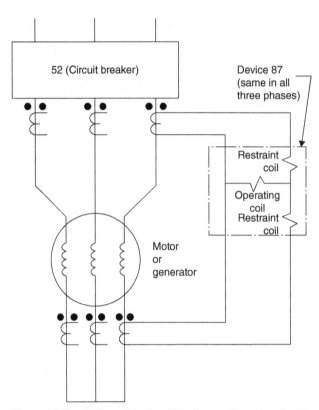

Figure 16.23   Differential relay (87). *Source*: Reproduced with permission of IEEE.

Negative sequence currents can be very damaging to the generator rotor. Negative sequence currents in the stator cause a rotating magnetic field in the generator that revolves in the opposite direction as the main field. This induces a current on the body of the generator rotor at twice the line frequency and this current has a tendency to travel along the conductive areas of the surface of the rotor forging and wedges. This can cause localize heating of the rotor body at points where there is greater resistance to this current flow (for example, between the rotor forging and retaining rings or in the areas of the flexible slots for a cylindrical machine). To protect the rotor from this damage, should negative sequence currents be detected in the stator currents, the protection system will trip the generator off-line thereby protecting the rotor forging.

As mentioned in Chapter 15, large generators use isophase bus between the generator termination leads and the generator step-up transformer to prevent a phase-to-phase short and use a neutral grounding transformer/resistor in the neutral-to-ground connection. Should one phase of the generator fault to ground at or near the high voltage termination of the generator, the phase-to-ground fault current is limited by this resistor to only a few amps limiting fault current in the machine and damage to the machine. An overvoltage relay (59N) is placed in parallel with the grounding resistor on the secondary side and/or a current transformer is placed in series with the grounding transformer on the primary side to feed an overcurrent device (51N). In the event of a phase-to-ground fault, the 59N will sense the voltage developed across the grounding resistor and the 51N will sense the current flowing in the neutral and these will act to trip the generator on the phase-to-ground fault.

Ground faults that occur inside the machine stator winding closer to the neutral side result in less ground current being developed and ground faults that occur inside the stator winding close to the neutral connection result in very little neutral current and very little voltage across the grounding resistor. As such, the 59N and 51N relays may not detect these faults. Because of this, a third harmonic undervoltage (27TN) relay that is tuned to the third harmonic of the generator frequency is placed across the grounding resistor. Under normal operations, the generator will generate a small amount of third harmonics currents (zero sequence currents) since the magnetic field is not perfectly balanced in the machine in the real-world application. Since these third order harmonic currents are in phase on all three phases, these currents flow in the neutral of the generator. These currents normally develop a voltage across the neutral grounding resistor at a frequency of three times the line frequency. Should a ground fault occur on one of the generator windings in the machine close to the neutral, the value of this third harmonic current is decreased. This results in a lower value of third harmonic voltage across the generator neutral. The third harmonic undervoltage relay 27TN monitors for this condition and, if it detects a reduction in this third harmonic voltage across the neutral grounding resistor, the 27TN actuates to trip the generator and protect the machine.

The main field of the generator does not have any intentional ground reference points. Both the positive and negative conductors are isolated from ground and the main field winding is insulated from the rotor body. As such, should a ground occur somewhere on the main field, initially there is no path for current and no fault current flows. However, this is a serious condition since, on the occurrence of a second ground at a different location in the machine, large currents will flow through the rotor body and can damage the rotor body (see negative sequence current discussion for more info on rotor body damage due to currents). Because of this, the field is occasionally monitored for grounds. If a ground is detected, the unit does not automatically trip, but an alarm is received and the unit is brought off-line as soon as safely possible to locate and repair the ground fault. On systems that use slip rings to connect the DC current to the main field, the ground detection circuit is typically on the excitation system. On brushless excitation systems, the main field does not come out on slip rings but is connected though diodes mounted on the rotor body. Therefore, an auxiliary set of slip rings is provided with a small set of brushes to connect the ground monitor equipment to the main field, as is shown in Figure 16.24.

Another event that can be damaging to a synchronous generator is loss of excitation. In loss of excitation, the field for the synchronous generator becomes de-energized causing the synchronous machine to operate as an induction machine. Now the generator speed can vary causing slip between the machine synchronous speed and actual speed. Just like an induction machine, this slip can induce currents in the rotor body (see negative sequence currents for description of adverse effect of currents in rotor body). To protect the machine in the event of a loss of excitation and protect the rotor body, the generator will trip on the loss of excitation condition.

Figure 16.24   Exciter field ground monitor brush assembly.

Additionally, should the field be reapplied when the generator breaker is closed, there is a very good chance the generator voltage will not be in phase with system voltage and the machine may be damaged due to the out-of-step condition due to the transient torque that will be developed in the generator.

GLOSSARY OF TERMS

- Capability Curve – A plot of generator real power (MW) and reactive power (Mvar) where real power is the horizontal axis of the plot and reactive power is the vertical axis of the plot.
- Coil Pitch – The coil pitch is the distance between the slot that the top half coil enters and the slot that the bottom half coil enters of an individual coil of an AC armature.
- Cylindrical Rotor – A rotor cylindrical in shape having parallel slots on it to place field windings. Faster rotating machines are constructed with cylindrical pole rotors.
- Eddy Currents – A localized electric current induced in a conductor by a varying magnetic field. These currents perform no useful work but result in I^2R losses and, as such, represent a loss in the generator.

- Flux Probe – Any instrument for measuring magnetic flux, usually by measuring the charge that flows through a coil when the flux changes.
- Full-Pitch Coil – A full-pitch coil is a coil design where the coil pitch is equal to the pole pitch.
- Gas-Intercooled Coils – Coils that have channels constructed inside the coils to allow for the gas inside the generator casing (typically air or hydrogen) to flow through the coil. The gas is used to remove heat from the coil.
- Generator – A machine that converts the mechanical energy delivered by the prime mover into electrical energy to be transmitted to the electrical system.
- Hysteresis Losses – Hysteresis losses occur due to the energy required to align up the magnetic materials in the iron each time the flux alternates from a positive orientation to a negative orientation and again from a negative orientation back to a positive orientation.
- Induction Generator – An induction generator is a type of electrical generator that is mechanically and electrically similar to a three-phase induction motor. Induction generators produce electrical power when their shaft is rotated faster than the synchronous frequency of the equivalent induction motor. Excitation must originate from the electrical system the generator is connected to.
- Indirectly Cooled Coils – Coils that are solid in construction and have ground wall insulation encapsulating the conductors. Heat is removed by the flow of gas inside the casing of the generator around the outside of the ground wall insulation.
- Overexciter Limiter (OEL) – The OEL compares the real and reactive power of the generator to the capability curve while the synchronous generator is operating in the overexcited region to ensure operation remains in the capability curve of the machine.
- Pole Pitch – The pole pitch is the distance measured on the circumference of the armature from the center of one pole to the center of the next pole.
- Rotor – The rotating portion of an electric generator or motor.
- Salient Pole Rotor – Salient pole type of rotor consists of large number of projected poles (salient poles) mounted on a magnetic wheel. Slower rotating machines are constructed with salient pole rotors.
- Stator – The stationary portion of an electric generator or motor.
- Stator Winding – A stator winding is the stationary winding in an electric generator or motor.
- Supersynchronous Speed – Speed above synchronous speed.
- Subsynchronous Speed – Speed below synchronous speed.
- Synchronous Generator – A synchronous generator is an AC generator having a DC exciter. Synchronous generators are used as stand-alone generators for emergency power and can also be paralleled with other synchronous generators and the utility system.

- Underexciter Limiter (UEL) – The UEL compares the real and reactive power of the generator to the capability curve while the synchronous generator is operating in the underexcited region to ensure operation remains in the capability curve of the machine.

- Voltage Regulator – The voltage regulator is the device that defines the amount of field current required for the desired amount of output voltage.

- Water-Intercooled Coils – Coils that have channels constructed inside the coils to allow for water to flow through the coil. The water is used to remove heat.

PROBLEMS

16.1 What is the speed of a synchronous four-pole generator operating at 60 Hz?

    A. 3600 rpm

    B. 1800 rpm

    C. 1200 rpm

    D. 1900 rpm

16.2 What is the frequency generated by a synchronous two-pole generator running at 3000 rpm?

    A. 50 Hz

    B. 60 Hz

    C. 100 Hz

    D. 120 Hz

16.3 The purpose of a flux probe is to

    A. detect stator bar shorts

    B. detect rotor bar shorts

    C. detect high eddy current loss in stator iron

    D. detect high vibration in journal bearing

16.4 A stator slot RTD is located where?

    A. Between top coil ground wall insulation and wedge

    B. Between top coil ground wall insulation and bottom coil ground wall insulation

    C. Between bottom coil ground wall insulation and slot bottom

    D. Between core laminations

16.5 Given a generator that has field resistance values given as 0.1516 ohms at 125°C and 0.1094 ohms at 25°C, determine the temperature of the field winding if the measured resistance of the field winding is 0.1225 ohms.

    A. 35°C

    B. 42°C

    C. 56°C

    D. 72°C

16.6   For a synchronous generator, real power delivered into the transmission system is controlled by the _____ between generator voltage and transmission system voltage, and reactive power is controlled by the _____ between generator voltage and transmission system voltage.

    A.   phase angle; voltage magnitude

    B.   voltage magnitude; phase angle

16.7   What is the power factor of a generator that is being operated at 100 kVA if the "real" part of the 100 kVA is 80 kW?

    A.   1.0

    B.   0.9

    C.   0.8

    D.   0.7

16.8   A three-phase, one-winding, two-pole synchronous generator contains 48 coils. Determine the number of coil groups in the machine.

RECOMMENDED READING

Electric Power Plant Design, Technical Manual TM 5-811-6, Department of the Army, USA, 1984.

Electrical Machines, Drives and Power Systems, 6th edition, Theodore Wildi, Prentice Hall, 2006.

The Engineering Handbook, 3rd edition, Richard C. Dorf (editor in chief), CRC Press, 2006.

IEEE 141-1993: IEEE Recommended Practice for Electric Power Distribution for Industrial Plants (IEEE Red Book).

IEEE 242-2001: IEEE Recommended Practice for Protection and Coordination of Industrial and Commercial Power Systems (IEEE Buff Book).

IEEE 67-2005: IEEE Guide for Operation and Maintenance of Turbine Generators.

IEEE 421.5-2005: IEEE Recommended Practice for Excitation System Models for Power System Stability Studies.

IEEE 421.1-2007: IEEE Standard Definitions for Excitation Systems for Synchronous Machines.

Industrial Power Distribution, 2nd edition, Ralph E. Fehr, III, Wiley-IEEE Press, 2016. ISBN: 978-1-119-06334-6.

Power Plant Engineering, Black & Veatch, edited by Larry Drbal, Kayla Westra, and Pat Boston, Chapman & Hall/Springer, 1996.

Standard Handbook of Powerplant Engineering, 2nd edition, Thomas C. Elliott, McGraw-Hill, 1998.

CHAPTER 17

MOTORS

GOALS

- To understand the basic design and construction of motors
- To understand some of the safety requirements when dealing with motor circuits and their associated loads
- To calculate the synchronous frequency and slip of an induction motor
- To calculate the mechanical power that a load draws
- To calculate the electrical power that a motor draws and the efficiency of the motor
- To calculate the required power factor capacitance needed to improve system power factor
- To understand the concept of minimum accelerating motor torque
- To identify the reduction in available motor mechanical torque for various types of induction motor reduced voltage starting methods
- To identify various types of motor cooling methods
- To be able to select an acceptable value of motor rated voltage for a given application
- To be able to calculate motor torque and speed as it is reflected through a gear or pulley system

---

ELECTRIC **MOTORS** convert electrical energy to mechanical energy to perform work. Motors are a very important element in the operation of the auxiliary systems that support power plant operations. For example, in nuclear power plants that are of a pressurized water reactor (PWR) design, the primary coolant is circulated between the reactor vessel and the steam generator by a reactor cooling pump (RCP). This is a critical function performed by electric motors. Therefore, it is helpful to understand motor applications and operational considerations.

The National Electrical Safety Code (NESC) requires some minimum safety precautions when dealing with motor circuits and the loads connected to the motors.

Energy Production Systems Engineering: An Introduction for Electrical Engineers to Electrical Power Generation Facilities, Systems, and Equipment, First Edition. Thomas H. Blair.

Separately excited DC motors and series DC motors, under the right conditions, may result in an overspeed condition which could generate a hazard to personnel in the vicinity of the motor and rotating equipment. To protect personnel in the area of the rotating equipment, NESC section 13 requires that all separately excited DC motors and series DC motors shall be provided with a speed-limiting device such as an overspeed trip circuit to limit the maximum speed of the machine to safe limits. Additionally, NESC requires that all motors that may start unexpectedly creating a hazard to personnel shall have the motor control circuit designed to prevent automatic starting of the motor after a power supply interruption. Upon power supply restoration, the operator must take some action to reset the control circuit and allow the motor to start. Lastly, just like the National Electric Code (NEC®), NESC requires that the motor control system be provided with some means of short circuit protection. Additionally, motors present moving mechanical equipment and all the safety comments in Chapter 7 on conveyors also apply to motor applications such as guarding moving parts to prevent personnel from becoming injured from rotating equipment.

Note that motor starters are designed to interrupt motor load current up to locked rotor current, but are not designed to interrupt fault current magnitudes in excess of motor locked rotor current. Therefore, the motor protection relay will trip the starter for current up to locked rotor current, but for fault currents in excess of motor locked rotor current, another device such as a circuit breaker or fuse is used to provide branch circuit protection.

The AC induction or synchronous motor is basically a rotating transformer with a primary and a secondary winding with a magnetic core that couples these two windings. While there are many types of construction of motors, in general, motors can be categorized as induction machines or synchronous machines. Most of this chapter is dedicated to the induction motor with some information on synchronous machines. For more information on synchronous motors, the reader can refer to Chapter 16 concerning generators. The chapter that covers generators addresses synchronous generators and, from a mechanical standpoint, synchronous motor and synchronous generator are very similar. The main thing that defines if the synchronous machine is a synchronous motor or a synchronous generator is the design of the direction of power flow for the equipment connected to the motor. If the equipment is a prime mover, then the synchronous machine is meant to operate as a generator and if the equipment is a load such as pump or fan, then the synchronous machine is meant to operate as a motor.

The induction motor is constructed with a stationary winding (the armature) that is installed in slots of a stationary core (stator). The other winding (the field) is installed on a core on the rotating element (rotor). The windings on the rotor are shorted at both ends of the rotor in an induction motor forming a "squirrel cage" (see discussion on wound rotor induction motors for an example of rotor that is not shorted at the ends of the rotor). AC voltage is applied to the stator winding. When a conductor moves in an area where there is magnetic flux, a voltage is induced on that conductor. This voltage is defined by Faraday's law. (Note in the following equations that values in bold text are vectors with properties of both magnitude and direction.)

$$\mathbf{EMF} = \ell(\mathbf{V} \times \mathbf{B})$$

where

EMF = voltage induced in a conductor of length *l* moving at velocity **V** through a magnetic flux density of **B**.

This induces a current in the stator winding which induces a rotating magnetic field inside the stator. With the motor shaft at standstill, the motor and stator perform much like a transformer. The lines of magnetic flux cut the conductors of the field circuit on the rotor which induces an electromagnetic force on the field conductor that is defined by

$$|emf| = n \, \varphi d/dt \tag{17.1}$$

emf = EMF induced in rotor circuit

n = number of turns on rotor circuit

dφ / *dt* = rate of change of flux

Since the ends of the field circuit are shorted together in a squirrel cage rotor, a current is induced in the field windings. This current inducted on the bars of the rotor puts a force on the rotor that causes it to start to rotate in the same direction of rotation as the rotating field applied by the stator. This force is defined by the Lorentz equation.

$$\mathbf{F} = \mathrm{I}(\ell \times \mathbf{B})$$

where

F = The force exerted on a conductor of length ℓ with current *I* flowing through the conductor and where the conductor is in a magnetic flux of density **B**.

or

$$|F| = B \, \ell \, i \tag{17.2}$$

F = force

B = magnetic flux density

ℓ = length of conductor

i = current flowing.

As the rotor increases in speed, the speed at which the lines of flux pass the rotor conductors is reduced. Looking at equation (17.1), if we reduce $d\varphi$ / dt, then we reduce emf induced in the rotor circuit. This reduces the current induced in the rotor circuit. Equation (17.2) tells us that a reduction in current results in a reduced force on the rotor.

Assuming the ideal case where there is nothing connected to the motor shaft and there are no losses in the motor, the rotor will increase in speed until the point when the shaft is spinning in synchronism with the magnetic field in the motor. This is known as the *synchronous speed* of the machine.

$$n_{\mathrm{s}} = 120 \times f/p \tag{17.3}$$

where

n_{s} = synchronous speed of machine (rpm)

f = frequency of voltage applied to stator (Hz)

p = number of *poles* in the machine

Example 17.1   Given a six-pole, 4000 V, three-phase, 60 Hz motor with a full-load speed of 1125 rpm, calculate the synchronous speed of the motor.

Solution: Using equation (17.3), we find the following.

$$n_s = 120 \times f/p$$
$$n_s = 120 \times 60 \text{ Hz}/6 \text{ pole}$$
$$n_s = 1200 \text{ rpm}$$

At this point, since the rotor conductors are in synchronism with the magnetic field, the rotor conductors are not cut by the flux in the motor and the value of $d\varphi/dt$ $= 0$. Therefore, the value of the emf from equation (17.1) is zero. Since there is no emf inducted in the rotor circuit, the current in the rotor circuit would be zero. Equation (17.2) would tell us that there is no force driving the rotor when the induction machine is rotating at synchronous speed. If there are no losses and no power is transferred, the motor would stay at this speed, known as the synchronous speed. In reality, there are losses in the motor such as windage losses that require a small amount of *torque*, so synchronous speed is never really achieved. Additionally, as load is applied to the shaft, this requires more torque from the rotor circuit, which requires more force. Since the force on the rotor is related to the value of the rate of change of flux ($d\varphi/dt$), this implies that the motor rotor must slow down to induce more current in the field winding. This brings up the term *slip*. The slip of the induction machine is defined by

$$s = [(n_s - n)/n_s] \times 100\% \qquad (17.4)$$

where

$s =$ slip of the machine (%)

$n_s =$ synchronous speed of the machine (rpm)

$n =$ operating speed of the machine (rpm)

Example 17.2   Given a six-pole, 4000 V, three-phase, 60 Hz motor with a full-load speed of 1125 rpm, calculate the slip of the motor when operating at full-load speed.

Solution: As we found from Example 17.1, the synchronous speed is,

$$n_s = 1200 \text{ rpm}$$

Given in the problem is the full-load speed of

$$n = 1125 \text{ rpm}$$

From equation (17.4), we can find the slip of the motor.

$$s = [(n_s - n)/n_s] \times 100\%$$
$$s = [(1200 \text{ rpm} - 1125 \text{ rpm})/1200 \text{ rpm}] \times 100\%$$
$$s = [(75 \text{ rpm})/1200 \text{ rpm}] \times 100\%$$
$$s = 6.25\% \text{ slip}$$

As mentioned above, as the induction machine's actual speed varies from the synchronous speed, the amount of force on the conductor of the rotor field varies. Since torque is the product of force and distance, the torque applied to the rotor shaft also varies as the speed of the induction machine varies. Therefore, the amount of torque available on the shaft of the machine is a function of the slip of the machine. The mechanical power available at the shaft of the motor is the product of the torque available on the shaft of the rotor and the speed of the machine and is calculated by

$$P_m = T\,n/5250 \tag{17.5}$$

where

P_m = mechanical power (HP)

T = torque (lb ft)

n = speed (rpm)

The factor of 5250 is necessary to convert the units on the right side of the equation which are (lb ft) × (rev/min) to units of *horsepower* (HP) on the left side of the equation. This is obtained knowing that one revolution is 2π radians and 1 HP = 33,000 lb ft/min. Utilizing unit conversion, we find

{1 (lb ft) × (rev/ min)} × {1 HP min /33,000 (lb ft)} × {2π radians/1 rev}
= 1/5250

Example 17.3   Given a six-pole, 4000 V, three-phase, 60 Hz motor that is turning at a speed of 1125 rpm and is delivering a torque of 750 lb ft, calculate the power (in HP) delivered to the load.

Solution: Using equation (17.5), we find the following.

$$P_m = T\,n/5250$$

$$P_m = (750 \text{ lb ft})1125 \text{ rpm}/5250$$

$$P_m = 161 \text{ HP}$$

As mentioned previously, the amount of torque the rotor develops is a function of the speed of the rotor and the synchronous speed of the machine. Depending on the construction of the machine, the amount of torque available for a certain value of slip can vary as shown in Figure 17.1 in a typical torque speed curve. Figure 17.1 and Table 17.1 provide guidance as to the various torque speed characteristic of both NEMA and IEC motors.

There are four distinct torque points that are of interest and they are shown in Figure 17.2. The *locked rotor torque* is the amount of torque the shaft can produce when the rotor is not moving. Once the rotor starts to move, as discussed above, the amount of force, and therefore torque, starts to reduce. At some point in the acceleration of the rotor, the amount of torque begins to increase with increasing speed. This point is known as the pull-up torque. At some point, the amount of torque available from the shaft reaches a peak value. This value is known as the *breakdown torque*. The last point is the *full-load torque* and is the amount of torque the rotor provides to the load when the motor is running at *full-load speed*.

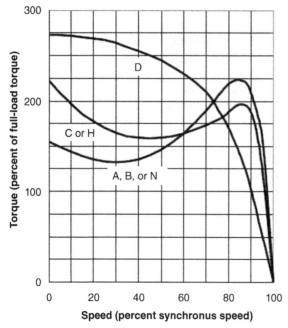

Figure 17.1 Torque speed curves for various NEMA and IEC design motors. *Source:* Reproduced with permission of National Electrical Manufacturers Association.

Motors absorb *real power* (kW) from the electrical system and convert that power to mechanical power at the shaft to perform some work such as driving a fan or pump. In addition, motors either absorb or deliver reactive power (kvar) from or to the electrical system. Induction machines always absorb reactive power from the electrical system since their field is produced from energy from the electrical system via the slip of the motor. Synchronous machines have an externally applied field and, as such, can either absorb or deliver reactive power to the electrical system. The relationship between real power, reactive power, and total power is shown in equation (17.6).

$$S = \sqrt{(P^2 + Q^2)} \tag{17.6}$$

where

S = total power (kVA)

P = real power (kW)

Q = reactive power (kvar)

The angle between the real power and the total power is defined in equation (17.7).

$$\theta = \cos^{-1}(P/S) \tag{17.7}$$

TABLE 17.1   Torque Speed Characteristics – Various NEMA and IEC Design Motors

Polyphase Characteristics	Locked Rotor Torque (percent-rated load torque)	Pull-Up Torque (percent-rated load torque)	Breakdown Torque (percent-rated load torque)	Locked Rotor Current (percent-rated load current)	Slip
Design A Normal locked rotor torque and high locked rotor current	70–275[a]	65–190[a]	175–300[a]	Not defined	0.5–5%
Design B Normal locked rotor torque and normal locked rotor current	70–275[a]	65–190[a]	175–300[a]	600–800	0.5–5%
Design C High locked rotor torque and normal locked rotor current	200–285[a]	140–195[a]	190–225[a]	600–800	1–5%
Design D High locked rotor torque and high slip	275	Not defined	275	600–800	≥5%
IEC design H High locked rotor torque and high locked rotor current	200–285[a]	140–195[a]	190–225[a]	800–1000	1–5%
IEC design N Normal locked rotor torque and high locked rotor current	75–190	60–140	160–200	800–1000	0.5–3%

Source: Reproduced with permission of National Electrical Manufacturers Association.
Note: These typical characteristics represent common usage of the motors – for further details, consult the specific performance standards for the complete requirements.
[a]Higher values are for motors having lower horsepower ratings.

where

θ = angle between real and total power vector (degree)

S = total power (kVA)

P = real power (kW)

The *power factor* is the cosine of the power angle θ or the ratio of the real power (kW) divided by the total power (kVA)

$$\text{pf} = \cos(\theta) = (P/S) \qquad (17.8)$$

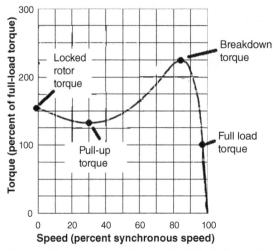

Figure 17.2 Torque speed curves for NEMA design B motor. *Source*: Reproduced with permission of National Electrical Manufacturers Association.

where

pf = power factor

S = total power (kVA)

P = real power (kW)

θ = phase angle between real and total power vector (degree)

For an induction motor, the current lags behind the voltage of the stator and reactive power is absorbed by the machine (Q is positive) and Figure 17.3 shows the power triangle for this configuration. Note here that the power angle Q is positive.

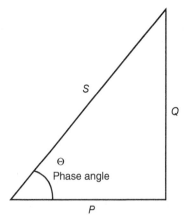

Figure 17.3 Power triangle for motor absorbing reactive power from the system. *Source*: Reproduced with permission of IEEE.

This appears as an inductive load to the electrical system. For a synchronous motor that is absorbing reactive power, this also looks like an induction motor from the standpoint of the power triangle and Figure 17.3 represents this situation.

In some installations of induction motors, power factor correction capacitors are added to the motor feeder to provide some capacitive reactance and reduce the amount of reactive power that the electrical system supplies to the motor feeder.

Example 17.4   Given a motor that is drawing 80 A of real power from an electrical system and is also drawing 60 A of reactive power from an electrical system, determine the power factor the motor presents to the electrical system.

Solution: Total current drawn is

$$I_{total} = \sqrt{(80^2 + 60^2)} = 100 \text{ A}$$
$$\text{pf} = (80 \text{ A}/100 \text{ A})$$
$$\text{pf} = 0.80$$

Example 17.5   Given a motor that is drawing 80 A of real power from an electrical system and is also drawing 60 A of reactive power from an electrical system. If we install power factor correction capacitors such that 60 A of reactive power is drawn from the capacitors and no reactive current is drawn from the line, determine the power factor the motor presents to the electrical system.

Solution: Total current drawn is

$$S = \sqrt{(80^2 + 0^2)} = 80 \text{ A}$$
$$\text{pf} = (80 \text{ A}/80 \text{ A})$$
$$\text{pf} = 1.00$$

So how do we determine the proper amount of capacitance to add to an induction motor to provide the desired about of power factor correction? When performing power factor correction, we need to ensure that the power factor is not corrected past the value of 1.0. This can result in overexcitation and can result in unstable operation of the motor. IEEE 141 and the EASA handbook provides a table to help determine the correct amount of power factor correction capacitors (in kvar) to add to the feeder circuit, determine first the original power factor of the motor feeder. Once this is determined, look up this value in the first column of the power factor correction of IEEE 141 and as is shown in Table 17.2.

Next, determine the new power factor desired and look up this point on the header across the top of the table. Find the value in the table for the intersection of the original power factor and the new power factor desired. This is the multiplier. Next, multiply the kW rating of the load by the multiplying factor to determine the amount of capacitance needed on the motor circuit to make the desired power factor correction (kvar) as shown in equation (17.9).

$$\text{kvar} = \text{kW} \times \text{multiplier} \tag{17.9}$$

TABLE 17.2 Chart of Power Factor Correction

Desired Power Factor (%)

Original power factor	0.80	0.81	0.82	0.83	0.84	0.85	0.86	0.87	0.88	0.89	0.90	0.91	0.92	0.93
0.50	0.982	1.008	1.034	1.060	1.086	1.112	1.139	1.165	1.192	1.220	1.248	1.276	1.306	1.337
0.52	0.893	0.919	0.945	0.971	0.997	1.023	1.050	1.076	1.103	1.131	1.159	1.187	1.217	1.248
0.54	0.809	0.835	0.861	0.887	0.913	0.939	0.966	0.992	1.019	1.047	1.075	1.103	1.133	1.164
0.56	0.730	0.756	0.782	0.808	0.834	0.860	0.887	0.913	0.940	0.968	0.996	1.024	1.054	1.085
0.58	0.655	0.681	0.707	0.733	0.759	0.785	0.812	0.838	0.865	0.893	0.921	0.949	0.979	1.010
0.60	0.583	0.609	0.635	0.661	0.687	0.713	0.740	0.766	0.793	0.821	0.840	0.877	0.907	0.938
0.62	0.516	0.542	0.568	0.594	0.620	0.646	0.673	0.699	0.726	0.754	0.782	0.810	0.840	0.871
0.64	0.451	0.474	0.503	0.529	0.555	0.581	0.608	0.634	0.661	0.689	0.717	0.745	0.775	0.806
0.66	0.388	0.414	0.440	0.466	0.492	0.518	0.545	0.571	0.598	0.626	0.654	0.682	0.712	0.743
0.68	0.328	0.354	0.380	0.406	0.432	0.458	0.485	0.511	0.538	0.566	0.594	0.622	0.652	0.683
0.70	0.270	0.296	0.322	0.348	0.374	0.400	0.427	0.453	0.480	0.508	0.536	0.564	0.594	0.625
0.72	0.214	0.240	0.266	0.292	0.318	0.344	0.371	0.397	0.424	0.452	0.480	0.508	0.538	0.569
0.74	0.159	0.185	0.211	0.237	0.263	0.289	0.316	0.342	0.369	0.397	0.425	0.453	0.483	0.514
0.76	0.105	0.131	0.157	0.183	0.209	0.235	0.262	0.288	0.315	0.343	0.371	0.399	0.429	0.460
0.78	0.052	0.078	0.104	0.130	0.156	0.182	0.209	0.235	0.262	0.290	0.318	0.346	0.376	0.407

Source: Reproduced with permission of IEEE.

where

kvar = rating of power factor correction

kW = rating of load real power

multiplier = factor from Table 17.2

Example 17.6   Given a motor that is drawing real power from the electrical system of 500 kW and is running at a power factor of 0.76. Using Table 17.2, calculate the amount of capacitance required (kvar) to improve the power factor to a new value of 0.93.

Solution: Find the original power factor of 0.76 on the first column of Table 17.2. Find the new power factor of 0.93 on the header of Table 17.2.

Find the multiplying factor where these two intersect. This is the multiplier.

$$\text{Multiplier} = 0.46$$

Next, to determine the amount of capacitance needed (kvar), multiply the real load of the motor by the multiplier utilizing equation (17.9).

$$\text{kvar} = \text{kW} \times \text{multiplier}$$

$$\text{kvar} = 500 \text{ kW} \times 0.46$$

The amount of capacitance needed is

$$\text{kvar} = 230 \text{ kvar}$$

The location on the motor feeder that the power factor correction capacitors are installed in is also important. There are three basic possible locations for the power factor correction as shown in Figure 17.4. Location (a) is downstream of the thermal overload relay (49) for the motor. Installing the capacitor at this locations results in a different current being drawn through the overload element than the motor pulls. Therefore, the overload relay is no longer a good approximate estimate of real motor temperature since it is not seeing the total amount of motor current. For this reason, the location (a) is not used unless the thermal device settings are adjusted to account for the difference between overload current and actual motor current. Additionally, if location (a) is utilized, NEC® section 460.9 requires that the rating or setting of the motor overload device must be based on the improved power factor of the motor circuit. However, motor feeder conductor must be sized without consideration of the improved power factor of the capacitor.

Location (c) is upstream of both the motor starter and the thermal overload device. This corrects the condition in location (a) where now the current flowing through the overload element is the same as motor current and, therefore, the overload is an accurate reflection of motor thermal condition. However, whether the motor starter is closed or open, the capacitor is always connected to the power source. Since the capacitor power factor correction is only required when the motor is energized, this may lead to an overexcited condition on the source side of the motor starter whenever the motor starter is open and the motor is de-energized. For this reason, location

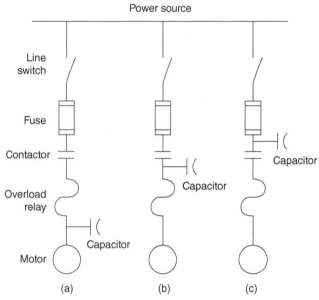

Figure 17.4 Power factor location for motor starter. *Source*: Reproduced with permission of IEEE.

(c) is not used unless the capacitor has its own means of isolation when the motor is not running.

Location (b) has the power factor correction capacitor between the thermal overload device and the motor starter. This is the ideal location as now the current flowing through the overload element is the same as motor current and, therefore, the overload is an accurate reflection of motor thermal condition. At the same time, the capacitor is downstream of the starter. This location automatically removes the power factor correction capacitor from the electrical system whenever the motor is de-energized and the need for the capacitive correction is not there.

Power factor correction capacitors should never be designed to adjust the power factor of the motor to unity. Capacitors may have values that vary in the order of +/− 15 percent of nameplate values. If the capacitance were larger than nominal, this could lead to an overexcitation condition causing excessive voltage at the motor and capacitor terminals. In practical application, the maximum amount of power factors correction added to a motor circuit is no more than 95% of the no-load kvar requirement of the motor. NEC® section 460 provides safe installation requirements for power factor correction capacitors. IEEE 141 provides a table shown in Table 17.3 for the suggested maximum allowable power factor correction capacitor values for various motor sizes and speeds.

For a synchronous motor that is delivering reactive power to the system, the value of reactive power (Q) is negative and the power angle θ is negative. This appears as a capacitor bank to the electrical system. Figure 17.5 shows the power triangle for a synchronous motor delivering reactive power to the electrical system.

TABLE 17.3 Maximum Capacitor Sizes for Direct Connection to the Terminals of an Induction Motor

Suggested maximum capacitor ratings—used for high-efficiency motors and older design (pre-"T-frame") motors

Number of Poles and Nominal Motor Speed (rpm)

| Induction motor horse-power rating | 2 | | 4 | | 6 | | 8 | | 10 | | 12 | |
| | 3600 rpm | | 1800 rpm | | 1200 rpm | | 900 rpm | | 720 rpm | | 600 rpm | |
	Capacitor kvar	Current reduction %	Capacitor kvar	Current reduction %	Capacitor kvar	Current reduction %	Capacitor kvar	Current reduction %	Capacitor kvar	Current reduction %	Capacitor kvar	Current reduction %
3	1.5	14	1.5	15	1.5	20	2	27	2.5	35	3	41
5	2	12	2	13	2	17	3	25	4	32	4	37
7.5	2.5	11	2.5	12	3	15	4	22	5	30	6	34
10	3	10	3	11	3	14	5	21	6	27	7.5	31
15	4	9	4	10	5	13	6	18	8	23	9	27
20	5	9	5	10	6	12	7.5	16	9	21	12.5	25
25	6	9	6	10	7.5	11	9	15	10	20	15	23
30	7	8	7	9	9	11	10	14	12.5	18	17.5	22
40	9	8	9	9	10	10	12.5	13	15	16	20	20
50	12.5	8	10	9	12.5	10	15	12	20	15	25	19
60	15	8	15	8	15	10	17.5	11	22.5	15	27.5	19
75	17.5	8	17.5	8	17.5	10	20	10	25	14	35	18
100	22.5	8	20	8	25	9	27.5	10	35	13	40	17
125	27.5	8	25	8	30	9	30	10	40	13	50	16
150	30	8	30	8	35	9	37.5	10	50	12	50	15
200	40	8	37.5	7	40	9	50	10	60	12	60	14
250	50	8	45	7	50	8	60	9	70	11	75	13
300	60	8	50	7	60	8	60	9	80	11	90	12
350	60	8	60	7	75	8	75	9	90	10	95	11
400	75	8	60	6	75	8	85	9	95	10	100	11
450	75	8	75	6	80	8	90	9	100	9	110	11
500	75	8	75	6	85	8	100	9	100	9	120	10

Source: Reproduced with permission of IEEE.
Note: For use with three-phase, 60 Hz, Design B motors (NEMA MG 1-1993) to raise full-load power factor to approximately 95%.

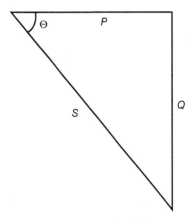

Figure 17.5   Power triangle for motor delivering reactive power to the system. *Source*: Reproduced with permission of IEEE.

An induction motor that is operating at a reduced load draws less real power (kW) from the electrical system. The reactive power associated with the inductance of the motor stator and rotor windings decreases, but the reactive power associated with core stored energy does not change due to the reactive energy of the core being proportional to the square of the applied voltage. Therefore, overall motor reactive power is reduced but to a less extent than the reduction in real power. Due to this, as the loading on an inductive motor is decreased; the power factor of that operating motor will become smaller. Figure 17.6 shows the response of a typical induction motor's power factor, the reactive power, and the total power drawn as the motor becomes unloaded. Since power factor is the ratio of real to total power, you can see as the motor becomes unloaded, the power factor decreases.

The ability of a synchronous motor to deliver reactive power to the electrical system is also a function of the load. For a motor that is operating at 100% of rated power (kW), the motor total power (kVA) is already at rated value and the reactive power the motor can tolerate is zero. As the synchronous motor becomes unloaded, the total power is reduced. This allows for the synchronous motor to be able to provide some reactive power to the electrical system. To maintain operation of the synchronous motor within its design capability, we need to maintain the total power draw (kVA) of the motor at the rating for the motor, so as the motor real power is increased, the reactive power capability of the machine is decreased. A special application of a synchronous motor is when the shaft is not connected to any mechanical load and the motor field current is controlled purely to provide reactive power to the system. This operation is known as a synchronous condenser or a rotating capacitor. Figure 17.7 shows the ability of a synchronous motor to provide reactive power to the electrical system as the real power demand of the motor is varied.

In order to understand the starting operation of the induction motor, it is helpful to look at the simplified model of this induction machine as is shown in Figure 17.8 and to look at some of the parameters of the machine as it starts from locked rotor condition to full speed (or loaded slip speed of the motor). The electrical side of the

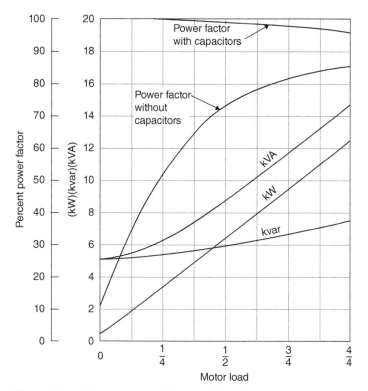

Figure 17.6   Induction motor kVA, kvar, kW and power factor in relation to load. *Source*: Reproduced with permission of IEEE.

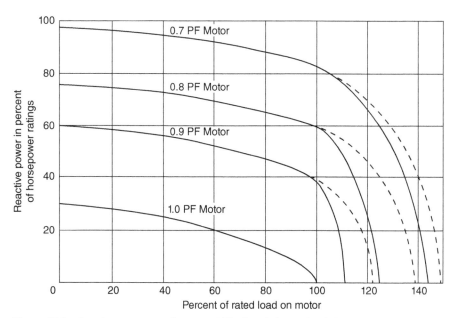

Figure 17.7   Synchronous motor kvar capability in relation to load. *Source*: Reproduced with permission of IEEE.

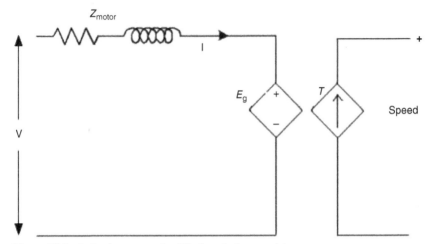

Figure 17.8 Induction motor simplified equivalent model.

motor can be simplified to an internal gap voltage or induced electromagnetic force (E_g) in series with an inductor and resistor (that account for stator and rotor resistance and inductance. The electrical power transferred to the rotor is the product of E_g and the current I. This is converted to mechanical power as the product of torque and speed. As a one-line diagram, this description is drawn as shown in equation (17.10). The important relationship is that the electrical power of the gap is equal to the mechanical power of the shaft.

$$\text{Power} = E_g \times I = T \times \text{Speed} \tag{17.10}$$

Note, this ignores some electrical and mechanical losses in the machine, but this ideal model provides us with some guidance for further discussion.

The value of the gap voltage is the product of the rotor resistance and rotor current and the slip of the machine and is defined in equation (17.11)

$$E_g = I_{rotor} \times R_{rotor}(1 - S)/S \tag{17.11}$$

If we plot the value of gap voltage as the machine increases in speed from the initial locked rotor condition to full speed, we see that it increases parabolically due to the effect slip has on the formula. At very low values of rotor speed, there is very little change in the gap voltage. Since E_g varies little, the machine can be looked at as a constant impedance device. That is to say, if the voltage applied to the machine is reduced, then, using Ohm's law and assuming constant impedance, the current is also reduced in proportion to the reduction of voltage. When the motor is close to full speed, for a small change in motor speed, there is a large change in gap voltage. The change in gap voltage is in the same direction as speed. Therefore, if motor rotor speed is reduced, the gap voltage is also substantially reduced. This increases the voltage drop across the motor inductance and resistance, increasing motor current. Therefore, near full speed, the motor acts as a constant power device in that, with a reduction of voltage applied to the motor, the motor slip increases, which reduces the

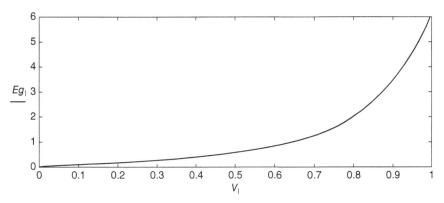

Figure 17.9   Induction motor E_g voltage or electromagnetic force versus speed curve.

gap voltage and increases the overall voltage drop across the motor inductance and resistance (Z_{motor}), which causes an increase in motor current. Figure 17.9 shows the typical relationship between motor internal or gap voltage and the rotational speed of the machine.

The stator current pulled during the acceleration of the motor is the product of the voltage across the inductance and resistance of the motor divided by the motor impedance as described in equation (17.12)

$$I = (V - E_g)/Z_{motor} \qquad (17.12)$$

Since at locked rotor condition, the value of the gap voltage is zero, the locked rotor current is defined as

$$I_{starting} = V/Z_{motor} \qquad (17.13)$$

As the motor rotor increases in speed, the value of E_g increases parabolically. This causes motor current to drop parabolically. This is the typical current speed curve for an induction motor. Figure 17.10 shows the typical relationship between

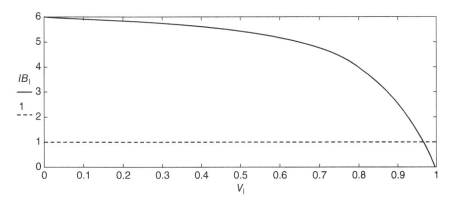

Figure 17.10   Induction motor current versus speed curve.

TABLE 17.4 Induction Motor Locked Rotor Currents

HP	Rated Voltage					
	200 V	230 V	460 V	575 V	2300 V	4000 V
0.5	23	20	10	8		
0.75	29	25	12	10		
1	34	30	15	12		
1.5	46	40	20	16		
2	57	50	25	20		
3	74	64	32	26		
5	106	92	46	37		
7.5	146	127	63	51		
10	186	162	81	65		
15	267	232	116	93		
20	333	290	145	116		
25	420	365	182	146		
30	500	435	217	174		
40	667	500	290	232		
50	834	725	362	290		
60	1000	870	435	348	87	50
75	1250	1085	542	434	100	62
100	1665	1450	725	580	145	83
125	2085	1815	907	726	181	104
150	2500	2170	1085	868	217	125
200	3335	2900	1450	1160	290	167
250	4200	3650	1825	1460	365	210
300	5060	4400	2200	1760	440	253
350	5860	5100	2550	2040	510	293
400	6670	5800	2900	2320	580	333
450	7470	6500	3250	2600	650	374
500	8340	7250	3625	2900	725	417

Source: Reproduced with permission of National Electrical Manufacturers Association.

motor stator current and the rational speed of the machine. Notice at locked rotor, the motor current is approximately six times full-load current (this is dependent on motor design).

Table 17.4 lists the typical locked rotor currents for various size and voltage rating motors.

For an across-the-line starting method, the applied voltage ideally does not change from locked rotor to full-speed condition of the motor. Since the total power drawn by the motor (kVA) is the product of motor stator voltage and motor current (as shown in Figure 17.10), the total power drawn versus speed curve is a reflection of the motor current versus speed curve as shown in equation (17.14) and reflected in Figure 17.11.

$$\text{kVA} = V_{\text{motor}} \times I_{\text{motor}} \times 1.73 \tag{17.14}$$

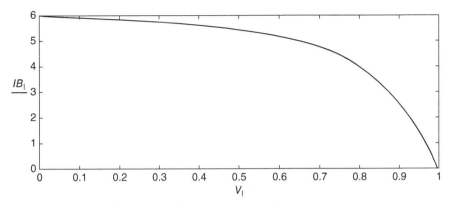

Figure 17.11   Induction motor total power versus speed curve.

Note that, if locked rotor current on the motor is approximately six times full-load current, then the locked rotor motor total power draw is approximately six times the total power draw of the motor at full load. This is assuming the ideal case with no voltage drop at the motor terminals.

Motor torque versus speed curves have been discussed previously and, as discussed, the motor design affects the motor torque speed curve. Knowing that, from a mechanical standpoint, mechanical power is the product of torque and speed, and we can take the motor's torque-speed curve and multiply that curve by the percent full speed at each point in the curve to derive the typical motor real power (kW) versus speed curve as shown in Figure 17.12.

Now, since we know that the motor power factor is the ratio of real power to total power and since we have previously defines the motor real power versus speed curve and the motor total power versus speed curve, we can multiply these two together to obtain the curve of motor power factor versus speed as shown in Figure 17.13. Note, just as in the discussion about motor load, as the motor rotor speed is reduced, the

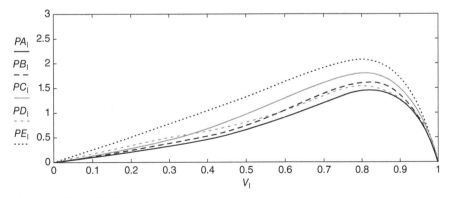

Figure 17.12   Induction motor real power versus speed curve.

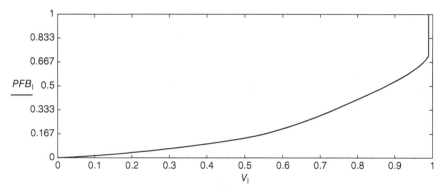

Figure 17.13   Induction power factor versus speed curve.

power factor of the induction motor is also reduced and, at locked rotor condition, the power factor is very small.

Some applications do not tolerate well either the peak mechanical torque a motor can provide when started across the line or the peak current of six or more times motor full-load current during an across-the-line start. When we need to reduce the current drawn by the motor and/or the torque available on the shaft of the motor during the acceleration of the motor, a reduced voltage starter is applied to the stator of the motor to achieve this reduction of motor current and/or motor shaft torque during the acceleration process. Some possible application issues are if, during the start a pump pumping fluid through a piping system, the sudden acceleration of the pump may generate a pressure wave in the piping system causing violent shaking and damage to the piping. This is called "water hammer." Another application issue is where the mechanical transmission is susceptible to early failure due to the peak torques the motor can produce during acceleration. Another application is due to a weak electrical distribution system where the starting current of the motor (six times full-load current or more) can cause severe voltage dips on the distribution system.

For the above applications, a starter is designed to reduce the applied voltage to the motor terminals. As discussed previously, when the motor is at standstill, the motor appears as a constant impedance device, so the reduction in motor voltage at any given value of slip speed will result in a proportional reduction of motor current. The torque produced in the rotor of the motor is proportional to the square of the current flowing in the machine. Therefore, with a reduction of motor voltage and subsequent proportional reduction of motor current, the torque available from the motor is reduced by the square of the reduction of voltage.

Sometimes, the question is asked if the application of reduced voltage starters reduces the heat generated in the motor during start, increases the heat generated in the motor during start, or has no difference on the heat generated in the motor during start as compared with an across-the-line starter. The answer is, "it depends."

Let us first look at the simplified case where there is no load on the motor shaft and there are negligible mechanical losses in the motor.

We know the torque of the motor is reduced in proportion to the square of the stator current. If we simplify the question and assume all motor losses are I^2R

losses, as we reduce the motor current, the motor losses are reduced by the square. However, this is not the whole story. These losses are in units of power and power is the rate of energy flow. Now let us look at what happens to the motor starting time with reduced current. The motor starting time is inversely proportional to the available torque on the shaft of the motor. Since the torque is reduced by the square of the reduction of current, the acceleration time is increased by the inverse of the square of the reduction of current. Now remember that energy is the product of power and time. With a reduced voltage start, if the power is reduced by the square of the current reduction but the acceleration time is increased by the inverse of the square of the reduction of current, then the overall energy dissipated in the rotor accelerating it from standstill to full speed is the same with a reduced voltage starter or across-the-line starter.

You can look at this from an energy standpoint and see the same result. With nothing connected to the shaft and no mechanical power drawn from the motor, when starting a motor, we are simply adding energy to the rotor to increase the kinetic energy of the rotating mass of the rotor. The amount of kinetic energy of the rotor is the same regardless of how the motor was started. In the above case, with no load on the motor, there is no significant difference in the heat generated in the motor regardless of the method of starting the motor.

Now let us look at the case where the shaft is connected to a mechanical load and as soon as the shaft starts rotating, power is absorbed by the mechanical load. The amount of accelerating torque in the motor is the difference between the available motor torque and the load demand torque. If we connect a load to motor during start, the load requires some torque to accelerate and this leaves less power (and torque) available to the motor for acceleration. Now, even though torque is reduced by the square of the reduction of motor current, the net available torque for acceleration is less than this amount because some of this torque is being used to drive the load. Therefore, the acceleration time will increase to a value more than by the inverse of the square of the reduction of current. Since the acceleration time has increased by more than the reduction of the I^2R losses in the machine, the application of a reduced voltage starter on a loaded machine may actually increase the heating in the machine. The amount of the increase in heating is a function of the mechanical load on the shaft as the machine accelerates to full speed.

There are various types of reduced voltage starters available. Table 17.5 lists these types of starters and their specific impact on the starting torque of the machine and the change in line current (note line current is not the same as motor current in the *autotransformer* case).

Now let us look at the motor torque versus speed curve and discuss the concept of accelerating torque during a motor start. With a motor uncoupled from its mechanical load, all the available motor mechanical torque at each point in the torque-speed curve is available to accelerate the rotor. However, when starting a motor and a coupled load, some of the motor available torque is used to drive the load. The load also has a torque-speed curve that defines the amount of torque the load needs to operate at each particular speed from zero speed to full speed. If we plot the motor torque-speed curve and the load torque-speed curve on the same graph and take the difference between the available motor mechanical torque and the load speed

TABLE 17.5   Starting Characteristics of Squirrel Cage Induction Motors

Starting Method	Voltage at Motor	Line Current	Motor Torque
Full-voltage value	100	100	100
Autotransformer			
80% tap	80	64[a]	64
65% tap	65	42[a]	42
50% tap	50	25[a]	25
Primary resistor typical rating	80	80	64
Primary reactor	80	80	64
80% tap	65	65	42
65% tap	50	50	25
50% tap			
Series-parallel	100	25	25
WYE-DELTA	100	33	33
Part-winding (1/2–1/2)			
2 to 12 Poles	100	70	50
14 and more Poles	100	50	50

Source: Reproduced with permission of Electrical Apparatus Service Association.
[a] Autotransformer magnetizing current not included. Magnetizing current is usually less than 25 percent of motor full-load current.

torque at each value of speed, we can define the accelerating torque available from the motor.

$$T_{\text{acceleration}} = T_{\text{motor}} - T_{\text{load}} \tag{17.15}$$

where

$T_{\text{acceleration}}$ = torque available for motor acceleration

T_{motor} = available motor mechanical torque

T_{load} = load required torque

As long as the magnitude of accelerating torque is positive, the motor will accelerate. When this value is negative the motor will decelerate. When this value is zero, the motor runs at this steady speed.

Figure 17.14 plots a standard NEMA B motor torque-speed curve and the load torque-speed curve for a typical variable torque load. The motor torque-speed curve is shown as the top curve and the load torque-speed curve is the bottom area that is highlighted in black. The area under the bottom curve represents the torque the load requires from the motor at any particular speed of operation. The area under the top curve represents the torque the motor has available at any particular speed of operation. The difference between the available motor mechanical torque and load required torque defines the available accelerating torque. This value is the area between these two curves and is shown in gray. The gray region defines the accelerating torque of the motor. This is the amount of torque available to the shaft of the motor for acceleration. The motor will come to a steady speed when these two curves meet. This is known as the operating point. From Figure 17.14, the operating point is about 95%

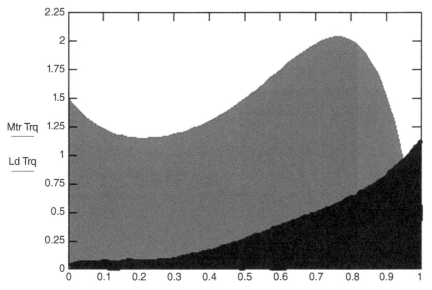

Figure 17.14   Induction motor which is started across the line – torque versus speed (gray area) curve compared with variable torque load versus speed (black area) curve.

of synchronous speed, so this motor has an operating value for slip of 5%. Note that throughout the speed range, the available motor mechanical torque always exceeds the load required torque as the motor is accelerating to operational speed. Therefore, throughout the speed range, the value of the accelerating torque remains positive.

Figure 17.15 plots a standard NEMA B motor torque-speed curve and the load torque-speed curve for a typical constant torque load. Notice as compared with the same motor driving a variable torque load, in this application, the difference between the available motor mechanical torque and load required torque, especially at lower values of speed is smaller. Therefore, the amount of accelerating torque for this motor is smaller and, since accelerating time is inversely proportional to accelerating torque, it will take this motor longer to accelerate the constant torque load as compared with the variable torque load (assuming the inertia of the load is the same). Note that throughout the speed range, the available motor mechanical torque always exceeds the load required torque. Therefore, throughout the speed range, the value of the accelerating torque remains positive.

Figure 17.16 plots a standard NEMA B motor torque-speed curve and the load torque-speed curve for a typical variable torque load. However, in this graph, the motor is being started with a reduced voltage starter with a current limit set to 350% of motor *full-load current* (FLA). As we discussed before, for any particular speed, the reduction in torque of the motor is the square of the reduction of voltage. Therefore, the motor torque-speed curve in the above figure shows this reduction in motor torque. Notice as compared with starting the motor across the line, in this application, the difference between the available motor mechanical torque and load required torque, especially at lower values of speed, is smaller. Therefore, the amount of accelerating

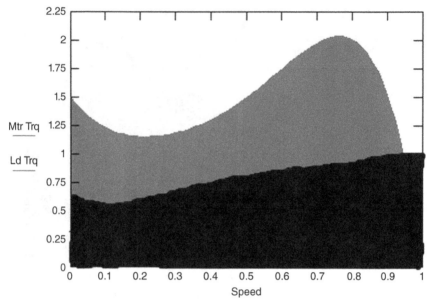

Figure 17.15   Induction motor which is started across the line – torque versus speed (gray area) curve compared with constant torque load versus speed (black area) curve.

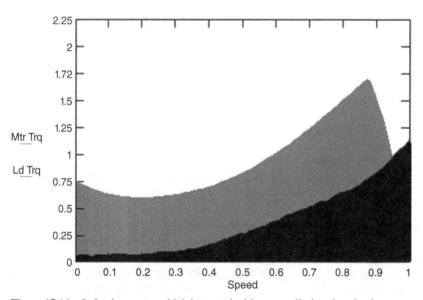

Figure 17.16   Induction motor which is started with current limit reduced voltage starter – torque versus speed (gray area) curve compared with variable torque load versus speed (black area) curve.

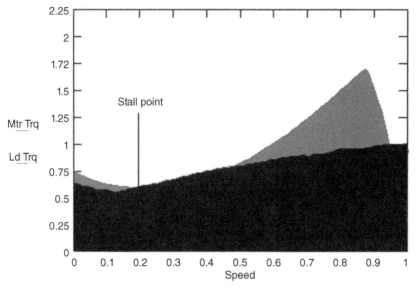

Figure 17.17   Induction motor which is started with current limit reduced voltage starter – torque versus speed (gray area) curve compared with constant torque load versus speed (black area) curve.

torque for this motor is smaller and, since accelerating time is inversely proportional to accelerating torque, it will take this motor longer to accelerate the variable torque load with a reduced voltage starter than if the motor were started across the line (ATL). Note that throughout the speed range, the available motor mechanical torque always exceeds the load required torque. Therefore, throughout the speed range, the value of the accelerating torque remains positive.

Figure 17.17 plots a standard NEMA B motor torque-speed curve and the load torque-speed curve for a typical constant torque load. As before, the motor is being started with a reduced voltage starter. However, now the load demand torque is higher due to fact that this is constant torque load. Notice what happens to the value of accelerating torque at a speed of 20%. At this point, the available motor mechanical torque drops below the value of the load required torque. At this point, the load required torque equals available motor mechanical torque and the value of accelerating torque is zero. Once the accelerating torque becomes zero or negative, the motor will not accelerate anymore and will stall at this point. Because of this issue, it is recommended that the engineer always compare the motor torque-speed curve (and consider any voltage drop or reduced voltage starting that may apply to the application) with the load torque-speed curve and ensure that at all speeds from zero speed to full speed that motor torque be greater than 110% of load torque.

Now, a word of caution. Some texts list a formula for average accelerating torque based on the *full-load torque* (FLT), *breakdown torque* (BDT), and *locked rotor torque* (LRT) values as shown in equation (17.16).

$$\text{Avg Acc Trq} = \{[(\text{FLT} + \text{BDT})/2] + \text{BDT} + \text{LRT}\}/3 \qquad (17.16)$$

where

> Avg Acc Trq = average accelerating torque
>
> FLT = full-load torque
>
> BDT = breakdown torque
>
> LRT = locked rotor torque

Note there is no term in this equation that accounts for load required torque so this estimation does not account for the load required torque. The real value of accelerating torque requires looking at the difference between the motor torque and load torque throughout the speed range, so caution should be exercised when using equation (17.16) for accelerating torque value.

REDUCED VOLTAGE STARTING METHODS

Wound Rotor Motor

The *wound rotor motor* has slip rings that allow connection of a stationary bank of external resistors to the moving rotor circuit. Figure 17.18 describes the wound rotor motor three-line diagram. By increasing the value of the external resistance that is connected to the rotor winding, the torque-speed curve of the motor is skewed such that, with additional resistance, the motor slips more, but the available torque from the motor is greater. These motors will first be started with a large value of rotor resistance and therefore available torque. As the motor accelerates, the external stationary rotor resistance is reduced by shunting out individual resistors with contactors. Once the motor reaches full speed, the rotor external resistance is completely shunted by the external contacts or a very minimal amount of resistance is left in the rotor depending on the machine steady-state torque requirements.

Autotransformer

As described above, as the voltage to the motor is reduced, for any given speed of motor, the available torque is reduced by the square of the reduction of the voltage.

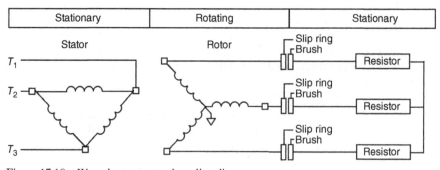

Figure 17.18   Wound rotor motor three-line diagram.

Therefore, one method of controlling torque is to control the voltage applied to the motor stator during the acceleration of the conveyor. The autotransformer has several taps and can adjust the voltage to the motor in discrete steps. The motor will be energized on the lowest tap setting of the autotransformer. As the motor accelerates, the taps will be increased on the autotransformer. At full speed, the motor will be placed directly in line with the electrical distribution system and the autotransformer will be removed from the circuit. One benefit of the autotransformer starter is that during reduced voltage start, the line current is reduced by the same ratio as the reduction of voltage on the motor side. For more information on voltage and current relationships for transformers, refer to Chapter 14 on electrical distribution.

Electronic Soft-Start

The electronic soft-start uses semiconductors to control the voltage to the motor. Figure 17.19 describes the electronic soft-start three-line diagram. The power semiconductor utilized in the electronic soft-start is most commonly two silicon controlled rectifiers (SCR) assembled anode to cathode (or reverse parallel). Since the soft-start has AC voltage applied to the source side and delivers AC voltage to the load side, the semiconductor must be able to block voltage of both polarities. A silicon-controlled rectifier is a semiconductor that can block voltage regardless if the anode is positive to the cathode or if the cathode is positive with respect to the anode which makes this semiconductor a natural choice for a soft-start application. Semiconductors configured as a transistor cannot tolerate reverse voltages from the collector to emitter. Therefore, transistors are utilized only when DC voltage is being converted to AC voltage where the transistor only must be able to block voltage of one polarity.

The semiconductor acts as an open switch to block the source voltage for the first part of the sinewave and as a closed switch to pass the source voltage to the load for the second part of the sine wave. The delay in the conduction of the semiconductor results in only part of the voltage sine wave being applied to the terminals of the motor which results in a reduced value of voltage at the motor. The waveforms on the line and load side of the soft-start are shown in Figure 17.19. The effect of an electronic soft-start is similar to the autotransformer from the motor side of the soft-start as it reduces the voltage to the motor to reduce the starting torque to the motor. The

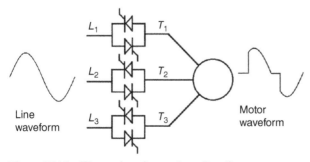

Figure 17.19   Electronic soft-start three-line diagram.

advantage of the electronic soft-start over the autotransformer is that the autotransformer only has certain discrete taps that can be selected. The soft-start can control the delay angle to any angle to get any value of voltage to the motor from zero voltage to full voltage. Additionally, modern soft-starts have torque regulation and speed regulation controls and can automatically vary voltage to the motor to regulate motor torque as the motor accelerates to full speed.

Power losses in a soft-start are mainly due to conduction losses and not switching losses. These conduction losses are the product of the anode to cathode current magnitude and the anode to cathode semiconductor voltage drop. Since the SCR is constructed of four layers of doped semiconductor material, the on state or conduction voltage drop is about 1.2 V. Once the motor has been started and accelerated to full speed, the soft-start can be bypassed utilizing a motor contactor, eliminating the losses and subsequent heat generated by the semiconductors.

Electronic Variable Frequency Drives

Variable frequency drives or variable speed drives convert the fixed *frequency*, fixed voltage input to a variable frequency variable voltage output. The benefit of the variable frequency drive over the soft-start is that, with the soft-start it can only adjust output voltage, but the frequency applied to the motor is fixed at line frequency. Therefore, the current during start still must rise above full-load current. Depending on the application, the current limit for a soft-start may be set as high as 300–400% of full-load current. Since the variable frequency drive outputs both a variable frequency and variable voltage, the motor can be accelerated with up to motor full-load torque by the variable frequency drive without ever exceeding full-load current of the motor. The variable frequency drive achieves this by slowly ramping up both the applied voltage and the applied frequency to the motor. The amount of current drawn by an induction motor is a function of the slip of the motor. Earlier in this chapter, in equation (17.4), we defined the slip of the motor as the difference between the motor synchronous speed (for the applied system frequency) and the actual speed of the motor or

$$s = [(n_s - n)/n_s] \times 100\%$$

By reducing the value of the frequency applied to the motor, we reduce the synchronous speed of the motor for that frequency and, thereby, we reduce the value of slip. This is how a variable frequency drive can accelerate a motor from zero speed to nominal nameplate speed without ever allowing the current to rise above the full-load current of the motor. Also, the torque control of a variable frequency drive is more constant than a soft-start for the same reason. Since motor torque is a function of slip and since the variable frequency drive can control the value of slip as the motor is accelerated, the variable frequency drive can better control the available motor torque during acceleration.

In the standard variable speed drive motor control application, the variable frequency drive will remain in the circuit in series with the motor winding both during start of the motor and while the motor is operating at full speed. Therefore, the associated losses of the power semiconductors will continue even when the motor is at full speed. Recent advances in variable frequency drives now allow some models

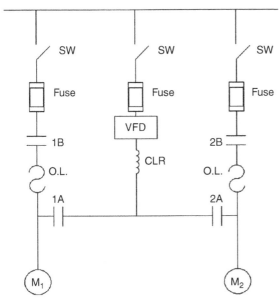

Figure 17.20   One variable frequency drive controlling multiple motors.

of variable frequency drive to synchronize the load voltage waveform to the source voltage waveform through a reactor, and then bypass the drive via a closed transition. This system is shown in Figure 17.20. In this process, the first motor M1 is started by closing contactor 1A. Then the variable speed drive or variable frequency drive (VFD) begins to accelerate motor M1. If system flow requires full speed of the pump connected to motor M1, the VFD will accelerate the motor M1 to full speed. Once the motor M1 reaches full speed, the bypass contactor 1B is closed. At this point, both contactor 1A and contactor 1B are closed. Since the output of the VFD is not a perfect sine wave, the current limiting reactor CLR is required to limit the current flow between the two sources during this short period. Once the VFD detects that the bypass contactor 1B is closed, the contactor 1A is opened and the output of the VFD is de-energized. Now the VFD is ready to accelerate the second motor using the same procedure.

In applications where the VFD is only used to accelerate the motor and we want to eliminate the losses associated with the VFD while the motor is operating at full speed, we can use this system to bypass the VFD once the motor reaches full speed. For more information on variable frequency drives and their applications, please see Chapter 18 which covers variable speed drives in detail.

Eddy Current Couplings

Eddy current couplings can also be used to control the mechanical torque provided to the conveyor system during acceleration. These devices are constructed of a stator and a rotor element that are magnetically coupled together. The field is controlled by the magnitude of current provided to the field of the eddy current coupling. The conveyor shaft and motor shaft never achieve the exact same speed as there is always some value of slip between the motor and conveyor sides of the eddy current coupling. The strength of the field applied to the coupling defines the amount of slip of the coupling. The power loss associated with the slip is dissipated at the coupling in the form of heat.

Advantages of eddy current couplings are

1. They require low power coil excitation.
2. They permit smooth, controlled starting.
3. The motor can be started and accelerated without connecting the load. On frequent start and stop applications the motor can run continuously while the load is coupled and uncoupled.
4. Variable speed can be obtained. However, in variable speed applications the additional slip creates more heat that must be dissipated.
5. They make possible the use of squirrel-cage motors and across-the-line starters.
6. A modified eddy current coupling can be used as a decelerating brake (not as a "holding brake," however).

The disadvantages of eddy current couplings are

1. They require additional drive space.
2. Water cooling must be provided for the larger sizes.
3. Generally, they are more expensive than a wound rotor motor and reduced voltage starting.

Fluid Couplings

Fluid couplings are similar to eddy current couplings, except the linkage between the motor shaft and conveyor shaft is provided by a mechanical fluid linkage in the fluid coupling instead of being provided by electromagnetic means of the field in the eddy current coupling. The fluid coupling consists of an impeller and a runner contained in an oil-filled housing. The impeller is connected to the motor shaft and the runner is connected to the conveyor shaft. The fluid couples the impeller and runner. Fluid couplings share similar advantages and disadvantages as the eddy current coupling, but there are two additional disadvantages to the fluid coupling. The fluid coupling tends to require more mechanical maintenance to maintain the device and, in application, they tend to create a substantial amount of mechanical noise during load acceleration.

Various standards are available to provide the user guidance to the selection and operation of motors in the generation station. One such standard is IEEE 841, IEEE Standard for the Petroleum and Chemical Industry-Severe Duty Totally Enclosed Fan-Cooled (TEFC) Squirrel Cage Induction Motors-Up to and Including 500 HP. While the title lists this as applying to the petroleum and chemical industry, it is commonly used for induction motors up to 500 HP installed in electric generator facilities due to the need for enhanced reliability of the motor in the utility application. While

IEEE 841 covers induction motors up to 500 HP, for motors larger than 500 HP, IEEE API 541 is used as a reference standard for motor specification. It covers minimum requirements and certain options to enhance the reliability of motors in the utility application. It also gives information for electrical and mechanical design features *insulation* systems, accessories, and tests.

For synchronous machines, IEEE API 546 – Brushless Synchronous Machines – 500 kVA and Larger is used for their specification. This standard covers the minimum requirements for form-wound and bar-wound brushless synchronous machines for use in petroleum, chemical, and other industrial applications. This standard includes synchronous motors and generators with both salient pole rotor designs (slower speed machines) as well as cylindrical pole rotor designs (higher speed machines).

The NEMA standard for both induction and synchronous machines is NEMA MG1 – Motors and Generators. This standard covers the minimum requirements and specifications of both induction and synchronous machines and covers machines up to 500 HP.

These standards address items such as standard dimensions for certain frames of motor, number of starts per hour a motor is required to be able to withstand, standard values for power and voltage ratings, enclosure definitions, bearing construction standards, vibration limits, and temperature rise requirements.

Motors manufactured in the United States have their power ratings in units of HP while motors manufactured outside the country have their power ratings in units of kW. Both of these are the mechanical power out available at the shaft and not the electrical power into the stator of the winding. The conversion between units of HP and units of kW is

$$kW = 0.746\,HP \qquad\qquad (17.17)$$

where

kW = rated shaft power in units of kW

HP = rated shaft power in units of HP

Motors manufactured in the United States according to NEMA standards also have a *service factor* rating. Motors manufactured to IEC standards do not carry a service factor rating. Therefore, IEC design motors rated in kW may be treated as having a service factor of 1.0 while motors rated in HP may have higher values of service factor. A common service factor for a NEMA-designed motor is 1.15. While the service factor is theoretically an amount of current the motor may be able to carry, it should never be depended upon for the design of the motor nor should it normally be used in the application. From NEMA MG1.43, "The service factor of an alternating current motor is a multiplier which, when applied to the rated HP, indicates a permissible HP loading which may be carried under conditions specified for service factor."

The insulation class of the motor defines the maximum temperature that the motor can safely be operated at without damaging the motor insulation and reducing the usable life of the motor. The various insulation classes for motors are shown in

TABLE 17.6   Insulation Class Definitions

Insulation Class – Class Defined by "Hot Spot"		Maximum Temperature Rise	
Class A	Class 105	105°C	221°F
Class E	Class 120	120°C	248°F
Class B	Class 130	130°C	266°F
Class F	Class 155	155°C	311°F
Class H	Class 180	180°C	356°F
Class N	Class 200	200°C	392°F

Source: Reproduced with permission of IEEE.

Table 17.6. Note the temperatures listed in Table 17.6 are the value of temperature rise from rated *ambient* temperature (40°C) to the temperature of the hot spot in the motor.

Operation of the motor above rating (either rated power or rated ambient temperature) can lead to exceeding the stator winding rated hot spot temperature for the insulation system. This will lead to accelerated degradation of the motor insulation and will shorten the motor operation life.

Motor *efficiency* values are very high compared with the losses in other plant equipment. Typical NEMA nominal efficiency ranges for motors vary from 91% to 95%. These values are at full load of the machine. Motor efficiency is reduced as the motor is unloaded below nominal power rating. Efficiency is the ratio of the shaft output power compared with the motor input power as defined in equation (17.18).

$$\text{Efficiency} = P_{out}/P_{in} \times 100\% \tag{17.18}$$

where

P_{out} = mechanical power out of shaft

P_{in} = electrical input power to motor

Motors are encased in enclosures and there are various designs of motor enclosure integrity. The best type of motor enclosure for the application is dependent on the type of environment that the motor is installed in. Table 17.7 shows a listing of motor enclosure types and their construction. where

TENV = totally enclosed, non-ventilated

TEFC = totally enclosed, fan-cooled

TEBC = totally enclosed, blower-cooled

TEWAC = totally enclosed, water-to-air-cooled

TEAAC = totally enclosed, air-to-air-cooled

WPII = weather-protected (two 90 degree turns in air path)

ODP = open drip proof

For environments that are clean and are not exposed to harsh environments, the WPII or ODP may be adequate design. Both of these designs allow outside air to circulate across the motor stator and rotor components for cooling and, as such, the

TABLE 17.7   Motor Enclosure Definitions

Typical Methods of Cooling (IC code)

NEMA	IC	Circuit Arrangement	Primary Coolant	Method of Movement	Secondary Coolant	Method of Movement
WP-I	IC	0	A	1		
WP-II	IC	0	A	1		
ODP	IC	0	A	1		
TEFC (guarded)	IC	4	A	1	A	1
TEBC	IC	4	A	1	A	6
TENV	IC	4	A	1	A	0
TEAO	IC	4	A	1	A	7

Source: Reproduced with permission of National Electrical Manufacturers Association.

Complete designation – full description (3 or 5 numerals or letters {numeral letter numeral [letter numeral]})

Simplified designation – 2 or 3 numerals or letters in the final position {numeral numeral numeral or numeral numeral letter}.

Example: The complete designation of TEFC guarded would be IC4A1A1; the simplified designation would be IC 411.

air should be relatively clean. The WPII differs from the ODP in that there are at least two 90 degree turns the outside air has to make before it reaches the stator to help prevent entrained material from coating the stator winding.

For environments that have contaminants in the air, the total enclosed motors, TENV, TEFC, TEBC, TEEWAC, or TEAAC motors are a better application. The TENV is totally enclosed. Air inside the motor casing is circulated around the motor stator and rotor and the case of the motor provides the heat sink for heat transfer to the outside environment. There is no forced airflow around the outside of the motor. The TEFC is similar except it has a blower on the outside of the motor casing mechanically connected to the shaft so that, when the motor is energized, the fan is turning forcing air to pass by the outside of the casing of the motor. The TEBC is similar in design but is a special application used for variable frequency applications. This motor has an external blower, but the blower is not connected to the motor shaft but is separately powered. For variable speed applications, if the blower were connected to the motor shaft, the blower would vary speed depending on motor shaft speed. Airflow across the motor would be reduced at lower shaft speeds which may lead to a reduction of heat transfer through the casing of the motor. To prevent this, a TEBC motor has a separately powered blower so the blower spins at full speed regardless of the speed of the shaft of the motor and ensures full cooling capability for the motor regardless of shaft speed. The TEWAC and TEAAC motors are larger sized motors where the case of the motor would not provide adequate cooling. These motors just like other totally enclosed motors have internal air circulating around stator and rotor to remove heat. However, since the motor case is no longer adequately sized to remove all the heat generated by the motor, this motor has a heat exchanger on top of the motor. The only difference between the TEAAC and TEWAC is the media that flows through the other side of the heat exchanger for the cooling of the motor. For the TEAAC, ambient airflows through the other side of the heat exchanger to remove the motor heat from the inside air of the motor. For the TEWAC, water flows through the other side of the heat exchanger to remove the motor heat from the inside air of the motor.

During running, the limiting thermal component to motor is the motor stator. During starting, the limiting thermal component to motor is the motor rotor. When starting an induction motor, the value of the slip is large. This induces large currents in the rotor at high frequency leading to increased I^2R losses in the rotor and the heat from these I^2R losses limits how long and how many times the induction motor may be started.

In a motor starter, there are two distinctly different functions that need to be accounted for, motor control and branch circuit protection. Circuit breakers are rated to interrupt fault current in their feeders, while motor contactors are not rated to interrupt fault current. Motor contactors are rated for more frequent operations than circuit breakers between maintenance cycles. When a circuit breaker is used as a method for starting a motor, then the circuit breaker protection relay can also be set up to trip the circuit breaker when it detects a short circuit on the motor feeder and it can provide the branch circuit protection. Since contactors are not rated to interrupt fault current, the motor starter protection relay is set up NOT to open the contactor on a short circuit on the motor feeder. It is only set up to open the contactor on motor thermal overload. Fuses or a breaker upstream of the contactor will be used for branch circuit protection to open up on the occurrence of a short circuit on the motor feeder. Failure to set up the protection system on a contactor to ignore a feeder short circuit can lead to the contactor trying to open under faulted conditions and this can and has resulted in catastrophic failure of the contactor.

As mentioned previously, if the voltage of the motor is reduced, then the available torque is reduced by the square of the reduction of voltage. If the torque is reduced, for the same inertial and connected load, the acceleration time increases. This must be taken into account when setting up the protection system for the induction motor. We need to ensure that, even with voltage sag during motor starting, the motor will successfully start. Many specifications and standards such as IEEE API541 will require the motor and load to be able to be successfully started with a source voltage at 80% of nominal motor value. Remember that torque changes as the square of voltage, so 80% of motor nominal voltage results in 64% of motor torque until the voltage recovers.

Synchronous motors are different from induction motors in that their field is externally powered and therefore the motor runs at synchronous speed. Just like synchronous generators, the field can be energized through slip rings and brushes or via a brushless exciter and a rotating diode wheel.

Unlike the synchronous generator which has a prime mover that is used to accelerate the synchronous generator mechanically, the synchronous motor is started by applying voltage to the stator of the machine without the DC current being applied to the field for excitation. Instead, a resistor is applied to the field of the synchronous motor to allow for a path for induced currents on the field windings to flow. Much like an induction motor, this allows for forces to push the rotor up to speed. However, depending on the design of the synchronous machine, these main field coils may not provide enough torque to successfully accelerate the machine. The typical synchronous motor will have damper bars installed around the rotor circuit. Damper bars are also known as "amortisseur windings." These damper bars are a form of squirrel cage just like the induction machine. When the synchronous motor is running at synchronous speed, these damper bars experience no current flow (because there is no slip) and are not effectively in the circuit. However, during start, these damper bars provide the required torque to accelerate the motor rotor to full speed. Additionally, during transients, these damper bars tend to smooth the response of the motor.

During operation of the synchronous motor, if the main field was lost, the motor stator remains energized, and the motor is still connected to a load, then the motor would start to behave like an induction machine and start to slip. This would induce currents in the rotor of the motor that can be damaging. Therefore, most synchronous motors will have some form of loss of excitation protection to de-energize the motor stator in the event the main field is lost.

Synchronous motors have the advantage that they result in more efficient utilization of power distribution resources. Synchronous motors can have leading power factor which means they can supply reactive power (kvar) to an electrical system. Most power systems will have a net lagging power factor. The addition of a synchronous motor to an electrical system can result in an improved power factor which results in a reduction of system current drawn for the same amount of real power delivered by the electrical distribution system. This has the effect of improving system efficiency. Synchronous motors have the disadvantages that they are more maintenance intensive due to their external power field circuit and brush maintenance. They are also more costly than a similarly rated induction motors.

Just as in synchronous generators, there are two rotor designs for synchronous motors. For high speed motors, the rotors are cylindrical. However, most synchronous motors are of lower speed design and lower speed motors utilize a salient pole design. In this design, the poles have the main field windings wrapped around them and these assembled poles are mounted on a spider that connects these poles to the motor shaft. Below are some photographs of a synchronous motor with eight poles.

In Figure 17.21, you can count the number of poles as eight. This is the non-drive end of the shaft. Looking at the end of the rotor, you can see the air mover

Damper bar or amortisseur winding

Coil

Fan

Rotor

Journal

Figure 17.21   Synchronous motor with eight poles and brushless exciter – main field.

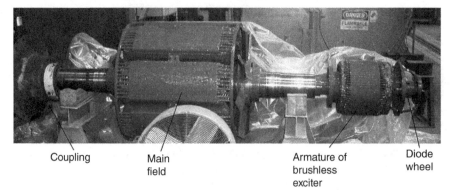

Coupling   Main field   Armature of brushless exciter   Diode wheel

Figure 17.22   Synchronous motor with eight poles and brushless exciter – main field and armature of brushless exciter.

connected to the shaft. As the rotor turns, the air mover forces air to circulate inside the motor casing to the heat exchanger. This rotor is from a TEAAC motor that uses an air-to-air heat exchanger on the top of the motor to remove the heat. Also, if you look at the pole faces, you can see copper bars that run along the length of the poles and are connected at both ends by a metal shorting ring. This is the damper bar mentioned previously and provides the torque necessary for starting of this motor as well as some dampening of the motor during transients.

In Figure 17.22, you see the same rotor, but now from a side view, where, on the left of the photo is the same motor main field and, on the same shaft on the right is the armature of the brushless exciter for this motor (the diode wheel has been removed for this photo but will be discussed later). This rotor uses a brushless exciter. From this side you can get a better view of the copper damper bars that are mounted in the face of the poles and see their connection to the shorting ring at both ends of the main field. At the far left you can see the coupling that makes the mechanical connection from the motor shaft (shown) to the load shaft (not shown). The load for this motor is an induced draft fan which for this application is a centrifugal fan. At the far right of the shaft you see two wires coming out from the hollow bore of the machine. These are the positive and negative leads that normally connect to the diode wheel and are connected to the main field on the left in the photo. You might ask why these leads enter the hollow center of the machine and transverse to the main field and then come back out of the shaft of the machine to connect the main field? The reason is that, when the shaft assembly is installed in the motor casing, there is a radial journal bearing that rides on the shaft of the motor between the main field and brushless exciter armature (in Figure 17.22, this location is about where the jack stand is on the right). Since this is a tight fit babbitted bearing, we cannot pass any electrical conductors on the shaft along this point and the only path to get from the diode wheel that is on the outside of the motor casing to the main field inside the motor casing is by going through the bore of the machine.

Figure 17.23 shows the motor stator winding and core installed in the motor case. This is a special motor with two separate windings, thus the complexity of the

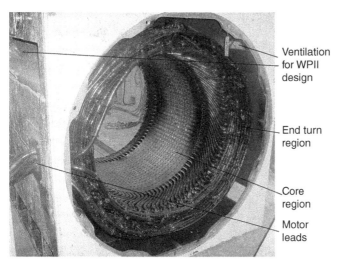

Figure 17.23   Synchronous motor with eight poles and brushless exciter – stator core and windings.

winding end turn area. This particular motor is fed from two different medium voltage LCI variable frequency drives. The reasoning for having two drives feeding one motor on two separate windings wound on one motor core is the need for enhanced reliability. By using two variable frequency drive channels fed from two different voltage sources feeding two motor windings on one motor, the loss of any one source, any one variable frequency drive channel, or the loss of a conductor to one of the two stators, do not cause this motor to become de-energized as it can be run off the single variable frequency drive channel and source that is unaffected.

Figure 17.24 is a close-up of the brushless exciter rotating armature from the previous photo in Figure 17.23. This armature has a field circuit that is mounted on the stationary portion of the motor casing and, as the motor shaft is rotated and current is applied to the stationary field, the armature you can see in Figure 17.24 rotates. This induces AC current in the windings of the armature. The three-phase leads leave the armature on the right of the photo and pass through a plate where the rotating diodes assembly (shown in Figure 17.26) mounts. You can also make out the DC leads that would normally lead the rotating diode assembly (Figure 17.26) and enter the bore of the shaft to feed the main field circuit.

Figure 17.25 shows the same brushless exciter armature, but now the rotating diode wheel has been mounted to the shaft and the three-phase armature leads have been connected to the rotating diode wheel. Note the three-phase connections are terminated on the back side of the diode wheel and small bus bars are used to get the armature three-phase current into the diode wheel.

Figure 17.26 shows the same rotating diode wheel but a front view. You can see the six diodes mounted on the aluminum rotating structure. Two diodes are connected to each armature lead from the AC brushless exciter armature winding and the aluminium rotating structure forms the positive and negative connections. In this photo,

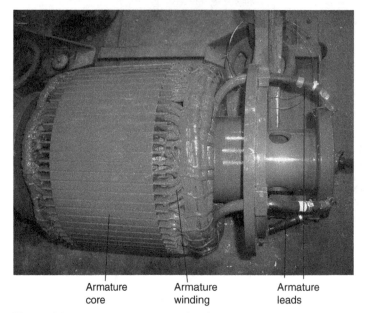

Armature   Armature   Armature
core   winding   leads

Figure 17.24   Synchronous motor with eight poles and brushless exciter – armature of brushless exciter.

Armature   Diode
leads   wheel

Figure 17.25   Synchronous motor with eight poles and brushless exciter – external of diode wheel.

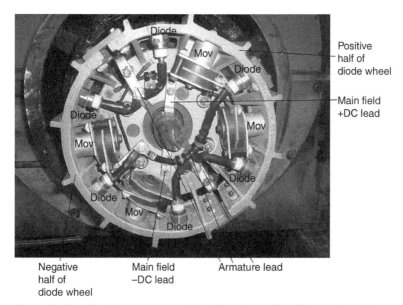

Figure 17.26   Synchronous motor with eight poles and brushless exciter – internal of diode wheel.

the conductors inside the shaft bore are connected to the two-bus connections (one on top and one on bottom) that connected the DC on the aluminum rotating structure to the main field through the internal shaft bore area. The electrical schematic of this rotating exciter circuit is shown in Figure 17.27.

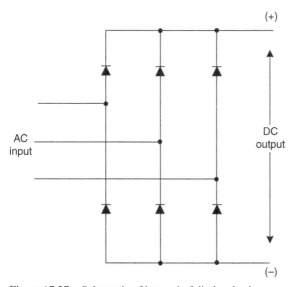

Figure 17.27   Schematic of internal of diode wheel.

TABLE 17.8 Motor Voltage Selection

Voltage Selection	Motor (HP)	Static
480 V	(1–200)	(1–200 kVA)
4160 V	(250–2000)	(250–2000 kVA)
13.8 kV	(2000–20,000)	(>2 MVA)

Source: Reproduced with permission of Tampa Electric Company.

When would a synchronous motor be specified over an induction motor? High speed motors (two- or four-pole) are almost always induction motors. Also, low power rated motors (500 HP or below) are almost always induction motors. The induction machine provides adequate service for a lower initial cost. When the application begins to get higher in power and lower in speed, this is where synchronous motors may be more commonly applied to utility applications. When the power of the motor is very large, the need to support the electrical system with reactive power becomes more important. In these applications, the ability of the synchronous motor to provide reactive vars to the electrical distribution system may be very attractive for the over electrical system design. Additionally, when the operational speed of the machine becomes very slow (16 poles or more), the efficiencies of the synchronous motor over the low speed induction machine may make the synchronous machine a more attractive choice.

The choice of motor nominal voltage depends on the power rating of the connected load. This varies depending on what voltage levels are available at a particular electrical distribution system, and what the plant standards are. However, for common applications, Table 17.8 shows the typical motor voltages specified for various power ratings of motors.

Now we will address motor bearings. Motor bearings are designed to maintain shaft position while still maintaining a low friction surface to allow for shaft rotation while minimizing friction losses. Some bearings are designed to maintain the radial position of the shaft and these are known as radial bearings. Other bearings are designed to maintain the axial position of the shaft and these are known as axial thrust or just "thrust" bearings.

Both radial and thrust bearings may be constructed as antifriction or journal bearings. Antifriction bearings use a race and roller assembly to isolate the bearing casing from the shaft surface area. For antifriction bearings, the typical lubrication is grease.

Antifriction bearings are used in smaller power rated motors and lower speed motor applications. Since these bearings use grease for their lubrication, operation of this motor at very low speeds is not a concern from the aspect of bearing wear and antifriction bearing are well suited for variable speed applications of motors.

The second type of bearing is the journal bearing. In this type of bearing, the surface of the shaft forms the race of the bearing and the casing of the bearing has a softer metal inserted in it called babbitt material. Oil is the lubrication for the journal bearing and oil is injected into the bearing housing. As the shaft spins, the speed of the rotor surface as it rotates by the babbitt produces a thin hydrodynamic oil film

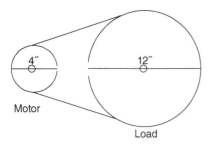

Figure 17.28   Motor torque and load torque application.

layer. This oil film provides the support for the shaft while at the same time providing the low friction surface to allow the shaft to freely rotate. Some applications utilize external pumps and oil coolers to supply the pressure or force for the oil injection into the bearing surface. In this application, the speed of shaft rotation is not an issue and this bearing would work well even in a variable speed application. However, some journal bearings are not externally supplied, but use an internal oil ring to force oil around the bearing surface. Since this oil ring uses shaft rotation to ensure adequate oil dispersion, the speed of the motor may be an issue in this type of bearing design. An "oil lift" pump is provided for no/low speed operation such that, if the shaft drops below some minimum speed where the self-lubrication is inadequate, then the external pump will run automatically to provide the required pressure to force the oil needed into the bearing surface. When a variable speed drive is used to drive a motor with babbited bearings, the bearings should be provided with lift oil pumps to provide adequate oil flow to the bearings during low speed operation.

Lastly, we need to address the difference between motor torque and load torque. When the motor is directly coupled to the shaft of the load and the motor shaft and load shaft both spin mechanically together at the same speed, then the load and motor torque are the same value. However, if the motor is coupled to the load by some type of transmission, where the motor shaft speed and load shaft speed are different from each other, then motor torque and load torque are not the same. Take for example the application shown in Figure 17.28 where the motor has a 4 in. diameter pulley connected to its shaft and the load has a 12 in. pulley connected to its shaft. The two pulleys are connected together by a common belt connected as shown in Figure 17.28. If the load requires X lb torque at a load speed of Y rpm, what is the speed of the motor shaft and what is the torque provided at the motor shaft?

First let us assume that this is an ideal machine. Assume there are no losses in the machine and that input power into the transmission system equals output power out of the transmission system.

$$P_{in} = P_{out} \qquad (17.19)$$

where

P_{in} = Power into the gear reducer (HP)

P_{out} = Power out of the gear reducer (HP)

Since power is the product of torque and speed, we can rewrite equation (17.19) in terms of torque and speed.

$$T_{in}\omega_{in} = T_{out}\omega_{out} \tag{17.20}$$

where

T_{in} = torque into the gear reducer (lb ft)

T_{out} = torque out of the gear reducer (lb ft)

ω_{in} = speed of shaft into the gear reducer (rpm)

ω_{out} = speed of shaft out of the gear reducer (rpm)

We can now rearrange equation (17.20) as shown in equation (17.21) to see the relationship of torque out of the gear reducer to the torque into the gear reducer as a function of the speed of the input to the gear reducer over the speed of the output of the gear reducer.

$$T_{out}/T_{in} = \omega_{in}/\omega_{out} \tag{17.21}$$

where

T_{in} = torque into the gear reducer (lb ft)

T_{out} = torque out of the gear reducer (lb ft)

ω_{in} = speed of shaft into the gear reducer (rpm)

ω_{out} = speed of shaft out of the gear reducer (rpm)

Now we just need to determine the rotation speed of the input shaft and the rotational speed of the output shaft to determine the relationship of torque to speed for each end of the mechanical transmission system.

The circumference of the pulley is given by

$$C = 2\pi r = \pi d \tag{17.22}$$

where

C = circumference of pulley

π = constant of 3.14

r = radius of pulley

d = diameter of pulley

The linear surface speed of the pulley is the product of the rotational speed of the pulley and the circumference of the pulley as shown in equation (17.22a)

$$V = C \times \omega \tag{17.22a}$$

V = linear velocity of circumference of the pulley

C = circumference of the pulley

ω = rotational speed of the pulley

Now we can write equation (17.23) in terms of equation (17.22a) to find the relationship between angular speed and the radius or diameter of the pulley.

$$V = 2\,\pi\,r\,\omega = \pi\,d\,\omega \qquad (17.23)$$

Realizing that, with the belt connecting the two pulleys, the linear speed of the surface of both pulleys is equal, we can state that

$$V_{in} = V_{out} \qquad (17.23a)$$

Using equation (17.22), we now find the relationship between angular velocity of both pulleys and the radius.

$$(2\,\pi\,r\,\omega)_{in} = (2\,\pi\,r\,\omega)_{out}$$

Removing the constants we find

$$r_{in}\omega_{in} = r_{out}\omega_{out} \qquad (17.24)$$

Rearranging we find that

$$r_{out}/r_{in} = \omega_{in}/\omega_{out} \qquad (17.25)$$

Now we know the relationship between angular speed and the radius of the pulley. The smaller pulley will have the higher rotational speed for the same linear surface velocity. We can use equation (17.25) and equation (17.21) to find the relationship between radius and torque

$$T_{out}/T_{in} = \omega_{in}/\omega_{out} = r_{out}/r_{in} \text{ or}$$

$$T_{out}/T_{in} = r_{out}/r_{in} \qquad (17.26)$$

Now we know that the smaller pulley will have the lower amount of torque for the same linear surface velocity.

Similarly, using 17.22, we now find the relationship between angular velocity of both pulleys and the diameter.

$$(\pi\,d\,\omega)_{in} = (\pi\,d\,\omega)_{out}$$

Removing the constants, we find

$$d_{in}\omega_{in} = d_{out}\omega_{out} \qquad (17.27)$$

Rearranging we find that

$$d_{out}/d_{in} = \omega_{in}/\omega_{out} \qquad (17.28)$$

Now, we know the relationship between angular speed and the diameter of the pulley. The smaller pulley will have the higher rotational speed for the same linear surface velocity. We can use equation (17.25) and equation (17.21) to find the relationship between diameter and torque

$$T_{out}/T_{in} = \omega_{in}/\omega_{out} = d_{out}/d_{in} \text{ or}$$

$$T_{out}/T_{in} = d_{out}/d_{in} \qquad (17.29)$$

Now we know that the smaller pulley will have the lower amount of torque for the same linear surface velocity.

Most electrical distribution systems do not operate exactly at the voltage level of the motor nameplate. For example, a common motor nominal voltage would be 460 V, three-phase, 60 Hz. However, the distribution system voltage may run higher or lower than exactly 460 V. If the voltage is raised above nominal voltage and/or if frequency is reduced below nominal such that the voltage to frequency ratio increases, the motor core operates closer to the saturation region. If the motor core becomes saturated, then excessive heating losses (due to hysteresis losses) will cause damage to the motor core. If the system voltage is less than the nominal voltage, then the motor, (being a constant power load) will compensate for the reduced voltage by pulling more current. This excess current can cause operation of thermal overload protection devices or, if overload protection is not properly set, this can lead to thermal failure of the stator winding due to excessive stator currents. So how much can the voltage vary without damaging the motor? NEMA MG1 states that motors shall be designed to operate successfully under running conditions at rated load with a variation in the voltage or the frequency up to the following limits

a. ±10% of rated voltage, with rated frequency for induction motors.
b. ±5% of rated frequency, with rated voltage.
c. A combined variation in voltage and frequency of 10% of the rated value (sum of absolute values), provided the frequency variation does not exceed ±5% of rated frequency.

So what types of motor applications are there in a power generation facility and what are the special considerations for the various applications? Below is a listing of some of the common motor applications in a generation facility along with a brief discussion of some special considerations for each application.

Coal-fired power plants have the coal delivered by truck, rail, or barge. With truck and rail delivery systems, there is some means of dumping the coal from the car or trailer. Either the car provides for doors to dump from under the car or, in some locations, the car is turned over to dump the coal from the car. To rotate a car full of coal requires substantial amount of torque until the coal begins to drop out of the car. High torque applications such as unloading rail cars utilize motors with special rotors designed for the high slip, high torque requirements of these applications. Mills and pulverizers are used to process the coal before it is injected into the boiler. These motors tend to be very low speed motors due to the very low speed operation of the mills or pulverizers that they are driving and, as such, tend to be constructed with a large number of poles. This construction makes the motor casing larger for the same power rating than a comparable high speed motor.

The coal is transported using conveyors. In the event the conveyors stop with coal (or other material on them), it will require a substantial amount of torque to start the conveyor belt when it is loaded with coal. The design of the conveyor motor should look at the instance when the conveyor starts loaded to ensure it can success-fully start the conveyor in this condition due to the higher mechanical starting torque that the conveyor requires when loaded with material. Boiler feed pumps are used to increase the boiler feedwater pressure to the values needed to send the feedwater to the boiler. These are large horse power, high speed pumps and, as such, require vibration monitoring to ensure that the mechanical operation of the pump is adequate for long-term service.

There are also many fans in the typical power plant such as the forced draft (FD) fan, the induced draft (ID) fan, the gas recirculation fan, the booster fan, and the primary air fan. All these fans are large HP and large inertial fans. Due to the size and inertia of the fans it is typical that it may take a long time to accelerate the fan to full speed and the motor must be designed to successfully accelerate the inertia of the fan, even with a depressed voltage source due to the worst-case design voltage drop in the feeder.

Circulating water pumps are a special application. These are vertical motors connected to axial flow pumps. Circulating water pumps have an axial thrust load associated with them due both to the vertical orientation of the motor and the axial flow pump that is connected to the motor. Also, these circulating water pump motors are very low speed motors so they are physically large motors. The pumps are axial flow pumps instead of radial flow pumps. Due to this design, if flow were restricted, load on the motor would increase instead of decrease so the starting of these pumps is unlike that of the typical radial pump. In the axial flow pump, the flow path must be proven (all valves proven open) and in some instances, the pump housing is primed before the pump motor is energized to prevent damage to the pump motor during a start.

GLOSSARY OF TERMS

- Alternator – A synchronous machine used to convert mechanical power into alternating current electric power.

- Ambient Temperature – The temperature of the surrounding cooling media. Commonly known as room temperature when the air is the cooling media in contact with the equipment.

- Autotransformer – A device that converts a fixed frequency fixed voltage input into a fixed frequency variable voltage output in several discrete steps. The autotransformer is used to reduce the voltage applied to a motor during the starting process.

- Breakdown Torque (BDT) – The maximum torque that an AC motor will develop with rated voltage applied at rated frequency without an abrupt drop in speed. Also termed pull-out torque or maximum torque.

- Constant Horsepower Load – Term used to describe a load where the power is constant for any given speed. In this application, torque is reduced as speed is increased and torque is increased as speed is reduced.

- Constant Torque Load – Term used to describe a load where the torque is constant for any given speed. In this application, power is reduced as speed is reduced and power is increased as speed is increased.

- Efficiency – The ratio of output power divided by the input power.

- Electronic Soft Start – A device that converts a fixed frequency and fixed voltage input to a fixed frequency and variable voltage output. The electronic soft-start is used to reduce the voltage applied to a motor during the starting process.

- Eddy Current Coupling – A device that contains a fixed speed rotor and an adjustable speed rotor separated by a small air gap. A direct current in a field coil produces a magnetic field that determines the torque transmitted from the input rotor to the output rotor.

- Fluid Coupling – The device connected between a motor and shaft that mechanically couples the two devices together using a fluid media.

- Frequency – A measurement of the number of cycles in a second and is given in units of hertz (Hz).

- Full-Load Current or Full-Load Amps (FLA) – The current required for any electrical machine to produce its rated output or perform its rated function.

- Full-Load Speed – The speed at which any rotating machine produces its rated output.

- Full-Load Torque (FLT) – The torque required to produce rated power at full-load speed.

- Horsepower (HP) – A unit of measurement of power. One horsepower equals 33,000 foot-pounds of work per minute (550 ft lb per sec) or 746 watts. In motors, the nominal HP refers to the shaft power of the motor, not the electrical power drawn by the motor.

- Insulation – Non-conducting materials separating the current-carrying parts of an electric machine from each other or from adjacent conducting material at a different potential.

- Kilowatt (kW) – A unit of measurement of power. One kilowatt equals 1.34 horsepower which equals 44,235 foot-pounds of work per minute (737 ft lb per sec). In motors, the nominal kW refers to the shaft power of the motor, not the electrical power drawn by the motor.

- Locked Rotor Current or Locked Rotor Amps (LRA) – Steady-state current taken from the line with the rotor of a motor with a rotor at standstill and at rated voltage and frequency.

- Locked Rotor Torque (LRT) – The minimum torque that a motor will develop at standstill, with rated voltage applied at rated frequency.

- Motor – A rotating machine that converts electrical power into mechanical power.

- Poles – The magnetic poles set up inside an electric motor or generator by the placement and connection of the windings.

- Power Factor (pf) – Ratio of real power or watts to total power or VA of an electrical circuit.

- Resistance Temperature Detector (RTD) – A device used for temperature sensing consisting of a wire coil or deposited film of pure metal for which the change in resistance is a known function of temperature.

- Rotor – The rotating element of any motor or generator. For an AC machine, the field is on the rotor. For a DC machine, the armature is on the rotor.
- Service Factor (sf) – The service factor of an alternating current motor is a multiplier which, when applied to the rated HP, indicates a permissible HP loading which may be carried under conditions specified for service factor.
- Slip – The difference between synchronous and operating speeds, compared to synchronous speed.
- Stator – The stationary element of any motor or generator. For an AC machine, the armature is on the stator. For a DC machine, the field is on the stator.
- Synchronous Speed – The speed of the rotating machine element of an AC motor that matches the speed of the rotating magnetic field created by the armature winding.
- Torque – The rotating force produced by a motor. The units of torque may be expressed as pound-foot, pound-inch (English system), or newton-meter (metric system).
- Variable Frequency Drive (VFD or VSD) – Device that converts a fixed frequency and fixed voltage input to a variable frequency and variable voltage output.
- Variable Torque Load – Term used to describe a load where the torque is reduced at reduced speeds and increased at increased speeds. In this application, power is reduced by the square or cube (depending on load type) as speed is reduced and power is increased by the square or cube (depending on load type) as speed is increased.
- Wound Rotor Motor – A motor with the rotor circuit wired out to slip rings. Varying the resistance connected to the slip rings varies the motor torque speed curve.

PROBLEMS

17.1 For a four-pole, synchronous motor that has 60 Hz, 460 V applied to the stator, what is the synchronous speed of the motor?

   A. 3600 rpm

   B. 1800 rpm

   C. 1200 rpm

   D. 900 rpm

17.2 For a four-pole motor with 60 Hz applied to stator, runs at full load at 1725 rpm, what is the percent slip of the motor at full load?

   A. 2.33% slip

   B. 3.17% slip

   C. 4.17% slip

   D. 6.33% slip

17.3   Given a load that has a maximum steady state torque requirement of 35 lb ft at a rotational speed of 900 rpm, what is the mechanical power requirement of this application in horsepower (HP)?

   A.   2 HP

   B.   4 HP

   C.   6 HP

   D.   8 HP

17.4   Given an induction motor that has 10 HP shaft power at a shaft rotational speed of 1100 rpm, what is the available torque output of the motor at the shaft?

   A.   11.9 lb ft

   B.   23.86 lb ft

   C.   47.73 lb ft

   D.   95.45 lb ft

17.5   Using the same motor application as described in Problem 17.4, if between the motor and the final load a 2:1 gear reducer is utilized, the load shaft speed is half of motor shaft speed (i.e., load shaft speed is 550 rpm), and assuming a perfectly efficient gear reducer (i.e., shaft power into gear reducer = shaft power out of gear reducer), what is the torque available on the load shaft downstream of the gear reducer? (*Hint*: use same formula as used in Problem 17.4, but use power and rpm available on load shaft.)

   A.   11.9 lb ft

   B.   23.86 lb ft

   C.   47.73 lb ft

   D.   95.45 lb ft

17.6   The method of reduced voltage starting where the line current is reduced proportional to the square of the reduction of voltage is

   A.   autotransformer

   B.   primary resistor

   C.   primary reactor

17.7   The transmission system shown in Figure 17.27 has a pulley with a diameter of 12″ on the load shaft and a pulley with a diameter of 4″ on the motor shaft. If the torque the load is requiring is 60 lb ft, what is torque on the motor shaft? (Assume an ideal transmission system where $P_{in} = P_{out}$).

   A.   15 lb ft

   B.   20 lb ft

   C.   30 lb ft

   D.   60 lb ft

17.8   The transmission system shown in Figure 17.27 has a pulley with a diameter of 12″ on the load shaft and a pulley with a diameter of 4″ on the motor shaft. If the rotational speed of the load is 380 rpm, what is rotational speed of the motor shaft? (Assume an ideal transmission system where $P_{in} = P_{out}$).

   A.   380 rpm

   B.   520 rpm

C. 1140 rpm

D. 2280 rpm

17.9 Using Table 17.1, find the capacitor rating required to improve the power factor of a 250 kW load from 0.70 to 0.90.

A. 95 kvar

B. 123 kvar

C. 134 kvar

D. 154 kvar

RECOMMENDED READING

Electric Power Plant Design, Technical Manual TM 5-811-6, Department of the Army, USA, 1984.

Electrical Machines, Drives and Power Systems, 6th edition, Theodore Wildi, Prentice Hall, 2006.

The Engineering Handbook, 3rd edition, Richard C. Dorf, (editor in chief), CRC Press, 2006.

IEEE API 546-1993: Brushless Synchronous Machines – 500 kVA and Larger.

IEEE API 541-1997: Form-Wound Squirrel Cage Induction Motors–250 Horsepower and Larger.

IEEE 141-1993: IEEE Recommended Practice for Electric Power Distribution for Industrial Plants (IEEE Red Book).

IEEE 841-1994: IEEE Standard for the Petroleum and Chemical Industry–Severe Duty Totally Enclosed Fan-Cooled (TEFC) Squirrel Cage Induction Motors–Up to and Including 500 HP.

IEEE 1068-2010: IEEE Recommended Practice for the Repair and Rewinding of Motors for the Petroleum and Chemical Industry.

Industrial Power Distribution, 2nd edition, Ralph E. Fehr, III, Wiley-IEEE Press, 2016. ISBN: 978-1-119-06334-6.

NEMA MG1-2010 – Motors and Generators.

Power Plant Engineering, Black & Veatch, edited by Larry Drbal, Kayla Westra, and Pat Boston, Chapman & Hall/Springer, 1996.

Standard Handbook of Powerplant Engineering, 2nd edition, Thomas C. Elliott, McGraw-Hill, 1998.

CHAPTER 18

VARIABLE FREQUENCY DRIVE SYSTEMS

GOALS

- To understand the basic design and construction of variable frequency drives (VFD)
- To calculate the synchronous frequency and slip (in percent and rpm) of an induction motor fed from a VFD
- To describe the difference and benefits/limitations between a voltage source drive and a current source drive
- To be able to define the maximum safe speed of an induction motor based on NEMA standards
- To describe the difference between a constant-power load, a constant-torque load, and a variable-torque load
- To calculate the pulse number for a given rectifier design and determine the phase shift required for the source transformer

VARIABLE FREQUENCY *drives* (VFD) and variable speed drives (VSD) are used to control the speed and/or torque of induction and synchronous motors at the power generation facility when speed control or torque control is required for the application. Their use varies from small pumps pumping water from one location to another for utilization to control flow or pressure of the water, to very large fans used to control the flow of air for combustion through the main furnace.

The types of VFD technologies available are varied and each has its own unique advantages and disadvantages. While details of construction vary from manufacturer to manufacturer, the three main types of VFD are the voltage source inverter (VSI), the current source inverter (CSI) and the pulse width modulated inverter (PWM).

A little clarification on terminology is in order. A *rectifier* converts AC voltage to DC voltage. An *inverter* converts DC voltage back to an AC voltage. The variable speed drive is sometimes called an inverter, but this is not accurate, as the typical variable speed drive consists of a rectifier section for conversion of AC source power

Energy Production Systems Engineering: An Introduction for Electrical Engineers to Electrical Power Generation Facilities, Systems, and Equipment, First Edition. Thomas H. Blair.
© 2017 by The Institute of Electrical and Electronics Engineers, Inc. Published 2017 by John Wiley & Sons, Inc.

to DC power, some type of filter section, and an inverter section for the conversion of the DC power back to an AC power source that can have its voltage and frequency output controlled to control the speed and available torque of a motor.

The synchronous speed of rotation of an AC motor is the product of the applied frequency and the inverse of the number of poles of the motor.

$$n_s = 120f/p \qquad (18.1)$$

n_s = synchronous speed of machine (rpm)

f = frequency applied to motor stator (Hz)

p = number of poles of motor

For a synchronous motor that has an external field applied to the machine, this machine runs at synchronous speed. For an induction machine, to produce torque, the rotor must slip (see Chapter 17 on motors for a more detailed discussion). When the motor stator is energized from rated voltage and frequency, the slip is defined in percent.

$$s = [(n_s - n)/n_s] \times 100\% \qquad (18.2)$$

where

s = slip of the machine (%)

n_s = synchronous speed of machine (rpm)

n = the operating speed of machine (rpm)

For full-load operation of an induction motor, the slip is defined as

$$s = [(n_{nl} - n_{fl})/n_{nl}] \times 100\% \qquad (18.3)$$

where

s = slip of the machine (%)

n_{nl} = synchronous speed of machine (rpm)

n_{fl} = the operating speed of machine (rpm)

However, when applying variable frequency drives to induction motors, it is more helpful to use the slip of the motor in rpm and not percentage. The slip of the motor, in rpm, is defined in equation (18.4).

$$s = (n_{nl} - n_{fl}) \qquad (18.4)$$

where

s = slip of the machine (rpm)

n_{nl} = synchronous speed of machine (rpm)

n_{fl} = operating speed of machine (rpm)

The reason we define the slip of the motor in a variable speed drive application by equation (18.4) is that, for the rotor to develop full torque, the rotor speed in rpm must be a certain number of revolutions per minute slower than the synchronous speed of the machine in rpm. This sets up a certain current and frequency in the rotor bars. If the stator is connected to a VFD and only half of the motor stator rated frequency is applied to the stator, then the synchronous speed of the motor is half of the nameplate synchronous speed. However, for the induction motor to develop rated torque, the rotor must slip (in rpm) the same amount as the rated slip of the motor at full frequency. This leads to a higher value of slip in percentage (equation 18.3) as the applied frequency to the stator is reduced. This is the reason why, when talking about an induction motor and its full-load speed when the motor is driven by a VFD, it is more helpful to use slip in rpm. The following example will expound on this concept.

Example 18.1   Given a motor rated at 460 V, 60 Hz, four-pole machine with a full-load speed of 1700 rpm, and given that the motor has 460 V, 60 Hz applied to the stator and if fully loaded, calculate the following.

   a. Calculate the synchronous speed of the motor.

   b. Calculate the slip of the motor in rpm.

   c. Calculate the slip of the motor in percent.

Solution: First, we need to calculate the synchronous speed with 60 Hz applied to the stator using equation (18.1).

$$n = 120 f/p$$
$$n = 120 \times 60 \text{ Hz}/4 \text{ poles}$$

   a. $n = 1800$ rpm

   The given full-load speed with 60 Hz applied is 1700 rpm. The slip of the motor in rpm is calculated using equation (18.4).

$$s = (n_{\text{nl}} - n_{\text{fl}})$$
$$s = (1800 \text{ rpm} - 1700 \text{ rpm})$$

   b. $s = 100$ rpm

   Now that we have the slip in rpm, we just need to normalize to the synchronous speed in rpm to determine slip in percentage using equation (18.3).

$$s = [(n_{\text{nl}} - n_{\text{fl}})/n_{\text{nl}}] \times 100\%$$
$$s = [(1800 \text{ rpm} - 1700 \text{ rpm})/1800 \text{ rpm}] \times 100\%$$
$$s = 0.0556 \times 100\%$$

   c. $s = 5.56\%$

Example 18.2   Given a motor rated at 460 V, 60 Hz, four-pole machine with a full-load speed of 1700 rpm when operated at rated voltage and frequency is now energized from a VFD. The output of the VFD is 230 V, 30 Hz. If the motor is delivering rated torque to the load (i.e., slip in rpm is the same as in the rated frequency case), calculate the following.

   a. Calculate the synchronous speed of the motor at 30 Hz.

   b. Calculate the slip of the motor in rpm.

   c. Calculate the slip of the motor in percent.

Solution: First, we need to calculate the synchronous speed with 30 Hz applied to the stator using equation (18.1).

$$n = 120f/p$$
$$n = 120 \times 30 \text{ Hz}/4 \text{ poles}$$

a. $n = 900$ rpm

We are given that the motor is delivering rated torque to the load. For the motor to provide rated torque to the load, the motor must slip the same amount (in rpm) at this reduced speed as it does at rated speed. This value was calculated in Example 18.1(b) using equation (18.4).

$$s = (n_{nl} - n_{fl})$$
$$s = (1800 \text{ rpm} - 1700 \text{ rpm})$$

b. $s = 100$ rpm

Now that we have the slip in rpm, we just need to normalize to the synchronous speed in rpm at the new applied frequency of 30 Hz to determine slip in percentage using equation (18.3).

$$s = [(n_{nl} - n_{fl})/n_{nl}] \times 100\%$$
$$s = [100 \text{ rpm}/900 \text{ rpm}] \times 100\%$$
$$s = 0.1111 \times 100\%$$

c. $s = 11.11\%$

As Examples 18.1 and 18.2 show, for rated torque, the motor slip in rpm is the same, but the percent slip changes because this is normalized to synchronous speed which changes when the motor-applied frequency changes. In Chapter 17 on motors, we discuss how the applied voltage also affects the torque of the machine and the discussion above assumes that the voltage-to-frequency ratio of the motor remains constant so that the voltage applied to the motor does not affect torque of the motor. Realize, by adjusting the voltage-to-frequency ratio, you can boost or buck actual motor torque available at a particular slip of the motor as long as we stay within the volts-to-hertz maximum tolerance of the motor as defined by NEMA MG1.

So how does the application of a VFD to an induction motor effect the motor torque-speed curve. It does not deform the curve but simply shifts the curve to the left (when the applied frequency is less than stator nominal) or to the right (when the applied frequency is greater than stator nominal). Figure 18.1 shows a typical motor torque speed curve with applied frequencies of 15 Hz, 30 Hz, 45 Hz, and 60 Hz.

The *voltage source inverter* (VSI) consists of a diode rectifier section that converts the electrical system AC voltage to a DC voltage. There is a capacitive filter in the DC bus to reduce the ripple from the AC rectification and provide a smoother DC voltage to the inverter section. This capacitive section presents a DC voltage source to the inverter section and this is where the term voltage source inverter is derived from. Then, there is an inverter section that converts this DC voltage back to an AC voltage of varying frequency and magnitude. With an inverter section with six transistor outputs, there are six different combinations of which transistor is on and which

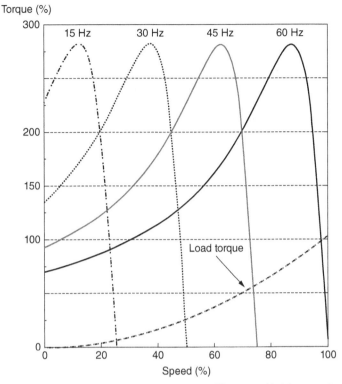

Torque (%)

Speed (%)

Figure 18.1   Motor torque-speed curve for different applied frequencies. *Source*:
Reproduced with permission of IEEE.

transistor is off. This provides a six-level or six-step output when the phase-to-phase
voltage is measured. Since the motor is highly inductive, the current waveform of the
motor stator is a smoother representation of the applied stator voltage and appears
as a sine wave. Operation of the VSI drive at very low stator frequencies and resultant very slow speeds sometimes can result in cogging of the motor due to the step
change in the output of the inverter section and the typical speed range of a VSI is ten
to one (or 60 Hz to 6 Hz). Motor cogging is the appearance of a motor rotor jumping
from pole to pole. The motor shaft moves quickly from one pole to the next, then
pauses, until the electromagnetic field in the motor stator changes, causing the rotor
to suddenly move again to the next pole.

The pulse width modulated (PWM) drive, as shown in Figure 18.2, is similar
in construction to the VSI in that the drive consists of a diode rectifier section that
converts the electrical system AC voltage to a DC voltage.

There is a capacitive filter in the DC bus to reduce the ripple from the AC
rectification and provide a smoother DC voltage to the inverter section. After the
filter section, there is an inverter section that converts this DC voltage back to an AC
voltage of varying frequency and magnitude. While the construction of the VSI and
PWM drive is similar, it is the gating pattern of the inverter section that differentiates

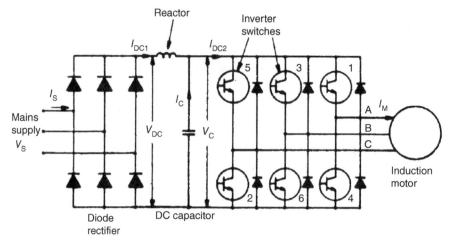

Figure 18.2   PWM VSD configuration. *Source*: Reproduced with permission of IEEE.

the VSI from the PWM drive. In the VSI drive, there are six switching events in one cycle which results in a six-step waveform voltage output of the drive. The PWM drive has the same six distinct combinations but each combination is switched on and off repetitively at varying pulse widths and a very high value of switching frequency (in the kHz range). This rapid switching results in an output AC voltage waveform of the inverter section that more closely represents a sinusoidal AC voltage waveform. Because of this, the typical PWM drive has a smoother output current waveform as shown in Figure 18.3 and does not suffer from the low speed cogging that sometimes occurs on the VSI drive.

The *current source inverter* (CSI) that is shown in Figure 18.4 consists of an SCR-controlled diode rectifier section that converts the electrical system AC current to a variable DC bus current. Silicon controlled rectifiers (SCR) are used to control the delay angle of the rectifier, thereby controlling the amount of current flowing into the DC bus. While the voltage source inverters like the PWM inverter and VSI use DC capacitors in their DC bus section to filter voltage, the CSI uses inductors in their

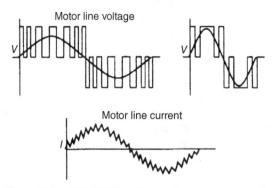

Figure 18.3   PWM VSD voltage and current outputs. *Source*: Reproduced with permission of IEEE.

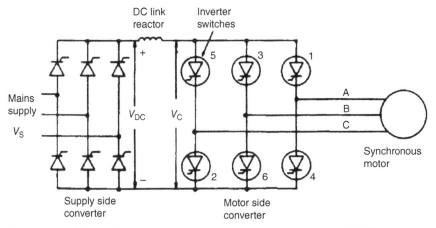

Figure 18.4 LCI configuration. *Source*: Reproduced with permission of IEEE.

DC bus to filter current. The inductor presents a filtered current source to the inverter section, thus the name current source inverter.

The constant current source from the filter of the LCI is then converted from DC current back to AC current in the inverter section which uses SCRs as the switching device. Figure 18.5 shows three SCRs compressed between heatsinks out of a 4000 V LCI drive.

Figure 18.5 LCI VSD SCR stack assembly. *Source*: Reproduced with permission of IEEE.

The way an SCR works is that, once gated, it is the reversal of current across the SCR anode and cathode that commutates the SCR off. Since the DC bus presents a constant current source, the energy for inverter SCR *commutation* comes from the motor connected to the LCI drive. Because the motor has to be able to provide this energy back to the drive for commutation, the only types of motors that can be used on an LCI drive are synchronous motors. The PWM and VSI drive can be used on either synchronous or induction type motors. Also, since the LCI drive uses motor energy for the commutation of the inverter section (it uses line energy for commutation of the rectifier section), the motor power rating connected to the LCI must be close to the value of the LCI nominal power rating. The PWM and VSI drives can be tested without a motor connected to the load, but an LCI drive cannot be fully tested without a motor close to nominal ratings connected to the drive. (There is a mode called crow bar test where the output inverter section is fully gated and the rectifier section can be tested, but to test the inverter section requires a motor rated for the LCI power rating connected to the load leads of the LCI.) For the LCI, the power factor of the motor is reflected back to the source, where for the PWM and VSI, the current waveform of the source does not reflect the current waveform of the motor load, so the power factor is not reflected back to the source. Much like the VSI drive, at very low speeds, the switching between devices is very slow and, in some instances very low speed operation of a motor with a CSI may present motor cogging.

Comparing the three basic technologies available, the VSI and PWM drive, being voltage source drives provide better protection to open circuit outputs whereas the CSI, being a current source drive, provides better protection to short circuit outputs. The VSI and PWM also being voltage source drives can handle an undersized motor and can be tested without the connection of any motor on the load leads of the drive, but will not tolerate an oversized motor well. The CSI being a current source drive can handle an oversized motor using the inherent current limiting feature of the CSI, but cannot handle an undersized motor and cannot have its inverter section tested with no motor connected.

The diodes of the PWM and VSI present a two-quadrant operating device such that power only flows from the AC side of the rectifier to the DC side of the rectifier and cannot flow in the other direction. Therefore, the PWM and VSI drives that have simple diode rectifier sections cannot regenerate. *Regeneration* is the process where, when the motor load starts driving the motor, the direction of power delivery at the motor stator changes and the motor becomes a generator and wants to deliver power back to the energy source which is through the VFD. The inverter section can pass power in either direction and will allow this energy into the VFD DC bus. However, in the PWM and VSI, the use of the diode bridge prevents this power from being delivered to the energy supply system. To compensate for this, in regenerative applications, there is a switch (transistor) and resistor across the DC bus of the PWM or VSI drive such that, when energy from the motor is regenerated back to the drive, this energy is dissipated locally as heat in the resistor (called *dynamic braking*). The CSI rectifier section can deliver power in both directions. From a technical aspect, there is no difference between the LCI rectifier section and the LCI inverter section except the point of gating of the SCR. When the motor becomes a generator, the "inverter" section of the LCI becomes a rectifier and allows energy flow from the motor to the DC bus, the voltage polarity of the DC bus changes (but current flow direction remains the same) so the energy flow in the DC bus changes, and the "rectifier" section becomes an inverter to transfer this energy in the DC bus back to the electrical system that the LCI is connected to.

Of the three technologies, the PWM has the smoothest performance at very low speeds. There are other drive technologies available but used much less frequently. One possible application is the need to drive very high torque loads at very low speeds. As mentioned previously, of the three common drive topologies, the CSI and VSI would not be good choices due to the possible low speed cogging that can occur. The PWM would be better choice, but may have issues with heat generated in motor due to high switching frequency. For motors that spend their entire service life at very low speeds, one specialty drive is the cycloconverter. This uses phase angle control to provide very low frequencies to motors to drive then at very low speeds smoothly and provide adequate high torque. A common application at a power plant for a cycloconverter would be in the dumping of rail cars where the rail car is turned over to dump the car. The speed the car is turned is very slow and the initial torque required to get the fully loaded car turning is large.

Another common application of variable speed drives in a power generation facility is for the control of flow of air from a fan. For example, in a coal plant, the forced draft (FD) fan controls the airflow through the boiler furnace section to regulate furnace pressure. As the combustion demand is increased (for more steam flow), the air that enters the furnace section is increased to support the combustion process and provide more air mass. There are two main mechanical methods to control airflow, by the use of inlet vanes or outlet dampers, and one electrical method of controlling the airflow, and that is with the use of a variable speed drive. (Note there are other methods that can be used, but are less common. For example, if the fan is an axial flow fan, then the blade pitch can be used to control the airflow.)

At rated flow, all of the three methods draw about the same amount of power from the motor. However, at reduced flow rates, the power drawn by the motor varies. The reason for this comes from the concept that the mechanical power than the fan draws is a function of the differential pressure across the fan and the rate of airflow through the fan. For the same reduction of airflow, using an outlet damper to restrict the airflow results in increased discharge pressure on the fan. This increases the differential pressure across the fan so, even though the flow of air through the fan is reduced which reduces the mechanical power drawn by the fan, the increase in the differential pressure across the fan has the opposite effect and tries to drive up the amount of mechanical power drawn from the motor. The other mechanical method of controlling airflow through a boiler is by adjustment of the inlet guide vanes on a centrifugal fan. The mechanics of this are similar except now the vanes that restrict airflow are on the suction side of the fan. The inlet guide vanes are arranged such that they provide some initial swirl of air to the vortex of the fan which increases efficiency of the fan. However, closing the inlet guide vanes does reduce the inlet pressure of the fan, thereby increasing the differential pressure across the fan. Therefore, when using inlet guide vane control of a centrifugal fan, when the inlet guide vanes are closed, the flow is reduced which has the tendency to reduce the power drawn by the fan, but the differential pressure across the fan is still increased due to the lower suction pressure at the vortex of the fan and this tries to increase power demand. Of the two mechanical methods of flow control, the inlet guide vane results in a lower power requirement of the fan at lower values of flow than does the outlet damper control method.

The third common method is to use a variable speed drive to control the frequency to the motor stator and, thereby, control the speed of the motor which controls the speed of the fan. As we lower the speed of the motor and fan assembly on a centrifugal fan, the flow responds in proportion to the speed. In the ideal case (if there are no other fans in series or parallel), the differential pressure is the square of the speed and, since power is the product of flow and differential pressure, the power is the cube of the change in speed. (See Chapter 8 on fans for more information on the centrifugal fan laws). A word of caution here is warranted. In the typical boiler airflow path, there are usually multiple motors. There are two forced draft fans in parallel and, in series with these, there are two induced draft (ID) fans in parallel or booster fans in parallel. In these applications, the change in differential pressure across the fan may not follow the "square of the speed" rule due to compensation from other fans in the system.

Now, as the flow is decreased, the differential pressure across the fan is also decreased and the power required by the fan at lower speeds is substantially less than it would be had the flow control method been either outlet damper or inlet vane control. The VFD itself is 95% efficient or more depending on the design of the drive, so there are some losses associated with the operation of the VFD that do not exist when using outlet damper or inlet vane control, but the losses in the VFD are much smaller than the losses associated with the two mechanical control methods. Therefore, VFD control results in a much lower power demand from the electrical system than the two mechanical methods of flow control. The disadvantage of the VFD is complexity and initial cost. Using a damper drive to control the position of either an outlet damper in inlet guide vane is much less expensive on initial capital costs and much simpler on the control system than implementing a variable speed drive to control motor speed. The advantage of the VFD is the long-term energy savings that is possible with a VFD in addition to the reduction of mechanical maintenance that comes with dampers and inlet guide vanes.

The chart in Figure 18.6 shows the typical power demand of a fan using each of the three methods of flow control described above.

The ability of a VFD to provide output frequencies above the motor nominal frequency and motor speeds above the nominal speed of the motor is another advantage of the VFD, but must be used with caution. The first obstacle to how much overspeed we can drive a motor is the mechanical limitation of the motor bearings and rotating members. As we increase the rotational speed of the motor, the centrifugal force increases as the square of the rotational speed.

$$F = m r \omega^2 \qquad (18.5)$$

F = centrifugal force

m = mass of object

r = radius the object from geometric center

ω = angular velocity of the object

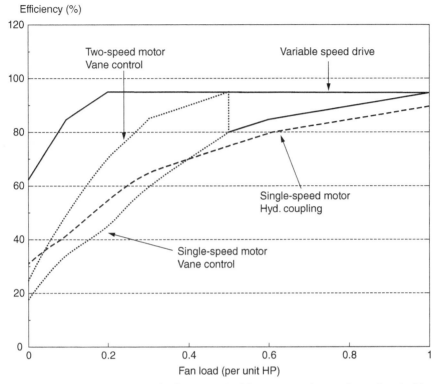

Figure 18.6   Efficiency curves for flow control of fan or pump. *Source*: Reproduced with permission of IEEE.

At some point, we can exceed the maximum force that the material and construction of the motor can tolerate. NEMA provides guidance through their recommendations on maximum safe speed a typical motor can tolerate, as shown in Table 18.1.

From Table 18.1 you can see that, for small-power rated motors, the percentage overspeed is greater (because the radius of the motor rotor is smaller and the mass of the motor is smaller) than for large- power rated motors. For example, from Table 18.1, looking at a 1 HP, 3600 rpm motor, the maximum speed is 7200 rpm. This is a 200% increase in speed. However, the same speed rated motor (3600 rpm) for a 50 HP motor, has a maximum speed of 4500 rpm. This is a 125% increase in speed.

Also, you can see for motors with a lower number of poles (or higher rated nominal synchronous speed), the percentage of overspeed at any particular HP rating is less than or equal to an equivalently power-rated motor that has more poles (i.e., lower nominal synchronous speed rating). For example, from Table 18.1, you can see that a 10 HP, two-pole, 3600 rpm motor has a maximum speed of 5400 rpm. This is an increase of 150%. A 10 HP, six-pole, 1200 rpm motor has a maximum speed of 2400 rpm. This is an increase of 200%.

Additionally, for the standard configuration, as long as we keep the ratio of motor voltage to motor-applied frequency approximately constant, the amount of

TABLE 18.1   Maximum Safe Operating Speed for Standard Squirrel Cage Motors

Horsepower	Synchronous Speed at 60 Hz		
	3600	1800	1200
1/4	7200	3600	2400
1/3	7200	3600	2400
1/2	7200	3600	2400
3/4	7200	3600	2400
1	7200	3600	2400
1.5	7200	3600	2400
2	7200	3600	2400
3	7200	3600	2400
5	7200	3600	2400
7.5	5400	3600	2400
10	5400	3600	2400
15	5400	3600	2400
20	5400	3600	2400
25	5400	2700	2400
30	5400	2700	2400
40	4500	2700	2400
50	4500	2700	2400
60	3600	2700	2400
75	3600	2700	2400
100	3600	2700	1800
125	3600	2700	1800
150	3600	2700	1800
200	3600	2300	1800
250	3600	2300	1800
300	3600	2300	1800
350	3600	1800	1800
400	3600	1800	—
450	3600	1800	—
500	3600	1800	—

Source: Reproduced with permission of National Electrical Manufacturers Association.

torque available from the motor is approximately constant. Once we reach nominal motor frequency and nominal motor voltage, if we increase frequency to the stator of the motor above this value, to keep motor torque constant, we would have to increase the voltage to the motor windings to keep the volts-to-hertz ratio constant. Depending on the winding insulation of the motor, this may be possible or the insulation may limit us to keeping the voltage applied to the motor stator at the nominal voltage value. If we keep the voltage on the stator constant as we increase the frequency to the motor, then the volts-to-hertz ratio becomes less. This has the same effect on the motor as reducing the voltage to the motor stator at rated frequency as the available torque of the motor is reduced by the square of the applied voltage. In the variable frequency drive application, as we increase the frequency above nominal frequency and keep voltage applied to the motor stator at nominal voltage, the volts-to-hertz ratio decreases inversely in proportion to the ratio of the applied frequency to nominal motor frequency. The available motor torque varies as the square of the volts-to-hertz ratio of the machine.

Example 18.3   Given a motor rated at 460 V, 60 Hz, 1800 rpm, 10 HP motor. This motor has a VFD supplying voltage and frequency to the motor stator and the maximum value of voltage that the VFD can provide to the motor is 460 V. The maximum frequency the VFD can supply to the motor is 120 Hz. Calculate the following.

  a. Calculate nominal torque available from the motor at 1800 rpm.

  b. Calculate the nominal volts-to-hertz ratio that this torque is based on.

  c. Calculate the volts-to-hertz ratio for the motor when it is at 10% overspeed or 1980 rpm (assume applied frequency of 66 Hz) knowing that the applied stator voltage does not change from 460 V.

  d. Calculate torque available from the motor at 10% overspeed or 1980 rpm (assume applied frequency of 66 Hz) knowing that the applied stator voltage does not change from 460 V.

  e. Calculate the percentage change in torque normalized to the nominal torque of the machine.

Solution:

  a. Using the formula from Chapter 17
         Equation (17.5) provides us the relationship between power and torque.

$$P = Tn/5250$$

  where

   P = power (HP)

   T = torque (lb ft)

   n = speed (rpm)

  Rearranging for torque, we find the following.

$$T = P \times 5250/n$$
$$T = 10 \text{ HP} \times 5250/1800 \text{ rpm}$$
$$T = 29.167 \text{ lb-ft}$$

  b. Next, we find the volts-to-hertz ratio that this torque is based on. Since stator voltage is 460 V and stator frequency is 60 Hz, the volts-to-hertz ratio is

$$\text{Volts-to-hertz ratio} = 460 \text{ V}/60 \text{ Hz}$$
$$\text{Volts-to-hertz ratio} = 7.67 \text{ V}/\text{Hz}$$

  c. Assuming the new applied frequency is 66 Hz, and the applied motor stator voltage is still 460 V, we find the new volts-to-hertz ratio to be

$$\text{Volts-to-hertz ratio} = 460 \text{ V}/66 \text{ Hz}$$
$$\text{Volts-to-hertz ratio} = 6.97 \text{ V}/\text{Hz}$$

Notice this is a 10% reduction of volts-to-hertz ratio which is the same percentage as the increase in overspeed of the motor.

d. The new available torque is the nominal torque times the square of the change in the volts-to-hertz ratio

$$T_{new} = T_{old}(V/Hz_{new}/V/Hz_{old})^2$$
$$T_{new} = 29.167 \text{ lb ft}(6.97 \text{ V}/Hz/7.67 \text{ V}/Hz)^2$$
$$T_{new} = 29.167 \text{ lb ft}(0.9091)^2$$
$$T_{new} = 29.167 \text{ lb ft}(0.82645)$$
$$T_{new} = 24.1 \text{ lb ft}$$

e. The percentage change in the torque is the new torque value normalized to the nominal torque value.

$$\text{Change in torque} = (T_{new}/T_{old})100\%$$
$$\text{Change in torque} = (24.1 \text{ lb ft}/29.167 \text{ lb ft})100\%$$
$$\text{Change in torque} = 82.6\%$$

Note this is the same result as if we just calculated the square of the change in volts-to-hertz ratio or

$$\text{Change in torque} = (V/Hz_{new}/V/Hz_{old})^2 100\%$$
$$\text{Change in torque} = 82.6\%$$

For most motor designs, there is a fan on the shaft of the machine to circulate air internally inside the enclosure of the motor. This fan forces air around the rotor and stator windings to remove heat. As the motor slows down in rotation, the speed of the fan is reduced and the circulation of air inside the casing of the motor is reduced. Due to this phenomenon at very low speeds of operation, some motor types have a reduction in torque available from the machine. The resultant torque-speed curve for a motor on a VFD with the limits of torque at overspeed and very low speed operation is shown in Figure 18.7.

Now we will discuss the relationship of speed to both torque and power and this entire discussion is based on the definition that mechanical power is the product of the torque demanded by the load and the rotational shaft speed of the load. This was defined back in Chapter 17 by equation (17.5).

$$P = Tn/5250$$

where

P = power demanded by load (HP)

T = torque demanded by the load (lb ft)

n = speed of the load shaft (rpm)

The first type of load is a *constant-power* load as shown in Figure 18.8. With this load, as the speed of the load is changed, the power that the load demands remains

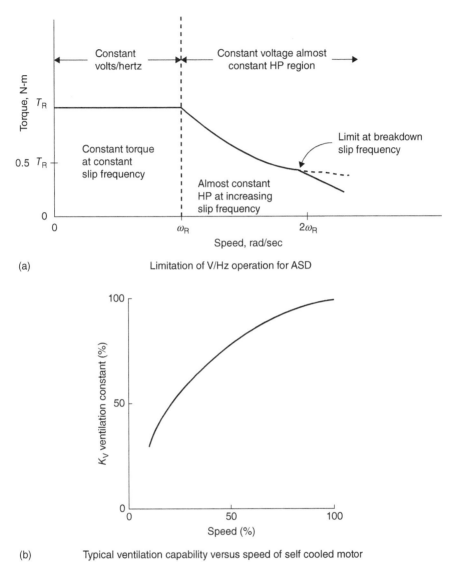

(a) Limitation of V/Hz operation for ASD

(b) Typical ventilation capability versus speed of self cooled motor

Figure 18.7 Torque-speed curve of standard motor showing high speed derate area (a) and low speed derate area (b). *Source*: Reproduced with permission of IEEE.

approximately constant. Since the value of power (P) remains constant, referring to equation (17.5), as we reduce the speed of the load, the torque that the load demands increases and as we increase the speed of the load, the torque that the load demands decreases. An example of a constant-power load would be machining applications such as a punch press. For a VFD to provide constant power, the volts-to-hertz ratio is varied inversely to the change in speed to vary torque so that, as speed is changed power remains constant.

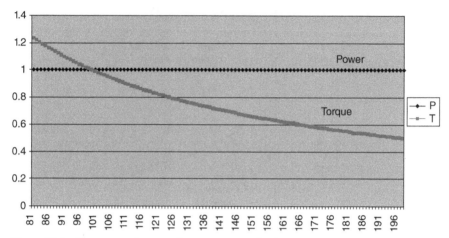

Figure 18.8   Torque-speed curve of constant-power load.

The second type of load is a *constant-torque* load as shown in Figure 18.9. With this load, as the speed of the load is changed, the torque that the load demands remains approximately constant. Since the value of torque (T) remains constant, referring to equation (17.5), as we reduce the speed of the load, the power that the load demands decrease and as we increase the speed of the load, the power that the load demands increases. An example of a constant-torque load would be a conveyor or auger. For a VFD to provide constant torque, the volts-to-hertz ratio is maintained constant. This ensures that torque available remains constant as the load speed is changed.

The third type of load is a linear *variable-torque* load as shown in Figure 18.10. With this load, as the speed of the load is changed, the torque that the load demands changes in proportion to the change in speed. Since the value of torque (T) changes linearly with the change in speed, referring to equation (17.5), as we reduce the speed

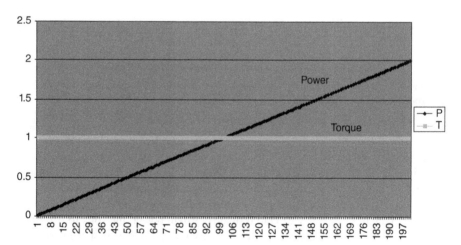

Figure 18.9   Torque-speed curve of constant-torque load.

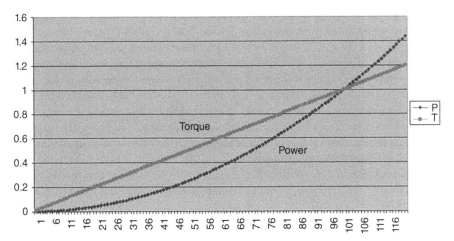

Figure 18.10   Torque-speed curve of variable (square)-torque load.

of the load, the power that the load demands decreases as the square of the change in speed and as we increase the speed of the load, the power that the load demands increases as the square of the change in speed. An example of a linear variable-torque load would be a positive displacement pump or a mixer. For a VFD to provide a linearly variable torque, the volts-to-hertz ratio is reduced linearly to the change in output frequency of the VFD. This ensures that torque available remains adequate for the load while ensuring the optimal efficiency of the motor by not providing more voltage on the stator than necessary for the torque the load is demanding at that speed.

The fourth type of load is a square variable-torque load as shown in Figure 18.11. With this load, as the speed of the load is changed, the torque that the load demands changes as the square of the change in speed. Since the value of torque

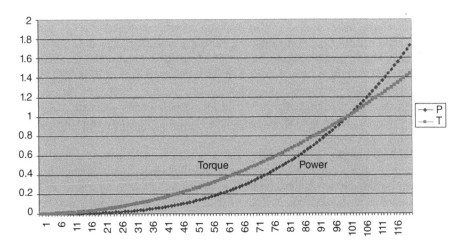

Figure 18.11   Torque-speed curve of variable (cube)-torque load.

(*T*) changes as the square of the change in speed, referring to equation (17.5), as we reduce the speed of the load, the power that the load demands decreases as the cube of the change in speed and as we increase the speed of the load, the power that the load demands increases as the cube of the change in speed. As stated in Chapter 8 covering fans and Chapter 9 covering pumps, with constant diameter, as the speed of a centrifugal load is changed, the flow changes proportionally, the pressure drop changes by the square and the power requirement changes by the cube. From an electrical viewpoint, as the speed of the load changes, the torque changes by the square of the change in speed and the power changes by the cube (which is required since the electrical power of motor must equal the sum of the mechanical power of the load and any losses in the motor and pump). These loads are very common in the power generation station as most centrifugal fans and pumps are square variable-torque loads. For a VFD to provide a square variable torque, the voltage-to-frequency ratio is reduced by the square to the change in output frequency of the VFD. This ensures that torque available remains adequate for the load while ensuring the optimal efficiency of the motor by not providing more voltage on the stator than necessary for the torque the load is demanding at that speed.

The above discussion that relates to the voltage to frequency ratio is the ideal case and assumes an ideal motor. In reality, for constant torque as the output frequency of the drive decreases, there is a slight increase in the voltage to frequency ratio to compensate for winding losses internal to the motor.

HARMONICS

A nonlinear load is defined as any load where impedance changes with the applied voltage. The changing impedance means that the current drawn by the non-linear load will not be sinusoidal even when it is connected to a sinusoidal voltage. Any nonlinear load will generate current harmonics on the source side of the load the magnitude and characteristic *harmonic* frequencies of which will be dependent on the type of *converter* seen by the system as well as the loading on the converter. These harmonic currents are a result of the non-sinusoidal current waveform that is drawn from the system during the conversion from AC to DC. Most variable frequency drives utilize either diode or SCR-type rectifiers for their source side conversion from AC to DC. For a voltage source drive, this rectifier is feeding a capacitor bank. The current only flows from the line through the rectifier to the DC capacitor bank when the AC instantaneous voltage exceeds the DC voltage. This occurs at the phase-to-phase voltage peaks of the AC waveform.

Harmonic currents can have several adverse effects on the electrical distribution system feeding the drive. Some of the effects are the possible overheating of distribution equipment due to nonlinear current waveforms and increased copper and iron losses in motors and transformers. These can also lead to possible increased neutral currents (if harmonics happen to be triplen of the fundamental frequency).

IEEE 519 provides guidance to the limits on the amount of harmonics that energy users can safety supply to the utility or distribution system and these values are defined at the point of common coupling (PCC) between the utility system and the user. Several techniques are used to reduce the harmonics that nonlinear loads such as variable speed drives provide to a distribution system. One solution is the use of line-side reactors or filters. Reactors represent an inductive impedance between the drive and the distribution system. For the same drive load, they tend to reduce the peak of the current draw and widen the current pulse, thereby reducing the magnitude of the harmonics involved, but the harmonics are still present. Another alternative to the reactor is the use of an inductor/capacitor. A filter that is constructed of a combination of an inductor and a capacitor may be used to more effectively reduce the harmonics on the source side. Adding a DC link (inductor in the DC bus) can also provide similar benefits. In some applications, a drive isolation transformer is installed on the source side to filter harmonics. The drive isolation transformer provides additional inductance just as a reactor would, but in addition, most drive isolation transformers are configured delta-wye. This allows for isolation of all triplen harmonics currents to the drive side of the transformer. Any triplen harmonics will circulate in the delta and not flow back into the source side power distribution system.

Alternately, some drive topologies will use multiple rectifier sections for the AC to DC conversion and will have a discrete transformer winding feeding each rectifier. This tends to spread out the current draw from the distribution system into a greater number of pulses, each with a smaller magnitude of peak current, thereby reducing the overall harmonic values generated. Each winding must be shifted a certain number of degrees from the previous winding. To determine the number of degree of phase shift, we first need to identify the number of pulses the overall rectifier section has. The number of pulses of the rectifier section is the product of the number of pulses for any one rectifier multiplied by the number of rectifiers as shown in equation (18.6).

$$P_{\text{total}} = n \times P_{\text{rectifier}} \qquad (18.6)$$

where

P_{total} = total number of pulses for all rectifiers

$P_{\text{rectifier}}$ = number of pulses for any individual rectifier

n = number of rectifiers

Example 18.4   Given a variable speed drive with six rectifiers on the front end of the drive, each rectifier is a six-pulse rectifier; what is the total pulse number of the rectifier system?

Solution: Using equation (18.6), we find the total pulse number of the rectifier to be as follows.

$$P_{\text{total}} = n \times P_{\text{rectifier}}$$

$$P_{\text{total}} = 6 \times 6$$

$$P_{\text{total}} = 36$$

Once we know the number of pulses for the entire rectifier assembly, we can determine the phase angle shift between windings using equation (18.7).

$$\text{Phase shift} = 360° / P_{total} \tag{18.7}$$

Example 18.5 Given a variable speed drive with six rectifiers on the front end of the drive, each rectifier is a six-pulse rectifier; what is the incremental phase shift between source side windings?

Solution: Using equation (18.7), we find the total pulse number of the rectifier to be as follows.

$$\text{Phase shift} = 360° / P_{total}$$

From Example 18.4, we found P_{total} to be 36. Therefore, the phase shift required is as follows.

$$\text{Phase shift} = 360° / 36$$

$$\text{Phase shift} = 10°$$

Table 18.2 summarizes the required phase shift of the source for the number of three-phase rectifiers for a particular drive configuration.

So how do we develop the correct phase shift between the voltage vectors for designs with more than one rectifier? From Chapter 15 on transformers, we know that a delta-wye configured transformer will provide us with a 30 degree phase shift from the delta side to the wye side where wye-wye and delta-delta transformers provide no phase shift from primary to secondary. Therefore, for the two rectifier design that requires a 30 degree phase shift, if we feed the 12-pulse (two rectifier assembly) drive input from a delta-delta transformer on one bridge and the other rectifier assembly from a delta-wye transformer on the other bridge; we will obtain the needed 30 degree phase shift between phase inputs of the two bridges. For higher order pulse count rectifier assemblies, we require phase shifts less than 30 degrees as shown in Table 18.2. The method utilized to achieve this is to utilize part of the winding from the delta-delta transformer with no phase shift and part of the winding from the delta-wye transformer with a 30 degree phase shift to obtain the desired phase shift and resultant voltage magnitude. These two voltages need to sum to a magnitude that we need for the rectifier required input voltage magnitude and also these two voltages

TABLE 18.2   Required Phase Shift for Number of Rectifier Pulses

No. of Rectifiers	Total Pulse #	Phase Shift
1	6	60
2	12	30
3	18	20
4	24	15
5	30	12
6	36	10

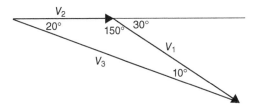

Figure 18.12   Voltage vectors for 18-pulse rectifiers: second rectifier.

need to provide the phase shift required to evenly distribute the current pulses at the waveform peak. Let us look at the "three rectifier" application, for example. For a drive that consists of three six-pulse rectifiers, the total number of pulses on the source side is

$$3 \text{ rectifiers} \times 6 \text{ pulses/rectifier} = 18 \text{ pulse total}$$

Using 360 degrees divided by 18 pulses gives us a required phase shift of

$$360 \text{ degrees}/18 \text{ pulses} = 20 \text{ degrees/pulse}$$

Figure 18.12 shows us our arrangement. The voltage from winding one (V_1) is from the secondary side of the delta-wye transformer and the secondary side is shifted from the primary side by −30 degrees. The voltage from winding two (V_2) is from the secondary side of the delta-delta (or wye-wye) transformer that presents no phase shift from primary to secondary. We know that the angle between V_2 and our actual voltage we wish to provide to our 18-pulse rectifier needs to be −20 degrees. With this information we derived the arrangement of voltage vectors in Figure 18.12. Now we simply need to identify the magnitude of the voltage vector V_1 and V_2 that provides a voltage vector V_3 with a magnitude of V and an angle of −20 degrees.

Using the law of sines, we can identify the relationship between the magnitude of vector V_2 and V_1 as

$$V_2/\sin(10) = V_1/\sin(20)$$

Rearranging, we find the relationship between V_2 and V_1 to be

$$V_2 = V_1 \times [\sin(10)/\sin(20)] = V_1 \times [0.173648/0.34202]$$

$$V_2 = 0.5077 \times V_1$$

We can again use the law of sines to identify the relationship between the magnitudes of vector V_3 and V_2 as

$$V_3/\sin(150) = V_2/\sin(10)$$

Rearranging, we find the relationship between V_3 and V_2 to be

$$V_3 = V_2 \times [\sin(150)/\sin(10)] = V_2 \times [0.5/0.173648]$$

$$V_3 = 2.87939 \times V_2$$

TABLE 18.3 Required Voltage Magnitudes and Angles for Number of Rectifier Pulses

	V_1		V_2		V_3	
Pulse #	Magnitude (per unit)	Angle (degrees)	Magnitude (per unit)	Angle (degrees)	Magnitude (per unit)	Angle (degrees)
18	0.684	−30	0.3473	0	1	−20
24	0.5176	−30	0.5176	0	1	−15
30	0.416	−30	0.618	0	1	−12
36	0.3473	−30	0.684	0	1	−10

Rearranging for V_2, we find that

$$V_2 = (1/2.8974) \times V_3 \text{ or}$$

$$V_2 = 0.3473 \times V_3$$

Now we can combine the relationship we found earlier between V_1 and V_2 to derive the relationship between V_3 and V_1 to be

$$V_3 = 2.87939 \times V_2 = 2.87939 \times (0.5077 \times V_1)$$

$$V_3 = 1.4619 \times V_1$$

Rearranging for V_1, we find that

$$V_1 = (1/1.4619) \times V_3 \text{ or}$$

$$V_1 = 0.684 \times V_3$$

Therefore, to obtain a secondary voltage of magnitude V and an angle of −20 degrees, we add vector V_1 (0.6798 × V) which is at an angle of −30 degrees and vector V_2 (0.3416 × V) which is at an angle of 0 degrees to obtain vector V_3 (1.0 × V) at an angle of −20 degrees. Table 18.3 summarizes the calculation for the first phase shift requirement for various rectifier pulse designs.

However, we are not yet done. We will feed our first rectifier with voltage vector of 0 degrees, and our second rectifier voltage vector will be at a voltage of magnitude V and an angle of −20 degrees. How about the third rectifier of our 18-pulse design? The third rectifier needs another 20 degree phase shift. The simplest method is to shift the third vector by +20 degrees. Notice that since this is between zero and +30 degrees, vector V_1 is now the vector from the wye transformer that is shifted +30 degrees. Figure 18.13 shows this relationship.

Figure 18.13 shows us our arrangement. The voltage from winding one (V_1) is from the secondary side of the delta-wye transformer and the secondary side is shifted from the primary side by +30 degrees. The voltage from winding two (V_2) is from the secondary side of the delta-delta (or wye-wye) transformer that presents no phase shift from primary to secondary. We know that the angle between V_2 and our actual voltage we wish to provide to our 18-pulse rectifier needs to be +20 degrees. With this information we derived the arrangement of voltage vectors in Figure 18.12. Now we simply need to identify the magnitude of the voltage vector V_1 and V_2 that provides a voltage vector V_3 with a magnitude of V and an angle of +20 degrees.

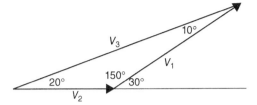

Figure 18.13   Voltage vectors for 18-pulse rectifiers: third rectifier.

Using the law of sines, we can identify the relationship between the magnitude of vector V_2 and V_1 as

$$V_2/\sin(10) = V_1/\sin(20)$$

Rearranging, we find the relationship between V_2 and V_1 to be

$$V_2 = V_1 \times [\sin(10)/\sin(20)] = V_1 \times [0.173648/0.34202]$$

$$V_2 = 0.5077 \times V_1$$

We can again use the law of sines to identify the relationship between the magnitudes of vector V_3 and V_2 as

$$V_3/\sin(150) = V_2/\sin(10)$$

Rearranging, we find the relationship between V_3 and V_2 to be

$$V_3 = V_2 \times [\sin(150)/\sin(10)] = V_2 \times (0.5/0.173648)$$

$$V_3 = 2.87939 \times V_2$$

Rearranging for V_2, we find that

$$V_2 = (1/2.8974) \times V_3 \text{ or}$$

$$V_2 = 0.3473 \times V_3$$

Now we can combine the relationship we found earlier between V_1 and V_2 to derive the relationship between V_3 and V_1 to be

$$V_3 = 2.87939 \times V_2 = 2.87939 \times (0.5077 \times V_1)$$

$$V_3 = 1.4619 \times V_1$$

Rearranging for V_1, we find that

$$V_1 = (1/1.4619) \times V_3 \text{ or}$$

$$V_1 = 0.684 \times V_3$$

Therefore, to obtain a secondary voltage of magnitude V and an angle of +20 degrees, we add vector V_1 ($0.6798 \times V$) which is at an angle of +30 degrees and vector V_2 ($0.3416 \times V$) which is at an angle of 0 degrees to obtain vector V_3 ($1.0 \times V$) at an angle of +20 degrees.

Now we build our transformer secondary to develop these phase shifts. For this application, we will assume that the transformer primary is connected in a delta and

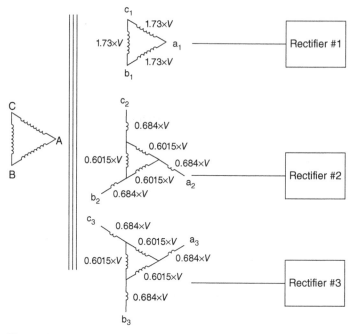

Figure 18.14   Arrangement of 18-pulse drive isolation transformer for ±20 degree phase shift.

we have a secondary winding that is connected in delta (no phase shift as in V_2) as well as a secondary winding connected in a wye (30-degree phase shift of V_1). The voltage vector V_1 is derived from the wye winding and is a magnitude of

$$V_1 = 0.684 \times V_3$$

Lastly, we need to configure the winding 2 for the voltage of V_2. Now we have one last item to discuss. The above phasor calculations only represented the single-line equivalent circuit which represents the voltage of a wye-connected winding. To convert this to our delta-connected secondary winding voltage V_2, we need to perform a wye-to-delta conversion by multiplying the V_2 value by 1.732. Therefore, our value of winding voltage for the secondary delta-connected transformer is

$$V_2 = 1.732 \times 0.3473 \times V_3 = 0.6015 \times V_3$$

The transformer construction is shown in Figure 18.14.

In summary, our first rectifier will receive A phase voltage at a magnitude of V volts and an angle of 0 degrees. Our second rectifier will receive an A phase voltage at a magnitude of V volts and an angle of -20 degrees and our third rectifier will receive an A phase voltage at a magnitude of V volts and an angle of $+20$ degrees. The calculations are done in a similar fashion for higher pulse number rectifier systems.

GLOSSARY OF TERMS

- Adjustable-Speed Drive – An electric drive designed to provide easily operable means for speed adjustment of a motor within a specified speed range.
- Base Speed – The lowest speed obtained at rated load and rated voltage at the temperature rise specified in the rating.
- Channel – A single path for transmitting electric signals, usually in distinction from other parallel paths.
- Commutation – The transfer of the current from one converter switching branch to another.
- Converter – An operative unit for electronic power conversion, comprising one or more electronic switching devices and any associated components, such as transformers, filters, commutation aids, controls and auxiliaries.
- Current Limiting – Is an overload protection mechanism that limits the maximum rms output current to a preset value, and automatically restores the output when the overload is removed.
- Dynamic Braking – A system of electric braking in which the excited machine is disconnected from the supply system and connected as a generator, the energy being dissipated in the winding and, if necessary, in a separate resistor.
- Efficiency – The ratio of load power to the total line power including the contribution of harmonics and auxiliary equipment.
- Harmonics – A sinusoidal component of a periodic wave or quantity having a frequency that is an integral multiple of the fundamental frequency.
- Inverter – A machine, device or system that changes direct-current power to alternating-current power.
- Motor Base Rating – When a motor is applied to an inverter, the motor base rating shall be the nameplate horsepower, voltage, frequency, speed, and torque.
- Pulse – The number of repetitive similar waveforms in the DC voltage output of the converter per cycle of alternating input voltage.
- Regenerative Braking – A form of dynamic braking in which the kinetic energy of the motor and driven machinery is returned to the power supply system by means of the drives input converter.
- Synchronous Speed – The motor speed in rpm defined by frequency and number of poles, $n = 120 f / p$, where p is the number of poles.

PROBLEMS

18.1 A variable frequency drive adjusts applied _____ to a motor stator to control the speed of the motor and adjusts applied _____ to a motor to control the available motor torque.

    A.   voltage; frequency

    B.   frequency; voltage

18.2 Given a motor rated at 460 V, 60 Hz, eight-pole machine with a full-load speed of 850 rpm, and given that the motor has 460 V, 60 Hz applied to the stator and if fully loaded, calculate the following.

A. Synchronous speed of the motor.

B. Slip of the motor in rpm.

C. Slip of the motor in percent.

18.3 Given a motor rated at 460 V, 60 Hz, eight-pole machine with a full-load speed of 850 rpm when operated at rated voltage and frequency is now energized from a VFD. The output of the VFD is 230 V, 30 Hz. If the motor is delivering rated torque to the load (i.e., slip in rpm is the same as in the rated frequency case), calculate the following.

A. Synchronous speed of the motor at 30 Hz.

B. Slip of the motor in rpm.

C. Slip of the motor in percent.

18.4 Given a motor rated at 460 V, 60 Hz, 900 rpm, 10 HP motor. This motor has a VFD supplying voltage and frequency to the motor stator and the maximum value of voltage that the VFD can provide to the motor is 460 V. The maximum frequency the VFD can supply to the motor is 120 Hz. Calculate the following.

A. Nominal torque available from the motor at 900 rpm.

B. Nominal volts-to-hertz ratio that this torque is based on.

C. Volts-to-hertz ratio for the motor when it is at 10% overspeed or 990 rpm (assume applied frequency of 66 Hz) knowing that the applied stator voltage does not change from 460 V.

D. Torque available from the motor at 10% overspeed or 990 rpm (assume applied frequency of 66 Hz) knowing that the applied stator voltage does not change from 460 V.

E. Percentage change in torque normalized to the nominal torque of the machine.

18.5 Given a variable speed drive with three rectifiers on the front end of the drive, each rectifier is a six-pulse rectifier, what is the total pulse number of the rectifier system and what is the phase shift required between transformer secondary windings?

RECOMMENDED READING

Electric Power Plant Design, Technical Manual TM 5-811-6, Department of the Army, USA, 1984.

Electrical Machines, Drives and Power Systems, 6th edition, Theodore Wildi, Prentice Hall, 2006.

The Engineering Handbook, 3rd edition, Richard C. Dorf, (editor in chief), CRC Press, 2006.

IEEE 958-2003: Guide for Application of AC Adjustable-Speed Drives for Electric Power Generating Stations.

IEEE 1566-2015: Draft Standard for Performance of Adjustable Speed AC Drives Rated 375 kW (500 HP) and Larger.

Industrial Power Distribution, 2nd edition, Ralph E. Fehr, III, Wiley-IEEE Press, 2016. ISBN: 978-1-119-06334-6.

NEMA Standards: Application Guide for AC Adjustable Speed Drive Systems – Industrial Control and Systems: Adjustable Speed Drives, IEEE PCIC-2001-7.

Power Plant Engineering, Black & Veatch, edited by Larry Drbal, Kayla Westra, and Pat Boston, Chapman & Hall/Springer, 1996.

Standard Handbook of Powerplant Engineering, 2nd edition, Thomas C. Elliott, McGraw-Hill, 1998.

CHAPTER 19

SWITCHGEAR

GOALS

- To understand the basic design and construction of industrial low voltage and medium voltage breakers and switchgear
- To understand key safety interlocks associated with switchgear
- To be able to draw and troubleshoot the typical close and trip circuits for medium voltage breakers
- To identify the correct IEEE standard that defines the requirements for a specific type of switchgear

---

CIRCUIT BREAKERS and switches are used to provide isolation points for the electrical system in the plant. Circuit breakers can be operated for control purposes to isolate equipment in the station for operation and maintenance and they can also be operated for protection purposes to isolate equipment in the event a fault is detected in the equipment. The circuit breaker is designed to open (de-energize) and close (energize) the feeder under load conditions. It is also designed and constructed to interrupt fault currents during a fault event successfully. Note, not all switching devices are designed to interrupt fault currents. Switches may or may not be rated to interrupt fault current, or even load current. The design of the switch will determine the ability of the switch to interrupt load. Motor starters are an example of equipment not rated for isolation of faults while they are intended to interrupt normal load.

Circuit breakers are designed with a mechanical device to provide the mechanical force to both close and open or trip the circuit breaker. This is a safety feature to ensure the breaker can be operated even with the loss of control power. Most breaker designs utilize a spring(s) on a camshaft as the driving force for closing and tripping of the breaker while some store the energy to close and trip the breaker in a storage device like an electrolytic capacitor. Low voltage power circuit breakers on 480 V or less substations, have their overcurrent protective functions powered by the current flowing through the breaker and the trip element is integral with the circuit breaker. During a fault condition, even with the loss of control power to the low voltage power circuit breaker, the overcurrent relay for the circuit breaker maintains power from the

Energy Production Systems Engineering: An Introduction for Electrical Engineers to Electrical Power Generation Facilities, Systems, and Equipment, First Edition. Thomas H. Blair.
© 2017 by The Institute of Electrical and Electronics Engineers, Inc. Published 2017 by John Wiley & Sons, Inc.

Figure 19.1   13.8 kV switchgear lineup.

current flowing through the breaker and will act to trip the breaker on an overcurrent condition.

Medium voltage breakers are not designed in the same manner. The protection relays for medium voltage circuit breakers are not integrated into the circuit breaker nor are they powered from the current of the primary circuit. The protection relays are provided externally (see Figure 19.1 where relaying is mounted in cell door and not on the breaker). As such medium voltage circuit breakers depend on control power being available to accurately and reliably trip or open the breaker in the event of a fault. Since the availability of control power is critical to the protective function of a medium voltage circuit breaker, the control power source is extremely reliable. The most reliable source in a utility electric generation station is a DC source from a station battery system. Even on a loss of all ac power in the power plant, the battery voltage is maintained and the breakers are able to provide their circuit protective functions. While the protective relay in medium voltage applications requires control power, the typical medium voltage breaker is closed and opened via mechanical springs in the breaker and there is a manual close and trip button on the face of the breaker along with a flag indicating breaker status. Figure 19.2 shows the manual close button (C), the manual trip button (A), the flag indicating breaker open or closed (B), and the flag indicating charging spring charged or discharged (D).

Figure 19.3 shows the mechanical linkage for a typical medium voltage power circuit breaker. Item (B) is the trip spring and item (J) is the close spring. These spring store the mechanical energy to close and trip the circuit breaker.

When the breaker opens (de-energizes) the feeder, and arc is drawn across the breaker contacts. The gap between contacts along with the insulating atmosphere between contacts provides the insulation to stop current flow. For breakers that use air between the contacts, an arc will be drawn across the contacts as they part. The size of this arc is a function of the current interrupted. Under fault conditions the arc size can

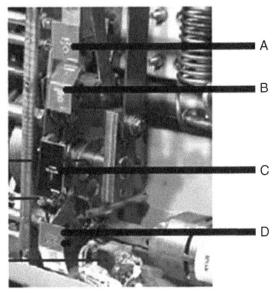

Figure 19.2   Manual close and trip buttons and charge and position flags on circuit breaker. *Source*: Reproduced with permission of Powell Industries.

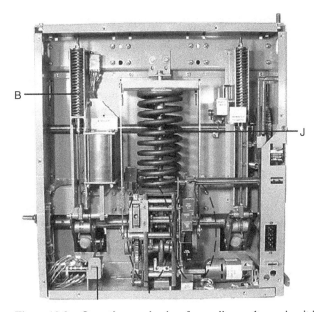

Figure 19.3   Operating mechanism for medium voltage circuit breaker. *Source*: Reproduced with permission of Powell Industries.

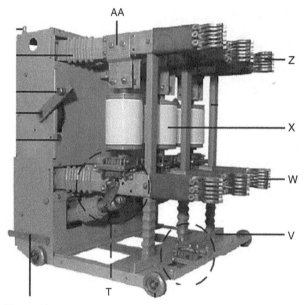

Figure 19.4   Vacuum bottle assembly of medium voltage circuit breaker. *Source*: Reproduced with permission of Powell Industries.

be substantial. For air breakers, there is an arc chute and small arc contacts associated with the main contacts. The main contacts open first and then the arc contacts begin to open. This allows the arc to occur at the arc contact area and not the main contact area. The arc chutes allow the hot gas of the arc to rise and it elongates the arc length allowing the arc to cool and extinguish. Additionally, there may be a *puffer* which is basically a piston in a cylinder. The action of the breaker opening forces the piston through the cylinder and injects air into the arc, blowing the arc up into the arc chutes. This is more intended for opening the air circuit breaker under very light load where the heat and force from the arc is not adequate alone to push the arc into the arc chute.

Other types of insulating media are available such as oil, vacuum bottles, or sulfur hexafluoride (SF_6) gas. See Figure 19.4 which shows the vacuum bottle assembly on a 13.8 kV, 1200 A circuit breaker. Item X of Figure 19.4 is the vacuum bottles inside the medium voltage breaker. Looking at Figure 19.4 we can identify other important components in the medium voltage power circuit breaker. Items Z and W are the main power stabs that connect the breaker to the power bus of the *switchgear* when the breaker is racked in. The power contacts are inside the vacuum bottle (X). The top side of the bottle is fixed in position by the bus at item AA. The connection at item T moves up when the breaker is closed and down when the breaker is opened. A very important contact is the ground brush contact shown at item (V). This provides a ground connection for the metal components of the breaker. This ensures personnel safety when making contact with the power circuit breaker.

Power circuit breakers that are draw-out type are designed to be withdrawn from the switchgear bus without de-energization of the switchgear bus. With the

Figure 19.5   13.8 kV switchgear breaker compartment.

power circuit breakers are removed from the switchgear cell, shutters will drop in front of the bus connections to remove the shock hazard associated with the energized bus that the circuit breaker connects to. Figure 19.5 shows a cubicle with the power circuit breaker removed from the cubicle. In this photo, the shutters are forced up so we can see the bus connections, but, the shutters would be in front of the bus connections. (This was done during initial checkout of the switchgear and the bus behind the stabs was de-energized. This would not be a safe operation were the bus stabs energized behind the shutter.) For this compartment, the source is on the bottom three stabs (labeled bus) and the load is on the upper three stabs. There is not a standard for which side (top or bottom) is line and which side is load. The equipment construction drawings and system drawings should always be consulted to understand which side of the breaker is the source side and which side is the load side when working on the breaker and depending the breaker as a point of isolation (see lockout/tag out in the electrical safety section in chapter 1). You can see in Figure 19.5 that the top three bus stabs have current transformers mounted around the bushings. These current transformers feed the protective relays that control the circuit breaker and provide load information to the computer system for the magnitude and phase angle of the current out of this feeder.

In the bottom right corner of the back of the cubicle you can see the secondary contact block. When the breaker is racked into its cubicle, secondary stabs make at this block that allow for control signal connections to the breaker for electrical trip and close functions as well as auxiliary contacts on the breaker indicating the breaker status (open or closed). In addition to the controls on the breaker, there are two switches in the cell. One switch, called a housing limit switch or truck operated contact (TOC) is activated when the breaker is racked into the bus. A second switch,

Figure 19.6   Aftermath of an arc flash in a 13.8 kV breaker.

called *mechanism operated contact* (MOC) is activated when the breaker is closed and deactivated when the breaker is open.

While the contacts of the circuit breaker are designed to interrupt load or fault current, the stabs of the breaker are not. There is no mechanism to extinguish an electric arc should the breaker be racked in or out with the breaker closed and current flowing through the breaker. As a safety feature, there is also a mechanical mechanism that trips the breaker open or *trip free* should the breaker start to be racked either in or out with the breaker closed. Many designs have it such that the racking or levering device cannot be inserted into the breaker to rack the breaker without tripping the breaker open. Were the breaker allowed to be withdrawn or inserted with the breaker contacts closed, an arc would be drawn at the bus stabs. Without the benefit of a means to reliably extinguish the arc at the point of connection/disconnection at the power stabs, this would lead to a catastrophic failure at the stab assembly of the breaker with potentially series results as shown in Figure 19.6.

Like all electrical equipment, breakers are design with insulating systems and must be able to withstand both the normal power system voltages as well as transient voltages the breaker may see during operation. Breaker and switch insulation systems undergo similar insulation tests as described in the section for transformers. The specific insulation levels of power circuit breakers for low voltage AC power circuit breakers are shown in Table 19.1. The specific insulation levels of power circuit breakers for low voltage DC power circuit breakers are shown in Table 19.2. The specific insulation levels of power circuit breakers for medium voltage metal-clad AC power circuit breakers are shown in Table 19.3. These values are for OEM testing when switchgear is new. When normal frequency insulation withstand tests are

TABLE 19.1   Voltage and Insulation Levels for AC LV Switchgear

| Rated Maximum Voltage (rms) | Insulation Levels | |
	Normal Frequency Withstand (rms)	Reference DC Withstand
254/508/635	2.2	3.1

Source: Reproduced with permission of IEEE.

TABLE 19.2   Voltage and Insulation Levels for DC LV Switchgear

| Rated Maximum Voltage (rms) | Insulation Levels | |
	Normal Frequency Withstand (rms)	Reference dc Withstand
300/325	2.2	3.1
800[a]	3.7	5.2
1200	4.6	6.5
1600	5.4	7.6
3200	8.8	12.4

Source: Reproduced with permission of IEEE.
[a]The value will be updated pending on the next revision.

to be made on LV switchgear after installation in the field, the switchgear shall not be tested at greater than 75% of the test values given in Tables 19.1, 19.2, and 19.3.

This is a good place to point out that, for low voltage switchgear, the standard that applies is IEEE C37.20.1. In this standard, low voltage switchgear is metal enclosed, not metal clad. For medium voltage switchgear, the standard that applies is IEEE C37.20.2. In this standard, the medium voltage switchgear is metal clad. The primary difference between is the construction between compartments, especially in the horizontal bus feed through the switchgear. For metal enclosed, there is no requirement to have barriers between different compartments. For metal-clad switchgear, this switchgear is required to have barriers between compartments. *Metal-clad switchgear* by definition always meets the requirements of *metal-enclosed switchgear*. However, due to the issue of barrier requirements, not all metal-enclosed switchgear can be classified as metal-clad switchgear. In a power generation facility, whenever possible, it is recommended to use metal-clad switchgear as the barriers tend to minimize the spread of damage should a fault occur inside the switchgear. See glossary of terms section of this chapter for more information on the construction of metal-enclosed and metal-clad switchgear.

For breakers connected to large motors, upon tripping of the normal source of power to the bus, the motors tend to continue to supply voltage to the load side of the breaker as they start to slow down. As the motors slow down, the magnitude of the

TABLE 19.3   Preferred Voltage and Insulation Levels for Metal-Clad (MC) Switchgear

| Rated Maximum Voltage (kV rms) | Insulation Levels | |
	Power Frequency Withstand (kV rms)	Lightning Impulse Withstand (BIL) (kV peak)
4.76	19	60
8.25	36	95
15	36	95
27	60	125
38	80	150

Source: Reproduced with permission of IEEE.

voltage and frequency of the voltage they induced in the bus becomes smaller. If an alternate source is to be closed in to re-energize the bus, this alternate source should only be allowed to closed and energize the motors when the motor voltage and the alternate source voltage are within certain tolerances. Closing the alternate source breaker with the difference in potential across the alternate source breaker can lead to torque spikes on the rotating electrical machines that can lead to mechanical damage to the machine. Motor bus transfer systems monitor the bus voltage, normal source voltage and alternate source voltage for magnitude, frequency, and phase angle and ensure during a transfer that the transfer to the alternate load will not cause damage to the rotating machinery.

Figure 19.7 shows the typical close and charging motor control circuit for a power circuit breaker. Table 19.4 defines some of the functions of the contacts in the control schematics of Figures 19.7 and 19.8.

The function of the charging motor (M) is to compress the main closing spring which is the mechanical stored energy mechanism. The closing spring provides the necessary energy to close the circuit breaker and the tripping spring provides the necessary energy to trip or open the circuit breaker. The motor cutoff switch (LS) provides an electrical break in the control circuit supplying the charging motor (M)

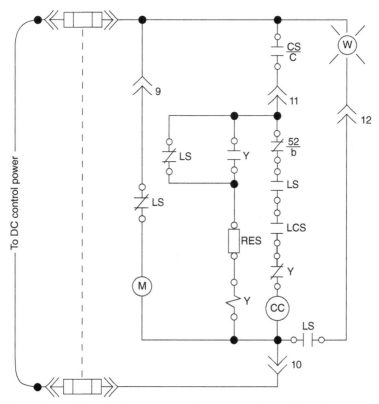

Figure 19.7 Circuit breaker close circuit schematic. *Source*: Reproduced with permission of Powell Industries.

TABLE 19.4   Device Descriptions for Figures 19.7 and 19.8

LS = Spring charge limit switch shown with breaker closing springs discharged

M = Breaker closing springs charging motor

52/b = Breaker normally closed auxiliary contact

52/a = Breaker normally open auxiliary contact

Y = Anti-pump relay

LCS = Latch check switch

PR = Protective relay

CS/C = Control switch close contact

CS/T = Control switch open contact

TC = Close coil

CC = Trip coil

Source: Reproduced with permission of Powell Industries.

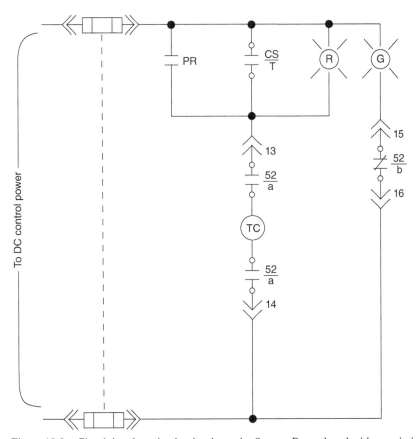

Figure 19.8   Circuit breaker trip circuit schematic. *Source*: Reproduced with permission of Powell Industries.

when the main closing spring is fully charged and the stored energy mechanism is ready for a closing operation. The anti-pump relay (Y) provides a logic function for the control circuit which prevents a continuous electrical close signal from causing the circuit breaker to continuously re-close after a trip signal. Electrical operation of the breaker is achieved by the use of solenoids which trigger the close and trip springs to close or open the breaker depending on which solenoid is triggered.

With the charging spring discharged, the spring charge limit switch (LS) is closed between the charging motor (M) and the secondary stab pin 9. This applies DC voltage to the charging motor and runs the charging motor until the closing springs become charged. The closing springs being charge will energize the LS contact. This opens the contact between the charging motor and the secondary stab pin 9 de-energizing the spring charging motor (M). As soon as the breaker is tripped and then reclosed, the closing spring will become discharged and the LS contact between the secondary stab pin 9 and the charging motor will close automatically recharging the closing spring.

With the breaker open, the contact 52/b is closed. The 52/b contact is an auxiliary contact that mirrors breaker status. The 52/b contact is closed with the breaker open and the 52/b contact is open when the breaker is closed. With the closing spring charged, the normally open spring charged limit switch (LS) contact below the 52/b contact is closed. This is a normally open contact off the LS mechanism. This contact is only closed when the closing spring is charged verifying that there is mechanical force ready to close the breaker. Downstream of the 52/b contact is the latch check switch (LCS). The latch check switch allows the circuit breaker to be used for instantaneous reclosing. The switch ensures that after a breaker trip the mechanical mechanism is reset and ready for a reclose before allowing the instantaneous re-closure. Downstream of the latch check switch (LCS) is a normally closed contact from the anti-pump relay (Y).

The anti-pump relay (Y) acts as a one-shot device. Looking at Figure 19.7, you can see that the anti-pump relay is driven by the close signal on stab pin 11 and the position of the charging spring limit switch normally closed contact. Before the breaker is closed, the anti-pump relay is not yet energized as the charging spring limit switch is open. Once the breaker closes, the closing spring discharges. This closes the normally closed charging motor limit switch LS which energizes the anti-pump relay coil (Y). The Y relay seals itself in with the Y relay normally open contact, in parallel with the LS normally closed contact. The normally closed contact from the Y relay prevents the close coil from being re-energized until the anti-pump relay Y resets. What resets the Y relay is the removal of the close command from the control switch contact CS/C.

Basically the anti-pump relay ensures that the close coil is energized only once for each contact closure of the control switch close contact CS/C. This ensures that, when the breaker is closed, if the breaker is closed into a fault and trips back open, the breaker will not reclose until the close command is removed and reasserted. This prevents the breaker from cycling open and closed into a fault and prevents failure of the breaker.

With the anti-pump relay de-energized, the normally closed contact from the Y relay is closed. When the control switch close contact (CS/C) is closed, this places voltage on stab pin 11. Assuming the 52/b contact, the LS contact, the LCS contact, and the Y contact are all closed, then the close coil (CC) is energized. When the breaker closes, the 52/b contact opens automatically removing power from the close coil.

Figure 19.8 shows the typical trip control circuit. Table 19.4 defines some of the functions of the contacts in the control schematics of Figure 19.8. There are auxiliary 52/a breaker auxiliary contacts in series with the trip coil of the breaker such that the trip coil is only energized when breaker is closed and needs to be opened or tripped. This prevents damaging of the trip coil should the trip signal remain on the breaker trip coil after the breaker opens and the trip coil no longer needs to be energized but the control switch or protective relay contact is still closed. The trip control coil will energize from either the control switch trip contact (CS/T) closing or any protection relay contacts (PR) closing. The green light is fed from the breaker normally closed auxiliary contact (52/b) and as such, whenever the breaker is opened, the 52/b contact is closed and the green light is energized indicating that the breaker is open. Notice that the red light is not only fed from two 52/a contacts in series but is also fed through the breaker trip coil between the two 52/a contacts. This is done for the following reason. The red light will be energized whenever the breaker is closed (from the 52/a contact) AND when there is continuity through the trip coil. If you walked up to this breaker and you see the green light on and the red light off, you know that the breaker is open and you have control power available to close the breaker. If you walked up to this breaker and you see the red light on and the green light off, you know the breaker is closed, you have control power available to trip the breaker, and you have continuity through the breaker trip coil, verifying the integrity of the trip coil. So what does it mean if you walk up to the breaker and neither the red light nor the green light is on? This means one of two things. Either we have opened our fuse to the trip circuit of the breaker, or the breaker is closed but the trip coil has failed and is open. This is a very important condition to be aware of and alarm on so that action can be taken to correct the issue. With the breaker closed and the trip coil open, a contact closure from the control switch trip contact (CS/T) or the protective relay (PR) will NOT open or trip the breaker. This means we have lost any protection this breaker may have been providing (overcurrent, differential, etc.) and this should be resolved immediately.

This is a very good place to explain a word of caution about the location of the red light. Notice that when the breaker is closed, the red light is illuminated by allowing current to flow through the trip coil. To prevent nuisance tripping of the breaker, the resistance that the red light presents must be much higher than the resistance that the trip coil presents to prevent this current from activating the trip coil. Said in another way, the current required to illuminate the red light must be substantially lesser than the minimum current required to activate the trip coil. The same discussion pertains to any device connected in parallel with the trip contacts. One common application of modern distributed control systems (DCS) is to parallel a voltage input to the DCS with the trip contacts. The input resistance of this DCS input must be substantially greater than the resistance of the trip coil to prevent nuisance tripping of the breaker. Additionally, when the breaker is critical to plant operations, it is a common practice to place two DCS voltage inputs that are in series with each other in parallel with the trip contact. This is done so that, in the event one of the two cards fails shorted, it does not cause a nuisance trip of the breaker.

IEEE standards define some of the requirements for construction of the *switchgear assembly*. The IEEE standards are broken up into the type of switchgear that is under consideration. For low voltage power circuit breakers, IEEE standard C37.20.1 applies. For metal-clad (MC) switchgear IEEE C37.20.2 applies. For metal-enclosed interrupter (MEI) IEEE C37.20.3 applies. Unless the customer specifies other arrangements, the phase arrangement on a three-phase assembled switchgear bus and connection are set up as phase A, phase B, and phase C from front to back, or top to bottom, or left to right as viewed from the front of the switchgear. The polarities on DC assembled switchgear buses and connections are set up as positive, neutral, negative from front to back, or top to bottom, or left to right as viewed from the front of the switchgear. While this is the standard defined by IEEE, switchgear is custom built to customer specifications, so the user should always refer to specific switchgear construction drawings for details on the phase orientation of the switchgear. One common exception to IEEE standard phase configuration is in older generation stations. In some older stations, A-B-C phase rotation was west to east and south to north instead of left to right and top to bottom.

Below is a description of some of the auxiliary contacts in a breaker, starter, or cubicle. A form "a" contact is a normally open contact while a form "b" contact is a normally closed contact. A form "c" contact has a normally open and normally closed contact with one side being common. Below are some specific standard nomenclatures for auxiliary contacts along with their description.

52/a is open when the breaker is in the de-energized or non-operated position. This is an auxiliary contact mounted directly on the breaker indicating status of the breaker.

52/b is closed when the device is in the de-energized or non-operated position. This is an auxiliary contact mounted directly on the breaker indicating status of the breaker.

52/aa is open when the operating mechanism of the main device is in the de-energized or non-operated position. This is an auxiliary contact mounted directly on the breaker indicating status of the breaker operating mechanism. This is also known as an early out contact as it is derived from the operating mechanism and not the breaker status itself and as such operates sooner than the 52/a contact.

52/bb is closed when the operating mechanism of the main device is in the de-energized or non-operated position. This is an auxiliary contact mounted directly on the breaker indicating status of the breaker operating mechanism. This is also known as an early out contact as it is derived from the operating mechanism and not the breaker status itself and as such operates sooner than the 52/b contact.

52TOC/a is open when the circuit breaker is not in the connected position. The TOC switch is mounted in the cubicle, not on the circuit breaker. TOC stands for *truck operated contact*.

52TOC/b is closed when the circuit breaker is not in the connected position. The TOC switch is mounted in the cubicle, not on the circuit breaker. TOC stands for *truck operated contact*.

52MOC/a is open when the circuit breaker is open. The MOC switch is mounted in the cubicle, not on the circuit breaker. MOC stands for *mechanism operated contact*.

52MOC/b is closed when the circuit breaker is open. The MOC switch is mounted in the cubicle, not on the circuit breaker. MOC stands for *mechanism operated contact*.

When coils on devices such as breakers and control relays are connected to a dc supply and de-energized and are not disconnected from both the positive and negative supply leads from the control power, these coils are arranged such that the positive supply is isolated from the relay leaving the negative supply connected. This is done to minimize the possibility of corrosion of the relay over long term service. When coils on devices such as breakers and control relays are connected to an ac supply and de-energized and are not disconnected from both the hot and neutral supply leads from the control power, these coils are arranged such that the hot supply is isolated from the relay leaving the neutral supply connected. This is done to prevent inadvertent energization of the coil should a ground occur in the control system.

For safe and reliable operation of the breaker, there are several mechanical interlocks that are used. To prevent moving the circuit breaker to or from the connected position when the circuit breaker is in the closed position, a trip free contact is installed on the circuit breaker. When the breaker is moved away from its fully inserted or fully disconnected position, the trip free contact will force the breaker open ensuring that the breaker is not racked in or out of a cubicle with the breaker

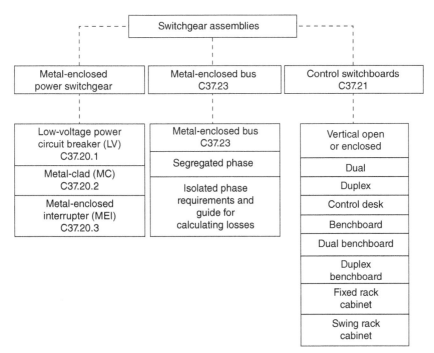

Figure 19.9 Application of IEEE Standards to Switchgear. *Source*: Reproduced with permission of IEEE.

closed. This also prevents closing the circuit breaker unless the primary disconnecting devices are in full contact or are separated by a safe distance. Also, circuit breakers are equipped with stored energy mechanisms that are designed to prevent the release of the stored energy unless the mechanism has been fully charged. Personnel are protected from the effects of accidental discharge of the stored energy by several mechanical interlocks. This is commonly done by discharging the mechanical springs of the breaker when the breaker is racked out. This way, once the breaker is racked out, any mechanical stored energy in the breaker has been discharged.

For more information on IEEE standards for specific types of switchgear the reader is encouraged to research the IEEE standard for the specific type of switchgear in question. Figure 19.9 shows which standard applies to which type of switchgear.

GLOSSARY OF TERMS

- Auxiliary Compartment – That portion of the switchgear assembly that is assigned to the housing of auxiliary equipment.
- Circuit Breaker Compartment – That portion of a switchgear assembly that contains one circuit breaker and the associated primary conductors and secondary control connection devices.
- Mechanism-Operated Contact (MOC) – A circuit breaker mechanism operated auxiliary switch that is mounted in the stationary housing and is operated by linkage that cooperates with the circuit breaker mechanism.
- Metal-Enclosed Low-Voltage Power Circuit-Breaker Switchgear (LV) – A low voltage switchgear where the breakers are contained inside individual grounded metal compartments. This switchgear includes the following equipment.
  - Low voltage power circuit breakers.
  - Bare bus or insulated bus and connections.
  - Instrument and control power transformers.
  - Instruments, meters, and relays.
  - Control wiring and accessory devices.
- Metal-Clad (MC) Switchgear – Metal-enclosed power switchgear that is characterized by the following additional features:
  - The main switching and interrupting device is of the removable or draw-out type arranged with a mechanism for moving it physically between connected and disconnected positions and equipped with self-aligning and self-coupling primary disconnecting devices and secondary disconnecting devices.
  - Major parts of the primary circuit, that is, the circuit switching or interrupting devices, buses, voltage transformers, and control power transformers, are completely enclosed by grounded metal barriers that have no intentional openings between compartments. This includes a metal barrier in front of, or a part of, the circuit interrupting device to ensure that, when in the connected position, no primary circuit components are exposed by the opening of a door.

- ○ All live parts are enclosed within grounded metal compartments.
- ○ Automatic shutters that cover primary circuit elements when the removable element is in the disconnected, test, or removed position.
- ○ Primary bus conductors and connections are covered with insulating material throughout.
- ○ Mechanical interlocks are provided for proper operating sequence under normal operating conditions.
- ○ Instruments, meters, relays, secondary control devices, and their wiring are isolated by grounded metal barriers from all primary circuit elements.
- Metal-Enclosed Power Switchgear (ME) – A switchgear assembly completely enclosed on all sides and top with sheet metal (except for ventilating openings and inspection windows) containing primary power circuit switching or interrupting devices, or both, with buses and connections.
- Switchgear – A general term covering switching and interrupting devices and their combination with associated control, instruments, metering, protective, and regulating devices.
- Switchgear Assembly – A general term used for equipment used for switching, interrupting, control, instrumentation, metering, protective, and regulating devices.
- Truck-Operated Contact (TOC) – A circuit breaker truck operated auxiliary switch that is mounted in the compartment of a removable circuit breaker and is operated by the circuit breaker frame.

PROBLEMS

19.1 True or false, metal-clad switchgear can always also qualify as metal-enclosed switchgear?

19.2 The IEEE number that identifies a normally open contact that is driven from the cell-mounted switch that indicates the breaker is fully racked-in is which of the following contacts?

- **A.** 52/a
- **B.** 52/b
- **C.** 52TOC/a
- **D.** 52TOC/b
- **E.** 52MOC/a
- **F.** 52MOC/b

19.3 The IEEE number that identifies a normally closed contact that is driven from the cell-mounted switch that indicates the breaker is closed is which of the following contacts?

- **A.** 52/a
- **B.** 52/b
- **C.** 52TOC/a

    D. 52TOC/b

    E. 52MOC/a

    F. 52MOC/b

19.4 The IEEE number that identifies a normally open contact that is mounted on the breaker and indicates the breaker is open or closed is which of the following contacts?

    A. 52/a

    B. 52/b

    C. 52TOC/a

    D. 52TOC/b

    E. 52MOC/a

    F. 52MOC/b

19.5 Given a trip circuit as shown in Figure 19.8, if the plant operator walks up to the circuit breaker and the red light is on and the green light is off, what is the status of the breaker?

    A. Racked in and closed.

    B. Racked in and open.

    C. Racked out and closed.

    D. Racked out and open.

19.6 True or false, a motor starter typically is designed to interrupt fault current on the feeder going to the motor?

RECOMMENDED READING

C37.20.1-1993: IEEE Standard for Metal-Enclosed Low-Voltage Power Circuit Breaker Switchgear.

C37.013-1997: IEEE Standard for AC High-Voltage Generator Circuit Breakers Rated on a Symmetrical Current Basis.

C37.04-1999: IEEE Standard Rating Structure for AC High-Voltage Circuit Breakers.

C37.14-1999: IEEE Standard for Low-Voltage DC Power Circuit Breakers Used in Enclosures.

C37.20.2-1999: IEEE Standard for Metal-Clad Switchgear.

C37.20.3-2001: IEEE Standard for Metal-Enclosed Interrupter Switchgear.

C37.20.4-2001: IEEE Standard for Indoor AC Switches (1 kV–38 kV) for Use in Metal-Enclosed Switchgear.

Electric Power Plant Design, Technical Manual TM 5-811-6, Department of the Army, USA, 1984.

Electrical Machines, Drives and Power Systems, 6th edition, Theodore Wildi, Prentice Hall, 2006.

The Engineering Handbook, 3rd edition, Richard C. Dorf, (editor in chief), CRC Press, 2006.

Industrial Power Distribution, 2nd Edition, Ralph E. Fehr III, Wiley-IEEE Press, 2016. ISBN: 978-1-119-06334-6.

Power Plant Engineering, Black & Veatch, edited by Larry Drbal, Kayla Westra, and Pat Boston, Chapman & Hall/Springer, 1996.

Standard Handbook of Powerplant Engineering, 2nd edition, Thomas C. Elliott, McGraw-Hill, 1998.

CHAPTER 20

BATTERY/VITAL BUS SYSTEMS

GOALS

- To understand the basic design and applications of DC batteries, battery chargers, inverters, and vital bus distribution equipment
- To understand key safety requirements when installing and operating batteries and associated vital bus equipment
- To describe the basic chemistry involved in a lead acid battery and the effects that temperature has on battery life
- To understand the minimum maintenance requirements as defined by IEEE 450
- To calculate the minimum battery capacity required for a particular DC system load as defined by IEEE 485
- To calculate the minimum ampere rating of the battery charger

---

THE **STATION** battery in the generation station is one of the most critical pieces of equipment related to the reliable and safe operation of the power plant. It provides power to critical equipment such as emergency lube oil pumps, hydrogen seal oil pumps, distributed control system, boiler control, and burner management system as well as other critical systems. Both the initial design and the maintenance of the battery system are of primary important to the plant electrical engineer. Several standards provide guidance to the design, maintenance, and installation of battery systems. IEEE standard 485 is used to size the battery for the loads they will be asked to support. IEEE 484 is used for the proper installation of battery systems in generating stations. IEEE 450 provides guidance to the maintenance and testing of battery systems in the power plant. When performing battery maintenance, only use non-sparking tools and ensure that the employee wears the required PPE for the task such as chemical resistant apron and a chemically resistant face shield. Also, ensure that the area is well-ventilated and no explosive gases have collected in the area.

For both safety and reliability, large lead acid battery installations are typically ungrounded. The battery charger is designed to detect an inadvertent ground on the DC system and alarm when a ground occurs. This allows for operation of the DC

Energy Production Systems Engineering: An Introduction for Electrical Engineers to Electrical Power Generation Facilities, Systems, and Equipment, First Edition. Thomas H. Blair.

system to continue while the ground is located and isolated. While detection of a ground is easily achieved in the battery charger, the location of the ground has typically involved isolating various sections of the DC system until the grounded circuit is isolated and the ground is removed from the battery. Recent advances in technology have resulted in an online method of detection of battery grounds. Test equipment known as a battery ground fault tracer is now available. This equipment is connected between the DC bus that is grounded and system ground. The test equipment injects a low frequency, low magnitude AC current into the DC system. A clamp on ammeter that is tuned to the low frequency AC waveform is used at the DC distribution panel to sense the location of the ground fault. The AC current will only flow from the DC grounded bus to system ground on the one feeder that contains the short from the DC bus to ground. Figure 20.1 shows how a ground fault tracer works.

A "battery" is a group of electrochemical cells that store energy in chemical form. Energy can be transferred to the battery from the electrical system it is

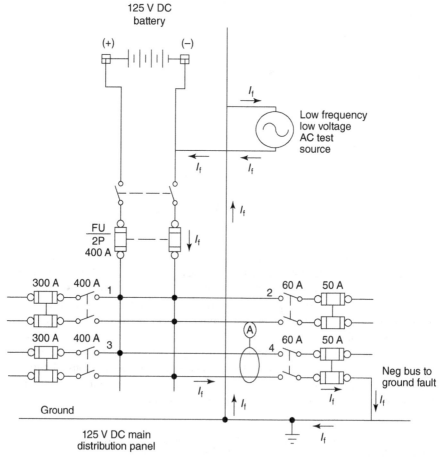

Figure 20.1   Battery ground fault tracer operation.

connected to (*charging*) and can be transferred from the battery to the electrical system it is connected to (*discharging*). The most common battery in the power plant is the wet cell lead acid battery and this will be the type of cell discussed in this text. There are other types of cells in service depending on the application. The attraction of the wet cell lead acid battery is the ability to maintain the battery and the availability of larger amp hour (Ah) cell ratings in the wet cell lead acid battery design. Most power stations have requirements to operate substantially large DC motors in an emergency condition for an extended period of time such as the hydrogen seal oil and turbine lube oil emergency pumps. The wet cell lead acid battery high amp hour rating lends itself well to this application. Depending on the design of the plant, the battery may be called upon to run the DC oil pumps for up to several hours before the unit can be safely stopped and the hydrogen purged from the generator casing.

Each cell in a lead acid battery provides a minimum of 2.17 V DC. The actual range of nominal cell voltages depends on the material that the battery is constructed from, the specific gravity of the electrolyte in the cell, and whether the battery charger is on float setting or equalizing setting. Typical values for ranges of cell voltages for antimony and calcium lead acid batteries are shown below.

Nominal Specific Gravity	Antimony	Calcium
1.250	2.17–2.30	2.21–2.30
1.215	2.15–2.20	2.17–2.26

For a minimum battery voltage of 130 Vdc, using the cell minimum voltage value of 2.17 Vdc, the number of cells required to provide this are 60 cells (60 cells × 2.17 V DC/cell = 130.2 Vdc). The cells are connected in series such that the voltage of each cell is summed and the amp hour rating of any one cell is the amp hour rating of the battery. The voltage at end of service will depend on what the minimum voltage rating of the equipment serviced by the load is. For example, if the battery is connected to a control system that has a range of DC supply voltage tolerable between 115 and 140 Vdc, then the battery will adequately supply that load until the overall battery voltage drops to the 115 Vdc value.

For some plants with very large DC motors, a second battery may be provided with twice as many cells (120 cells) all connected in series for a battery voltage of 260 Vdc. Using a motor with a larger voltage rating reduces the current drawn by the motor and thereby reduces the hour requirement of the individual cells.

In some applications, more amp hours are needed than what one string of cells can provide. In this instance, there may be several chains of cells, each with the individual cell amp hour rating, but these chains are connected in parallel thereby increasing the amp hour rating of the overall battery to a value larger than one individual cell amp hour rating.

Sulfuric acid (H_2SO_4) is a strong acid and, even before charging or discharging, the sulfuric acid will break down into a hydrogen ion (H^+) and sulfate ion (HSO_4^-).

$$H_2SO_4 \rightarrow H^+ + HSO_4^-$$

During the discharge cycle, the reaction between the metal lead (Pb) in the anode plate and the sulfate ion (HSO_4^-) produce lead sulfur ($PbSO_4$), hydrogen ions (H^+), and two excess electrons (e^-). Since electrons are produced at the anode plate, this is the negative plate of the cell. The reaction is chemically defined by equation (20.1).

$$Pb + HSO_4^- \rightarrow PbSO_4 + H^+ + 2e^- \qquad (20.1)$$

During the charge cycle, the reaction between lead dioxide (PbO_2), the available hydrogen ions (H^+), and the ionized sulfate ion (HSO_4^-) at the cathode plate and some water molecules and some donated excess electrons producing sulfate ($PbSO_4$) and water (H_2O). Since electrons are absorbed by the cathode, the cathode is the positively charged plate. The reaction is chemically defined by equation (20.2):

$$PbO_2 + HSO_4^- + 3H^+ + 2e^- \rightarrow PbSO_4 + 2H_2O \qquad (20.2)$$

When batteries are being discharged, electrons flow from the anode ($-$) to the cathode ($+$). In this text book, we will define the direction of current flow using hole flow as the direction of current flow. Restated, we consider that current flows from the positive terminal to the negative terminal. This means that when the battery is discharging, conventional current flow is from the cathode ($+$) to the anode ($-$) on the circuit outside the battery. This means that inside the battery, during discharge, current flow is from the anode ($-$) to the cathode ($+$). When batteries are being charged, conventional current flow is from the anode ($-$) to the cathode ($+$) on the circuit outside the battery. This means that inside the battery, during discharge, current flow is from the cathode ($+$) to the anode ($-$). Realize that "electron flow" would be the opposite of "hole" flow.

The ability of the cells to support the chemical reaction is a function of the health of the electrolyte and plates in the cell and the temperature of the cell. As the cell temperature is increased above nominal, the cell plate life is shortened. Table 20.1 is a chart of cell temperatures and the resultant expected life of the battery due to elevated temperatures. Note that for a cell, the reference temperature is 77°F (nominal); the expected life of a cell is 20 years.

To ensure that all cells of the battery will be reliable, the specific gravity of the cells should be monitored. The expected specific gravity of the electrolyte varies with

TABLE 20.1 Expected Life of Battery Cell as Function of Ambient Temperature

Average Temperature Degrees Fahrenheit	Anatomy Years	Calcium Years
107°F	6 yrs	5 yrs
92°F	12 yrs	10 yrs
77°F	20 yrs	20 yrs
62°F	22 yrs	22 yrs
47°F	25 yrs	25 yrs

Source: Reproduced with permission of IEEE.

TABLE 20.2   Recommended Cell Temperature

Calcium	Antimony	T_{avg} (annual) (°F)	$T_{cell\,max}$ (monthly) (°F)
1.215	1.215	77.5	90
1.250	1.250	72	85

Source: Reproduced with permission of IEEE.

the temperature of the cell. Typical specific gravity values and maximum annual and monthly cell temperatures are provided in Table 20.2.

Both to ensure adequate cooling and to prevent the collection of hydrogen gas in the battery compartment, the battery compartment is well-ventilated. Hydrogen gas has a specific density of about 10% that of air. As such, if hydrogen gas is vented from the cell, it has a tendency to rise. Ventilation systems are designed to aid in the dispersion of hydrogen gases at the top of the battery compartment to ensure an explosive concentration of hydrogen gas does not collect. The production of hydrogen gas is more commonly a result of overcharging the cells than from normal charge and discharge events.

In the standard *flooded cell*, there are gray plates and brown or black plates. The negative terminal (anode) connects to the gray plates. The positive terminal (cathode) connected to the brown or black plates. There is one more negative (gray) plate than positive (black) plate in each cell. Bubbling or boiling of the electrolyte is a potential indication of an overcharging condition and should be investigated. Another indication of a potential issue is if the positive plates (brown or black normally) take on a color of the negative plates (gray). If this should occur, this is an indication of sulfation of the plate and indicates that the plate's ability to support the chemical reaction needed is limited or nonexistent. Another indication of an issue with a cell is indicated by white bands that develop across the plate. This occurs during periods of high current flow and can lead to shorted cells. A shorted cell will result in slightly increased voltage on the other cells and higher levels of charging current for the same charger applied voltage. The higher current may lead to other cells developing these same shorts and, in very short time, there may be several failed cells in the battery. Additionally, the higher charging currents can cause the electrolyte to boil off. With a loss of electrolyte, the battery basically looks like an open and will not provide current if needed by the system. Therefore, maintaining proper electrolyte level is critical to proper battery operation.

To ensure batteries are properly maintained, when the voltage of any cell drops below a value of 2.13 V, the voltage from the charger is increased from the floating voltage to a slightly higher value of the equalizing charge setting. This forces current to flow back into batteries and tends to balance out the individual cell voltages. A *float voltage* is a voltage from a battery charger that is applied to a battery to maintain it in a fully charged condition during normal operation. An *equalizing voltage* is a voltage, which is higher than the float voltage, from a battery charger that is applied to a battery to correct inequalities in cell specific gravity or cell voltage among battery cells. The float voltage is always slightly less than the equalizing voltage. Note that equalizing does not apply to all battery technologies, but does apply to lead acid. Consult the OEM maintenance document to determine battery charging levels.

Since battery systems are so critical to plant operation, their maintenance is of great importance. The following is a description of the minimum steps to be taken to maintain a battery on a monthly basis.

Check/record float voltage

Check/record appearance

Check/record electrolyte levels

Check/record leaks

Check/record terminal condition

Check/record ambient temperature

Every 3 months, in addition to the monthly maintenance, the following should be done to ensure the reliable operation of the battery.

Check/record specific gravity of each cell

Check/record voltage of each cell and total battery voltage

Check/record temperature of electrolyte in representative cells

Annually, in addition to the above, the following should be performed on the battery.

Check/record detailed visual inspection of each cell

Check/record cell-to-cell connection resistance; compare with baseline

When the battery is installed there is an initial battery *capacity test* (called an *acceptance test*) to ensure that the battery has been designed, constructed, and installed properly and will perform as needed if called upon. The acceptance test should meet a specific discharge rate and be for a duration relating to the manufacturers rating or to the purchase specifications requirements. Batteries may have less than *rated capacity* when delivered. Unless 100% capacity upon delivery is specified, initial capacity can be as low as 90% of the rated. Under normal operating conditions, capacity should rise to at least rated capacity in normal service after several years of float operation.

At the end of the warranty period, this same acceptance test will be done to ensure there are no issues with the battery before the warranty runs out. After this acceptance test, the batteries are tested once every 5 years (called a *performance test*). The first performance test should occur within the first 2 years of battery life. This is typically done just before the battery warranty expires to ensure any issues discovered can be resolved under warranty. As long as the battery exceeds 85% of capacity of initial test, further performance testing is done once every 5 years. Annual performance tests of battery capacity should be made on any battery that shows signs of degradation or has reached 85% of the service life expected for the application. Degradation is indicated when the battery capacity drops more than 10% from its capacity on the previous performance test, or is below 90% of the manufacturers rating. If the battery has reached 85% of service life, delivers a capacity of 100% or greater of the manufacturer's rated capacity, and has shown no signs of degradation,

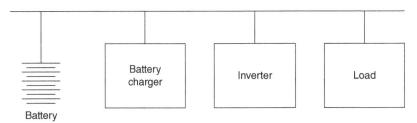

Figure 20.2   Simplified DC distribution system.

performance testing at 2-year intervals is acceptable until the battery shows signs of degradation. Voltage polarity reversal of a cell is also a good indicator for further investigation into the need for individual cell replacement.

In most power stations, the station battery is tied to a common bus. Figure 20.2 shows a normal DC distribution system that incorporates the battery, battery charger, inverter, and various DC loads on one common bus. One or more battery chargers, one or more inverters, and other DC loads are all connected to this common bus.

Under normal conditions, the battery charger is fed from an AC source and rectifies the AC source to DC. The charger applies this DC to the common bus. The charger provides current to the inverter and other station DC loads and provides a small trickle charge to the battery. During maintenance, the voltage output of the battery charger is raised slightly to the equalizing voltage value to increase the current to the battery, but the battery charger is normally providing load current to the inverter and load. During testing of the battery, the battery is disconnected from this DC bus and the charger provides the DC power the inverter and load. Only during a loss of AC power, when the battery charger would become de-energized, does the battery support the inverter and DC load.

The typical values for the float voltage (per cell) for antimony cells is 2.15–2.19 V/cell and the typical values for the float voltage (per cell) for calcium cells is 2.17–2.26 V/cell. The typical values for the equalizing voltage (per cell) for antimony cells is 2.24–2.36 V/cell and the typical values for the equalizing voltage (per cell) for calcium cells is 2.30–2.48 V/cell.

Normally, the battery charger is sized to provide current for all the loads connected to the DC system and, in addition, the necessary charge to recharge the battery in a certain amount of time. As such, the sequence to design the proper capacity or size of a battery is to first define the total DC load on the DC system. Once the load is defined and the time duration of the load, the battery capacity is designed. Lastly, once the DC load and battery capacity are known, the battery charger is designed to provide the DC current necessary to support the plant load and enough current to maintain a float charge on the station battery.

DESIGN OF BATTERY SYSTEMS (DC SYSTEM LOAD AND BATTERY CAPACITY)

The first step in the design of the DC system is to identify all the necessary DC loads and define any future spare capacity that needs to be accounted for in the design of the system. Once these loads are defined, the second step is to design the battery capacity necessary to support these loads for the duration necessary in the event of a loss of battery charger to the DC system. In a power generation station, there are many loads that required DC power in the event of a loss of AC power to the battery chargers. Some loads are continuous and some are only momentary. Typical continuous loads include items such as lighting, motor running currents, inverter fed loads, indicating lights, and annunciator loads. Typical momentary loads are motor starting currents, switchgear operations (opening and closing of circuit breakers), and field flashing of emergency generators. One such example is the need for a backup DC lubrication oil pump. The generator and large motors have shaft-driven oil pumps, but on a loss of AC power to the station, as the shaft of the machines slow down, the shaft-driven pumps will not provide the lubricating oil necessary to the machine bearings. This oil is needed to both provide lubrication to the journal bearings to prevent wiping of the bearings and the oil also provides a means for heat removal from the bearing surface. Lubrication must be maintained until the shaft has completely stopped and the temperatures of the bearings have been reduced to reasonable levels. Another example is the need for seal oil for generators that contain hydrogen for the cooling media. This seal oil is supplied by a pump coupled to an AC motor. On a loss of AC power, a backup DC motor and pump must be started immediately, to maintain seal oil pressure greater than the pressure of the hydrogen gas in the generator. Additional loads also include various breaker close and trip operations, various control system functions (such as operating valves) to place the system in a stable shutdown state. The process to shut down a system and the necessary DC current and time duration is defined. An example of such loads and durations is shown in Table 20.3.

Once the necessary DC load and times are defined, the next step is to calculate the capacity of the battery to be connected to the DC system. Since various battery manufacturers and various battery designs will have varying capacity, this process tends to be an interactive process. An initial estimate is done using a certain battery manufacturer's datasheet for a specific series of battery with an initial estimate of cell size (the number of plates in the cell). The construction of the cell is such that the

TABLE 20.3 Sample Load List for Battery Capacity Calculation

Description	Load in Amperes			
	Min 1	Min 2–60	Min 61–119	Min 120
Emergency Lube Oil Pump	150	150	150	150
Emergency Seal Oil Pump	105	70	34	47
Breaker Trip	15	0	0	15
Breaker Close	10	0	0	10
Inverter Load	30	30	30	30
Total DC Load	310	250	214	252

Source: Reproduced with permission of IEEE.

number of negative plates in a cell is always one more than the number of positive plates in a cell. This leads to equation (20.3) that defines the total number of plates as a function of the number of positive plates in a cell.

$$\text{Total \# of plates} = 1 + (2 \times \text{\# of positive plates}) \tag{20.3}$$

As the number of plates in a cell increases, the total surface area of the cell plates increases, thereby increasing the amp hour rating of each cell. For a battery that is constructed of one set of cells in series, the amp hour rating of each cell is the amp hour rating of the overall battery.

The IEEE 485 process to define necessary battery capacity can be done either utilizing the amp hour rating (for a certain discharge rate) of a battery cell or it can be done using the current per positive plate rating (again for a certain discharge rate) of a battery cell. The process looks at each *period* of discharge and the largest value of amp hours or number of positive plates is selected from the evaluation of all periods. Lastly, correction factors are added for temperature corrections for battery system ambient temperature, aging factor, and any additional design margin desired for future DC system growth.

- Temperature Correction – As temperature decreases the capacity of a cell decreases (and vice versa as the temperature increases). Manufacturers quote cell capacity at a given temperature and appropriate correction factors should be used for other temperatures.

- Aging Factor – Battery performance is relatively stable throughout its life, dropping rapidly towards the end. To ensure the battery can meet the design requirements throughout its life, the IEEE 485 standard suggests the initial capacity should be 125% of the design capacity.

- Design Margin – To cater for unexpected circumstances (increased loads, poor maintenance, recent discharge, etc.) it is common to allow a design margin of 10–15%.

As an example, given the load data in Table 20.3, we first need to settle in on a battery manufacturer and series of battery so that we can calculate the necessary battery capacity. Once the cell series is defined, the battery cell OEM will provide a datasheet that shows the current the type of cell can provide for various durations for various values of end-of-cell voltage (see calculation of minimum number of cells for importance of the value of end-of-cell voltage).

The IEEE 485 describes two methods of evaluating battery cell capacity. The term R_t is the number of amperes each plate can supply for t min, at 25°C, and a certain cell electrolyte specific gravity value to a defined minimum cell voltage. The other term used K_t is the ratio of amp hour capacity, at a standard time rate, at 25°C and to a defined minimum voltage which can be delivered for t min. (Note that due to units used, while R_t is inversely proportional to K_t, they are not equal).

We will use R_t in our evaluation, but either method is valid.

The value R_t is calculated by the following formula.

$$R_t = \text{ampacity (for given value of discharge time)}/\text{\# of positive plates} \tag{20.4}$$

TABLE 20.4   Battery Discharge Rates

Time	8 hr	2 hr	1 hr	30 min	1 min
Amperes	123.3	313	478	679	981
R_t	8.22	20.86667	31.86667	45.26667	65.4

Source: Reproduced with permission of IEEE.

For example, if we chose an OEM battery that has 15 positive plates that can provide 123.3 A for 8 hr, then to calculate the 8 hr R_t value, we simply divide the current supplied by the number of positive plates in the cell. For this battery, there are 15 positive plates, so the value for R_t for an 8 hr discharge is

$R_t(8 \text{ hr}) = 123.3$ A (from OEM)/15 positive plates $= 8.22$ A/positive plate

We can calculate the R_t values for all other times for this 15 positive plate cell once we have discharge rate information from the battery OEM. Table 20.4 is an example of the calculation of these values defining R_t values for various discharge times and the equations are shown below Table 20.4.

R_t (2 hr) $= 313$ A (from OEM) / 15 positive plates $= 20.87$ A/positive plate

R_t (1 hr) $= 478$ A (from OEM) / 15 positive plates $= 31.87$ A/positive plate

R_t (30 min) $= 679$ A (from OEM) / 15 positive plates $= 45.27$ A/positive plate

R_t (1 min) $= 981$ A (from OEM) / 15 positive plates $= 65.4$ A/positive plate

Next, we develop our analysis of each discharge period. For each period, if the next period total current is greater than the previous, we do not need to calculate the lower of the two and can skip this calculation and move to the next period with higher current. We will show this with the following example.

The first discharge period (as can be seen from the DC load list of Table 20.3) is 310 A for the first minute. We enter into column 1 the first period. In column 2, we enter the current during the first period. From the table, our value for current is 310 A. In column 3, we enter the increase/decrease of load from the previous period. Since we went from 0 A in the previous period to 310 A during this period, the increase is 310 A. In column 4, we list the duration of this period. For our example, this is 1 min. For column 5, we list the total time from the end of the total analysis period to the beginning of this period (summation of this period and all later periods). Since there are no other periods for our first section, this is simply the value of 1 min. Next, in column 6, we enter the calculated value of battery capacity in terms of R_t that was developed in Table 20.4. For this section the time is 1 min, so from Table 20.4, the 1 min value of R_t is 65.4 and this value is entered into column 6. Column 7 is the quotient of the dividend value in column 3 divided by the divisor value in column 6. If the resultant is positive, this value goes on to the left and if the resultant is negative, this value goes on to the right. For this first period, the value of required positive plates for this section is

$$+310 \text{ A}/65.4 = 4.74 \text{ plates}$$

TABLE 20.5A   First Period Calculation in Tabular Format

1	2	3	4	5	6	7	
					Capacity at T		
	Load	Changes	Duration of	End of	min rate. (6 A)	Required section	
	(A) $[A_1,$	in Load	Period	Section	R_t (A/positive	# positive plates =	Sum (7A
Period	A_2, etc.]	(A)	(min)	(min)	plate)	(3)/(6)	+ 7B)
Section 1: First Period Only – If A_2 is greater than A_1, go to section 2.							No
1	310	310	1	1	65.4	4.740061	4.740061
Total							4.740061

Source: Reproduced with permission of IEEE.

Lastly, we sum the values in the left and subtract any values on the right. For this first period, we find the total to be 4.74 plates. This first section is summarized in Table 20.5a.

Looking back at the sample load profile of Table 20.3, we see that the load changes from period 1 to period 2 from 310 A for 1 min to 250 A for the following 59 min. This is represented in the calculation as one current of 310 A for a period of 60 min and −60 A for the time 2–60 minutes. Since the total current at the end of period 2 is less than period 1, we continue this calculation. The data is summarized in Table 20.5b.

Looking back at the sample load profile of Table 20.3, for the next period, current is further reduced from a value of 250 A to a value of 214 A. This is represented in the calculation as a current of 310 A for a period of 119 min and −60 A for the time 2–119 min and a further −36 A for the time 61–119 min. However, since the current in period 4 increases from 214 A to 252 A, we can skip this calculation and proceed to the last period. This is due to fact that, since load current increases in period 4 above that of period 3, the amperes per plate will always be greater from the last period with higher current and there is no need to calculate the third period.

For the last period, current is increased from a value of 214 A to 252 A. This is represented in the calculation as a current of 310 A for a period of 120 min and −60 A for the time 2–120 min and a further −36 A for the time 61 min–120 min and a final increase in current of +38 A for the period 119–120 min. A summary of all these calculations is shown in Table 20.5c.

As can been seen in Table 20.5c, the required number of positive plates from section 1 was 4.74, the required number of positive plates from section 2 was 7.85, the

TABLE 20.5B   Second Period Calculation in Tabular Format

Section 2: Second Period Only – If A_3 is greater than A_2, go to section 3.							No
1	310	310	1	60	31.86667	9.728033	9.728033
2	250	−60	59	59	31.86667		−1.88285	−1.88285
Total							7.845188

Source: Reproduced with permission of IEEE.

TABLE 20.5C All Calculations in Tabular Format

1	2	3	4	5	6	7	
Period	Load (A) (A_1, A_2, etc.)	Changes in Load (A)	Duration of Period (min)	End of Section (min)	Capacity at T min rate. (6A) R_t (A/positive plate)	Required section # positive plates = (3)/(6)	SUM (7A + 7B)
Section 1: First Period Only – If A_2 is greater than A_1, go to section 2							
1	310	310	1	1	65.4	4.740061	No 4.740061
Total						4.740061	4.740061
Section 2: Second Period Only – If A_3 is greater than A_2, go to section 3							
1	310	310	1	60	31.86667	9.728033	No 9.728033
2	250	–60	59	59	31.86667	–1.88285	–1.88285
Total							7.845188
Section 3: Third Period Only – If A_4 is greater than A_3, go to section 4							
1							
2							
3							
Total							Yes
Section 4: Fourth Period Only – If A_5 is greater than A_4, go to section 5							
1	310	310	1	120	20.86667	14.85623	No 14.85623
2	250	–60	59	119	20.86667	–2.8754	–2.8754
3	214	–36	59	60	31.86667	–1.12971	–1.12971
4	252	38	1	1	65.4	0.58104	0.58104
Total							11.43216

Source: Reproduced with permission of IEEE.

required number of positive plates from section 3 we did not need to calculate since section 4 had higher current, and the required number of positive plates from section 4 was 11.43. We select the maximum values of plate quantities in each of the section and find in this example that the *minimum* number of positive plates to meet the defined loading of Table 20.2 for this example is 11.43 (*Note*: we cannot have 0.43 plates, but will correct this after we take into account any adjustments). Now in addition to this we can add a correction factor for battery temperature, a correction factor for battery aging and lastly a correction factor for design margin (for future growth of DC system). For this example, a factor of 1.2 is multiplied to the value of 11.43 to derive a MINIMUM number of positive plates for our battery to be 13.7 positive plates. If the next available number of positive plates in a commercially available type battery is 15 positive plates, we would select the battery with 15 positive plates.

Lastly, if we found the number of positive plates required was different than our assumption we made when we selected our battery data sheet at the beginning of the calculation, we would need to go back to our original design criteria and select the new number of plates we have found and recalculate the battery load data for the correct size battery cell as the number of plates effects the value of our capacity factor (R_t) in Table 20.4 that we used for our calculations. If we used a different plate design, then we need to recalculate based on the number of plates from the above calculation and perform the analysis one more time. It is important to ensure that the values of battery cell capacity R_t are derived from the same OEM and cell size (i.e., number of positive plates) that we recommend at the end of the analysis since the value of capacity per positive plate (R_t) does change slightly between cell sizes.

Now that we have determined the cell capacity we need for our application, lastly, we need to calculate the number of cells in our battery. There must be enough cells connected in series to provide the minimum needed DC system voltage when the battery cells are completely discharged, but there cannot be too many cells that would cause the maximum DC system voltage to exceed the connected equipment rating when the battery cells are at equalizing charge voltage. The maximum number of cells in a battery is defined by the maximum allowable battery voltage that the equipment connected to the DC system can tolerate when the batteries are being charged in equalizing mode and the cell voltage during this charge. The equation for calculation of the number of battery cells in a battery system based on the maximum allowable battery voltage that the equipment connected to the DC system can tolerate is defined in equation (20.5).

$$\text{Maximum number of cells} = V_{\text{max}}/V_{\text{cell\_max}} \tag{20.5}$$

where

V_{max} = maximum allowable battery voltage

$V_{\text{cell\_max}}$ = cell voltage needed for equalizing charge

The minimum number of cells in a battery is defined by the minimum allowable battery voltage that the equipment connected to the DC system can function reliably on when the batteries are at the end of discharge and the value of the cell voltage at end of discharge. The equation for calculation of the number of battery cells in a battery system based on the minimum allowable battery voltage that the equipment connected to the DC system can reliably tolerate is defined in equation (20.6).

$$\text{Minimum number of cells} = V_{min}/V_{cell\_min} \tag{20.6}$$

where

V_{min} = minimum allowable battery voltage

V_{cell_min} = cell voltage at discharged state

The total number of cells in a battery must fall between these two values for reliable operation.

DESIGN OF BATTERY SYSTEMS (BATTERY CHARGER)

Once the DC system load and battery capacity is designed and specified, then the battery charger requirements can be designed. The minimum current rating of the battery must provide the current necessary to recharge a depleted battery in a certain amount of time defined by the project. It must supply the required DC connected load and, at the same time, recharge the battery in the number of hours specified. This is accomplished by utilizing equation (20.7).

$$A = ((AHR \times 1.10)/T + L) \tag{20.7}$$

where

A = ampere rating of the charger (but not less than 20% of 8 hr battery discharge rate)

AHR = ampere hours removed from battery

1.10 = conversion factor for lead acid cells

T = maximum number of hours for battery recharge

L = continuous load of DC system (A)

The battery charger output must be derated for altitude, when it is installed in locations above 3300 ft (1000 m), and for temperature, when the ambient temperature exceeds 50°C. The battery charger manufacturer can provide these derating factors.

Example 20.1   Given a battery is sized such that it will be discharged at a rate of 15 A for 8 hr, we want to design the charger to recharge this battery in a period of 12 hr while still feeding the DC-connected load of 4.5 A. The charger will be installed in an environment that is less than 50°C ambient temperature and less than 3300 ft elevation. What is the minimum required ampacity of the charger?

Solution: From the stated problem and utilizing equation (20.7), we find our variables to be

AHR = 15 A × 8 hr = 120 Ah

T = 12 hr

L = 4.5 A

Using equation (20.7), we find the minimum ampacity for the battery charger to be

$A = ((\text{AHR} \times 1.10) \, / \, T + L)$

$A = ((120 \text{ Ah} \times 1.10) \, / \, 12 \text{ hr} + 4.5)$

$A = 15.5 \text{ A}$

The next largest available commercially available charger size should be ordered.

GLOSSARY OF TERMS

- Acceptance Test – A constant-current or constant-power capacity test made on a new battery to confirm that it meets specifications or manufacturer's ratings.
- Battery Duty Cycle – the load currents a battery is expected to supply for a specified time period.
- Capacity Test – A discharge of a battery at a constant current or constant power to a specified terminal voltage.
- Critical Period – That portion of the duty cycle that is the most severe, or the specified time period of the battery duty cycle that is most severe.
- Duty Cycle – The loads a battery is expected to supply for specified time periods while maintaining a minimum specified voltage.
- Equalizing Voltage – The voltage, higher than float, applied to a battery to correct inequalities among battery cells (voltage or specific gravity).
- Float Voltage – The voltage applied to a battery to maintain it in a fully charged condition during normal operation.
- Flooded Cell – A cell in which the products of electrolysis and evaporation are allowed to escape to the atmosphere as they are generated. These batteries are also referred to as "vented."
- Performance Test – A constant-current or constant-power capacity test made on a battery after it has been in service, to detect any change in the capacity.
- Period – An interval of time in the battery duty cycle during which the current is assumed to be constant for the purposes of cell sizing calculations.
- Rated Capacity (lead acid) – The capacity assigned to a cell by its manufacturer for a given discharge rate, at a specified electrolyte temperature and specific gravity, to a given end-of-discharge voltage.
- Service Test – A test in the as "found condition" of the battery's capability to satisfy the battery duty cycle.
- Terminal Connection – Connections made between cells or at the positive and negative terminals of the battery, which may include terminal plates, cables with lugs, and connectors.

PROBLEMS

20.1 Ventilated lead acid batteries should be replaced if, during discharge test, the capacity of the battery drops below what percentage of original capacity?

A. 90%

B. 85%

C. 80%

D. 70%

20.2 True or false, under float operation the battery provides current to the load and the battery charger.

20.3 True or false, the negative plates in a healthy flooded cell are gray in color.

20.4 Given a battery that is rated for a life expectancy of 20 years with a maximum ambient temperature of 77°F, what is the life expectancy of the battery if the ambient is maintained at 62°F?

20.5 When performing maintenance on a battery, what is the minimum frequency of recording the charger float voltage?

A. Monthly

B. Quarterly

C. Annually

D. Every 5 years

20.6 Given a battery is sized such that it will be discharged at a rate of 30 A for 8 hr. We want to design the charger to recharge this battery in a period of 16 hr while still feeding the DC connected load of 10 A. The charger will be installed in an environment that is less than 50°C ambient temperature and less than 3300 ft elevation. What is the minimum required ampacity of the charger?

RECOMMENDED READING

Electric Power Plant Design, Technical Manual TM 5-811-6, Department of the Army, USA, 1984.

The Engineering Handbook, 3rd edition, Richard C. Dorf, (editor in chief), CRC Press, 2006.

IEEE Power Engineering Society, IEEE 946-2004: IEEE Recommended Practice for the Design of DC Auxiliary Power Systems for Generating Stations, 2005.

IEEE Power Systems Committee, Orange Book IEEE 446-1987: IEEE Recommended Practice for Emergency and Standby Power Systems for Industrial and Commercial Applications, 1992.

IEEE 450-2010: Recommended Practice for Maintenance, Testing, and Replacement of Large Lead Storage Batteries for Generating Stations and Substations.

IEEE 484-2002: Recommended Practice for Installation Design and Installation of Large Lead Storage Batteries for Generating Stations and Substations.

IEEE 485-2010: Recommended Practice for Sizing Large Lead Storage Batteries for Generating Stations and Substations.

Industrial Power Distribution, 2nd Edition, Ralph E. Fehr III, Wiley-IEEE Press, 2016. ISBN: 978-1-119-06334-6.

Power Plant Engineering, Black & Veatch, edited by Larry Drbal, Kayla Westra, and Pat Boston, Chapman & Hall/Springer, 1996.

Standard Handbook of Powerplant Engineering, 2nd edition, Thomas C. Elliott, McGraw-Hill, 1998.

CHAPTER 21

GROUND SYSTEM

GOALS

- To understand the four fundamental functions of the energy production station ground system
- To understand the purpose of the neutral ground system, the equipment ground system, and the safety ground system
- To calculate the amount of voltage across a ground detection relay open delta for an ungrounded system
- To calculate the required neutral resistance required to limit fault current to a certain value for a resistance grounded system
- To calculate the *ground grid* resistance, touch potentials and step potentials as defined by IEEE 80
- To calculate the inductance of the ground loop

GROUNDING OF an electrical power distribution system is concerned specifically with the design of an intentional electrical connection of the power system to ground or earth. The grounding system has four distinct functions in the power generation station.

- First, the ground system is required to provide a low impedance ground-fault current return path in order to activate the protection and clear or alarm the ground fault as soon as possible.
- Second, the ground system is also designed to limit the voltages on station structures and accessible equipment to safe levels.
- Third, the ground system may be required to provide shielding of electrical noise for instrumentation.
- Lastly, the ground system should provide a low impedance path for protection of plant equipment from lightning strikes.

To provide all these functions, the ground system can be broken up into three distinct systems:

Energy Production Systems Engineering: An Introduction for Electrical Engineers to Electrical Power Generation Facilities, Systems, and Equipment, First Edition. Thomas H. Blair.

- the neutral ground system
- the equipment ground system
- the safety ground system

The *neutral ground* system is intended to establish the ground reference of the electrical system. The neutral ground connection is usually made to the neutral point of the supply equipment. The neutral ground system may or may not carry current under normal circumstances depending on the system design. The *safety ground* system is intended to protect personnel from injury and equipment from damage. These parts do not normally carry current, except in the event of a phase-to-ground fault. The *equipment ground* system ensures a low impedance return path for ground current during a phase-to-ground fault and ensures that touch and step potential values are within those allowed.

There are four basic types of neutral grounding systems, each with its own distinct advantages and disadvantages.

UNGROUNDED SYSTEM

The first type of *system grounding* is the ungrounded system. The ungrounded system does not have any intentional electrical system connections to ground or earth potential. This system is used when the desire for a highly reliable electrical system is needed. Upon the first phase-to-ground fault in the electrical system, only a small charging current flows through the ground (due to distributed capacitance of the phase conductors). No damage occurs to the grounded equipment (assuming the insulation of the system is properly rated and installed). This system provides time for the phase-to-ground fault to be located and isolated. One of the disadvantages of this system is the method of determining the location of the phase-to-ground fault. With no intentional path for ground current to flow, the entire electrical system potential (voltage-to-ground) is shifted. There is essentially no residual current. Therefore, there is no automatic means to determine the exact location where the phase-to-ground faults exists. The normal method of finding the location of a phase-to-ground fault in an ungrounded system is by the systematic isolation of feeder circuits by opening their isolating devices one at a time and seeing if the ground is cleared. Due to the reliability of this system and the immunity to circuit tripping on the first phase-to-ground fault occurrence, this system is utilized in many power generator stations, especially older units. It can also be found in many military applications when reliability is critical. Another disadvantage of the ungrounded system is the magnitude of the transient voltage that can be impressed on the insulation system. With an ungrounded system, the transient voltage can be large. As the impedance of the connection between the electrical system and ground is lowered, the peak voltage that an insulation system must withstand is reduced.

The typical ground-fault detection relay for an ungrounded system is the use of a three-phase transformer that is connected wye on the system side and broken delta on the secondary side. This is shown in Figure 21.1.

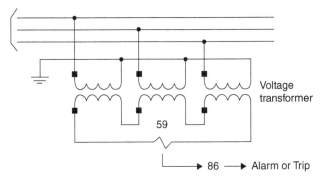

Figure 21.1   Ground detection relay (59N) for a three-phase ungrounded system. *Source*: Reproduced with permission of IEEE.

A resistor and a relay coil are placed across the broken delta secondary of a three-phase transformer. With a balanced distribution system with balanced cable capacitance on all three phases and with no faults, the earth connection is at phase-to-phase voltage / 1.732 and each phase voltage is separated by 120 degrees. The vector diagram is shown in Figure 21.2. Since the voltage magnitudes are balanced and they are all displaced by 120 degrees, the voltage across the relay 59N is nearly zero.

Should a phase-to-earth fault occur, the voltage magnitude on that one phase-to-earth becomes zero or negligible and the phase-to-ground voltage on the other two ungrounded phases becomes at or close to the phase-to-phase value of voltages. The phase angle between all three voltages is still 120 degrees. The vectors for this situation are shown in Figure 21.3.

This presents a voltage across the relay and activates the relay indicating the grounded condition. For a solidly grounded fault, the magnitude of the voltage across the broken delta approaches the phase-to-phase voltage of the primary multiplied by the turns ratio of the transformer.

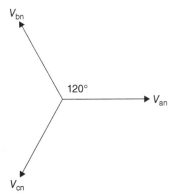

Figure 21.2   Balanced phase-to-ground voltages of an ungrounded system (no phase-to-ground fault).

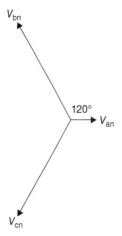

Figure 21.3   Unbalanced phase-to-ground voltages of an ungrounded system (phase A-to-ground fault).

Example 21.1   For a 480 V three-phase, three-wire ungrounded system, where the ground detection relay (59N) is fed from a 4:1 transformer, what is the value of voltage across the broken delta during the solidly grounded fault on the 480 V side?

Solution: Select any one phase on the primary to fault. For example, let phase B-to-ground fault. In this condition, the phase-to-ground voltages found on the primary of the transformer are as shown below.

$$V_{an} = 480\angle0°$$
$$V_{bn} = 0\angle - 120°$$
$$V_{cn} = 480\angle120°$$

The resultant voltage on the secondary side, knowing that the transformer turns ratio is 480:120, we find secondary values to be

$$V_{ab} = 120\angle0°$$
$$V_{bc} = 0\angle - 120°$$
$$V_{ca} = 120\angle120°$$

The voltage across the resistor is the summation of these three voltages.

$$V_{resistor} = V_{ab} + V_{bc} + V_{ca}$$
$$V_{resistor} = 120\angle0° + 0\angle - 120° + 120\angle120°$$

Performing vector analysis, the resultant voltage vector for the voltage across the resistor is

$$V_{resistor} = 120[0.5 + j0.866]$$
$$V_{resistor} = 120\angle60°$$

RESISTANCE GROUNDED SYSTEM

The second type of electrical system grounding is the resistance grounded system. Resistance grounded systems use a resistance between the source transformer neutral connection and earth. This resistor is sized for a specific amount of fault current in the event a phase-to-ground fault occurs. The formula to calculate the required resistance to obtain a certain phase-to-ground fault current is shown in equation (21.1).

$$R_n = V_{ln}/I_{fault} \tag{21.1}$$

where

R_n = neutral grounding resistor resistance (Ω)

V_{ln} = phase-to-neutral voltage of the transformer (V)

I_{fault} = amount of phase-to-ground fault current desired (A)

The value of the line-to-neutral voltage can be found by using equation (21.2).

$$V_{ln} = V_{ll}/1.732 \tag{21.2}$$

where

V_{ln} = phase-to-neutral voltage of the transformer (V)

V_{ll} = phase-to-phase voltage of the transformer (V)

Example 21.2   We wish to design a three-phase, four-wire, 4160 V (phase-to-phase) electrical distribution grounding system such that a solid phase-to-ground fault will be limited to 400 A. What size of resistor between the transformer neutral and ground is required to limit current to this value?

Solution: First, we need to find the value of the line-to-neutral voltage using equation (21.2).

$$V_{ln} = V_{ll}/1.732$$
$$V_{ln} = 4160 \text{ V}/1.732$$
$$V_{ln} = 2402 \text{ V}$$

Now, we can calculate the required resistance by using equation (21.1)

$$R_n = V_{ln}/I_{fault}$$
$$R_n = 2402 \text{ V}/400 \text{ A}$$
$$R_n = 6 \ \Omega$$

There are two types of resistance grounded systems, high resistance grounded systems and low resistance grounded systems. High resistance grounded systems are used on low voltage systems and occasionally for system voltages up to 4160 V three-phase. Much like the ungrounded system, when the intention is to not immediately clear the phase-to-ground fault, but to allow for time for the phase-to-ground fault to be isolated and repaired, the high impedance system is used. The resistor in the neutral-to-ground connection of the transformer is sized to limit bolted phase-to-ground fault current to values 5 A or less. Unlike the ungrounded system where once a phase-to-ground fault occurs, the troubleshooting consists of opening individual circuits on the distribution system until the ground is removed to locate the fault, the high resistance ground provides a residual current of up to 5 A (depending on the impedance of the phase-to-ground fault at the point of the fault). Wrapping a clamp on ammeter around the phase conductors at each load, will indicate which circuit has the phase-to-ground fault without de-energizing the circuit by sensing the zero sequence currents. As such, many newer power plants utilize the high resistance grounded system for their electrical distribution. Permanently installed ground-fault detection systems for switchgear and motor control centers (MCCs) such as the one shown in Figure 21.4 are available and are used in high resistance grounded system that can display where the ground has occurred to simplify troubleshooting.

For system voltages of 4160 V and above, these systems are typically low resistance grounded systems. The fault current is designed to be substantial on a solid phase-to-ground fault. This is done to ensure that the protective relaying will function immediately to remove the fault by de-energizing the circuit where the phase-to-ground fault has occurred even if there is substantial impedance in the phase-to-ground fault path. The neutral to ground resistor is sized for values for bolted phase-to-ground fault currents of between 100 A and 2000 A with the most common values being 400 A and 1000 A. The tradeoff in the selection of the resistor size is the security of ensuring that, during a ground fault with an impedance in the ground-fault path, the relaying will be able to detect the ground fault and trip the correct circuit protection device to isolate the ground fault, while keeping ground-fault current as small as possible to minimize the equipment damage that occurs at the site of the phase-to-ground fault.

REACTANCE GROUNDED SYSTEM

The third type of electrical system grounding is the reactance grounded system. In this system, a reactance is placed between the transformer neutral and ground connection. In the event of a fault, the reactance in the neutral substantial reduces the ground-fault current to values between 100 and 800 A. Either an inductive reactance or a Peterson coil can be used in the neutral to reduce the value of ground-fault currents. The value of the inductive reactance in the neutral is typically chosen at a value slightly larger than the system-distributed capacitance to maintain the overall reactance in the ground path inductive which has the effect of reducing the potential transient overvoltage that the system might experience during an arcing fault. A system in which the inductive current is slightly larger than the capacitive earth fault current is overcompensated. A system in which the induced earth fault current is slightly smaller than the capacitive earth fault current is undercompensated. To get substantial reduction of the transient overvoltages during a fault, the reactance must be sized to provide a substantial amount of phase-to-ground fault current, which allows for substantial

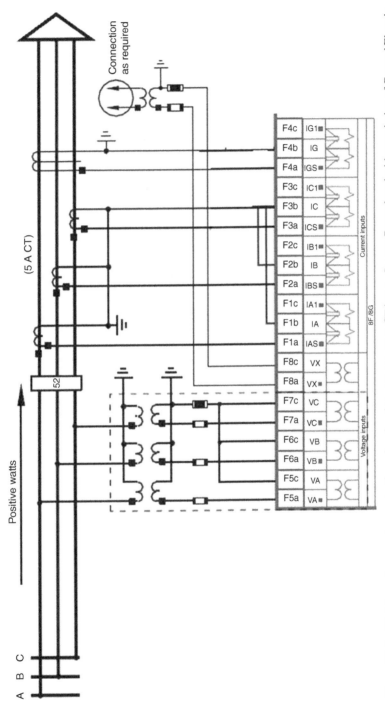

Figure 21.4 High impedance system protection relay showing zero sequence CT input. *Source:* Reproduced with permission of General Electric Company.

equipment damage and also potential personal injury. Therefore, reactance grounded systems are rarely used in power plants.

SOLIDLY GROUNDED SYSTEM

The fourth type of grounding system is the solidly grounded system where the neutral of the source transformer is directly connected to system ground. In this design, system ground loop impedance and the impedance of the fault are the only limitations to current during a phase-to-ground fault. The protection system senses the large currents that flow from phase-to-ground during a ground fault event and immediately removes the feeder with the phase-to-ground fault. This arrangement results in a less reliable electrical system than the ungrounded or high impedance grounded systems, since it removes the circuit with the ground fault immediately and does not allow for equipment to remain energized to provide operations time to find and correct phase-to-ground fault and shut down affected equipment in a control fashion. For facilities under the jurisdiction of the National Electrical Code® (NEC®) installation requirements, a low voltage system is required to be solidly grounded. However, for power plant facilities that fall under the National Electrical Safety Code (NESC) installation requirements, the lack of reliability of this system makes its use on power plant facilities utilized only in limited areas where reliability is not an issue or areas not directly associated with the generation, transmission, or distribution functions of the utility.

The grounding system affects the cable selection process for the medium voltage electrical distribution system. There are three basic levels of conductor insulation for medium voltage cables, 100%, 133%, and 173%, that are required to be used depending on which grounding system is utilized and the expected time to clear the ground fault.

- For solidly grounded systems where the phase-to-ground fault is immediately removed from the energized electrical distribution system, the minimum insulation level required is the 100% level.
- For impedance grounded systems where the fault will not initially be removed but will be removed within 1 hr, the 133% insulation level should be used as minimum.
- For impedance grounded systems where the fault will not initially be removed and may remain on the electrical distribution system for periods exceeding 1 hr, the 173% insulation level should be used as minimum.
- Ungrounded systems require the use of a 173% insulation level cable system.

In addition to the grounding or earthing system design as discussed above, all electrically conductive equipment in a power plant facility is bonded both to the local ground grid system as well as to the *grounding conductor* routed with the power cables in the cable tray or conduit.

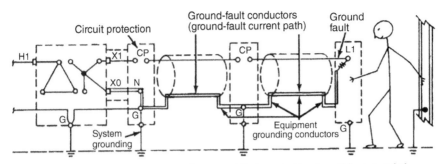

Figure 21.5   Grounding arrangement for ground-fault protection in solidly grounded systems. *Source*: Reproduced with permission of IEEE.

The function of the *bonding* connections between conductive structures and equipment and the ground system is to ensure the touch and step potentials are minimized to safe levels. The system is designed to maintain the touch potential between two conductive surfaces at a minimum potential as well as the step potential between a conductive surface and the walking surface as shown in Figure 21.5 (for solidly grounded systems), Figure 21.6 (for resistance grounded systems), and Figure 21.7 (for ungrounded systems).

IEEE 80: Guide for Safety in AC Substation Grounding provides engineering guidance with the design calculations for bonding of equipment at a utility generation station for personnel safety. The formula for calculating the resistance of a ground grid is defined by IEEE 80 as

$$R_g = \frac{\rho}{4}\sqrt{\frac{\pi}{A}} + \frac{\rho}{L} \tag{21.3}$$

where

ρ = resistivity of the soil (Ωm)

A = the area of the grid (m^2)

L = total length of the buried conductors (m)

Refer IEEE 80.

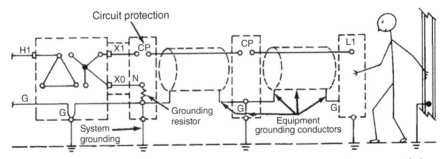

Figure 21.6   Grounding arrangement for ground-fault protection in resistance grounded systems. *Source*: Reproduced with permission of IEEE.

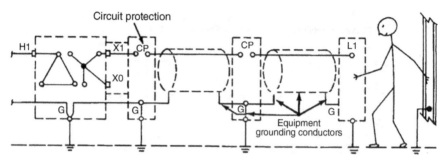

Figure 21.7   Grounding arrangement for ground-fault protection in ungrounded systems.
Source: Reproduced with permission of IEEE.

Example 21.3   A power plant ground grid design with a total length of buried bare copper conductor of 40 meters formed in such a fashion to provide a grid area 100 m², is buried in soil with a resistivity of the soil is 50 Ωm. Calculate the expected ground resistance of this ground grid.

Solution: Using equation (21.3), we find the ground grid resistance to be as follows.

$$R_g = \frac{\rho}{4}\sqrt{\frac{\pi}{A}} + \frac{\rho}{L}$$

$$R_g = \frac{50\ \Omega m}{4}\sqrt{\frac{\pi}{100\ m^2}} + \frac{50\ \Omega m}{40\ m}$$

$$R_g = 0(2.215\ \text{ohms}) + (1.25\ \text{ohms}) = 3.465\ \text{ohms}$$

From IEEE 80, the limits to the maximum touch and step potentials allows in the design of the ground grid system is a function of the fault clearing time and the soil surface layer resistivity as shown in Table 21.1.

Example 21.4   What is the step and touch voltage potential limits for a system that will clear a fault in six cycles (based on a 60 Hz system) if the surface layer resistivity is found to be 500 Ωm? Please use Table 21.1 for this calculation.

Solution: First, we need to identify the clearing time in seconds since the problem gave us clearing time in cycles. This is calculated utilizing equation (21.4).

$$\text{Clearing time (sec)} = \text{clearing time (cycles)}/f \qquad (21.4)$$

$$\text{Clearing time (sec)} = 6\ \text{cycles}/60\ \text{Hz} = 0.1\ \text{sec}$$

From Table 21.1, for a clearing time of 0.1 sec and a surface layer resistivity of 500 Ωm, we find the maximum step potential to be 1162 V, while the maximum touch potential is 513.8 V.

Example 21.4 is simplified by the fact that the surface layer resistivity was given, but in practical application, the actual surface layer resistivity is calculated. In power generation stations, rock or some other material with a high resistivity is

TABLE 21.1   Step and Touch Potential Limits from IEEE 80

Surface Layer Resistivity	Fault Clearing Time					
	0.100 sec		0.200 sec		0.300 sec	
	Step Voltage	Touch Voltage	Step Voltage	Touch Voltage	Step Voltage	Touch Voltage
[Ωm]	[V]	[V]	[V]	[V]	[V]	[V]
None	985.0	469.5	763.0	363.7	645.9	307.9
500	1162.0	513.8	900.1	398.0	762.0	336.9
1000	1802.3	673.9	1396.1	522.0	1181.9	441.9
1500	2422.8	829.0	1876.7	642.1	1588.8	543.6
2000	3036.8	982.5	2352.4	761.1	1991.5	644.3
2500	3647.9	1135.3	2825.8	879.4	2392.2	744.5
3000	4257.5	1287.7	3298.0	997.5	2792.0	844.4
3500	4866.2	1439.8	3769.5	1115.3	3191.2	944.2
4000	5474.3	1591.9	4240.6	1233.1	3590.0	1043.9
4500	6082.0	1743.8	4711.3	1350.8	3988.5	1143.5

*Source: Reproduced with permission of IEEE.

used which increases the resistance between the person and earth ground potential. This has the effect of increasing the allowable tolerable touch and step potentials for a person. The formulas for the maximum allowable touch and step potentials for persons with (50 kg) and (70 kg) weights are shown below. In this text, assume a value of 1.0 for C_s (h_s, K). This is a reasonable estimate if no surface material is applied to increase the surface material resistance. If surface material is used, then refer to IEEE 80 or IEEE 665 for determination of the correct value for C_s (h_s, K).

It is beneficial to use a surface material with high resistivity to allow for larger values of allowable touch and step potentials as this reduces the requirements of the ground grid design required to meet these values.

The touch and step potential limits for a person weighing 110 lb (50 kg) are calculated utilizing equations (21.5) and (21.6).

$$E_{step50} = [1000 + 6C_s(h_s, K)\rho_s][0.116/\sqrt{(t_s)}] \tag{21.5}$$

$$E_{touch50} = [1000 + 1.5C_s(h_s, K)\rho_s][0.116/\sqrt{(t_s)}] \tag{21.6}$$

where

1000 is the value of body resistance (ohms)

1.5 is the factor for the resistance of two feet in parallel

6 is the factor for the resistance of two feet in series

C_s (h_s, K) is 1 if there is no protective surface later (refer IEEE 665 or IEEE 80 if surface layer is protective layer of rock.)

ρ_s is the wet resistivity of the surface rock

t_s is the shock duration (sec)

Example 21.5   What is the step and touch voltage potential limits for a system that will clear a fault in six cycles (based on a 60 Hz system) if the surface layer resistivity is found to be 1000 Ωm? Assume that there is no protective surface layer in this application. Please use equations (21.5) and (21.6) for these calculations. Assume a weight of 110 lb (50 kg).

Solution: Given that there is not protective surface layer in this application, we can use equation (21.5) to calculate the maximum step potential voltage for a person that weighs 110 lb.

$$E_{step50} = [1000 + 6C_s(h_s, K)\rho_s][0.116/\sqrt{(t_s)}]$$

$$E_{step50} = [1000 + 6(1)1000 \, \Omega m][0.116/\sqrt{0.1}]$$

$$E_{step50} = [1000 + 6000][0.03668]$$

$$E_{step50} = 257 \text{ V}$$

The maximum touch potential voltage for a person that weighs 110 lb can be found using equation (21.6).

$$E_{touch50} = [1000 + 1.5C_s(h_s, K)\rho_s][0.116/\sqrt{(t_s)}]$$

$$E_{touch50} = [1000 + 1.5(1)1000 \, \Omega m][0.116/\sqrt{0.1}]$$

$$E_{touch50} = [1000 + 1500][0.03668]$$

$$E_{touch50} = 92 \text{ V}$$

The touch and step potential limits for a person weighing 154 lb (70 kg) are described in equations (21.7) and (21.8).

$$E_{step70} = [1000 + 6C_s(h_s, K)\rho_s][0.157/\sqrt{(t_s)}] \tag{21.7}$$

$$E_{touch70} = [1000 + 1.5C_s(h_s, K)\rho_s][0.157/\sqrt{(t_s)}] \tag{21.8}$$

Example 21.6   What is the step and touch voltage potential limits for a system that will clear a fault in six cycles (based on a 60 Hz system) if the surface layer resistivity is found to be 1000 Ωm? Assume that there is no protective surface layer in this application. Please use equations (21.7) and (21.8) for these calculations. Assume a weight of 154 lb (70 kg).

Solution: Given that there is not protective surface layer in this application, we can use equation (21.7) to calculate the maximum step potential voltage for a person that weighs 154 lb.

$$E_{step70} = [1000 + 6C_s(h_s, K)\rho_s][0.157/\sqrt{(t_s)}]$$

$$E_{step70} = [1000 + 6(1)1000 \, \Omega m][0.157/\sqrt{0.1}]$$

$$E_{step70} = [1000 + 6000][0.04965]$$

$$E_{step70} = 348 \text{ V}$$

The maximum touch potential voltage for a person that weighs 154 lb can be found using equation (21.8).

$$E_{\text{touch}70} = [1000 + 1.5C_s(h_s, K)\rho_s][0.157/\sqrt{(t_s)}]$$
$$E_{\text{touch}70} = [1000 + 1.5(1)1000\,\Omega\text{m}][0.157/\sqrt{0.1}$$
$$E_{\text{touch}70} = [1000 + 1500][0.04965]$$
$$E_{\text{touch}70} = 124 \text{ V}$$

The function of the grounding conductor that is routed from the power source to the utilization equipment is to minimize the ground potential rise between the source and the load during a phase-to-ground fault. The inductance of the path that ground current flows is a logarithmic function related to the distance the ground path is in relation to the phase conductors. The inductance of the ground path is defined in equation (21.9).

$$L = 4 \times 10^{-7} \ln(D/r')\,\text{H/m} \tag{21.9}$$

where

L = inductance of the loop between the phase conductors and the ground return path (H/m)

D = distance between phase conductor and ground path

r' = geometric mean radius (GMR) of conductor

Looking at this equation, if we minimize the distance between the phase conductor and the ground return path, we minimize the inductance of the ground loop. By minimizing the inductance of the ground loop, we minimize the impedance of the ground loop since the impedance of the ground loop is given by the equation in (21.10).

$$jX_g = j2\pi fL \; (\Omega/m) \tag{21.10}$$

where

X_g = ground path impedance (Ω/m)

f = system frequency (Hz)

L = inductance of the ground path. (H/m)

Example 21.7 Given a ground conductor that is 4/0 conductor with an effective radius of 11.684 mm and a distance between the ground conductor and phase conductor if 1 m (1000 mm), calculate the inductance of the ground loop in H/m and the impedance of the ground loop in ohms per meter to a phase-to-ground fault. Assume a system frequency of 60 Hz.

Solution: Using equation (21.9) we find the inductance of the loop per linear meter to be the following.

$$L = 4 \times 10^{-7} \ln(D/r')\,\text{H/m}$$
$$L = 4 \times 10^{-7} \ln(1000 \text{ mm}/11.684 \text{ mm})\,\text{H/m}$$
$$L = 4 \times 10^{-7}\ln(85.6)\,\text{H/m}$$

$$L = 4 \times 10^{-7}(4.45) \text{ H/m}$$
$$L = 1.8 \times 10^{-6} \text{ H/m}$$
$$L = 1.8 \text{ mH/m}$$

Using equation (21.10) we find the inductive reactance of the loop per linear meter to be the following.

$$jX_g = j2\pi fL \ (\Omega/\text{m})$$
$$jX_g = j2\pi(60 \text{ Hz})(1.8 \times 10^{-6} \text{ H/m})(\Omega/\text{m})$$
$$jX_g = j2\pi(60 \text{ Hz})(1.8 \times 10^{-6} \text{ H/m})(\Omega/\text{m})$$
$$jX_g = j0.000671 \ (\Omega/\text{m})$$
$$jX_g = j671 \ (\text{m}\Omega/\text{m})$$

Example 21.8 Given a ground conductor that is 4/0 conductor with an effective radius of 11.684 mm and a distance between the ground conductor and phase conductor if 1 in (25.4 mm), calculate the inductance of the ground loop in H/m and the impedance of the ground loop in ohms/m to a phase-to-ground fault. Assume a system frequency of 60 Hz.

Solution: Using equation (21.9) we find the inductance of the loop per linear meter to be the following.

$$L = 4 \times 10^{-7}\ln(D/r') \text{ H/m}$$
$$L = 4 \times 10^{-7}\ln(25.4 \text{ mm}/11.684 \text{ mm}) \text{ H/m}$$
$$L = 4 \times 10^{-7}\ln(2.174) \text{ H/m}$$
$$L = 4 \times 10^{-7}(0.777) \text{ H/m}$$
$$L = 0.3 \times 10^{-6} \text{ H/m}$$
$$L = 0.3 \text{ mH/m}$$

Using equation (21.10), we find the inductive reactance of the loop per linear meter to be the following.

$$jX_g = j2\pi fL \ (\Omega/\text{m})$$
$$jX_g = j2\pi \ (60 \text{ Hz})(0.3 \times 10^{-6} \text{ H/m})(\Omega/\text{m})$$
$$jX_g = j2\pi \ (60 \text{ Hz})(0.3 \times 10^{-6} \text{ H/m})(\Omega/\text{m})$$
$$jX_g = j0.000117 \ (\Omega/\text{m})$$
$$jX_g = j117 \ (\text{m}\Omega/\text{m})$$

By minimizing the impedance of the ground loop, for the same magnitude of fault current, we minimize the potential rise from source to load during a fault and, thereby, minimize the ground potential rise from source to load.

To achieve a good connection to ground, multiple grounding electrodes or grounding rods may be utilized. The distance between the rods should be the sum of the length of the ground rods.

$$\text{Distance between rods} = L_{r1} + L_{r2} \tag{21.11}$$

Example 21.9 If we are to use two ground rods with a length of 8 ft for our ground connection to our ground grid, what is the minimum distance between ground rods recommended?

Solution: Using equation (21.11), we find the correct distance between the ground rods to be as follows.

$$\text{Distance between rods} = L_{r1} + L_{r2}$$
$$\text{Distance between rods} = 8 \text{ ft} + 8 \text{ ft}$$
$$\text{Distance between rods} = 16 \text{ ft}$$

The maximum allowable ground resistance at generation facilities is substantially smaller than for industrial plants or residential locations. This is due to fact that the source impedance is smaller at generation facilities so the fault currents that can occur at generation stations are substantially higher than values found in most industrial locations. Additionally, due to very low values of zero sequence impedance, it is not uncommon that, at the utility site, the phase-to-ground fault currents may equal or exceed the three phase fault currents. In most industrial locations, the phase-to-ground fault current is only a fraction of the three phase fault values.

The recommended maximum allowable ground resistance for *generating stations* and substations is 1 Ω. For comparison, the recommended maximum allowable ground resistance at industrial facilities is 5 Ω and the recommended maximum allowable ground resistance at residential facilities per the NEC® is 25 Ω.

GLOSSARY OF TERMS

- Bonding – The permanent joining of metallic parts to form an electrically conductive path that will ensure electrical continuity and the capacity to conduct safely any current likely to be imposed.
- Bonding Jumper – A reliable conductor to ensure the required electrical conductivity between metal parts that need to be electrically connected.
- Equipment Grounding Conductor – The conductor used to connect the noncurrent-carrying metal parts of equipment, raceways, and other enclosures to the service equipment, the service power source(s) ground, or both.
- Four-Wire System – A three-phase system consisting of three-phase conductors and a neutral conductor.
- Generating (Energy Production) Station – A plant wherein electric energy is produced by conversion from some other form of energy (for example, chemical, nuclear, solar, mechanical, or hydraulic) by means of suitable apparatus. This includes all generating station auxiliaries and other associated equipment required for the operation of the plant. Not included are stations producing power exclusively for use with communications systems.
- Grid Mesh – Any one of the open spaces enclosed by the grounding grid conductors.

- Ground Grid – A system of horizontal ground electrodes that consists of a number of interconnected bare conductors buried in the earth, providing a common ground for electrical devices or metallic structures, usually in one specific location. *Note*: Grids buried horizontally near the surface of the earth are also effective in controlling the surface potential gradients. A typical grid usually is supplemented by a number of ground rods and may be further connected to auxiliary ground electrodes to lower its resistance with respect to remote earth.

- Ground Well – A hole with a diameter greater than an inserted ground rod, drilled to a specified depth, and backfilled with a highly conductive material. The backfill will be in intimate contact with the earth.

- Grounding Conductor – A conductor used to connect equipment or the grounded circuit of a wiring system to a grounding electrode or electrodes (i.e., ground grid).

- Neutral Ground – An intentional ground applied to the neutral conductor or neutral point of a circuit, transformer, machine, apparatus, or system.

- Overhead Ground Wire – A grounded, bare conductor suspended horizontally between supporting rods or masts to provide protection from lightning strikes for structures, equipment, or suspended conductors within the *zone of protection* created by the combination of the masts and the overhead ground wire.

- Safety Ground – The connection between a grounding system and metallic parts that are not usually energized but that may become live due to a fault or an accident; often referred to as equipment or frame ground.

- Substation – An enclosed assemblage of equipment (e.g., switches, circuit breakers, buses, and transformers) under the control of qualified persons, through which electric energy is passed for the purposes of switching or modifying its characteristics.

- System Ground – The connection between a grounding system and a point of an electric circuit (for example, a neutral point).

- Transferred Voltage – That voltage between points of contact, hand to foot or feet, where the grounded surface touched is intentionally grounded at a remote point (or unintentionally touching at a remote point a conductor connected to the station ground system). Here the voltage rise encountered due to ground fault conditions may equal or exceed the ground potential rise of the ground grid discharging the fault current (and not a fraction of this total as is encountered in the usual touch contact).

- Zone of Protection – The adjacent space provided by a grounded air terminal, mast, or overhead ground wire that is protected against most direct lightning strikes.

PROBLEMS

21.1 A power plant ground grid design with a total length of buried bare copper conductor of 4 m formed in such a fashion to provide a grid area 1 m² is buried in soil which has a resistivity 100 Ωm. Calculate the expected ground resistance of this ground grid.

21.2   What is the step and touch voltage potential limits for a system that will clear a fault in 12 cycles (based on a 60 Hz system) if the surface layer resistivity is found to be 2000 Ωm? Assume that there is no protective surface layer in this application. Please use equations (21.5) and (21.6) for these calculations. Assume a weight of 110 lb (50 kg).

21.3   What is the step and touch voltage potential limits for a system that will clear a fault in 12 cycles (based on a 60 Hz system) if the surface layer resistivity is found to be 2000 Ωm? Assume that there is no protective surface layer in this application. Please use equations (21.7) and (21.8) for these calculations. Assume a weight of 154 lb (70 kg).

21.4   Given a ground conductor that is 1/0 conductor with an effective radius of 8.252 mm and a distance between the ground conductor and phase conductor is 1 m, calculate the inductance of the ground loop in henneries per meter and the impedance of the ground loop in ohms per meter to a phase-to-ground fault. Assume a system frequency of 60 Hz.

21.5   If we are to use two ground rods with a length of 10 ft for our ground connection to our ground grid, what is the distance between ground rods recommended?

RECOMMENDED READING

ANSI C2-1993, National Electrical Safety Code.

Electric Power Plant Design, Technical Manual TM 5-811-6, Department of the Army, USA, 1984.

Electrical Machines, Drives and Power Systems, 6th edition, Theodore Wildi, Prentice Hall, 2006.

The Engineering Handbook, 3rd edition, Richard C. Dorf, (editor in chief), CRC Press, 2006.

Emerald Book IEEE 1100-2005: IEEE Recommended Practice for Powering and Grounding Electronic Equipment.

Green Book IEEE 142-2007: IEEE Recommended Practice for Grounding of Industrial and Commercial Power Systems.

IEEE 80-2000: IEEE Guide for Safety in AC Substation Grounding.

IEEE 81-2012: IEEE Guide for Measuring Earth Resistivity, Ground Impedance, and Earth Surface Potentials of a Ground System (Part 1).

IEEE 487-1992: IEEE Recommended Practice for the Protection of Wire-Line Communication Facilities Serving Electric Power Stations (ANSI).

IEEE 665-1995: IEEE Guide for Generating Station Grounding.

IEEE 666-1991: IEEE Design Guide for Electric Power Service Systems for Generating Stations.

IEEE 1050-1989: IEEE Guide for Instrumentation and Control Equipment Grounding in Generating Stations.

IEEE C37.101-2006: IEEE Guide for Generator Ground Protection.

IEEE C62.92-2005 (series): IEEE Guide for the Application of Neutral Grounding in Electrical Utility Systems.

Industrial Power Distribution, 2nd edition, Ralph E. Fehr, III, Wiley-IEEE Press, 2016. ISBN: 978-1-119-06334-6.

NFPA 70®-2014 – National Electrical Code®, National Fire Protection Association.

NFPA 77-1993, Static Electricity.

NFPA 780-1992, Lightning Protection Code.

Power Plant Engineering, Black & Veatch, edited by Larry Drbal, Kayla Westra, and Pat Boston, Chapman & Hall / Springer, 1996.

Red Book IEEE 141-1993: IEEE Recommended Practice for Electric Power Distribution for Industrial Plants, 1993.

Standard Handbook of Powerplant Engineering, 2nd edition, Thomas C. Elliott, McGraw-Hill, 1998.

CHAPTER 22

ELECTRICAL SYSTEM PROTECTION AND COORDINATION

GOALS

- To understand the fundamental concepts of reliability, security, selectivity, speed, simplicity, and economics as they pertain to proper electrical system protection and coordination
- To define the most common protective relays functions utilized in an energy production facility distribution system
- To calculate the relay activation time, given a standard time-current curve (TCC) and value of relay pickup and delay setting
- To understand the application of subtransient, transient, and steady state reactance values
- To calculate the equivalent three-phase wye impedance values, given a three-phase delta circuit or vice versa
- To perform per unit calculations
- To identify correct and incorrect coordination characteristics on a TCC plot

THE PROTECTION system provides for the oversight of the distribution system and its purpose is to remove faulted equipment from the electrical distribution system while keeping unaffected equipment energized. Therefore, the function of protective relaying has two distinct components.

The first function of the electrical protection system is to reliably remove a faulted piece of equipment from an electrical system as quickly as possible to minimize the damage to the equipment during the fault condition and the incident energy level at the faulted location to minimize the personal injury hazard associated with the incident energy level at a faulted piece of equipment. The second function of the

Energy Production Systems Engineering: An Introduction for Electrical Engineers to Electrical Power Generation Facilities, Systems, and Equipment, First Edition. Thomas H. Blair.
© 2017 by The Institute of Electrical and Electronics Engineers, Inc. Published 2017 by John Wiley & Sons, Inc.

electrical protection system is *not* to remove equipment that is not faulted to ensure the reliable operation of the electrical distribution system. Said briefly, the electrical protection system function is to trip when fault condition is present AND not to trip when fault condition not present. Based on the above functions of the electrical protection system, several concepts are developed. Protective relaying is always a balancing act between these concepts.

1. **Reliability** – relay system should trip when fault exists in protective zone
2. **Security** – relay system should only trip when fault exists in protective zone
3. **Selectivity** – relay system should trip minimum equipment to remove fault
4. **Speed** – relay system should remove a fault fast to minimize damage and minimize personnel hazard due to incident energy level
5. **Simplicity** – minimum amount of equipment—maximize reliability
6. **Economics** – reasonable cost

The most common protective relays in the electrical distribution system of a power plant facility are the following relays:

27 = undervoltage relay

49 = thermal relay

50 = instantaneous overcurrent relay

51 = time delay overcurrent relay

59 = overvoltage relay

81U = underfrequency relay

81O = overfrequency relay

87 = differential current relay

See Annex B for a complete list of IEEE device numbers and functions.

The *undervoltage relay* (27), as shown in Figure 22.1, receives an input from a PT input to monitor system voltage. Upon detection of a loss of potential, this relay activates to perform some protective function or possibly to close an alternate source of power to restore voltage to the de-energized bus. See the section on motor bus transfer in Chapter 17 for more details on re-energization of a de-energized bus that has motors and the precautions needed. Another common function of an undervoltage relay is to automatically open any mechanically latching motor controls. This is done so that, when the bus is re-energized, the motors that are fed from latching motor controllers do not start up in an uncontrolled fashion due to the fact that the motor control device remains closed during the bus undervoltage.

The *thermal relay* (49) shown in Figure 22.2 receives a current input from a current transformer CT secondary to monitor the current flowing to a piece of equipment. Since in most electrical equipment, the temperature rise is due to I^2R losses, the overload can estimate the temperature of the electrical equipment from a measurement of the current fed to the equipment. Frequently, the final load of a circuit with an overload relay is a motor. For an overload condition where the thermal limits of the motor are exceeded, the thermal relay operates a control element such as a motor starter to remove power from the motor and protects the motor from thermal damage.

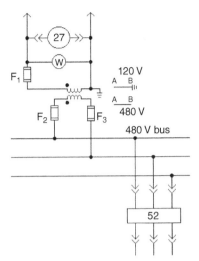

Figure 22.1   Device 27, undervoltage relay. *Source*: Reproduced with permission of Tampa Electric Company. Reproduction is forbidden without the express consent of Tampa Electric Company.

The *instantaneous overcurrent relay* (50) shown in Figure 22.3 receives an input from a current transformer CT to monitor the current flowing to a feeder. If current exceeds the set point of the relay it acts instantaneously to de-energize the feeder. For coordination purposes, it is assumed that the maximum delay due to relay activation time is one cycle and this is added to the operating time of the final control

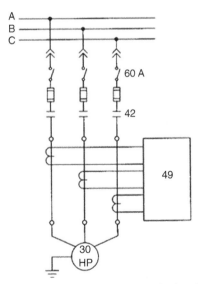

Figure 22.2   Device 49, thermal relay. *Source*: Reproduced with permission of Tampa Electric Company. Reproduction is forbidden without the express consent of Tampa Electric Company.

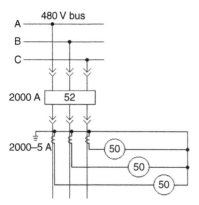

Figure 22.3   Device 50, instantaneous overcurrent relay. *Source*: Reproduced with permission of Tampa Electric Company. Reproduction is forbidden without the express consent of Tampa Electric Company.

element such as a power circuit breaker. For the instantaneous overcurrent relay, there is one basic adjustment. The user can adjust the current required to cause the relay to activate.

For motor circuits the instantaneous overcurrent relay should not activate on a motor overload condition, including locked rotor during startup. The instantaneous relay pickup should be set at a value in excess of the peak motor locked rotor current. It is possible for the motor to draw 173% of nameplate locked rotor current for the first cycle due to DC offsets that may occur depending on when in the voltage waveform, the motor starter contactor closes. IEEE C37.96: IEEE Guide for AC Motor Protection provides guidance for the proper considerations of motor circuit protection. If the inrush current is not accurately known, then an additional 10–25% is usually added as a safety factor when settings are calculated. Finally, a relay tolerance (usually 10%) is then added to arrive at the final setting. It has been the author's experience that a range of 200–250% of motor locked rotor current is a secure and reliable value for the pickup setting of the instantaneous relay in motor protection applications.

The time delay overcurrent relay (51) shown in Figure 22.4 receives an input from a current transformer CT to monitor the current flowing to a feeder. If the current exceeds the set point of the relay for a given amount of time, it actuates to de-energize the feeder. The time for a relay to activate is dependent on the amount of current flowing and can be derived from the relay time-current curve provided by the manufacturer of the relay. Figure 22.5 shows a typical TCC curve for an *inverse time overcurrent relay*.

Notice from Figure 22.5, there are two basic adjustments to the device 51 relay. The user can adjust the current required to start the relay timing. This is called the *pickup setting*. The second setting, called the *time dial* setting, adjusts the amount of time it takes once the current in the relay exceeds the pickup level until the relay trip contact activates. To achieve coordination with other devices in the circuit, a different characteristic relay protection curve may be required. IEEE defines several standard characteristic curves inverse, very inverse, and extremely inverse and three of these curves are shown in Figure 22.6.

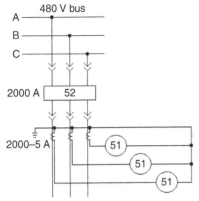

Figure 22.4 Device 51, time delay overcurrent relay. *Source*: Reproduced with permission of Tampa Electric Company. Reproduction is forbidden without the express consent of Tampa Electric Company.

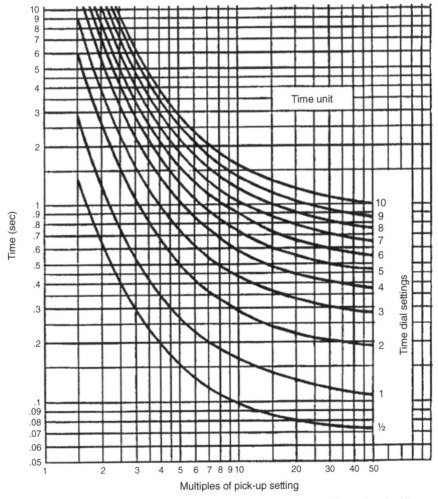

Figure 22.5 Time-current curve (TCC) for time delay relay. *Source*: Reproduced with permission of IEEE.

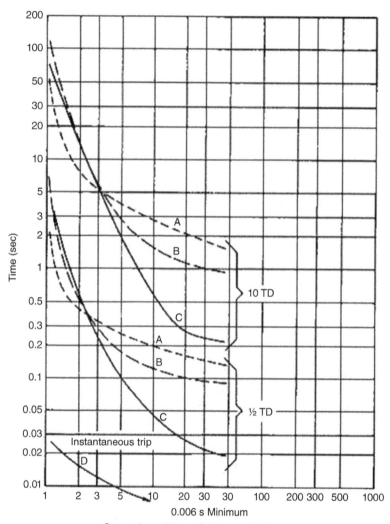

Figure 22.6 Typical relay time-current characteristics. *Source*: Reproduced with permission of IEEE.

A special application of the device 51 time delay overcurrent relay is in the sensing of ground faults on electrical systems as is shown in Figure 22.7. For high impedance systems, ground faults that have low values of current associated with them that may not be detected by the phase overcurrent relays due to the high primary current rating of the current transformer. In this instance, a current transformer with a

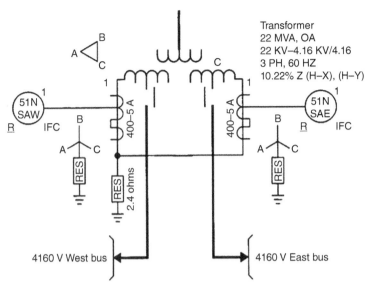

Figure 22.7 Device 51N, neutral-ground fault time delay overcurrent relay. *Source*: Reproduced with permission of Tampa Electric Company. Reproduction is forbidden without the express consent of Tampa Electric Company.

lower primary current rating can be placed in the neutral of the transformer feeding the electrical system and a time delay overcurrent relay can be used to detect this condition and de-energize the electrical system or set an alarm an annunciator in the control room in the event of a phase-to-ground fault. Another method to detect a ground fault would be the use of an *overvoltage relay* device 59G across the neutral grounding resistor. The following section describes the function of the overvoltage relay.

The overvoltage relay (59) receives an input form a potential transformer PT input to monitor feeder voltage as is shown in Figure 22.8. If the set point is exceeded, the relay trips the breaker to de-energize the feeder and protect the electrical equipment on the feeder from the overvoltage condition.

The frequency relay (*overfrequency* (81O) and *underfrequency* (81U)) that is shown in Figure 22.9 receives an input form a potential transformer PT input to monitor feeder frequency. One application of a device 87U relay is to use it along with a device 59 relay to provide volts-to-hertz protection for electrical machines with iron cores. Examples of possible applications are motors, generators, and transformers. The electrical machine cores are susceptible to saturation and damage if the ratio of applied voltage to applied frequency exceeds an established design limit. Should the frequency relay detect this condition, the relay trips the breaker to de-energize the feeder and protect the electrical equipment on the feeder from the adverse frequency condition. Another application of the frequency relay is for use on electrical distribution systems that are fed from generators and not paralleled with an electric utility system. The over- and underfrequency relays can provide operating personnel an alarm to indicate potential issues with the generator governor control system, allowing the personnel time to investigate and correct the cause.

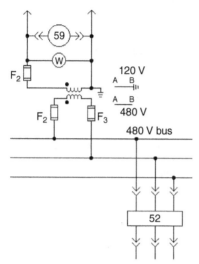

Figure 22.8 Device 59, overvoltage relay. *Source*: Reproduced with permission of Tampa Electric Company. Reproduction is forbidden without the express consent of Tampa Electric Company.

The *differential relay* (87) receives an input from a CT on all sides of a piece of equipment for each phase. The relay compares the current flowing into the equipment on one side of a winding with the current flowing out of the equipment on the other winding on each phase. If the current entering the equipment matches the current leaving the equipment, then the currents are balanced and no fault exists on the equipment. In Figure 22.10, in this condition, the value of the restraint current I_1 is equal to the value of the restraint current I_2 and the value of the operate current I_0 is zero.

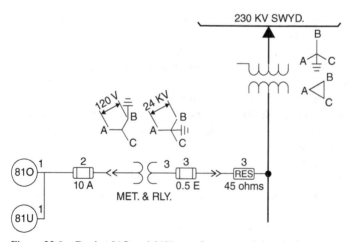

Figure 22.9 Device 81O and 81U, overfrequency and underfrequency relay. *Source*: Reproduced with permission of Tampa Electric Company. Reproduction is forbidden without the express consent of Tampa Electric Company.

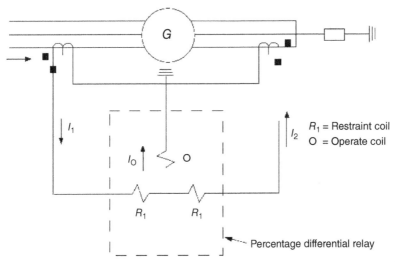

Figure 22.10　Device 87, differential relay. *Source*: Reproduced with permission of IEEE.

If a fault occurs in the equipment, then the amount of current entering the equipment will not match the amount of current leaving the equipment. The differential relay detects the current flowing into the equipment and not returning on the other side of the winding and trips the source to the equipment. In Figure 22.10, in this condition, the value of the operate current I_0 is the difference between the restraint current I_1 and the restraint current I_2. When the ratio of the operate current I_0 to restraint current I_1 exceeds a set amount, the relay will operate. Differential relays will ignore faults external to the machine assuming the CT does not undergo saturation due to the high magnitude of external fault currents. Special care needs to be taken with the orientation of the polarity of the current transformers and the OEM manual should be consulted for correct orientation of the polarity of the current transformers.

To determine the types of protective relays needed for protection of an electrical system and the settings for the protective relays, a short-circuit study must be completed of the electrical system. The short-circuit study also provides information needed for selecting adequate momentary ratings and interrupt ratings of the equipment in the electrical system. Part of the job of the electrical engineer at a power generation facility is to keep the system documentation up to date with changes and this includes the plant short-circuit analysis.

Generators and motors can be modeled as ideal voltage sources with series impedance values that define the available short-circuit current the generator can produce as is shown in Figure 22.11.

As discussed in Chapter 16 on generators, the value of the reactance of the synchronous generator (and synchronous motor) is not a steady value but varies from a low value at the start of a fault on the electric machine to a larger value at a time several seconds into the fault.

X''_d = subtransient reactance ($0 < t < 0.2$ sec)

X'_d = transient reactance ($0.2 < t < (0.5–2.0$ sec))

X_d = synchronous reactance (($0.5–2.0$ sec) $< t$), steady state

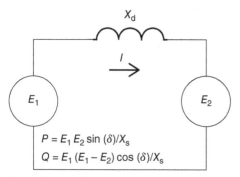

Figure 22.11   Simplified one line diagram of generator (E_1), generator reactance (X_d) and system (E_2).

Some generator data sheets will provide two values for generator subtransient reactance based on test conditions. The value X''_{dv} is the generator subtransient reactance at rated voltage. When the machine is at rated voltage and the machines leads undergo a fault condition, the magnitudes of stator currents are high. Therefore, the core is under saturation and the value of X''_{dv} is small. The value X''_{di} is the generator subtransient reactance at rated current. To maintain the machine at rated current, the voltage is suppressed. Therefore, the core in not under saturation during this condition. Therefore, the value of X''_{di} is larger than the value of X''_{dv}. To obtain conservative results, the value X''_{dv} should be used in the short-circuit calculations. A typical generator decrement curve that shows the effect of this changing value of reactance of time is shown in Figure 22.12.

Induction machines do provide an initial value of fault current for the first few cycles of a fault on the induction machines terminals, but after about six cycles, the fault current contribution of induction machines decays to zero. This is because the induction machine does not have an externally supplied main field. The induction machine field collapses just a few cycles after the fault and the machine voltage decays accordingly. Because of this, induction machines are provided with subtransient reactance values that define the value of the initial fault current the machine can provide, but they do not have values for transient or steady state reactance.

The utility connection typically does not have a value for subtransient or transient reactance, but is only described by a steady-state or synchronous reactance value. The utility connection can be simplified by being represented by an infinite voltage source with fixed impedance defined by the steady-state impedance.

For low voltage systems, both the resistance and reactance of the system is included in the simplified model as the resistance can be a substantial part of the overall system impedance. Also, a fault on a low voltage system may have some arc resistance associated with it. To determine maximum short-circuit values though, the resistance of the arc is assumed to be zero. However, for medium and high voltage systems, the resistance of the system is a very small part of the overall impedance and these values can be ignored to obtain conservative values of short-circuit currents.

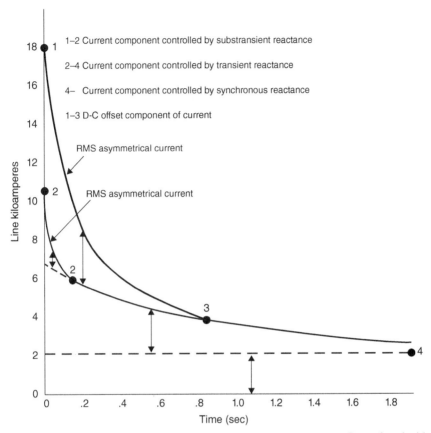

Figure 22.12 Typical synchronous generator decrement curve. *Source*: Reproduced with permission of IEEE.

The procedure for performing a basic short-circuit study is as follows:

Step 1: Prepare system diagrams

Step 2: Collect and convert impedance data

Step 3: Combine impedances

Step 4: Calculate short-circuit current

In a three-phase system, sometimes delta-connected loads need to be converted to equivalent wye-connected loads to determine the single-phase equivalent circuit for the model. The formula to convert a delta-connected load to an equivalent wye-connected load is shown in Figure 22.13 and described as follows.

$$a = (BC)/(A + B + C) \qquad (22.1)$$

$$b = (AC)/(A + B + C) \qquad (22.2)$$

$$c = (AB)/(A + B + C) \qquad (22.3)$$

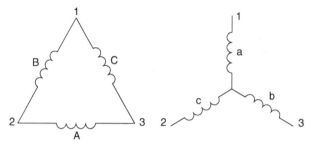

Figure 22.13 Delta-to-wye conversion. *Source*: Reproduced with permission of IEEE.

For completeness, the formula to convert a wye-connected load to an equivalent delta-connected load is shown in Figure 22.14.

$$A = (bc/a) + b + c \qquad (22.4)$$
$$B = (ac/b) + a + c \qquad (22.5)$$
$$C = (ab/c) + a + b \qquad (22.6)$$

Once the electrical system has been modeled as an equivalent one-line system, the values are converted to *per unit* values. This simplifies the model by elimination of the turns ratio of the transformers in the system and reduces these devices to simple impedances and makes the calculations of the short-circuit study simpler (and less prone to errors). Two values are assumed at one point in the circuit as base values. For this instance, the base power rating (base kVA) and the base voltage (base kV) are assumed. These values do not have to be the nominal values of the electrical system, but typically they are. This is done so that the per unit values of power and voltage wind up being 1.0 which are the nominal values at one point in the circuit.

Once the base kVA and base kV values are chosen, the remaining values of base current and base impedance can be calculated. The equations that follow are one method to calculate the values of base current and base impedance. There are several forms to these equations that the reader may come across in other literatures

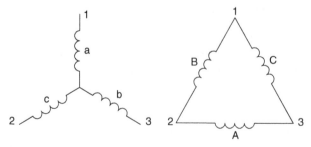

Figure 22.14 Wye-to-delta conversion. *Source*: Reproduced with permission of IEEE.

which are equally valid to utilize. To calculate the value of the base current, given the assumed values of base power and base voltage, use the following formula:

$$I_{base} = MVA_{base}(1{,}000{,}000)/(1.732V_{base}) \qquad (22.7)$$

where

I_{base} = base line current (A)

MVA_{base} = base three-phase total power (MVA)

V_{base} = base voltage phase-to-phase (V)

To calculate the value of the base impedance, given the assumed values of base power and base voltage, use the following formula:

$$Z_{base} = V_{base}/(1.732I_{base}) \qquad (22.8)$$

where

Z_{base} = base impedance (ohms)

I_{base} = base line current (A)

V_{base} = base voltage phase-to-phase (V)

Now that the base value of impedance is determined, the impedance values in ohms in the simplified one-line diagram can be converted to per unit as follows:

$$Z_{pu} = Z_{actual}(\Omega)/Z_{base} \qquad (22.9)$$

where

Z_{pu} = actual impedance value (per unit)

Z_{actual} = actual impedance value (ohms)

Z_{base} = base impedance (ohms)

After the system has been modeled and the fault currents determined, the relay settings can be developed to achieve the primary function of the protective circuit that is to trip the primary control device such as a power circuit breaker when fault conditions are present AND do not trip when fault conditions are not present. To ensure that only the faulted equipment is removed from service, we plot the relay and breaker clearing times on a log/log paper and look at the clearing times for various levels of fault current. This is called a TCC or time-current curve. This plot has values of current on the horizontal axis and values of time on the vertical axis. The relays and associated interrupting devices that are electrically closer to the loads should trip faster than the relays and associated interrupting devices that are electrically closer to the source.

For example, in the TCC shown in Figure 22.15, the one-line diagram on the left shows us that the circuit breaker CB_1 is electrically closer to the source and the circuit breaker CB_2 is electrically closer to the load. Plotting the trip times for both CB_1 and CB_2, we want to ensure that the curve for CB_1 (source side breaker) lies above the curve for CB_2 for all values of fault current possible from the electrical system up to the available short-circuit current. The TCC that is shown in Figure 22.15 shows

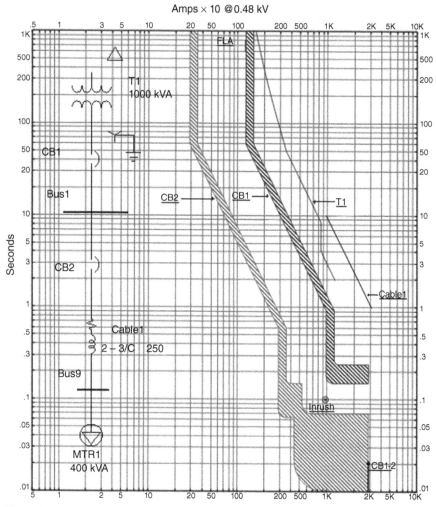

Figure 22.15   Correct TCC Coordination.

properly coordinated breakers. If a fault occurs on the motor MTR_1, for all possible values of current, CB_2 will open before CB_1. This will remove power from the faulted motor but maintain bus 1 energized to allow any other equipment that is fed from bus 1 remain energized and in service.

An example of improperly coordinated relay or breaker settings is shown in Figure 22.16. The curves for CB_1 and CB_2 are coordinated for lower fault currents of values 1000 A or below. However, for fault current values between 1000 A and 3000 A, breaker CT_1 will open in about 200 ms while CB_2 will take 400 ms to open. If this system experienced a fault at the motor MTR_1 and the faulted current value was greater than 1000 A, then breaker CB_1 would open first not only de-energizing the faulted motor but also bus 1. Any equipment connected to bus 1 that is not faulted

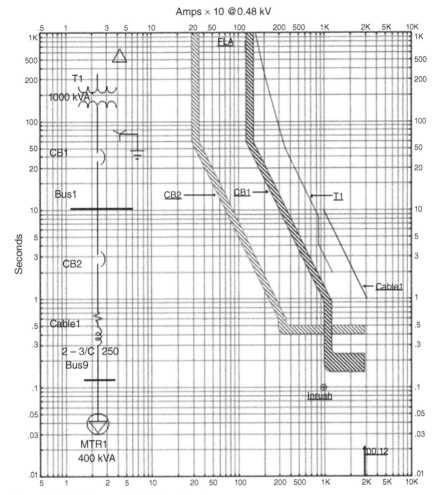

Figure 22.16   Incorrect TCC Coordination.

would also be de-energized. This is an example of improper coordination of the protection system.

GLOSSARY OF TERMS

- Differential Relay (87) – A current-operated relay that receives current from all sides of a piece of protected equipment and compares the currents on all sides of each phase. It produces a trip signal whenever the ratio of the operate current to restraint current exceeds a set amount in the relay.

- Instantaneous Overcurrent Relay (50) – A current-operated relay that produces an instantaneous trip signal when the magnitude of current exceeds a preset value.

- Inverse-Time Overcurrent Relay (51) – A current-operated relay that produces an inverse time-current characteristic by integrating a function of current with respect to time. Pickup current is the current at which integration starts positively and the relay produces an output when the integral reaches a predetermined value.

- Overfrequency Relay (81O) – A frequency-operated relay that produces a trip signal when the magnitude of frequency exceeds a preset value.

- Overvoltage Relay (59) – A voltage-operated relay that produces a trip signal when the magnitude of voltage exceeds a preset value.

- Thermal Overload Relay (49) – A current-operated relay that receives a current signal. The relay converts this current signal into an estimated value of temperature in the protected equipment. The thermal relay produces a trip signal whenever the estimated value of temperature exceeds the set amount in the relay.

- Underfrequency Relay (81U) – A frequency-operated relay that produces a trip signal when the magnitude of frequency drops below a preset value.

- Undervoltage Relay (27) – A voltage-operated relay that produces a trip signal when the magnitude of voltage drops below a preset value.

- Time Dial – The time dial is the control that determines the value of the integral at which the trip output is actuated, and hence controls the time scale of the time-current characteristic produced by the relay. In the induction-type relay, the time dial sets the distance the disk must travel, which is the integral of the velocity with respect to time.

PROBLEMS

22.1 As defined by IEEE, device 87 is a (an)

    A. instantaneous overcurrent relay

    B. time delay overcurrent relay

    C. differential overcurrent relay

    D. lockout device

22.2 Given a load configuration below where the load is configured in a delta and each resistance phase-to-phase is 30 ohms, what is the equivalent line-to-neutral resistance value?

    A. 90 ohms

    B. 10 ohms

    C. 30 ohms

    D. 3 ohms

22.3 Given a system and using 480 V as your base voltage (V_{base}) and using 10 MVA as your base power (S_{base}), what is the base current (I_{base}) for system analysis?

    A. $I_{base} = 12$ MA

B. $I_{base} = 12$ kA

C. $I_{base} = 12$ A

D. $I_{base} = 12$ mA

22.4 3. For the time delay overcurrent relay characteristics that are shown in the figure below, what is the estimated time for the relay to trip for a current that is 10 times pickup if the relay is set to time dial 7?

A. 1.7 sec

B. 1.0 sec

C. 0.5 sec

D. 0.1 sec

RECOMMENDED READING

A.E. Kennelly, "Equivalence of triangles and three-pointed stars in conducting networks," *Electrical World and Engineer*, vol. 34, pp. 413–414, 1899.

ANSI C2-1990: National Electrical Safety Code 1.

Electric Power Plant Design, Technical Manual TM 5-811-6, Department of the Army, USA, 1984.

Department of the Army, TM 5-811-14, Coordinated Power System Protection, 2007.

Electrical Machines, Drives and Power Systems, 6th edition, Theodore Wildi, Prentice Hall, 2006.

The Engineering Handbook, 3rd edition, Richard C. Dorf (editor in chief), CRC Press, 2006.

IEEE 315-1975: Graphic Symbols for Electrical and Electronics Diagrams.

IEEE C37.5-1979: Guide for Calculation of Fault Current for Application of AC High-Voltage Circuit Breakers Rated on a Total Current Basis.

IEEE C37.97-1979: Guide for Protective Relay Applications to Power System Buses.

IEEE 141-1993: Recommended Practice for Electric Power Distribution for Industrial Plants (IEEE Red Book).

IEEE 446-1995: Recommended Practice for Emergency and Standby Power for Industrial and Commercial Applications (IEEE Orange Book).

IEEE 80-2000: Guide for Safety in AC Substation Grounding.

IEEE 242-2001: Recommended Practice for Protection and Coordination of Industrial and Commercial Power Systems (IEEE Buff Book).

IEEE C37.95-2002 IEEE Guide for Protective Relaying of Utility-Consumer Interconnections.

IEEE C37.90-2005: Relays and Relay Systems Associated with Electric Power Apparatus.

IEEE 493-2007: Recommended Practice for the Design of Reliable Industrial and Commercial Power Systems (IEEE Gold Book).

IEEE C37.2-2008: Standard Electrical Power System Device Function Numbers.

IEEE C37.91-2008: Guide for Protective Relay Applications to Power Transformers.

IEEE C37.96-2012: Guide for AC Motor Protection.

Industrial Power Distribution, 2nd Edition, Ralph E. Fehr III, Wiley-IEEE Press, 2016. ISBN: 978-1-119-06334-6.

NFPA 70® – National Electrical Code®, National Fire Protection Association, 2014.

Power Plant Engineering, Black & Veatch, edited by Larry Drbal, Kayla Westra, and Pat Boston, Chapman & Hall/Springer, 1996.

Standard Handbook of Powerplant Engineering, 2nd edition, Thomas C. Elliott, McGraw-Hill, 1998.

CHAPTER **23**

CONTROL SYSTEMS

GOALS

- To understand the fundamental four functions that a control system performs
- To understand the basic concepts behind boiler burner management control systems
- To understand the difference between a tangentially fired and wall-fired boiler
- To understand concepts of turbine governor control systems
- To understand the various types of feedback and feedforward control systems
- To be aware of standard color backgrounds for the graphics displayed on an operator HMI
- To be aware of industry standard alarm sequences and colors

WHERE THE plant field instrumentation system is the nerve center for the power generation facility, the control system is the brain of the power generation facility. The control system takes inputs from the field instrumentation system, performs necessary calculations and conversions, displays the status of the field equipment to the operations personnel and will provide commands to actuators in the field to control equipment as needed to ensure a safe and reliable power plant operation. Any control system performs a minimum of four functions.

- First, it must measure the field physical parameter that is trying to be regulated such as pressure, temperature, or level.
- Second, it must compare that feedback physical value with some desired set-point for that field value.
- Third, it must calculate the error between the desired set-point for the physical parameter and the actual measured value of the physical parameter.
- Last, it must provide an output that drives some activating device to correct the system and bring the actual physical value closer to the desired set-point value.

Energy Production Systems Engineering: An Introduction for Electrical Engineers to Electrical Power Generation Facilities, Systems, and Equipment, First Edition. Thomas H. Blair.
© 2017 by The Institute of Electrical and Electronics Engineers, Inc. Published 2017 by John Wiley & Sons, Inc.

These four steps constitute a closed-loop control feedback system. There are many types of control systems in the modern power generation facility and this chapter will cover some of the more common ones in regards to the safe and reliable operation of the power generation facility.

One of the most important control systems in the fossil fuel plant is the *boiler control system* and *burner management system*. This system provides two main functions. It maintains a constant steam flow or constant steam pressure (depending on the method of steam turbine control) under varying load conditions through the proper input of fuel and air mixture to the main combustors. The amount of air has to be adequate to ensure enough oxygen is present to ensure complete combustion of the fuel while, at the same time, ensuring that the amount of excess air (excess oxygen) is controlled to control the formation of NO_x and SO_x molecules and ensure adequate efficiency. If the fuel/air mixture is rich (more fuel than available air), then the fuel will not completely combust. This leads to the release of partial combustion byproducts like carbon monoxide (CO). CO is in itself combustible and this would lead to combustibles existing in the exhaust duct with presents an explosive danger in the downstream equipment. In addition, the loss of combustibles out of the stack represents an increase in the heat rate of the power plant (increased heat rate is undesirable). This would mean that, for the same amount of steam pressure or flow, we would need to provide additional fuel and this increases the BTU/kWh or $/kWh for the same conditions as compared to the one having adequate oxygen for complete combustion.

The second function of the combustion control and burner management system is to ensure the safe and efficient operation of the boiler throughout the boiler's load range. The document that is the primary guidance for combustion control and burner management controls is NFPA 85: Boiler and combustion systems hazards code. This document describes the minimum controls to ensure that the furnace is operated in a safe and reliable state for various boiler conditions such as initial light off, normal operation and safe shutdown. Some of the basic concepts of NFPA 85 will be introduced in the following section.

There are four fundamental methods of controlling the combustion system depending on what the feedback signal(s) are and what the control system response is to the feedback signals. The first method of control is the *series pressure control* with airflow being the primary control element and fuel flow the secondary control element. In the series pressure control system with airflow as primary control element, the feedback signal is the steam pressure. As steam pressure changes, the burner control system initially changes the airflow to the furnace. When the control system senses the change in the airflow, the control system then adjusts fuel flow to match the new requirement of airflow. This system is fast to respond for an increase in load, but slow to respond during a reduction in load due to the requirement to ensure adequate air (oxygen) is supplied to the furnace section during the transient.

Let us look at the case where the load increases. The increase in load results in an initial reduction in steam pressure. This reduction in steam pressure is sensed by the burner management control system and the burner management controls increase the amount of air supplied to the furnace, but initially, the amount of fuel has not yet changed. This makes the fuel/air mixture in the furnace more lean with a greater amount of excess air available. Since the air content has increased, the fuel flow can increase quickly to catch up to the additional amount of air without the danger of developing a condition where we do not have adequate air sufficient for complete combustion.

Now, let us look at the case where load decreases. The decrease in load results in an initial increase in steam pressure. This increase in steam pressure is sensed by the burner management control system and the burner management controls decrease the amount of air supplied to the furnace, but initially, the amount of fuel has not yet changed. This makes the fuel/air mixture in the furnace more rich with a lesser amount of excess air available. This reduction in airflow with the same fuel flow carries with it the risk of allowing the condition of insufficient air to be in the furnace for the existing flow of fuel. Therefore, the airflow is changed slowly to allow time for the airflow feedback to start reducing the amount of fuel flow in the combustion chamber. This is the reason that the *series pressure control with airflow being the primary control element responds quickly to a sudden increase in load demand but responds slowly to a sudden decrease in load demand.*

The second method of control is the *series pressure control* with fuel flow being the primary control element and airflow the secondary control element. In the series pressure control system with fuel flow as primary control element, the feedback signal is still the steam pressure, but now as steam pressure changes, the burner control system initially changes the fuel flow to the furnace. When the control system senses the change in the fuel flow, the control system then adjusts airflow to match the new requirement of fuel flow. This system is slow responding for an increase in load, and fast responding during a reduction in load due to the requirement to ensure adequate air (oxygen) is supplied to the furnace section during the transient.

Let us look at the case where load increases. The increase in load results in an initial reduction in steam pressure. This reduction in steam pressure is sensed by the burner management control system and the burner management controls increases the amount of fuel supplied to the furnace, but initially, the amount of air has not yet changed. This makes the fuel/air mixture in the furnace more rich with a lesser amount of excess air available. This runs the risk of momentarily having an oxygen-starved environment in the combustion zone, so the rate at which the fuel is increased must be slow enough to ensure that the increase in air follows at a rate to ensure adequate oxygen in the combustion zone during the transient.

Now, let us look at the case where load decreases. The decrease in load results in an initial increase in steam pressure. This increase in steam pressure is sensed by the burner management control system and the burner management controls decrease the amount of fuel supplied to the furnace, but initially, the amount of air has not yet changed. This makes the fuel/air mixture in the furnace more lean with a more amount of excess air available. This reduction in fuel flow with the same airflow allows for more excess air to be available in the combustion zone. There is no risk of having an oxygen starved combustion zone so this system can respond rapidly to reductions in load changes. This is the reason that the *series pressure control with fuel flow being the primary control element responds quickly to a sudden decrease in load demand but responds slowly to a sudden increase in load demand.*

As we can see with either of the systems described above, the control system is limited in one direction to its response to transients. A third method of control that removes this restriction is the *parallel pressure control* system. In this control system, the feedback is still main steam pressure, but now a load change will directly control both airflow and fuel flow without the requirement of waiting for either fuel or air feedback to change before changing the other parameters as was done in the series pressure control systems.

These three systems can be used with fuels that have a fairly constant energy content per pound of fuel (BTU/lbm) such that the energy content can be depended upon for transient control. As the fuel command is changed, the control system is depending on the fact that, if the BTU/lbm content of the fuel is constant, a change in the lbm of the fuel delivered will provide a certain amount of energy change (BTU). This is true for fuels such as natural gas. However, in fuels such as oil and coal, the type of fuel oil or coal can dramatically affect the BTU/lbm content of the fuel. Therefore, the control system for this type of fuel needs both steam pressure and steam flow for feedback signals since the steam pressure (at given temperature) defines the enthalpy of the steam and the steam flow defines the mass flow rate, so the product of these will tell the control system the rate at which energy is being delivered by the control system to the turbine and the burner management control system can adjust the fuel/airflow to ensure adequate BTU is available from the combustion zone of the furnace.

This leads us to the fourth type of burner management control system. The fourth control system which is used when the fuel content BTU/lbm can vary is the *series/parallel pressure flow*. In this system, steam pressure is monitored and provides a feedback signal for the controls for fuel flow. The steam flow is monitored and provides a feedback signal for the airflow controls.

So what constitutes the final element for control of fuel? That depends on the type of technology used for the power conversion technology. For nuclear power, the reactivity of the reactor controls the power level of the power plant. Depending on the design of the reactor, the final control element may be the position of the control rods which control the reactivity around the nuclear fuel. (Please see Chapter 6 on nuclear power for other elements such as moderator temperature as means for effecting reactivity of the main reactor in a nuclear power plant.) In coal-fired boilers, the coal is pulverized in the mills and the primary air fans provide the air for transportation of the pulverized coal from the mills to the burners. The damper on the primary air fan or the coal feeder that feeds the coal to the mill may be fuel control device as this controls the amount of transportation air through the mill and coal to the mill and, thereby, the amount of pulverized coal that enters the burner. In oil-fired units (as well as gas-fired units), there is a fuel valve that controls the flow of oil or gas to the combustion section and this is the final control element for fuel. For hydro plants, depending on the design of the turbine section, this may have either gates that act much like valves that control the amount of water through the turbine or, alternately, some hydro plants have variable pitch blades on the rotor of the turbine and the position of this blade is changed to change the amount of power delivered to the turbine.

What constitutes the final element for the control of air? The forced draft fan (FD) provides air to the furnace to ensure adequate air for complete combustion. The flow of air through the FD fan is the parameter that is controlled. There may be inlet guide vanes on the FD fan to provide control of airflow, or there may be an outlet damper to provide control of the airflow. For newer units, the motor of the FD fan may be connected to a solid-state variable frequency drive to control the speed of the motor and, therefore, the flow of air through the fan. In addition, in the combustion section of the furnace, there are dampers at the inlet of each burner section and the dampers help ensure that the flow through each combustion section is adequate so that we have complete combustion at each burner. Note that tangentially fired boilers do not have individual combustion zones, but one main flame. However, tangentially fired boilers also have secondary air dampers on tilting controls to control the amount and direction of the airflow into the combustion zone.

As discussed previously, a major function of the combustion control system is to ensure proper balance of air to fuel mixture during both steady state and transient conditions. Optimizing the airflow for a certain amount of fuel flow ensures there is sufficient oxygen for complete combustion while ensuring that the excess oxygen is minimized.

We have already discussed the issue of too little oxygen, and the concerns with incomplete combustion. The concerns with too much oxygen are several. First, with too much air, the excess mass is heated in the furnace and this excess mass leaves the furnace section with the energy that it gains inside the furnace. This leads to a reduced efficiency as it is energy lost out the stack. Additionally, providing too much air (and therefore oxygen) leads to available oxygen and increased formation of NO_x byproducts which have to be removed from the gas stream in post-combustion equipment.

So how do we know if we are providing the adequate amount of air (oxygen) to the combustion chamber? The two most commonly used parameters to monitor the amount of oxygen provided to the combustion section is the amount of oxygen (O_2) in the economizer section (exhaust of the combustion section of the boiler) or the amount of carbon monoxide (CO). If multiple fuels are used in the furnace section, then almost always, oxygen is used to monitor as the amount of carbon monoxide varies greatly depending on the fuel used. The advantages of using carbon monoxide for the feedback of adequate air in the combustion chamber is that the amount of CO is unaffected by air in leakage whereas the amount of O_2 is greatly affected by air in leakage. Also, the operating point for CO has a very narrow range of operation which makes control much easier.

The function of the burner management system is to provide for the safe startup, operation, and shutdown of the fuel supply system to the boiler and consist of various control loops. The pre-fire purge control loop purges the air in the boiler before initial light off to ensure that any combustibles in the boiler or duct system have been completely purged before the application of any ignition energy source to the combustion zone. This ensures that the initial light off is controlled and prevents an explosion of combustible gasses in the boiler section or ductwork. The ignition control loop monitors for permissive and verification of igniter flame. In the event that flame detection is lost, the ignition control system sends a trip signal to the master fuel controller to remove fuel from the combustion section and stop fuel from being admitted to the combustion chamber with no ignition source is present. The master fuel trip (MFT) control loop provides permissive for startup, operation, and shutdown. It will close the final element to remove the fuel stream from the combustion zone should any unsafe condition present itself. The post fire purge control loop ensures that, after

TABLE 23.1 NFPA 85 Section Applications

NFPA 85 Section	Application Description
8501	Single-burner boiler operations
8502	Multiple-burner boiler furnaces
8503	Pulverized fuel systems
8504	Atmospheric fluidized bed boiler operation
8505	Stoker operation
8506	Heat recovery steam generators

Source: Reproduced with permission from NFPA 85, *Boiler and Combustion Systems Hazards Code*, Copyright © 2015, National Fire Protection Association. This is not the complete and official position of the NFPA on the referenced subject, which is represented only by the standard in its entirety. The student may download a free copy of the NFPA 497 standard at: http://www.nfpa.org/codes-and-standards/document-information-pages?mode=code&code=85

boiler shutdown, the air system through the boiler remains active long enough to ensure that all combustibles have been purged from the boiler and the duct work so that we do not set up an area where unintended combustion could take place.

NFPA 85: Boiler and combustion systems hazards code provides specific guidance for the safe operation of various types of boilers and steam generators, depending on the type of unit utilized. Various sections of NFPA apply depending on the type of boiler under consideration. Table 23.1 lists the various sections of NFPA 85 and which type of boiler they apply.

The ignition control loop monitors for the existence of a flame in the combustion zone and, on a loss of flame detection, will remove the fuel source from the combustion chamber via a master fuel trip. Wall-fired burners have individual combustion zones, one at each burner. Therefore, they have individual flame monitors, one at each combustion zone. For tangentially fired boilers, during startup, each burner has its own combustion zone. However, after startup, there is only one main flame in the center of the boiler as shown in Figure 23.1. Therefore, during startup of a tangentially fired boiler, there is flame detection at each burner. Once this unit is above

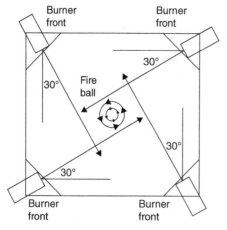

Figure 23.1   Tangentially fired furnace.

about 30% load, it transitions to one main flame in the center of the boiler and the individual flame monitors must be able to detect a flame in the center of the boiler. There are various technologies available for use for flame detection. An ionic flame rod monitors the conductivity to detect the existence of the flame whereas the optic sensor monitors for either infrared radiation (IR) (typical for coal burners) or ultraviolet radiation (UV) (typical for gas burners).

In Chapter 14 on electrical system, we discussed the excitation control systems available as this controls the amount of reactive power or vars delivered or absorbed from the electrical system. For control of the amount of real power or watts delivered to the electrical system, the prime mover main governor controls the amount of real power delivered to the generator shaft which, in turn, controls the amount of watts (real power) the generator delivers to the electrical system. The governor controls operating parameters of the prime mover depending on the type of technology of the prime mover. For example, for the steam turbine, the governor controls the position of the governor valves to control the rate of main steam flow through the turbine (for constant pressure control) or controls the position of the governor valves to control main steam header pressure (for constant flow control). The control system drives an actuator in the field to control the final element.

The above discussion is a simplified model and only applies to an electrical system with only one source, that of the generator being discussed. In the electrical generator and prime mover assembly, conservation of energy requires that under steady-state conditions, the amount of power into a system is equal to the amount of power out of a system. For the ideal generator and prime mover, this means that at steady state conditions (speed is not changing), the amount of power into the prime mover must equal the amount of electrical power out of the generator (plus any losses). If the amount of power into the prime mover is greater than the amount of electrical power out of the generator, then the excess amount of power results in a positive value of accelerating torque and the speed of the assembly will start to increase. If the amount of electrical power out of the generator is greater than the amount of amount power into the prime mover, then the imbalance of power results in a negative value of accelerating torque and the speed of the assembly will start to decrease.

Given a prime mover and generator assembly that is in steady-state operation, when a power transient occurs (change on load torque), the amount of power provided by the prime mover has not yet changed, but the power (and thereby torque) required from the generator has. This causes a mismatch between the load torque and supply torque and the speed of the machine begins to change. The governor control system senses the change of speed. This is not instantaneous and there is some slight lag in this feedback. The governor output changes to drive the modulating device. This changes the power delivered to the prime mover (again, this is not instantaneous but there is some slight lag while the torque to the prime mover changes). This rebalances the load torque with the supply torque. This process continues until the load and supply torques are balanced again and the machine is at steady state. The steps that the governor control system undergoes to compensate for this change in load are summarized below.

1. Change of load torque
2. Change of speed (lag)

3. Change in governor output

4. Change in modulating device

5. Change in developed torque (lag)

6. Change in speed (go to item 3)

The above discussion described just a few of the many control systems in the electric utility generation station. Sensors in the field act as the nervous system for the power generation facility and provide the needed information to the control system for the controls to monitor the condition of the plant and to make correct control decisions. This information that is fed back to the control system is compared with set points and the difference between the ideal value (*set-point*) and the actual value (*feedback*) is known as the *error* signal. Control systems can also look at the rate of change of the error signal to see how quickly the error is increasing or decreasing and they can also look at the product of the amount of time and size of the error to determine the amount of error that is contained in the error signal. Each of these calculations has a potential purpose in the defining the response of the control system to an error between set point and feedback.

The three main types of control calculations are the *proportional* (percent), *integral* (reset), and the *differential* (derivative) calculations or *PID*. The first type of control calculation is the proportional (percent) control calculation. The proportional control calculation is a function of the product of the error signal and some predefined amount of gain. The equation for determination of the output of a proportional controller is

$$P_e = (K_p)E \tag{23.1}$$

where

P_e = output of the percent controller

E = error

K_p = proportional gain

This output is directly proportional to the error between the set-point and feedback signal at the instant of the last sample.

The second type of control calculation is the integral (reset) control calculation is a function of the product of the error signal and the length of time that the error signal has been in existence. It not only considers the latest sample of the error signal but looks back in time to see what the error signal has been. The longer the error signal has been in existence or the greater in magnitude the error signal is, the greater the output of the integral controller response to correct the command signal. The equation for the determination of the integral (reset) controller is

$$P_i = (1/T_i) \int Edt \tag{23.2}$$

where

P_i = output of the integral controller

T_i = integral (or reset) time

E = error

The third type of control calculation is the derivative or differential calculation. This is the product of the time base and the rate of change of the error signal. It defines the slope or rate at which the error signal is growing or decaying. The equation for determination of the derivative or differential calculation is

$$P_d = (T_d)(dE/dt) \tag{23.3}$$

where

P_d = output of derivative controller

T_d = rate (or derivative) time

dE/dt = rate of change of error

The final output of the PID controller is the summation of the three different control calculations as shown in equation (23.4);

$$P = P_e + P_i + P_d \tag{23.4}$$

where

P = output of the controller to the modulating device

P_e = output of the percent controller

P_i = output of the integral controller

P_d = output of derivative controller

Now that we have discussed some basics of control systems and the types of calculations used in the control system, we can start looking at complete control loops. Complete control loops are comprised of the element that provides the feedback signal to the control system, the controller that compares the feedback signal to the set point, and the command signal that is delivered to a final element to manipulate a parameter to regulated the feedback signal to the set point. When designing a control loop, it is the author's experience that it is best to work backwards from the final control element through the controller and back to the instrumentation providing signals for the control system to determine the fail-safe direction of the final control element and therefore, the direction of feedback and feedforward signals as the following examples will show.

The first type of control loop is the *feedback modulating control loop*. An example of this is the main turbine lube oil cooling system as shown in Figure 23.2.

The reader is encouraged to reference common system identification and equipment identification terminology as described in Annex C and Annex D and also ISA standard instrumentation terminology as described in Chapter 24 for an understanding of how the instrumentation and equipment numbers are derived in this example. In the main turbine lube oil heat exchanger (1-TML-E-001), heat is transferred from the main turbine lube oil system (TML) to the chilled water system (CDS). Warm turbine lube oil enters the heat exchanger and gives off its thermal energy to the chilled water system to allow for a well-regulated main turbine oil temperature out of the cooler that is then supplied to the main turbine components. To regulate the main turbine oil outlet temperature, we could either regulate oil flow or regulate chilled

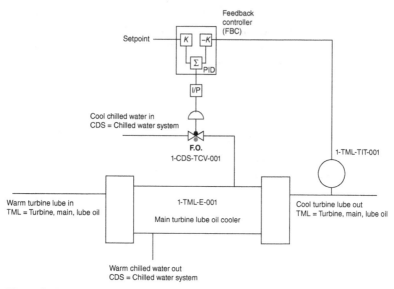

Figure 23.2   Modulating feedback control loop.

water flow. However, we never want to starve the main turbine lube oil system for a supply of oil, so we elect to place a valve (the final control element) in the chilled water system. This valve has been given an equipment number of 1-CDS-TCV-001. Note from the equipment number that this valve is in the chilled water system (CDS) and is designated as a temperature control valve (TCV). Even though this valve is controlling flow of water, the value or parameter being controlled by the valve is the output temperature of the oil from the heat exchanger. Therefore, this is a temperature control valve, not a flow control valve.

From the symbol of the valve and knowing ISA standards, we know that the actuator is an air-operated actuator. Now that we have decided where to place our final control element, we want to evaluate the "fail safe" position of this valve. Should the valve fail close, then the main turbine oil will have no cooling in the heat exchanger, resulting in higher temperatures of the lube oil supplied to the turbine which in turn results in lower oil viscosity and a potential of loss of lubrication to the turbine system. Should the valve fail open, then the main turbine oil will have more cooling in the heat exchanger than the design resulting in lower temperatures of the lube oil supplied to the turbine resulting in higher oil viscosity but without the potential of loss of lubrication to the turbine system. Of the two failure modes, we would rather have the valve fail fully open and provide too much cooling to the turbine lube oil than have the valve fail fully closed and have insufficient cooling for the turbine lube oil. Therefore, we will select an air-operated valve that will fail in the open position. To achieve this, our valve actuator will utilize air to close the valve and utilize a spring to open the valve. In Figure 23.2, this is indicated below the valve by the term F.O. (fail open.) Now that we have designed the correct location and valve actuator type, next we work upstream to the energy source that drives the valve. In this case, this is an air-actuated valve. As we described before, the valve actuator will utilize air to close the valve and utilize a spring to open the valve. Therefore, on a loss of instrument air pressure, this valve will be driven open by the spring in the actuator and fail in the open position as desired. The next failure mode we must evaluate is a loss of 4–20 mA signal to the I/P controller. The failure mode of the current loop will present 0 mA (or possibly 4 mA) to the I/P controller, so we want to design the control such that when the command signal to the I/P controller is at 4 mA, the valve will be fully open. We now have defined the required input to output response of our I/P controller in that, when the control loop current is at 4 mA, the output will be to provide minimum instrument air to the valve actuator and when the control loop current is 20 mA, the output will be to provide maximum instrument air to the valve actuator.

Next, we have to define the type of control system we will use to regulate the desired variable (in this case, cool-turbine lube oil outlet temperature) and the type of instrumentation required to implement our control system. As mentioned above, we will first learn about the basic feedback modulating control loop by using this system as our control loop and then find ways to modify this basic control system to add anticipatory signals to enhance regulation.

In the feedback modulating control loop, our goal is to regulate the temperature of the lube oil out of the lube oil heat exchanger and therefore we make this our feedback signal. The controlling parameter (cooling water flow) is the amount of cooling water we supply to the heat exchanger. The feedback parameter (oil out temperature) is monitored with a temperature indicating transmitter (1-TML-TIT-001) and this signal is sent to a feedback PID controller (FBC). This controller compares the feedback signal with a set point and develops an error signal. This error signal is then sent to a PID block where it is conditioned as needed for the application. Then an output signal is developed by the controller and delivered to the I/P converter to convert the electrical signal to an air signal which is then delivered to the valve actuator for the valve that regulates the cooling water flow to the heat exchanger. Adjustment is then made to the valve position to ensure the amount of cooling water flow is adequate for the temperature desired. For the air actuator we have chosen, we know that 4 mA out of the controller drives the valve 1-CDS-AOV-001 open and 20 mA out of the controller drives the valve 1-CDS-AOV-001 closed. Although scalable, the typical setup for a temperature-indicating transmitter is to increase the analog out signal with increasing temperature. Since we want the command signal to the I/P section to reduce, opening the valve for an increasing value of lube oil temperature, we multiply the analog feedback signal by a negative constant $(-K)$ so, when it sums with the set point, it reduces the output signal to the I/P, thereby opening the valve and increasing the flow of cooling water to the heat exchanger.

The *feedback modulating control loop* is adequate for most control systems and is very commonly used. However, let us look at the response of the control loop to a system transient. From thermodynamics, we remember that for a system to be in steady state, the amount of energy into a system must be equal to the amount of energy out of a system. We also remember energy is enthalpy (a function of temperature for sensible heat transfer) and flow. Therefore, in a steady state condition, input oil and water temperature and flow are steady. What happens when one of these parameters changes, causing the input energy and output energy to not be equal (transient condition)? The feedback modulating control loop does not monitor input flow or temperature of either the oil or water system. During a transient, either or both temperature and/or flow changes to the oil and/or water system. Now, since energy into the heat exchanger does not equal energy out of the heat exchanger, the output temperature will begin to change. If input energy is greater than output energy, for example, if the warm turbine oil flow increases, then the output temperature will begin to rise. The standard *feedback modulating control loop* must wait to see the increase in output temperature before the control system responds to open the valve (1-CDS-AOV-001) thereby increasing cooling water flow to regain energy balance in the heat exchanger and bringing our discharge temperature back to set point. However, if we were to monitor the flow of the oil into the heat exchanger and bring this signal into our control system, we would be able to anticipate the change in the energy balance of the heat exchanger. Knowing that increasing flow of warm oil into the heat exchanger is increasing the energy into the heat exchanger from the turbine oil system, we know we must begin to open the valve (1-CDS-AOV-001) thereby increasing cooling water flow to maintain the energy balance in the heat exchanger which means we must reduce our command signal to the I/P controller, so we multiply the feedforward analog signal by $(-K)$ to invert the direction of control to be as we need it for this system. This is known as *feedforwarding* signal. Figure 23.3 shows the control loop for a modulating feedback control loop with feedforward anticipatory signal.

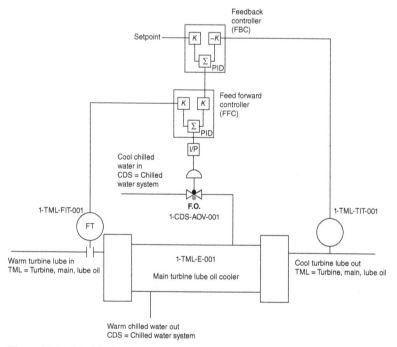

Figure 23.3   Modulating feedback control loop with feedforward anticipatory signal in the lube oil loop.

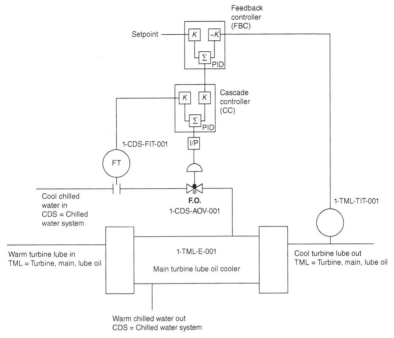

Figure 23.4   Modulating feedback control loop with feedforward anticipatory signal in cooling water loop.

Adding a feedforward anticipatory signal to a feedback control loop increases the system response to transients in the lube oil system being regulated (in this instance main turbine lube oil). What if we experience a transient in the cooling water system? The feedforward system of Figure 23.3 would not detect any change in cooling water flow. However, we can use the same feedforward control system but if we now monitor cooling water flow, we can anticipate changes in the cooling water system flow. Figure 23.4 shows a system that monitors cooling water flow. Note for this system, where we now monitor cooling water, an increase in cooling water flow will be indicated by an increase in our feedforward signal and the correct response of the control valve to this situation is to turn the temperature control valve in the close direction which means that our command to our I/P controller must increase, so we multiply the incoming feedforward signal by a value of (K).

Since both a change in the incoming oil stream and incoming water stream to the heat exchanger may cause a change in output oil temperature, we may want to have an anticipatory signal from each stream. We can combine multiple feedback signals utilizing the cascade loop as shown in Figure 23.5. In this system, the temperature controller output is summed with the feedforward signal from the flow in the lube oil system. This output is then sent to the flow controller that is cascaded with the temperature controller. The cooling water flow signal is sent to the flow controller as feedback and the error in the flow controller is sent to the cooling water flow control valve (1-CDS-AOV-001) to control the rate of cooling water flow through the feedwater heater.

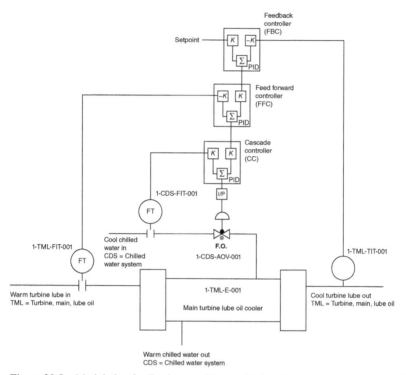

Figure 23.5   Modulating feedback control loop with feedforward anticipatory signal and a cascade loop.

Next, we will discuss the modern control room and control system. Pre 1980s, signals were sent to a control room and analog displays provided operations with the operating condition of the plant. Post 1980s, more commonly a *distributed control system* (DCS) is utilized to accept feedback of plant instrumentation signals, storage of historical data, calculation of automatic functions, and controls and provides the operations personnel with an interface to provide commands to the control system. This control system will include several drops, each having processors that run independently, but communicate information back and forth between controllers or drops. Each controller contains the ability to perform the control calculations described above and provide command signals to equipment to regulate control loops.

Figure 23.6 shows a typical distribution of controllers in a power generation facility.

Note that each drop has a primary and a secondary or backup controller such that if one controller fails, the backup will take over and keep the control system functioning. This also allows for updates to the control logic while still keeping the control system online. The backup controller will be updated with new logic, then the backup controller will be switched to be the primary controller and the other controller will then be updated with logic change. (Note, please consult with specific DCS instruction literature for procedure to update controllers as procedures vary between manufacturers). Additionally, the controller will have redundant power supplies to ensure

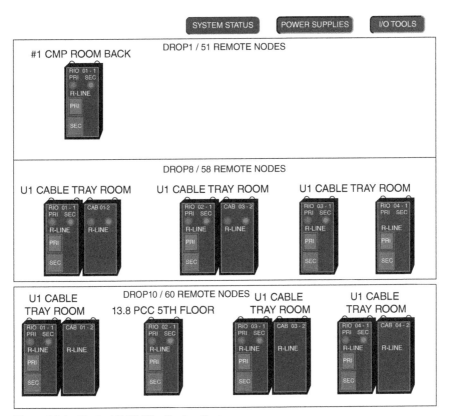

Figure 23.6   Typical DCS Distribution System.

reliability of the control system. The DCS also provides for the interface between operations personnel and the control system with *human machine interface* (HMI) screens. These screens show the system one-line diagram or PID diagram in a fashion to allow a quick understanding of the current status of the unit. HMIs are not designed to allow for logic changes, but only set point changes and to allow operations to start and stop equipment. For engineering logic changes, the control system provides a different interface called an engineering workstation. This is a computer that provides the required software needed to interrogate the ladder logic of the controllers, update the logic scheme, and update processors in the remote I/O drops as described above.

Now, we will discuss some of the conventions used in the graphical representation of information on HMI screens of the control system and common conventions used for alarming of abnormal conditions in the power generating facilities. Normally, the graphics the operators see are either PID representations of the mechanical and instrumentation system or one-line representations of the electrical system. Figure 23.7 displays a typical graphic of an electrical system and Figure 23.8 displays a typical graphic for a mechanical system. ANSI/ISA-5.1-2009 Instrumentation Symbols and Identification, provides detailed direction for symbols and colors standardized for the industry.

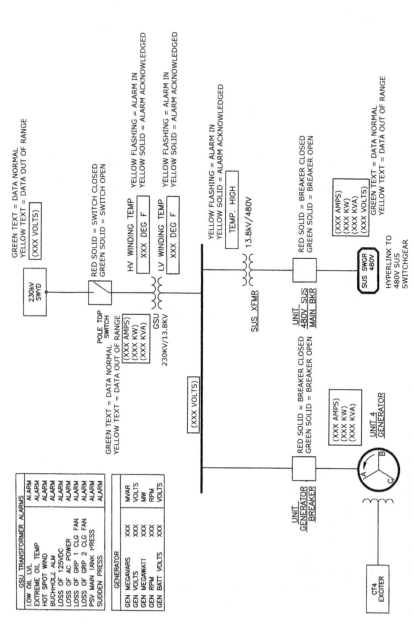

Figure 23.7 Typical DCS Graphic Display, Electrical.

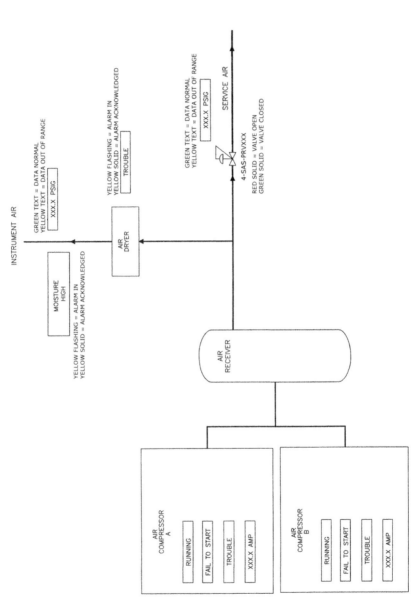

Figure 23.8 Typical DCS Graphic Display Mechanical.

TABLE 23.2   Standard Control System Graphic Color Code System

Color	Description of Use
Green (steady)	Status: off, open circuit, closed valve, data normal
Yellow (flashing)	Breaker or starter tripped
Light red (steady)	Status: on, closed circuit, open valve
Yellow (steady)	Alarm acknowledged
Bright red (flashing)	Data: alarm unacknowledged

Source: ISA-5.5-1985, Graphic Symbols for Process Displays. Reprinted with permission of the International Society of Automation, www.isa.org/standards.

The standard color background for the graphics displayed on an operator HMI screen is black. Table 23.2 defines the various other colors used on the HMI screen and what their meaning is to the operator that is interpreting this information.

Let us look at some standard HMI graphics and discuss what the colors and symbols mean. First, we will look at an electrical system. Typically, the electrical system is displayed on the operators screen in an electrical one-line format. The snapshot screen in Figure 23.7 shows the 13.8 kV electrical distribution system for a 60 MW combustion turbine unit. The station service feed is on the right and the generator is on the left. The transmission system is at the top of the figure. When breakers are shown in green, this indicates that these breakers are open but ready for service. If these breakers were tripped, then these symbols would also have a flashing yellow box around them. Similarly, data from field instrumentation that is not out of design values is shown in green, while data from field instruments that is out of normal design values is shown in yellow. As an example, in Figure 23.7, values such as voltage, current, and real and reactive power of the system are displayed in green. Should the values become unavailable, the text changes to yellow indicating the alarm condition. Alarms may be shown graphically alongside the equipment they are associated with but additionally there will typically be an alarm summary table as shown in Figure 23.7.

There are some pieces of information that are so important to the operation of the plant that they are displayed on every graphic so that, regardless of which control screen the operator is on, these data are in front of the operator at all times. Some examples of critical plant parameters are listed below.

MW = megawatt or real power that the generator is delivering to the system (MW).

THROT PRS = pressure at the governor valves of the steam turbine (psig).

FW FLOW = feedwater system flow to the boiler (KPPH or 1000 lb/hr)

STM FLOW = steam system flow from the main steam section of the boiler (KPPH or 1000 lb/hr)

AIR FLOW = airflow through furnace side of main boiler (%)

FUEL FLOW = fuel flow to the furnace of the main boiler (%)

OPACITY = opacity of exhaust gas from the stack of the unit (%)

Similarly, mechanical systems are shown in their PID (piping and instrumentation diagram) simplified format. Figure 23.8 is an example of a compressed air supply system for the 60 MW unit we are using for our example. The compressed air is supplied in this example by two 100% redundant air compressors as shown on the left of Figure 23.8. For each compressor, when the compressor is energized, this is indicated by a red "running" text box and when the compressor is de-energized, this is indicated by a green "stopped" text box. The valve labeled 4-SAS-PRV-XXX is a pressure regulating valve (PRV) associated with the service air system (SAS) for unit 4. Note that in this example, the air compressors and air receiver are the supply system for both the instrument air system (IAS) and the service air system (SAS). The main difference between these two air systems is the equipment that is serviced by these systems. The instrument air system is dedicated to feeding valve air actuators and instrumentation with narrow airlines that would be affected by moisture in the compressed air system. The instrument air system contains an air dryer (as is shown in Figure 23.8) to reduce the dew point of the air being sent to the instrument air system. The service air system, on the other hand, supplies equipment such as purge air for explosion proof enclosures where moisture or dew point of the compressed air is not as critical to the operation of the equipment.

One item to stress is the colors of red and green. Red means energized, on, running, valve open, motor on, etc. When you think of red, think that the equipment is not safe to work on. Green means de-energized, off, stopped, valve closed, motor off, etc. When you think of green, think that the equipment is safe to work on. The reason we mention this and highlight it is that sometimes these can cause confusion when you transition from an electrical system to a mechanical system. A red breaker is closed, but a red valve is open. A green breaker is open, but a green valve is closed. If you do not think of red and green in terms of open or closed, but if you think of red and green as safe and not safe, then the electrical and mechanical graphical representation of red and green share symmetry.

The last item to discuss would be the topic of alarming on abnormal conditions. Alarms are meant to notify plant operations to abnormal conditions in the plant processes and systems to give them an opportunity to address, troubleshoot, and resolve these abnormal conditions before they effect the safe and reliable operation of the power generation facility. Because the control system is trying to quickly and effectively communicate important information to the operations department of the facility, it is critical to maintain consistency of alarm information presentation to the operations personnel.

The standard alarm sequence is shown in Table 23.3. Initially, the process condition is within normal boundaries and the sequence state is normal. There is no visual or audible indication of any alarm. The sequence state changes to the alarm state when the process condition first varies out of the normal range. In this state, the visual indication turns on and flashes and the audible indication begins to annunciate the alarm condition to gain the attention of the operations personnel. Once the operations personnel read and understand the alarm condition, they acknowledge the condition on the HMI graphics. This moves the sequence state to acknowledge. The visual indication becomes steady on signal and the audible indication becomes silent. The annunciator will remain in this condition until the alarm condition is corrected.

TABLE 23.3 Alarm Sequence

Process Condition	Sequence State	Visual Indication	Audible Indication
Normal	Normal	Off	Silent
Abnormal	Alarm	Flashing	Audible
Abnormal	Acknowledge	On	Silent
Normal	Acknowledge	Off	Silent

Source: ISA-18.1 (R2004), Annunciator Sequences and Specifications. Reprinted with permission of the International Society of Automation, www.isa.org/standards

Once the alarm condition is corrected, the sequence state moves to the fourth state and the visual indication turns back off and the system is ready for the next condition that becomes abnormal.

GLOSSARY OF TERMS

- Boiler Control System – A safety system for power generation companies that controls the boiler pressure during steady state and transient operation within safe levels.

- Burner Management System – A safety system for power generation companies that enables the safe start up, operation, and shut down of the burner system for a boiler or steam generator.

- Distributed Control System (DCS) – A system comprised of software, hardware, cabling, sensors, and activators that is used to control and monitor equipment wherein control elements are distributed throughout the system.

- Graphic User Interface (GUI) – A GUI is a form of HMI where the system information is displayed graphically.

- Human Machine Interface (HMI) – An HMI is a software application that presents information to an operator or user about the state of a process, and to accept and implement the operators control instructions.

- Parallel Pressure Control System – Control logic where airflow and fuel flow are both primary control elements.

- PID Controller – A proportional–integral–derivative controller (PID controller) is a generic control loop feedback mechanism (controller) widely used in industrial control.

- Series/Parallel Pressure Flow – Control logic where airflow is the primary control element for steam flow and fuel flow is the primary control element for steam pressure.

- Series Pressure Control (Fuel) – Control logic where fuel flow is the primary control element and airflow is the secondary control element.

PROBLEMS

23.1   Utilizing conventional color codes, a motor-operated valve that is the color "red" on the HMI indicated that the valve is

    A.   fully open

    B.   mid-travel (i.e., not fully open and not fully closed)

    C.   fully closed

23.2   Utilizing conventional color codes, a pump energized by a three-phase motor that is the color "red" on the HMI indicated that the pump is

    A.   energized and operating

    B.   in a faulted condition

    C.   de-energized

23.3   The formula in equation (23.2) given below represents the _____ function of a PID controller.

$$P_i = (1/T_i) \int E dt$$

    A.   percent

    B.   integral

    C.   differential

23.4   The control system shown below is an example of a _____ System.

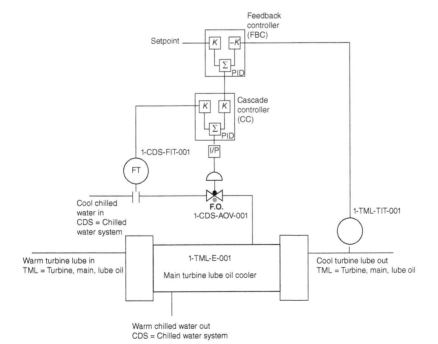

    A.   Feedforward – lube oil signal

    B.   Feedforward – cooling water signal

    C.   Feedforward and cascade

23.5   What standard addresses boiler control and burner management system safety requirements?

    A.   NEC®

    B.   NESC

    C.   NFPA 70E

    D.   NFPA 85

RECOMMENDED READING

Electrical Machines, Drives and Power Systems, 6th edition, Theodore Wildi, Prentice Hall, 2006.

The Engineering Handbook, 3rd edition, Richard C. Dorf (editor in chief), CRC Press, 2006.

Fundamentals of Process Control Theory, 3rd edition, Paul W. Murrill, ISA, 2000.

IEEE C37.1-2007: Definition, Specification, and Analysis of Systems used for Supervisory Control, Data Acquisition, and Automatic Control.

IEEE C37.21-1985: IEEE Standard for Control Switchboards.

IEEE 101-1987: IEEE Application Guide for Distributed Digital Control and Monitoring for Power Plants.

IEEE 1050-1996: IEEE Guide for Instrumentation Control Equipment Grounding in Generating Stations.

ISA5.1-2009: Instrumentation Symbols and Identification, International Society of Automation.

ISA18-1979 (R2004), International Society of Automation, Instrument Signals and Alarms.

NFPA85-2015 – Boiler and combustion systems hazards code.

Power Plant Engineering, Black & Veatch, edited by Larry Drbal, Kayla Westra, and Pat Boston, Chapman & Hall/Springer, 1996.

Standard Handbook of Powerplant Engineering, 2nd edition, Thomas C. Elliott, McGraw-Hill, 1998.

T.A. Hughes, *Measurement and Control Basics*, 4th edition, International Society of Automation (ISA), 2007.

CHAPTER 24

INSTRUMENTS AND METERS

GOALS

- To understand the concepts of accuracy and precision
- To understand industry standard color coding of thermocouple extension cables
- To design a series-connected thermocouple loop to add and subtract two temperature signals
- To calculate the expected resistance of an RTD of a certain material and temperature coefficient for a given ambient temperature
- To calculate the flow through an orifice given an orifice size and differential head
- To calculate the flow through a pitot tube given the material density and differential head
- To convert between units of measurement of pressure and temperature
- To understand the concept of boiler shrink and swell during boiler transient steam flow conditions
- To calculate the amount of pressure head causing displacement of a liquid in a tube
- To understand the construction and applications of current transformers and potential transformers
- To be able to determine the amount of energy that an electromechanical kilowatt-hour meter senses
- To be able to identify the purpose and type of plant instrument from the instrument identification as defined by ISA 5.1
- To be able to identify the purpose and type of plant equipment for a typical plant system and equipment identification table
- To be able to identify the various types of engineering system drawings utilized in an electric power generation station and the drawing symbols utilized

Energy Production Systems Engineering: An Introduction for Electrical Engineers to Electrical Power Generation Facilities, Systems, and Equipment, First Edition. Thomas H. Blair.
© 2017 by The Institute of Electrical and Electronics Engineers, Inc. Published 2017 by John Wiley & Sons, Inc.

WHILE THE DCS system receives information on the status of equipment in the facility and makes control decisions to operate this equipment, the instrumentation forms the nervous system of the power generation facility to feed valid data back to the control system. The instrumentation system monitors various physical parameters in the power plant and provides either local or remote indication of these parameters to the operational staff to assist in the reliable and safe operation of the power station. Additionally, many systems in the modern power station are automated through a distributed control system (DCS). This control system relies on the accurate and precise feedback from the instrumentation system on the status of the plant physical parameters. Many parameters such as pressure, temperature, flow, level, vibration, oxygen content (O_2), carbon monoxide content (CO), NO_x content, opacity, pH (acid or base value of a fluid), conductivity, and other parameters are monitored and fed back to the control system of the power plant.

Above, we mentioned that the control system requires an accurate and precise feedback from the instrumentation system of a signal that represents the physical condition of the plant equipment systems and processes. At this point, it is worth defining these two terms and their meaning to the control system. The *accuracy* of a measurement of a physical parameter is the degree of conformity of the measured quantity to the actual physical parameter measured. The *precision* of a series of measurements is a measure of the repeatability of the measured quantity.

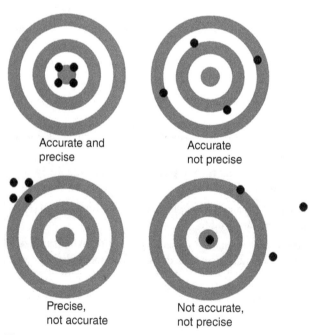

Figure 24.1   Comparison of accuracy to precision.

Accuracy is a measure of how close the value displayed or transmitted by the field instrument represents the actual parameter measured. If the actual pressure in the system is 1200 psig and the display or signal transmitted back to the control system indicates a pressure of 1200 psig, then this system is very accurate. If the actual pressure in the system is 1200 psig and the display or signal transmitted back to the control system indicates a pressure of 1 psig, then this system is not very accurate (and some calibration is in order).

Precision, on the other hand, is a measure of how closely a group of samples agrees with each other. It is not a measure of how close a sample is to the measured parameter. For example, if a field instrument samples pressure 10 times and the 10 values fed back to the control system are all the same value, regardless of whether the measured parameter agrees with the sample values, this is a very precise system. On the other hand, if the 10 values are all wildly different, then this is not a very precise system.

Figure 24.1 shows this concept graphically.

Example 24.1   Given the sample data below determine the following.

a. Is the system accurately reflecting the physical parameter measured in the field?

b. Looking at the group of feedback signals, is the system precise?

Sample #	Sample Value (psig)	Physical Parameter (psig)
1	1	1200
2	1200	1200
3	50	1200
4	900	1200
5	20	1200
6	750	1200
7	60	1200
8	0	1200
9	500	1200
10	2	1200
Average	348	

Solution: Since the 10 different values vary greatly, the system is not precise. Since the average value does not closely represent the measured parameter, this is not accurate.

Example 24.2   Given the sample data below determine the following.

a. Is the system accurately reflecting the physical parameter measured in the field?

b. Looking at the group of feedback signals, is the system precise?

Sample #	Sample Value (psig)	Physical Parameter (psig)
1	1	1200
2	670	1200
3	50	1200
4	700	1200
5	20	1200
6	10000	1200
7	60	1200
8	0	1200
9	500	1200
10	2	1200
Average	1200	

Solution: Since the 10 different values vary greatly, the system is not precise. Since the average value closely represents the measured parameter, this is accurate.

Example 24.3 Given the sample data below determine the following.

a. Is the system accurately reflecting the physical parameter measured in the field?

b. Looking at the group of feedback signals, is the system precise?

Sample #	Sample Value (psig)	Physical Parameter (psig)
1	1	1200
2	1	1200
3	1	1200
4	1	1200
5	1	1200
6	1	1200
7	1	1200
8	1	1200
9	1	1200
10	1	1200
Average	1	

Solution: Since the 10 different values do not very between samples, the system is precise. Since the average value does not closely represent the measured parameter, this is not accurate.

Example 24.4 Given the sample data below determine the following.

a. Is the system accurately reflecting the physical parameter measured in the field?

b. Looking at the group of feedback signals, is the system precise?

Sample #	Sample Value (psig)	Physical Parameter (psig)
1	1200	1200
2	1200	1200
3	1200	1200
4	1200	1200
5	1200	1200
6	1200	1200
7	1200	1200
8	1200	1200
9	1200	1200
10	1200	1200
Average	1200	

Solution: Since the 10 different values do not very between samples, the system is precise. Since the average value closely represents the measured parameter, this is accurate.

TEMPERATURE

There are many sensors in use in the power generation facility to monitor temperature. Temperatures on bearings, motor windings, gas streams, steam streams, hotwell fluid, and the like are all monitored. The device to monitor the various systems may be different depending on the needs of the physical system being monitored.

One simple device is the *bimetallic thermometer*. This device uses dissimilar metals that utilizes the property of differential thermal growth of the two dissimilar metals to move a needle. The needle position indicates the temperature of the measured media. Bimetallic thermometers can monitor temperatures from −200°F to +1000°F. The advantage of the bimetallic thermometer is the low capital cost of the device, but it is limited in its use due to the fact that it is used for local indication only as one has to read the needle on the thermometer to obtain information on the temperature. Also, due to their construction, the calibration of bimetallic thermometers is sensitive to mechanical shocks.

Another device that is used to measure the temperature of a system in the field is the *thermocouple*. This device is also constructed of dissimilar metals and the two metals form a junction at the end where the temperature is to be monitored. The dissimilarity generates current from one metal to the other and the amount of current generated is a function of the temperature of the dissimilar metal junction. The amount of current and therefore the temperature range of the thermocouple are defined by the materials that make up the dissimilar metal junction. Table 24.1 defines the standard types of thermocouples, their applicable temperature ranges, and the materials they are constructed from. It also shows the standard color code for thermocouple conductors as defined by ANSI MC 96.1.

TABLE 24.1   Thermocouple Color Codes and Information According to ISA MC 96.1

ANSI	Color Code		Alloy Combination		Temperature
Code	(+) Lead	(−) Lead	(+) Lead	(−) Lead	Range (°C)
J	White	Red	Fe	Cu-Ni	−210–1200
K	Yellow	Red	Ni-Cr	Ni-Al	−270–1372
T	Blue	Red	Cu	Cu-Ni	−270–400
E	Purple	Red	Ni-Cr	Cu-Ni	−270–1000
N	Orange	Red	Ni-Cr-Si	Ni-Si-Mg	−270–1300

Source: ANSI/ISA-MC 96.1 (1982), Temperature Measurement Thermocouples. Reprinted with permission of the International Society of Automation, www.isa.org/standards

Table 24.2 does the same but for thermocouple extension cables as defined by IEC standards.

The author thanks the International Electrotechnical Commission (IEC) for permission to reproduce Information from its International Standards. All such extracts are copyright of IEC, Geneva, Switzerland. All rights reserved. Further information on the IEC is available from www.iec.ch. IEC has no responsibility for the placement and context in which the extracts and contents are reproduced by the author, nor is IEC in any way responsible for the other content or accuracy therein.

One of the advantages of the thermocouple over the bimetallic thermometer is the ability to transmit the signal to a distant device for indication or control. In most utility environments, the small current generated by the thermocouple cannot be reliably transmitted over great distances. So, to boost the small signal of the thermocouple, the thermocouple is connected to a local device that amplifies the signal to one of the standard controls signal levels used in the plant instrumentation system. Typical signal levels are 0–1 mA, 4–20 mA, 0–5 Vdc, or 0–10 Vdc. This signal can then be transmitted longer distances reliably.

Thermocouples are used to measure the temperature of bearings used to support motors, fans, and pumps. When used on insulated bearings, the thermocouple must

TABLE 24.2   Thermocouple Color Codes and Information

IEC	(+) Lead	Alloy Combination		Max Op Temp (°C)
Code	Color Code	(+) Lead	(−) Lead	Steel Sheath (^{18}Cr, ^{8}Ni)
J	Black	Fe	Cu-Ni	750
K	Green	Ni-Cr	Ni-Al	800
T	Brown	Cu	Cu-Ni	400
E	Violet	Ni-Cr	Cu-Ni	800
N	Pink	Ni-Cr-Si	Ni-Si	800

Source: IEC 60584-3 ed.2.0, Thermocouples, Part 3: Extension and compensating cables – Tolerances and identification system, Copyright © 2007 IEC Geneva, Switzerland, www.iec.ch, and IEC 61515:1995, Mineral insulated thermocouple cables and thermocouples, Copyright © 1995 IEC Geneva, Switzerland, www.iec.ch. Reproduced with permission of the International Electrotechnical Commission, www.iec.ch

be insulated between the dissimilar junction of the thermocouple and the bearing housing to ensure that the thermocouple does not provide an unintentional ground for the bearing housing. Another application for the thermocouple is in the monitoring of the temperature of steam flow. In this application, the steam is under very high pressure (values are around 2500 psig), so we cannot simply drill a hole in the steam pipe and slip the thermocouple into the stream of steam. In instances like this, a hole is cut into the steam pipe and a *thermowell* is inserted in the steam stream. This thermowell rises to the temperature of the steam flowing past it. The thermocouple is then inserted in the thermowell and makes contact with the bottom of the thermowell. In practicality, the thermocouple is not sensing the temperature of the steam directly but sensing the temperature of the thermowell. This temperature is then sent back to the control system.

In some instances, we are less concerned with the absolute temperature of a certain location and more interested in the difference of temperatures between two different locations or the summation of the temperatures of two different locations. Thermocouples can be connected in series to either add or subtract from the two different signals for the summation or difference between two systems.

For example, in Figure 24.2, the transducer receives input from both thermocouple T_1 and thermocouple T_2. When determining if the connection is additive or subtractive, first assign polarity marks at each device. Then start at the any point in the loop and sum the voltage drops in the loop. Solve for the voltage at the receiver and this will tell you if the thermocouples are connected such that their temperatures are additive or subtractive. Using Figure 24.2, we have assigned polarities in the figure. Let us start at the positive lead of T_1 and go clockwise around the loop. Doing this, we derive the following equation

$$+T_{sensor} + T_2 - T_1 = 0$$

Next, we rearrange the equation to isolate the sensor on one side

$$T_{sensor} = T_1 - T_2$$

This tells us that, with this configuration the sensor, we will detect the difference in temperature between thermocouple T_1 and thermocouple T_2.

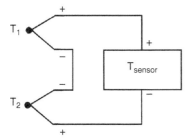

Figure 24.2   Series connection for thermocouple system.

Another device used to detect temperatures of equipment is the *Resistance Temperature Detector* (RTD). This device is made of a conductive material and has a specific temperature coefficient. As the temperature of the material changes, the resistance of the element changes linearly.

$$R_t = R_0[1 + \alpha(t - t_0)] \tag{24.1}$$

where

R_t = resistance of the RTD at actual temperature (Ω)

R_0 = resistance of the RTD at a standard temperature t_0 (Ω)

t = actual temperature of the RTD (°C)

t_0 = reference temperature of the RTD (°C)

α = temperature coefficient of the material used in the RTD (°C^{-1})

We can simplify equation (24.1) to the form of a linear equation $y = mx + b$ and the equation becomes

$$R_t/R_0 = \alpha\Delta t + 1 \tag{24.2}$$

where

R_t = resistance of the RTD at actual temperature (Ω)

R_0 = resistance of the RTD at a standard temperature t_0 (Ω)

α = temperature coefficient of the material used in the RTD (°C^{-1})

$$\Delta t = t - t_0 \tag{24.3}$$

where

t = actual temperature of the RTD (°C)

t_0 = reference temperature of the RTD (°C)

From equation (24.2) we can see that the relationship between resistance of the RTD and the temperature of the RTD is a linear function of temperature of the RTD.

Since the RTD is connected to the other end device via conductors, these conductors have some amount of resistance that is in series with the RTD and this would prevent the direct reading of the RTD value as the total loop resistance is

$$R_{total} = R_{rtd} + 2R_{conductor} \tag{24.4}$$

where

R_{total} = total loop resistance (Ω)

R_{rtd} = resistance of RTD (Ω)

$R_{conductor}$ = conductor resistance between the RTD and the sensing device (Ω)

To remove the resistance of the loop from the value sensed by the sensor, a third (and sometimes fourth) wire is brought back from the RTD back to the sensor on a third input. This senses the resistance of the loop ($2R_{conductor}$) and the sensor subtracts this resistance from the value detected on the input to obtain the value of the RTD resistance.

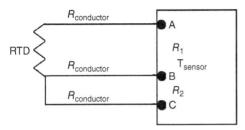

Figure 24.3 RTD connection.

Using Figure 24.3 as reference, the sensor measures two resistances. The first resistance, R_1, is a measurement of the RTD resistance and twice the conductor resistance or

$$R_1 = R_{rtd} + 2\,R_{conductor} \tag{24.5}$$

The second resistance, R_2 is a measurement of the loop resistance or

$$R_2 = 2\,R_{conductor} \tag{24.6}$$

The sensor then subtracts the two resistances to determine the actual value of the RTD resistance.

$$R_{sensor} = R_1 - R_2$$
$$R_{sensor} = (R_{rtd} + 2\,R_{conductor}) - (2\,R_{conductor})$$
$$R_{sensor} = R_{rtd} \tag{24.7}$$

The value of the reference resistance R_0 is a function of the material used. Three common materials used and their reference temperatures at a temperature of 25°C are shown in Table 24.3.

The value of the temperature coefficient (α) is defined by the standard organization that the RTD is constructed to. There are two main standards, an ANSI standard and an IEC/DIN standard. For example, the IEC/DIN standard temperature coefficient for Platinum is 0.00385 $\Omega/(\Omega°C)$ while the ANSI standard temperature coefficient for Platinum is 0.003926 $\Omega/(\Omega°C)$. Both RTDs will read the same at the reference temperature but will vary at actual temperature since their temperature coefficients are different. Care should be exercised in the specification of RTDs to be used at the generation facility to maintain consistency of which standard is to be used.

For measurement of gas flow in a power plant, one typical device is an acoustic pyrometer. This device has a sound generator and receiver on opposite sides of the duct. The speed with which the sound passes across the media is a function of the

TABLE 24.3 Common RTD Resistance Reference Values

Copper (10 Ω)

Nickel (25 Ω)

Platinum (100 Ω)

density of the gas and, therefore, a function of the temperature of the gas. By detecting the amount of time it takes for a sound wave to transverse from transmitter to receiver, one can calculate the temperature of the gas.

Similarly, temperature can be calculated by the use of light. The instrument used is a fiber optic sensor. An intrinsic fiber optic sensor performs the temperature measurement, whereas an extrinsic fiber optic sensor has an external transducer that converts the light information to an electrical signal for determination of temperature. Optical sensors measure the amount of thermal radiation of the light emitting source to calculate the temperature of the media in question.

FLOW

There are several end elements that can be used to sense the amount of flow, each with their own advantages and disadvantages. The most cost-effective element is the *orifice flow meter* which is the simple differential pressure sensor monitoring two sides of an orifice plate in a stream of fluid.

The amount of flow across an *orifice* is proportional to the cross-sectional area of restriction and proportional to the square root of the differential pressure across the element. Said another way, the differential pressure across the orifice plate is the square of the flow. When the ratio of the diameter of the orifice to the diameter of the pipe is less than 0.3, equation (24.8) defines the relationship among flow, differential pressure across the orifice, and the diameter of the orifice.

$$Q = 19.636 \times C \times d_1^2 \times h^{1/2} \tag{24.8}$$

where

Q = flow (gpm)

C = orifice coefficient (see Figure 24.4 for values for water)

d_1 = diameter of orifice or nozzle opening (in.)

h = differential head at orifice (feet of liquid)

Figure 24.4 gives typical values for the orifice coefficient for water media.

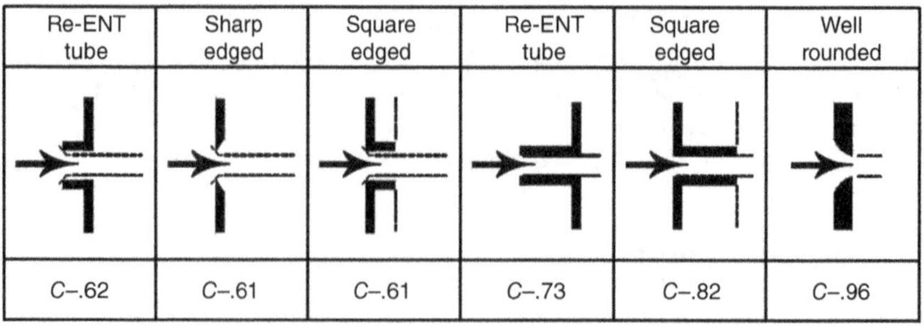

Re-ENT tube	Sharp edged	Square edged	Re-ENT tube	Square edged	Well rounded
C–.62	C–.61	C–.61	C–.73	C–.82	C–.96

Figure 24.4   Typical orifice coefficients for water media. *Source*: Reproduced with permission of Tampa Electric Company. Reproduction is forbidden without the express consent of Tampa Electric Company.

Remembering that power loss across a valve or orifice is the product of the flow and differential pressure across the valve or orifice, and since differential pressure is the square of the flow, we can see that the power loss across the orifice is a cube function of flow. Due to this, while the orifice plate is one of the less costly methods of measuring flow from an initial installation cost, the losses associate with the element will be large for higher values of flow. The tab on the orifice plate contains information regarding its construction. Items such as the equipment number of the plate, the diameter ratio, and the pipeline size are all stamped on the tab of the orifice plate which extends to the outside of the pipe flange where the information can be read.

Other methods are available that have lower losses associate with the element and therefore are more efficient. The *flow nozzle* is one such device. This element, just like the orifice plate causes a pressure drop across the element and this pressure drop is an indication of flow. However, the flow nozzle is constructed much like the nozzle in a turbine where the cross-sectional area is reduced, velocity is increased and pressure is reduced but to a lesser extent than with the orifice at the same flow rate. At the end of the flow nozzle the diameter suddenly increases to the pipe diameter and this has some losses associated with it. Since the pressure drop is less for the same flow, the power losses associated with the flow nozzle are less than with the orifice plate. Since the power loss is less, the efficiency of the flow nozzle is higher than with the orifice plate. The flow nozzle also reduces flow turbulence around the element. The disadvantage of the flow nozzle over the orifice plate is that it has a higher initial capital cost.

Another device used to measure flow is the *venturi flow meter* which utilizes the *venturi effect* to measure flow. In this device, the cross-sectional area is reduced gradually similar to the flow nozzle, but at the discharge the cross-sectional area is increased gradually back to the ID of the pipe. Since we do not have the sudden increase in cross-sectional area at the discharge of the flow element, the venturi tube is more efficient and lower loss device than the flow nozzle.

Another device used to measure flow is the *pitot tube*. This device has two penetrations to the pipe containing the media that we are measuring flow. One penetration has its tube perpendicular to the direction of flow of the media. As such, this penetration only monitors the static pressure of the media. The other penetration has its tube parallel to the direction of flow and, as such, is monitoring both the static pressure and the dynamic pressure of the media due to the velocity of the media. The sensing elements subtracts the pressure read from the perpendicular tube (static) from the pressure read from the parallel tube (static + dynamic). The resultant pressure is the pressure from the dynamic flow of the media only. The flow can be calculated from the value of the dynamic pressure. For a pitot tube with a velocity probe coefficient of 1.0, equation (24.9) approximates the relationship between fluid flow (V), the velocity pressure (VP) and the density of the media (d).

$$V = (\text{VP}/d)^{1/2} \tag{24.9}$$

where

V = velocity (fpm)

VP = velocity or dynamic pressure (in. H_2O)

d = density (lb/ft^3)

Looking for flow of solid or particulate material, a rotating paddle with a switch can be used to indicate whether flow exists in a stream of solid material. Another type of indication of flow of solid material is the coal feeder. This usually has two modes of control called "volumetric" and "gravimetric." In volumetric mode, the material is dumped on a conveyor belt and the material is formed to a certain height and width. Knowing the conveyor speed, we can calculate the "volume" of the material that is delivered, thus the term volumetric mode. In gravimetric mode, the material is again dumped on a conveyor belt but a scale weighs the material. Knowing the weight and the speed of the conveyor, we can calculate the "mass" or gravity of the material transported and this is known as the "gravimetric" mode.

There are other flow elements used in the power generation facility, too many to cover in detail here. However, some of the elements you will come across are

- Magnetic flow meters
- Vortex flow meters
- Coriolis meters
- Ultrasonic meters
- Positive displacement meters

PRESSURE

The pressures of various systems are monitored and transmitted back to the control room and control system to ensure reliable plant operations. Before we discuss the various final elements that are used to monitor pressure, we must first talk about units and reference points. In electrical circuits, the voltage at any node in a circuit takes on meaning only when referenced to another node in the circuit. For example, when we say a motor has 480 V applied to the stator, this is the voltage on one phase of the motor as referred to another phase of the motor. Similarly, when we say we have a certain pressure at a certain node in a piping system, we need to refer that pressure value to some reference point. The two most common reference points are atmospheric pressure and absolute vacuum.

When a pressure is in reference to absolute vacuum, the units used are pounds per square inch absolute (psia). When a pressure is in reference to atmospheric, the units used are pounds per square inch gauge (psig). Sometimes, we may want to measure the pressure when the pressure is between absolute zero and atmospheric. This is more commonly called measuring the vacuum of a system. At sea level, the value of atmospheric pressure is 14.7 psia or 0 psig. Percentage of vacuum is a measurement of how far from atmospheric pressure the pressure is. For example, atmospheric pressure is 0% vacuum. Pure vacuum is 100% vacuum. Half way between atmospheric pressure and a pure vacuum is 50% vacuum. There are various units used when discussing measurement of vacuum. Table 24.4 shows the conversion from one system to another.

Example 24.5 Given a vacuum measurement of 50% vacuum, what is the equivalent vacuum measured in units of psia?

TABLE 24.4   Unit Comparison for Pressure Measurements

% Vacuum	Torr	Micron	psia	Inches Mercury Absolute	Inches Mercury Gauge	kPa
0	760.000	760,000.00	14.70	29.92	0.0000	101.4
1.3	750.120	750,120.00	14.51	29.53	0.3890	100.0818
1.9	745.560	745,560.00	14.42	29.35	0.5685	99.4734
7.9	699.960	699,960.00	13.54	27.56	2.3637	93.3894
21	600.400	600,400.00	11.61	24.64	6.2832	80.106
34	501.600	501,600.00	9.70	19.75	10.17	66.924
47	402.800	402,800.00	7.79	15.86	14.06	53.742
50	380.000	380,000.00	7.35	14.96	14.96	50.7
61	296.400	296,400.00	5.73	11.67	18.25	39.546
74	197.600	197,600.00	3.82	7.78	22.14	26.364
87	98.800	98,800.00	1.91	3.89	26.03	13.182
88	91.200	91,200.00	1.76	3.59	26.33	12.168
89.5	79.800	79,800.00	1.54	3.14	26.78	10.647
90.8	69.920	69,920.00	1.35	2.75	27.17	9.3288
92.1	60.040	60,040.00	1.16	2.36	27.56	8.0106
93	53.200	53,200.00	1.03	2.09	27.83	7.098
93.5	49.400	49,400.00	0.96	1.94	27.98	6.591
94.8	39.520	39,520.00	0.76	1.56	28.36	5.2728
96.1	29.640	29,640.00	0.57	1.17	28.75	3.9546
96.6	25.840	25,840.00	0.50	1.02	28.90	3.4476
97.4	19.760	19,760.00	0.38	0.7779	29.14	2.6364
98.7	9.880	9,880.00	0.19	0.3890	29.53	1.3182
99	7.600	7,600.00	0.15	0.2992	29.62	1.014
99.9	0.760	760.00	0.01	0.0299	29.89	0.1014
99.9	0.760	760.00	0.01	0.0299	29.89	0.1014
99.99	0.076	76.00	0.00	0.0030	29.92	0.01014
99.999	0.008	7.60	0.00	0.0003	29.92	0.001014
100	0.000	0.00	0.00	0.0000	29.92	0

Solution: From Table 24.4, the pressure at 50% vacuum is 7.35 psia. This could have also been calculated as follows:

$$P = 14.7 \text{ psia} \times (100\% - 50\%)/100\% = 7.35 \text{ psia}$$

So what type of final element is used to measure pressure? The simplest type is the *Bourdon tube*. This device is a tube formed into a curve. The measured static pressure is admitted to the inside of the tube and acts to straighten out the tube. The greater the static pressure is, the more the tube tends to straighten out. The coiled tube is connected to an indicator, so as the tube straightens out, the dial moves indicating the static pressure inside the Bourdon Tube. The advantage of this device is its simplicity and low capital cost. The disadvantage of this device is it is limited to local indication of pressure only.

A more flexible type of final element is the *bellows sensor*. This device uses a moving shaft that is connected to a flexible bellows assembly. On the one side of the bellows assembly is the media being monitored for pressure and on the other side of the bellows is a spring providing the opposing force to the pressure on the other side of the bellows. As the static pressure increases, the force on the one side of the bellows increases, which compresses the spring more and moves the shaft. This shaft can be either connected to a local indicator or be connected to a device to convert the position of the moving shaft to an electrical signal for remote indication.

Another type of final element used to sense pressure is the *capacitive sensor*. This device is constructed with a movable electrode in the center and two fixed electrodes on the outside of the device. Between the movable and fixed electrodes is a piece of dielectric material. This assembly forms a capacitance. The value of the capacitance is determined by the distance between electrodes as shown in equation (24.10).

$$C = \varepsilon A/D \qquad (24.10)$$

where

C = capacitance

ε = permittivity of dielectric material

A = surface area of the electrode

D = distance between inner and outer electrode

LEVEL

The simplest method of measuring level is the dip stick where a stick is dipped inside a tank and then removed to indicate the level in the tank. This method is very simple and has a very low capital cost, but the value can only be read locally. Site glasses or gauge glasses require less labor, but can only be read locally. Around the late 1980s and early 1990s, before the advent of more modern instrumentation that could tolerate high values of pressure and temperatures, the steam drum on the boiler would have a local site glass and a camera would be installed to view the sight glass remotely. More modern devices are now available that can tolerate the high pressures and temperatures of the steam drum and other locations and convert the pressure signal to an electrical signal to be transmitted to the control room or control system.

One of the challenges of using devices that tap off the main tank is to ensure that, in the sample piping, the temperature of the media being monitored is the same temperature as the media in the main tank. This is due to the fact that as the temperature of the media changes, the specific density of the media changes. As it becomes cooler, it becomes denser. Therefore, if we allow the media in the site glass or external tubing to become cooler, the fluid in the sight glass will be denser. Since the sight glass is connected to the main tank at the top and bottom entry points, the pressure at these points is the same in the tank and in the sight glass. Since the media in the site glass is cooler and denser, it takes less height of the fluid in the sight glass to see the same differential pressure from top tap to bottom tap. Therefore, with the media in the sight glass cooler, the level indicated in the cooler would indicate lower than the actual level in the tank. This is critical in certain applications such as feedwater heater level controls. On the steam side of the feedwater heater, we maintain level in the heat exchanger as low as possible. This allows for the heat transfer on the steam side to occur in the latent heat transfer phase instead of the sensible heat transfer phase. This provides for more energy transfer from the steam side to the feedwater side. As the level on the steam side of the feedwater heat exchanger increases and tubes begin to become covered by condensate, the heat transfer at these tubes is no longer using latent heat transfer and the efficiency of the feedwater heater is decreased, which decreases the overall plant efficiency heat rate. (See more detail in chapter 2 on thermodynamics and the steam cycle). In addition to having the level too high affecting the heat rate of the plant, there is also the danger that, if the feedwater heater completely fills up, then condensate from the steam side of the feedwater heater could be carried back to the steam turbine causing water impingement on the turbine blades and catastrophic failure of the steam turbine. Therefore, it is important that the level control systems in the plant receive accurate values for levels in the plant.

Manometric gauges measure a reference head or pressure to the measured head or pressure of the media of concern. A common application of a manometric gauge is the boiler drum level where the reference head is the bottom of the steam drum and the measured head is the top of the steam drum. The steam drum level is a function of the amount of steam leaving the steam drum and the amount of feedwater that enters the drum. We control the feedwater rate to control level in the steam drum.

There are some challenges to boiler drum level control. During transients, the drum level responds initially in one way, but then recovers and responds in the other direction for final direction. This has to do with the nucleate boiling that occurs in the water wall tubes of the boiler. At steady state, the rate of nucleate boiling is steady. Let us evaluate what occurs when we experience a sudden load increase. This results in a sudden increase in steam flow. This initially lowers the pressure in the water wall tubes, which increases the nucleate boiling in the water wall. Since gas takes up more volume than fluid, this decreases the density of the steam water mixture in the water wall tubes, and forces the level of the steam drum higher. One's first reaction would be to reduce the feedwater rate into the steam drum to reduce the level. However, remember that what started this is the increase in steam flow from the drum. If the feedwater feed rate did not change, drum level would quickly start to lower due to the increase in steam demand. Therefore, with a sudden increase in steam flow from the drum, the correct response is the increase the feedwater feed rate despite the fact that, initially the level in the steam drum increased. This is known as boiler swell.

Now let us evaluate what occurs when we get a sudden load decrease. This results in a sudden decrease in steam flow. This initially raises the pressure in the water wall tubes, which decreases the nucleate boiling in the water wall. Since gas takes up more volume than fluid, this increases the density of the steam-water mixture in the water wall tubes, and forces the level of the steam drum lower. One's first reaction would be to increase the feedwater rate into the steam drum to raise level back up. However, remember that what started this is the decrease in steam flow from the drum. If feedwater feed rate did not change, drum level would quickly start to rise due to the decrease in steam demand. Therefore, with a sudden decrease in steam flow from

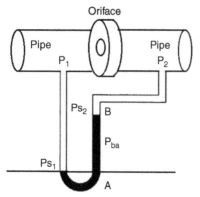

Figure 24.5 Simplified manometer.

the drum, the correct response is the decrease the feedwater feed rate despite the fact that initially the level in the steam drum decreases. This is known as boiler shrink.

A simple *manometer* is shown in Figure 24.5 where the manometer is installed across an orifice plate (see section on flow final elements above). Looking at Figure 24.5, we see the fluid in the port P_2 is physically higher in height and at point P_1. Let us evaluate this static condition to see which port's pressure is greater, P_1 or P_2.

On port P_1, the height of the fluid in the port is at a lower height than the fluid in port P_2. Let us draw a line at the height of the fluid in port P_1 shown in Figure 24.5. Realize at a static condition, the pressure at the two ports at the same height must be the same. Therefore, the pressure at the surface of the fluid in port P_1 (will call this pressure Ps_1) must be equal to the pressure in port P_2 at point A. We will call this pressure P_a. Therefore, we can write the following equation,

$$Ps_1 = P_a$$

Note that the pressure of the surface of the port P_2 is higher in elevation and is located at point B. There is a certain amount of head pressure associated with the amount of fluid in port P_2 between points A and B. This head we will call P_{ba}. The pressure at point A (P_a) is the sum of the pressure at the surface of port P_2 (Ps_2) and the head of the fluid from point B to point A (P_{ba}). We can write this as

$$P_a = Ps_2 + P_{ba}$$

Now, if we combine the above two equations we find,

$$Ps_1 = P_a$$
$$Ps_1 = (Ps_2 + P_{ba})$$

For the above to be true, the pressure at port P_1 must be greater than the pressure at port 2, by the amount of head between point B and point A (P_{ba}). You can see from the above that the measurement of the difference in height of the fluid between the two ports is proportional the differential pressure across the orifice.

Other types of level indication are switches used in conductive medias. When the level rises in a tank and makes contact with a sensor, the sensor can detect the conductivity of the media and transmit this change in conductivity to a remote location. Note this is simple device, but only provides a discrete value and not analog value. For medias that contain non-conductive media, float switches can be used to indicate level reaching a certain point. Float switches may be of the free moving type or the displacer type. The free moving type float remains out of the media being measured and floats on top of the media. As the level changes the float passes by switches located at several elevations and the position of the float, and thereby tank level is indicated by which switch is activated. The displacer float stays in place and is connected to a spring. As the level increases, the media applies more head pressure to the float and the float transmits this head pressure to the spring and moves the spring and connected shaft by a small amount. The displacement of the float and spring assembly provides indication of the head applied to the float and, therefore, level in the tank.

To enhance the reliability of digital-level indication, there will be two probes to detect level. One indicator provides a low (or high)-level alarm and second indicator provides low-low (or high-high) level alarm. This adds to the reliability of the signal as the failure of one probe does not result in the complete loss of high- or low-level indication.

The amount of pressure or hydraulic head on a pressure transmitter exposed to a certain amount of fluid at a potential height is given by equation (24.11).

$$\text{Hydraulic head } P = \rho \, g \, h \qquad (24.11)$$

where

P = Hydraulic head

ρ = density of fluid

g = gravitational constant

h = height of fluid

If we have the fluid height and specific gravity, we can simplify this formula to calculate for us the height of the fluid based on an indicated pressure as described in equation (24.12).

$$H \,(\text{in.}) = [P(\text{psig})/\text{SG}] \times 27.678 \qquad (24.12)$$

where

P = differential pressure (psig)

H = height of fluid (in.)

SG = specific gravity of the fluid under consideration

The value of 27.678 is a conversion factor to convert units of pounds force per square inch to inches of fluid. This conversion factor comes from the fact that 2.31 ft (or 27.678 in.) of water at a specific gravity of 1.0 will result in a differential pressure from the top of the water column to the bottom of the water column of 1 pound per square inch differential.

For tanks open to atmosphere, the pressure is in unit psig where the reference pressure is the atmospheric pressure at the top of the tank. For closed tanks, the pressure is a differential indication where the pressure in the top of the tank is the reference pressure.

Example 24.6   Given a tank that is open to atmosphere on top of the tank with a fluid that has a specific gravity of 1.0, if the pressure gauge indicates a pressure of 1 psig, what is the height of the fluid in the tank in inches?

Solution: Using equation (24.12), we find the following.

$$H(\text{inc.}) = [P(\text{psig})/\text{SG}] \times 27.678$$
$$H(\text{in.}) = [1 \text{ psig}/1.0] \times 27.678$$
$$H(\text{in.}) = 27.678 \text{ in.}$$

Example 24.7   A water storage tank is vented to atmosphere. The tank is located at sea level and contains 100,000 gal of 80°F water (specific gravity of 1.0). A pressure gauge at the bottom of the tank reads 10 psig. What is the approximate water level in the tank?

Solution: Using equation (24.12), we find the following.

$$H(\text{in.}) = [P(\text{psig})/\text{SG}] \times 27.678$$
$$H(\text{in.}) = [10 \text{ psig}/1.0] \times 27.678$$
$$H(\text{in.}) = 276.78 \text{ in. or } 23 \text{ ft}$$

In addition to the level detectors described above, other technologies are available that do not require contact with the media involved. Some of these technologies are radio frequency (RF)-level sensors that use radio waves to detect level, optic sensors that have a transmitter of light source and a receiver to detect level, microwave level sensors which use a microwave transmitter and receiver, ultrasonic level sensors that use sound wave transmitters and receivers, and radiation sensor that uses a gamma source and a radiation detector. Ultrasonic, microwave, and radiation technologies are commonly applied when the material is nonconductive such as with many solids. Some of these materials at the generation station include items such as coal, fly-ash, and limestone.

Other Plant Instrumentation

Potential transformers (PT), as shown in Figure 24.6, are used to monitor voltages in the power plant. Typical power plants utilize voltages in values from 480 V up to 500,000 V. These voltages are too high to be sent directly to plant instrumentation systems. PTs are used to reduce the magnitude of the voltage sent to the control system. The typical value for secondary voltage is 120 Vac.

Similarly, current transformers (CT), as shown in Figure 24.7, are used to monitor currents in the power plant. Typical power plants have very large loads. Currents in excess of 1000 A are typical. These values of current are too high to be sent directly

Figure 24.6   Potential transformer. *Source*: Reproduced with permission of Eaton Cutler-Hammer.

Figure 24.7   Current transformer.

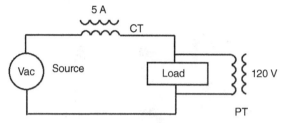

Figure 24.8   Current transformer and potential transformer one line diagram.

to plant instrumentation systems. CTs are used to reduce the magnitude of the current sent to the control system. The typical value for secondary current is 5 A or 1 A.

While we talk about current transformers (CT) and potential transformers (PT) as if there were two different devices, they are both transformers. A CT could be used as a PT and a PT can be used as a CT if the turns ratio and core cross-sectional area are adequate for the application. The actual difference between a CT and a PT is the application of the transformer as shown in Figure 24.8.

The PT is applied in parallel with the source or the load that we have interest in monitoring. The CT is applied in series with the source or the load that we have an interest in monitoring. Since the PT is in parallel, the output voltage is the input voltage times the turns ratio of the potential transformer. If you open the secondary of a PT, then the secondary impedance of the PT becomes infinite. The impedance reflected to the primary is the square of the turns ratio times the secondary resistance. With an open secondary, the PT presents an open circuit to the voltage monitored. As such, the secondary of a PT circuit can be open-circuited safely. If you short-circuit the secondary of a PT, then the secondary impedance of the PT becomes zero. The impedance reflected to the primary is the square of the turns ratio times the secondary resistance. With a shorted secondary, the PT presents a short circuit to the voltage monitored. As such, excessive currents can be developed and therefore, the secondary of a *PT circuit cannot be shorted circuited safely*.

Since the CT is in series with the load or source, the output current is the input current times the turns ratio of the current transformer. If you short-circuit the secondary of a CT, then the secondary impedance of the CT becomes zero. The impedance reflected to the primary is the square of the turns ratio times the secondary resistance. With a shorted secondary, the CT presents a short circuit to the primary circuit. Since the CT is in series with the load, the secondary current will be the primary current times the square of the turns ratio and, therefore, the CT can be safely short-circuited on the secondary side.

If you open the secondary of a CT, then the secondary impedance of the CT becomes infinite. The impedance reflected to the primary is the square of the turns ratio times the secondary resistance. With an open secondary, the CT presents an open circuit to the primary circuit. This will look like an open on the primary side. This will cause the voltage on the secondary side of the CT to increase to very large values. As such, the secondary of a *CT circuit cannot be open-circuited safely*. Additionally, since it is unsafe to open-circuit the secondary of the current transformer, fuses are

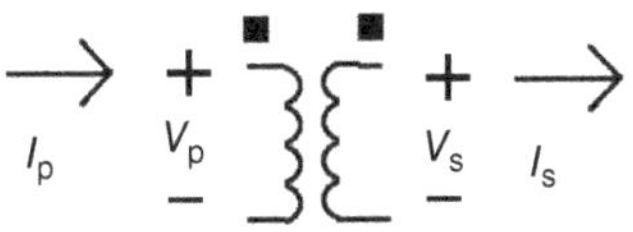

Figure 24.9 Transformer polarity.

not used in the secondary circuit of a current transformer. Since a short circuit on a PT secondary circuit results in unsafe currents, all ungrounded conductors on the secondary of a PT are fused to protect the transformer from a short on the secondary circuit.

When looking at the polarity of a transformer, there are polarity marks showing the polarity of the voltage and current waveforms. Using Figure 24.9, the voltage polarity of the primary voltage (V_p) matches the voltage polarity of the secondary voltage (V_s). Also the current flowing into the polarity mark of the primary I_p is the current flowing out of the polarity mark of the secondary I_s.

The accuracy of a current transformer CT used in a metering application is usually stated as a percent at rated burden. This leads to a "B" rating for a CT. The standard burden (B = 1) for a CT with a 1 A secondary is 1 VA ($I^2 \times R = (1\ A)^2 \times 1\ W = 1\ VA$). The standard burden (B = 1) for a CT with a 5 A secondary ($I^2 \times R = (5\ A)^2 \times 1\ W = 25\ VA$) is 25 VA.

A listing of standard current transformer burdens for CTs with a secondary of 5 A is given in Table 24.5.

Example 24.8   Given, a CT rated for 5 A secondary, that has an accuracy of 0.3 at B = 0.1, determine the secondary burden in units of VA that the accuracy value is defined at.

Solution: Since the CT has a secondary rated current of 5 A, the burden of B = 0.1 means the VA secondary load on the CT for the accuracy is $0.1 \times 25\ VA = 2.5\ VA$. (This can also be obtained directly from Table 24.5).

Current transformers used in protective relaying applications utilize a different system to define the accuracy. The reason that there is a separate system for defining the accuracy of CTs used in a protective circuit versus a system for defining the accuracy of CTs used in a metering circuit is the application of the CT. For the metering

TABLE 24.5   Standard Current Transformer Burdens

Type	Max VA Burden	Max Ext Impedance (ohms)
B = 0.1	2.5	0.1
B = 0.2	5.0	0.2
B = 0.5	12.5	0.5
B = 1.0	25.0	1.0
B = 2.0	50.0	2.0

Source: Reproduced with permission of IEEE.

circuit, the CTs' function is to accurately reflect the current of the circuit under normal loads. Therefore, the CT accuracy rating is based on nominal secondary current and the maximum VA that is connected to the secondary of the current transformer. For the protective relay circuit, the CTs' function is to accurately reflect currents to the protective relay during a fault. As such, the system for defining the accuracy of CTs in protective relay circuits define the accuracy of the CT at a value of secondary current magnitudes higher than nominal and based on a certain amount of load on the secondary circuit.

There are several classes of current transformers for protective relay circuits. Class C or K covers current transformers in which the leakage flux in the core does not have an appreciable effect on the ratio within the limit of current and burden. A majority of current transformers in service at a generation station will be the Class C current transformer. Class K classification is a special case of the Class C transformer that has a saturation point or a knee-point voltage of at least 70% of the secondary terminal voltage rating (IEEE C57.15).

Class T current transformers are transformers where the leakage flux has an appreciable effect on the ratio within the specified limits (IEEE C57.15). These transformer accuracies are tested due to this appreciable effect to ensure accuracy, thus the T rating for "tested."

The transformer accuracy is given as a value as PPXVVV, where PP is the maximum percent error that the transformer will present at VVV secondary volts at 20 times secondary current. The value X is the class of current transformer as discussed above. For multitap current transformers, the accuracy is based on the maximum number of turns of the current transformer and the burden varies directly with the tap of the transformer. For example, a Class C400 multitap current transformer with a nominal turns ratio of 1000:5 CT, when used on the 500 A tap, has new class of C200 as calculated in equation (24.13).

$$C_{tap} = C_{rated}(TAP_{actual}/TAP_{maximum}) \qquad (24.13)$$

where

C_{tap} = secondary voltage maximum at 20 times secondary current for actual transformer tap

C_{rated} = secondary voltage maximum at 20 times secondary current for nominal turns ratio

TAP_{actual} = actual tap setting

$TAP_{maximum}$ = nominal primary current rating

Example 24.9 For a Class C400 multitap current transformer with a nominal turns ratio of 1000:5 CT, when used on the 500 A tap, determine the accuracy class of the transformer on the 500 A tap.

Solution: Using equation (24.13) we find the following.

$$C_{tap} = C_{rated}(TAP_{actual}/TAP_{maximum})$$
$$C400 \times (500/1000) = C200$$

TABLE 24.6   Standard Accuracy Class Ratings for Current Transformers in Metal-Clad Switchgear

Ratio	B0.1	B0.2	B0.5	B0.9[b]	B1.8[b]	Relaying Accuracy[c]
50:5	1.2	2.4[d]				C or T10
75:5	1.2	2.4[d]				C or T10
100:5	1.2	2.4[d]	—	—		C or T10
150:5	0.6	1.2	2.4[d]			C or T20
200:5	0.6	1.2	2.4[d]	—	—	C or T20
300:5	0.6	1.2	2.4[d]	2.4[d]		C or T20
400:5	0.3	0.6	1.2	1.2	2.4[d]	C or T50
600:5	0.3	0.3	0.3	1.2	2.4[d]	C or T50
800:5	0.3	0.3	0.3	0.6	1.2	C or T50
1200:5	0.3	0.3	0.3	0.3	0.3	C100
1500:5	0.3	0.3	0.3	0.3	0.3	C100
2000:5	0.3	0.3	0.3	0.3	0.3	C100
3000:5	0.3	0.3	0.3	0.3	0.3	C100
4000:5	0.3	0.3	0.3	0.3	0.3	C100

Source: Reproduced with permission of IEEE Std C57.13.
[b] These were formerly standard burdens of B1.0 and B2.0, respectively, which now are classified as relaying burdens in IEEE Std C57.13.
[c] These accuracies may not be sufficient for proper relaying performance under all conditions. To enable proper relaying performance, the user should make an analysis of CT performance considering the relaying requirements for the specific short circuit currents and secondary circuit impedances (see 8.7.1 of C57.13).
[d] This metering accuracy is not in IEEE Std C57.13.

For installation in metal-clad switchgear, the standard current transformer accuracies for metering and relaying are listed in Table 24.6. These are the minimum required accuracies defined by IEEE and if higher burden capability is required by the application, the switchgear purchaser needs to define the required burdens necessary for the CTs to be used.

For protective circuits where the currents can exceed several times nominal value, the burden of the conductors between the protective device and the current transformer may be a significant portion of the secondary burden. NEC® requires CT secondary conductors to be at least AWG #10 or larger. Also, NEC® requires one side of the secondary circuit to be solidly grounded providing a reference plane for the rest of the circuit.

A transformer, be it a voltage or current transformer, functions by the rate of change of flux. The voltage induced on the secondary of a transformer is defined by

$$e = n(d\varphi/dt) \tag{24.14}$$

where

e = induced voltage (V)

n = number of turns of the transformer

$d\varphi/dt$ = rate of change of flux with time (Wb/s)

The flux flows through the magnetic core of the transformer. As the magnitude of current flowing through the primary winding of the transformer changes, the amount of flux changes accordingly as long as we remain in the linear region of the core B-H curve. If the flux density increases beyond the linear region of the B-H curve, saturation of the core occurs. With further increases in current, the flux in the core remains constant at this saturation point. Upon saturation, even though the magnitude of flux is large, the rate of change of flux is zero. Therefore, the voltage induced on the secondary becomes zero, until the current, and thereby the flux of the core, is reduced below saturation.

PTs and CTs are used to indicate voltage and current, but they are also used to feed power and energy meters as shown in Figure 24.10.

An electromechanical kilowatt-hour (kWh) meter has both voltage and current inputs and drives a disk based on the in-phase portion of current and voltage (i.e., the amount of real power). Figure 24.11 shows the typical electro-mechanical kWh meter arrangement.

The kWh register records the revolutions of the meter disk in terms of electrical energy that the meter has detected. The register usually has five dials. The register

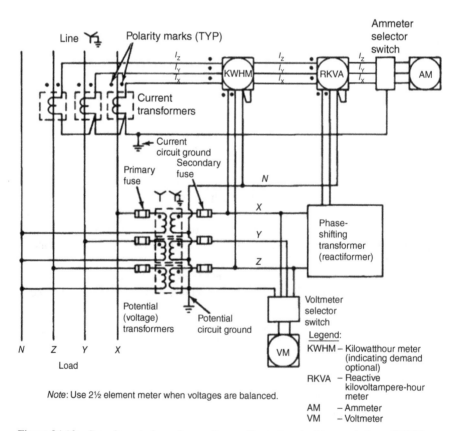

Figure 24.10 Sample metering scheme. *Source*: Reproduced with permission of IEEE.

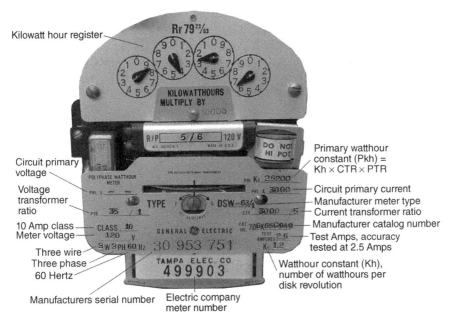

Figure 24.11 Typical electro-mechanical kWh demand meter nameplate. *Source*:
Reproduced with permission of Tampa Electric Company. Reproduction is forbidden without
the express consent of Tampa Electric Company.

reading must be multiplied by the billing multiplier in order to convert the reading to
kWh.

The billing multiplier is the number by which the register reading is multiplied
to obtain kW, kWh or kQh. The billing multiplier on a specific meter is directly proportional to the current transformer (CT) and potential transformer (PT) ratios. For
example, given a meter fed by a current transformer with a ratio of 400:5 or a CTR of
80:1 and fed by a potential transformer with a ratio of 288:120V or a PTR of 2.4:1,
the billing multiplier is the product of the PTR and CTR or 80 × 2.4 or 192. For a
kWh register reading of 02543 and a billing multiplier of 192 (from the meter face),
the following calculation is made:

$$2543 \text{ kWh} \times 192 = 488,256 \text{ kWh}$$

The demand reset interval indicator is used to indicate the relative point of the
30-min demand interval. It resets itself at the end of each interval. Electronic demand
meters do not have this feature. The demand register records the peak kW demand
usage over a specific time period. The register reading must be multiplied by the
billing multiplier in order to convert the reading to kW. The disk constant (K_h) is the
number of Wh represented by one revolution of the mechanical meter disk or by one
electronic (disk) revolution in a solid-state meter. "Number of wires" is defined as
the number of wires or conductors that comprise the service. A single-phase service
will have either two wires for 120 V service or three wires for 120/240 V service.
Three-phase services will have *four wires* for delta-connected services (240/120 V

and 480/240 V) and Y-connected services (208Y/120 V and 480Y/277 V). Meter voltage is the voltage at which the voltage coil of the meter is energized. Typical meter voltages are 120, 208, 240, 277, or 480 V.

The kWh meter integrates the real power delivered (or absorbed) over time and measures the amount of energy that is delivered (or absorbed). By definition, power is the rate at which energy flows or

$$P = E/t \tag{24.15}$$

where

> P = power (kW)
>
> E = energy (J)
>
> t = time (sec)

In a kWh meter, to determine the power that the meter is indicating, count the seconds for the meter to make a given number of revolutions of the disk and then use the following formula to determine the power in kW.

$$P = 3.6 \times r \times K_h \times M/t \tag{24.16}$$

where

> P = power (kW)
>
> r = number of revolutions observed
>
> K_h = meter disk constant (Wh/rev)
>
> t = time (sec)
>
> M = billing multiplier

The billing multiplier is the product of the potential transformer ratio (PTR) and the current transformer ratio (CTR).

$$M = PTR \times CTR \tag{24.17}$$

where

> PTR = potential transformer ratio
>
> CTR = current transformer ratio

kWh meters are classified as to the maximum safe amount of current they can monitor. The meter class is the designation of the meter's maximum continuous load range in amperes. Self-contained meters are class 60, 100, 200, 320, or 480. Instrument transformer-rated meters are class 10 or 20. Table 24.7 is a listing of the common classes along with the maximum steady state current each can safely monitor.

The number of PTs and CTs required to drive a kWh meter is defined by the circuit that is monitored. A single-phase, two-wire system requires only one PT and one CT to accurately determine the energy transferred through the system. For a three-phase, four-wire system, three PTs and three CTs are required to determine the energy transferred through the system. A stator is an electromagnetic element energized by a voltage coil and one or more current coils. A stator provides the driving torque for the meter disk. The number of stators required for proper metering is determined from

TABLE 24.7   Common Classes of Energy Meters

Meter Class	Maximum Safe Current (A)
Class 10	10
Class 20	20
Class 100	100
Class 200	200
Class 320	320

Source: Reproduced with permission of IEEE.

Blondel's theorem, which states that "in a system of n conductors, $n-1$ meter elements (stators), properly connected, will measure the power or energy taken." Table 24.8 shows the various possible system configurations and the required number of PTs and CTs for the energy meter.

Solid-state kWh meters are available with a combination of functions such as kWh, kvarh, kQh, and kVAh. Also available are bi-directional Wh meters and varh meters in a single device. An advantage of these meters is that one set of input connections meets the requirements for any configuration. Solid-state meters are also useful where the customer is using solid-state switching devices on his load, such as silicon-controlled rectifiers (SCRs), which may create harmonics on the power delivered at the meter. A conventional induction disk kWh meter cannot accurately measure such loads.

In addition to being able to determine the amount of real power and energy absorbed from a system or delivered to a system, many systems monitor the reactive power delivered or absorbed from an electrical system. Reactive power delivered to an electrical system tends to support the voltage of the system and reactive power absorbed from an electrical system tends to suppress the voltage of the system. To measure the amount of reactive energy, kvarh meters are used. Internally, this is similar in construction to the kWh meter. However, a potential transformer is applied between the PT and the voltage input to this meter to shift the potential 90 electrical degrees. Therefore, while internally the meter is still determining real power, due to the 90-degree shift, from a system viewpoint, the quantity represents the reactive power of the system.

TABLE 24.8   Metering Requirements for PTs and CTs

Service Voltage	Stators	CT	PT	Load Restriction
1-phase, 2-wire	1	1	1	None
1-phase, 3-wire	1	2	1	None
1-phase, 3-wire	2	2	2	None
1-phase, 3-wire (wye)	2	2	2	None
3-phase, 3-wire (delta)	2	2	2	None
3-phase, 4-wire (wye)	$2^1/_2$	3	2	Balanced
3-phase, 4-wire (wye)	3	3	3	None
3-phase, 4-wire (delta)	3	3	3	None
3-phase, 4-wire (delta)	2	3	2	Balanced mid tap

Source: Reproduced with permission of IEEE.

One last item of caution is mentioned for the reader. All sockets used for commercial and industrial installations are normally equipped with a bypass switch. The bypass is a device that can be operated to bypass the load from the meter without interrupting the customer and the bypass switch releases the meter for removal. The function of the bypass switch is to allow for removal of the meter when the meter is not under load. In many applications, the bypass is not rated for meter full load. In many instances, the bypass switch is only rated for about half of the continuous load. Before utilizing a bypass switch, ensure that the connected load is less than the bypass switch's rated load to ensure the bypass can be operated safely. Additionally, whenever removing or installing a meter into a socket, always follow safe work practices by wearing the required PPE for the task.

INSTRUMENT IDENTIFICATION STANDARDS

A common industry standard that is used to document instrumentation identification is the International Society of Automation Standard ISA 5.1. This standard defines instrumentation identification based on the function of the instrument, not the physical construction of the instrument. For example, a differential pressure transmitter used with an orifice plate to indicate the flow of material in a pipe is not called a differential pressure transmitter but is called a flow transmitter. ISA 5.1 uses the first letter of the instrument identification to describe the measured or initiating variable and not the manipulated or final variable. For example, let us look at a storage tank that has a supply line with a valve in the supply line. The valve is used to adjust or manipulate the flow into the tank based on the measured value of level in the tank. While the valve actually controls the flow rate into the tank, the initiating value for the control system for the valve is the level in the tank and, as such, the valve is designated as a level valve and not a flow valve. The format for instrumentation ID will have three letters (up to five if modifying letters are used) followed by a dash, then a number and another letter.

- The first letter defines the measured or initiating variable and may have a modifier associated with it to provide more information on the initiating variable.
- The second letter defines the type of instrument, whether it has a read out, or a passive function such as local display or alarm.
- The third letter defines the output function such as transmitter and may have a modifier associated with it to provide more specific on the type of output of the instrument.

Each instrument system in a generation facility has a unique "loop" number and this is the number that follows the dash. Lastly a final letter is placed on the instrument number. This is a sequential number that is assigned to the instrument to ensure the ID of the instrument is unique. The format is summarized below.

$$ABC - 1D$$

where

A = Measured or Initiating Variable (May have modifier)

B = Readout or Passive Function

C = Output Function (May have modifier)

1 = Loop number

D = Sequence letter

Table 24.9 shows the description of the first three letters of the instrument ID.

Example 24.10   You locate in instrument in the field with a tag. On the tag is the instrument number of PIT-098D. From this instrument ID, determine the following.

a. What is the initiating process variable that the instrument is monitoring?

b. Is there local indication on this device?

c. Does this device transmit a signal back to a remote location?

d. What is the loop number this device is associated with?

Solution: The instrument ID from Table 24.9 is a pressure (signified by P) indicating (signified by I) transmitter (signified by T).

a. The initiating process is pressure (P)

b. The instrument has local indication (I)

c. The instrument is a transmitter that transits its value to remote location (T)

d. Loop number is loop 098.

We have discussed instrument identification systems in a power generation facility which is well defined by ISA standards but how about non-instrument equipment identification systems in a generation facility. We would like to have identification systems that can quickly identify both the system that the equipment is associated with as well as the type of equipment identified by the equipment ID number. Throughout the industry, there are variations to the specific methods of identifying equipment usually dependent on the engineering firm that designed the original system, but all these systems have one thing in common. In all systems, the equipment identification number identifies the unit number, the type of equipment, and the system the equipment is associated with. IEEE 505 provides a very basic standard nomenclature system for generating station electrical power systems. This standard recommends breaking the equipment numbering system down to unit number, the equipment function (or function in the system), and then further to the equipment type. For example, a unit-connected station service transformer should be identified by the unit number, the function of the transformer in the system (provides station service power when the unit is online), and the type of equipment (a transformer). This equipment would then be known as *1-station service-transformer*.

However, the above would make a very long equipment number and, in an effort to shorten equipment numbers and still have meaningful nomenclature, IEEE 505

TABLE 24.9   Instrument Identification Letters

	First Letter (measured or initiating variable)	Variable Modifier	Second Letter (readout or passive function)	Third Letter (output/active function)	Function Modifier
A	Analysis		Alarm		
B	Burner, combustion		User choice	User choice	User choice
C	Conductivity			Control	
D	Density or S.G.	Differential			Deviation
E	Voltage		Sensor, primary element		
F	Flow, flow rate	Ratio (fraction)			
G	Gauging		Glass, gauge, viewing		
H	Hand				High
I	Current		Indicate		
J	Power		Scan		
K	Time	Time rate of change		Control station	
L	Level		Light (pilot)		Low
M	Moisture (humidity)				Middle (intermediate)
N	User choice		User choice	User choice	User choice
O	User choice		Orifice (restriction)		Open
P	Pressure (vacuum)		Point (Test connection)		
Q	Quantity (event)	Integrate (totalize)	Integrate (totalize)		
R	Radiation		Record (print)		
S	Speed (Frequency)	Safety		Switch	
T	Temperature			Transmit	
U	Multivariable		Multifunction	Multifunction	
V	Vibration, mechanical analysis			Valve, damper, louver	
W	Weight (Force)		Well, probe		
X	Unclassified	X-axis	Accessory devices, unclassified	Unclassified	Unclassified
Y	Event, state, presence	Y-axis		Auxiliary device, relay (computer)	
Z	Position, dimension	Z-axis, safety instrumented system		Drive, actuator	

Source: ANSI/ISA-5.1-2009, Instrumentation Symbols and Identification, Table 4.1: Identification letters. Reprinted with permission of the International Society of Automation, www.isa.org/standards

provides common abbreviations used in the generator industry. For example, in IEEE 505, a transformer is abbreviated as "xfmr," the function of station service is abbreviated "sta svce." Utilizing these common abbreviations, the above example would now be *1-sta svce-xfmr*.

This achieves the three goals of identifying the type of equipment, the system the equipment is associated with, and the unit that the equipment is associated with. While the general concepts presented in IEEE 505 are utilized in generation engineering, the details vary between both utilities and the A&E firms that create the structure of the plant nomenclature system. Most energy production facilities take the concepts in IEEE 505 and further reduce the length of the abbreviations to minimize the length of the equipment ID while still maintaining the usefulness of the equipment ID in describing both the type of equipment and the function of the equipment. Different utilities will have unique standards, but below is one example of an equipment identification system that is currently in use in many coal and natural gas plants.

The typical numbering system will consist of a number identifying both the system identifier and the equipment identifier as shown below:

U-SSS-EEE###

where

U = unit number – number of the generator that equipment supports

SSS = three digit system or function identifier as listed in Annex C

EEE = three digit equipment identifier as listed in Annex D

= sequential number to ensure equipment identifier is unique

Annex C lists standard system identification codes and Annex D lists standard equipment identification codes used in the power generation industry.

Going back to ISA standards, we find common symbols of plant instruments and these are shown in Figure 24.12 through Figure 24.17. Figure 24.12 lists general instrumentation device and function symbols utilized on instrumentation drawings. The symbol used communicates to the reader both the type of system the instrument is associated with (by the outside part of the figure) and the location of the instrument (by the inside part of the figure). In regard to the outside of the symbol, for devices and signals that are computer-generated, the outside of the symbol is a hexagon. For discrete instruments, the outside of the symbol is a circle. For process control generated signals, the outside symbol is a square. Where the device is associated with a safety system, there will also be a diamond shape inside the figure.

In regard to the inside of the symbol, for devices located remotely in the field, there are no internal horizontal lines in the symbol. For devices located in the front of a main or central panel, there is one solid horizontal line. For devices located in the rear of a main or central panel, there is one dashed horizontal line. For devices located in the front of a secondary panel, there are two solid horizontal lines. For devices located in the rear of a secondary panel, there are two dashed horizontal lines. Figure 23.12 provides examples of these symbols.

Figure 24.13 describes in more detail some of the symbols utilized to describe primary instrumentation that monitors physical parameters in the field and provides

Shared display, shared control		C	D	Location and accessibility
A	**B**			
Primary choice or basic process control system	Alternate choice or safety instrumented system	Computer systems and software	Discrete	Location and accessibility
				• Located in field. • Not panel, cabinet, or console mounted. • Visible at field location. • Normally operator accessible.
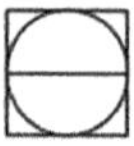	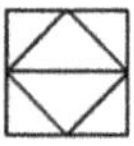	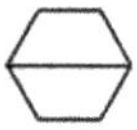	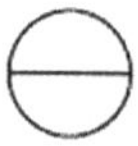	• Located in or on front of central or main panel or console. • Visible on front of panel or on video display. • Normally operator accessible at panel front or console.
		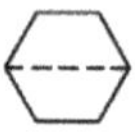	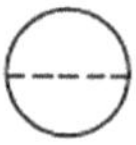	• Located in rear of central or main panel. • Located in cabinet behind panel. • Not visible on front of panel or on video display. • Not normally operator accessible at panel or console.
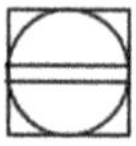	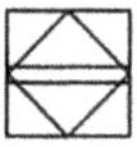	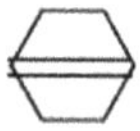	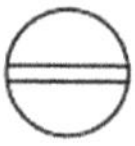	• Located in or on front of secondary or local panel or console. • Visible on front of panel or on video display. • Normally operator accessible at panel front or console.
		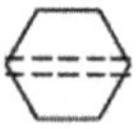	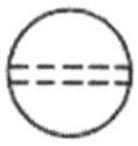	• Located in rear of secondary or local panel. • Located in field cabinet. • Not visible on front of panel or on video display. • Not normally operator accessible at panel or console.

Figure 24.12 General instrument or function symbols. *Source*: ANSI/ISA-5.1-2009, Instrumentation Symbols and Identification, Instrumentation device and function symbols. Reprinted with permission of the International Society of Automation, www.isa.org/standards

feedback to the control system. The symbol is based on the type of instrument and the primary measured value that is being monitored.

Figure 24.14 describes in more detail some of the symbols utilized to describe secondary instrumentation such as local gages and meters. These symbols are based on the type of instrument and the primary measured value that is being monitored.

The lines of interconnection between instrumentation symbols communicate to the reader of the drawing both the type of connection and links to other

	Symbol	Description
Analysis		• Conductivity, moisture, etc. • Single element sensing probe.
Burner		• Ultraviolet flame detector. • Television flame monitor.
Flow		• Generic orifice plate. • Restriction orifice.
Flow	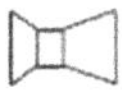	• Venturi tube.
Flow		• Flow nozzle.
Flow		• Standard pitot tube.
Flow		• Averaging pitot tube.
Flow		• Turbine flowmeter. • Propeller flowmeter.
Level	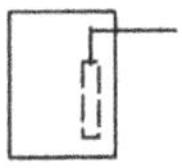	• Displacer internally mounted in vessel.
Level	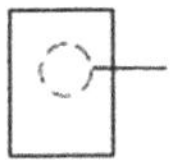	• Ball float internally mounted in vessel. • May be installed through top of vessel.
Level	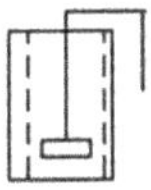	• Float with guide wires. • Location of readout should be noted, at grade, at top, or accessible from a ladder. • Guide wires may be omitted.
Level	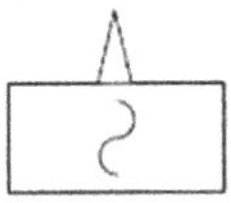	• Radar.
Pressure	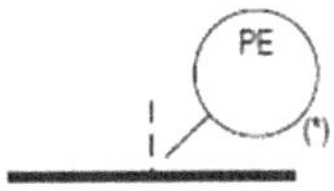 PE (*)	• Strain gage or other electronic type sensor. • Notation (*) should be used to identify type of element. • Bubble may be omitted if connected to another instrument.
Temperature	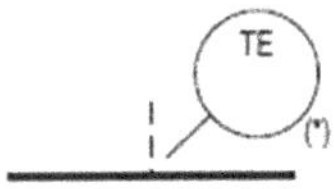 TE (*)	• Generic element without thermowell. • Notation (*) should be used to identify type of element. • Bubble may be omitted if connected to another instrument.

Figure 24.13   Measurement symbols: primary elements. *Source*: ANSI/ISA-5.1-2009, Instrumentation Symbols and Identification, Measurement symbols: primary elements, Reprinted with permission of the International Society of Automation, www.isa.org/standards

Symbol (4)	Description
Flow — (FG) —	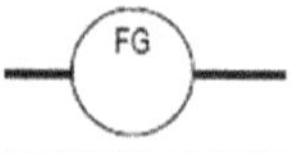 • Sight glass.
Level [ ] (LG)	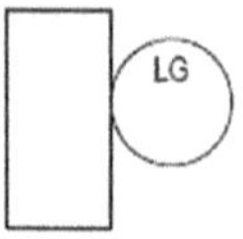 • Gage integrally mounted on vessel. • Sight glass.
Level [ ] (LG)	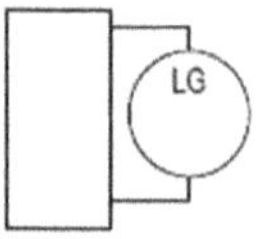 • Gage glass externally mounted on vessel or standpipe. • Multiple gages may be shown as one bubble or one bubble for each section.
Pressure (PG)	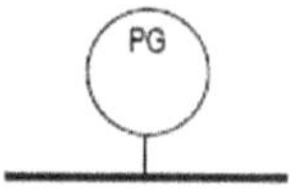 • Pressure gage.
Temperature (TG)	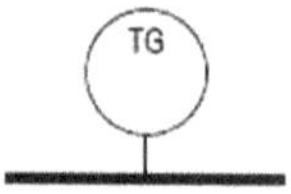 • Thermometer.

Figure 24.14 Measurement symbols: secondary instruments. *Source*: ANSI/ISA-5.1-2009, Instrumentation Symbols and Identification, Measurement symbols: secondary instruments. Reprinted with permission of the International Society of Automation, www.isa.org/standards

drawings as defined by ISA 5.1 Table 5.3.2 and a sample of these symbols is provided in Figure 24.15 below.

Some field instrumentation does not feed the plant control system but is directly connected to a local field actuator to regulate a parameter. One example of this is a pressure-regulating valve that senses pressure on one side of the valve and the valve actuator controls the position of the valve to maintain this pressure at the set point. Figure 24.16 shows some of these self-regulating devices and the standard symbols used to identify these on plant drawings.

Just as each type of instrument and valve has a unique symbol on a plant drawing, the actuator has a unique symbol as well. Valves may have motor-type actuators on them, or air-operated actuators, or may be manually operated valves. The symbols for some of the most common types of field actuators are shown in Figure 24.17.

Figure 24.17 describes the types of valve actuator symbols found on instrumentation drawings but sometimes we need to be able to communicate the failsafe position of the valve actuator. This is usually done with arrows indicating the direction that the actuator travels when power (electric, pneumatic, hydraulic, etc.) is removed from the actuator. Figure 24.18 describes the symbols used to identify the failsafe position of field actuators.

The failsafe position of a valve is critical in many applications and the engineering designing the system should verify that loss of any source of power drives the valve in the failsafe position. For example, if we have spring diaphragm actuator on a valve, and the air is admitted via a solenoid valve, we should ensure that the

Symbol	Application
IA  ———————	• IA stands for instrument air • IA may be replaced by PA [plant air], NS [nitrogen}, or GS [any gas supply]. • Indicate supply pressure as required, e.g., PA-70 kPa, NS-150 psig, etc.
ES ———————	• Instrument electric power supply. • Indicate voltage and type as required, e.g. ES-220 Vac. • ES may be replaced by 24 Vdc, 120 Vac, etc.
HS  ———————	• Instrument hydraulic power supply. • Indicate pressure as required, e.g., HS-70 psig.
——/——/—	• Undefined signal. • Use for process flow diagrams. • Use for discussions or diagrams where type of signal is not of concern.
 ——//——//—	• Pneumatic signal, continuously variable or binary.
 — — — — — — — —	• Electronic or electrical continuously variable or binary signal. • Functional diagram binary signal.
 ———————	• Functional diagram continuously variable signal. • Electrical schematic ladder diagram signal and power rails.
 ——L——L—	• Hydraulic signal.
(4) —— o —— o ——	• Communication link and system bus, between devices and functions of a shared display, shared control system. • DCS, PLC, or PC communication link and system bus.
(5)  —●——●—	• Communication link or bus connecting two or more independent microprocessor or computer-based systems. • DCS-to-DCS, DCS-to-PLC, PLC-to-PC, DCS-to-Fieldbus, etc. connections.
 —◎——◎—	• Mechanical link or connection.
(3) a) 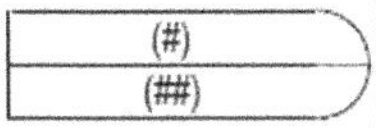 (#) (##) a) 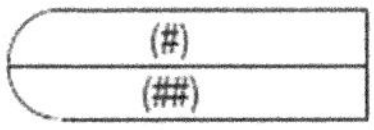 (#) (##) b) 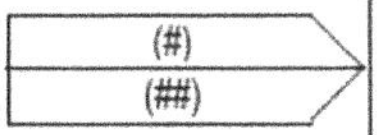 (#) (##) b) 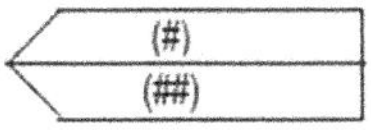 (#) (##)	• Drawing-to-drawing signal connector, signal flow from left to right. • (#) = Instrument tag number sending or receiving signal. • (##) = Drawing or sheet number receiving or sending signal.
 (*)|———	• Signal input to logic diagram. • (*) = Input description, source, or instrument tag number.
 ———|(*)	• Signal output from logic diagram. • (*) = Output description, destination, or instrument tag number.

Figure 24.15   Line symbols. *Source*: ANSI/ISA-5.1-2009, Instrumentation Symbols and Identification, Line symbols: instrument to instrument equipment connections. Reprinted with permission of the International Society of Automation, www.isa.org/standards

Symbol	Description
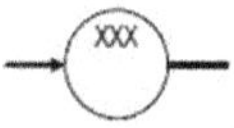	• Automatic flow regulator. • XXX = FCV without indicator. • XXX = FICV with integral indicator.
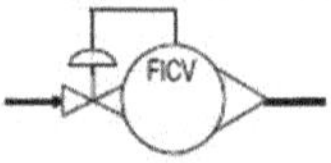	• Constant flow regulator.
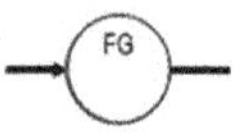	• Flow sight glass. • Type shall be noted if more than one type used.
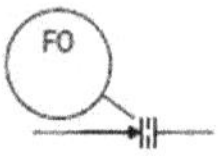	• Generic flow restriction. • Single stage orifice plate as shown. • Note required for multi-stage or capillary tube types.
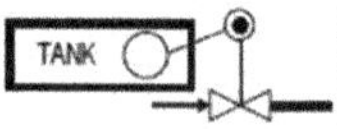	• Level regulator. • Ball float and mechanical linkage.
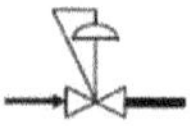	• Backpressure regulator. • Internal pressure tap.
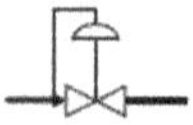	• Backpressure regulator. • External pressure tap.
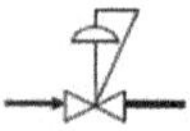	• Pressure-reducing regulator. • Internal pressure tap.
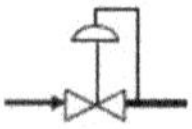	• Pressure-reducing regulator. • External pressure tap.
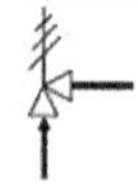	• Generic pressure safety valve. • Pressure relief valve.
	• Generic vacuum safety valve. • Vacuum relief valve.
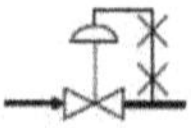	• Temperature regulator. • Filled thermal system.
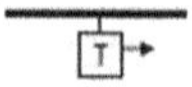	• Generic moisture trap. • Steam trap. • Note required for other trap types.

Figure 24.16   Self-Actuated final control element symbols. *Source*: ANSI/ISA-5.1-2009, Instrumentation Symbols and Identification, Self-actuated final control element symbol. Reprinted with permission of the International Society of Automation, www.isa.org/standards

Symbol	Description
	• Generic actuator. • Spring-diaphragm actuator.
	• Spring-diaphragm actuator with positioner.
	• Pressure-balanced diaphragm actuator.
	• Linear piston actuator. • Single acting spring opposed • Double acting.
	• Linear piston actuator with positioner.
 (M)	• Rotary motor operated actuator. • Electric, pneumatic, or hydraulic. • Linear or rotary action.
 [S]	• Modulating solenoid actuator. • Solenoid actuator for process on-off valve.
	• Actuator with side-mounted handwheel.
	• Actuator with top-mounted handwheel.
	• Manual actuator. • Hand actuator.
 (E/H)	• Electrohydraulic linear or rotary actuator.
[S]	• Automatic reset on-off solenoid actuator. • Non-latching on-off solenoid actuator.

Figure 24.17 Final control element actuator symbols. *Source*: ANSI/ISA-5.1-2009, Instrumentation Symbols and Identification, Final control element actuator symbols. Reprinted with permission of the International Society of Automation, www.isa.org/standards

Method A (1) (10)	Method B (1) (10)	Definition
 FO		• Fail to open position.
 FC		• Fail to closed position.
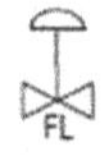 FL		• Fail locked in last position.
 FL/DO		• Fail at last position. • Drift open.
 FL/DC		• Fail at last position. • Drift closed.

Figure 24.18 Control valve failure and de-energized position indications. *Source*: ANSI/ISA-5.1-2009, Instrumentation Symbols and Identification, Control valve failure and de-energized position indications. Reprinted with permission of the International Society of Automation, www.isa.org/standards

valve travels to the failsafe position (open or closed) on either a loss of air pressure, or a loss of control power to the solenoid controlling the valve.

Example 24.11 You locate a piece of equipment in the field with a tag. On the tag is the equipment number 2EGS-PNL001. From this equipment ID, determine the following.

a. What generating unit is this equipment associated with?
b. What type of system is this equipment connected to?
c. What type of equipment is this?
d. What is the sequence number of this piece of equipment?

Solution:

a. The first number is the equipment ID is 2, so this is associated with the number 2 generator in the facility.
b. The system code is EGS which, from Annex C, stands for generator standby system.
c. The equipment code is PNL which, from Annex D, stands for instrument panel
d. The sequence number is 001.

TABLE 24.10   Drawing Types Utilized in the Electric Generation Industry

PFD	Process flow diagram
MSG	Material selection guides
UFD	Utility flow diagrams
PID	Piping and instrument drawings
KLD	Instrument loop diagrams
KCD	Instrument configuration diagrams
KSK	Functional control diagrams
LSK	Logic diagrams

Source: Reproduced with permission of Tampa Electric Company. Reproduction is forbidden without the express consent of Tampa Electric Company.

So this is an instrument panel associated with the generator standby system on unit 2.

Drawings and documentation are the key methods for communicating information about the various systems in the power generation facility. The ability of the plant engineer to quickly and accurately read the various types of drawings in the power generation facility is critical to the successful transfer of information. There are many types of drawings in the typical power generation facility used to adequately document information. Table 24.10 provides a listing of the types of documentation and the acronyms used in the power generation industry to document instrumentation systems.

The instrumentation on these drawings is identified by standard symbols and terminology. The type of circle used to identify an instrument help to identify the location of the instrument. An instrument identified by a circle with no lines through the diameter is an instrument located in the field. An instrument identified by a circle with two solid lines through the circle is an instrument located at a local control panel in the field. An instrument identified by a circle with two dashed lines through the diameter is an instrument located behind a control panel. An instrument identified by a circle with one dashed line through the diameter is an instrument mounted in a common instrument rack room. An instrument identified by a circle with one solid line through the diameter is an instrument mounted in the control room.

A hexagonal figure with a solid line through the center is a computer or DCS system calculated value and not a physical instrument in the field (we will see this used in a KLD or loop drawing later in this chapter). A circle with a box around it and a solid line through it is an HMI (human machine interface) display.

The type of valve in a PID can be identified by the drawing symbol used for the valve. Figure 24.19 shows the standard symbol for various types of valves in the facility. (For more information on the types of valves utilized in a generation facility, see Chapter 25 on valves).

Figure 24.20 shows a typical PID drawing where we pull all these standard drawing and equipment numbering systems together to allow us to communicate the system and equipment details in a drawing. The valve that controls the flow of fluid into tank 1-CCC-TK-901 is controlled by valve 1-CCC-LV-900. From Annex C, the system is the component cooling water – common system (CCC), and this is associated with unit 1. For the storage tank (1-CCC-TK-901), we go to Annex D for the equipment ID and find that this is a tank (TK) and the sequence number is 901. Notice

Symbol	Description
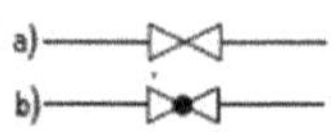 a) ———▷◁——— b) ———▷◀————	• Generic two-way valve. • Straight globe valve. • Gate valve.
	• Generic two-way angle valve. • Angle globe valve. • Safety angle valve.
	• Generic three-way valve. • Three-way globe valve. • Arrow indicates failure or unactuated flow path.
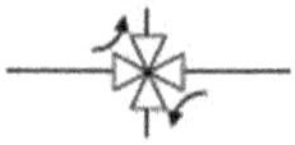	• Generic four-way valve. • Four-way four-ported plug or ball valve. • Arrows indicates failure or unactuated flow paths.
	• Butterfly valve.
	• Ball valve.
	• Permanent magnet variable speed coupling.
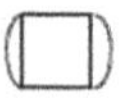	• Electric motor.
	• Generic damper. • Generic louver.
	• Parallel blade damper. • Parallel blade louver.
	• Opposed blade damper. • Opposed blade louver.

Figure 24.19 Final control element symbols. *Source*: ANSI/ISA-5.1-2009, Instrumentation Symbols and Identification, Final control element symbols. Reprinted with permission of the International Society of Automation, www.isa.org/standards

that the type of valve is not a flow control valve (FC) by ISA standards but a level control valve (LV). While this valve is controlling the amount of flow into the tank, this valve is not called a flow control valve, but a level control valve. The name of the valve is identified by the final element that is providing the controlling signal. If we start at valve 1-CCC-LV-900 and follow the command signal upstream, we find that the valve is driven by air and a current to air transducer 1-CCC-LY-900. This in turn is driven from a level controller 1-CCC-LC-900, which receives its feedback signal from both 1-CCC-LT-900A and 1-CCC-LT-900B through a controller 1-CCC-LY-900A that selects which feedback signal to utilize in the controller 1-CCC-LC-900. Notice that for instrumentation equipment types, we normally refer back to ISA standards and not Annex D (equipment system codes).

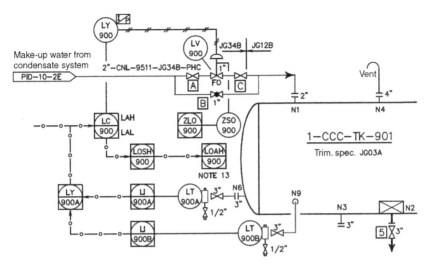

Figure 24.20   Typical PID diagram showing piping and instrumentation. *Source*:
Reproduced with permission of Tampa Electric Company. Reproduction is forbidden without
the express consent of Tampa Electric Company.

In addition to the PID drawings, these symbols are utilized on other types of
drawings designed to provide required information in a different format. One example
is the instrument loop diagram (KLD). Figure 24.21 shows a typical instrument loop
diagram. These are designed to show the complete signal path for one loop from the
field instrument(s) back to the control system and is an important tool in troubleshooting and verification of instrument loops. In the KLD drawing shown in Figure 24.21,
on the left, you see the field installed instrument PSL-688. From the ISA 5.1, we know
that this is a pressure switch that is activated on low pressure. This switch is located
in equipment 1-IAS-IPNL-603. The system code for this equipment is IAS which
from Annex C is the Instrument Air System. The equipment code for this equipment
is IPNL which from Annex D is tell us this is an instrument panel. We then see this
switch is connected to a terminal block in the panel 1-WTS-CAB-621. The system
code for this equipment is WTS which from Annex C is the water treating system.
The equipment code for this equipment is CAB which from Annex D tells us this is
a Cabinet. The drawing itself tells us this is terminated on terminal block TB-2 pins
33 and 34. So we know this termination point is in a cabinet in the water treating
system area and it is terminated on TB-2 pins 33 and 34. From there, the pressure
switch wires are terminated in equipment 1-WTS-CAB-601. Again this is a cabinet
in the water-treating system area. The instrument loop drawing also shows us that
this signal is then converted to a calculated point (identified by the dashed line) with
identification number PAL-688. Looking back to the ISA 5.1 information, we see this
is a pressure alarm low (PAL) that is driven from the device PSL-688 in the field.

Understanding and committing to memory the above symbols and some of the
ISA 5.1 identifiers will make reading drawings quicker and more informative for the
power generation facility engineer and is a critical skill necessary to be an efficient
resource for reliable operations.

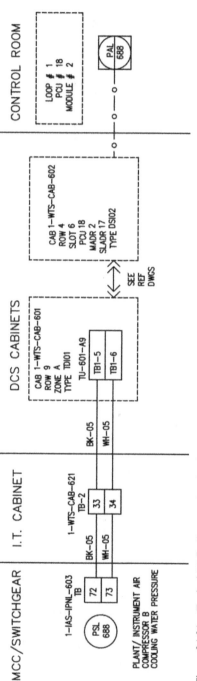

Figure 24.21   Typical KLD diagram showing instrument loop. *Source:* Reproduced with permission of Tampa Electric Company. Copyright 2016 Tampa Electric Company. Reproduction is forbidden without the express consent of Tampa Electric Company.

GLOSSARY OF TERMS

- Accuracy – Degree of conformity of measured quantity to actual parameter.
- Manometer – An instrument for measuring the pressure acting on a column of fluid.
- Orifice Flow Meter – A device that monitors pressure difference across a flow restriction plate with a hole or orifice in the plate and calculates flow based on certain parameters.
- Pitot Tube – An open-ended right-angled tube pointing in opposition to the flow of a fluid and used to measure pressure.
- Precision – Measure of repeatability of measured quantity for same actual parameter.
- Resistance Temperature Detector (RTD) – Temperature sensors that exploit the predictable change in electrical resistance of some materials with changing temperature.
- Thermocouple – A thermoelectric device for measuring temperature, consisting of two wires of different metals connected at two points, a voltage being developed between the two junctions in proportion to the temperature difference.
- Venturi Flow Meter – A device that monitors pressure difference across a flow restriction plate with a hole or orifice in the plate and calculates flow based on certain parameters.
- Venturi effect – the reduction in fluid pressure that results when a fluid flows through a constricted section of pipe.

PROBLEMS

24.1 Fill in the blanks: "A current transformer with a rating of 10C400 that has a 5 A rated secondary, can develop up to _____ V on secondary circuit at 20 times rated secondary current (100 A) and not exceed _____ accuracy."

    A. 120 V, 10%

    B. 400 V, 10%

    C. 120 V, 5%

    D. 400 V, 5%

24.2 What are at least four physical parameters that are needed to be monitored in a power plant to provide adequate information for process control?

    _____

    _____

    _____

    _____

24.3 You need to replace a thermocouple in a plant process and no information is known on the device. However, when locating the device you see that the conductor color on the positive connection is yellow and the conductor color on the negative connection is red. Knowing that the plant was built to ANSI standards, what is the type thermocouple you need to use to replace the existing device?

    A. Type J

    B. Type K

    C. Type T

    D. Type E

24.4 You need to replace a thermocouple in a plant process and no information is known on the device. However, when locating the device, you see that the conductor color on the positive connection is green and the conductor color on the negative connection is white. Knowing that the plant was built to IEC standards, what is the type thermocouple you need to use to replace the existing device?

    A. Type J

    B. Type K

    C. Type T

    D. Type E

24.5 A pressure of 30 psia approximately equals _____.

    A. 30.0 psig

    B. 15.3 psig

    C. 14.7 psig

    D. 0.0 psig

It can also be found by subtracting atmospheric pressure of 14.7 psia from 30 psia.

24.6 With the thermocouple connection shown below, what is the measured temperature (assume cold junction temperature compensation at the measuring end)?

    A. $T_{\text{meas}} = T_1 + T_2$

    B. $T_{\text{meas}} = T_1 - T_2$

    C. $T_{\text{meas}} = T_2 - T_1$

    D. $T_{\text{meas}} = -T_1 - T_2$

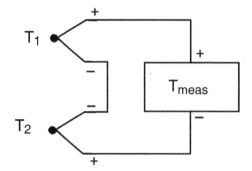

24.7 The typical reference resistance of a copper RTD is?

    A. 10 ohms

    B. 25 ohms

    C. 100 ohms

    D. 200 ohms

24.8 Refer to the drawing of a differential pressure manometer (see figure below). A differential pressure manometer is installed across an orifice in a ventilation duct. With the ventilation conditions as shown, the pressure at P_1 is _____ than P_2, and airflow is from _____.

    A. greater; left to right

    B. greater; right to left

    C. less; left to right

    D. less; right to left

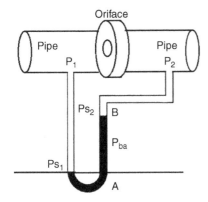

24.9 A water storage tank is vented to atmosphere. The tank is located at sea level and contains 100,000 gal of 80°F water (specific gravity of 1.0). A pressure gauge at the bottom of the tank reads 5.6 psig. What is the approximate water level in the tank?

    A. 13 ft

    B. 17 ft

    C. 21 ft

    D. 25 ft

24.10 If a tank contains 30 ft of water at 60°F and the top of tank is open to atmosphere and the reference standoff pipe contains 30 ft of water at the same temperature and is also open to the atmosphere, what is the approximate D/P sensed by the detector?

    A. Greater than zero (tank-side pressure greater than standoff pipe side)

    B. Zero

    C. Less than zero (tank-side pressure less than standoff pipe side)

24.11 If a tank contains 30 ft of water at 60°F and the top of tank is open to the atmosphere and the reference standoff pipe contains no water (0 ft) and is also open to the atmosphere, what is the approximate D/P sensed by the detector?

A. Greater than zero (tank-side pressure greater than standoff pipe side)

B. Zero

C. Less than zero (tank-side pressure less than standoff pipe side)

24.12 If a tank contains 30 ft of water at 600°F and the top of tank is open to the atmosphere and the reference standoff pipe contains 30 ft of water but at a temperature of 60°F and the standpipe is also open to the atmosphere, what is the approximate D/P sensed by the detector?

A. Greater than zero (tank-side pressure greater than standoff pipe side)

B. Zero

C. Less than zero (tank-side pressure less than standoff pipe side)

24.13 If flow is doubled through an orifice plate, what is the increase in differential pressure across the flow plate?

A. New differential pressure is the same as the original differential pressure

B. New differential pressure is two times the original differential pressure

C. New differential pressure is four times the original differential pressure

D. New differential pressure is eight times the original differential pressure

24.14 A transducer is installed on a tank that is scaled to send a 4–20 mA signal proportional to a tank level of 4–20 ft (such that 4 ft = 4 mA and 20 ft = 20 mA). At five different times, when the tank level was physically measured at 12 ft, the ideal output of the transmitter should have been 12 mA, but actual output was 9 mA all five times samples were tested. The question is, is this transducer accurate, precise, both, or neither?

A. Very accurate but not very precise

B. Not very accurate but very Precise

C. Both accurate and precise

D. Neither accurate nor precise

RECOMMENDED READING

Electrical Machines, Drives and Power Systems, 6th edition, Theodore Wildi, Prentice Hall, 2006.

The Engineering Handbook, 3rd edition, Richard C. Dorf (editor in chief), CRC Press, 2006.

Fundamentals of Process Control Theory, 3rd edition, Paul W. Murrill, ISA, 2000.

IEC 60584-3 ed.2.0, Thermocouples – Part 3: Extension and Compensating Cables – Tolerances and Identification System.

IEC 61515:1995, Mineral Insulated Thermocouple Cables and Thermocouples.

IEEE 141: IEEE Recommended Practice for Electric Power Distribution for Industrial Plants (IEEE Red Book), 1993.

IEEE 505-1977: IEEE Standard Nomenclature for Generating Station Electric Power Systems.

Industrial Power Distribution, 2nd edition, Ralph E. Fehr, III, Wiley-IEEE Press, 2016. ISBN: 978-1-119-06334-6.

ISA5.1-2009, International Society of Automation, Instrumentation Symbols and Identification.

Power Plant Engineering, Black & Veatch, edited by Larry Drbal, Kayla Westra, and Pat Boston, Chapman & Hall/Springer, 1996.

Standard Handbook of Powerplant Engineering, 2nd edition, Thomas C. Elliott, McGraw-Hill, 1998.

T.A. Hughes, *Measurement and Control Basics*, 4th edition, International Society of Automation (ISA), 2007.

CHAPTER 25

VALVES AND ACTUATORS

GOALS

- To understand the various types of isolation and control valves utilized in an electric generation facility
- To understand the difference between a valve, a valve actuator, and a valve positioner
- To understand the purpose and identify the failsafe position of a valve actuator
- To calculate the increase in torque from the actuator motor to the valve shaft when coupled through a gear reducer
- To calculate the amount of head loss that a valve presents to the system given the amount of flow, valve diameter and resistance coefficient of the valve
- To identify the operation of a pneumatic valve manifold and types of operators for various manifold designs

---

*V**ALVES* **ARE** used in a facility for both control and isolation purposes. The valve is the final control device that controls the flow of material through the piping system. It must have some means to move the valve members to control the state of the valve. For valves that are used only for isolation, they are driven fully open or fully closed and verification that the valve is on its seat is adequate feedback. These valves are known as isolation valves. In other applications, the valve regulates a parameter such as flow, level, pressure, etc. In such a control application, the valve may modulate somewhere between fully open and fully closed. These valves are knows as *control valves*. A control valve is a device capable of modulating flow at varying degrees between minimal flow and full capacity in response to a signal from an external control device. In control applications, the position of the valve is critical for the control loop and the *valve actuator* may have a *valve positioner* that obtains a *feedback signal* on the actual valve position and compares it with the desired position and sends a command to the valve actuator depending on the error between the feedback and desired position so the *actuator* can move the valve into the correct position.

Energy Production Systems Engineering: An Introduction for Electrical Engineers to Electrical Power Generation Facilities, Systems, and Equipment, First Edition. Thomas H. Blair.

The simplest valve actuators are manual operators such as *hand wheels* or chains. The constraint with manual operators is that the person controlling the valve position must be local to the valve operator. Many valves in power stations are not easily accessible. Take for example the inner containment isolation valves for the main steam lines for a boiling water reactor (BWR). These may be physically located inside the containment vessel and may not be accessible during operation of the reactor, but must be controlled by some remote means.

For valves where remote control is required, the three main modes of control for the valve position are by a *pneumatic valve actuator*, a *hydraulic actuator*, or a *motor-driven actuator*.

There are various types of valves utilized in industry and the type of valve defines the type of actuator required to control the valve position. For example, *butterfly valves* or *ball valves* have a movement from fully closed to fully open position of 90 degrees. The actuator for these valves only moves the valve $1/4$ of a full rotation to move from fully closed to fully open and is known as a "partial turn actuator." On the other hand, a gate valve attached to a stem requires multiple complete turns to get the valve from the fully open position to the fully closed position as this valve element must travel the complete diameter of the valve to change state. As such, the actuator for the gate valve must make many turns to move the valve from the fully open position to the fully closed position. This type of actuator is known as a "multi turn actuator."

Another type of valve actuator is a linear actuator. Most linear actuators are pneumatic diaphragm actuators. A typical air-actuated diaphragm control valve is shown in Figure 25.1. Their design is such that there is a diaphragm inside a sealed cylinder. There are two types of pneumatic diaphragm actuators depending on if the failsafe (loss of air signal) is fully closed, fully open, or to remain in current position. Note that the design of the valve shown in Figure 25.1 is such that the high pressure side of the valve assists the force of the spring and pushes the valve disk against the seat. This tends to help seal the valve disk against the seat.

For a valve that must travel fully open or fully closed on loss of air pressure, one side of the diaphragm has spring pressure on it. The other side of the diaphragm is constructed with an air chamber. The side that has the spring is chosen such that on loss of air pressure, the valve moves to the safe position (fully open or fully closed) as defined by the application of the valve. As air is admitted to the air chamber, the air pressure applies a force on the diaphragm that opposes the force of the spring and moves the valve stem.

As a side note, the control signal is set up such that a loss of the control signal will result in the valve traveling the same direction as if there was a loss of air pressure. Many *valve positioners* utilize an I/P controller. The term "I/P controller" represents a current "*I*" to pressure "*P*" controller. At 4 mA, the amount of air pressure is minimal and at 20 mA, the amount of air pressure is maximal. If the electrical control signal fails and the control signal is driven to a minimum value (4 mA), air pressure will also be driven to a minimal value and the valve will fail in the desired failsafe direction of spring pressure.

For a valve that must remain in its existing position on loss of air pressure, there is no spring in the valve actuator, but air is admitted to both sides of the diaphragm and the air pressure is balanced on each side to position the valve in the desired position.

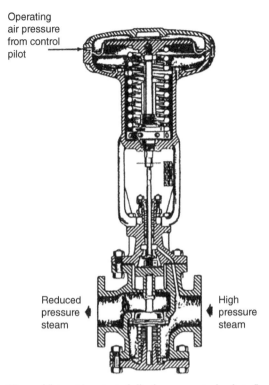

Operating
air pressure
from control
pilot

Reduced
pressure
steam

High
pressure
steam

Figure 25.1   Air actuated diaphragm control valve. *Source*: Reproduced with permission of
U.S. Navy Training Manual.

On loss of air pressure, neither side of the valve actuator sees any change in pressure and the valve remains in its existing position. The failed position of the valve is shown below the valve symbol. In Figure 25.2, this valve is a "failed closed" valve as designated by the symbol FC under the valve.

The hydraulic valve actuator uses hydraulic fluid instead of air for the driving force to move the valve to the correct position. It is constructed of a cylinder that has hydraulic chambers on each side and ports at each chamber to allow for the admission or expulsion of hydraulic fluid from the chamber. The pressure of the hydraulic fluid on the surface of the piston provides the driving force to move the valve to the desired position. Figure 25.3 shows the symbol for a valve with a hydraulic actuator.

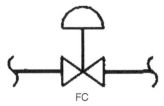

FC

Figure 25.2   Symbol for a pneumatic actuated valve. *Source*: ANSI/ISA-5.1-2009,
Instrumentation Symbols and Identification. Reprinted with permission of the International
Society of Automation, www.isa.org/standards

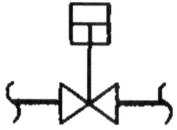

Figure 25.3   Symbol for a hydraulic actuated valve. *Source*: ANSI/ISA-5.1-2009, Instrumentation Symbols and Identification. Reprinted with permission of the International Society of Automation, www.isa.org/standards

Another valve actuator that allows the valve to remain in its existing position on a loss of command signal to the actuator is the motor valve actuator. Depending on the application, AC or DC motors may be used as the driving force for the valve actuator. The actuator will have both a limit switch to indicate valve fully open and valve fully closed as well as torque switches to ensure the valve is fully seated into the valve body. Figure 25.4 shows the symbol for a motor-actuated valve.

The motor is geared to the actuator to provide high torque, low speed operation. Looking back in Chapter 17 on motors, we discussed that, with the ideal gear reducer, power into the gear reduced is equal to the power out of the gear reduced.

$$P_{in} = P_{out} \tag{25.1}$$

where

P_{in} = power into the gear reducer (HP)

P_{out} = power out of the gear reducer (HP)

Since power is the product of torque and speed, we can rewrite equation (25.1) in terms of torque and speed.

$$T_{in}\omega_{in} = T_{out}\omega_{out} \tag{25.2}$$

where

T_{in} = torque into the gear reducer (lb ft)

T_{out} = torque out of the gear reducer (lb ft)

ω_{in} = speed of shaft into the gear reducer (rpm)

ω_{out} = speed of shaft out of the gear reducer (rpm)

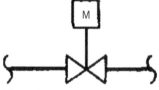

Figure 25.4   Symbol for a motor-actuated valve. *Source*: ANSI/ISA-5.1-2009, Instrumentation Symbols and Identification. Reprinted with permission of the International Society of Automation, www.isa.org/standards

We can now rearrange equation (25.2) to see the relationship of torque out of the gear reduced to the torque into the gear reducer as a function of the speed of the input to the gear reducer over the speed of the output of the gear reducer.

$$T_{out}/T_{in} = \omega_{in}/\omega_{out} \tag{25.3}$$

Therefore, the ratio of the torque increase in a gearbox of a motor-actuated valve is inversely proportional to the speed ratio between the motor and the final driving element.

The motor-operated valve is available with a manual operator and a clutch lever that switches control of the valve actuator from the motor (normal position) to the manual operator (manual position). This ensures that the manual operator (a hand wheel) does not move when the motor is driving the actuator as this could present a personnel safety hazard unless the hand wheel was guarded.

A positioner is a device installed on a valve actuator to ensure that the valve is at a correct position of opening or closing as per the control signal. An I/P converter only sends the opening/closing request to valve but cannot confirm its position. The positioner senses the valve actual position through a feedback link connected to the valve assembly. The positioner compares the feedback signal to the set point and calculates an error signal. The positioner will send this error signal to the actuator to move the valve until the error signal is zero indicating the valve is in the position that the command signal is asking for.

VALVE TYPES

There are many types of valves used in the power generation facility. Each valve design has unique advantages and disadvantages that define the best application for that particular valve design. Following is a description of the most common type of valves used in the utility industry along with their traits and common applications.

BALL VALVE

The ball valve as the name indicates is shaped like a ball with an orifice drilled through it to control the amount of flow through the valve. When the orifice is parallel to the flow of fluid, the valve is fully open and when the orifice is perpendicular to the flow of the fluid, the valve is fully closed. The ball valve can be used in mid-travel to act as a restriction to flow. This changes the effective orifice size as seen by the fluid. However, the ability to control flow via the *globe valve* is much better than the ball valve, so for flow control applications, the globe valve is more commonly chosen, but for low cost flow control where accuracy is not critical, the ball valve provides a good compromise between cost and accuracy. This valve only requires a $\frac{1}{4}$ turn or 90-degree turn to move from the fully open to the fully closed position.

The actuator is aligned such that when the actuator indicator is in parallel to the flow of the fluid, the valve is fully open and when the actuator indicator is

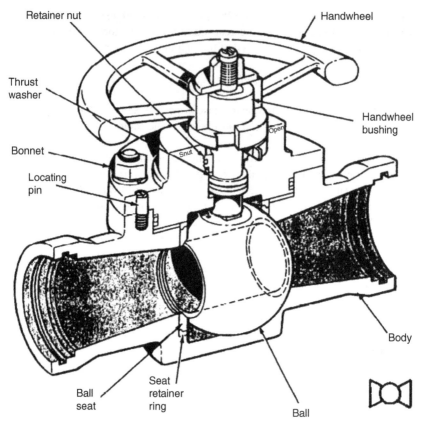

Figure 25.5   Ball valve cross section and schematic symbol. *Source*: Reproduced with permission of U.S. Navy Training Manual.

perpendicular to the flow of the fluid, the valve is fully closed. This provides external indication of the internal valve position to the operator controlling the valve position.

The ball valves ability to provide a good seat to fully closed position is superior to that of the gate or globe valve and as such, ball valves are used in applications where the need to ensure minimum valve leaking in the fully closed position is critical to the operation.

Figure 25.5 shows the arrangement and schematic symbol for a ball valve.

The ball check valve is a type of check valve where the ball does not have an orifice in the disk. The position of the ball in the valve body and the space between the ball and the valve seat provide the control of flow. When flow tries to run in the opposite direction, dynamic fluid pressure forces the ball against the seat of the valve body thus preventing flow in the opposite direction.

The butterfly valve is constructed of a disk and a seat in the valve body. Its operation is similar to the ball valve in that when the disk is parallel to the flow of fluid, the valve is fully open and when the disk is perpendicular to the flow of the fluid, the valve is fully closed. The butterfly valve can be used in mid travel to act as a restriction to flow. This changes the effective size of the restriction to fluid flow. However, the

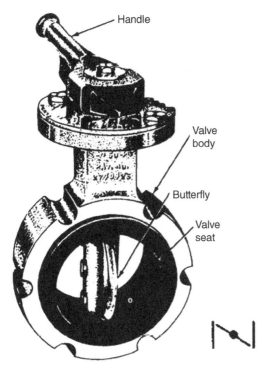

Handle

Valve
body

Butterfly

Valve
seat

Figure 25.6   Butterfly valve cross section and schematic symbol. *Source*: Reproduced with permission of U.S. Navy Training Manual.

ability to control flow via the globe valve is much better than the butterfly valve, so for control applications, the globe valve is more commonly chosen. Like the ball valve, this valve only requires a $1/4$ turn or 90-degree turn to move from the fully open to the fully closed position. Since the operating device of the butterfly valve contains less material than the ball valve, they tend to be less expensive and lighter in weight than a similarly sized ball valve. When fully open, the butterfly valve disk is still within the fluid in the valve and, as such, there is still some pressure drop and associated losses with the butterfly valve even when fully open. The same is not true for the ball valve as the orifice in the ball valve is about the same diameter as the ID of the valve body and, when the ball valve is fully open, the pressure drop across it is minimal, with minimal valve losses. Remember that the losses for flow restrictions such as valves an orifices is proportional the flow and the pressure drop across the restriction. As such, for the same amount of flow, if the pressure drop increases, then the power losses due to the restriction increase as well. Figure 25.6 shows the arrangement and schematic symbol for a butterfly valve.

CHECK VALVE

The check valve is a device that allows flow in one direction but blocks flow in the opposite direction. It is the mechanical equivalent of the diode in an electrical

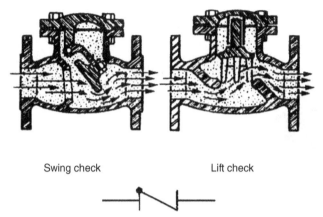

Swing check   Lift check

Figure 25.7   Check valve types and schematic symbol. *Source*: Reproduced with permission of U.S. Navy.

circuit that allows electron flow in only one direction. Check valves have many designs depending on the needs of the application, but commonly there is a disk inside a valve body. When dynamic fluid pressure is from the forward direction, the dynamic pressure forces the valve disk away from the valve seat, opening the valve and allowing the fluid to pass. When dynamic fluid pressure is from the reverse direction, the dynamic pressure forces the valve disk toward the valve seat, closing the valve and preventing the fluid from passing. The discharge of a centrifugal pump, where there are multiple pumps in parallel is a very common application for the check valve. With one pump running and another pump in parallel not running, without the check valve and if the pump that is off is not valved out of the system, the running pump discharge head would tend to force fluid back through the pump that is not running, and the overall efficiency of the running pump would be reduced. To prevent this, the pump in parallel that is off, either must be valved out or, alternately if a check valve is installed in the discharge of the pump that is off, the dynamic head from the running pump discharge would force the check valve of the pump that is off to the closed position preventing this backflow from occurring. Figure 25.7 shows the arrangement and schematic symbol for a check valve.

GATE VALVE

The construction of a gate vale is such that a disk is raised or lowered into the valve body increasing or decreasing the space between the disk and valve body thereby allowing more or less flow through the valve. The flow path through the valve is a nonlinear function and the pressure losses for a gate valve are very high. Also, due to the design of the gate valve, a partially opened gate valve may vibrate due to dynamic interactions between the disk and the fluid flowing past the disk. Therefore, the most common application for the gate valve is only for a valve used for isolation where the two positions of the valve will be fully open or fully closed. The gate valve is almost

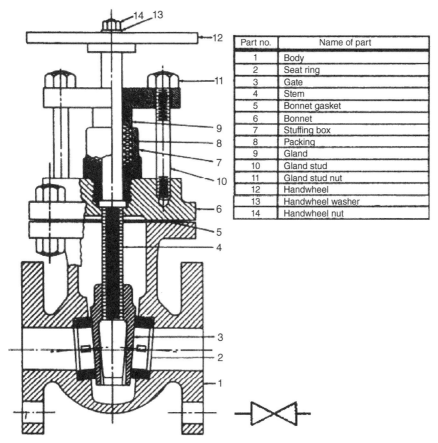

Part no.	Name of part
1	Body
2	Seat ring
3	Gate
4	Stem
5	Bonnet gasket
6	Bonnet
7	Stuffing box
8	Packing
9	Gland
10	Gland stud
11	Gland stud nut
12	Handwheel
13	Handwheel washer
14	Handwheel nut

Figure 25.8   Gate valve cross section and schematic symbol. *Source*: Reproduced with permission of U.S. Navy Training Manual.

never used for control or modulating applications. When the valve is fully open, the disk is raised entirely out of the valve body. Therefore, in the fully open position, the gate valve has minimal pressure drop losses associated with it.

While the gate always raises or lowers, depending on the construction of the valve stem, the stem may rise with the gate or the stem may be a non-rising type where the gate rises on the stem and the stem remains at the same level. Figure 25.8 shows the arrangement and schematic symbol for a gate valve.

GLOBE VALVE

As mentioned in the discussion of the ball and butterfly valve, the globe valve is usually the choice when the valve must control or modulate the fluid flow through the valve body. The construction of the globe valve is a movable plug or plunger that meets up with a seat inside the valve body. The internals of the standard globe valve

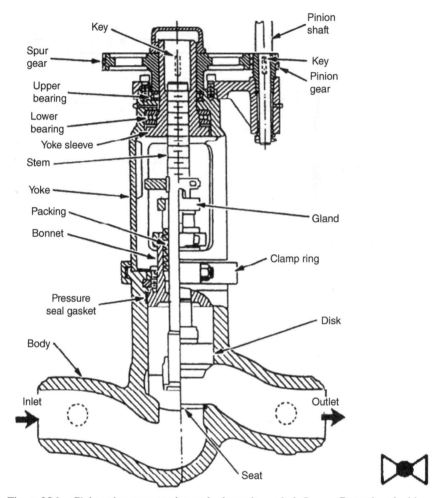

Figure 25.9 Globe valve cross section and schematic symbol. *Source*: Reproduced with permission of U.S. Navy Training Manual.

are such that the fluid must make a 90-degree turn, flow through the valve seat, and then make another 90-degree turn to leave the valve body. These two turns present a pressure drop even with the valve is fully open. Therefore, even when the globe valve is fully open, there is a power loss associated with the pressure drop across this valve.

The excellent control and regulating features of the globe valve make it the most common choice for applications requiring throttling or control of fluid flow. Figure 25.9 shows the arrangement and schematic symbol for a globe valve.

RELIEF VALVE

The relief valve is a very important valve in the safe and reliable operation of the power generation facility. These valves monitor system pressure (or vacuum) and

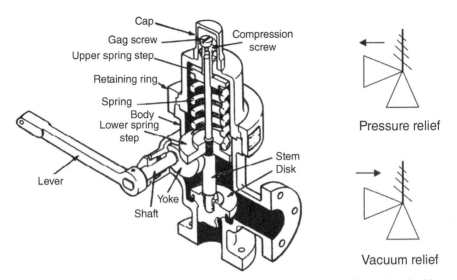

Pressure relief

Vacuum relief

Figure 25.10   Relief valve cross section and schematic symbol. *Source*: Reproduced with permission of U.S. Navy Training Manual.

open when the maximum amount of pressure (or vacuum) is reached. These valves are critical to the safety of the system as they are designed to be the "weak" point of a pressure vessel or system. Should pressures exceed design values; the damage to the system can be catastrophic. These relief valves will open and release the pressure (or vacuum) and then reseat when the sensed pressure (or vacuum) drops back to the reseat pressure of the valve. The blowdown is usually stated as a percentage of set pressure (or vacuum) and refers to how much the pressure needs to drop (or vacuum needs to rise) before the valve resets, closing the path for release of pressure (or vacuum). The blowdown can vary from roughly 1–20%. A typical relief valve is shown in Figure 25.10.

Many relief valves vent to atmosphere when the release of the fluid does not constitute a hazardous environment. When the material in the system may present a hazard, the location for release must address this hazard. For example, the pressure relief valve for the hydrogen system that supports the generator is at a very high level in the plant and the location of the end of this pipe is a classified area. Pipes from the hydrogen relief valve will be piped to one of the highest locations (e.g., the turbine roof) and vent to an area that is well ventilated. Hydrogen has a specific density of about 10% that of air and, as such will quickly rise and disperse.

VALVE LOSSES

Just as pumps present an energy source to a mechanical system when there is flow and differential pressure across the pump, valves present a loss of energy to a mechanical system due to the flow through the valve and the differential pressure across the valve. The head loss (HL) across the valve can be calculated by equation (25.4) and the power losses (P_v) associated with the flow and differential pressure drop across the valve can be calculated by equation (25.5).

$$HL = (k\,Q^2)/(D^4\;385) \tag{25.4}$$

where

   HL = head loss (ft)

   k = resistance coefficient

   Q = flow (gpm)

   D = diameter (in.)

The value of 385 is required as it accounts for the acceleration of gravity (32 ft/sec^2) and also converts 12 in. to 1 ft or $12 \times 32.2 = 385$

$$P_v = \rho\, g\, HL\, Q \tag{25.5}$$

where

   P_v = valve loss (W)

   ρ = fluid density (kg/m^3)

   g = standard acceleration of gravity (9.80665 m/sec^2)

   HL = head loss across the valve (m)

   Q = flow rate through the valve (m^3/sec)

Example 25.1   Given a resistance coefficient of 3.85, a flow through a valve of 1000 gal/min, a valve diameter of 10 in., determine the head loss across this valve and the associated power loss due to losses in this valve given a fluid density of 10 kg/m^3?

Solution: First, we need to calculate the head loss of the valve. To do this, we can use equation (25.4)

   $HL = (kQ^2) / (D^4 \times 385)$

   $HL = (3.85) (1000\ \text{gpm})^2) / ((10\ \text{in.})^4\ 385)$

   $HL = 1$ ft

Next, we need to calculate the power loss in the valve, but to do this, we need to convert from standard to metric units as shown below.

   $HL = 1\ \text{ft} \times (0.3408\ \text{m} / 1\ \text{ft}) = 0.3408\ \text{m}$

   $Q = 1000\ \text{gpm} \times (0.0631\ \text{m}^3/\text{sec}/1000\ \text{gpm}) = 0.0631\ \text{m}^3/\text{sec}$

Now, we can use equation (25.5) to calculate the losses in this valve using the assumed fluid density of 1 kg/m^3.

   $P_v = \rho g HL Q$

   $P_v = (10\ \text{kg/m}^3)(9.80665\ \text{m/sec}^2)(0.3408\ \text{m})(0.0631\ \text{m}^3/\text{sec})\ \text{W}$

   $P_v = (10\ \text{kg/m}^3)(9.80665\ \text{m/sec}^2)(0.3408\ \text{m})(0.0631\ \text{m}^3/\text{sec})\ \text{W}$

   $P_v = 2.1\ \text{W}$

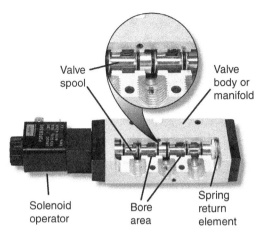

Valve spool

Valve body or manifold

Solenoid operator

Bore area

Spring return element

Figure 25.11   Pneumatic pilot valve module assembly. *Source*: Reproduced with permission of Parker Pneumatic Division.

Pneumatic Actuator Pilot Valve Manifold Controls

For pneumatically controlled valves that are not in an application that requires modulation control, the pneumatic control system will drive the compressed air to or exhaust the compressed air from to the pneumatic actuating cylinder that operates the main valve to the fully open or the fully close position in the direction required. The "valve" is formed by the close fit of a spool piece inside a bore area of a valve body or manifold as is shown in Figure 25.11. Actuators connected to the ends of the spool piece drive the spool piece to the position required to connect the ports as needed. The device that directs the flow path of this compressed air is a *pneumatic pilot valve module*.

Depending on the design of the controls this may be a two- to five-port device. The number of ports simply defines the number of orifices that the valve body contains. Most (but not all) manufacturers label the valve ports per ISO (International Standards Organization) symbology or NFPA (National Fluid Power Association) symbology. ISO (International Standards Organization) symbology defines port 1 as pressure inlet, even-numbered ports (2 and 4) as working ports and odd-numbered ports (3 and 5) as exhaust ports. NFPA (National Fluid Power Association) symbology defines port P as pressure inlet, port A and port B as working ports and port EA and EB (or in the case where there is only one exhaust port, port E) as exhaust ports.

Figure 25.12 shows the typical port configuration symbolic representation for two- to five-valve port designs for ISO (International Standards Organization) symbology and four- to five-port designs for NFPA (National Fluid Power Association) symbology. When two ports are drawn connected by an arrow, the ports are tied together and the direction of pneumatic fluid flow (commonly instrument air) is indicated by the direction of the arrow. When a port is not connected to any other port, the port symbol is not connected with an arrow. For example, all the ports and, in Figure 25.12, are schematically shown as blocked.

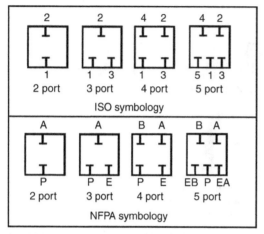

Figure 25.12   Pneumatic valve manifold port symbol.

From Figure 25.12, we can see that the two port manifold has a supply port (1 or P) on the bottom and a cylinder working port (2 or A) on top. The three port manifold has a supply port (1 or P) on the bottom left, an exhaust port (3, E or EA) on the bottom right and one cylinder working port (2 or A) on top. The four-port manifold has a supply port (1 or P) on the bottom left, an exhaust port (3 or E) on the bottom right, one cylinder port (2 or A) on the top right and another cylinder port (4 or B) on the top left. The five-port manifold has a supply port (1 or P) on the bottom center, one exhaust port on the bottom left (3 or EA), another exhaust port (5 or EB) on the bottom right, one cylinder port (2 or A) on the top right and another cylinder port (4 or B) on the top left.

Now that we have an understanding of how we symbolically represent the number of ports on a pneumatic pilot valve module, we will discuss how we symbolically represent the number of positions that a pneumatic valve manifold can have and the ports connected in each position. The number of positions is shown by the number of blocks in the pneumatic valve symbol. For example, in Figure 25.13, we show a two-port valve that shows two position blocks. In the left position block, both the supply port (1) on the bottom and the cylinder working port (2) on the top are closed. In the right position block, the supply port (1) on the bottom is connected to the cylinder working port (2) on the top and the arrow in the position block shows the direction of instrument airflow from high pressure source to valve actuator cylinder. Note that

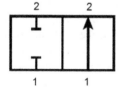

Figure 25.13   Pneumatic valve manifold position symbol.

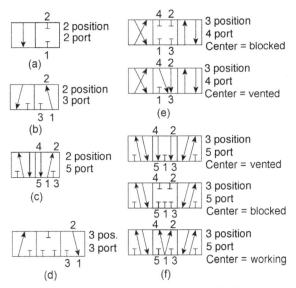

Figure 25.14 Typical pneumatic valve manifold position symbols. *Source*: Reproduced with permission of Parker Pneumatic Division.

flow only occurs when pressure is available at the supply port (1). One of the first things to check when a pneumatic valve is not supplying pressure on a working port is to verify that pressure is available at the supply port (1).

Now, we can combine these to show some very common pneumatic valve position and port configurations as shown in Figure 25.14.

A two-port, two-position directional valve is shown in Figure 25.14a. This configuration consists of two ports connected to each other with passages that are either connected or disconnected depending on the position of the spool piece inside the valve body bore. In one position, the supply port (1) is connected to the working port (2) and the flow path is open. In second position, the supply port (1) and the working port (2) are blocked and the flow path is blocked. This configuration serves as an on/off function and is used for isolation purposes.

A three-port, two-position directional valve is shown in Figure 25.14b. This configuration consists of three ports connected through passages in the valve body. In one position, the supply port (1) is connected to the working port (2) and the valve supplies air pressure to the working port. In second position, the working port (2) and the exhaust port (3) are connected together and the valve exhausts air pressure from the working port. This configuration can be used in two applications. The first application is a pressure selector valve, where the device connected to the working port will have its pressure supplied by either the pressure port or exhaust port depending on the position of the spool. The second application is a diverter valve where the pressure source is connected to the working port (2) and the air can be diverted to either port 1 or port 3.

A five-port, two-position directional valve is shown in Figure 25.14c. This configuration consists of five ports connected through passages in the valve body. In one position, the supply port (1) is connected to the first working port (2) and the second working port (4) is connected to the exhaust port (5). In the second position, the supply port (1) is connected to the second working port (4) and the first working port (2) is connected to the exhaust port (3). The most common application of this type of configuration is for air actuated valve control.

A three-port, three-position directional valve is shown in Figure 25.14d. This configuration consists of three ports connected through passages in the valve body. In the first position, the supply port (1) is connected to the working port (2) and the flow path is open and the exhaust port (3) is blocked. In the second position, all three ports are blocked. In the third position, the supply port (1) is blocked and the working port (2) and the exhaust port (3) are connected. This configuration serves as an on/off function similar to the two-port, two-position valve, except this valve has a middle position where all ports are blocked for isolation.

A four-port, three-position directional valve is shown in Figure 25.14e. This configuration consists of four ports connected through passages in the valve body. In the first position, port (1) is connected to port (4) and port (2) is connected to port (3). In the third position, port (1) is connected to port (2) and port (4) is connected to port (3). There are two variations of this design as shown in Figure 25.14e depending on the path presented by the second position. In the second position, all four ports may be blocked which is needed when we need isolation of ports in the second position, or ports (2) and (4) may be connected to port (3) allow for a common exhaust path from ports (2) and (4) to the exhaust port.

A five-port, three-position directional valve is shown in Figure 25.14f. This configuration consists of five ports connected through passages in the valve body. In the first position, the pressure port (1) is connected to the first working port (2) and second working port (4) is connected to the second exhaust port (5). In the third position, the pressure port (1) is connected to the second working port (4) and the first working port (2) is connected to the first exhaust port (3). There are three variations of this design as shown in Figure 25.14f depending on the path presented by the second position. The first variation is to have the second position designed such that the first working port (2) is connected to the first exhaust port (3) and the second working port (4) connected to the second exhaust port (5). This allows both working ports to be vented to the exhaust in the second position. A second variation is to have all five ports blocked. This allows for isolation of all ports in the second position. The third variation is to have the pressure port (1) connected to both working ports (2 and 4) and the exhaust ports blocked. This allows for both working ports to obtain supply pressure in the second position.

Now that we understand valve construction and symbology, we can now turn our focus to the devices that actuate the pneumatic pilot valve module. The most common types are manual via a handle or button on the manifold, remotely by use of an electrical solenoid, or automatically by an internal spring and these are symbolically shown in Figure 25.15. Other types of actuators are by the use of a level switch, a foot pedal switch, a proximity switch, or a hydraulic actuated switch. Just as ports have standard symbology and identifiers, operators also have standard symbology and identifiers. In a five-port valve, ISO refers to operator 12 as a spring operator and 14 as a manual operator. For the following discussion, we will assume that we have an operator 12 (a spring operator) on the right side of Figure 25.14c and we have

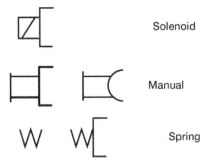

Solenoid

Manual

Spring

Figure 25.15   Typical pneumatic valve manifold position symbols. *Source*: Reproduced with permission of Parker Pneumatic Division.

an operator 14 (manual operator) on the left side of Figure 25.14c. When operator 12 (on right of Figure 25.14c) is in control of the valve spool, the pressure port (1) is connected to the first working port (2), the second working port (4) is connected to the second exhaust port (5), and the first exhaust port (3) is blocked. Conversely, when operator 14 (on left of Figure 25.14c) is in control of the valve spool, the pressure port (1) is connected to the second working port (4), the first working port (2) is connected to the first exhaust port (3), and the second exhaust port (5) is blocked.

Symbolically, when the actuator is energized (such as in the case of the solenoid), the piston is "pushed away" from the symbol. For example, for the two-position, three-port, manually actuated (14), spring return (12), pneumatic pilot valve module that is shown in Figure 25.16, when the manual actuator (14) shown on the left of Figure 25.16 is activated, the source compressed air available at port 1 is connected to the cylinder connected to port 2 as is shown by the symbology next to the manual operator (14). When the manual actuator (14) is released, the spring (12) returns the pneumatic valve manifold to its default position where the compressed air that is connected to port 1 is blocked and the cylinder that is connected to port 2 is exhausted to port 3 as is shown by the symbology next to the spring operator (12).

Lastly, let us describe how to decide which pneumatic valve manifold configuration is needed for which type of application. We will use the following example to explain the design process from the final control element (the valve) back to the pilot control device (the pilot pneumatic valve manifold).

Example 25.2   Design a pneumatic valve control system for an actuator that is connected to a knife gate valve where the valve is closed when the stem is driven down and the valve is opened when the stem is driven up. For this application, we have defined the failsafe mode for this particular application to be with the valve open (the

14   2   2   12

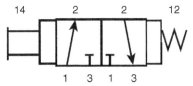

1   3   1   3

Figure 25.16   Example for manually operated, spring return, pneumatic valve manifold position symbol.

failsafe mode is application- and valve-design specific). The valve is to be operated manually from the pilot pneumatic valve manifold to close the valve and, when the manual operator is released, the valve is to automatically open.

Solution: For this application, we have defined the failsafe mode for this particular application to be with the valve open (this is application and valve-design specific). We also defined the direction of actuation for the main valve such that to close the valve, we drive the stem down and to open the valve, we drive the stem up. Since the failsafe mode for the valve is with the valve open, this means that the pneumatic valve-actuating cylinder must drive the valve stem up when air is removed from the cylinder. Therefore, we would select a pneumatic valve-actuating cylinder that has a spring to force the valve stem up with no air supplied and we would have the actuator drive the valve stem down when air is supplied to the valve-actuating cylinder.

Next, we need to select the correct pneumatic pilot valve module. From the design description, the valve is to be operated manually from the pilot pneumatic valve manifold to close the valve and, when the manual operator is released, the valve is to automatically open. Therefore, when we manually depress the actuator on the pilot pneumatic valve manifold, this should admit air into the main valve pneumatic actuating cylinder. When we release the manual actuator on the pilot pneumatic valve manifold, a spring in the pilot pneumatic valve manifold should return the manifold to a position that exhausts the main valve pneumatic actuating cylinder to atmosphere, allowing the main valve pneumatic actuating cylinder spring to raise the valve shaft opening the valve.

Figure 25.17 achieves this design.

Looking at Figure 25.17, when the pneumatic valve manifold is manually depressed, the air supply port 1 is connected to the valve cylinder port 2 and the valve actuator has air pressure applied to the top of the diaphragm. In this position, the pneumatic valve manifold port 3 is isolated. This applies air pressure to the top of the valve actuator and drives the valve stem down to close the gate valve. This is shown in Figure 25.17.

When the manual operator on the pneumatic valve manifold is released, the spring in the pneumatic valve manifold will return the pneumatic valve manifold to this normal position where the air supply port 1 is isolated and the cylinder supply port 2 is connected to the exhaust port 3. This removes air pressure from the top of the valve actuator by exhausting this chamber to atmosphere and the spring in the valve

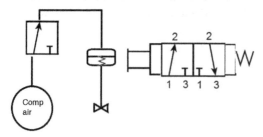

Figure 25.17   Example for pneumatic valve control system: air to close.

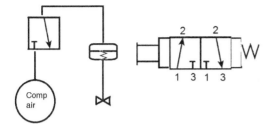

Figure 25.18   Example for pneumatic valve control system: spring return to open.

actuator drives the valve stem up to open the gate valve. This is shown in Figure 25.18. Additionally, the failsafe position of the valve is with the main valve shaft raised and the main valve open should air pressure be lost, as required by our system description.

The selection of the correct type of valve for the application, the correct type of valve actuator (and positioner when used) and the correct type of pneumatic pilot valve manifold is all part of the design of the system. Consideration to the failsafe mode for the valve is critical to the design. The selection of the valve actuator and the pilot pneumatic valve manifold should always be made to ensure the final control element (the valve) fails in the safest state possible.

GLOSSARY OF TERMS

- Actuator – A device connected to the valve that allows remote movement of the valve. It can be an air, fluid or electrically powered device.
- Air-To-Close Valve – A valve where an increase in air pressure to the actuator is required to cause the valve to close. This is another way of saying the valve is fail open or normally open.
- Air-to-Open Valve – A valve where an increase in air pressure to the actuator is required to cause the valve to open. This is another way of saying the valve is fail closed or normally closed.
- Air Valve – A valve that is used to control airflow. This flow is generally small.
- Ball Valve – A valve design which uses a spherical ball as the closing element. Closure is achieved by turning the ball.
- Butterfly Valve – A quarter-turn valve design which includes a circular body. It has a rotary motion disk closure member, which is pivotally supported by its stem, allowing the disk to rotate 90 degrees to open and close the valve.
- Control Valve – Also known as the final control element. A power-operated device used to modify the fluid flow rate in a process control system.
- Feedback Signal – The return signal that results from a measurement of the directly controlled variable.
- Feedforward Signal – The return signal that results from a measurement of an anticipatory element to the control system.

- Gate Valves – A multi-turn valve which as a gate-like disk and two seats to close the valve. This valve is used in fully open/ fully closed applications and the gate disk moves linearly, perpendicular to the direction of flow.

- Globe Valve – A valve with a linear motion, push-pull stem, whose one or more ports and body are distinguished by a globular-shaped cavity around the port region.

- Hand Wheel – A manual override device used to stroke a valve or limit its travel.

- Motor Valve Actuator – A motor-operated device that is used to actuate or move a valve.

- Pneumatic Valve Actuator – A pneumatic operated device that is used to actuate or move a valve.

- Positioner – A device used to position a valve with regard to a signal. The positioner compares the input signal with a mechanical feedback link from the actuator. It then produces the force necessary to move the actuator output until the mechanical output position feedback corresponds with the pneumatic signal value.

- Valve – A device by which the flow of liquid, gas, or loose bulk material may be started, stopped, or regulated by a movable part that opens, shuts, or partially obstructs one or more ports or passageways.

- Valve Actuator – A device that actuates or moves a valve.

- Valve Positioner – A mechanical device used to position something.

PROBLEMS

25.1 What is the purpose of a check valve?

    A. Regulate the rate of flow through the discharge pipe.

    B. Prevent flow of fluid in reverse direction.

    C. Provide overpressure protection for piping system.

    D. Prevent clogging of the suction line.

25.2 What is the purpose of a valve positioner?

    A. Provide the driving force to actuate the valve.

    B. Monitor feedback signal of valve position and drive the valve actuator until the set point and feedback values agree.

    C. Provide overpressure protection for piping system.

    D. Prevent flow of fluid in reverse direction.

25.3 What is the purpose of a relief valve?

    A. Regulate the rate of flow through the discharge pipe.

    B. Prevent flow of fluid in reverse direction.

    C.   Provide overpressure protection for piping system.

    D.   Prevent clogging of the suction line.

25.4   Given a motor that is running at ten times the speed of the valve actuator where the motor is providing 10 ft lb torque to the gear reducer, determine the amount of torque that the valve actuator is providing to the valve stem.

25.5   Given a resistance coefficient of 3.85, a flow through a valve of 5000 gal/min, a valve diameter of 12 in., determine the head loss across this valve and the associated power loss due to losses in this valve given a fluid density of 50 kg/m^3?

25.6   Design a pneumatic valve control system for an actuator that is connected to a globe valve where the valve is closed when the stem is driven down and the valve is opened when the stem is driven up. For this application, we have defined the failsafe mode for this particular application to be with the valve closed (the failsafe mode is application and valve-design specific). The valve is to be operated remotely by a solenoid attached to the pilot pneumatic valve manifold to open the valve and, when the solenoid is de-energized, the valve is to automatically close.

RECOMMENDED READING

Electrical Machines, Drives and Power Systems, 6th edition, Theodore Wildi, Prentice Hall, 2006.

The Engineering Handbook, 3rd edition, Richard C. Dorf (editor in chief), CRC Press, 2006.

Power Plant Engineering, Black & Veatch, edited by Larry Drbal, Kayla Westra, and Pat Boston, Chapman & Hall / Springer, 1996.

Parker Hannifin Corporation, Technical Data Catalog, www.parker.com/pneumatics, 2016.

Standard Handbook of Powerplant Engineering, 2nd edition, Thomas C. Elliott, McGraw-Hill, 1998.

US Navy Engineman 3rd Class Training Manual, Naval Education and Training Professional Development and Technology Center, 2003.

US Navy Training Manual Machinist Mate 2nd class, Naval Education and Training Professional Development and Technology Center, 2003.

CHAPTER 26

EMISSION CONTROL SYSTEMS

> **GOALS**
>
> - To understand the various types of emissions in the air stream from an electric generation facility and identify types of emission control systems utilized to limit the magnitude of emissions
> - To identify the major components of an electrostatic precipitator and describe the function and operation of those components
> - To describe the basic chemistry behind fuel NO_x formation and identify the methods utilized to limit the amount of NO_x in emissions gas streams
> - To identify the major components of a selective catalytic reduction (SCR) system and describe the chemistry behind the SCR process
> - To describe the basic chemistry behind SO_2 formation and identify the methods utilized to limit the amount of SO_2 in emissions gas streams
> - To identify the major components of a flue gas desulfurization (FGD) process and describe the chemistry behind the (FGD) process
> - To describe the function of the continuous emission monitoring system (CEMS)
> - To understand the effects of carbon dioxide (CO_2) on the environment and describe the five basic areas of effort being developed to control the amount of CO_2 emissions from energy production facilities

$\mathbf{T}$**HE OPERATION** of power plants to convert chemical, mechanical, thermal, or other energy resources to electrical energy for the transmission of energy to end users involves some production of byproducts. Operational permitting requirements place limits on both the quantity and quality of these emissions of byproducts. This chapter will review some of the restrictions on power generation facility operating emissions and the technologies used to control these emissions. All aspects of the power plant operation are subject to regulation of emissions including any water streams from the plant (including rainwater runoff), air emissions, and solid waste emissions.

Energy Production Systems Engineering: An Introduction for Electrical Engineers to Electrical Power Generation Facilities, Systems, and Equipment, First Edition. Thomas H. Blair.

There are various concerns with the air emissions of a power plant which include particulate emissions, nitrogen oxide emission, and sulfur oxide emissions. *Continuous emissions monitoring systems* (CEMS) are used to continuously monitor the air emissions from a power generation facility to ensure compliance with permitting requirements.

PARTICULATE EMISSION CONTROL

There are several sources of particulate emissions. The most common emission source is the non-combustibles from fossil fuel. For the typical coal plant, some of the noncombustible material reaches a melting temperature and forms a slag. This slag collects at the bottom of the furnace section. It solidifies in the slag tank at the bottom of the boiler. At the bottom of the slag tank, it is ground to a smaller size and this generates *bottom ash*. The bottom ash is then sluiced from the tank (using water for the transport media) to an external storage facility. At this point, it is de-watered (process to remove the water from the bottom ash material), and then collected to sell or landfill. Bottom ash is used in industry for various processes such as sand-blasting material.

Other noncombustible materials which are lighter in nature become entrained in the gas flow through the furnace and are transported through the exhaust ductwork. The first collection point for these particles is the first 90-degree turn out of the economizer section of the boiler. Hoppers are installed at this location to collect the heavier particles that fall out of the gas stream at this point due to the 90-degree turn in the airflow path. These non-combustibles are known as *economizer ash*. The lighter particles of ash will remain in the air stream past these bends in the ductwork and have to be removed using another means. This is known as fly ash. *Fly ash* is removed by either using *bag houses* which are mechanical filters to capture the particles and allow the gas to pass through or by the use of *electrostatic precipitators*. Smaller units will use the bag house technology and larger units will use the electrostatic precipitator technology.

Figure 26.1 shows the typical arrangement of an *electrostatic precipitator*. The diameter of the ductwork entering the electrostatic precipitator increases dramatically in size. Much as we learned in fluid dynamics, as we increase the cross-sectional area of the ductwork, the same amount of mass is now flowing through a larger diameter path, leading to a much lower linear velocity of the gas stream inside the electrostatic precipitator. Just this action of reducing the linear speed of the gas stream allows some of the particles to begin to drop out of the gas stream. To help assist in the removal of particulates from the gas stream, there are TR (transformer/rectifier) sets that step up and rectify the station service voltage (480 Vac, three-phase) to a very high dc voltage (25,000 Vdc to 125,000 Vdc) which is used to feed electrodes in the precipitator. The transformer and rectifier are enclosed in the same enclosure. The positive terminal is grounded and connected to collection electrodes or plates in the electrostatic precipitator. The negative terminal is insulated and connected to the discharge electrodes which are pipes or wires.

Both the collection electrodes or plates and the discharge electrodes or wires are suspended from the top of the precipitator. These are energized from the TR sets

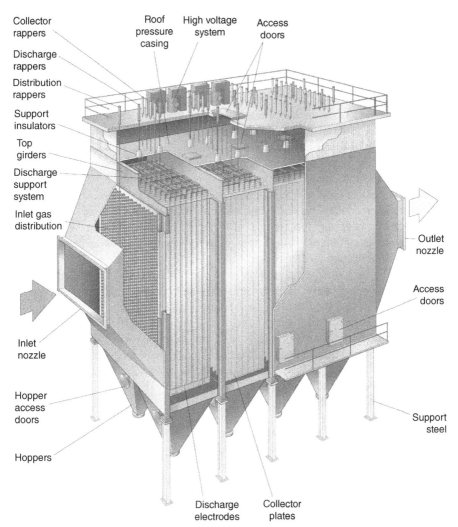

Collector rappers
Discharge rappers
Distribution rappers
Support insulators
Top girders
Discharge support system
Inlet gas distribution
Inlet nozzle
Hopper access doors
Hoppers
Roof pressure casing
High voltage system
Access doors
Outlet nozzle
Access doors
Support steel
Discharge electrodes
Collector plates

Figure 26.1   Dry electrostatic precipitator. *Source*: Reproduced with permission of Babcock & Wilcox.

that are located on the top of the precipitator. As ash flows through the electrostatic precipitator, the high electrostatic field causes the ash to become positively charged. Once positively charged, the ash becomes attracted to the collecting plate and will collect on the plate. As this process continues, the plate will become covered by charged ash particles and the removal efficiency is reduced. To clean off the plate, rappers are installed on top of the precipitators and mechanically vibrate the collecting plates such that the ash falls from these plates into hoppers at the bottom of the electrostatic precipitator. The rappers are basically heavy metal cores with a coil surrounding the core. When rapping is needed, the coil of the rapper is momentarily energized which lifts the core inside the rapper. Then the coil is de-energized and the core in the rapper falls due to the force of gravity and strikes the collection plate. This vibrates the collection plate. Other technologies such as vibrators or hammers can be used as well to vibrate the collection plates to remove the ash from the collection plates.

Under the hoppers at the bottom of the electrostatic precipitator as well as under the hoppers on the discharge side of the economizer section, there is an ash transportation system (sometimes by a belt conveyor, but more commonly by air, since ash is transported easily via airflow) to transport this ash to the final processing location. This ash is sold to vendors that utilize this ash in the production of other materials such as roofing material and block manufacturing. In some circumstances where the properties of the ash prevent its use in other applications, this ash is stored in landfills.

Depending on the type of fuel used (low sulfur fuels), the moisture content of the ash, and even the temperature of the gas stream, some of the ash in the gas stream may be less conductive and therefore does not easily obtain a charge from the electrostatic precipitator. In this event, sulfur in the form of SO_3 may be injected in small quantities after the furnace but before the electrostatic precipitator to increase the sulfur content of the ash. This improves the conductivity of the ash and results in a higher ash collection efficiency.

Another technology used to remove fly ash from the gas stream is the use of fabric filters or bag houses. These use mechanical filtration to remove the fly ash from the gas stream. The filters are sewn into cylindrical tubes or "bags." During the normal operating cycle of the bag house, the flue gas enters the bottom of the bag house in the center of the filter bags. The particulates are collected in the inside of the filter bags and the clean flue gas leaves the bag house at the top of the bag house. As the ash collects in the filter bags, there is more restriction to the flow of air and the ash must be removed from the bags.

During the cleaning cycle, air is forced in the opposite direction as furnace exhaust gas flow to break free particulates which drop into the bottom of the bag house where the particulate is collected in an ash hopper. The fly ash is collected for future transport to the processing facility.

There are other technologies used, to a lesser frequency. Just as in the case where the economizer ash collected at the first 90-degree turn out of the economizer section, we can use the same mechanical force to remove the lighter particles, but the force must be increased to remove the lighter fly ash. *Cyclone collectors* use centrifugal force to separate the fly ash from gas stream. The flue gas enters the cyclone collector and is forced at high velocity through whirl vanes. The high centrifugal force forces the fly ash to collect on the outside of the collector. The cleaned flue gas is exhausted out of the top of the cyclone collector.

Another type of technology used to collect fly ash is the wet venturi scrubber. This device uses liquid to capture the fly ash. The flue gas enters the wet venturi scrubber and accelerates in the venture where water droplets are admitted to collect the ash. The clean flue gas exits the wet venturi scrubber and the ash slurry is transported to a de-watering facility much like the bottom ash as described earlier.

NITROGEN OXIDES EMISSIONS CONTROL

Air that is supplied to the furnace has many constituents. The two largest constituents by volume are nitrogen (N_2) at about 78% and oxygen (O_2) at about 21%. For combustion, we only need the oxygen and the nitrogen is not part of the combustion reaction (see chapter 4 for details of the combustion process). Nitrogen that is entrained in the air supplied to the furnace can lead to the formation of NO$_x$ particles. This is known as *thermal NO$_x$ formation* and constitutes about 25% of the total amount of NO$_x$ generated during the combustion process. This process is affected by the temperature at which complete combustion occurs.

In addition to the nitrogen in the air, some fuels also contain elemental nitrogen in the fuel. Nitrogen that is entrained in the fuel supplied to the furnace is known as *fuel NO$_x$ formation* and constitutes about 75% of the total amount of NO$_x$ generated during the combustion process.

The nitrogen in the air and the nitrogen in the fuel react during the combustion process to form two molecules: nitrogen monoxide (NO) and nitrogen dioxide (NO$_2$). nitrogen monoxide (NO) reacts with oxygen in the air to form nitrogen dioxide (NO$_2$). Equation (26.1) shows the reaction equation for this chemical reaction.

$$2NO + O_2 \rightarrow 2NO_2 \qquad (26.1)$$

Nitrogen dioxide (NO$_2$), when released into the environment, has the ability to react with moisture in the air (H$_2$O) to form nitric acid (HNO$_3$). Equation (26.2) shows the reaction equation for this chemical reaction.

$$3NO_2 + H_2O \rightarrow 2HNO_3 + NO \qquad (26.2)$$

The formation of nitric acid can lead to changes in pH balance of the surface and ground waters in the local environment. Therefore, the release of NO$_x$ is regulated. NO$_x$ is a common term to include both molecules, nitrogen monoxide (NO) and nitrogen dioxide (NO$_2$). This is commonly known as "acid rain." About 25% of the amount of "acid rain" is originated from the reaction with NO$_2$ and H$_2$O.

NO$_x$ can be controlled either at the point where it is generated (i.e., in the combustion process) or in the gas stream by removing the NO and NO$_2$ molecules (or both may be required depending on permitting limitations).

COMBUSTION CONTROL OF NO$_x$

Combustion control of NO$_x$ formation is achieved by one of four basic methods. We can control and reduce the amount of NO$_x$ formation by either reducing the temperature at which complete combustion takes place, by reducing the oxygen concentration such that there is a reduced quantity of oxygen molecules available to generate NO$_x$, by reducing the time the oxygen-rich air is exposed to the higher temperatures of the furnace, and by using low NO$_x$ burner systems. The quantity of thermal NO$_x$ molecules generated is a function of the temperature of the flame where complete combustion occurs. For cooler flame temperatures, less NO$_x$ is formed and for hotter flame temperatures, more NO$_x$ is formed. Also, controlling the amount of oxygen supplied to the combustion zone controls the amount of NO$_x$ formation. To generate NO$_x$ we need excess oxygen not consumed in the combustion process. Typical low NO$_x$ burners operate on the concept that, if most of the combustion occurs in an oxygen-starved environment, then this restricts the formation of NO$_x$. Therefore, these burners have several "zones" of combustion and the oxygen in each zone is carefully controlled to ensure that we form minimal amount of NO$_x$ but at the same time, adequate oxygen is supplied for combustion, especially at the last stage where temperatures are reduced to ensure complete combustion.

The low NO_x burner allows for the control of airflow to the combustion zones and sets up multiple combustion zones. The zone closer to the burner tip is at higher temperature and oxygen availability is restricted. The zone farthest from the burner tip is at lower temperature and this is where final complete combustion occurs. Since complete combustion is occurring at a lower temperature, this design results in less thermal NO_x formation due to the cooler temperature of the gas where complete combustion takes place.

On a tangentially fired boiler, the burners on located at the corners of the boiler. The flame location where complete combustion takes place is not at the burner tip for this design but basically one fireball is maintained in the center of the boiler. By controlling the zone where complete combustion takes place NO_x emissions can be reduced by this design of furnace. The burners are tilting and can adjust the elevation in the boiler where the fireball is located. Additionally there are LNCFS systems (low NO_x concentric firing system) that have the ability to direct auxiliary secondary air around the fireball thereby cooling it where complete combustion takes place. The lower temperature for complete combustion results in reduced NO_x formation.

In a combustion turbine, NO_x is formed in the combustion zone just before the gas enters the turbine blades. The main method of pre-combustion NO_x control in a combustion turbine is by controlling the flame temperature. Various methods are used in combustion turbines to reduce the flame temperature. One of the most common methods is by injection of steam or water to the air stream entering the combustion section. This will cool the flame temperature and result in a reduced amount of NO_x generated. Injecting water or steam into the air stream also has the effect of adding mass to the airflow thereby increasing the efficiency of the turbine, in addition to a reduction of NO_x. Another method to cool the flame is by injection of nitrogen (N_2) into the combustion section.

POST-COMBUSTION CONTROL OF NO_x

For post-combustion control of NO_x, the main method for removing NO_x is by using a *selective catalytic reduction* (SCR) system, as shown in Figure 26.2. In this system, ammonia is injected into the gas stream just before a catalyst. SCR catalysts are made of ceramic materials such as titanium oxide. This ceramic material contains an active catalytic component which is some type of oxide of a base metal such as vanadium, molybdenum, or tungsten. The ammonia (NH_4) in the presence of the NO_x molecules and a catalyst forms elemental nitrogen (N_2) and water vapor (H_2O).

The desired SCR chemical reaction to obtain elemental nitrogen (N_2) and water vapor (H_2O) is shown in equations (26.3) and (26.4). These are exothermic reactions which means that energy is released in the form of heat in the reaction.

$$4NO + 4NH_3 + O_2 - (\text{Catalyst}) \rightarrow 4N_2 + 6H_2O + \text{Heat} \qquad (26.3)$$
$$2NO_2 + 4NH_3 + O_2 - (\text{Catalyst}) \rightarrow 3N_2 + 6H_2O + \text{Heat} \qquad (26.4)$$

There are some potential undesirable reactions that can occur in the SCR system if sulfur is present in the flue gas stream. The first is known as oxidation of the sulfur

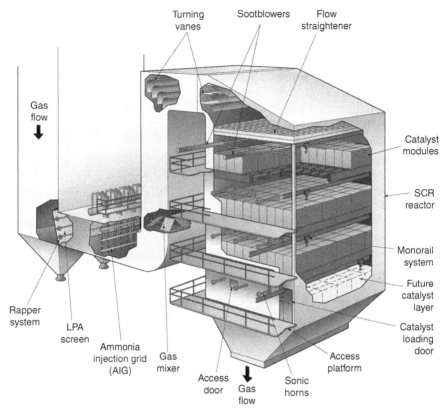

Figure 26.2   Selective catalytic reduction (SCR) system. *Source*: Reproduced with permission of Babcock & Wilcox.

molecules where sulfur dioxide (SO$_2$) oxidation generates sulfur trioxide (SO$_3$). As is described in the next section on sulfur-removal, the sulfur trioxide (SO$_3$) will react with moisture (H$_2$O) in the air to crease sulfuric acid (H$_2$SO$_4$). The sulfuric acid is corrosive to the catalyst and ductwork system and has the potential to cause acid rain if it is released in the atmosphere. The chemical reaction that causes this formation of sulfur trioxide (SO$_3$) is shown in equation (26.5).

$$2SO_2 + O_2 \rightarrow 2SO_3 (SO_2 \text{ oxidation}) \qquad (26.5)$$

The second reaction is known as ammonium bisulfate (NH$_4$HSO$_4$) formation. The third reaction is known as ammonium sulfate (2NH$_4$HSO$_4$) formation. Ammonium bisulfate and ammonium sulfate are salts that will deposit and clog downstream equipment in the ductwork. Cleanup of this material requires blasting of the coated equipment. The three most common materials used for this blasting are water, beads, or CO$_2$ (dry ice).

The SO$_2$ oxidation reaction increases significantly at temperatures above 700°F. The formation of ammonium bisulfate and ammonium sulfate increases significantly at temperatures below 570°F. This leaves an optimal operating temperature range for the flue gas in the SCR between 570°F and 700°F. However, to ensure that moisture is not an issue with the catalyst, the actual operating range is 650–700°F. To minimize the SO_2 oxidation reaction rate, minimize the formation of salts, and to minimize the moisture of the flue gas in the catalyst, the flue gas in the SCR is maintained between 650°F and 700°F.

The amount of ammonia injected into the flue gas stream is controlled to allow just enough for the catalyst to react with the flue gas NO_x, but not an excess amount such that some ammonia remains in the flue gas after the SCR. The measure of the excess ammonia past the SCR structure is known as *ammonia slip*. Ammonia remaining in the gas stream after the SCR can be absorbed by the flyash in the gas stream and this can affect the quality (and resell ability) of the ash collected downstream of the SCR in equipment such as the bag house or electrostatic precipitator.

The ammonia used for injection into the gas stream upstream of the SCR may be delivered in two forms. The first form is pure ammonia which is known as *anhydrous ammonia*. This is 100% ammonia (NH_3). The term anhydrous means "no water." The other form is aqueous ammonia. This is approximately 25% ammonia (NH_3), and 75% water (H_2O). The term aqueous means "with water."

SULFUR DIOXIDE EMISSIONS CONTROL (SCRUBBER)

Many fuels that are used for the combustion process may contain elemental sulfur (S). Some have higher amounts than others. For example, depending on the grade of fuel oil, fuel oil has much higher amounts of elemental sulfur than other fuels such as natural gas. Sulfur is a combustible material and has energy content described in Chapter 4 on combustion in this textbook. Any sulfur that is not completely combusted may react in the combustion process to sulfur dioxide (SO_2). If this is allowed to exit with the flue gas, it will react with oxygen (O_2) first to create sulfur trioxide (SO_3).

$$2SO_2 + O_2 \rightarrow 2SO_3 \qquad (26.6)$$

The sulfur trioxide (SO_3) will react with moisture (H_2O) in the air to crease sulfuric acid (H_2SO_4).

$$SO_3 + H_2O \rightarrow H_2SO_4 \qquad (26.7)$$

Just like the formation of nitric acid, the formation and release into the environment of sulfuric acid can lead to changes in pH balance of the surface and ground waters in the local environment. Therefore, the release of sulfur dioxide (SO_2) is regulated. It is estimated that about 75% of the amount of "acid rain" is originated from the reaction with SO_3 and H_2O.

Like NO_x control that has both pre-combustion and post-combustion controls, the control of SO_2 has two methods of control. The pre-combustion control is to use low sulfur containing fuels. The post-combustion control can use either a dry or wet process for control. The post-combustion process of removing sulfur oxides (SO_x) is known as the *flue gas desulfurization* (FGD) system. There are two main locations for flue gas desulfurization (FGD) systems based on if the processes are a dry-type or wet-type process. It is also commonly called the "scrubbing" process in that the process scrubs the flue gas of the SO_2 in the gas stream. The dry process is known as dry furnace sorbent injection (DFSI) and occurs on the source side of the electrostatic precipitator of bag house. The wet process occurs after the electrostatic precipitator of bag house and just before the stack. Following is a description of the dry process.

Dry furnace sorbent injection (DFSI) is the main dry-type process used to remove SO_x from the flue gas. Dry sorbent desulfurization systems are used to remove sulfur dioxide (SO_2) from flue gas at fossil fuel-fired electric generating stations. In a dry sorbent injection system, limestone, dolomite, lime, or hydrated lime is injected into the boiler or into flue gas ahead of the bag house or electrostatic precipitator. The SO_2 reacts with the dry sorbent, and the dry reaction products are collected, along with fly ash, in the bag house or electrostatic precipitator. The fly ash collected is sold for use as feedstock to cement plants or for stabilization of earthen structures. The DFSI system is a dry injection system and, as such, no water is consumed, and no wastewater is produced. The flue gas is not cooled nor saturated with water, so reheating of desulfurized flue gas is not required. Plants with wet desulfurization systems have moisture in their duct system and stack downstream of the FGD system and, as such, the ductwork and stack liner must be constructed to ensure that the moisture does not cause corrosion of these systems. With the DFSI system, no moisture is added to the flue gas, so these concerns are not present. If this is a retrofit to an existing plant with a dry stack, the stack cannot be modified to handle the wet gases of the wet scrubbing system and a new stack must be built and the old dry stack demolished. This can be a substantial cost to the project and the selection of a dry scrubbing system can avoid these potential costs.

In the DFSI system, limestone forms calcium oxide (CaO) (calcination) and reacts with SO_2 and oxygen to form calcium sulfate $CaSO_4$ (sulfation). Following are equations of reaction depending on if limestone, dolomite, lime, or hydrated lime is utilized as reagents. If limestone is used as the reagent, the calcination reaction is given in equation (26.8).

$$CaCO_3 \rightarrow CaO + CO_2 \qquad (26.8)$$

The sulfation reaction is given in equation (26.9).

$$2CaO + 2SO_2 + O_2 \rightarrow 2CaSO_4 \qquad (26.9)$$

If dolomitic limestone (where concentration of magnesium carbonate or MgCO3 is greater than 20%) is used as the reagent, the calcination reaction is given in equation (26.10).

$$CaCO_3 + MgCO_3 \rightarrow CaO + MgO + 2CO_2 \qquad (26.10)$$

The sulfation reaction is given in equation (26.11).

$$2CaO + 2MgO + 2SO_2 + O_2 \rightarrow 2CaSO_4 + 2MgO \qquad (26.11)$$

If quicklime is used as the reagent, there is no calcination reaction as the quicklime is already in form for the sulfaction reaction and the sulfaction reaction is given in equation (26.12).

$$2CaO + 2SO_2 + O_2 \rightarrow 2CaSO_4 \qquad (26.12)$$

If hydrated lime is used as the reagent, the calcination reaction is given in equation (26.13).

$$Ca(OH)_2 \rightarrow CaO + H_2O \qquad (26.13)$$

The sulfation reaction is given in equation (26.14).

$$2CaO + 2SO_2 + O_2 \rightarrow 2CaSO_4 \qquad (26.14)$$

If hydrated dolomitic lime (dolomitic hydroxide) is used as the reagent, the calcination reaction is given in equation (26.15).

$$Ca(OH)_2 + Mg(OH)_2 \rightarrow CaO + MgO + 2H_2O \qquad (26.15)$$

The sulfation reaction is given in equation (26.16).

$$2CaO + 2MgO + 2SO_2 + O_2 \rightarrow 2CaSO_4 + 2MgO \qquad (26.16)$$

The wet scrubbing system occurs after the electrostatic precipitator or bag house, just upstream of the stack. In this process, water that contains the reagent is sprayed into the gas stream as a fine mist and the reagent in the water reacts with the SO_x molecules. There are four technologies available that are listed below.

1. Forced oxidized wet limestone
2. Magnesium enhanced wet lime
3. Seawater
4. Ammonium sulfate

Of these four, the forced oxidized wet limestone process is the most frequently used and will be the only one discussed in this text book.

The forced oxidized wet limestone process is described in Figure 26.3 and starts with a delivery and storage of raw limestone. When needed for scrubbing operations, the limestone is first processed by pulverizing the limestone to minimize its size. Grinding the limestone performs two functions. It allows for enhanced solubility of the limestone when mixed with the slurry. This allows for easier transport of the slurry to the reaction tank as well as enhanced reaction of the limestone with the SO_x in the gas stream when it is sprayed in the reaction tank and then mixed with water to form limestone slurry. This slurry is then injected into the reaction tank. This slurry is cycled from the bottom of the reaction tank to the spray nozzles at the top of the tank where the limestone slurry reacts with the SO_x in the flue gas that flows through the reaction tank. There is a mixer in the bottom of the tank where the limestone slurry is stored to keep the limestone in solution with the slurry. The reaction of the limestone slurry with the flue gas that contains the SO_x molecules reacts to generate gypsum. The gypsum limestone slurry mix is transported to cylinders that separate the gypsum from the limestone. The unused limestone is returned to the reaction tank and the gypsum is sent to filters or presses for dewatering. Once the water is removed from the gypsum, the gypsum is stored for sale to customers that will use the gypsum for manufacture of other materials such as wall board.

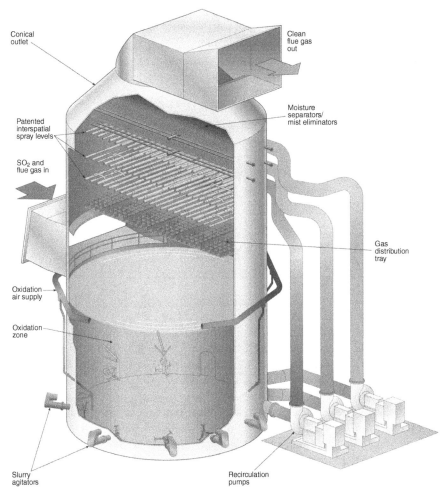

Conical outlet

Clean flue gas out

Moisture separators/ mist eliminators

Patented interspatial spray levels

SO$_2$ and flue gas in

Gas distribution tray

Oxidation air supply

Oxidation zone

Slurry agitators

Recirculation pumps

Figure 26.3   Forced oxidized wet limestone process. *Source*: Reproduced with permission of Babcock & Wilcox.

CONTINUOUS EMISSION MONITORING SYSTEM (CEMS)

So how do we know what the opacity (particulate quantity) of a flue gas stream leaving a stack is or what the SO$_x$ quantity of a flue gas stream leaving a stack is? Most power generation facilities use a continuous emission monitoring system (CEMS) to monitor the condition of the flue gas leaving a stack. A continuous emission monitoring system is the name for the all the equipment necessary to monitor, record and display the status of emissions from a gas stream. CEMS are required under some of the EPA regulations for continual compliance determinations. The individual subparts of the EPA rules specify the reference methods that are used to substantiate the accuracy and precision of the CEMS. Performance specifications are used for evaluating the acceptability of the CEMS at the time of or soon after installation and whenever specified in the regulations. Quality assurance procedures in Appendix F to 40 CFR 60 are used to evaluate the effectiveness of quality control (QC) and quality assurance (QA) procedures and the quality of data produced by any CEMS that is used for determining compliance with the emission standards on a continuous basis as specified in the applicable regulation.

In situ systems monitor the flue gas at the conditions present in the stack at the monitoring location. *Extractive systems* are located remotely from the stack and they draw a gas sample to a remote location for analysis. For the analysis of opacity, the amount of visible light that passes through the gas stream is analyzed and the opacity value is given in percent. A higher opacity value indicates more particulate in the gas stream as observed by more restriction to light.

CARBON DIOXIDE (CO_2) AND GREENHOUSE GAS EMISSION CONTROL

In the process of combustion, when carbon based molecules are part of the combustion process, one of the primary molecules that results is carbon dioxide (CO_2). Please refer to Chapter 4 on combustion for more information on the chemical process of organic molecule (carbon molecule) combustion. Carbon dioxide has been identified as one of several greenhouse gases. A *greenhouse gas* is a gas that contributes to the greenhouse effect by absorbing infrared radiation, for example, carbon dioxide and chlorofluorocarbons. The *greenhouse effect* is the trapping of the sun's warmth in a planet's lower atmosphere due to the greater transparency of the atmosphere to visible radiation from the sun than to infrared radiation emitted from the planet's surface. Since the transparency of the atmosphere is greater for solar energy arriving to the earth than for energy reflected by the earth's surface, some of this energy is retained in the atmosphere. Temperature is the measure of the average molecular kinetic energy of a substance. Since energy can be neither created nor destroyed, the imbalance between energy received into the atmosphere by solar energy and the energy released by the atmosphere results in a net increase in energy in the atmosphere. This net increase in energy in the atmosphere results in elevated temperatures of the earth's atmosphere. The primary greenhouse gases in the earth's atmosphere are water vapor, carbon dioxide, methane, nitrous oxide, and ozone.

A change in the composition of the greenhouse gases in the atmosphere will cause a change in the energy balance equation between energy received by the atmosphere and energy released by the atmosphere. This change in the balance of energy would result in a change in the average molecular kinetic energy of the atmosphere, which, over time will cause a change in atmospheric temperatures. This is also known as *global warming* or *climate change*.

While there are many natural sources for the greenhouse gases listed above, human activities have been identified as causing a significant increase in the atmospheric concentration of carbon dioxide. The primary source for this human source of carbon dioxide is the emissions from carbon-based fuels such as petroleum, coal, oil, natural gas, and wood. Looking at Chapter 4 on the basic combustion process, we see that the electric generation industry is a primary consumer of fossil fuel products and thereby a primary source of emissions of carbon dioxide.

On May 13, 2010, the U.S. Environmental Protection Agency (EPA) issued a final rule that addressed greenhouse gas emissions from stationary sources under the Clean Air Act (CAA) permitting programs. This final rule sets thresholds for greenhouse gas (GHG) emissions that define when permits under the New Source Review Prevention of Significant Deterioration (PSD) and title V Operating Permit programs are required for new and existing industrial facilities. The EPA regulations now apply to any single source emitter that emits at least 100,000 tons/year. So what size generation is covered under the EPA regulations? We know that for a coal-fired power plant, a typical heat rate is about 10,000 BTU/kWh. Also, we know from the Chapter 4 on combustion that the combustion of 1 lb of carbon releases about 14,000 BTU and we know that the combustion of 1 lb of carbon releases about 3.67 lb of CO$_2$. Given that 2000 lb is 1 ton, we can then approximate what size single source generation station may be covered by the EPA limit of 100,000 tons/year qualification. The EPA limits can be converted to the amount of energy a station provides in a year (approximately) as shown below.

$$\frac{(100,000 \text{ tons CO}_2)}{yr} \times \frac{(2000 \text{ lb})}{ton} \times \frac{(1 \text{ lb C})}{3.667 \text{ lb CO}_2} \times \frac{(14,000 \text{ BTU})}{lb \text{ C}} \times \frac{(1 \text{ kWh})}{10,000 \text{ BTU}}$$
$$= \frac{76,363,629 \text{ kWh}}{yr}$$

If we average this energy delivery over 1 year, we find that any generation station in excess of about 8700 kW or about 8.7 MW may be covered by the EPA regulations regarding greenhouse gas emissions. This is calculated below.

$$(76,363,629 \text{ kWh/yr}) \times (1 \text{ yr/}8760 \text{ hr}) = 8717 \text{ kW} = 8.7 \text{ MW}$$

Each stationary source that holds a clean air permit will have the new requirements defined in their permit. The Clean Air Permit will define which best available control technology (BACT) will be utilized to limit the release of greenhouse gasses.

Control of carbon dioxide emissions can be broken down into five areas of effort. The first area of effort is the study is the carbon capture and sequestration (CCS) of carbon dioxide. The second area of effort is focused on treatment of the byproduct of carbon dioxide after combustion to break up the carbon dioxide molecule into its fundamental elements of carbon and oxygen. The third area of effort is the replacement of existing fossil fuel generation resources with non-fossil fuel generation resources. The fourth area of effort is the improvement in fossil fuel process heat rates. The fifth and newest area of effort is carbon capture during the combustion process.

Effort 1: Carbon Capture and Sequestration (CCS)

Carbon capture and sequestration (CCS) involves two processes. The first process captures the carbon dioxide and the second process stores the carbon dioxide. The

CO_2 concentration in the flue gas ranges from 3% for natural gas to about 15% for pulverized coal.

Carbon dioxide can be captured either pre-combustion or post-combustion. Post-combustion carbon dioxide capture involves a process to remove or "scrub" the CO_2 from the flue gas stream after combustion but before the exhaust gasses leave the process. The post-combustion process used to capture CO_2 called "amine scrubbing" is well known and used in the chemical processing industry for stripping other molecules such as hydrogen sulfide (H_2S) from gas streams. It comprises a chemical solvent in an aqueous solution of some type of alkylamines. Figure 26.4 describes this scrubbing process for removal of carbon dioxide. The gas to be cleaned enters the process at location 1 of Figure 26.4 by a flue gas blower at location 2 and enters the absorption column at location 4. The "lean" amine is delivered to an absorber tower where it reacts with the exhaust flue gas to strip the CO_2 from the flue gas resulting in cleaned flue gas leaving the system at location 5. Prior to entering the absorption tower, any useful heat in the flue gas is recovered and the flue gas is cooled to absorption temperature in a flue gas cooler at location 3. The CO_2 dissolves into the amine solution. Once the amine solution has captured CO_2, it is known as "rich" amine. This rich amine is then pumped to a stripper tower at location 8 through a heat exchanger at location 6 that raises the temperature of the rich amine solution. The absorption/desorption process is governed by the temperature difference of absorber tower and the desorber tower. It is necessity to cool the flue gas before it enters the absorber which is achieved at location 3.

In the desorption tower at location 8 the dissolved CO_2 is removed from the "rich" amine solution. CO_2 stripping in desorber requires a substantial amount of energy in the form of heat to be introduced through the heat exchanger (location 10) which is heated by plant process steam. This additional energy required to operate this system results a reduced efficiency of the system. This "lean" amine is then resent back to the absorber tower at location 4 through the heat exchanger at location 6 to reduce the lean amine temperature and begin this process anew. The exhaust gas that is rich in CO_2 entering the absorber tower at location 1 is called "sour gas" and the exhaust gas that leaves the absorber tower at location 5 with the CO_2 gas removed is called "sweet gas." At location 10, the CO_2 captured in the desorption tower is heated typically by a steam source in a heat exchanger and also compressed to a useful pressure. The energy consumed by the system is mostly consumed at location 10. The location of the solvent reclaiming system is very dependent on the type of system utilized. The system shown in Figure 26.4 utilizes a non-solvent amine solution. Non-solvent amine solutions do not require the higher temperatures of solvent amine solutions. Therefore, for non-solvent amine-based systems, the solvent reclaimer is normally located on the cooler side of the amine loop between the heat exchanger at location (6) and the absorber tower at location (4). Solvent amine based systems which utilize a volatile amine system normally will have their amine solution reclaimer, located on the warmer side of the amine loop, is normally located between the heat exchanger at location (6) and the desorption tower at location (8).

In pre-combustion carbon dioxide capture, the fuel is treated prior to combustion to separate the elemental components of the fuel. The carbon is converted to carbon dioxide and this CO_2 stream is then moved to the second part of the process

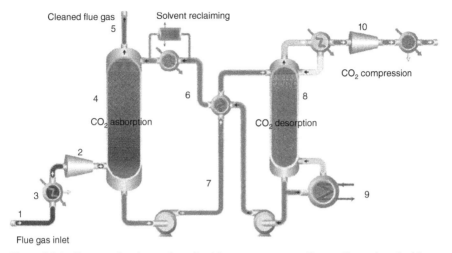

Figure 26.4 Post-combustion carbon dioxide capture process. *Source*: Reproduced with permission of Siemens Power and Gas Division.

called carbon sequestration. In chapter 4 we learned about coal gasification. This is one process where prior to combustion, the fossil fuel is processed into the fundamental elements that make up the fuel. In this process, the coal is milled and then mixed with water and delivered to a preliminary combustion chamber called a "gasifier." Pure oxygen is admitted to the gasifier and the coal slurry and oxygen (O_2) undergo a reaction to release the elements carbon monoxide (CO) and hydrogen (H_2). The chemical reaction that generates the carbon monoxide (CO) and hydrogen (H_2) molecules is shown in equation (26.17). First, the carbon and oxygen are combusted to generate carbon dioxide and release energy in the form of heat.

$$C + O_2 \rightarrow CO_2 + \text{Heat} \tag{26.17}$$

The thermal energy or heat generated in the above reaction is then used to convert carbon dioxide into carbon monoxide as shown in equation (26.18).

$$C + CO_2 + \text{Heat} \rightarrow 2CO \tag{26.18}$$

Additionally, the thermal energy or heat generated in the above reaction is also used to convert carbon in the coal and water in the slurry mixture into hydrogen and carbon monoxide as shown in equation (26.19).

$$C + H_2O + \text{Heat} \rightarrow H_2 + CO \tag{26.19}$$

In the typical gasification process, since the carbon monoxide and hydrogen both have energy of combustion available, these are delivered to the combustion turbine. However, in the CCS process, the carbon monoxide is further reacted with the pure oxygen stream to create carbon dioxide as shown in the reaction described by equation (26.20).

$$2CO + O_2 \rightarrow 2CO_2 + \text{Heat} \tag{26.20}$$

The carbon dioxide is now delivered to the sequestration process and the hydrogen is delivered to the combustion turbine. In the combustion section of the turbine, we now only combust hydrogen and the emission from hydrogen combustion is water as shown in the chemical equation defined by equation (26.21).

$$2H_2 + O_2 \rightarrow 2H_2O + Heat \tag{26.21}$$

Now that we have identified methods to capture carbon dioxide, we discuss the second part of the process of CCS, that of carbon sequestration. *Carbon sequestration* describes long-term storage of carbon dioxide or other forms of carbon to either mitigate or defer global warming. The presence of carbon dioxide in the atmosphere increases the amount of energy stored in the atmosphere leading to average elevated global temperatures. If we were to capture the carbon dioxide released during combustion and store this in a location that would not allow for its release into the atmosphere, this would reduce the amount of carbon dioxide in the atmosphere and thereby reduce the greenhouse effect. There are two basic types of carbon sequestration processes currently under development. One process, known as geo-sequestration involves storage of gaseous CO_2 in deep geologic formations such as oil fields, gas fields, saline formations, unutilized coal seams, and saline-filled basalt formations. The carbon dioxide is pressurized and pumped into these deep formations and then the formation is sealed to prevent release of the carbon dioxide. Another proposed method of storing carbon dioxide is by allowing CO_2 to chemically reacting with various metal oxides which produces a solid form of carbonate. This is a naturally occurring reaction and is one of several methods that nature uses to remove CO_2 from the atmosphere. However, the natural reaction between CO_2 and metal oxides is a slow process. In order to utilize this technology for storage of the quantities of carbon dioxide that are emitted from power generation stations, we would need to accelerate the reaction and research is ongoing along this line. The benefit of using a carbonate form of storage for CO_2 versus geo-sequestration is the stability of the storage media. Once CO_2 has been converted to a solid form of carbonate, it has a lower energy level than CO_2 and therefore, this is a stable long-term form of storage. One of the concerns about geo-sequestration is the ability to store the carbon dioxide under pressure for long periods of time. If a leak were to develop and the carbon dioxide under pressure was able to escape into the atmosphere, in addition to the greenhouse effect, it would cause, a large concentration of carbon dioxide released in a short amount of time could present an asphyxiation hazard to animals and humans in the area of the release. This potential hazard is not just theoretical. In nature, there are several lakes located on areas of volcanic activity. These lakes naturally store carbon dioxide in layers at the bottom of the lake. Lake Nyos is one such lake that is located in the Northwest Region of Cameroon. On August 21, 1986, possibly as the result of a landslide, Lake Nyos suddenly emitted a large cloud of CO_2, which suffocated 1700 people and 3500 livestock in nearby towns and villages. If geo-sequestration is to be utilized, then the concern for the asphyxiation hazard will have to be addressed.

Additionally, the major energy consumer of this process is in the compression and purification unit (location 10 of Figure 26.4) for the processing of flue gases into dense phase CO_2 ready for transportation, use and/or storage. Research is still underway to find more energy efficient methods of compressing and purifying the CO$_2$ gas stream thereby improving the cycle efficiency.

Effort 2: Carbon Dioxide Reaction into Carbon and Oxygen

Another area of research is to chemically break up the carbon dioxide molecule back into its elemental components of carbon and oxygen, neither of which presents the concern for greenhouse effects. Prior to discussion technologies to achieve this reaction, let us review the energy of combustion for carbon (see Chapter 4 on combustion reactions for more detail).

When carbon is combusted to create carbon dioxide, the maximum amount of energy released is 14,093 BTU/lbm. And, the chemical reaction is defined in equation (26.22).

$$C + O_2 \rightarrow CO_2 + \text{Heat (14,093 BTU/lbm)} \tag{26.22}$$

To break up CO$_2$ into its elements of carbon and oxygen is a two-step process. The first process involves adding energy to the carbon dioxide molecule to break it up into carbon monoxide and one oxygen atom. The amount of energy to achieve this reaction is 4347 BTU/lbm. The chemical reaction is shown in equation (26.23).

$$CO_2 + \text{Heat (4743 BTU/lbm)} \rightarrow CO + (1/2)\,O_2 \tag{26.23}$$

Then, additional energy is provided to further break up the carbon monoxide molecule into elemental carbon and oxygen. The amount of energy to achieve this reaction is 9753 BTU/lbm. The chemical reaction is shown in equation (26.24).

$$CO + \text{Heat (9,753 BTU/lbm)} \rightarrow C + (1/2)\,O_2 \tag{26.24}$$

If we total the amount of energies of the above reaction equations, we find that the net amount of energy is approximately zero (law of conservation of energy). Therefore, to convert CO$_2$ back into the elemental components of carbon and oxygen requires the same amount of energy we obtained from the combustion of oxygen and carbon. If the energy source we use was derived from a carbon-based fuel, then we generate as much carbon dioxide in the generation of that energy as we could convert in the final reaction. However, this does not address efficiencies of the energy conversion process. We learned earlier in Chapter 2 covering thermodynamics that no thermodynamic process is 100% efficient. Therefore, in reality, if the energy source we use was derived from a carbon-based fuel, then we generate more carbon dioxide in the generation of that energy as we could convert in the final reaction and we would have a net increase in the amount of CO$_2$ at the end of the process. For this reason, if this technology were to be used, then the source of the energy must be a non-carbon-based source to ensure that our result is a net reduction in the amount of carbon dioxide. Research is underway utilizing non-fossil fuel sources of energy to accomplish this reaction. Keeping in mind that no energy conversion process is 100% efficient, brings up the question, is it more efficient to combust fossil fuels thus creating carbon dioxide, and then utilize a second energy conversion process with its inefficiencies to break the carbon dioxide back into carbon and oxygen, or would it be more efficient to just not combust the fossil fuel in the first place and utilize the non-fossil fuel energy source as our source for electrical energy. Because the latter is a more efficient conversion of energy from one form (chemical, solar, geothermal, tidal, etc.) to electrical energy, the third method for reducing greenhouse gases is the replacement of fossil fuel generation sources with non-fossil fuel generation sources.

Effort 3: Replacement of Fossil Fuel Generation with Non-Fossil Fuel Generation

Back in the beginning of this text, we explained that energy production is really a process of energy conversion. There are a multitude of energy sources available to be converted to electrical energy for transportation to the end user of this energy. The basic function of the electric utility is to provide for its customers inexpensive and reliable energy. Fossil fuels have been utilized largely due to their ability to deliver on both the cost and reliability requirements. For example, coal can easily be stored and made available regardless of the availability of transportation and the cost per BTU for fossil fuels have been historically low as compared with other technologies. However, the concern for increased costs due to regulation of carbon dioxide emissions may soon make fossil fuels less attractive in regards to cost.

There are many potential energy sources that can be utilized instead of fossil fuels. If we replace the energy source with non-carbon-containing energy sources, then we would reduce the amount of carbon dioxide that is emitted into the air per BTU of electrical energy delivered. There are many technologies that have been available for many years. Geothermal energy is where thermal energy from the earth's crust is utilized to drive a steam cycle. Solar energy can be converted to electrical energy in one of two forms, either thermal energy conversion or radiated energy conversion (see Chapter 28 on solar energy for more information on these processes). Wind energy can be captured by a wind turbine and used to drive an electric generator to convert this kinetic energy into electrical energy. Nuclear energy utilizes the heat of decay of unstable elements as they decay into lower energy stable elements or nuclides. This heat is used to drive a water based thermal cycle to drive a steam generator which eventually drives a generator. Another resource is hydroelectric energy where the potential energy contained in a head of water is converted to electrical energy. Other sources of non-fossil fuel-based energy include tidal energy where the kinetic energy of the tidal flows is captured to drive an electric generator.

Getting back to the basic function of the electric utility, the primary function is to provide inexpensive and reliable energy for its customers. These two functions need to be addressed in the utilization of alternative sources of energy. For example, solar energy is only available during periods of sunlight and wind energy is only available when atmospheric conditions are advantageous to the development of wind. While these forms of energy cannot be directly stored for conversion to electrical energy later, there are methods to convert this energy into other forms of energy that can later be converted to electrical energy when needed. For example, solar energy can be converted to electrical energy and this energy can be utilized to pump water into a dam converting this electrical energy into potential energy. Later, when sunlight is not available, this potential energy then can be converted back to electrical energy in a hydroelectric generation station. However, no energy conversion process is 100% efficient. As we saw in Chapter 2 on thermodynamics, as the efficiency of the conversion process is reduced, the cost for the energy increases. The process of conversion of some of these energy sources into other forms of energy for storage can reduce the overall efficiency of the process thereby increasing the cost for the energy resource.

Effort 4: Improved Efficiency

The fourth area of potential reduction of greenhouse gas emissions is the area of improved fossil fuel process efficiency (or reduction in heat rate). Heat rate is defined as the amount of energy into a process for the amount of energy out of a process. The typical unit for heat rate is BTU/kWh where BTU defines the amount of energy into the process and kWh defines the amount of electrical energy out of a process. For combustion of a fossil fuel, the complete combustion of 1 lb of carbon releases a maximum energy value of 14,093 BTU. If we improve the efficiency of the energy conversion process, this results in a lower value of heat rate or a lower value of BTU/kWh. Stated another way, for a certain amount of energy in kWh out of the process, a process with an improved efficiency will require less BTU into the process to deliver the given amount of electrical energy kWh out of the process. By requiring less input energy in BTU, this means that less carbon is needed to deliver the electrical energy needed. By combusting less carbon in a process with a greater efficiency, we release less carbon dioxide into the atmosphere for the same amount of electrical energy delivered.

Effort 5: Carbon Capture During Combustion (Oxyfuel Combustion)

There is a fairly new field of research is known as oxyfuel combustion. In this process, much like gasification, the combustion gas is concentrated oxygen (but the purity of the oxygen concentration for oxyfuel combustion is less stringent then for gasification). The concentrated oxygen is combined or mixed with recycled flue gases from the economizer section of the boiler. The flue gases contain carbon dioxide (CO$_2$) and also some portions of un-combusted fuel in the form of carbon monoxide (CO) and other minor elements. Recycling this flue gas back through the combustion chamber and mixing it with the concentrated oxygen allows for the carbon monoxide to be completely combusted into carbon dioxide (CO$_2$). This improves the percentage of carbon that is completely combusted resulting in more energy released per pound mass of carbon combusted. Additionally, this process results in a more concentrated stream of CO$_2$ in the exhaust stream and a reduced amount of nitrogen in the exhaust stream (since nitrogen is removed in the process of generating oxygen for the combustion gas). While pre-combustion and post-combustion processes require substantial amounts of energy in their processes (thereby possibly reducing the efficiency of the process), oxyfuel combustion offers the potential to allow for concentration of CO$_2$ (thereby increasing the removal efficiency) with substantially reduced energy input requirements to the process and resulting in less of an impact to overall net plant heat rate.

Remaining research for the oxyfuel combustion process includes refining and optimizing the two major energy users of this process; namely the air separation unit (ASU) for the production of oxygen, and a compression and purification unit (CPU) for the processing of flue gases into dense phase CO_2 ready for transportation, use and/or storage.

GLOSSARY OF TERMS

- Baghouse – A facility that removes fly ash from the flue gas by the use of fabric filter bags.

- Continuous Emission Monitoring System (CEMS) – The total equipment necessary for the determination of a gas or particulate matter concentration or emission rate using pollutant analyzer measurements and a conversion equation, graph, or computer program to produce results in units of the applicable emission limitation or standard.

- Cyclone – The cone-shaped air-cleaning apparatus which operates by centrifugal separation that is used in particle-collecting and fine-grinding operations.

- Electrostatic Precipitator (ESP) – Collection of coal combustion fly ash requires the application of an electrostatic charge to the fly ash, which then is collected on grouped plates in a series of hoppers. Fly ash collected in different hoppers may have differing particle size and chemical composition, depending on the distance of the hopper from the combustor. The ESP ash may also be collected as a composite.

- Flue Gas Desulfurization (FGD) – Removal of the sulfur gases from the flue gases, using a high calcium sorbent such as lime or limestone. The three primary types of FGD processes commonly used by utilities are wet scrubbers, dry scrubbers, and sorbent injection.

- Fuel NO_x – NO_x generated where the source of nitrogen is contained in the fuel used in the combustion process.

- Greenhouse Gas – A gas that contributes to the greenhouse effect by absorbing infrared radiation, for example, carbon dioxide and chlorofluorocarbons.

- Greenhouse Effect – The trapping of the sun's warmth in a planet's lower atmosphere due to the greater transparency of the atmosphere to visible radiation from the sun than to infrared radiation emitted from the planet's surface.

- Selective Catalytic Reduction (SCR) – Is a means of converting nitrogen oxides, also referred to as NO_x with the aid of a catalyst into elemental nitrogen, N_2, and water, H_2O. A gaseous reductant, anhydrous ammonia, aqueous ammonia or urea, is added to a stream of flue or exhaust gas and is absorbed onto a catalyst. Carbon dioxide, CO_2 is a reaction product when urea is used as the reductant.

- Thermal NO_x – Is the NO_x generated where the source of nitrogen is the air used in the combustion process.

PROBLEMS

26.1 Which equation below shows the formation of nitric acid (acid rain)?

A. $NO + \frac{1}{2} O_2 \rightarrow NO_2$

B. $3NO_2 + H_2O \rightarrow 2HNO_2 + NO$

C. $4NO + 4NH_3 + O_2$ –(Catalyst) $\rightarrow 4N_2 + 6H_2O + Heat$

D. $2NO_2 + 4NH_3 + O_2$ –(Catalyst) $\rightarrow 3N_2 + 6H_2O + Heat$

26.2 True or False: thermal NO_x is generated from nitrogen contained in the fuel that is provided to the combustion process.

26.3 For wet scrubbing of exhaust gas to remove SO_2, which equation below shows the sulfation reaction if the reagent used is limestone?

A. $CaO + SO_2 + \frac{1}{2} O_2 \rightarrow CaSO_4$

B. $CaO + MgO + SO_2 + \frac{1}{2} O_2 \rightarrow CaSO_4 + MgO$

C. $CaO + SO_2 + \frac{1}{2} O_2 \rightarrow CaSO_4$

D. $CaO + SO_2 + \frac{1}{2} O_2 \rightarrow CaSO_4$

26.4 For wet scrubbing of exhaust gas to remove SO_2, which equation below shows the sulfation reaction if the reagent used is dolomitic limestone?

A. $CaO + SO_2 + \frac{1}{2} O_2 \rightarrow CaSO_4$

B. $CaO + MgO + SO_2 + \frac{1}{2} O_2 \rightarrow CaSO_4 + MgO$

C. $CaO + SO_2 + \frac{1}{2} O_2 \rightarrow CaSO_4$

D. $CaO + SO_2 + \frac{1}{2} O_2 \rightarrow CaSO_4$

26.5 For wet scrubbing of exhaust gas to remove SO_2, which equation below shows the sulfation reaction if the reagent used is quicklime?

A. $CaO + SO_2 + \frac{1}{2} O_2 \rightarrow CaSO_4$

B. $CaO + MgO + SO_2 + \frac{1}{2} O_2 \rightarrow CaSO_4 + MgO$

C. $CaO + SO_2 + \frac{1}{2} O_2 \rightarrow CaSO_4$

D. $CaO + SO_2 + \frac{1}{2} O_2 \rightarrow CaSO_4$

26.6 For wet scrubbing of exhaust gas to remove SO_2, which equation below shows the sulfation reaction if the reagent used is hydrated lime?

A. $CaO + SO_2 + \frac{1}{2} O_2 \rightarrow CaSO_4$

B. $CaO + MgO + SO_2 + \frac{1}{2} O_2 \rightarrow CaSO_4 + MgO$

C. $CaO + SO_2 + \frac{1}{2} O_2 \rightarrow CaSO_4$

D. $CaO + SO_2 + \frac{1}{2} O_2 \rightarrow CaSO_4$

RECOMMENDED READING

Electrical Machines, Drives and Power Systems, 6th edition, Theodore Wildi, Prentice Hall, 2006.

Efficiency of CO₂ Dissociation in a Radio-Frequency Discharge, Laura F. Spencer, and Alec D. Gallimore, Plasma Chem Plasma Process, Springer Science+Business Media, LLC 2010, DOI: 10.1007/s11090-010-9273-0.

Environmental Protection Agency EPA web site, "Continuous Emission Monitoring - Information, Guidance, etc.," http://www.epa.gov/ttnemc01/cem.html, 2016.

NOx Emission Reduction by Furnace Cleanliness and Combustion Management, H. Randy Carter, and Wayne K. Larson, Diamond Power Specialty Company. 1993.

Power Plant Engineering, Black & Veatch, edited by Larry Drbal, Kayla Westra, and Pat Boston, Chapman & Hall / Springer, 1996.

Standard Handbook of Powerplant Engineering, 2nd edition, Thomas C. Elliott, McGraw-Hill, 1998.

U.S., Technical Manual, TM 5-815-1/AFR 19-6, Air Pollution Control Systems for Boilers and Incinerators, Department of the Army and the Air Force, USA, 1988.

CHAPTER 27

WATER TREATMENT

GOALS

- To understand the various types of water streams utilized in an electric generation facility
- To identify the different types of processes utilized in water treatment systems utilized for control of water chemistry
- To perform unit conversion calculations
- To perform dosage and removal efficiency calculations
- To perform pump flow, power and efficiency design calculations

---

POWER PLANTS use water in many systems for many different functions. The quality of the water directly impacts the ability of the system to function reliably. For example, in the power plant thermal cycle (i.e., boiler), the hardness of the water is critical to the heat transfer capability of the tubes in the boiler. As water flows through the tubes in the boiler, calcium and magnesium can deposit out and bake on the water wall tube surfaces. Deposition of these materials onto the surface of the boiler tubes increases the thermal resistance of the tube surface, reducing the heat transfer capabilities of the tubes in the boiler. To prevent issues such as these, the quality of the water is well-regulated through several processes.

Treatment of the water systems at the power plant utilizes various processes or series of processes depending on the type of material being removed from the water system and for the service that it will be placed in. The processes include the following: *sedimentation, clarification, filtration, softening, oxidation, degasification,* and *demineralization.* Demineralization can further be broken down into the processes used to remove minerals and ions and these three processes are *evaporation, ion exchange,* and *reverse osmosis.* Following is a description of each of these processes.

Oxidation is a process whereby we convert soluble gas to solids by forcing oxygen into a water system. The oxidation process uses either chemical addition and/or a process of aeration where the water is forced to mix with air to increase the surface area and increase the amount of oxygen contained in the water.

Energy Production Systems Engineering: An Introduction for Electrical Engineers to Electrical Power Generation Facilities, Systems, and Equipment, First Edition. Thomas H. Blair.
© 2017 by The Institute of Electrical and Electronics Engineers, Inc. Published 2017 by John Wiley & Sons, Inc.

Sedimentation is a process whereby the water is stored in a certain area, called a settling pond, where the suspended non-soluble materials will settle out of the water stream. These materials settle to the bottom of the storage area and this sludge is pumped out or dredged out occasionally and sent for treatment. The suction is taken from the top of the storage area for the next step in the water treatment process which is clarification.

Clarification is similar to sedimentation, but includes the addition of certain chemicals, that allow the non-soluble material to coagulate. Once this material coagulates, it more easily settles out of the water source. *Flocculation* is a process that follows clarification where the water is slowly stirred. This process uses the chemicals added during the clarification process and the agitation of the flocculation process to further increase the particle size and provides the ability to remove more suspended matter from the water stream. Following Flocculation is the *filtration* process where these coagulated particles are filtered out of the water stream via a mechanical filter process and the water is sent to the next step in the water treatment process.

For some water streams, such as fire water system, the water quality at this stage may be adequate for water treatment to get reliable service from the system. However, some systems such as boiler water system, requires further processing of the water to remove unwanted soluble and ions from the water before it is used in the power plant system. This is known as *demineralization* process and includes processes such as ion exchange, evaporation, and reverse osmosis.

Ion exchange utilizes resin beds (typically two) to replace unwanted ions with less objectionable ions. *Cation exchanges* will exchange positively charged ions in the water stream and *anion exchanges* will exchange negatively charged ions in the water. Since the resin beds have only a limited number of ions for exchange with the water system and, periodically, we need to "regenerate" the resin bed to replenish the ions. Also, the resin bed is sensitive to contamination from suspended and non-suspended solids in the water stream and, as such, ion exchange follows *filtration*.

Evaporation utilizes conversion of water to steam to separate the pure water from the unwanted ions and soluble material. The water is evaporated in an evaporation or auxiliary boiler and the steam is transported to a heat exchanger where it is recondensed. The unwanted materials called brine in the boiler are pumped from the boiler for disposal. A typical evaporation system is shown in Figure 27.1.

The *reverse osmosis* (RO) process uses pressure difference across a semipermeable membrane to separate the water stream from the unwanted materials. The pressure difference across the membrane separates the incoming stream into two discharge streams, one known as the concentrate and one known as the permeate. The permeate stream is the purified water stream that is sent to the plant process system. The concentrate system contains the unwanted impurities that are discharged for disposal.

Lastly, we discuss the importance of pH which is a measure of the acid or base concentration of a liquid. As water (H_2O) undergoes chemical reactions, it may break into its constituents of either hydrogen ions (H^+) or hydroxyl ions (OH^-). Tests can be performed that measure the pH of a fluid. A fluid with the same amount of hydrogen

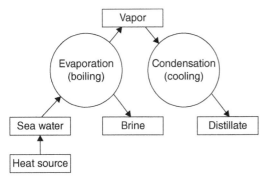

Figure 27.1   Typical evaporation distillation process. *Source*: Reproduced with permission of U.S. Navy Training Manual.

ions (H^+) and hydroxyl ions (OH^-) has a pH value of 7.0. A fluid with more hydrogen ions (H^+) than hydroxyl ions (OH^-) has a pH value less than 7.0 and is known as an acid. A fluid with less hydrogen ions (H^+) than hydroxyl ions (OH^-) has a pH value greater than 7.0 and is known as a base. The pH scale varies from 0.0 to 14.0. The pH will affect the corrosion rate of metals like those that make up the boiler tubes. Figure 27.2 shows the corrosion rate on boiler tube metal for various values of pH.

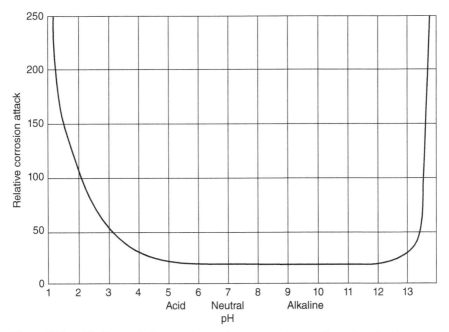

Figure 27.2   pH effect on boiler metal corrosion process. *Source*: Reproduced with permission of U.S. Navy Training Manual.

The reader will note that, in order to minimize boiler tube corrosion, the ideal range of pH is between 6.0 and 10.0.

Water Treatment Calculations

The ability to perform various calculations regarding the treatment of wastewater is critical to the generation utility engineer. An example is to calculate the dosage of a chemical for treatment of the process water. If this is not performed correctly, the water that comes into contact with plant equipment could damage plant equipment and lead to expensive and time consuming repairs. In the following section, we cover some of the basic water treatment calculations that are performed in the water treatment plant of a utility generation station.

FLOW

In the instrumentation section, we discussed the primary elements utilized to measure flow. The typical units of flow are million gallons per day (mgd), gallons per minute (gpm), and cubic feet per second (cfs). To convert from MGD to GPM, utilize equation (27.1).

$$GPM = MGD \times 694.4 \text{ (gpm)} \tag{27.1}$$

where

GPM = flow (gpm)

MGD = flow (mgd)

694.4 converts units from MGD to GPM

To convert from GPM to MGD, use equation (27.2).

$$MGD = GPM/694.4 \text{ (mgd)} \tag{27.2}$$

where

MGD = flow (mgd)

GPM = flow (gpm)

694.4 converts units from MGD to GPM

694.4 is needed to convert units from millions of gallons per day to gallons per minute as shown below.

(1 million gal/min) × (1,000,000 gal/1 million gal) × (1 day/1440 min) = 694.4

For the conversion from cubic feet per second to gallons per minute or million gallons per day, we need to determine the gallons in one cubic foot of fluid. To convert from gallons to cubic feet of volume, we can use the conversion that 7.48052 gal of water takes up approximately one cubic foot of volume. Using this relationship, we can convert from CFS to GPM using equation (27.3).

$$GPM = CFS \times 448.8 \text{ (gpm)} \tag{27.3}$$

where

GPM = flow (gpm)

CFS = flow (cfs)

448.8 converts units of GPM to CFS

To convert from GPM to CFS, utilizing equation (27.4).

$$CFS = GPM/448.8 \text{ (cfs)} \tag{27.4}$$

where

CFS = flow (cfs)

GPM = flow (gpm)

448.8 converts units of GPM to CFS

448.8 is needed to convert units from cubic feet per second to gallons per minute as shown below.

$$(1 \text{ ft}^3/\text{sec}) \times (7.48052 \text{ gal}/\text{ft}^3) \times (60 \text{ sec}/1 \text{ min}) = 448.8$$

To convert from MGD to CFS, utilize equation (27.5).

$$CFS = MGD \times 1.547 \text{ (cfs)} \tag{27.5}$$

where

CFS = flow (cfs)

MGD = flow (mgd)

1.547 is needed to convert units from million gallons per day to cubic feet per second as shown below.

$$\frac{(1 \text{ Mgal/min}) \times (1{,}000{,}000 \text{ gal}/1 \text{ Mgal}) \times (1 \text{ day}/1440 \text{ min})}{(1 \text{ ft}^3/\text{sec}) \times (7.48052 \text{ gal}/\text{ft}^3) \times (60 \text{ sec}/1 \text{ min})} = 1.547$$

To convert from CFS to MGD, utilize equation (27.6).

$$MGD = CFS/1.547 \text{ (mgd)} \tag{27.6}$$

where

MGD = flow (mgd)

CFS = flow (cfs)

Example 27.1   A system averages an influent flow of 2.5 mgd. What is the average flow rate in gallons per minute that flows into the system?

Solution: Utilizing equation (27.1) we find

GPM = MGD × 694.4 (gpm)

GPM = 2.5 × 694.4 (gpm)

GPM = 1736 (gpm)

AREAS AND VOLUMES

The next area of calculations are surface areas or cross-sectional areas and containment volumes. To calculate the cross-sectional area of a pipe or tank, we can use either the radius or diameter to calculate the cross-sectional area (pipes and tanks are defined by units of diameter, not radius). We can utilize either equation (27.7) or (27.8) to determine the cross-sectional area depending on if we are given the radius or the diameter of the object.

Use equation (27.7) to calculate the cross-sectional area given a radius.

$$A = 3.1416 \times R^2 \ (\text{ft}^2) \tag{27.7}$$

where

$A = $ area (ft^2)

$R = $ radius (ft)

3.1416 is the approximate value of π

Use equation (27.8) to calculate the cross-sectional area given a diameter.

$$A = 0.7854 \times D^2 \ (\text{ft}^2) \tag{27.8}$$

where

$A = $ area (ft^2)

$D = $ diameter (ft)

0.7854 is the approximate value of $\pi/4$

VOLUME

The volume of a rectangular tank is the product of the length, width, and height of the tank. We can use equation (27.9) to determine the volume of a rectangular tank.

$$V = 0.7854 \times L \times W \times H \ (\text{ft}^3) \tag{27.9}$$

where

$V = $ volume (ft^3)

$L = $ length (ft)

$W = $ width (ft)

$H = $ height (ft)

To convert any volume from gallons to cubic feet of volume, we can use the conversion that 7.48052 gal of water takes up approximately one cubic foot of volume.

$$V(\text{gal}) = 7.48052 \times V \ (\text{ft}^3) \tag{27.10}$$

where

$V = $ volume (ft^3)

$V = $ volume (gal)

The volume of a circular tank is the product of the cross-sectional area of the base of the tank and the height of the tank. We can use equation (27.11) or equation (27.12) to determine the volume of a cylindrical tank.

$$V = 3.1416 \times R^2 \times H \ (\text{ft}^3) \tag{27.11}$$

where

$V =$ volume (ft^3)

$R =$ radius (ft)

$H =$ height (ft)

or

$$V = 0.7854 \times D^2 \times H \ (\text{ft}^3) \tag{27.12}$$

where

$V =$ volume (ft^3)

$D =$ diameter (ft)

$H =$ height (ft)

Example 27.2   A circular demineralized water storage tank has a diameter of 50 ft and a usable height of 10 ft. What is the usable volume of the tank in cubic feet? What is the usable volume of the tank in gallons?

Solution: Utilizing equation (27.12), we find

$V = 0.7854 \times D^2 \times H \ (\text{ft}^3)$

$V = 0.7854 \times 50^2 \times 10 \ (\text{ft}^3)$

$V = 19{,}365 \ (\text{ft}^3)$

Utilizing equation (27.10), we convert from cubic feet to gallons.

$V \ (\text{gal}) = 7.48052 \times V \ (\text{ft}^3)$

$V \ (\text{gal}) = 7.48052 \times 19365 \ (\text{ft}^3)$

$V \ (\text{gal}) = 144{,}860 \ \text{gal}$

To calculate the flow of a material if only the linear velocity of the material and dimensions of the pipe are known, if we multiply the cross-sectional area by the linear velocity, we will obtain volumetric velocity.

$$\text{Flow} = \text{velocity} \times \text{area} \tag{27.13}$$

where

Flow $=$ volumetric flow rate (ft^3/s)

velocity $=$ linear velocity of material (ft/s)

area $=$ cross-sectional area of pipe (ft^2)

DETENTION TIME

The detention time is the length of time for one gallon of water to pass through an enclosed volume such as a tank or a pond. The detention time is simply the total capacity of the tank or pond in gallons divided by the flow rate through the volume.

$$\text{Detention time} = 24 \times V(\text{gal})/(1000000 \times \text{MGD}) \text{ (hr)} \qquad (27.14)$$

where

$V = \text{volume (ft}^3)$

$\text{MGD} = \text{flow (mgd)}$

The following example will show a calculation of detention time.

Example 27.3 A 100,000 gal tank received 0.5 mgd of flow. What is the detention time of the fluid in the tank?

Solution: Utilizing equation (27.14), we calculate the detention time as shown below.

Detention time = $24 \times V$ (gal) / (1,000,000 × MGD) (hr)

Detention time = $24 \times 100,000 / (1,000,000 \times 0.5)$ (hr)

Detention time = 4.8 (hr)

DOSAGE

During the water treatment process, various stages will require the injection of chemicals into the process water to treat the water. Chemical doses are measured in either parts per million (ppm) or milligrams per liter (mg/L). A dose of one part per million is defined as one unit of chemical in units of weight, to 1,000,000 units of process water in units of weight. The conversion between parts per million and milligrams per liter is

$$1 \text{ ppm} = \text{mg/L}$$

Since one liter of water weighs approximately 8.34 lb, the conversion between the volume of process water and the weight of process water is

$$1 \text{ gallon} = 8.34 \text{ lb}$$

The dosage is the rate of addition of the chemical to treat a certain amount of process water (in units of mgd) and can be found using equation (27.15).

$$\text{Dosage} = \text{Dose} \times \text{MGD} \times 8.34 \text{ (lb/day)} \qquad (27.15)$$

where

Dosage = rate of chemical injection to be added to the process water (lb/day)

Dose = amount of chemical in weight needed to treat the process water (ppm or mg/L)

MGD = flow (mgd)

Example 27.4   How many pounds per day of chlorine are needed to provide a dosage of 3.0 mg/L to a water stream that flows at 2.0 mgd?

Solution:

Utilizing equation (27.15), we calculate the dosage as follows:

Dosage = Dose × MGD × 8.34 (lb/day)

Dosage = 3.0 × 2.0 × 8.34 (lb/day)

Dosage = 50 (lb/day)

PROCESS REMOVAL EFFICIENCY

The process removal efficiency is a measure of the ability of a process to remove a material from a solution. It is a calculated as the difference between the flow of material into a process or influent and the flow of material out of a process or effluent normalized to the value of the influent of the process. The equation to calculate the process removal efficiency is

$$\eta\text{-process} = (\text{Influent} - \text{Effluent})/\text{Influent} \qquad (27.16)$$

where

η-process = process removal efficiency (unitless)

Influent = amount of material entering a process (ppm or mg/L)

Effluent = amount of material leaving a process (ppm or mg/L)

Example 27.5   A clarifier tank has an influent of suspended solids of 500 mg/L and an effluent flow of suspended solids of 100 mg/L. What is the removal efficiency of this process for suspended solids?

Solution: Utilizing equation (27.16), we calculate the removal efficiency as

η-process = (influent − effluent) / Influent

η-process = (500 mg/L − 100 mg/L) / 500 mg/L

η-process = 0.8 or 80%

PUMP CALCULATIONS

One of the most common machines in the water treatment area is the pump. To move water through the various processes of the water treatment facility requires the addition of energy to the water stream. This is achieved by the use of a pump. The function of the pump is to convert the mechanical energy in the shaft of the pump to kinetic energy of the water in the pump. The amount of energy provided to the process water is the product of the pressure rise across the pump from suction to discharge of the pump and the flow of water through the pump. The equation to calculate the power a pump transfers to the water is

$$P_w = TDH \times Flow/3960 \tag{27.17}$$

where

P_w = amount of power transferred to the water (HP)

TDH = total dynamic head (pressure) of the pump (ft)

Flow = flow of water through the pump (gal/min)

3960 is required to convert units of gal/min/ft to units of HP

1 horsepower (HP) = 746 W = 0.746 kW = 3960 gal/min/ft

The pump needs a source of energy to obtain the mechanical energy in the shaft of the pump. The pump is connected to an electric motor. The function of the electric motor is to convert electrical energy in the windings of the motor to mechanical energy in the shaft of the motor. If equipment were ideal with no losses associated with their operation, the amount of electrical power (P_e) into the motor would be equal to the amount of mechanical power (P_m) on the shaft and this would be equal to the amount of power transferred to the water (P_w). However, the operation of a pump has associated with it some losses (P_p-loss) of energy due to friction between the pump and the water being pumped. Pump efficiency (η-pump) is the ratio of the mechanical power (P_m) into the pump and the amount of power transferred to the water (P_w) as shown in Figure 27.3 and described by equation (27.18).

$$\eta\text{-pump} = P_w/P_m \tag{27.18}$$

where

η-pump = efficiency of the pump

P_w = amount of power transferred to the water (HP)

P_m = amount of mechanical power delivered to the shaft of the pump (HP)

Similarly, the operation of a motor has associated with it some losses of energy due to friction in the motor bearings and windage losses and also some electrical losses in the windings of the motor (P_m-loss). Motor efficiency (η-motor) is the ratio

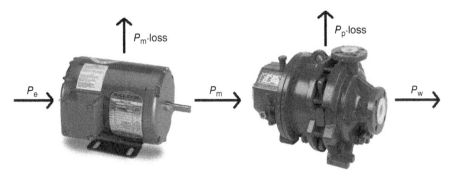

Figure 27.3   Pump efficiency.

of the electrical power (P_e) delivered to the motor winding and the amount of mechanical power (P_m) that is transferred to the shaft of the motor.

$$\eta\text{-motor} = P_m/P_e \qquad (27.19)$$

where

η-motor = efficiency of the motor

P_m = amount of mechanical power delivered to the shaft of the pump (HP)

P_e = amount of electrical power delivered to the motor (HP)

Electrical power is not measured in units of HP, but in units of kW. One HP is equal to 0.746 kW. Therefore, we can rewrite equation (27.19) in terms of electrical power in kW to derive equation (27.20).

$$\eta\text{-motor} = 0.746 \times P_m/P_e \qquad (27.20)$$

where

η-motor = efficiency of the motor

P_m = amount of mechanical power delivered to the shaft of the pump (HP)

P_e = amount of electrical power delivered to the motor (kW)

0.746 converts power from units of HP to units of kW

By combining equations (27.18) and (27.20), we can derive the relationship between power delivered to the water by the pump and the electrical power delivered to the motor windings.

$$P_w = (\eta\text{-pump} \times P_m) \qquad (27.21)$$

where

η-pump = efficiency of the pump

P_w = amount of power transferred to the water (HP)

P_m = amount of mechanical power delivered to the shaft of the pump (HP) and

$$P_m = (\eta\text{-motor} \times P_e/0.746) \tag{27.22}$$

where

η-motor = efficiency of the motor

P_m = amount of mechanical power delivered to the shaft of the pump (HP)

P_e = amount of electrical power delivered to the motor (kW)

Therefore

$$P_w = (\eta\text{-pump} \times \eta\text{-motor} \times P_e/0.746) \tag{27.23}$$

where

η-motor = efficiency of the motor

P_w = amount of power transferred to the water (HP)

P_e = amount of electrical power delivered to the motor (kW)

To determine the cost for the electrical energy to run a pump for a certain amount of time, we multiply the average electrical power that a pump draws, with the percentage of time the pump runs during that period of time and multiply by the cost of energy.

Example 27.6   Given that a pump supplies a water system with a total dynamic head of 238 ft at a flow rate of 500 gpm, determine the water horsepower (P_w) that the pump delivers.

Solution: Utilizing equation (27.17), we calculate the water horsepower to be

P_w = TDH × Flow / 3960

P_w = 238 × 500 / 3960

P_w = 30 HP

Example 27.7   Given that a pump supplies 30 HP of power to a water system and that the pump efficiency is 85% and the motor efficiency is 92%, determine the electrical power (kW) required to provide 30 HP water power.

Solution: Using equation (27.23) and rearranging for P_e, we find:

$P_e = 0.746 \times P_w / (\eta\text{-pump} \times \eta\text{-motor})$

$P_e = 0.746 \times 30 / (0.85 \times 0.92)$

$P_e = 28.62$ kW

Example 27.8   If a pump that draws 28.62 kW of electrical power runs for 12 hr/day and the electrical rate for power is 0.12 \$/kWh, how much does it cost to run the pump for 1 year (365 days)?

Solution: The cost to run the pump is the product of the cost per kWh for the energy, the amount of time the motor is energized, the average electrical power the motor draws, and the time period under consideration.

$$\text{Cost} = P_e \times \text{time energized} \times \text{cost per kWh} \times \text{period of operation}$$
$$\text{Cost} = 28.62 \text{ kW} \times (12/24) \times 0.12 \text{ \$/kWh} \times 365 \text{ days} \times (24 \text{ hr/day})$$
$$\text{Cost} = \$15,043$$

GLOSSARY OF TERMS

- Anion Ion Exchange – Exchange negatively charge ions in the water stream.
- Cation Ion Exchanges – Exchange positively charged ions in the water stream.
- Clarification – The process where certain chemicals are added to the water that allow the non-soluble material to coagulate. Once this material coagulates, it more easily settles out of the water source.
- Demineralization – The process of removal of the minerals and mineral salts from a water stream.
- Evaporation – The process of converting water to steam to separate pure water from the unwanted ions and soluble materials.
- Filtration – The process of passing water through a filter media to remove suspected material from the water stream.
- Flocculation – The process where the water is slowly stirred.
- Ion Exchange – Process utilizing resin beds to replace unwanted ions with less objectionable ions.
- Oxidation – The process whereby we convert soluble gas to solids by forcing oxygen into a water system.
- Reverse Osmosis (RO) – A process that uses pressure difference across a semipermeable membrane to separate the water stream from the unwanted materials.
- Sedimentation – The process whereby the water is stored in a certain area where the suspended non-soluble materials will settle out of the water stream.
- Softening – The process of removing dissolved impurities by a chemical reaction with certain materials to convert the soluble to a form that is more easily removed.

PROBLEMS

27.1 Which process is known as the process of adding certain chemicals to the water to allow non-soluble material to coagulate?

  A. Clarification

  B. Evaporation

C. Flocculation

D. Sedimentation

27.2 Which process is known as the process of converting water to steam to separate the pure water from the unwanted ions and soluble materials?

A. Clarification

B. Evaporation

C. Flocculation

D. Sedimentation

27.3 Which process is known as the process where the water is slowly stirred?

A. Clarification

B. Evaporation

C. Flocculation

D. Sedimentation

27.4 Which process is known as the process whereby the water is stored in a certain area where the suspended non-soluble materials will settle out of the water stream?

A. Clarification

B. Evaporation

C. Flocculation

D. Sedimentation

27.5 A system averages an influent flow of 10 mgd. What is the average flow rate in gallons per minute that flows into the system?

27.6 A circular demineralized water storage tank has a diameter of 40 ft and a usable height of 30 ft. What is the usable volume of the tank in cubic feet? What is the usable volume of the tank in gallons?

27.7 A 200,000 gal tank received 1.5 mgd of flow. What is the detention time of the fluid in the tank?

27.8 How many pounds per day of chlorine are needed to provide a dosage of 15.0 mg/L to a water stream that flows at 3.0 mgd?

27.9 A clarifier tank has an influent of suspended solids of 500 mg/L and an effluent flow of suspended solids of 100 mg/L. What is the removal efficiency of this process for suspended solids?

27.10 Given that a pump supplies a water system with a total dynamic head of 238 ft at a flow rate of 2000 gpm, determine the water horsepower (P_w) that the pump delivers.

27.11 Given that a pump supplies 120 HP of power to a water system and given that the pump efficiency is 80% and the motor efficiency is 93%, determine the electrical power (kW) required to provide 120 HP water power.

27.12 If a pump that draws 120.3 kW of electrical power runs for 20 hr/day and the electrical rate for power is $0.15 kWh, how much does it cost to run the pump for 1 year (365 days)?

RECOMMENDED READING

Electrical Machines, Drives and Power Systems, 6th edition, Theodore Wildi, Prentice Hall, 2006.

The Engineering Handbook, 3rd edition, Richard C. Dorf (editor in chief), CRC Press, 2006.

EPA, Primer for Municipal Wastewater Treatment Systems, EPA 832-R-04-0, http://water.epa.gov/aboutow/owm/upload/2005_08_19_primer.pdf

http://www.ragsdaleandassociates.com/WastewaterSystemOperatorsManual/Chapter%2017%20-%20Mathematics.pdf

http://www.epa.state.oh.us/Portals/28/documents/opcert/WWformulas.pdf

http://www.tennessee.gov/assets/entities/environment/attachments/study_1104.pdf

Power Plant Engineering, Black & Veatch, edited by Larry Drbal, Kayla Westra, and Pat Boston, Chapman & Hall / Springer, 1996.

Standard Handbook of Powerplant Engineering, 2nd edition, Thomas C. Elliott, McGraw-Hill, 1998.

US Navy Training Manual Machinist Mate 2nd class, Naval Education and Training Professional Development and Technology Center, 2003.

CHAPTER 28

SOLAR AND WIND ENERGY

GOALS

- To understand the basic construction of a typical wind turbine
- To calculate the amount of available power from a wind turbine power curve for a given wind speed
- To understand the four basic types and construction of a thermal solar energy generator facility
- To understand the basic construction of a photovoltaic solar energy facility
- To understand some of the basic safety installation requirements for solar installations
- To be able to draw the basic single-line diagram of a PV array utilized in a utility distribution application

---

WIND ENERGY

Wind is the natural movement of the air due to differences in atmospheric temperatures and pressures. It is produced by the uneven heating of the earth's surface by the sun. Since the earth's surface is made of various land and water formations, it absorbs the sun's radiation unevenly. Additionally, due to the curvature of the earth, some areas of the earth receive a greater amount of solar energy per square meter of surface area than other areas of the earth. As the surface of the earth is heated unevenly, the atmosphere is also heated unevenly. This creates a differential thermal energy level in various locations in the atmosphere. This difference of energy levels causes movements of currents in the atmosphere, which is the definition of wind. This flow of air represents a potential source of kinetic energy. To capture this energy and convert it into a different form of energy, namely electrical energy, a wind turbine is utilized.

The wind turbine converts the kinetic energy of the wind into mechanical energy at the shaft of the turbine. The shaft of the turbine in turn is connected to a generator where this mechanical energy is converter to electrical energy. For utility operations, the generator is required to operate at a substantially faster speed than the

Energy Production Systems Engineering: An Introduction for Electrical Engineers to Electrical Power Generation Facilities, Systems, and Equipment, First Edition. Thomas H. Blair.
© 2017 by The Institute of Electrical and Electronics Engineers, Inc. Published 2017 by John Wiley & Sons, Inc.

shaft of the rotor connected to the wind blade. The wind turbine turns at speeds of 30–400 rpm depending on rotor design. The typical wind turbine AC generator is a two-pole machine with a synchronous rotational speed of 3600 rpm or a four-pole machine with a synchronous rotational speed of 1800 rpm. To match the needed rotational speed of the generator with the operational rotational speed of the turbine rotor, a gearbox is incorporated. The most common type of utility wind turbine utilizes a blade with a lift design. This is similar in design to the axial flow fan where the pitch of the blades on the rotor can be adjusted as needed for control. Much like the aerodynamics of an airplane wing, the design of the wind turbine blade is such that the area of the leeward side of the turbine blade is larger than the windward side. This creates a vacuum or "lift" on the leeward side of the turbine blade improving upon the efficiency of the energy conversion process from kinetic energy of the wind to the mechanical energy of the shaft of the turbine. The specific wind turbine design determines the actual acceptable values for wind speed.

The power curves for various types of wind turbines are different, depending on the design of the machine. For very low wind speeds below the minimum operational speed of the machine, the wind turbine is not able to convert an adequate amount of energy to drive the generator. Under these conditions, the wind turbine rotor is locked by the disc brake to prevent movement of the turbine blades. Above the minimum operational speed of the machine, the break is released and the turbine blades drive the generator. The faster the wind speed is, the greater the amount of energy that is generated by the generator. In the area of operation between minimum operational speed and the speed that provides rated power at maximum blade pitch, the pitch of the turbine blades remains fixed for maximum energy transfer between the wind and the turbine. The relationship between power and wind speed is a cubic function. Above the speed that provides rated power at maximum blade pitch, the generator operates at its rated output power capability. When the speed of the wind exceeds this speed, then the pitch of the turbine blades is adjusted to maintain the power transfer from the wind turbine to the generator at the rating of the generator. For wind speeds in excess of the generator design speed (this is called cut-out speed), the wind turbine is shut down to prevent failure of the structure. The blade pitch is adjusted such that most of the wind does not impinge on the blade and the wind turbine rotor is locked by the disc brake to prevent rotation.

As we mentioned above, in the operational area between the cut-in wind velocity and the wind velocity that causes rated power, the ideal amount of power out of the generator is a function of the cube of wind speed. The equation that defines the ideal maximum amount of available mechanical power is shown in equation (28.1).

$$P_{max} = 0.5 \times \rho \times A \times V^3 \tag{28.1}$$

where

ρ = density of air (kg/m^3)

A = swept area = $3.14 \times R^2$ (m^2) (R is the radius of the sweep area)

V = wind velocity (m/s)

The density of air is affected by the atmospheric pressure, atmospheric temperature, and relative humidity of the atmosphere. For air at a pressure of 760 mm Hg, temperature of 20°C and relative humidity of 40%, the density is 1.2 kg/m³. As the atmospheric pressure increases, the air density increases. As atmospheric temperature increases, the air density decreases. As relative humidity increases, the air density decreases. From equation (28.1), we can see that as density increases, available power increases up to the rating of the generator. Therefore, increased air pressure, decreased air temperature, and decreased relative humidity will all lead to increases in maximum power output of the machine.

Example 28.1   Given an air density of 1.2 kg/m³, a blade radius of 10 m, and a wind velocity of 22.3 miles/hr (10 m/s), determine the theoretical maximum mechanical power that can be transferred by the wind turbine.

Solution: First, we have to determine the swept area. The radius of the turbine blade was given as 10 m, so the area of the swept area is

$$A = 3.14 \times R^2 \text{ m}^2$$
$$A = 3.14 \times 10^2 \text{ m}^2$$
$$A = 3.14 \times 100 = 314 \text{ m}^2$$

Utilizing equation (28.1) for maximum mechanical power as a function of wind speed, we find

$$P_{max} = 0.5 \times \rho \times A \times V^3$$
$$P_{max} = 0.5 \times (1.2 \text{ kg/m}^3) \times (314 \text{ m}^2) \times (10 \text{ m/s})^3$$
$$P_{max} = 188495 \text{ kg m}^2/\text{s}^3 = 188.5 \text{ kW or } 188.5 \text{ kJ/s}$$

Equation (28.1) defines the theoretical maximum amount of power in the kinetic energy of the wind that is available. As the blade presents resistance to wind flow, the wind through the turbine blade slows down. Since the maximum amount of power transfer is based on the entire amount of kinetic energy of the wind being transferred to the turbine blades (as shown in equation 28.1 for P_{max}), if all of this power were transferred to the turbine, this would imply that the remaining kinetic energy of the wind is zero, but that would imply that there is no wind. This is not an achievable condition. We define the turbine blade efficiency (η) as the ratio of the amount of mechanical power derived from the turbine (P_{mech}) to the maximum amount of available wind power to the turbine (P_{max}). The turbine blade efficiency is defined by equation (28.2).

$$\eta = P_{mech}/P_{max} \tag{28.2}$$

where

η = turbine blade efficiency

P_{mech} = mechanical power derived from the turbine

P_{max} = maximum amount of available wind power to the turbine

Example 28.2   Given the same parameters that were used in Example 28.1, if the turbine blade efficiency is 30%, determine the mechanical power available to the turbine shaft.

Solution: From example 28.1, we found $P_{max} = 188.5$ kW. Using the definition of turbine blade efficiency, we find that the mechanical power available is
    Utilizing equation (28.2), we calculate the mechanical power available to be

$$\eta = P_{mech}/P_{max}$$
$$P_{mech} = \eta \times P_{max}$$
$$P_{mech} = 0.3 \times 188.5 \text{ kW}$$
$$P_{mech} = 56.6 \text{ kW}$$

The generator may be a DC generator or an AC generator. For the DC generator (smaller rated units), the output of the DC generator is fed to an *inverter* and filter combination that is used to invert the DC voltage to AC voltage and synchronize the AC voltage to the grid. For larger applications, an AC induction generator is utilized. AC induction machines are preferred over DC generators for larger applications due to the reduced frequency of maintenance of the AC machine over the DC machine. The generator is mounted at the top of the supporting structure inside the canopy and this makes maintenance more challenging in this application. For more details on the design and operation of AC induction machines for wind applications, the reader is directed to Chapter 16 covering generator and, specifically the section regarding AC inductor machine operation in the sub-synchronous and super-synchronous modes.

THERMAL SOLAR ENERGY

Concentrating solar power (CSP) systems receive solar energy and concentrate this energy into a collector. One benefit of thermal solar energy over photovoltaic solar energy is the concept of energy storage. In photovoltaic (PV) solar energy, when the source of solar radiation is inhibited (such as by cloud cover), the power output of the *PV cell* is reduced immediately (unless some energy storage device such as a battery is utilized). Unlike PV solar energy, thermal solar energy stores energy in thermal form in a media and this has the advantage of allowing some tolerance of the generation system to momentary reductions in solar energy available.
    Solar thermal electric energy generation concentrates the solar radiation from the sun and this concentrated radiant energy is converted to thermal energy. This thermal energy is then utilized to drive a heat engine. The working fluid that is heated by the concentrated sunlight can be a liquid or a gas depending on the technology utilized. This heat engine converts this thermal energy to mechanical energy. The mechanical energy is then delivered to a generator that converts this mechanical energy into electrical energy.
    Thermal solar technology is categorized by the type of reflector or collector utilized in the collection of solar radiation. There are several types of utility grade thermal solar energy plants currently in use or under development.

PARABOLIC TROUGH SOLAR FIELD TECHNOLOGY

Parabolic trough solar field technology utilizes parabolic trough collectors. The primary building block for these collectors is the *single parabolic trough solar module*. These collectors are constructed into a series of modules called a *parabolic trough solar array*, as shown in Figure 28.1. Each array contains a single axis tracking control system to position the parabolic trough solar collector array. The design of each array is such that the solar energy is reflected by the reflector and concentrated on the absorber tube. In the absorber tube, the radiant solar energy is converted to thermal energy in the media inside the absorber tube. Multiple modules are connected in series for form an array and multiple arrays are connected in parallel to form the *solar field*. The parabolic arrays track the angle of the solar energy being received by the array to maximize the efficiency of the solar collection system. Each module has three main components. The first part of the solar module is known as the *concentrator structure*. This is the skeleton of the parabolic trough solar collector and its main function is to support the reflector and absorber tube. In addition to supporting the weight of the reflector and absorber tube, the concentrator structure provides the support to allow for the movement of the reflector to maintain alignment with the solar energy. The second component of the solar module is the *reflector* (or mirror). This is also the most visible component when looking at a solar array field. The reflectors are formed in the shape of a parabola, and the focus point for this parabola is the absorber tube. The third component of the solar module is the *absorber tube* or heat collection element.

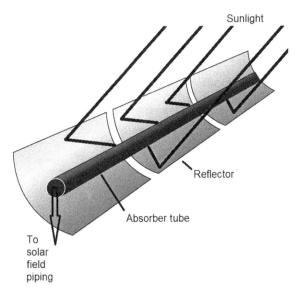

Figure 28.1   Parabolic trough solar collector assembly. *Source*: Reproduced with permission of National Renewable Energy Laboratory (NREL), http://www.nrel.gov/docs/fy11osti/48895.pdf.

The solar energy reflected off the mirrors is concentrated on these absorber tubes, heating up the tubes. This heat is then transferred to a heat transfer fluid that circulates inside these tubes. One typical type of heat transfer fluid is oil. In this type of application, the oil is heated to around 400–500°C. Some new research has been done with molten nitrate salts as an alternate media to oil. There are several benefits of molten nitrate salt over oil as a thermal energy transfer media. Salts are less expensive, denser and have a higher energy density than oil. Additionally, oils present a potential environmental hazard that is difficult to clean up when a spill occurs. Salts on the other hand solidify when a leak occurs, so they can be cleaned up relatively easily. Oils present a fire hazard, while salts do not. Due to the salts higher energy density, the need for expansion tanks and heat exchanger surface area is reduced which has the potential to reduce the initial material capital costs for a salt-based plant. Lastly, salts do not degrade over time like oil does. This reduces the operational cost of replacement of the heat transfer media that occurs with an oil system.

The connections between collector tubes are a special area of concern during installation and maintenance. As the collection tubes heat and cool, these tubes expand and contract. This expansion and contraction is accommodated for in the connection area for the collector tubes between modules. These flexible connections should be monitored to ensure the integrity of the connection. Over time, the process of expansion and contraction can place increased stress on these connections and these tend to be an area of enhanced focus for leaks.

Once the fluid is heated in the collector tubes, this heated fluid is delivered to a heat exchanger to transfer this thermal energy to a steam and water cycle. The steam is then expanded across a turbine to convert the thermal energy in the steam to mechanical energy. The turbine is connected to a generator where this mechanical energy is converted to electrical energy for distribution. A typical process diagram for a parabolic trough solar power plant is shown in Figure 28.2.

Diagram 2. Parabolic trough solar power plant

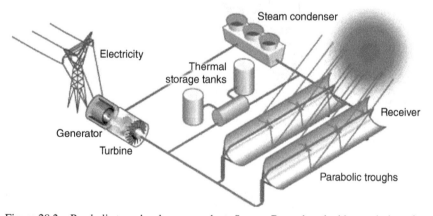

Figure 28.2 Parabolic trough solar power plant. *Source*: Reproduced with permission of United States Department of Labor, Bureau of Labor Statistics, http://www.bls.gov/

Figure 28.3   Solar power tower power plant. *Source*: Reproduced with permission of U.S. Department of Energy, Waste Isolation Pilot Plant, http://www.wipp.energy.gov/science/energy/powertower.htm.

SOLAR POWER TOWERS

The solar power tower power plant is also known as the central power tower plant or heliostat power plant. As shown in Figure 28.3, the solar power tower uses an array of flat but movable mirrors that are called heliostats to direct solar energy onto one central point or receiver mounted on top of the solar power tower. In the receiver on the power tower, a fluid received this solar energy and converts it to thermal energy. Earlier designs of the power tower utilized water as the transfer media and transferred this energy to a steam turbine. Current designs utilize molten nitrate salt because this has a greater thermal energy storage capability and higher heat transfer capability than that of water and steam.

In the molten nitrate salt design, the salt solution is delivered to a steam generator where the thermal energy from the molten nitrate salt is transferred to a water and steam cycle. In the water and steam design, water is boiled and the steam is delivered to a steam turbine which converts this thermal energy to mechanical energy. The turbine is connected to a generator where this mechanical energy is converted to electrical energy for distribution. A typical process diagram for steam-based solar power tower power plant is shown in Figure 28.4.

Diagram 3. Solar power tower plant

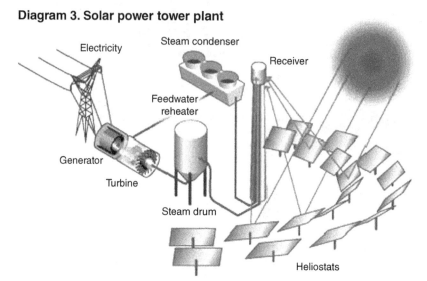

Figure 28.4 Solar power tower plant. *Source*: Reproduced with permission of United States Department of Labor, Bureau of Labor Statistics, http://www.bls.gov/.

DISH SYSTEM

A relatively new technology that utilizes at its heart a very old technology is the dish system. The dish system contains reflectors arranged in a parabolic shape very much like that of a satellite dish. This parabolic shape directs the solar energy onto a central receiver. The central receiver is a Stirling engine which converts the thermal energy to mechanical energy. This process is achieved by the expansion of the fluid in the Stirling engine which in turn drives a piston engine. The basic concept of the Stirling engine was developed by Robert Stirling in 1816. The Stirling engine is also known as the external combustion engine. Unlike the internal combustion engine where combustion takes place in the engine cylinder, the Stirling engine makes use of external heat exchangers. The combustion, or in this case, the elevation of the molten nitrate salt temperature takes place in a separate closed loop. Heat exchangers are used to transfer this thermal energy to the close loop that feeds the Sterling engine. In the Stirling engine cylinder, this fluid is allowed to expand converting this thermal energy into mechanical energy on the shaft of the engine. In the Stirling engine, the working gas is generally compressed in the colder portion of the engine and expanded in the hotter portion resulting in a net conversion of heat into work. The generator is connected to the shaft of this engine and converts this mechanical power to electrical power. A typical process diagram for the dish system is shown in Figure 28.5.

PHOTOVOLTAIC SOLAR ENERGY

Photovoltaic cells are specialized semiconductor diodes that convert solar energy into electrical energy via the photovoltaic effect. The photovoltaic effect is the basic

Diagram 4. Solar dish power plant

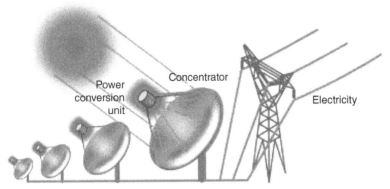

Figure 28.5   Solar dish power plant. *Source*: Reproduced with permission of United States Department of Labor, Bureau of Labor Statistics, http://www.bls.gov/.

process through which solar cells convert radiant energy from the sun into electrical energy. Figure 28.6 depicts this basic process. Each *solar cell* is constructed of two pieces of silicon that make up the semiconductor materials.

While there are many designs available, boron and phosphorous are one combination of common doping materials. For this type of semiconductor, the p-layer semiconductor material is doped with boron. Boron is a group IIIA element with the atomic number 5 and, by inspection on the periodic table (see Table 4.3), we see that boron has three available outer shell electrons. As such, the boron-doped p-semiconductor layer attracts electrons. The n-layer semiconductor material is doped with phosphorous. Phosphorous is a group VA element with the atomic number 15

Diagram 1. The photovoltaic effect

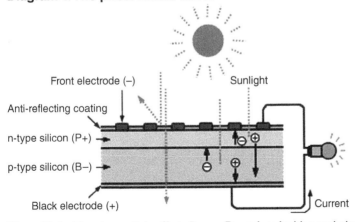

Figure 28.6   The photovoltaic effect. *Source*: Reproduced with permission of United States Department of Labor, Bureau of Labor Statistics, http://www.bls.gov/.

and, by inspection on the periodic table (see Table 4.3), we see that phosphorous has five available outer shell electrons. As such, the phosphorous-doped semiconductor layer gives up electrons. When sunlight strikes the semiconductor materials, some of the photons in sunlight are absorbed by the semiconductor layers and this absorption adds radiant energy to the semiconductor layer. The semiconductor layer converts this radiant energy into electrical energy by releasing electrons from the semiconductor material.

Individual solar cells are grouped into solar modules. Solar modules are defined as a complete, environmentally protected unit capable of producing DC power when exposed to sunlight. The output of each PV module is around 50 Vdc. Several modules are then arranged in parallel to for a *solar panel*. These solar panels are grouped sets of modules that are prewired to form a field installable component. Solar panels are then arranged electrically in series into a *solar array* (sometimes called a PV string or PV array) to increase the DC output voltage of the cell. The largest PV arrays can generate up to 1100 Vdc. Then multiple solar arrays are connected together in parallel at a combiner box or collector box. The combiner or collector box is simply a junction box where a group of PV arrays are connected to one electrical point. The conductors between the solar array and the combiner box are known as *photovoltaic source circuits* and the conductors downstream of the combiner box are known as *photovoltaic output circuits*.

Photovoltaic source circuits are DC circuits and the sizing of the conductors is covered under NEC® 690.8 while the sizing of overcurrent protection devices for these circuits is covered under NEC® 690.9. For sizing of the conductors, NEC® requires that the conductor be rated for 125% of the maximum output current the solar array can output at the peak of solar irradiation, even if the inverter connected to the system is rated for less than this current. PV module voltage will vary with ambient temperatures. As the ambient temperature decreases, the PV module output voltage increases. NEC® requires that the output-connected electrical equipment be rated for 125% of the PV rated output voltage. This factor of 125% of nominal voltage leads us to a requirement to size conductors at 125% of nominal current just for the condition where voltage is 125% of nominal voltage. In addition to this factor, NEC® considers this value to be the maximum continuous current and, as such, it requires a second 125% factor for a total increase in rating of 125% × 125% = 156.25%. For example, if a solar array has a rated output of 2.4 kW at 240 Vdc, then the rated current of the array is 2.4 kW / 240 Vdc = 10 Adc. The conductor must be rated for 156.25% × 10 A or 15.6 A. The solar array by design is current limiting. However NEC® 690.9 requires overcurrent protection where there are multiple sources that could exceed the rating of the conductor in the event of a short circuit. Locations are at combiner boxes where multiple photovoltaic strings are connected in parallel. NEC® requires a branch circuit protection of 125% of the DC rating of the equipment. For example, for a solar array rated at 10 Adc, NEC® requires a 10 Adc × 125% = 12.5 Adc fuse or breaker. The next available standard size fuse would be 15 Adc. Note that NEC® requires this branch circuit protection to be rated to interrupt DC current and also, as is the case in the photovoltaic source circuits, the interrupting device must be listed to protect the circuit when the source is on either side of the protective device. Note this ampacity rating does not take into account any additional de-ratings due to ambient conditions or any de-ratings due to cable routing in tray or conduit and the NEC® should be consulted for any additional cable ampacity sizing requirements due to those considerations. Cables should be rated for 90°C operation and their ampacity based on 70°C operation unless actual conditions dictate more severe deration.

Just as in the case of the PV array, inverters are inherently current limiting. Therefore, if this was a standalone system with only one inverter, then NEC® would not require overcurrent protection on the photovoltaic output circuits. However, typical utility systems have multiple inverters paralleled and feeding the utility distribution system through one point of common coupling. In the case where multiple inverts are paralleled, NEC® requires the use of an overcurrent protection device at points in the photovoltaic output circuits where multiple sources could exceed the rating of the conductors. This requires an overcurrent protective device on the output of each inverter. Additionally, where the single point of common coupling occurs between the inverter output and the utility system, the utility system can provide a substantial amount of short-circuit current and NEC® again requires the use of an overcurrent protective device in this location. The point of connection between the inverter outputs and the utility system is covered under NEC® section 705. NEC® requires an overcurrent protective device that is rated for 125% of the nominal rated current of the circuit being protected. There are four methods for determination of cable ampacity, but the most conservative is to size the conductor for 125% of the nominal rated current of the system at the connection point between the utility and the PV installation. Just as in the case of the photovoltaic input circuits, where the overcurrent protective device has sources on both sides of the device, NEC® requires the overcurrent protective device to be listed for use in such an application.

There must be a main disconnecting means in the photovoltaic output circuit that is located within 6 ft of the combiner box to provide a means for isolation of the DC voltage from the combiner box and the inverter. NEC® requires ground fault protection on the photovoltaic circuits. DC current is more difficult to interrupt that AC since DC has no natural current zero crossing. A disconnecting means must be provided for all equipment to isolate each piece of equipment from all sources and, where multiple disconnecting means are needed to isolate a piece of equipment from multiple sources, those disconnects must be physically grouped together and identified in regards to their function. Where the open-circuit DC voltage exceeds 600 V, each pole of the DC system must be physically separated from the other pole including when these conductors pass through conduit, junction boxes, and disconnect devices. This boundary must be maintained for the entire photovoltaic output circuit.

Since the output of the PV array is *direct current* (DC) and the *electrical grid* is alternating current (AC), the output of the PV array needs to be converted to AC. This is the function of the PV inverter which converts the DC voltage from the PV array to AC voltage. Additionally, the PV inverter controls the rms value of output voltage and the phase angle of the output voltage to the inverter transformer to ensure synchronization with the electrical grid and control reactive power flow to the electrical grid. Current inverter technology can accept up to 1100 Vdc. The standard inverter output voltage values are 480 Vac and 690 Vac. While the inverter output is 480 Vac and 690 Vac, the typical utility distribution voltages are well in excess of 1000 V ac. The inverter transformer steps the voltage level up from the inverter output to the voltage level required by the collector bus. The inverter power rating tends to be the limitation on the maximum number of PV cells that can feed one inverter. The largest size inverter currently is 1500 kVA with a common size being 500 kVA. Inverter technology is rapidly developing and the author expects larger rated inverters to be commercially available in the near future. On a PV generation facility that has an output rating in excess of a standard inverter output rating, there will be multiple PV inverters connected in parallel to a *collector bus* to supply the required kVA rating of the PV generation facility.

In order to parallel inverter outputs, each PV inverter will have an isolation transformer called the *inverter transformer*. The inverter transformer primarily is used to step up the voltage, but it also presents an impedance between the inverter output voltage which is a pulse width-modulated output waveform and the collector bus voltage which is sinusoidal. An additional benefit of the inverter transformer is that it allows for phase angle control between the inverter output and the collector bus. This allows for control of the reactive power flow as it allows the inverter to adjust the phase angle of the voltage output from the inverter section as necessary to control reactive power flow out of the inverter. As a third benefit, the inverter transformer provides electrical isolation of the inverter circuit, other inverter circuits connected in parallel, and the collector bus. A ground fault on one circuit will not be reflected across the inverter transformer. Inverter transformers are rated for the non-sinusoidal currents they will see from the inverter section (see Chapter 15 covering transformer applications for non-sinusoidal currents).

For larger utility distribution applications, there will be one transformer called a *substation collector transformer* that is the interface between the common collector bus and the electric utility distribution system. This arrangement for a utility application is shown in the block diagram of Figure 28.7. This arrangement gives us one point of connection (point of common coupling) to the distribution system. This also

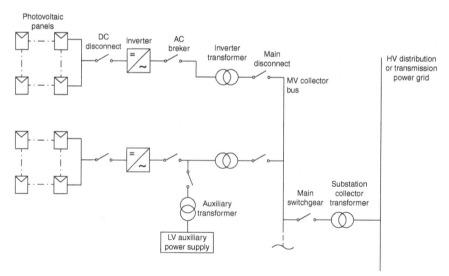

Figure 28.7 Typical equipment block diagram for PV array for utility distribution application. *Source*: Reproduced with permission of IEEE.

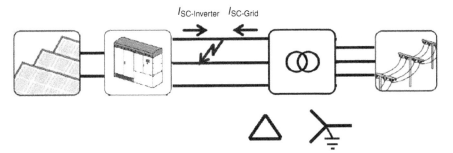

Figure 28.8   Evaluation of phase-to-phase fault on collector bus. *Source*: Reproduced with permission of IEEE.

has the benefit of providing a total of two points of electrical isolation between the utility distribution system and the DC electrical system provided by the PV cells. This can be of benefit in the event of a ground fault on one of these systems as the transformer would tend to isolate this ground fault. Additionally, this gives us a single point for the installation of circuit protection between the collector bus and the utility distribution grid.

A circuit breaker with protection relays will be utilized at this point of common coupling. The breaker provides the capability to connect and disconnect from the electrical grid under normal operations as well as the ability to disconnect during abnormal operations such as an electrical system fault. The protection scheme for the substation collector transformer must be coordinated with the utility distribution system. For example, in Figure 28.8, if a phase-to-phase fault occurs on the collector bus and if the substation collector transformer is arranged as a delta-connected winding on the collector bus and a solidly grounded wye connection on the transmission side, then a phase-to-phase fault on the collector bus would appear as a single phase-to-ground fault on the distribution system and the magnitude of the current would be limited by the impedance of the substation collector transformer. The inverter's inherent current limit feature will limit the fault current from the inverter, but the distribution system will feed this fault. The protection scheme should be designed to identify this type of fault and isolate the fault from the distribution system.

While the inverter may have a maximum nominal output rating, this rating is limited primarily by two factors, the environmental temperature and the solar radiance. *Solar irradiance* is the power per unit area produced by the sun in the form of electromagnetic radiation. Solar irradiance may be limited by atmospheric optical properties such as cloud cover and particulate concentration in the atmosphere. Additionally, solar irradiance is affected the by the change of the seasons. Depending on the location of the solar facility, the angle, the sun and, thereby, the amount of solar radiation per square meter changes between Spring, Summer, Fall, and Winter seasons. Since the inverter utilizes transistors for conversion of the DC voltage from the PV array to AC voltage for the inverter transformer, these transistors have maximum current ratings that decrease rapidly with elevated ambient temperatures. Figure 28.9 shows a typical deration of an inverter due to elevated ambient temperatures.

During daylight hours, the inverter receives DC voltage from the PV arrays and can provide both active and reactive power, but during the night, the inverter output is

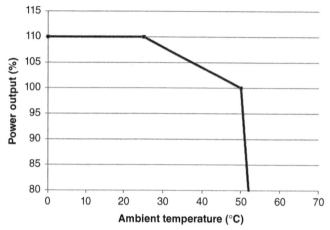

Figure 28.9   Inverter power output as a function of the ambient temperature.
Source: Reproduced with permission of IEEE.

limited to reactive power control only, depending on the inverter configuration. This results in a typical inverter transformer load profile as shown in Figure 28.10.

- Curve #1 represents DC power from the PV cells as a percentage of nominal inverter power.
- Curve #2 represents the transformer active load as a percentage of nominal inverter power.
- Curve #3 represents the ambient temperature (°C)
- Curve #4 represents the transformer reactive load as a percentage of nominal load.

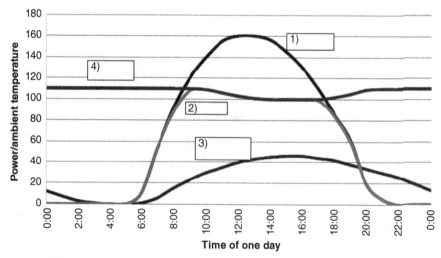

Figure 28.10   Inverter transformer load profile. *Source*: Reproduced with permission of IEEE.

The ratio of the peak power available from the PV field (peak of curve 1 in Figure 28.10) to the nominal power rating of the inverter is known as the PV-to-inverter ratio. Current designs have typical values of PV-to-inverter ratios of 1.15 to 1.3.

Notice that the daytime limitation when present is due to ambient temperatures, whereas the evening limitation is due lack of sun exposure to the PV cells. Some PV applications utilize battery systems to store excess energy in the daytime and these can be used for short periods in the evening to extend the operation of the PV electrical generation facility. IEEE C57.159 provides further guidance to the details of the proper interconnection between a distributed photovoltaic (DPV) power generation system and the electrical distribution grid.

There are several advantages to photovoltaic solar energy. Unlike fossil fuel plants that utilize large rotating equipment that produce a substantial amount of noise, solar grid arrays do not utilize such equipment and they are much quieter operationally. This makes locating solar farms near residential areas less of a nuisance than a conventional generation facility. In addition to being quiet, the PV solar grid array can be constructed in smaller kVA ratings and distributed along the existing utility distribution system, thereby providing support to the local electrical grid, potentially increasing reliability without having to upgrade the transmission system. Additionally, the solar grid arrays do not emit pollutants via water and air streams to the environments as other technologies can. If designed and installed correctly, PV solar grid arrays require substantially less maintenance than traditional generation stations, which would reduce operations and maintenance costs.

There are some disadvantages to PV solar energy. The manufacturing of PV cells requires the use of chemicals and some of these chemicals are known to be toxic, like cadmium and arsenic. As long as PV cells are handled correctly and, at their end of life, if they are disposed of properly or recycled, this concern can be managed. Another disadvantage to PV solar energy is the initial cost for installation. Currently, the installed cost for a PV solar generation station exceeds the installed cost for other types of generation facilities. However, over time the cost for PV cell manufacturing and PV solar generation station installation is rapidly decreasing due to improvements in manufacturing processes and photocell solar efficiencies are increasing. If this trend continues, PV cells may quickly become competitive with other generation technologies.

The third and most challenging limitation to PV solar energy technology is the availability of the energy source. Generation stations do not really generate energy. Rather, they are energy conversion facilities. For us to deliver electrical energy to the grid, we must have some reservoir of energy in a different form to allow us to convert that form of energy to electrical energy for distribution. Solar PV generation stations depend on the availability of sunlight as the original form of energy to be converted to electrical energy. During periods of low sunlight or at night, PV generation stations do not have solar energy as a resource to generate electricity from. We are back to that fundamental law which states, "energy cannot be created nor destroyed, only altered in form." If we are to store solar energy, we must develop some media to allow for this storage. One such possibility in use is the storage of this energy in chemical form as in a battery bank. When used, the amp hour rating of existing battery designs is not intended to provide full capacity throughout the night, but only to provide several hours of capacity during evening hours when demand is still high.

Despite the few challenges that PV solar energy presents, the popularity of solar energy is increasing. Since 2008, US installations have grown 17-fold from 1.2 gigawatts (GW) to an estimated 20 GW in 2015. As the installation costs decrease, it is anticipated that PV solar electric generation will gain a greater market share of the electric utility industry. Current research is underway to improve the efficiency of the solar cell. Current efficiencies range from between 14% and 17%. As efficiency increases, a greater amount of electrical energy can be obtained from the same square foot of exposed semiconductor material thereby reducing the payback period for the capital costs to install.

Solar cell efficiency is the ratio of the electrical output of a solar cell to the incident energy in the form of sunlight. Solar efficiency is calculated by dividing a cell's power output at its maximum power point (P_m) by the irradiance or input light (G) and the surface area of the solar cell (A_c).

$$\text{Solar efficiency} = P_m/(G \times A_c) \qquad (28.3)$$

where

> P_m = maximum power point (W)
>
> G = irradiance (input light) (W/m^2)
>
> A_c = surface area of the solar cell (m^2)

GLOSSARY OF TERMS

- Ampere (A) or Amp – The unit for the electric current; the flow of electrons. One amp is one coulomb passing in one second. One amp is produced by an electric force of one volt acting across a resistance of one ohm.

- Array – Any number of photovoltaic modules connected together to provide a single electrical output. Arrays are often designed to produce significant amounts of electricity.

- Cell – The basic unit of a photovoltaic panel or battery.

- Direct Current (DC) – Electric current in which electrons flow in one direction only. Opposite of alternating current.

- Electrical Grid – An integrated system of electricity distribution, usually covering a large area.

- Inverter – Device that converts DC electricity into AC.

- Photovoltaic Array – An interconnected system of PV modules that function as a single electricity-producing unit. The modules are assembled as a discrete structure, with common support or mounting. In smaller systems, an array can consist of a single module.

- Photovoltaic Cell – The smallest semiconductor element within a PV module to perform the immediate conversion of light into electrical energy (dc voltage and current).

- Photovoltaic (PV) Source Circuits – The conductors of the DC system between the solar arrays and the combiner or collector box.
- Photovoltaic (PV) Output Circuits – The conductors of the DC system between the combiner or collector box and the inverter.
- Solar Module – A complete, environmentally protected unit consisting of multiple photovoltaic cells that is capable of producing DC power when exposed to sunlight.
- Solar Panel – A group of solar modules that are grouped and prewired to form a field installable component.
- Solar Cell Efficiency – The ratio of the electrical output of a solar cell to the incident energy available on the surface of the solar cell in the form of sunlight.

PROBLEMS

28.1   In a solar cell doped with boron and phosphorous, which layer is the n-layer semiconductor?

   A.   Boron-doped semiconductor

   B.   Phosphorous-doped semiconductor

28.2   Given an air density of 1.2 kg/m^3, a blade radius of 15 m, and a wind velocity of 22.3 mph (10 m/sec), determine the theoretical maximum mechanical power that can be transferred by the wind turbine.

28.3   Given the same parameters that were used in Problem 28.2, if the turbine blade efficiency is 30%, determine the mechanical power available to the turbine shaft.

RECOMMENDED READING

Federal Energy Regulatory Commission, www.ferc.gov/

http://www.energy.siemens.com/hq/pool/hq/power-generation/renewables/wind-power/wind/wind%20turbines/E50001-W310-A103-V6-4A00_WS_SWT_3_6_107_US.pdf

IEEE C57.105-1978: Guide for Application of Transformer Connections in Three-Phase Distribution Systems

IEEE C57.159 (Draft standard): Guide on Transformers for Application in Distributed Photovoltaic (DPV) Power Generation Systems

IEEE 937, 2007: IEEE Recommended Practice for Installation and Maintenance of Lead-Acid Batteries for Photovoltaic Systems

IEEE 1661, 2007: Guide for Test and Evaluation of Lead-Acid Batteries Used in Photovoltaic (PV) Hybrid Power Systems

IEEE 2030-2011: Guide for Smart Grid Interoperability of Energy Technology and Information Technology Operation with the Electric Power System (EPS), and End-Use Applications and Loads

IEEE 1547, 2015: Standard for Interconnecting Distributed Resources with Electric Power Systems

National Renewable Energy Laboratory, NREL, http://www.nrel.gov/docs/fy11osti/48895.pdf

NEC® – National Electrical Code®, 2014.

Siemens Wind Turbine Wind Turbine SWT-3.6-107,

United State Department of Labor, Bureau of Labor Statistics, http://www.bls.gov/green/solar_power/

US DOE, Office of Energy Efficiency & Renewable Energy, www.doe.gov

ANNEXES

ANNEX A   NEMA ENCLOSURE TYPES

(NEMA 250-2003, Reprinted with permission of National Electrical Manufacturers Association)

Type 1 Enclosures constructed for indoor use to provide a degree of protection to personnel against access to hazardous parts and to provide a degree of protection of the equipment inside the enclosure against ingress of solid foreign objects (falling dirt).

Type 2 Enclosures constructed for indoor use to provide a degree of protection to personnel against access to hazardous parts; to provide a degree of protection of the equipment inside the enclosure against ingress of solid foreign objects (falling dirt); and to provide a degree of protection with respect to harmful effects on the equipment due to the ingress of water (dripping and light splashing).

Type 3 Enclosures constructed for either indoor or outdoor use to provide a degree of protection to personnel against access to hazardous parts; to provide a degree of protection of the equipment inside the enclosure against ingress of solid foreign objects (falling dirt and windblown dust); to provide a degree of protection with respect to harmful effects on the equipment due to the ingress of water (rain, sleet, snow); and that will be undamaged by the external formation of ice on the enclosure.

Type 3R Enclosures constructed for either indoor or outdoor use to provide a degree of protection to personnel against access to hazardous parts; to provide a degree of protection of the equipment inside the enclosure against ingress of solid foreign objects (falling dirt); to provide a degree of protection with respect to harmful effects on the equipment due to the ingress of water (rain, sleet, snow); and that will be undamaged by the external formation of ice on the enclosure.

Type 3S Enclosures constructed for either indoor or outdoor use to provide a degree of protection to personnel against access to hazardous parts; to provide a degree of protection of the equipment inside the enclosure against ingress of solid foreign objects (falling dirt and windblown dust); to provide a degree of

Energy Production Systems Engineering: An Introduction for Electrical Engineers to Electrical Power Generation Facilities, Systems, and Equipment, First Edition. Thomas H. Blair.
© 2017 by The Institute of Electrical and Electronics Engineers, Inc. Published 2017 by John Wiley & Sons, Inc.

protection with respect to harmful effects on the equipment due to the ingress of water (rain, sleet, snow); and for which the external mechanism(s) remain operable when ice laden.

Type 3X Enclosures constructed for either indoor or outdoor use to provide a degree of protection to personnel against access to hazardous parts; to provide a degree of protection of the equipment inside the enclosure against ingress of solid foreign objects (falling dirt and windblown dust); to provide a degree of protection with respect to harmful effects on the equipment due to the ingress of water (rain, sleet, snow); that provides an additional level of protection against corrosion and that will be undamaged by the external formation of ice on the enclosure.

Type 3RX Enclosures constructed for either indoor or outdoor use to provide a degree of protection to personnel against access to hazardous parts; to provide a degree of protection of the equipment inside the enclosure against ingress of solid foreign objects (falling dirt); to provide a degree of protection with respect to harmful effects on the equipment due to the ingress of water (rain, sleet, snow); that will be undamaged by the external formation of ice on the enclosure that provides an additional level of protection against corrosion; and that will be undamaged by the external formation of ice on the enclosure.

Type 3SX Enclosures constructed for either indoor or outdoor use to provide a degree of protection to personnel against access to hazardous parts; to provide a degree of protection of the equipment inside the enclosure against ingress of solid foreign objects (falling dirt and windblown dust); to provide a degree of protection with respect to harmful effects on the equipment due to the ingress of water (rain, sleet, snow); that provides an additional level of protection against corrosion; and for which the external mechanism(s) remain operable when ice laden.

Type 4 Enclosures constructed for either indoor or outdoor use to provide a degree of protection to personnel against access to hazardous parts; to provide a degree of protection of the equipment inside the enclosure against ingress of solid foreign objects (falling dirt and windblown dust); to provide a degree of protection with respect to harmful effects on the equipment due to the ingress of water (rain, sleet, snow, splashing water, and hose-directed water); and that will be undamaged by the external formation of ice on the enclosure.

Type 4X Enclosures constructed for either indoor or outdoor use to provide a degree of protection to personnel against access to hazardous parts; to provide a degree of protection of the equipment inside the enclosure against ingress of solid foreign objects (windblown dust); to provide a degree of protection with respect to harmful effects on the equipment due to the ingress of water (rain, sleet, snow, splashing water, and hose-directed water); that provides an additional level of protection against corrosion; and that will be undamaged by the external formation of ice on the enclosure.

Type 5 Enclosures constructed for indoor use to provide a degree of protection to personnel against access to hazardous parts; to provide a degree of protection of the equipment inside the enclosure against ingress of solid foreign objects (falling dirt and settling airborne dust, lint, fibers, and filings); and to provide a degree of protection with respect to harmful effects on the equipment due to the ingress of water (dripping and light splashing).

Type 6 Enclosures constructed for either indoor or outdoor use to provide a degree of protection to personnel against access to hazardous parts; to provide a degree of protection of the equipment inside the enclosure against ingress of solid foreign objects (falling dirt); to provide a degree of protection with respect to harmful effects on the equipment due to the ingress of water (hose-directed water and the entry of water during occasional temporary submersion at a limited depth); and that will be undamaged by the external formation of ice on the enclosure.

Type 6P Enclosures constructed for either indoor or outdoor use to provide a degree of protection to personnel against access to hazardous parts; to provide a degree of protection of the equipment inside the enclosure against ingress of solid foreign objects (falling dirt); to provide a degree of protection with respect to harmful effects on the equipment due to the ingress of water (hose-directed water and the entry of water during prolonged submersion at a limited depth); that provides an additional level of protection against corrosion and that will be undamaged by the external formation of ice on the enclosure.

Type 12 Enclosures constructed (without knockouts) for indoor use to provide a degree of protection to personnel against access to hazardous parts; to provide a degree of protection of the equipment inside the enclosure against ingress of solid foreign objects (falling dirt and circulating dust, lint, fibers, and filings); and to provide a degree of protection with respect to harmful effects on the equipment due to the ingress of water (dripping and light splashing).

Type 12K Enclosures constructed (with knockouts) for indoor use to provide a degree of protection to personnel against access to hazardous parts; to provide a degree of protection of the equipment inside the enclosure against ingress of solid foreign objects (falling dirt and circulating dust, lint, fibers, and flyings); and to provide a degree of protection with respect to harmful effects on the equipment due to the ingress of water (dripping and light splashing).

Type 13 Enclosures constructed for indoor use to provide a degree of protection to personnel against access to hazardous parts; to provide a degree of protection of the equipment inside the enclosure against ingress of solid foreign objects (falling dirt and circulating dust, lint, fibers, and flyings); to provide a degree of protection with respect to harmful effects on the equipment due to the ingress of water (dripping and light splashing); and to provide a degree of protection against the spraying, splashing, and seepage of oil and non-corrosive coolants.

Type 7 Enclosures constructed for indoor use in hazardous (classified) locations classified as Class I, Division 1, Groups A, B, C, or D as defined in NFPA 70®.

Type 8 Enclosures constructed for either indoor or outdoor use in hazardous (classified) locations classified as Class I, Division 1, Groups A, B, C, and D as defined in NFPA 70®.

Type 9 Enclosures constructed for indoor use in hazardous (classified) locations classified as Class II, Division 1, Groups E, F, or G as defined in NFPA 70®.

Type 10 Enclosures constructed to meet the requirements of the Mine Safety and Health Administration, 30 CFR, Part 18.

ANNEX B   IEEE DEVICE NUMBERS AND FUNCTIONS

(*IEEE C37.2 Standard Electrical Power System Device Function Numbers, Acronyms and Contact Designations*)

The devices in switching equipment are referred to by numbers, according to the functions they perform. These numbers are based on a system which has been adopted as standard for automatic switchgear by IEEE. This system is used in connection diagrams, in instruction books, and in specifications.

Device No.	Function	Device No.	Function
1	Master element	49	Machine or transformer thermal relay
2	Time-delay starting or closing relay	50	Instantaneous overcurrent or rate-of-rise relay
3	Checking or interlocking relay	51	AC time overcurrent relay
4	Master contactor	52	AC circuit breaker
5	Stopping device	53	Exciter or DC generator relay
6	Starting circuit breaker	54	Turning gear engaging device
7	Rate-of-change relay	55	Power factor relay
8	Control power disconnecting device	56	Field application relay
9	Reversing device	57	Short-circuiting or grounding device
10	Unit sequence switch	58	Power rectifier misfire relay
11	Multifunction device	59	Overvoltage relay
12	Overspeed device	60	Voltage or current balance relay
13	Synchronous-speed device	61	Density switch or sensor
14	Underspeed device	62	Time delay stopping or opening relay
15	Speed, or frequency matching device	63	Liquid or gas pressure level
16	Data communications device	64	Ground protective relay
17	Shunting or discharge-switch	65	Governor
18	Accelerating or decelerating device	66	Notching or jogging device
19	Starting to running transition contactor	67	AC directional overcurrent relay

20	Electrically operated valve	68	Blocking or "out-of-step" relay
21	Distance relay	69	Permissive vontrol device
22	Equalizer circuit breaker	70	Electrically operated rheostat
23	Temperature control device	71	Liquid level switch
24	Volts per hertz relay	72	DC circuit breaker
25	Synchronizing or synchronism check	73	Load resistor contactor
26	Apparatus thermal device	74	Alarm relay
27	Undervoltage relay	75	Position-changing mechanism
28	Flame detector	76	DC overcurrent relay
29	Isolating contactor or switch	77	Telemetering device
30	Annunciator relay	78	Phase-angle measuring or out-of-step protective relay
31	Separate excitation device	79	AC reclosing relay
32	Directional power relay	80	Flow switch
33	Position switch	81	Frequency relay
34	Master sequence switch	82	DC reclosing relay
35	Brush-operating or slip-ring short-circuiting device	83	Automatic selective control or transfer relay
36	Polarity device	84	Operating mechanism
37	Undercurrent or underpower relay	85	Carrier or pilot-wire receiver relay
38	Bearing protective device	86	Locking-out relay
39	Mechanical condition monitor	87	Differential protective relay
40	Field (over/under excitation) relay	88	Auxiliary motor or motor generator
41	Field circuit breaker	89	Line switch
42	Running circuit breaker	90	Regulating device
43	Manual transfer or selector device	91	Voltage directional relay
44	Unit sequence starting relay	92	Voltage and power directional relay
45	Abnormal atmospheric condition monitor	93	Field-changing contactor
46	Reverse-phase-balance current relay	94	Tripping or trip-free relay
47	Phase-sequence voltage relay	95	Used only for specific applications
48	Incomplete sequence relay	96	Used only for specific applications
		97	Used only for specific applications
		98	Used only for specific applications
		99	Used only for specific applications

ANNEX C   COMMON SYSTEM CODES FOR POWER GENERATION FACILITIES

(Reprinted with permission of Tampa Electric Company. Reproduction is forbidden without the express consent of Tampa Electric Company)

System Code	System Description
AAH	Anhydrous ammonia system
ABA	Auxiliary boiler – combustion air
ABB	Auxiliary boiler – burner control
ABD	Auxiliary boiler – blowdown
ABF	Auxiliary boiler – feedwater and condensate
ABH	Auxiliary boiler – chemical feed
ABM	Auxiliary boiler – steam
ABS	Auxiliary boiler – system
ACS	Air-cooled condenser – system
ADS	Automatic depressurization – system
AFP	Fly ash precipitator – fluidizing air
AFS	Fly ash silo – fluidization air
AFW	Fly ash sluicing water
AHB	Ash handling – bottom ash disposal
AHC	Ash handling – chemical feed
AHF	Ash handling – fly ash disposal and/or reinjection
AHP	Ash handling – pyrite
AHW	Bottom ash handling system – sluicing water
AML	Lean amine
AMM	Ammonia additive system
AMR	Rich amine
AMS	Ambient temperature monitoring – system
ANN	Annunciator system
ARA	Air removal – auxiliary condenser
ARC	Air removal – main condenser
ASA	Aqueous system absorber loop – FGD
ASF	Auxiliary steam – fossil – system
ASN	Air separation unit, nitrogen
ASO	Air separation unit, oxygen
ASQ	Aqueous system quencher loop – FGD
ASP	Air feed to sulfur unit
ASR	Auxiliary steam – radioactive waste
ASS	Auxiliary steam – nuclear – system
ASU	Air separation unit
ATA	Absorber tower area – FGD
BAT	Battery
BCC	Bearing cooling – chemical feed
BCP	Bearing cooling – purification system

BCS	Bearing cooling – system
BDC	Blowdown – cooling tower
BDE	Blowdown effluent
BDG	Blowdown – steam generator (nuclear)
BLW	Black water handling
BRC	Brine condensate system
BRN	Brine system
BRS	Brine sewer
BWS	Battery – 24 V – station
BWY	Battery – 24 V – yard
BXS	Battery – 48 V – station
BXY	Battery – 48 V – yard
BYS	Battery – 125 V/250 V station
BYY	Battery – 125 V – yard
BZS	Battery – 250 V station
BZY	Battery – 250 V – yard
CAC	Chemical Feed – core aux cooling
CAS	Containment cleaning –system
CAU	Caustic
CBS	Conductor cooling – system
CCB	Chemical cleaning – boiler
CCC	Component cooling water – common
CCD	Component cooling – CNDS demineralizer – liquid waste
CCE	Component cooling – charging pump seal
CCF	Component cooling water – ESF area
CCH	Component cooling – chemical feed
CCI	Component cooling – safety injection pump seal
CCL	Component cooling – water (fossil)
CCN	Cooling water – reactor plant
CCP	Chemical cleaning – plant
CCR	Closed loop cooling water returns
CCS	Closed loop cooling water supply
CCW	Component cooling – chilled
CDS	Chilled water – system
CEC	Common elect equip – control RM complex
CES	Common electrical system – (control building)
CFS	Core flooding – system (B&W)
CGS	Vacuum cleaning – system
CHS	Chemical and volume control system
CKC	Core cooling – standby
CLB	Chemical cleaning – boiler
CLG	Chemical cleaning – steam generator (nuclear)
CLM	Chemical cleaning – main steam and feedwater (nuclear)
CLP	Chemical cleaning – preboiler and heater shield
CLS	Chemical cleaning – system

CMP	Containment press monitor system
CMS	Containment atmosphere monitoring – system
CNA	Condensate – auxiliary condensate
CNC	Condensate – chemical treatment
CND	Condensate – demineralizer
CNE	Condensate, drains – air-cooled condenser
CNH	Steam condensate – high pressure
CNL	Steam condensate – low pressure
CNM	Condensate – main condensate
CNO	Condensate pump lube oil – system
CNV	Converter
COC	Communication – carrier
COJ	Communication – maintenance
COL	Communication – telephone (leased line)
COM	Communication – microwave
COP	Communication – paging (public address)
COR	Communication – radio
COS	Communication – sound power
COT	Communication – telephone (inter-plant)
CPD	Containment atmospheric dilution
CPM	Containment hydrogen mixing
CPP	Containment purge
CPS	Reactor containment (inerting and purge)
CPX	Condensate polishing demineralizer
CRS	Cold reheat – system
CSH	Core spray – high pressure
CSL	Core spray – low pressure
CSS	Containment spray – system
CSY	Condensate from syngas
CTG	Cold gas clean-up tail gas
CTS	Reactor core temperature – system
CVS	Containment vacuum – system
CWA	Circulating water – condenser tube cleaning
CWB	Burner cooling water
CWS	Circulating water – system
CXC	Combustion air heater – condensate
CXF	Combustion air heater – hot air
CXG	Combustion air heater – glycol
CXW	Combustion air heater – water
CXZ	Combustion air heater – steam
CYC	Coal handling – dust collection
CYH	Coal handling yard (trip, trip conveyor, and bunker level)
CYS	Coal handling – suppression
DAR	Drains – amine recovery sewer
DAS	Drains (aerated) reactor plant – system

DCS	Distributed control system (DCS equipment only)
DEC	Drains – control area/building floor and equipment
DED	Drains – radioactive waste building
DEF	Equipment and floor drains
DEM	Chemical wastewater requiring PH adjustment
DER	Drains – reactor building equipment
DET	Drains – turbine building equipment
DEW	Drains – water treat area/building floor and equipment
DFA	Drains, floor – fuel and auxiliary building
DFB	Drains, floor – boiler plant
DFD	Drains, floor – diesel gen building
DFE	Drains, floor – service area
DFF	Drains – fuel and auxiliary building equipment
DFM	Drains, floor – miscellaneous buildings
DFN	Drains, floor turbine and reactor building
DFR	Drains, floor – reactor plant
DFT	Drains, floor – turbine plant
DFV	Drains, floor – solid waste building
DFW	Drains, floor – radioactive waste building
DGS	Drains (hydrogenated) reactor plant – system
DGV	Drains, valve stem
DHS	Decay heat removal – system (B&W)
DMS	Motor control center DC – system
DPS	Drains – powerhouse
DRB	Roof drainage – boiler area
DRC	Roof drainage – coal bunker
DRF	Drywell floor seal – system
DRP	Roof drainage – precipitator
DRS	Drywell cooling – system
DRT	Roof drainage – turbine
DSM	Drains – moisture separator (nuclear)
DSR	Drains – moisture separator reheater
DSS	Dust suppression system
DTF	Drains – flash tank
DTM	Drains – turbine plant miscellaneous
DUH	Station sump drainage and unwatering system
DWS	Domestic/potable water system
DWT	Domestic water treating
DXA	Dewatering system – subsurface
DXF	Dewatering – fishway
DXR	Dewatering – tailrace tunnel
DXT	Dewatering – turbine
EBS	Emergency boration – system
ECS	Environmental monitoring – system
EDA	Essential bus – AC supply

EDB	Essential bus – DC supply
EDS	Essential bus – system
EGA	Generator, standby – air compressor – diesel starting air
EGD	Generator, standby – combustion air damper control
EGE	Generator, standby – excitation
EGF	Generator, standby – fuel
EGO	Generator, standby – lube oil system
EGP	Generator, standby – protection (include ACB and instrument)
EGS	Generator, standby – system
EGT	Generator, standby – temperature
EHS	Motor control center – standby – system
EJA	Unit substation – standby – AC control and heater supply
EJB	Unit substation – standby – DC control supply
EJS	Unit Substation – standby system
ENA	Switchgear, standby, 4160 V – AC control and heater supply
ENB	Switchgear, standby, 4160 V – DC control supply
ENS	Switchgear, standby, 4160 V – system
EPA	Switchgear, standby, 13.8 kV – AC control and heater supply
EPB	Switchgear, standby, 13.8 kV – DC control supply
EPS	Switchgear, standby, 13.8 kV – system
ESR	Emergency sludge reclaim system – FGD
ESS	Extraction – system
EXA	Excitation – main generator – AC control and heater supply
EXB	Excitation – main generator – DC control supply
EXC	Excitation – main generator – cooling
EXG	Excitation – main generator – ground detector
EXP	Excitation – main generator – protection
EXS	Extraction – main generator – system (power circuits.)
EXV	Excitation – main generator – voltage regulator
FCS	Fuel-coal station (rappers, feeders, pulverizers, gates, etc.)
FDS	Final dewatering system – FGD
FGA	Flue gas absorber loop – FGD
FGB	Flue gas bypass (PRD) – FGD
FGC	Flue gas conditioning
FGI	Flue gas inlet – FGD
FGM	Flue gas mixing chamber – FGD
FGO	Flue gas outlet – FGD
FGQ	Flue gas quencher loop – FGD
FGS	Fuel – gas – system
FIN	Fines handling
FLT	Harmonic filter
FNR	Fuel – nuclear – refueling
FNS	Fuel, nuclear – fuel storage
FNT	Fuel – nuclear – transfer
FOA	Fuel – oil – auxiliary boiler (supply to)

FOC	Fuel – oil – additive system
FOF	Fuel – oil – engine driven fire pump
FOH	Fuel – oil – heaters
FOM	Fuel – oil – main boiler (supply to)
FOS	Forced oxidation system – FGD
FOY	Fuel – oil – yard facilities
FPA	Fire detection – alarm system
FPC	Fire protection – dry chemical
FPF	Fire protection – foam system
FPG	Fire protection – Halon
FPH	Fire protection – high pressure CO_2
FPL	Fire protection – low pressure CO_2
FPM	Fire protection – supervision system
FPP	Fire protection – fire pump (makeup pump, diesel, etc.)
FPS	Fire protection system – pulverizers
FPW	Fire protection – water
FSS	Fines sewer system
FWA	Feedwater – auxiliary feedwater
FWD	Steam – water dump – system
FWF	Feedwater – pump fluid drive
FWK	Feedwater – motor driven feedwater pump coupling
FWL	Feedwater – pump and driver oil systems
FWP	Feedwater and condensate pump – seals and leakoff
FWR	Feedwater – recirculation and balance drum leakoff
FWS	Feedwater – system
FWT	Feedwater – high pressure
FWV	Boiler feedwater system
GAS	Gasification
GDT	Gate – draft tube
GHM	Gypsum handling – maintenance
GHS	Gypsum handling – system
GJB	Generator – junction box
GMC	Generator – main stator cooling
GMH	Generator – main – hydrogen and CO_2
GML	Generator – main leads
GMO	Generator – main – seal oil
GMS	Generator – main system
GRW	Gray water handling
GSF	Gas – helium – fuel handling and purge
GSH	Gas – hydrogen – system (nuclear)
GSL	Gas – helium – system
GSN	Gas – nitrogen – system (nuclear)
GSO	Gas – oxygen – system (nuclear)
GSP	Gas – helium – PCRV seal and purge
GSR	Gas – helium – recovery

GSS	Grinding slurry sewer system
GTS	Gas treatment, standby – system
GWS	Gray water sewer
HAG	Hydraulic power – governor
HAS	Hydraulic power – system
HAV	Hydraulic power – gates/valves
HDH	Heater drain – high pressure
HDL	Heater drain – low pressure
HPS	High press injection – system (B&W)
HRS	Hot reheat – system
HTG	Syngas system
HTR	Sorbent regeneration gas system
HTS	Heat tracing – system
HVA	Chilled water – office
HVB	Ventilation – boiler plant
HVC	Air conditioning – control building – area
HVD	Gas heating – direct fired – system
HVE	Ventilation – service building
HVF	Ventilation – fuel building
HVG	Glycol heating – system
HVH	Hot water heating – system
HVI	Ventilation – auxiliary boiler room
HVJ	Ventilation – water treatment building
HVK	Chilled water – control building or area
HVL	Air conditioning – service building
HVM	Steam heating – system
HVN	Chilled water – ventilation
HVO	Air conditioning – office building
HVP	Ventilation – diesel gen building
HVQ	Ventilation – engineered safety feature building
HVR	Ventilation – reactor plant
HVS	Heating, ventilating, air cond. – system
HVT	Ventilation – turbine area
HVU	Ventilation – containment structure
HVV	Ventilation – main steam valve building
HVW	Ventilation – radioactive waste building
HVX	Refrigeration – system
HVY	Ventilation – yard structures
HVZ	Heating, ventilating, air cond condensing water – system
HWA	Ventilation – auxiliary building
HWE	Ventilation and cooling – switchgear building
HWH	Ventilation and cooling – warehouse
HWN	Ventilation – service water building
HWO	Heating, hot water – admin. building
HWP	Ventilation – precipitator building

HWS	Ventilation and cooling – security building
HWT	Ventilation and cooling – waste treatment building
HWW	Ventilation – service water
HWY	Ventilation and cooling – health physics building
IAC	Instrument air – containment
IAS	Instrument air system
ICS	Reactor core isolation cooling – system
IGS	Ignition gas – system
IHA	Information handling – annunciator
IHC	Information handling – computer
IHO	Information handling – oscillography
IHS	Information handling – system
IIS	Incore instrumentation system
IOS	Ignition oil – system
IRS	Indirect reheat system – FGD
ISA	Instrument system air – FGD
ISC	Isolation – containment
ISD	Isolation – drywell
ISM	Isolation – main steam
ISOL	Isolation transformer or switch
ISR	Isolation – reactor vessel
JAA	Adm. building – substructure all
JAB	Adm. building – superstructure all
JBA	Boiler room – substructure fossil
JBB	Boiler room – substructure fossil
JCA	Control room – substructure all
JCB	Control room – superstructure
JDA	Diesel gen. room – substructure
JDB	Diesel gen. room – superstructure
JFA	Fuel storage house – substructure
JFB	Fuel storage house – superstructure
JGA	Spillway gate control house – substructure
JGB	Spillway gate control house – superstructure
JHA	Cooling tower – substructure
JHB	Cooling tower – superstructure
JHC	Cooling tower, circ. water – substructure
JHD	Cooling tower, circ. water – superstructure
JHE	Cooling tower, serv. water – substructure
JHF	Cooling tower, serv. water – superstructure
JIA	Information building – substructure
JIB	Information building – superstructure
JJA	Waste disposal building – substructure
JJB	Waste disposal building – superstructure
JKA	Miscellaneous building – substructure
JKB	Miscellaneous building – superstructure

JLA	Tank area structures – substructure
JLB	Tank area structures – superstructure
JLC	Reactor plant tank area structure – substructure
JLD	Reactor plant tank area structure – superstructure
JMA	Water treatment building – substructure
JMB	Water treatment building – superstructure
JNA	Radioactive Waste building – substructure
JNB	Radioactive Waste building – superstructure
JOA	Office building – substructure
JOB	Office building – superstructure
JPA	Primary auxiliary building – substructure
JPB	Primary auxiliary building – superstructure
JQA	Decontamination building – substructure
JQB	Decontamination building – superstructure
JRA	Reactor building – substructure
JRB	Reactor building – superstructure
JSA	Gar. storage building – substructure
JSB	Gar. storage building – superstructure
JTA	Turbine room – substructure
JTB	Turbine room – superstructure
JUA	Vacuum priming building – substructure
JUB	Vacuum priming building – superstructure
JVA	EL equipment areas – switchgear – substructure
JVB	EL equipment areas – switchgear – superstructure
JVC	EL equipment areas – XFMR – substructure
JVD	EL equipment areas – XFMR – superstructure
JWA	Screen well and pump house – substructure
JWB	Screen well and pump house – superstructure
JXA	Auxiliary boiler room – substructure
JXB	Auxiliary boiler room – superstructure
JYA	Switchyard control house – substructure
JYB	Switchyard control house – superstructure
JZA	Service building – substructure
JZB	Service building – superstructure
KAA	Auxiliary generator house – substructure
KAB	Auxiliary generator house – superstructure
KBA	Boron evap. pump house – substructure
KBB	Boron evap. pump house – superstructure
KCA	Chlorine building – substructure
KCB	Chlorine building – superstructure
KDA	Discharge vacuum primary pump house – substructure
KDB	Discharge vacuum primary pump house – superstructure
KFA	Fuel oil pump house – substructure
KFB	Fuel oil pump house – superstructure
KGA	Gate house – substructure

KGB	Gate house – superstructure
KHA	Hyd. storage building – substructure
KHB	Hyd. storage building – superstructure
KIA	Intake structure – substructure
KIB	Intake structure – superstructure
KJA	Containment auxiliary structure – substructure
KJB	Containment auxiliary structure – superstructure
KKA	Containment spray pump area storage – substructure
KKB	Containment spray pump area storage – superstructure
KMA	Main steam shop – substructure
KMB	Main steam shop – superstructure
KNA	Fire pump house – substructure
KNB	Fire pump house – superstructure
KPA	Purge air equipment building – substructure
KPB	Purge air equipment building – superstructure
KRB	Radiation containment area storage building – superstructure
KSA	Stack. – substructure
KSB	Stack. – superstructure
KTA	Weather tower – substructure
KTB	Weather tower – superstructure
KWA	Warehouse – substructure
KWB	Warehouse – superstructure
LAA	Lighting – AC – administration building
LAB	Lighting – AC boiler area
LAC	Lighting – AC – control room (main)
LAD	Lighting – AC – diesel generator room
LAE	Lighting – AC – settling pond
LAF	Lighting – AC – fuel storage area
LAG	Lighting – AC – spillway gate containment house
LAH	Lighting – AC – cooling tower and pond area
LAI	Lighting – AC – information area
LAJ	Lighting – AC – waste disposal building
LAK	Lighting – AC – miscellaneous buildings
LAL	Lighting – AC – gallery area
LAM	Lighting – AC – water treatment building
LAN	Lighting – AC – Radioactive waste building
LAO	Lighting – AC – outdoor
LAP	Lighting, AC precipitator
LAP	Lighting – AC – primary auxiliary building
LAQ	Lighting – AC – decontamination building
LAR	Lighting – AC – reactor building
LAS	Lighting, AC scrubber
LAS	Lighting – AC – system (supply)
LAT	Lighting – AC – turbine area
LAU	Lighting – AC – vacuum priming building

LAV	Lighting – AC – warning
LAW	Lighting – AC – screen well and pump house
LAX	Lighting – AC – auxiliary boiler room
LAY	Lighting – AC – switchyard or substation (including control house)
LAZ	Lighting – AC – service building
LBA	Lighting – DC – administration building
LBB	Lighting – DC – boiler area
LBC	Lighting – DC – control room (main)
LBK	Lighting – DC – miscellaneous areas
LBP	Lighting – DC – primary auxiliary building
LBR	Lighting – DC – reactor building
LBS	Lighting – DC – system (supply)
LBT	Lighting – DC – turbine area
LBV	Lighting – DC – warning
LBZ	Lighting – DC – service building
LCS	Supplementary leak collection – system
LFC	Load frequency control – system
LHM	Limestone handling area maintenance equipment – FGD
LHS	Limestone handling system – FGD
LMS	Leakage monitoring system
LOS	Lube oil (treatment and storage only) system
LPN	Low pressure nitrogen system
LSS	Limestone slurry system – FGD
LSV	Leakage control, penetration valve
LTG	Lighting
LUS	Lubrication – system
LWC	Liquid waste – condensate demineralizer system
LWS	Liquid waste (radioactive) – system
MBA	Boiler, main – secondary air and flue gas
MBB	Boiler, main burner control
MBD	Boiler, main – drains to boiler blow off tank
MBE	Boiler, main – seal air
MBG	Boiler, main – ignition air
MBH	Boiler, main – chemical feed
MBI	Boiler, main – circulation seal injection
MBK	Boiler, main – SO_2 removal
MBL	Boiler, main – continuous blowdown, header vents and drains
MBM	Boiler, main – monitoring
MBO	Boiler, main – once through start-up – system
MBP	Boiler, main – protection (boiler trip)
MBR	Boiler, main – steam line blowout
MBR	Boiler, main – start-up drum system
MBS	Boiler, main system
MBT	Boiler, turbine start-up
MBW	Boiler, main – primary air

MBX	Boiler, main – scanner and television cooling air
MBZ	Boiler, main – soot blowing
MCB	Main control board
MCC	Motor control center
MDS	Motor-operated doors – system
MHB	Material handling (cranes, hoist, etc.) boiler area
MHF	Material handling (cranes, hoist, etc.) – fuel storage area
MHG	Material handling (cranes, hoist, etc.) – spillway gate and pump house
MHH	Material handling (cranes, hoist, etc.) – chlorine storage area
MHI	Material handling (cranes, hoist, etc.) – intake structure
MHJ	Material handling (cranes, hoist, etc.) – waste disposal building
MHK	Material handling (cranes, hoist, etc.) – misc. bldgs.
MHN	Material handling (cranes, hoist, etc.) – radioactive waste building
MHP	Material handling (cranes, hoist, etc.) – PRI auxiliary building
MHR	Material handling (cranes, hoist, etc.) – reactor building (containment)
MHS	Material handling (cranes, hoist, etc.) – system
MHT	Material handling (cranes, hoist, etc.) – turbine area
MHW	Material handling (cranes, hoist, etc.) – screen well and pump house
MHZ	Material handling (cranes, hoist, etc.) – service building
MMS	Meteorological monitoring system
MPN	Medium pressure nitrogen
MPP	Medium pressure purge water (uneconomized)
MSD	Main turbine steam drain system
MSE	Main turbine EXH steam to air-cooled condenser
MSI	Main steam isolation valve seals
MSR	Main steam – reheat (moist sep and reheat – nuclear)
MSS	Main steam system (fossil and nuclear)
MSV	Main stop valve
MTX	Main transformer including auxiliaries (coolers, LTC, alarms, etc.)
MUS	Makeup and purification – system (B&W)
MWS	Makeup water – system (raw water)
NBS	Nitrogen blanketing – system (fossil)
NGS	Miscellaneous distribution panels – all areas
NHS	Motor control center, normal – system
NIT	Nitrogen for use in combustion turbine
NJA	Unit substation – AC control and heater supply
NJB	Unit substation – DC – control supply
NJS	Unit substation – system
NLA	Switchgear, normal, 2400 V – AC control and heater supply
NLB	Switchgear, normal, 2400 V – DC control supply
NLS	Switchgear, normal, 2400 V – system
NME	Neutron monitoring – system
NMI	Neutron monitoring – intermediate range
NMP	Neutron monitoring – power range
NMS	Neutron monitoring – source range

NMT	Neutron monitoring – traverse incore probes
NNA	Switchgear – normal, 4160 V – AC control and heater supply
NNB	Switchgear – normal, 4160 V – DC control supply
NNS	Switchgear – normal, 4160 V system
NOA	Switchgear – normal, 6900 V – AC control and heater supply
NOB	Switchgear – normal, 6900 V – DC control supply
NOS	Switchgear – normal, 6900 V – system
NPA	Switchgear – normal, 13.8 kV – AC control and heater supply
NPB	Switchgear – normal, 13.8 kV – DC control supply
NPS	Switchgear – normal, 13.8 kV system
NSS	Neutron shield tank and cooling – system
NXS	Pipe displacement monitoring – system
OFC	Office and commissary equipment
OFG	Off gas
ORS	Organism return system
OWS	Oil-water separation system
OXS	Oxygen to sulfur recovery unit
OXY	Oxygen for use in gasifier
PBS	Sanitary sewerage facilities system
PCS	Protection, cathodic – system
PCW	Process water
PDS	Primary dewatering system – FGD
PES	Pressurization, emergency – system
PGS	Primary grade water – system
PHS	Personnel access door – system
POC	Power outlet – outlets for controlling devices
POG	Power outlet – general purpose, single phase
POI	Power outlet – indicating and recording devices
POP	Power outlet – general purpose, three phase
PRO	Propane
PTS	Purge treatment system – FGD
PWR	Power island
PWS	Potable water
QSS	Quench spray system (containment)
RVA	Relief or vent to atmosphere
RAD	Radiation monitoring – auxiliary boiler room
RBC	Reboiler – condensate return (fossil)
RBD	Reboiler drains auxiliary steam – system
RBF	Reboiler – feedwater (fossil)
RBH	Reboiler, chemical feed
RBM	Reboiler – steam (fossil)
RCA	Reactor coolant – auxiliary service
RCM	Reactor coolant – main circulator service
RCR	PCRV pressure relief – system
RCS	Reactor coolant (PRI loop) – system

RCW	Recycle water
RDC	Reactor rod drive – control
RDI	Reactor rod drive – instrumentation
RDS	Roof drain system
RDT	Reactor rod drive, test
RVF	Relief or vent to flare
RHB	Chemical feed system
RHS	Residual heat removal – system
RMA	Radiation monitoring – administration building
RMC	Radiation monitoring – control room
RMD	Radiation monitoring – diesel generator room
RMF	Radiation monitoring – fuel storage area
RMI	Radiation monitoring – information center
RMJ	Radiation monitoring – waste disposal building
RMK	Radiation monitoring – miscellaneous buildings
RMN	Radiation monitoring – radioactive waste building
RMP	Radiation monitoring – PRI auxiliary building
RMR	Radiation monitoring – reactor building
RMS	Radiation monitoring – system
RMT	Radiation monitoring – turbine building
RMW	Radiation monitoring – screen well and pump house
RMZ	Radiation monitoring – service building
RPM	Reactor protection system – M-G set
RPS	Reactor protection – system
RRC	Reactor water recirculation – M-G set coupling
RRG	Reactor water recirculation – M-G set generator and excitation
RRL	Reactor water recirculation – lube oil system
RRM	Reactor water recirculation – M-G set motor
RRP	Reactor water recirculation – protection – (electrical)
RSA	Relief to acid flare
RSS	Containment recirculation spray – system
RSY	Relief system
RTX	Reserve station service trans, including auxiliaries
RUB	Scrubbing
RVS	Reactor vessel – system
RWS	Return water system – FGD
SAD	Strong acid drain system
SAH	Service air – high pressure
SAS	Service air – system
SAT	Service air – tail water depression
SBA	Soot blowing air – system
SBD	Sodium bicarbonate and salt solid discharge system
SBS	Sodium bicarbonate and salt solid Supply System
SCA	Station control bus (nonvital) – AC
SCB	Station control bus (nonvital) – DC supply

SCC	Station control – bypass/in operation status indication
SCI	Station control bus (nonvital) – indication (breakers, valves, etc.)
SCM	Station control monitoring (vital) – indication
SCR	Selective catalytic reduction system
SCV	Station control bus (vital) – AC
SDS	Steam drain system (fossil)
SER	Suppression – reservoir evaporation
SES	Nuclear servicing equipment – system (other than refuel)
SFC	Spent fuel – Fuel pool cooling and cleanup (inc. pool and aux)
SFT	Fuel transfer – system
SGC	Steam generator water cleanup
SGF	Steam generator – chemical feed
SGL	Steam generator wet lay-up – system
SGR	Steam generator recirculation
SHP	High pressure steam (1650 PSIG)
SHS	Reserve shutdown actuation – system
SIH	Safety injection – high pressure
SIL	Safety injection – low pressure
SIS	Safety injection – system
SLA	Slurry additive
SLG	Slag handling
SLP	Low pressure steam (50 psig)
SLR	Slurry
SLS	Standby liquid control (poisoning) – system
SLW	Slurry water
SMP	Medium pressure steam (400 psig)
SNC	Syngas cooling
SNH	Steam condensate
SNS	Neutron shielding – system
SOR	Absorbing
SPB	Station protection – DC control supply
SPF	Station protection – reserve station service line
SPG	Station protection – generator
SPI	Station protection – generator, main transformer, station service bus
SPL	Station protection – generator line
SPM	Station protection – main transformer
SPR	Station protection – reserve station service transformer
SPS	Station protection – station service transformer
SPU	Station protection – unit
SPX	Station protection – auxiliary power transformer
SPY	Station protection yard DC control supply
SRG	Sulfur recovery unit tail gas
SRH	Sour water
SRR	Storm sewer – roof drainage
SRS	Storm sewer – system

SRW	Storm sewer – storm and waste water
SSA	Sampling system – fuel and auxiliary building
SSC	Sampling system – containment
SSF	Sampling system – failed fuel detection
SSM	Scrubber system area maintenance equipment – FGD
SSR	Sampling system – reactor plant
SSS	Sampling system – plant
SST	Sampling system – turbine plant (nuclear)
SSY	Sampling system
SSW	Sampling system – radioactive waste building
STA	Station ambient condition monitors
STC	Steam turbine condensate system
STP	Stripping
STX	Station service trans – normal including auxiliaries
SUL	Elemental sulfur
SUM	Supervisory control – multiplex
SVD	Miscellaneous SV discharge
SVH	Feedwater shell relief vents and drips
SVS	Steam vents – system
SVV	Steam vents – safety valves
SWC	Service water – chilled water
SWG	Switchgear
SWH	Service water
SWM	Service water – turbine plant
SWP	Service water – system (nuclear)
SWS	Service water – system
SWT	Service water – traveling screens – wash and disposal
SWV	Service water – spherical valve seal
SXS	Start-up test instrumentation – system
SYD	Synchronizing – diesel generator
SYG	Synchronizing – main generator
SYL	Synchronizing line
SYR	Synchronizing – reserve system
SYS	Synchronizing – station service
TCI	Turbine, combustion – supervisory instrumentation
TFA	Turbine, feed pump – alarms and trips
TFC	Turbine, feed pump – control
TFD	Turbine, feed pump – disconnect coupling
TFE	Turbine, feed pump – gland seal
TFG	Turbine, feed pump – turn gear
TFI	Turbine, feed pump – supervisory instrumentation
TFL	Turbine, feed pump – lube oil
TFM	Turbine, feed pump – steam and exhaust
TJB	Turbine junction box
TMA	Turbine, main – alarms and trips

TMB	Turbine, main control
TME	Turbine, main – gland seal and exhaust system
TMG	Turbine, main – turn gear
TMH	Turbine, main – hydraulic gear
TMI	Turbine, main, supervisory instrumentation
TMJ	Turbine, main – jacking oil
TML	Turbine, main, lube oil
TMR	Turbine, main – unit run back
TMS	Turbine, main system
TPW	Total process water (makeup) – FGD
TSW	Sanitary waste treatment system
TSY	Fuel gas (treated syngas)
USS	Unit substation
USY	Untreated syngas
VAS	Vents (aerated) – system (nuclear)
VBA	Vital bus – AC supply (to bus)
VBB	Vital bus – DC supply (to inverter)
VBN	Vital bus, normal
VBS	Vital bus system (incoming supply from inverter)
VNT	Vents – turbine and reactor building
VPS	Vacuum priming (water side) – system
VRS	Vents gaseous reactor plant – system
VRW	Vents – radioactive waste building equipment
VTP	Vents, turbine plant equipment
WCS	Reactor water cleanup – system
WDD	Demineralized water
WDS	Waste disposal system – FGD
WHM	Waste handling area maintenance equipment – FGD
WLT	Liquid waste – tritiated
WOS	Waste oil disposal – system
WOW	Oily water sewer
WPS	Welding power – system
WSC	Solid waste – condensate demineralizer system
WSR	Service water
WSS	Solid waste (radioactive) – system
WTA	Water treated – sulfuric acid
WTC	Water treating – chlorination
WTF	Water treating – upflow filter
WTH	Water treating – hypochlorite
WTL	Water treating – clarifier
WTP	Waste treatment – condensate polishing
WTS	Water treating – system
WTU	Water treatment – fossil station
WTW	Water treating – chemical waste
WUS	Warm up oil – system

WV	Service water – spherical valve seal
WWF	Waste water – fresh
YDK	Yardwork
YPB	13.8 kV switchyard/substation – bus diff/PT's/MET/oscillography
YPC	13.8 kV switchyard/substation – PCB control/aux/breaker fail/mod SW
YPF	13.8 kV switchyard/substation common breaker failure
YPL	13.8 kV switchyard/substation – line relay/met/mod switch/ground switch
YPS	13.8 kV switchyard/substation – synchronizing
YPT	13.8 kV switchyard/substation – transformer aux/tap changer
YPU	13.8 kV switchyard/substation – yardwork
YPV	13.8 kV switchyard/substation – substructure
YPW	13.8 kV switchyard/substation – superstructure
YRB	34 kV switchyard/substation – bus diff/PT's/met oscillography
YRC	34 kV switchyard/substation – PCB control/aux breaker fail
YRF	34 kV switchyard/substation common breaker failure
YRL	34 kV switchyard/substation – line relay/met/mod switch/ ground switch
YRS	34 kV switchyard/substation – synchronizing
YRT	34 kV switchyard/substation – transformer aux/tap changer
YRU	34 kV switchyard/substation – yardwork
YRV	34 kV switchyard/substation – substructure
YRW	34 kV switchyard/substation – superstructure
YTB	69 kV switchyard/substation – bus diff/PT's/met oscillography
YTC	69 kV switchyard/substation – PCB control/aux breaker fail/mod SW
YTF	69 kV switchyard/substation common breaker failure
YTL	69 kV switchyard/substation – line relay/met/mod switch/ground switch
YTS	69 kV switchyard/substation – synchronizing
YTT	69 kV switchyard/substation – trans aux/tap changer
YTU	69 kV switchyard/substation – yardwork
YTV	69 kV switchyard/substation – substructure
YTW	69 kV switchyard/substation – superstructure
YVB	138 kV switchyard/substation – bus diff/PT's/met/oscillography
YVC	138 kV switchyard/substation – PCB control/aux/breaker fail/mod SW
YVF	138 kV switchyard/substation common breaker failure
YVL	138 kV switchyard/substation – line relay/met/mod switch/ground switch
YVS	138 kV switchyard/substation – synchronizing
YVT	138 kV switchyard/substation – trans aux/tap changer
YVU	138 kV switchyard/substation – yardwork
YVV	138 kV switchyard/substation – substructure
YVW	138 kV switchyard/substation – superstructure
YWB	230 kV switchyard/substation – bus diff/PT's/met oscillography
YWC	230 kV switchyard/substation – PCB control/aux/breaker fail/mod SW
YWF	230 kV switchyard/substation common breaker failure
YWL	230 kV switchyard/substation – Line relay/met/mod switch/ground switch
YWS	230 kV switchyard/substation – synchronizing
YWT	230 kV switchyard/substation – trans aux/tap changer

YWU	230 kV switchyard/substation – yardwork
YWV	230 kV switchyard/substation – substructure
YWW	230 kV switchyard/substation – superstructure

ANNEX D   COMMON EQUIPMENT CODES FOR POWER GENERATION FACILITIES

(Reprinted with permission of Tampa Electric Company. Copyright 2016 Tampa Electric Company. Reproduction is forbidden without the express consent of Tampa Electric Company)

Equipment Code	Equipment Description
ACB	Circuit breaker, air
ACC	Accumulator
ACD	Air control device
ACH	Chamber, air
ACT	Actuator
ACU	Air conditioning unit
ACUS	Air conditioning unit, self-contained
AD	Anode
ADS	Adsorber
AF	Arrestor, flame
AGIT	Agitator
AHU	Air handling unit
AMPL	Amplifier
ANT	Antenna
AOD	Damper, air-operated
AOL	Louver, air-operated
AOV	Valve, air-operated
AP	Panel, access
APH	Preheater, air
ARV	Air release valve
ASP	Separator, air
B	Boiler
BAT	Battery
BD	Bus duct
BIN	Storage bin
BKR	Breaker
BLOW	Blower
BL	Blender
BO	Soil boring
BFD	Potential device, bushing
BTC	Test cabinet, breaker

BURN	Burner
BUS	Bus
C	Compressor
CAB	Cabinet
CAPB	Cabinet, penetration
CAP	Capacitor (electrical)
CB	Catch basin
CBV	Valve, continuous blowdown
CCT	Test, concrete compression
CH	Coil, heater
CHGR	Charger, battery
CHL	Chiller, liquid
CHUT	Chute
CIV	Combined intermediate valve
CLC	Coil, cooling
CLR	Clarifier
CN	Centrifuge
CND	Condenser (mechanical)
CNDT	Conditioner
CNSL	Console
CNV	Conveyor
CO	Cleanout
COL	Trayed or packed columns
COLL	Collector
COOL	Cooler
CP	Computer
CPD	Potential device, coupling type
CR	Crystallizer
CRH	Crusher
CRN	Crane or hoist
CSK	Cask
CUB	Cubicle, electrical
CUR	Condensing unit, refrigerant
CVV	Conveyance vehicle
CWS	Seal, water cavity
CYL	Cylinder
D	Drum
DA	Deaerator
DCS	Cooler, drain
DEH	Dehumidifier
DEMN	Demineralizer
DESH	Desuper-heater
DIFF	Diffuser
DIV	Diverter
DMP	Damper

DMPF	Damper, fire
DMST	Demister
DNF	Drain, floor
DNR	Drain, roof
DO	Door
DRA	Door, access
DRY	Dryer
DTH	Test hole, duct
DWC	Cooler, drinking water
E	Exchanger, heat
EA	Air fan exchangers
ED	Drain, equipment
EDU	Eductor
EFV	Valve, excess flow
EG	Engine generator
EJ	Joint, expansion
EL	Elevator
ENC	Encapsulation
ENG	Engine
EV	Evaporator
EX	Turbo-expander
EXC	Exciter
FCD	Cleanout, floor
FDR	Feeder
FHR	Reel, fire hose
DHY	Hydrant, fire
FL	Flange
FLI	Flange, insulating
FLS	Flange, spectacle
FLT	Filter
FLTM	Filter, motor-operated
FLTT	Filter, air, temporary
FLU	Filter train
FN	Fan
FO	Orifice, flow
FP	Connection, flow
FPV	Fire protection valve
FSU	Splitter unit, flow
FTG	Fitting, pipe or sheet metal duct
FUN	Funnel
FUR	Furnace
FX	Vanes, straightening
G	Generator
GA	Equipment, garage
Gate	Gate

GCR	Gas recycle unit, cryogenic
GDR	Grinder/grinding mill
GEAR	Gear, drive
GL	Generator leads
GLC	Cooler, generator leads
GRL	Grille
GRU	Gas removal unit
H	Heater
HD	Hood
HDS	Switch, disconnect, hand-operated
HDST	Hand or headset
HH	Hand-hole
HORN	Sonic horns
HOSE	Hose, all types
HR	Rack, hose
H/A	Switch, control, hand automatic
HS	Switch, hand
HSS	Switch, hand selector
HUM	Humidifier
HVU	Unit, heating and ventilating
IB	Inlet basin
ICV	Connection, vacuum cleaning
IND	Inductor
IOE	Exchanger, ION
IPB	Bus, isolated phase
IPNL	Panel, instrument
J	Ejector
JB	Box, junction
JIS	Jet impingement shield
JK	Jack, jack panel
KIE	Equipment, kitchen
KYBD	Keyboard
LAV	Lavatory
LDC	Load center
LFT	Lifting equipment
LH	Lock hopper
LNR	Liner
LTA	Arrestor, lightning
LTR	Rod, lightning
LTRY	Rod, lightning, switchyard
LTU	Tuner, line
LVR	Louver, non-powered
LVRE	Louver electrical
M	Motor, electric
MCB	Main control board

MCC	Motor control center
MCS	Motor control center, special
MCV	Main control valve, turbine
MD	Metal detector
MDS	Switch, disconnect, motor-operated
MFD	Manifold
MG	Motor generator
MH	Man hole
MIKE	Microphone
MIX	Mixer
MIXB	Box, mixing
MIXP	Mixer, forwarding
MOD	Damper, motor-operated
MOL	Louver, motor-operated
MOV	Valve, motor-operated
MS	Magnetic separator
MSP	Moisture separator
MST	Starter, magnetic
MSV	Main stop valve, turbine
NOZ	Nozzle
NRV	Valve, non-return
NSA	Assembly, neutron source
OCB	Circuit breaker, oil
ODR	Odorizer
OSC	Oscillograph
OV	Sight overflow
OWR	Oxygen and water removal unit
P	Pump
PB	Starter, manual
PH	Pipe hanger
PHD	Pothead
PIL	Pilings foundation
PKG	Package unit
PLAT	Platform
PLC	Pole, communications
PLP	Pole, power
PNL	Panel
PNLL	Panel, lighting electrical
PNLD	Panel, distribution electrical
PP	Piping
PRE	Pressurizer
PREC	Precipitator
PRR	Pipe rupture restraint (snubbers)
PSA	Pipe support anchor
PSG	Pipe support guide

PSR	Pipe support restraint
PSRH	Pipe support rod
PSS	Pipe support sliding
PSSH	Pipe support spring
PSSP	Pipe support suppressor
PSST	Pipe support strut
PULV	Pulverizer (coal)
PUR	Purifier
PWRS	Power supply
R	Register
RAK	Rack or storage stand
RBNR	Recombiner
RBRS	Radial shear bar assemblies
RBW	Weld reinforcing bar
RCP	Reactor containment penetration
RCPT	Power receptacle
RDMA	Rod drive mechanism, part length
RDMF	Rod drive mechanism, full length
RECT	Rectifier
REF	Refrigeration unit
REG	Regulator
REN	Regenerator
RES	Resistor
REV	Reactor vessel
RFV	Ventilator, roof, gravity
RHE	Rheostat
RIV	Reheat intercept valve, turbine
RLY	Relay
RODA	Reactor rod, part length
RODF	Reactor fod, full length
RODP	Reactor fod, poisoning
RSV	Reheat stop valve, turbine
RSY	Thermal oxidizer
RTB	Restricting tube
S	Unit substation
SAS	Sink, sample
SCA	Scale, weighing
SCAC	Self-cleaning air cleaner
SCB	Basket, collection, trash, screen well
SCL	Cooler, sample
SCR	Dual flow traveling screen
SER	Reservoir, evaporation suppression
SEW	Safety eye wash
SG	Steam generator
SH	Super-heater, electric or oil-fired

SHLD	Shielding
SHRD	Rod drive mechanism, air cooling shroud
SIL	Silencer (inlet or exhaust)
SILO	Silo
SIRN	Siren
SKD	Skid
SKIM	Skimmer
SL	Sling
SLU	Lift unit, sewerage
SMPT	Sample point
SO	Screened opening
SOT	Test, soil compaction
SOV	Valve, solenoid-operated
SP	Separator
SPEC	Piping special, water-cooled inlet blind
SPKR	Loud speaker
SRK	Rakes, screen well
SS	Sampling system
SSC	Screen, traveling
SSH	Shower
SSK	Sink, service
SSP	Surge suppressor, volume chamber
SSR	Rack, screen well
STAK	Stack or chimney
STG	Steam turbine generator
STR	Strainer, permanent
STRF	Strainer, system flush
STRT	Strainer, temporary
STU	Treatment unit, sewerage
SUMP	Sump
SUP	Support, equipment
SW	Switch
SWC	Switch, indication and control
SWG	Switchgear
SWS	Switch, safety
T	Turbine
TB	Box, terminal
TBK	Block, terminal
TBL	Lugs, terminal
TD	Temperature damper
TEL	Telephone
TFT	Tube, fuel transfer
TG	Turbine, gas
TH	Thermostat
TK	Tank

TL	Tool
TP	Temperature point
TRP	Trap
TRS	Switch, transfer
TVC	Television camera
TVM	Television monitor
TW	Thermowell
TWR	Tower
TX	Power controller, modulating
UC	Cooler, unit
UH	Heater, unit
UHE	Heater, unit, electric
URN	Urinal
US	Unit substation
UV	Unit ventilator
V	Valves (non-powered)
VAR	Reactive power
VEX	Valve, explosive
VT	Wave trap, electrical
WB	Baler (solid waste)
WC	Water closet
WCH	Cement hopper (solid waste)
WDR	Drum roller (solid waste)
WDW	Window
WF	Wash fountain
WH	Whistle
WVT	Wave trap, electrical
WW	Wireway (raceway)
X	Transformer
XA	Transformer auxiliaries
XC	Transformer current
XCA	Transformer current auxiliary
XD	Transformer distribution
XG	Transformer grounding
XL	Transformer lighting
XM	Transformer main
XNS	Transformer normal station service
XP	Transformer potential
XRC	Transformer regulating control
XS	Transformer station service
XSR	Transformer station reserve
Z	Penetration

ANNEX E   UNIT CONVERSION FACTORS

Useful Unit Conversion Factors

To Obtain	Multiply	By
area of circle	(diameter of circle)^2	0.7854
area of sphere	(diameter of sphere)^2	3.1416
atmospheres	cm of Hg @ 0°C	0.013158
atmospheres	ft of H_2O @ 39.2°F	0.029499
atmospheres	in. of Hg @ 32°F	0.033421
atmospheres	in. of H_2O @ 39.2°F	0.0024583
atmospheres	pounds per square foot	0.00047254
atmospheres	pounds per square inch	0.068046
BTU	ft lb	0.0012854
BTU	HP hr	2545.1
BTU	kWh	3413
BTU	Wh	3.413
BTU/hr	mech HP	2545.1
BTU/hr	kW	3413
BTU/hr	watts	3.4137
BTU/kWh	kg cal/kWh	3.9685
BTU/min	ft lb/min	0.0012854
BTU/min	mech HP	42.418
BTU/min	kW	56.896
BTU/sec	mech HP	0.70696
BTU/sec	kg cal/hr	0.0011024
BTU/sec	kW	0.94827
calories	ft lb	0.32389
calories	Wh	860.01
cal/gram	BTU/lb	0.55556
centimeters	inches	2.54
cm of Hg @ 0°C	atmospheres	76
cm of Hg @ 0°C	ft of H_2O @ 39.2°F	2.242
cm of Hg @ 0°C	grams/sq cm	0.07356
cm of Hg @ 0°C	in. of H_2O @ 4°C	0.1868
cm of Hg @ 0°C	lb/sq in.	5.1715
cm of Hg @ 0°C	lb/sq ft	0.035913
cm/sec	ft/min	0.508
cm/sec	ft/sec	30.48
cm of H_2O @ 39.2°F	atmospheres	1033.24
cm of H_2O @ 39.2°F	lb/sq in.	70.31
circumference of circle	diameter of circle	3.1416
cu ft	cu meters	35.314
cu ft	cu yards	27
cu ft	gal (USA, liq)	0.13368

cu ft	liters	0.03532
cu ft/min	cu meters/sec	2118.9
cu ft/min	gal (USA, liq) /sec	8.0192
cu ft/lb	cu meters/kg	16.02
cu ft/lb	liters/kg	0.01602
cu ft/sec	cu meters/min	0.5886
cu ft/sec	gal (USA, liq) /min	0.002228
cu ft/sec	liters/min	0.0005886
cu in.	cu centimeters	0.061023
cu in.	gal (USA, liq)	231
cu in.	liters	61.03
cu meters	cu ft	0.028317
cu meters	cu yards	0.7646
cu meters	gal (USA, liq)	0.0037854
cu meters	liters	0.001000028
cu meters/hr	gal/min	0.22712
cu meters/kg	cu ft/lb	0.062428
cu meters/min	cu ft/min	0.02832
cu meters/min	gal/sec	0.22712
cu meters/sec	gal/min	0.000063088
cu yards	cu meters	1.3079
degrees of arc	radians	57.296
feet	meters	3.281
ft of H_2O @ 39.2°F	atmospheres	33.899
ft of H_2O @ 39.2°F	cm of Hg @ 0°C	0.44604
ft of H_2O @ 39.2°F	in. of Hg @ 32°F	1.133
ft of H_2O @ 39.2°F	lb/sq ft	0.016018
ft of H_2O @ 39.2°F	lb/sq in.	2.3066
ft/min	cm/sec	1.9685
ft/min	km/hr	54.68
ft/min	miles (USA, statute)/hr	88
ft/sec	km/hr	0.9113
ft/sec	meters/sec	3.2808
ft/sec	miles (USA, statute)/hr	1.4667
ft/sec sec	km/hr sec	0.91133
ft/sec sec	meters/sec sec	3.2808
ft lb	BTU	778
ft lb	kg meters	7.233
ft lb	kWh	2655200
ft lb	mech HP hr	1980000
ft lb/min	BTU/min	778
ft lb/min	kg cal/min	3087.4
ft lb/min	kW	44254
ft lb/min	mech HP	33000
ft lb/sec	BTU/min	12.96

ft lb/sec	kW	737.56
ft lb/sec	mech HP	550
gal (USA, liq)	barrels (petroleum, USA)	42
gal (USA, liq)	cu ft	7.4805
gal (USA, liq)	cu meters	264.173
gal (USA, liq)	cu yards	202.2
gal (USA, liq)	gal (Imperial, liq)	1.201
gal (USA, liq)	liters	0.2642
gal (USA, liq)/min	cu ft/sec	448.83
gal (USA, liq)/min	cu meters/hr	4.4029
gal (USA, liq)/sec	cu ft/min	0.12468
gal (USA, liq)/sec	liters/min	0.0044028
grams	ounces	28.35
grams	pounds	453.5924
grams/cm	pounds/in.	178.579
grams/cu cm	lb/cu ft	0.016018
grams/cu cm	lb/cu in.	27.68
grams/cu cm	lb/gal	0.119826
inches	centimeters	0.3937
in. of Hg @ 32°F	atmospheres	29.921
in. of Hg @ 32°F	ft of H_2O @ 39.2°F	0.88265
in. of Hg @ 32°F	kg/sq meter	0.0028959
in. of Hg @ 32°F	lg/sq in.	2.036
in. of Hg @ 32°F	in. of H_2O @ 4°C	0.07355
in. of H_2O @ 4°C	in. of Hg @ 32°F	13.6
in. of H_2O @ 39.2°F	lb/sq in.	27.673
kg	pounds	0.45359
kg cal	BTU	0.252
kg cal	ft lb	0.00032389
kW	BTU/min	0.01758
kW	ft lb/min	0.00002259
kW	ft lb/sec	0.00135582
kW	mech HP	0.7457
kWh	BTU	0.000293
kWh	ft lb	3.766E-07
kWh	mech HP hr	0.7457
liters	gal (USA, liq)	3.78533
mech HP	BTU/hr	0.0003929
mech HP	BTU/min	0.023575
mech HP	ft lb/sec	0.0018182
mech HP	kW	1.341
mech HP hr	BTU	0.00039292
mech HP hr	ft lb	5.0505E-07
mech HP hr	kWh	1.341
meters	ft	0.3048

meters	in.	0.0254
meters	miles (USA, statute)	1609.34
meters/min	ft/min	0.3048
meters/min	miles (USA, statute)/hr	26.82
meters/sec	ft/sec	0.3048
meters/sec sec	ft/sec sec	0.3048
ounces	grams	0.035274
ounces (USA, liq)	gal (USA, liq)	128
pounds	grams	0.0022046
pounds/cu ft	pounds/gal	7.48
pounds/cu in.	grams/cu cm	0.036127
pounds/in.	grams/cm	0.0056
pounds/sq in.	atmospheres	14.696
pounds/sq in.	cm of Hg @ 0°C	0.19337
pounds/sq in.	ft of H_2O @ 39.2°F	0.43352
pounds/sq in.	in. of Hg @ 32°F	0.491
pounds/sq in.	in. of H_2O @ 39.2°F	0.0361
pounds/sq in.	kg/sq cm	14.223
pounds/sq ft	kg/sq meter	0.20481
pounds/sq in.	kg/sq meter	0.0014223
pounds/gal (USA, liq)	kg/liter	8.3452
pounds/gal (USA, liq)	pounds/cu ft	0.1337
pounds/gal (USA, liq)	pounds/cu in.	231
quarts (USA, liq)	cu cm	0.0010567
quarts (USA, liq)	cu in.	0.01732
quarts (USA, liq)	liters	1.057
radians	degrees of arc	0.017453
sq in.	circular inches	0.7854
sq in.	circular mils	7.854E-07
volume of sphere	(diameter of sphere)^3	0.5236
watts	BTU/sec	1054.8
yards	meters	1.0936

ANNEX F   SOLUTIONS TO PROBLEMS

Chapter 1

1.1 What are the four minimum steps to establish an electrically safe work condition?

   A. _____

   B. _____

   C. _____

   D. _____

Solution:

Isolate all sources of electrical energy

Apply Lockout/tag out device

Verify absence of voltage

Ground phase conductors where possibility of induced voltage exists

1.2 What activities below constitute energized electrical work?

A. Taking voltage readings using a contact voltage meter on a panel that has just been isolated from the source but not yet locked out

B. Racking out circuit breaker from energized bus

C. Removal of MCC buckets from energized bus

D. All the above

Solution:
D (All the above).

1.3 A task has been assigned to remove a breaker bucket from a 480 V three-phase MCC where there is a bolted fault current of 35 kA and the clearing time is 5 cycles, can you use the tables in NFPA 70E for the determination of the correct arc flash PPE for this task? If you can, what is the arc flash PPE level for this task?

Solution:
You can use the tables as this condition is covered in NFPA70E Table 130.7(C) (15)(a)(b) in the fourth row where the maximum fault current is 42 kA and the maximum clearing time is 20 cycles. The arc flash PPE Category is 4 and the arc flash boundary is 14 ft.

1.4 A task has been assigned to remove a breaker from a 125 Vdc switchboard panel where there is a bolted fault current of 2 kA and the clearing time is 5 cycles, can you use the tables in NFPA 70E for determination of the correct arc flash PPE for this task? If you can, what is the arc flash PPE level for this task?

Solution:
You can use the tables as this condition is covered in NFPA70E Table 130.7(C)(15)(B) in the second row where the maximum fault current is 4 kA and the maximum clearing time is 2 sec. The arc flash PPE Category is 1 and the arc flash boundary is 3 ft.

1.5 A task has been assigned to verify the phasing of a three-phase, 22 kV circuit where the bolted fault current is 15.1 kA and the clearing time is 5 cycles. What is the arc flash PPE level that is required for this task?

Solution:
You can use the NESC tables (Table 1.13 of this text) to determine the correct arc flash PPE level. Looking at the 15.1–25 kV row, and finding the next higher fault current level over 15.1 kA, we see that for a 4 cal system, the maximum tolerable clearing time is 4.4 cycles and for an 8 cal system, the maximum tolerable clearing time is 8.8 cycles. Since our task involves a clearing time of 5 cycles, we must use at minimum the 8 cal level arc flash PPE to perform this task.

1.6 Given the following activity, using Table 130.4(C) of NFPA70E what is the limited approach boundary and the restricted approach boundary in feet and inches? The activity is racking out a three-phase, 480 V line-to-line, power circuit breaker with doors open from an energized switchgear bus; therefore, the exposed bus is fixed and not movable.

Limited approach boundary _____

Restricted approach boundary _____

Solution:

Limited approach boundary = 3 ft, 6 in.

Restricted approach boundary = 1 ft, 0 in.

1.7 Given the following activity, using Table 130.4(D) of NFPA70E what is the limited approach boundary and the restricted approach boundary in feet and inches? The activity is to perform voltage testing of a 250 Vdc panel board circuit breaker with doors open from an energized switchgear bus, therefore the exposed bus is fixed and not movable.

Limited approach boundary _____

Restricted approach boundary _____

Solution:

Limited approach boundary = 3 ft, 6 in.

Restricted approach boundary = Avoid contact.

1.8 An area where flammable gases or vapors can exist under normal operating conditions is an example of

A. Class I, Division 1 location

B. Class I, Division 2 location

C. Class II, Division 1 location

D. Class II, Division 2 location

E. Class III, Division 1 location

F. Class III, Division 2 location

Solution:

A (Class I, Division 1 location).

1.9 Using Table 1.1 in text for a situation where there is a 480 V MCC assembly across from a concrete wall on opposite side of switchgear, what is the minimum aisle space required between the assembly and the concrete wall?

A. 3 ft

B. $3\frac{1}{2}$ ft

C. 4 ft

Solution:

B ($3\frac{1}{2}$ ft).

(*Note*: Concrete is considered a conductive surface.)

1.10 An energized electrical work permit is required to perform energized work. What are some of the items that should be listed in the energized electrical work permit? (Name at least five)

_____

_____

_____

_____

_____

Solution:

The answer is at least five of the items listed below.

1. A description of the circuit and equipment to be worked on and their location

2. Justification for why the work must be performed in an energized condition

3. A description of the safe work practices to be employed

4. Results of the shock hazard analysis

5. Determination of shock protection boundaries

6. Results of the flash hazard analysis

7. The flash protection boundary

8. The necessary personal protective equipment to safely perform the assigned task

9. Means employed to restrict the access of unqualified persons from the work area

10. Evidence of completion of a job briefing, including a discussion of any job-specific hazards

11. Energized work approval from authorizing or responsible management, safety officer, or owner

1.11 The motor controller that drives a coal pulverizer motor for a power generation facility has been deemed as a location where, under normal operations, coal dust will not be present suspended in air in sufficient quantities to form an explosive concentrations and due to housekeeping controls, the layer of dust will not be sufficient to prevent cooling of the motor; however, on a failure of one of the multiple seals in the area, coal dust can become suspended in the atmosphere in quantities sufficient to form an explosive concentration should those seals fail. What are the correct Class and Division for this hazardous area classification?

Solution:

This would be a Class 2 (combustible dust), Division 2 (not explosive under normal operation, but may form explosive mixture under abnormal operation).

Chapter 2

2.1 A combined cycle power plant utilizes two thermal cycles to drive the two generators. One generator is driven by a combustion turbine and the other generator is driven by a steam turbine. What are the names of the two thermal cycles that are referenced by the term "combined cycle?"

A. Rankine – Carnot

B. Rankine – Brayton

C. Carnot – Brayton

Solution:
B (Rankine – Brayton).

2.2 Complete this sentence. Overall power plant thermal efficiency will decrease if…

   A. the steam temperature entering the turbine is increased

   B. the temperature of the feedwater entering the steam generator is increased

   C. the amount of condensate depression (subcooling) in the main condenser is decreased

   D. the temperature of the steam at the turbine exhaust is increased

Solution:
D (the temperature of the steam at the turbine exhaust is increased).

2.3 The theoretical maximum efficiency of a steam cycle is given by equation (2.24). A power plant is operating with a stable steam generator pressure of 900 psia at saturated steam conditions with a saturation temperature of 532.02°F. What is the approximate theoretical maximum steam cycle efficiency this plant can achieve by establishing its main condenser vacuum at 1.0 psia at saturated steam conditions with a saturation temperature of 101.69°F? (*Hint*: Assume both the steam generator and condenser are at saturated steam conditions. Use ASME steam tables and assume saturated steam conditions in the steam generator and condenser to determine the temperature of the working fluid in the two states.)

   A. 35%

   B. 43%

   C. 57%

   D. 65%

Solution:
Using equation (2.24),

$$\text{Eff}_{th\,max} = (1 - T_{out}/T_{in}) \times 100\%$$

From saturated steam tables,
900 psia has a saturation temperature of 532.02°F
Converting to Rankine,

$$T_R = T_F + 460 = 532.02 + 460 = 992.02°R$$

1 psia has a saturation temperature of 101.74°F
Converting to Rankine,

$$T_R = T_F + 460 = 101.69 + 460 = 561.69°R$$
$$\text{Eff}_{th\,max} = (1 - (561.69°R/992.02°R)) \times 100\%$$
$$\text{Eff}_{th\,max} = 43.38\%$$

The answer is B (43%).

2.4 Main condenser pressure is 1.0 psia. During the cooling process in the condenser, the temperature of the low pressure turbine exhaust decreases to 100°F, at which time it is a…

   A. saturated liquid

   B. saturated vapor

C. subcooled liquid

D. superheated vapor

Solution:

The saturation temperature at 1.0 psia is 101.69°F. Since actual temperature is less than saturation temperature, the media is in subcooled form.

The answer is C (subcooled liquid).

2.5 A liquid is saturated with 0% quality. Assuming pressure remains constant, the addition of a small amount of heat will…

A. raise the liquid temperature above the boiling point

B. result in a subcooled liquid

C. result in vaporization of the liquid

D. result in a superheated liquid

Solution:

C (result in vaporization of the liquid).

2.6 Which one of the following is the approximate steam quality of a steam-water mixture at 250°F with an enthalpy of 1000 BTU/lbm?

A. 25%

B. 27%

C. 83%

D. 92%

Solution:

We can rearrange equation (2.19) for steam quality as shown below.

$$h = h_1 + h_{lv} \times (\%SQ/100\%)$$
$$\%SQ = 100\% \times (h - h_f)/h_{fg}$$

We find from the steam table that the enthalpy of 0% steam quality saturated steam at 250°F is 218.62 BTU/lbm. We find from the steam table that the enthalpy of 100% steam quality saturated steam at 250°F is 1164 BTU/lbm. Utilizing equation (2.17), the change in enthalpy from 0% quality to 100% quality is

$$h_{lv} = (h_v - h_1)$$
$$h_{lv} = (1164 - 218.62) \text{ BTU/lbm}$$
$$h_{lv} = 945.38 \text{ BTU/lbm}$$

Now we can rearrange for steam quality as shown below.

$$h = h_1 + h_{lv} \times (\%SQ/100\%)$$
$$\%SQ = (h - h_1)/h_{lv} \times 100\%$$
$$\%SQ = 100\% \times (1000 - 218.62)/945.38$$
$$\%SQ = 82.65\%$$

The answer is C (83%).

2.7 If a saturated vapor is at 205°F and has a steam quality of 90%, its specific enthalpy is approximately…

A. 173 BTU/lbm.

B. 271 BTU/lbm.

C.   1050 BTU/lbm.

D.   1147 BTU/lbm.

Solution:

Using equation (2.19), we can determine the enthalpy of a saturated water-steam mixture with a steam quality of 90%.

We find from the steam table that the enthalpy of 0% steam quality saturated steam at 205°F is 173.13 BTU/lbm. We find from the steam table that the enthalpy of 100% steam quality saturated steam at 205°F is 1147.6 BTU/lbm. Utilizing equation (2.17), the change in enthalpy from 0% quality to 100% quality is

$$h_{lv} = (h_v - h_l)$$
$$h_{lv} = (1147.6 - 173.13) \quad \text{BTU/lbm}$$
$$h_{lv} = 974.47 \text{ BTU/lbm}$$

Now we can determine the enthalpy at 90% steam quality utilizing equation (2.19).

$$h = h_l + h_{lv} \times (\%\text{SQ}/100\%)$$
$$h = 173.13 \text{ BTU/lbm} + 974.47 \text{ BTU/lbm} \times (90\%/100\%)$$
$$h = A.1050 \text{ BTU/lbm}$$

The answer is C (1050 BTU/lbm).

2.8   If steam pressure is 230 psia is at a temperature of 900°F, what is the approximate amount of superheat? (Given the saturation temperature of saturated steam at 230 psia is 393.71°F.)

A.   368°F

B.   393°F

C.   506°F

D.   535°F

Solution:

From saturated steam pressure table (or as given above) we see the saturation temperature associated with 230 psia is 393.71°F. The amount of superheat is simply the difference between the actual steam temperature and the saturation temperate of the steam at that pressure. Therefore,

$$\text{SH} = 900°\text{F} - 393.71°\text{F}$$
$$\text{SH} = 506.29°\text{F}$$

The answer is C (506°F).

2.9   Which one of the following is the approximate amount of thermal energy required to convert 2 lbm of water at 100°F and 100 psia to a saturated vapor at 100 psia? (Given the value of enthalpy of subcooled water at 100°F and 100 psia is 68.3 BTU/lbm and the value of enthalpy of saturated vapor at 100% steam quality, and 100 psia is 1187.5 BTU/lbm.)

A.   560 BTU

B.   1120 BTU

C.   2238 BTU

D.   3356 BTU

Solution:

The enthalpy of water at 100°F and 100 psia (subcooled liquid) is 68.3 BTU/lbm. The enthalpy of saturated steam at 100% steam quality at 100 psia is 1187.5 BTU/lbm. The difference is the energy needed to raise 1 lbm of material's energy level.

$$1187.5 \text{ BTU/lbm} - 68.3 \text{ BTU/lbm} = 1119.2 \text{ BTU/lbm}$$

Since we have 2 lbm of material, the amount of energy is the enthalpy times the mass or

$$E = h \times m = 1119.2 \text{ BTU/lbm} \times 2 \text{ lbm}$$
$$E = 2238 \text{ BTU.}$$

So the answer is C (2238 BTU).

2.10 In the basic heat cycle there are four processes, compression, expansion, evaporation, and condensation. Referring Figure 2.4, these four processes are represented by four lines. Which line presents the compression process of the basic steam cycle?

A. Line 1–2.

B. Line 2–3.

C. Line 3–4.

D. Line 4–1.

Solution:

D (Line 4–1).

2.11 In the basic heat cycle shown in Figure 2.4, the heat into the system (Q_{in}) is defined as which of these blocks?

A. Box 1, 2, 3, 4, 1.

B. Box 1, 2, 5, 6, 1.

C. Box 4, 3, 5, 6, 4.

Solution:

B (Box 1, 2, 5, 6, 1).

2.12 Given a temperature of 0°F, what is this temperature in the Rankine scale?

Solution:

Using equation (2.1), where $T_F = 0°F$

$$T_R = T_F + 460°F$$
$$T_R = 0°F + 460°F$$
$$T_R = 460°R.$$

2.13 Condensate depression is the process of...

A. removing condensate from turbine exhaust steam

B. spraying condensate into turbine exhaust steam

C. heating turbine exhaust steam above its saturation temperature

D. cooling turbine exhaust steam below its saturation temperature

Solution:

D (cooling turbine exhaust steam below its saturation temperature).

2.14 The law that states that any thermodynamic process is irreversible as the net entropy of a system and its surroundings always increases is which law?

A. The first law of thermodynamics

B. The second law of thermodynamics

C. Ohm's Law

D. Stuart's Law

Solution:
B (The second law of thermodynamics).

2.15 Given the following values for compressor inlet and outlet isentropic and non-isentropic enthalpies, determine the compressor efficiency.

$$h_1 = 250 \text{ BTU/lbm}$$
$$h_2 = 1250 \text{ BTU/lbm}$$
$$h_{2S} = 1000 \text{ BTU/lbm}$$

Solution:
Using equation (2.40), we find

$h_{comp} = (h_{2S} - h_1)/(h_2 - h_1)$
$h_{comp} = (1000 \text{ BTU/lbm} - 250 \text{ BTU/lbm})/(1250 \text{ BTU/lbm} - 250 \text{ BTU/lbm})$
$h_{comp} = 750 \text{ BTU/lbm}/1000 \text{ BTU/lbm}$
$h_{comp} = 0.75 \text{ or } 75\%.$

2.16 Given the following values for turbine inlet and outlet isentropic and non-isentropic enthalpies, determine the compressor efficiency.

$$h_3 = 3000 \text{ BTU/lbm}$$
$$h_4 = 1000 \text{ BTU/lbm}$$
$$h_{4S} = 700 \text{ BTU/lbm}$$

Solution:
Using equation (2.41) we find

$h_{comp} = (h_3 - h_4)/(h_3 - h_{4S})$
$h_{turb} = (3000 \text{ BTU/lbm} - 1000 \text{ BTU/lbm})/(3000 \text{ BTU/lbm} - 700 \text{ BTU/lbm})$
$h_{turb} = 2000 \text{ BTU/lbm}/2300 \text{ BTU/lbm}$
$h_{turb} = 0.87 \text{ or } 87\%.$

Chapter 3

3.1 In a balanced draft boiler, the _____ fan is used to control boiler pressure (vacuum).

A. forced draft

B. induced draft

C. primary air

D. gas recirculation

Solution:
B (induced draft).

3.2 In a balanced draft boiler, the _____ is used to control the fuel flow.

   A. primary air fan

   B. forced draft fan

   C. induced draft fan

   D. pulverizer

   Solution:

   A (primary air fan).

3.3 Given a wet bulb temperature of 55°F and a relative humidity ratio of 0%, determine the dry bulb temperature.

   Solution:

   Using Figure 3.14, find the value of 55°F wet bulb temperature on the diagonal lines in the figure. Then find the humidity curve of 0% (*Hint*: notice that 0% humidity corresponds to the horizontal axis of Figure 3.14). Find the intersection for these two lines and the vertical dry bulb temperature that is associated with these two points is found to be 95°F.

3.4 The furnace of the boiler is where _____.

   A. coal is ground or pulverized

   B. water is converted to steam

   C. fuel and air is mixed and combusted

   Solution:

   C (fuel and air is mixed and combusted).

3.5 Complete this sentence:

   In a heat recovery steam generator, the superheater approach temperature is the difference in temperature between the gas _____ the _____ section and the steam _____ the _____ section of the steam generator.

   A. entering, superheater, leaving, superheater

   B. leaving, superheater, entering, superheater

   C. entering, economizer, leaving, economizer

   D. leaving, economizer, entering, economizer

   E. entering, evaporator, leaving, evaporator

   F. leaving, evaporator, entering, evaporator

   Solution:

   A (entering, superheater, leaving, superheater). See Figure 3.4.

3.6 Complete this sentence:

   In a heat recovery steam generator, the economizer approach temperature is the difference in temperature between the gas _____ the _____ section and the steam _____ the _____ section of the steam generator.

   A. entering, superheater, leaving, superheater

   B. leaving, superheater, entering, superheater

   C. entering, economizer, leaving, economizer

   D. leaving, economizer, entering, economizer

E. entering, evaporator, leaving, evaporator

F. leaving, evaporator, entering, evaporator

Solution:
C (entering, economizer, leaving, economizer). See Figure 3.4.

3.7 What is the function of a classifier?

Solution:
The function of the classifier is to allow the finer particles of coal to leave the classifier in the coal pipes and to be transported to the furnace burners. The larger particles of coal that require further processing are separated in the classifier and returned to the mill for further processing.

3.8 Complete this sentence:

A Downcomer is a tube or pipe in a boiler through which _____ flows downward

A. primary air

B. coal

C. fluid or water

D. secondary air

Solution:
C (fluid or water).

Chapter 4

4.1 What is the amount of heat energy released in BTU when 10 lb of Hydrogen is completely combusted releasing water?

Solution:
From Table 4.1, the heat released per pound mass for the complete combustion of 1 lbm of carbon is 61,000 BTU/lbm. The total heat released is the product of the heat per pound mass multiplied by the mass of material combusted.

$$E = 61,000 \text{ BTU/lbm} \times 10 \text{ lbm}$$
$$E = 610,000 \text{ BTU/lbm}.$$

4.2 What is the molecular weight of pentane (C_5H_{12})?

Solution:
One molecule of pentane (C_5H_{12}) consists of five carbon (C_5) and 12 hydrogen (H_{12}) atoms. Five carbon (C_5) atoms have a molecular weight of $12 \times 5 = 60$. Twelve hydrogen atoms (H_{12}) have a total molecular weight of $12 \times 1 = 12$. The total molecular weight for a pentane (C_5H_{12}) molecule is the sum of the individual components or $60 + 12 = 72$ moles.

4.3 To completely combust 100 lb of hydrogen sulfide (H_2S) gas requires how much minimum oxygen in pounds?

Solution:
From Table 4.4, the ratio of pounds of oxygen to pounds of hydrogen sulfide (H_2S) is 1.41/1.

Multiplying the mass of fuel by the ratio of oxygen to fuel give us

Pounds oxygen = 1.41 lb of oxygen/1 lb of hydrogen sulfide × 100 lb of hydrogen sulfide

Pounds oxygen = 141 lb of oxygen

Therefore, to combust 100 lb of hydrogen sulfide (H_2S) gas requires, at minimum, 141 lb of oxygen.

4.4   To completely combust 100 lb of hydrogen sulfide (H_2S) gas requires how much minimum air in pounds?

Solution:
From Table 4.4, the ratio of pounds of air to pounds of hydrogen sulfide (H_2S) is 6.12/1. Multiplying the mass of fuel by the ratio of air to fuel give us

Pounds oxygen = 6.12 lb of air/1 lb of hydrogen sulfide × 100 lb of hydrogen sulfide

Pounds oxygen = 612 lb of air

Therefore, to combust 100 lb of hydrogen sulfide (C_5H_{12}) gas requires, at minimum, 612 lb of air.

4.5   Which ASTM number oil is also known as Bunker C oil?

Solution:
ASTM #6 fuel oil.

4.6   What is the temperature range required to preheat ASTM #5 fuel oil at to ensure adequate viscosity?

Solution:
Typical preheat is between 160°F and 220°F.

4.7   What thermodynamic cycle is utilized in an integrated gasification combined cycle plant?

A.   Brayton

B.   Rankine

C.   Carnot

D.   Both Brayton and Rankine

Solution:
The thermal energy from the gasifier drives a steam turbine using the Rankine thermodynamic cycle and the syn-gas is combusted in a combustion turbine using the Brayton thermodynamic cycle, so the correct answer is D (Both Brayton and Rankine).

Chapter 5

5.1   Using equation (5.1), calculate the power developed (in HP) by a hydraulic turbine if the net head of water is 250 ft, turbine discharge flow is 1000 ft³/sec, and turbine efficiency is 0.8?

Solution:

Using equation (5.1), we calculate the power as follows

$$P = (H \times \eta \times h)/8.81$$
$$P = (250 \text{ ft}) \times (1000 \text{ ft}^3/\text{sec}) \times (0.8)/8.81$$
$$P = 22{,}700 \text{ HP.}$$

5.2   If the power developed by a turbine with a head of 200 ft is 20,000 HP and the head is reduced to 100 ft, but the runner discharge diameter remains constant, what is the new HP developed by the hydraulic turbine?

Solution:

Using equation (5.2), we find the relationship

$$P \propto D^2 \, H^{1.5}$$

In the problem statement head is changed, but runner diameter is held constant. Therefore, D does not change and equation (5.2) simplifies to

$$P \propto H^{1.5}$$

Putting this in equation form we find

$$(P_1/P_2) = (H_1/H_2)^{1.5}$$

Solving for P_2

$$P_2 = P_1 \times (H_1/H_2)^{-3/2}$$
$$P_2 = 20{,}000 \text{ HP} \times (200 \text{ ft}/100 \text{ ft})^{-3/2}$$
$$P_2 = 7071 \text{ HP.}$$

5.3   If the speed developed by a turbine with a head of 200 ft is 360 rpm and the head is reduced to 100 ft, but the runner discharge diameter remains constant, what is the new speed developed by the hydraulic turbine?

Solution:

Using equation (5.3) we find the relationship

$$n \propto H^{0.5}/D$$

In the problem statement head is changed, but runner diameter is held constant. Therefore, D does not change and equation (5.2) simplifies to

$$n \propto H^{0.5}$$

Putting this in equation form, we find

$$(n_1/n_2) = (H_1/H_2)^{1/2}$$

Solving for n_2

$$n_2 = n_1 \times (H_1/H_2)^{-1/2}$$
$$n_2 = 360 \text{ rpm} \times (200 \text{ ft}/100 \text{ ft})^{-1/2}$$
$$n_2 = 255 \text{ rpm.}$$

5.4   If the flow through a turbine with a head of 200 ft is 1000 ft³/s and the head is reduced to 100 ft, but the runner discharge diameter remains constant, what is the new flow through the hydraulic turbine?

Solution:

Using equation (5.4), we find the relationship

$$Q \propto D^2 H^{1.5}$$

In the problem statement head is changed, but runner diameter is held constant. Therefore, D does not change and equation (5.2) simplifies to

$$Q \propto H^{1.5}$$

Putting this in equation form we find

$$(Q_1/Q_2) = (H_1/H_2)^{1/2}$$

Solving for Q_2

$$Q_2 = Q_1 \times (H_1/H_2)^{-1/2}$$
$$Q_2 = 1000 \ ft^3/sec \times (200 \ ft/100 \ ft)^{-1/2}$$
$$Q_2 = 707 \ ft^3/sec.$$

Chapter 6

6.1 You have four radioactive cookies – one an alpha emitter cookie, one a beta emitter cookie, one a gamma emitter cookie, and one neutron emitter cookie. You must eat one, hold one in your hand, put one in your pocket, and give the last one you throw away. Which cookie do you eat, which cookie do you hold in your hand, which cookie do you put in your pocket and which cookie do you throw away to minimize your radiation exposure?

Solution:

If you had four cookies, and you could eat one, put one in your pocket, hold one in your hand, and throw the other away, what would you do? This problem is used to help people understand the effects of different kinds of radiation. This has largely to do with each type of radiation's dose factor.

Gamma cookie: You would eat this one, because it does not cause much somatic (body tissue) damage.

Beta cookie: You would put this one in your pocket, as it will be shielded by one layer of clothing.

Alpha cookie: Hold this one in your hand, because it will be blocked by your first layer of skin.

Neutron cookie: Throw this one away due to the large somatic damage it can cause.

Eat the gamma cookie – little tissue damage, and hard to shield

Hold the alpha cookie in your hand – blocked by first layer of skin

Put the beta cookie in your pocket – shielded by clothing

Throw the neutron cookie away – high energy, high mass, most tissue damage.

6.2 Typically, in a fission reactor fuel cell, the cladding contains the fuel pellets and what type of material to improve the thermal conductivity between the fuel pellets and cladding?

A. Hydrogen

B. Helium

C. Oxygen

D. Nitrogen

Solution:
B (Helium).

6.3 A saturated steam-water mixture with an inlet steam quality of 60% is flowing through a moisture separator. The moisture separator is 100% efficient for removing moisture. How much moisture will be removed by the moisture separator from 50 lbm of the steam-water mixture?

    A. 10 lbm

    B. 20 lbm

    C. 30 lbm

    D. 40 lbm

Solution:
This relies on information provided in the thermodynamics background in Chapter 3. The fact that we have 50 lbm of steam-water mixture that is at a steam quality of 60%, means that 60% of the steam-water mixture is in the form of steam and 40% of the steam-water mixture is in the form of water. Therefore, the mass of steam is

$$\text{Mass steam} = 50 \text{ lbm} \times 60\% = 30 \text{ lbm}$$
$$\text{Mass water} = 50 \text{ lbm} \times 40\% = 20 \text{ lbm}$$

    If the moisture separator is 100% efficient, that means it removes 100% of the water contained in the mixture. Therefore, the mass of water removed is 20 lbm. The answer is B (20 lbm).

6.4 A reactor coolant system is being maintained at 1000 psia. A pressurizer safety relief valve is slowly discharging to a collection tank, which is maintained at 5 psig. Use the ideal assumption that, since the flow is so low, the process is a throttling process. Assuming 100% quality steam in the pressurizer vapor space, what is the approximate enthalpy of the fluid entering the tank?

    A. 1210 BTU/lbm

    B. 1193 BTU/lbm

    C. 1178 BTU/lbm

    D. 1156 BTU/lbm

Solution:
In the problem, it was given that, since flow is very small, we can assume that the process is throttling process which means there is no change in the enthalpy across the value. Looking up the enthalpy of the steam at 100% steam quality and 1000 psia in the pressurizer, we find the enthalpy is 1193 BTU/lbm. Since enthalpy does not change across the relief valve, the entropy of the steam in the collection tank is the same 1193 BTU/lbm.
The answer is B (1193 BTU/lbm).

6.5 The thermodynamic cycle efficiency of a nuclear power plant can be increased by…

    A. decreasing power from 100% to 25%

    B. removing a high-pressure feed water heater from service

C. lowering condenser vacuum from 29 in. to 25 in.

D. decreasing the amount of condensate depression (subcooling)

Solution:

Decreasing power, removal of an HP feedwater heater, or lowering condenser vacuum all have the effect of decreasing thermodynamic cycle efficiency. Decreasing the amount of condensate depression will increase the thermodynamic cycle efficiency of the nuclear plant.

The answer is D (decreasing the amount of condensate depression (subcooling)).

6.6 To achieve maximum overall nuclear power plant thermal efficiency, feed water should enter the steam generator (S/G) _____ and the pressure difference between the S/G and the condenser should be as _____ as possible.

A. as subcooled as practical; great

B. as subcooled as practical; small

C. close to saturation; great

D. close to saturation; small

Solution:

To maintain maximum overall thermal efficiency, we want the feedwater to the steam generator as close to saturation as possible (less subcooling) and we want to maximize the pressure difference between the condenser and the steam generator.

The answer is C (close to saturation; great).

6.7 A nuclear power plant is operating at 85% reactor power when the extraction steam to a high-pressure feedwater heater is isolated. After the transient, the operator returns reactor power to 85% and stabilizes the plant. Compared to conditions just prior to the transient, current main turbine generator output (kW) is…

A. higher because increased steam flow is causing the turbine to operate at a higher speed

B. lower because decreased steam flow is causing the turbine to operate at a lower speed

C. higher because plant thermal efficiency has increased

D. lower because plant thermal efficiency has decreased

Solution:

For a nuclear plant, the output is maintained at 60 Hz, which means that turbine speed does not change. Therefore, A and B are not correct answers.

The correct answer is D (lower because plant thermal efficiency has decreased) since the thermal efficiency of the thermodynamic process has decreased when a feedwater heater is taken out of service.

6.8 A pressurizer is operating in a saturated condition at 636°F. If a sudden pressurizer level decrease of 10% occurs, pressurizer pressure will _____ and pressurizer temperature will _____.

A. remain the same; decrease

B. remain the same; remain the same

C. decrease; decrease

D. decrease; remain the same

Solution:
When level in the pressurizer decreases, the same steam occupies more space, so the specific density of the steam decreases. Looking at the steam tables, this means the pressure will decrease and the pressurizer temperature will also decrease since this is in a saturated condition.

The answer is C (decrease; decrease).

6.9 Nuclear reactor fuel rods are normally charged with _____ gas to improve the heat transferred by _____ from the fuel pellets to the cladding.

　A.　helium; convection

　B.　helium; conduction

　C.　nitrogen; convection

　D.　nitrogen; conduction

Solution:
A (helium; convection).

6.10 If a nuclear reactor is operated within core thermal limits, then…

　A.　plant thermal efficiency is optimized.

　B.　fuel cladding integrity is ensured.

　C.　pressurized thermal shock will be prevented.

　D.　reactor vessel thermal stresses will be minimized.

Solution:
B (fuel cladding integrity is ensured).

6.11 Establishing natural circulation requires that a heat sink be _____ in elevation than a heat source and that a _____ difference exists between the heat sink and heat source.

　A.　lower; pressure

　B.　lower; temperature

　C.　higher; pressure

　D.　higher; temperature

Solution:
For natural circulation to occur, the heat source must be lower than the heat sink. The reason for this is that the heat source will increase the temperature of the fluid which reduces the specific density of the fluid causing the fluid to rise. The heat sink will decrease the temperature of the fluid which increases the specific density of the fluid causing the fluid to lower. Therefore the answer is D (higher; temperature).

6.12 For a reactor operating in a "critical" state, the reactivity (ρ) is _____ and the neutron multiplication factor (k) is _____

　A.　$\rho > 1, k > 0$

　B.　$\rho < 0, k < 1$

　C.　$\rho = 0, k = 1$

　D.　$\rho = 1, k = 0$

Solution:
From equation (6.4), (6.5), and (6.6), the answer can be found to be C ($\rho = 0, k = 1$).

6.13 The primary purpose of the "moderator" is
    A. primary heat extraction
    B. lubrication
    C. slow neutrons produced by fission
    D. shielding
    Solution:
    C (slow neutrons produced by fission).

6.14 What two types of reactor types utilize "light water" as both coolant and moderator?
    A. BWR and PWR
    B. PHWR and PTGR
    C. GCR and LMFBR
    Solution:
    A (BWR and PWR).

6.15 The purpose of a "pressurizer" in a PWR design is to
    A. control primary loop temperature and pressure
    B. control secondary loop temperature and pressure
    C. control primary loop flow and pressure
    D. control secondary loop flow and pressure
    Solution:
    A (control primary loop temperature and pressure).

6.16 The reactor type whose design provides for online refueling is
    A. PWR
    B. BWR
    C. PHWR
    D. HTGR
    Solution:
    C (PHWR).

6.17 What is the annual maximum permissible occupational radiation exposure for a worker in rem?
    A. 1 rem
    B. 5 rem
    C. 10 rem
    D. 15 rem
    Solution:
    From Table 6.2, the answer is D (15 rem).

Chapter 7

7.1 Given a belt conveyor where the diameter of the head and tail pulleys will be 18 in., determine the rotational speed of the head pulley necessary to drive the belt at a linear speed of 94 fpm.

Solution:
Utilizing equation (7.9), we find that the necessary rotation speed of the head pulley is

$$RPM = 3.819719 \times V/D_h$$
$$RPM = 3.819719 \times 94 \text{ fpm}/18 \text{ in.}$$
$$RPM = 20 \text{ rpm.}$$

7.2 Given a belt conveyor where the initial feed rate is 10 tons per hour (tph) at a belt speed of 20 lfpm that is pulling 10 kW of power, if we change the motor speed such that the new belt speed is 40 lfpm, the new feed rate.

Solution:
Utilizing equation (7.13),

$$Q_1/Q_2 = n_1/n_2$$

Rearranging

$$Q_2 = Q_1(n_2/n_1)$$
$$Q_2 = (10 \text{ tph})(40 \text{ lfpm})/(20 \text{ lfpm})$$
$$Q_2 = 20 \text{ tph.}$$

7.3 Given a belt conveyor where the initial feed rate is 10 tons per hour (tph) at a belt speed of 20 lfpm that is pulling 10 kW of power, if we change the motor speed such that the new belt speed is 40 lfpm, calculate the new power drawn at the new speed.

Solution:
Utilizing equation (7.14),

$$P_1/P_2 = n_1/n_2$$

Rearranging,

$$P_2 = P_1 n_2/n_1$$
$$P_2 = (10 \text{ kW})(40 \text{ lfpm})/(20 \text{ lfpm})$$
$$P_2 = 20 \text{ kW.}$$

7.4 Given a belt conveyor where the initial feed rate is 10 tons per hour (tph) at a belt speed of 20 lfpm, that is pulling 10 ft lbm of torque, if we change the motor speed such that the new belt speed is 40 lfpm, calculate the new value of torque delivered by the motor.

Solution:
By definition, a belt conveyor is a constant torque application. Therefore, the torque does not change with belt speed and the new value of torque is 10 ft lbm of torque.

7.5 Given a belt conveyor with a linear velocity of 10,000 ft/min and a belt tension of 500 lb, calculate the power required to run this conveyor at steady-state speed.

Solution:
Using equation (7.15), we find the power to be as follows.

$$P = (F \times V)/33,000$$
$$P = 500 \text{ lb } 10,000 \text{ fpm}/33,000$$
$$P = 152 \text{ HP.}$$

7.6 For a belt with an idler spacing of 3 ft and a combined belt and material weight of 500 lb/ft, calculate the tension in the belt for maximum sag of 3%.

Solution:
Using equation (7.18), we find the following.

$$T(3\% \text{ sag}) = 4.2 \, S_i \, W$$
$$T(3\% \text{ sag}) = 4.2(3 \text{ ft})(500 \text{ lb/ft})$$
$$T(3\% \text{ sag}) = 6300 \text{ lb}.$$

7.7 For a belt with an idler spacing of 3 ft and a combined belt and material weight of 500 lb/ft, calculate the tension in the belt for maximum sag of 2%.

Solution:
Using equation (7.19), we find the following.

$$T(2\% \text{ sag}) = 6.25 \, S_i \, W$$
$$T(2\% \text{ sag}) = 6.25(3 \text{ ft})(500 \text{ lb/ft})$$
$$T(2\% \text{ sag}) = 9375 \text{ lb}.$$

7.8 For a belt with an idler spacing of 3 ft and a combined belt and material weight of 500 lb/ft, calculate the tension in the belt for maximum sag of 1.5%.

Solution:
Using equation (7.20), we find the following.

$$T(1.5\% \text{ sag}) = 8.4 \, S_i \, W$$
$$T(1.5\% \text{ sag}) = 8.4(3 \text{ ft})(500 \text{ lb/ft})$$
$$T(1.5\% \text{ sag}) = 12600 \text{ lb}.$$

Chapter 8

8.1 Using equation (8.1), given that the needed volume of a fan application is 100,000 cfm and the pressure difference across the fan is 20 in. of water, the ideal air HP of the fan is most closely what value? For this problem, assume a compressibility factor of 1 and assume the fan is 100% efficient.

A. 210 Air HP

B. 315 Air HP

C. 385 Air HP

D. 410 Air HP

Solution:
Using equation (8.1), we find the following

$$\text{Air HP} = (k \times V \times H)/(6356 \times \text{eff})$$
$$\text{Air HP} = (1 \times 100{,}000 \text{ cfm} \times 20'' \text{ water})/(6356 \times 100\%)$$
$$\text{Air HP} = 315 \text{ Air HP}.$$

The answer is B (315 Air HP).

8.2 Given the centrifugal fan laws in Chapter 8, for a given fan size, system resistance, and air density, as fan speed is reduced by a factor of two, fan power is reduced by a factor of

A. 2

B. 4

C. 8

D. 16

E. 1.414 (square root of 2)

Solution:
When speed is changed, power varies to cube of speed. Since speed is reduced by $\frac{1}{2}$, the power requirement is reduced by $(\frac{1}{2})^3 = 1/8$th. This can be found using equation (8.4).

$$HP_1/HP_2 \propto (N_1/N_2)^3$$

The answer is C (8).

8.3 Given the centrifugal fan laws in Chapter 8, for a given fan size, system resistance and air density, as fan speed is reduced by a factor of two, fan pressure is reduced by a factor of

    A. 2

    B. 4

    C. 8

    D. 16

    E. 1.414 (square root of 2)

Solution:
When speed is changed, the differential pressure across the fan varies to square of speed. Since speed is reduced by $\frac{1}{2}$, the differential pressure is reduced by $(\frac{1}{2})^2 = 1/4$th. This can be found using equation (8.3).

$$P_1/P_2 \propto (N_1/N_2)^2$$

The answer is B (4).

8.4 Given the centrifugal fan laws in Chapter 8, for a given fan size, system resistance and air density, as fan speed is reduced by a factor of two, fan flow is reduced by a factor of

    A. 2

    B. 4

    C. 8

    D. 16

    E. 1.414 (square root of 2)

Solution:
When speed is changed, the flow through a fan through the fan varies proportionally to speed as is described in equation (8.2). Therefore the flow is $\frac{1}{2}$ of the original flow.

$$F_1/F_2 \propto N_1/N_2$$

The answer is A (2).

8.5 Which fan design has the highest efficiency?

    A. Radial tipped

    B. Forward curved

    C. Backward curved

    D. Airfoil

Solution:
From Table 8.1, the fan blade with the highest efficiency is Airfoil. The answer is D.

8.6   Which fan design has the highest tolerance for erosion?

   A.   Radial tipped

   B.   Forward curved

   C.   Backward curved

   D.   Airfoil

   Solution:

   From Table 8.1, the fan blade with the highest tolerance for erosion is Radial tipped. The answer is A.

8.7   For an axial flow fan, what is the most common method of controlling airflow?

   A.   Discharge damper

   B.   Inlet damper

   C.   Inlet guide vanes

   D.   Controlling blade pitch

   Solution:

   D (Controlling blade pitch).

Chapter 9

9.1   For a positive displacement pump, if the speed of the pump is doubled which doubles the flow through the pump, what is the effect on pump power assuming the head across the pump is constant?

   Solution:

   Using equation (9.3), we see power is proportional to flow and head.

$$P_m = \rho g H Q / \eta p$$

Or

$$P \propto H Q$$

   Putting this in equation form we find

$$(P_1/P_2) = (H_1/H_2)(Q_1/Q_2)$$

   Rearranging,

$$(P_2/P_1) = (H_2/H_1)(Q_2/Q_1)$$

   If head is maintained constant and if flow is doubled, then

$$(P_2/P_1) = (1/1)(2/1)$$
$$(P_2/P_1) = 2 \text{ or the power requirement will double.}$$

9.2   Given a centrifugal pump where the initial flow is 100 gpm and the initial speed is 1800 rpm, if we change the pump speed to a speed of 900 rpm, based on the assumption that impeller diameter does not change, calculate the value of the new flow from the pump.

   Solution:

   Using equation (9.10),

$$Q_1/Q_2 \propto (n_1/n_2) \text{ and } (d_1/d_2)$$

Rearranging

$$Q_2 = Q_1[(n_2/n_1)(d_2/d_1)]$$

Since diameter does not change in this problem, this simplifies to

$$Q_2 = Q_2(n_2/n_1)$$
$$Q_2 = (100 \text{ gpm})(900 \text{ rpm})/(1800 \text{ rpm})$$
$$Q_2 = 50 \text{ gpm}.$$

9.3 Given a centrifugal pump where the head is 100 ft and the initial speed is 1800 rpm, if we change the pump speed to a speed of 900 rpm, based on the assumption that impeller diameter does not change, calculate the value of the new head developed by the pump.

Solution:
Using equation (9.11)

$$H_1/H_2 \propto (n_1/n_2)^2 \text{ and } (d_1/d_2)^2$$

Rearranging

$$H_2 = H_1[(n_2/n_1)^2(d_2/d_1)^2]$$

Since diameter does not change in this problem, this simplifies to

$$H_2 = H_1(n_2/n_1)^2$$
$$H_2 = (100 \text{ ft})((900 \text{ rpm})/(1800 \text{ rpm}))^2$$
$$H_2 = 25 \text{ ft}.$$

9.4 Given a centrifugal pump where the initial power is 500 bhp and the initial speed is 1800 rpm; if we change the pump speed to a speed of 900 rpm, based on the assumption that impeller diameter does not change, calculate the value of the new power drawn from the pump.

Solution:
Using equation (9.12)

$$P_1/P_2 \propto (n_1/n_2)^3 \text{ and } (d_1/d_2)^3$$

Rearranging

$$P_2 = P_1[(n_2/n_1)^3(d_2/d_1)^3]$$

Since diameter does not change in this problem, this simplifies to

$$P_2 = P_1(n_2/n_1)^3$$
$$P_2 = (5 \text{ BHP})((900 \text{ rpm})/(1800 \text{ rpm}))^3$$
$$P_2 = 62.5 \text{ BHP}.$$

9.5 We are given an application where we need to pump water at a velocity of 100 m/sec. The temperature of the water is at 20°C. When water is at 20°C (68°F), it has a density or specific weight of 998 N/m³ (62.303 lb/ft³) and a saturation vapor pressure of 2333 Pa (0.339 psia).

Solution:
Using equation (9.8), we calculate the available net positive suction head as follows.

$$\text{NPSHA} = (V^2/2g) + (p/\text{SW}) - (p_v/\text{SW})$$
$$\text{NPSHA} = ((100 \text{ m/sec})^2/2(9.80665 \text{ m/sec}^2)) + (10 \text{ Pa}/998 \text{ kg/m}^3)$$
$$\quad -(2333 \text{ Pa}/998 \text{ kg/m}^3)$$
$$\text{NPSHA} = 507.5 \text{ m}.$$

Chapter 10

10.1 What is the term for water that is removed from a condenser that uses evaporation in cooling towers as part of the cooling process to prevent the buildup of suspended solids in the water stream?

Solution:
Blow down.

10.2 The condenser vacuum is measured at 29.2 in. of mercury. What is the desired temperature of the condensate in the hotwell to ensure we do not create cavitation in the condensate pump, but we do not adversely affect the thermal efficiency of the steam cycle?

Solution:
Ideally, the amount of subcooling should be about 2°F. From Table 10.1, at a condenser vacuum of 29.2 in. of mercury, the corresponding saturation temperature is 72°F. Adding 2°F subcooling to the saturation temperature results in an ideal hotwell temperature of 70°F.

10.3 What is the term for the minimum pressure required at the suction port of the pump to keep the pump from cavitating?

Solution:
Net positive suction head required (NPSHR).

10.4 List at least three safety procedures when operating a condenser.

Solution:
Any three of the following

1. Always be on the alert for air leaks in the vacuum system
2. Shut down steam turbine if a loss of vacuum occurs along with a hot or flooded condenser
3. During shutdowns, vent the water chest.
4. Keep the saltwater side of operating condensers free of air.
5. Test the condensate for any increases in salinity.
6. Never bring an open flame to a freshly opened saltwater side of a condenser.
7. When making entry to a condenser, follow OSHA confined space requirements.

Chapter 11

11.1 A steam turbine stage, the design where pressure drops only across the fixed nozzle and not across the moving buckets is

A. reaction turbine

B. impulse turbine

Solution:
B (impulse turbine).

11.2 A turbine installation where there are two LP turbines connected on a common shaft driving one generator is known as a

A. tandem compound turbine

B. cross-compound turbine

Solution:
A (tandem compound turbine).

11.3 The valve that is used to control steam flow to the HP section turbine for control during full-load operation is known as the

   A.   governor valve

   B.   intercept valve

   C.   stop valve

   D.   throttle valve

Solution:
A (governor valve).

11.4 The purpose of the turbine turning gear is to rotate the turbine during shutdown and startup to

   A.   prevent bowing of turbine rotor due to weight of rotor

   B.   build up oil film in journal bearings

   C.   prevent bowing of turbine rotor due to uneven heating

   D.   check for rotor rotation direction

Solution:
C (prevent bowing of turbine rotor due to uneven heating).

11.5 The type of turbine control that refers to the sequential opening of the steam control (governor) valve (partial arc admission) is

   A.   throttle control

   B.   governing control

   C.   variable pressure control

Solution:
B (governing control).

11.6 The pump that takes suction from the deaeration tank and provides water to the boiler is the

   A.   condensate pump

   B.   boiler recirculation pump

   C.   boiler feed pump

   D.   cooling water pump

Solution:
C (boiler feed pump).

11.7 The device that converts the thermal energy of steam into mechanical energy on a shaft is

   A.   boiler

   B.   steam generator

   C.   steam turbine

   D.   electric generator

Solution:
C (steam turbine).

11.8 The type of control method where the boiler or steam generator main steam pressure is varied and the control valves are left in a fixed open position is known as

    A. throttling control

    B. governing control

    C. variable pressure control

    Solution:

    C (variable pressure control).

11.9 Using Figure 11.13 for typical vibration analysis, on a turbine running at 3600 rpm, we indicate a displacement of 0.10 mils. What is the most likely condition of the bearings?

    A. Extremely smooth

    B. Very smooth

    C. Smooth

    D. Very good

    E. Good

    F. Fair

    G. Slightly rough

    H. Rough

    I. Very rough

    Solution:

    D (Very good) or C (Smooth) are correct.

11.10 Using Figure 11.13 for typical vibration analysis, on a turbine running at 3600 rpm, we indicate a displacement of 3.0 mils. What is the most likely condition of the bearings?

    A. Extremely smooth

    B. Very smooth

    C. Smooth

    D. Very good

    E. Good

    F. Fair

    G. Slightly rough

    H. Rough

    I. Very rough

    Solution:

    I (Very rough) or H (Rough) are correct.

11.11 Using Figure 11.13 or Table 11.2 for typical vibration analysis, on a turbine running at 1800 rpm, we indicate a displacement of 0.4 mils. What is the most likely condition of the bearings?

    A. Extremely smooth

    B. Very smooth

    C. Smooth

D.   Very good

E.   Good

F.   Fair

G.   Slightly rough

H.   Rough

I.   Very rough

Solution:

Looking at Figure 11.13 it is between Good and Very good so E. Good or D. Very good is acceptable.

Looking at Table 11.2, we see that the value of 0.4 is the "Very good" range for 1800 rpm which is it is just to the "Very good" side, so D (Very good) would be the preferred answer. 0.2080 mils peak to peak and 0.4159 mils peak to peak.

11.12   What is the vibration in units of inches per second of a piece of equipment rotating at 720 rpm given that the displacement is measured as total displacement of 2 mils peak to peak?

A.   0.7536 in./sec

B.   0.04578 in./sec

C.   0.07536 in./sec

D.   0.4578 in./sec

Solution:

First, we must convert or units of measurement of displacement from mils peak to peak to inches peak to peak. Knowing that 1000 mils = 1 in., we find the displacement to be

$$D = 2 \text{ mils peak to peak} \times (1 \text{ in.}/1000 \text{ mil})$$
$$D = 0.002 \text{ in. peak to peak}$$

Now we utilize equation (11.2) to find the velocity of vibration to be

$$V = \pi \times F \times D$$
$$V = \pi \times 720 \text{ cycles/min} \times 0.002 \text{ in. peak to peak}$$
$$V = 4.5216 \text{ in./min}$$

However, the standard units for velocity are in inches per second, so converting to standard units we find

$$V = (4.5216 \text{ in./ min}) \times (1 \text{ min /60 sec}) = 0.07536 \text{ in./ sec}.$$

The answer is C (0.07536 in./sec).

Chapter 12

12.1   What are the four parts of the Brayton combustion cycle?

_____

_____

_____

_____

Solution:

Compression

Combustion

Expansion

Exhaust

12.2   What are the six major components of a combustion turbine?

_____

_____

_____

_____

_____

_____

Solution:

Air inlet

Compressor

Combustion system

Turbine

Exhaust

Support systems

12.3   Given that the plant generator output is 400 MW (or 400,000 kW), the fuel used has a heating value of 10,000 BTU/lbm, and the fuel is consumed by the plant at a rate of 200,000 lbm/hr, determine the plant heat rate.

Solution:

Using equation (12.1), we find

$$HR = (FCR\ FHV)/GO$$
$$HR = (200{,}000\ lbm/hr \times 10{,}000\ BTU/lbm)/400{,}000\ kW$$
$$HR = 5000\ BTU/kWh.$$

12.4   What is the function of the air inlet assembly?

Solution:

The function of the air inlet assembly is to condition the air being supplied to the compressor stage of the system to ensure the quality and cleanliness of the inlet air is adequate for the compression stage.

12.5   What is the function of the combustion section?

Solution:

The function of the combustion section is to raise the temperature of the incoming air stream supplied from the compressor section to achieve the combustion process of the Brayton cycle.

12.6   What is the function of the turbine section?

Solution:

The function of the turbine section is to expand the gas stream and convert the energy contained in the gas stream to mechanical energy delivered to the shaft of the turbine.

12.7 What is the function of the recuperator?

Solution:
A recuperator captures waste heat in the turbine exhaust system to preheat the compressor discharge air before it enters the combustion chamber.

12.8 What is pre-combustion method to minimize NO_x formation?

Solution:
Pre-combustion NO_x control is done by reducing the flame temperature where complete combustion occurs. This is done by injection of either steam or elemental nitrogen (N_2).

12.9 What is post-combustion method to minimize NO_x formation?

Solution:
Post-combustion NO_x control is by chemical reaction with NO and NO_2 at a selective catalytic reduction system (SCR).

Chapter 13

13.1 For a two-stroke gas or oil reciprocating engine it takes how many shaft revolutions to complete one full cycle?

A. One

B. Two

C. Three

D. Four

Solution:
It takes two piston strokes or one shaft revolution to complete one full cycle.
The answer is A (One).

13.2 For a four-stroke gas or oil reciprocating engine it takes how many shaft revolutions to complete one full cycle?

A. One

B. Two

C. Three

D. Four

Solution:
It takes four piston strokes or two shaft revolutions to complete one full cycle.
The answer is B (Two).

13.3 What is the total displacement of an eight-cylinder engine where each cylinder displaces a volume of 15 in.^{3}?

Solution:
Using equation (13.6), we find the total engine displacement to be

$$D = \text{disp/cylinder} \times N$$
$$D = 15 \text{ (in.}^3) \times 8$$
$$D = 120 \text{ in.}^3$$

13.4 What is the ideal power available from the combusted gas for a four-cycle, eight-cylinder engine with a total engine displacement of 120 in.^{3} that operates at 2000 rpm and has an average pressure that the combusted gas places on the piston during the expansion process of 165 (lbf/in.^{2})?

Solution:

Given the following values from the problem statement,

$K = 2$ for a four-cycle engine.

$V = 2000$ rpm

$D = 120$ (in.^{3})

$P_{avg} = 165$ (lbf/in.^{2})?

Using equation (13.7), we find the ideal power available from the combusted gas to be

$$P_{in} = (V \times D \times P_{avg})/(K \times 396,000)$$
$$P_{in} = (2000 \text{ rpm}) \times (120 \text{ in.}^3) \times (165 \text{ lbf/in.}^2)/(2 \times 396,000)$$
$$P_{in} = 50 \text{ HP.}$$

13.5 What is the efficiency of a reciprocating engine that has a value for ideal power available from the combusted gas of 50 HP and a measured available shaft power of 30 HP?

Solution:

Using equation (13.6), we find the efficiency of this engine to be

$$\eta\text{- engine} = P_{out}/P_{in}$$
$$\eta\text{- engine} = 30 \text{ HP}/50 \text{ HP}$$
$$\eta\text{- engine} = 0.6 \text{ or } 60\%.$$

13.6 What is the volume of the cylinder at the bottom of the cylinder travel if the volume of the cylinder at the top of the cylinder travel is 2 in.^{3} and the compression ratio (CR) is 10?

Solution:

Using equation (13.1), we rearrange for volume of the cylinder at the bottom of the cycle and find this volume to be

$$CR = V_6/V_3$$
$$V_6 = CR \times V_3$$
$$V_6 = 10 \times 2 \text{ in.}^3$$
$$V_6 = 20 \text{ in.}^3$$

13.7 What is the displacement of the cylinder at the bottom of the cylinder travel if the volume of the cylinder at the top of the cylinder travel is 2 in.^{3} and the compression ratio (CR) is 10?

Solution:

From Problem 13.6, we found the volume of the cylinder at the bottom of travel to be

$$V_6 = 20 \text{ in.}^3$$

Using equation (13.2), we find the displacement per cylinder to be

$$\text{disp/cylinder} = V_6 - V_3$$
$$\text{disp/cylinder} = 20 \text{ in.}^3 - 2 \text{ in.}^3$$
$$\text{disp/cylinder} = 18 \text{ in.}^3$$

Chapter 14

14.1 The system configuration shown below is an example of what type of plant distribution system?

  A. Primary selective system

  B. Secondary selective system

  C. Loop system

  D. Radial system

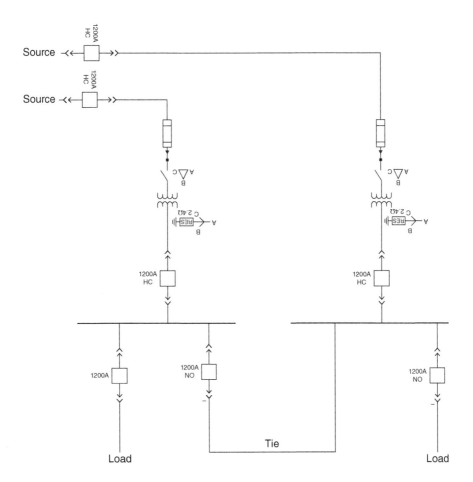

Solution:

B (Secondary selective system).

14.2 In the figure below, which figure depicts "segregated phase bus duct"?

  A. Arrangement A

  B. Arrangement B

  C. Arrangement C

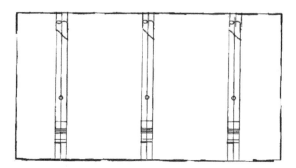

Arrangement A

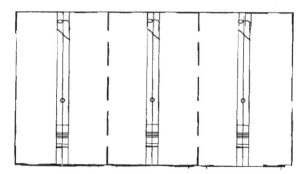

Arrangement B

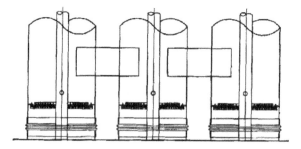

Arrangement C

Solution:

B (Arrangement B).

14.3 For a straight pull of 200′, pulling a cable weighing 1 lb/ft, what is the pulling tension assuming coefficient of friction to be 0.5?

A. 50 lb

B. 100 lb

C. 150 lb

D. 200 lb

Solution:
Using equation (14.4), we find the tension to be

$$T = W \times 0.5$$
$$T = (1 \text{ lb/ft}) \times (200 \text{ ft}) \times 0.5$$
$$T = 100 \text{ lb}$$

The answer is B (100 lb).

14.4 What is the pull tension for a cable pull through a 90 degree bend where the tension at the bend inlet is 100 lb assuming coefficient of friction to be 0.5?

A. 59 lb

B. 159 lb

C. 219 lb

D. 319 lb

Solution:
Using equation (14.5), we find the tension to be

$$T_c = T_1 \times e^{(0.5 \times a)}$$
$$T_c = 100 \text{ lb} \times e^{(0.5 \times (90/57.3))}$$
$$T_c = 219 \text{ lb}$$

The answer is C (219 lb).

14.5 A control cable is being pulled into 30 ft of 3″ trade size conduit. See figure below for physical layout. We have a choice for the direction of the cable pull as shown by part (A) or part (B) of the figure below. Side wall bearing pressure (SWBP) of the cable is 500 lbf/ft and coefficient of friction is not known but assumed to be no greater than 0.5. For this problem, use Table 14.14 for cable data. For both the cable pull direction shown in part (A) and the cable pull direction shown in part (B), determine the following.

a. The total effective conduit length

b. The total degrees of bend

c. The maximum allowable effective conduit length for cable specified

d. The maximum pulling tension for application

e. Is this cable pull allowable from the direction indicated?

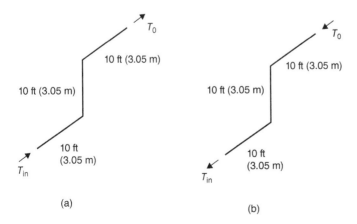

(a)

(b)

Solution:

Part (A):

For the cable pull direction shown in part (A), to determine item (a) and (b), we first need to set up a table to list each pull, the actual length and in what direction so we can convert the actual conduit length to the effective conduit length. The first pull is horizontal, so the actual conduit length (20 ft) and effective length are the same. The second pull is vertical/up, so the effective length is twice the actual length (40 ft). The third pull is horizontal so the actual conduit length (20 ft) and effective length are the same. Then we sum up the total degrees of bends (to be used when we look up the maximum allowable effective length in the IEEE 1185 tables) and the total effective length of the conduit. The total effective length is found to be 80 ft (a). See the table below for summary. The total number of bends is 180° (b) from the table below.

Section Type	Angle (degree)	Actual Length (ft)	Effective Length (ft)
Straight horizontal		20	20
Bend up	90		
Vertical/up		20	40
Bend down	90		
Straight horizontal		20	20
End of pull (totals)	180		40

From the problem description, we were given that we are using a 3 in. conduit trade size and pulling a cable with a maximum side wall bearing pressure of 500 lbf/ft. With this information, we can find the maximum effective conduit length for a pull with 180° total bends in table A1a of IEEE 1185 (see Table 14.14) and find the maximum allowable effective conduit length to be 62 ft (c) and the maximum pulling tension to be 478 lbf (d). Since the total effective length was 80 ft and the maximum allowable effective length is 62 ft, this pull is not allowable (e) and will damage the cable.

Part (B):

For the cable pull direction shown in part (B), to determine item (a) and (b), we first need to set up a table to list each pull, the actual length and in what direction so we can convert the actual conduit length to the effective conduit length. The first pull is horizontal, so the actual conduit length (20 ft) and effective length are the same. The second pull is vertical/down, so the effective length is aero (0 ft). The third pull is horizontal so the actual conduit length (20 ft) and effective length are the same. Then we sum up the total degrees of bends (to be used when we look up the maximum allowable effective length in the IEEE 1185 tables) and the total effective length of the conduit. The total effective length is found to be 40 ft (a). See the table below for summary. The total number of bends is 180° (b) from the table below.

Section Type	Angle (degree)	Actual Length (ft)	Effective Length (ft)
Straight horizontal		20	20
Bend down	90		
Vertical down		20	0
Bend Up	90		
Straight horizontal		20	20
End of pull (totals)	180		40

From the problem description we were given that we are using a 3 in. conduit trade size and pulling a cable with a maximum side wall bearing pressure of 500 lbf/ft. With this information, we can find the maximum effective conduit length for a pull with 180° total bends in table A1a of IEEE 1185 (see Table 14.14) and find the maximum allowable effective conduit length to be 62 ft (c) and the maximum pulling tension to be 478 lbf (d). Since the total effective length was 40 ft and the maximum allowable effective length is 62 ft, this pull is allowable (e) and will damage the cable.

14.6 We have a cable that we are pulling with a tension of 20 lb running through a pulley making a 180 degree bend. This is shown in the figure below on the right. That pulley is then connected to another pulley which is shown on the figure below on the left. This second pulley is attached to the wall. Calculate the force that the pulley on the left will impose on the wall from the cable tension of 20 lb.

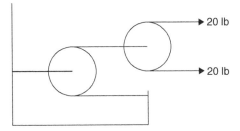

Solution:

Looking at the figure below, we have two sources of 20 lb of force in the positive X direction on the right pulley so the total force of the right pulley on the left pulley is 40 lb in the negative X direction. The left pulley has two forces, each of 40 lb pulling in the positive X direction, so the left pulley places onto the wall a net force of 80 lb. Therefore, the net force on the wall is 80 lb (see figure below).

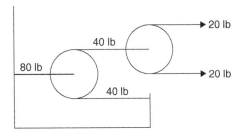

Chapter 15

15.1 For a 28,800 VA, single-phase transformer that is rated 14,400 V on the primary and 120 V on the secondary where the secondary winding is connected to a 1 ohm load, determine the following.

A. Turns ratio (n)

B. Rated primary current

C. Rated secondary current

D. The impedance seen by the circuit connected to the primary of the transformer.

Solution:

A. Turns ratio (n) can be found utilizing equation (15.3)

$$V_P/V_S = n$$
$$n = 14{,}400 \text{ V}/120 \text{ V}$$
$$n = 120$$

B. Rated primary current

$$S_P = V_P \times I_P$$
$$I_P = S_P \times V_P$$
$$I_P = 28{,}800 \text{ VA} \times 14{,}400 \text{ V}$$
$$I_P = 2 \text{ A}$$

C. Rated secondary current

$$S_S = V_S \times I_S$$
$$I_S = S_S \times V_S$$
$$I_S = 28{,}800 \text{ VA} \times 120 \text{ V}$$
$$I_S = 240 \text{ A}$$

D. The impedance seen by the circuit connected to the primary of the transformer due to the 1 ohm resistor connected to the secondary circuit can be found utilizing equation (15.5)

$$Z_P/Z_S = n^2$$
$$Z_P = n^2 \times Z_S$$
$$Z_P = (120)^2 \times 1 \text{ ohm}$$
$$Z_P = 14{,}400 \text{ ohms.}$$

15.2 A generator step-up transformer consists of three single-phase transformers that are under test to determine their single-line equivalent circuit values of core resistance and core reactance. A no-load test is performed on the transformer and, when rated primary potential of 36,000 V is applied to the primary of the GSU, the measured value for real power loss was 6000 W and the measured value for primary current was 3 A. Determine the effective values for transformer core resistance (R_c) and core reactance (X_m).

Solution:
Utilizing equation (15.10), we find the core resistance to be

$$R_c = V_p^2/P_{nl}$$
$$R_c = (36000)^2/6000 \text{ W}$$
$$R_c = 216{,}000 \text{ ohms}$$

Utilizing equation (15.11), we find the core reactance to be

$$X_m = V_p^2/\{(V_p \times I_p)^2 - P_{nl}^2\}^{1/2}$$
$$X_m = 36000^2/\{(36000 \times 3)^2 - 6000^2\}^{1/2}$$
$$X_m = 12019 \text{ ohms.}$$

15.3 A generator step-up transformer consists of three single-phase transformers that are under test to determine their single-line equivalent circuit values of core resistance and core reactance. A full-load test is performed on the transformer and, when rated secondary current is drawn on the shorted secondary of the transformer, the measured primary potential was found to be 2000 V, the measured value for real power loss was 90,000 W and the measured value for the primary current was found to be 300 A.

Determine the effective values for transformer primary and secondary winding resistance $(R_p + R_s')$ and primary and secondary winding reactance $(X_p + X_s')$.

Solution:
Utilizing equation (15.12), we find the primary and secondary winding resistance to be

$$(R_p + R_s') = P_{fl}/I_p^2$$
$$(R_p + R_s') = 90000/(300)^2$$
$$(R_p + R_s') = 1.0 \text{ ohms.}$$

Utilizing equation (15.13), we find the primary and secondary winding reactance to be

$$(X_p + X_s') = V_p^2/\{(V_p \times I_p)^2 - P_{fl}^2\}^{1/2}$$
$$(X_p + X_s') = 2000^2/\{(2000 \times 300)^2 - 90,000^2\}^{1/2}$$
$$(X_p + X_s') = 6.74 \text{ ohms.}$$

15.4 What should be the winding hot spot alarm on a dry-type station service transformer be that was constructed with a class 180 insulation system and measures the average winding temperature using an RTD?

Solution:
From Table 15.6, for a class 180 insulation system, for a transformer that has an average temperature rise monitored by resistance, the alarm should be set at or below 115°C above a 30°C ambient.

15.5 What is should be the winding hot spot alarm on a dry-type station service transformer be that was constructed with a class 180 insulation system and measures the actual hot spot temperature rise?

Solution:
From Table 15.6, for a class 180 insulation system, for a transformer that monitors the actual hot spot temperature rise, the alarm should be set at or below 140°C above a 30°C ambient.

15.6 What is the standard BIL level for a transformer winding that will be connected to an electrical system with a nominal voltage of 15.0 kV?

Solution:
From Table 15.5, for a nominal system voltage of 15.0 kV, the standard BIL is 60 kV.

15.7 A transformer with a primary winding inductance of 10 mH is connected to an electrical system with a nominal voltage of 4.16 kV. A vacuum breaker is used to isolate the transformer in the event of an electrical fault. During an electrical fault, the breaker interrupted 30,000 A in a time of 10 ms. What is the potential peak voltage developed across the transformer winding in this instance?

Solution:
Utilizing equation (15.9), we find the potential voltage developed to be

$$V_1 = L(di/dt)$$
$$V_1 = (0.010 \text{ H}) \times (30,000 \text{ A})/(0.010 \text{ sec})$$
$$V_1 = 30,000 \text{ V.}$$

Chapter 16

16.1 What is the speed of a synchronous four-pole generator operating at 60 Hz?

A. 3600 rpm

B. 1800 rpm

C.   1200 rpm

D.   1900 rpm

Solution:

Using equation (16.1), we find,

$$f = n \times p/120$$

Rearranging for speed we find,

$$n = f \times 120/p$$
$$n = 60 \times 120/4$$
$$n = 1800$$

The answer is B (1800 rpm).

16.2   What is the frequency generated by a synchronous two-pole generator running at 3000 rpm?

A.   50 Hz

B.   60 Hz

C.   100 Hz

D.   120 Hz

Solution:

Using equation (16.1), we find,

$$f = n \times p/120$$
$$f = 3000 \times 2/120$$
$$f = 50 \text{ Hz}$$

The answer is A (50 Hz).

16.3   The purpose of a flux probe is to:

A.   Detect stator bar shorts

B.   Detect rotor bar shorts

C.   Detect high eddy current loss in stator iron

D.   Detect high vibration in journal bearing

Solution:

B (Detect rotor bar shorts).

16.4   A stator slot RTD is located where?

A.   Between top coil ground wall insulation and wedge

B.   Between top coil ground wall insulation and bottom coil ground wall insulation

C.   Between bottom coil ground wall insulation and slot bottom

D.   Between core laminations

Solution:

B (Between top coil ground wall insulation and bottom coil ground wall insulation).

16.5   Given a generator that has field resistance values given as 0.1516 ohms at 125°C and 0.1094 ohms at 25°C, determine the temperature of the field winding if the measured resistance of the field winding is 0.1225 ohms.

A.   35°C

B.   42°C

C. 56°C

D. 72°C

Solution:

Using equation (16.9), we find the following.

$$M = (R_h - R_c)/(T_h - T_c)$$
$$M = 0.000422 \text{ ohm/°C}$$
$$T_2 = [(Rt_2 - Rt_1)/M] + T_1$$
$$T_2 = [(Rt_2 - 0.1094)/0.000422] + 25°C$$
$$T_2 = [(0.1225 - 0.1094)/0.000422] + 25°C$$
$$T_2 = 56°C$$

The answer is C (56°C).

16.6 For a synchronous generator, real power delivered into the transmission system is con-trolled by the _____ between generator voltage and transmission system volt-age, and reactive power is controlled by the _____ between generator voltage and transmission system voltage.

A. phase angle, voltage magnitude

B. voltage magnitude, phase angle

Solution:

A (phase angle, voltage magnitude).

16.7 What is the power factor of a generator that is being operated at 100 kVA and the "real" part of the 100 kVA is 80 kW?

A. 1.0

B. 0.9

C. 0.8

D. 0.7

Solution:

Using equation (16.7), we find the following.

$$MVA = MW/pf$$

Rearranging equation (16.6) for power factor, we find the following.

$$pf = kW/kVA$$
$$pf = 80 \text{ kW}/100 \text{ kVA}$$
$$pf = 0.8$$

The answer is C (0.8).

16.8 A three-phase, one-winding, two-pole synchronous generator contains 48 coils. Deter-mine the number of coil groups in the machine.

Solution:

Coil groups = (Number of coils) / (Number of phases × Number of windings × Number of poles)

Coil groups = (48 coils) / (3 phases × 1 winding × 2 poles)

Coil groups = 8 coils per group.

Chapter 17

17.1   For a four-pole, synchronous motor that has 60 Hz, 460 V applied to the stator, what is the synchronous speed of the motor?

  A.   3600 rpm

  B.   1800 rpm

  C.   1200 rpm

  D.   900 rpm

Solution:

Using equation (17.3) we find the following.

$$n_s = 120 \times f/p$$
$$n_s = 120 \times 60/4 = 1800 \text{ rpm}$$
$$n_s = 1800 \text{ rpm.}$$

The answer is B (1800 rpm).

17.2   For a four-pole motor with 60 Hz applied to stator, runs at full load at 1725 rpm, what is the percent slip of the motor at full load?

  A.   2.33% slip

  B.   3.17% slip

  C.   4.17% slip

  D.   6.33% slip

Solution:

Using equation (17.4), we find the following.

$$s = [(n_s - n)/n_s] \times 100\%$$
$$s = [(1800 \text{ rpm} - 1725 \text{ rpm})/1800 \text{ rpm}] \times 100\%$$
$$s = 4.17\%.$$

The answer is C (4.17%).

17.3   Given a load that has a maximum steady-state torque requirement of 35 lb ft at a rotational speed of 900 rpm, what is the mechanical power requirement of this application in horsepower (HP)?

  A.   2 HP

  B.   4 HP

  C.   6 HP

  D.   8 HP

Solution:

Using equation (17.5), we find the following.

$$P = Tn/5250$$
$$P = (35 \text{ lb ft})900 \text{ rpm}/5250$$
$$P = 6 \text{ HP}$$

The answer is C (6 HP).

17.4   Given an induction motor that has 10 HP shaft power at a shaft rotational speed of 1100 rpm, what is the available torque output of the motor at the shaft?

A.   11.9 lb ft

B.   23.86 lb ft

C.   47.73 lb ft

D.   95.45 lb ft

Solution:
Using equation (17.5), we find the following.

$$P = Tn/5250$$

Rearranging for torque, we find the following.

$$T = 5250\,P/n$$
$$T = 5250\ 10\ \text{HP}/1100\ \text{rpm}$$
$$T = 47.73\ \text{lb ft}$$

The answer is C (47.73 lb ft).

17.5   Using the same motor application as described in Problem 17.4, if between the motor and the final load a 2:1 gear reduced is utilized, the load shaft speed is half of motor shaft speed (i.e., load shaft speed is 550 rpm), and assuming a perfectly efficient gear reducer (i.e., shaft power into gear reduced = shaft power out of gear reducer), What is the torque available on the load shaft downstream of the gear reducer? (*Hint*: Use same formula as used in Problem 17.4, but use power and rpm available on load shaft)

A.   11.9 lb ft

B.   23.86 lb ft

C.   47.73 lb ft

D.   95.45 lb ft

Solution:
Using equation (17.5), we find the following.

$$P = Tn/5250$$

Rearranging for torque, we find the following.

$$T = 5250\,P/n$$
$$T = 5250\ 10\ \text{HP}/550\ \text{rpm}$$
$$T = 95.45\ \text{lb ft}$$

The answer is D (95.45 lb ft).

17.6   The method of reduced voltage starting where the line current is reduced proportional to the square of the reduction of voltage is

A.   Autotransformer

B.   Primary resistor

C.   Primary reactor

Solution:
Referencing Table 17.5, the answer is A (Autotransformer).

17.7   The transmission system shown in Figure 17.27 has a pulley with a diameter of 12″ on the load shaft and a pulley with a diameter of 4″ on the motor shaft. If the torque the load is requiring is 60 lb ft, what is the torque on the motor shaft? (Assume an ideal transmission system where $P_{in} = P_{out}$.)

   A.   15 lb ft

   B.   20 lb ft

   C.   30 lb ft

   D.   60 lb ft

Solution:
Using equation (17.29), we find the following.

$$T_{out}/T_{in} = d_{out}/d_{in}$$

Rearranging, we find the following.

$$T_{out} = T_{in} \times d_{out}/d_{in}$$
$$T_{out} = 60 \text{ lb ft} \times 4″/12″$$
$$T_{out} = 20 \text{ lb ft}$$

The answer is B (20 lb ft).

17.8   The transmission system shown in Figure 17.27 has a pulley with a diameter of 12″ on the load shaft and a pulley with a diameter of 4″ on the motor shaft. If the rotational speed of the load is 380 rpm, what is the rotational speed of the motor shaft? (Assume an ideal transmission system where $P_{in} = P_{out}$).

   A.   380 rpm

   B.   520 rpm

   C.   1140 rpm

   D.   2280 rpm

Solution:
Using equation (17.28), we find the following.

$$d_{out}/d_{in} = \omega_{in}/\omega_{out}$$

Rearranging, we find the following.

$$\omega_{out} = \omega_{in} \times d_{in}/d_{out}$$
$$\omega_{out} = 380 \text{ rpm} \times (12″/4″)$$
$$\omega_{out} = 1140 \text{ rpm}$$

The answer is C (1140 rpm).

17.9   Using Table 17.2, find the capacitor rating required to improve the power factor of a 250 kW load from 0.70 to 0.90.

   A.   95 kvar

   B.   123 kvar

   C.   134 kvar

   D.   154 kvar

Solution:
Using Table 17.1, with original pf = 0.7 and desired pf = 0.9, the multiplier is 0.536.

Using equation (17.9), we now find the amount of kvar of capacitance we need to correct from a power factor is 0.7 to a power factor of 0.9.

$$kvar = kW \times multiplier$$
$$kvar = 250 \times 0.536$$
$$kvar = 134 \, kvar$$

The answer is C (134 kvar).

Chapter 18

18.1   A variable frequency drive adjusts applied _____ to a motor stator to control the speed of the motor and adjusts applied _____ to a motor to control the available motor torque.

   A.   voltage; frequency

   B.   frequency; voltage

Solution:
The answer is B as the frequency defines the speed of the motor (as shown in equation 18.1) while the volts-to-frequency ratio define the torque of the machine at a certain slip (in rpm).

18.2   Given a motor rated at 460 V, 60 Hz, eight-pole machine with a full-load speed of 850 rpm, and given that the motor has 460 V, 60 Hz applied to the stator and if fully loaded, calculate the following.

   A.   The synchronous speed of the motor

   B.   The slip of the motor in rpm

   C.   The slip of the motor in percent

Solution:
First, we need to calculate the synchronous speed with 60 Hz applied to the stator using equation (18.1).

$$n = 120 f/p$$
$$n = 120 \times 60 \, Hz/8 \, poles$$

   A. $n = 900$ rpm

The given full-load speed with 60 Hz applied is 1700 rpm. The slip of the motor in rpm is calculated using equation (18.4).

$$s = (n_{nl} - n_{fl})$$
$$s = (900 \, rpm - 850 \, rpm)$$

   B. $s = 50$ rpm

Now that we have the slip in rpm, we just need to normalize to the synchronous speed in rpm to determine slip in percentage using equation (18.3).

$$s = [(n_{nl} - n_{fl})/n_{nl}] \times 100\%$$
$$s = [(900 \, rpm - 850 \, rpm)/900 \, rpm] \times 100\%$$
$$s = 0.0556 \times 100\%$$

   C. $s = 5.56\%$.

18.3 We will control a motor with a variable frequency drive (VFD). The motor is rated at 460 V, 60 Hz, eight-pole machine with a full-load speed of 850 rpm. The output of the VFD is 230 V, 30 Hz. If the motor is delivering rated torque to the load (i.e., slip in rpm is the same as in the rated frequency case), calculate the following.

A. Calculate the synchronous speed of the motor at 30 Hz

B. Calculate the slip of the motor in rpm

C. Calculate the slip of the motor in percent

Solution:

First, we need to calculate the synchronous speed with 30 Hz applied to the stator using equation (18.1).

$$n = 120\,f/p$$
$$n = 120 \; 30 \text{ Hz}/8 \text{ poles}$$

A. $n = 450$ rpm

We are given that the motor is delivering rated torque to the load. For the motor to provide rated torque to the load, the motor must slip the same amount (in rpm) at this reduced speed as it does at rated speed. This value was calculated in Example 26.1(b) using equation (18.4).

$$s = (n_{nl} - n_{fl})$$
$$s = (450 \text{ rpm} - 400 \text{ rpm})$$

B. $s = 50$ rpm

Now that we have the slip in rpm, we just need to normalize to the synchronous speed in rpm at the new applied frequency of 30 Hz to determine slip in percentage using equation (18.3).

$$s = [(n_{nl} - n_{fl})/n_{nl}] \times 100\%$$
$$s = [50 \text{ rpm}/450 \text{ rpm}] \times 100\%$$
$$s = 0.1111 \times 100\%$$

C. $s = 11.11\%$.

18.4 Given a motor rated at 460 V, 60 Hz, 900 rpm, 10 HP. This motor has a VFD supplying voltage and frequency to the motor stator and the maximum value of voltage that the VFD can provide to the motor is 460 V. The maximum frequency the VFD can supply to the motor is 120 Hz. Calculate the following.

A. Calculate nominal torque available from the motor at 900 rpm.

B. Calculate the nominal volts-to-hertz ratio that this torque is based on.

C. Calculate the volts-to-hertz ratio for the motor when it is at 10% overspeed or 990 rpm (assume applied frequency of 66 Hz) knowing that the applied stator voltage does not change from 460 V.

D. Calculate torque available from the motor at 10% overspeed or 990 rpm (assume applied frequency of 66 Hz) knowing that the applied stator voltage does not change from 460 V.

E. Calculate the percentage change in torque normalized to the nominal torque of the machine.

Solution:

A. Using equation (17.5) from Chapter 17,

$$P = T\,n/5250$$

where
$P = $ power (HP)
$T = $ torque (lb ft)
$n = $ speed (rpm)
    Rearranging for torque, we find the following.

$$T = P\,5250/n$$
$$T = 10\ \text{HP}\ 5250/900\ \text{rpm}$$
$$T = 58.33\ \text{lb ft}$$

B. Next, we find the volts-to-hertz ratio that this torque is based on. Since stator voltage is 460 V and stator frequency is 60 Hz, the volts-to-hertz ratio is

$$\text{Volts-to-hertz ratio} = 480\ \text{V}/60\ \text{Hz}$$
$$\text{Volts-to-hertz ratio} = 7.67\ \text{V}/\text{Hz}$$

C. Assuming the new applied frequency is 66 Hz, and the applied motor stator voltage is still 460 V, we find the new volts-to-hertz ratio to be

$$\text{Volts-to-hertz ratio} = 460\ \text{V}/66\ \text{Hz}$$
$$\text{Volts-to-hertz ratio} = 6.97\ \text{V}/\text{Hz}$$

    Notice this is a 10% reduction of volts-to-hertz ratio which is the same percentage as the increase in overspeed of the motor.

D. The new available torque is the nominal torque times the square of the change in the volts-to-hertz ratio

$$T_{new} = T_{old}(\text{V}/\text{Hz}_{new}/\text{V}/\text{Hz}_{old})^2$$
$$T_{new} = 58.33\ \text{lb ft}\ (6.97\ \text{V}/\text{Hz}/7.67\ \text{V}/\text{Hz})^2$$
$$T_{new} = 58.33\ \text{lb ft}\ 0.9091^2$$
$$T_{new} = 58.33\ \text{lb ft}\ 0.82645$$
$$T_{new} = 48.2\ \text{lb ft}$$

E. The percentage change in the torque is the new torque value normalized to the nominal torque value

$$\text{Change in torque} = (T_{new}/T_{old})100\%$$
$$\text{Change in torque} = (48.2\ \text{lb ft}/58.33\ \text{lb ft})100\%$$
$$\text{Change in torque} = 82.6\%$$

    Note this is the same result as if we just calculated the square of the change in volts-to-hertz ratio or

$$\text{Change in torque} = (\text{V}/\text{Hz}_{new}/\text{V}/\text{Hz}_{old})^2 100\%$$
$$\text{Change in torque} = 82.6\%.$$

18.5 Given a variable speed drive with three rectifiers on the front end of the drive, each rectifier is a six-pulse rectifier, what is the total pulse number of the rectifier system and what is the phase shift required between transformer secondary windings?

Solution:
Using equation (18.6), we find the total pulse number of the rectifier to be as follows.

$$P_{total} = n \times P_{rectifier}$$
$$P_{total} = 3 \times 6$$
$$P_{total} = 18$$

Using equation (18.7), we find the total pulse number of the rectifier to be as follows.

$$\text{Phase shift} = 360 \text{ degrees}/P_{total}$$
$$\text{Phase shift} = 360 \text{ degrees}/18$$
$$\text{Phase shift} = 20 \text{ degrees}.$$

Chapter 19

19.1   True or false, metal-clad switchgear can always also qualify as metal enclosed switchgear?

Solution:
From text in Chapter 18, "Metal-clad switchgear by definition always meets the requirements of metal enclosed switchgear. However, due to the issue of barrier requirements, not all metal enclosed switchgear can be classified as metal-clad switchgear."
   Therefore, the answer to the question is true.

19.2   The IEEE number that identifies a normally open contact that is driven from the cell-mounted switch that indicates the breaker is fully racked in is which of the following contacts?

A.   52/a

B.   52/b

C.   52TOC/a

D.   52TOC/b

E.   52MOC/a

F.   52MOC/b

Solution:
From text, the answer is C (52TOC/a).

19.3   The IEEE number that identifies a normally closed contact that is driven from the cell-mounted switch that indicates the breaker is closed is which of the following contacts?

A.   52/a

B.   52/b

C.   52TOC/a

D.   52TOC/b

E.   52MOC/a

F.   52MOC/b

Solution:
From text, the answer is F (52MOC/b).

19.4 The IEEE number that identifies a normally open contact that is mounted on the breaker and indicates the breaker is open or closed is which of the following contacts?

    A. 52/a

    B. 52/b

    C. 52TOC/a

    D. 52TOC/b

    E. 52MOC/a

    F. 52MOC/b

Solution:

From text, answer is A (52/a).

19.5 Given a trip circuit as shown in Figure 19.8, if the plant operator walks up to the circuit breaker and the red light is on and the green light is off, what is the status of the breaker?

    A. Racked in and closed

    B. Racked in and open

    C. Racked out and closed

    D. Racked out and open

Solution:

Looking at Figure 19.8, we see that both the red and green lights are driven by contacts on the breaker through the secondary stabs of the breaker. Therefore, if either of the lights is on, then this implies the breaker is racked in. The red light comes from the 52/b contact on the breaker, so this light is on whenever the breaker is closed. Therefore, the answer is A (Racked in and closed).

19.6 True or false, A motor starter is designed to interrupt fault current on the feeder going to the motor?

Solution:

Circuit breakers and fuses are designed to interrupt fault current. However, motor starters alone are not designed to interrupt fault currents (but they can interrupt load current). A motor starter will incorporate a motor circuit protector, circuit breaker, or fuse to provide fault overcurrent protection to the starter. Therefore the answer is false.

Chapter 20

20.1 Ventilated lead acid batteries should be load tested annually or replaced if, during discharge test, the capacity of the battery drops below what percentage of original capacity?

    A. 90%

    B. 85%

    C. 80%

    D. 70%

Solution:

From text, the answer is B (85%).

20.2 True or false; under float operation, the battery provides current to the load and the battery charger.

Solution:
From text, the answer is false. Under float operation, the battery charger provides current to both the battery and the load.

20.3 True of false, the negative plates in a healthy flooded cell are gray in color.

Solution:
From text, the answer is true. The negative terminal (anode) connects to the gray plates. The positive terminal (cathode) connects to the brown or black plates.

20.4 Given a battery (Anatomy) that is rated for a life expectancy of 20 years with a maximum ambient temperature of 77°F, what is the life expectancy of the battery if the ambient temperature is maintained at 62°F?

Solution:
From Table 20.1, the new lift expectancy with the lower ambient temperature is about 22 years.

20.5 When performing maintenance on a battery, what is the minimum frequency of recording the charger float voltage?

A. Monthly

B. Quarterly

C. Annually

D. Every 5 years

Solution:
From text, the minimum frequency of the following maintenance is quarterly. The answer is B (Quarterly).

20.6 Given a battery is sized such that it will be discharged at a rate of 30 A for 8 hr. We want to design the charger to recharge this battery in a period of 16 hr while still feeding the DC-connected load of 10 A. The charger will be installed in an environment that is less than 50°C ambient temperature and less than 3300 ft elevation. What is the minimum required ampacity of the charger?

Solution:
From the stated problem and utilizing equation we find our variables to be

$$AHR = 30\,A \times 8\,hr = 240\,A\,hr$$
$$T = 16\,hr$$
$$L = 10\,A$$

Using equation (20.7), we find the minimum ampacity for the battery charger to be

$$A = ((AHR \times 1.10)/T + L)$$
$$A = ((240\,AHr \times 1.10)/16\,Hr + 10\,A)$$
$$A = 26.5\,A$$

The next largest commercially available charger size should be ordered.

Chapter 21

21.1   A power plant ground grid design with a total length of buried bare copper conductor of 4 m formed in such a fashion to provide a grid area 1 m². Is buried in soil with a resistivity of the soil is 100 Ωm. Calculate the expected ground resistance of this ground grid.

Solution:
Using equation (21.3), we find the ground grid resistance to be as follows.

$$R = \frac{\rho}{4}\sqrt{\frac{\pi}{A}} + \frac{\rho}{L}$$

$$R = \frac{100 \ \Omega m}{4}\sqrt{\frac{\pi}{1 \ m^2}} + \frac{250 \ \Omega m}{100 \ m}$$

$$R = (25 \ \Omega)(1.77) + (2.5 \ \Omega) = 3.465 \ \Omega$$

$$R = (44.3 \ \Omega) + (2.5 \ \Omega) = 3.465 \ \Omega$$

$$R = 46.8 \ \Omega.$$

21.2   What is the step- and touch-voltage potential limits for a system that will clear a fault in 12 cycles (based on a 60 Hz system) if the surface layer resistivity is found to be 2000 Ωm? Assume that there is no protective surface layer in this application. Please use equations (21.5) and (21.6) for these calculations. Assume a weight of 110 lb (50 kg).

Solution:
Given that there is not protective surface layer in this application, we can use equation (21.5) to calculate the maximum step-potential voltage for a person that weighs 110 lb.

$$E_{step50} = [1000 + 6C_s(h_s, K)\rho_s][0.116/\sqrt{(t_s)}]$$

$$E_{step50} = [1000 + 6(1) \ 2000 \ \Omega m][0.116/\sqrt{0.2}]$$

$$E_{step50} = [1000 + 12000][0.05188]$$

$$E_{step50} = 674 \ V$$

The maximum touch-potential voltage for a person that weighs 110 lb can be found using equation (21.6) as shown below.

$$E_{touch50} = [1000 + 1.5C_s(h_s, K)\rho_s][0.116/\sqrt{(t_s)}]$$

$$E_{touch50} = [1000 + 1.5(1) \ 2000 \ \Omega m][0.116/\sqrt{0.2}]$$

$$E_{touch50} = [1000 + 3000][0.05188]$$

$$E_{touch50} = 208 \ V.$$

21.3   What is the step- and touch-voltage potential limits for a system that will clear a fault in 12 cycles (based on a 60 Hz system) if the surface layer resistivity is found to be 2000 Ωm? Assume that there is no protective surface layer in this application. Please use equations (21.7) and (21.8) for these calculations. Assume a weight of 154 lb (70 kg).

Solution:
Given that there is no protective surface layer in this application, we can use equation (21.7) to calculate the maximum step-potential voltage for a person that weighs 154 lb.

$$E_{step70} = [1000 + 6\,C_s(h_s, K)\rho_s][0.157/\sqrt{(t_s)}]$$
$$E_{step70} = [1000 + 6(1)\,2000\;\Omega m][0.157/\sqrt{0.2}]$$
$$E_{step70} = [1000 + 12000][0.0702]$$
$$E_{step70} = 913\;V$$

The maximum touch-potential voltage for a person that weighs 154 lb can be found using equation (21.8) as shown below.

$$E_{touch70} = [1000 + 1.5\,C_s(h_s, K)\rho_s][0.157/\sqrt{(t_s)}]$$
$$E_{touch70} = [1000 + 1.5\,(1)\,2000\;\Omega m][0.157/\sqrt{0.2}]$$
$$E_{touch70} = [1000 + 3000][0.0702]$$
$$E_{touch70} = 281\;V.$$

21.4 Given a ground conductor that is 1/0 conductor with an effective radius of 8.252 mm and with a distance of 1 m between the ground conductor and phase conductor, calculate the inductance of the ground loop in henries per meter and the impedance of the ground loop in ohms per meter to a phase-to-ground fault. Assume a system frequency of 60 Hz.

Solution:
Using equation (21.9) we find the inductance of the loop per linear meter to be the following.

$$L = 4 \times 10^{-7}\ln{(D/r')}\quad H/m$$
$$L = 4 \times 10^{-7}\ln{(1000\;mm/8.252\;mm)}\quad H/m$$
$$L = 4 \times 10^{-7}\ln{(121.19)}\quad H/m$$
$$L = 4 \times 10^{-7}(4.8)\quad H/m$$
$$L = 1.9 \times 10^{-6}\quad H/m$$
$$L = 1.9\quad \mu H/m$$

Using equation (21.10) we find the inductive reactance of the loop per linear meter to be the following.

$$j \times g = j2pfL(\Omega/m)$$
$$j \times g = j2p(60\;Hz)(1.9 \times 10^{-6}\;H/m)(\Omega/m)$$
$$j \times g = j2p(60\;Hz)(1.9 \times 10^{-6}\;H/m)(\Omega/m)$$
$$j \times g = j0.000723\;(\Omega/m)$$
$$j \times g = j723\;\mu\Omega/m.$$

21.5 If we are to use two ground rods with a length of 10 ft for our ground connection to our ground grid, what is the recommended distance between ground rods?

Solution:
Using equation (21.11) we find the correct distance between the ground rods to be as follows.

$$\text{Distance between rods} = L_{r1} + L_{r2}$$
$$\text{Distance between rods} = 10\;ft + 10\;ft$$
$$\text{Distance between rods} = 20\;ft.$$

Chapter 22

22.1 As defined by IEEE, device 87 is a (an)

    A. instantaneous overcurrent relay

    B. time delay overcurrent relay

    C. differential overcurrent relay

    D. lockout device

Solution:
Referring to the table in Annex B, the answer is C (differential overcurrent relay).

22.2 Given a load configuration below where the load is configured in a delta and each resistance phase-to-phase is 30 ohms, what is the equivalent line-to-neutral resistance value?

    A. 90 ohms

    B. 10 ohms

    C. 30 ohms

    D. 3 ohms

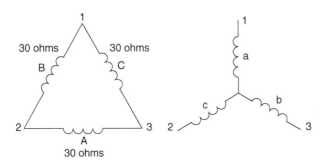

Solution:
Using equations (22.1), (22.2), and (22.3), we find the following.

$$a = (B\ C)/(A + B + C)$$
$$b = (A\ C)/(A + B + C)$$
$$c = (A\ B)/(A + B + C)$$
$$a = (30 \times 30)/(30 + 30 + 30) = 10 \text{ ohms}$$
$$b = (30 \times 30)/(30 + 30 + 30) = 10 \text{ ohms}$$
$$c = (30 \times 30)/(30 + 30 + 30) = 10 \text{ ohms}$$

The answer is B (10 ohms).

22.3 Given a system and using 480 V as your base voltage (V_{base}) and using 10 MVA as your base power (S_{base}), what is the base current (I_{base}) for system analysis?

    A. $I_{base} = 12$ MA

    B. $I_{base} = 12$ kA

    C. $I_{base} = 12$ A

    D. $I_{base} = 12$ mA

Solution:

Since we are given base voltage in V and base power in MVA, we can use equation (22.7) as shown below to determine the base current.

$$I_{base} = MVA_{base}(1,000,000)/[1.732\ V_{base}]$$
$$I_{base} = 10\ MVA(1,000,000)/[1.732\ (480)]$$
$$I_{base} = 12,028\ amps$$
$$I_{base} = 12\ kA$$

The answer is B ($I_{base} = 12$ kA).

22.4 For the time delay overcurrent relay characteristics that are shown in the figure below, what is the estimated time for the relay to trip for a current that is 10 times pickup if the relay is set to time dial 7?

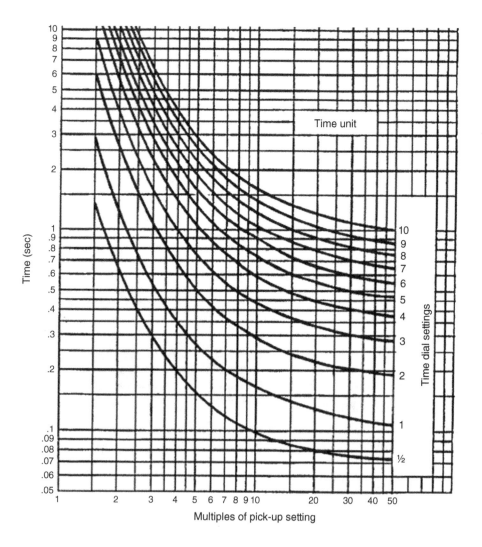

A. 1.7 sec

B. 1.0 sec

C. 0.5 sec

D. 0.1 sec

Solution:

We are given that the current is 10 times the pickup. If we plot this with the intersection of the line for a time dial of 7 sec as shown in the figure below, we find the trip time to be 1.0 sec. The answer is B (1.0 sec).

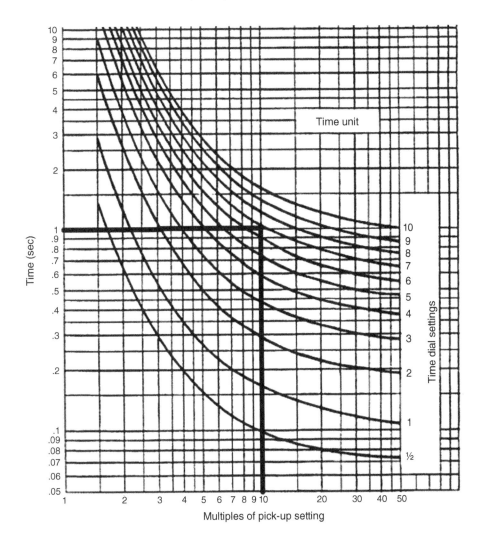

Chapter 23

23.1 Utilizing conventional color codes, a motor-operated valve that is the color "Red" on the HMI indicates that the valve is

    A. fully open

    B. mid travel (i.e., not fully open and not fully closed)

    C. fully closed

Solution:

Using Table 23.2, the color can be found to be that of fully open. The answer is A.

23.2 Utilizing conventional color codes, a pump energized by a three-phase motor that is the color "Red" on the HMI indicates that the pump is

    A. energized and operating

    B. in a faulted condition

    C. de-energized

Solution:

Using Table 23.2 the color can be found to be that of energized and operating. The answer is A.

23.3 The formula in equation (23.2) and given below represents the _____ function of a PID controller.

$$P_i = (1/T_i) \int Edt$$

    A. percent

    B. integral

    C. differential

Solution:

Referring to equation (23.2), the answer is B (integral).

23.4 The control system shown below is an example of a _____ system.

    A. feedforward – lube oil signal

    B. feedforward – cooling water signal

    C. feedforward and cascade

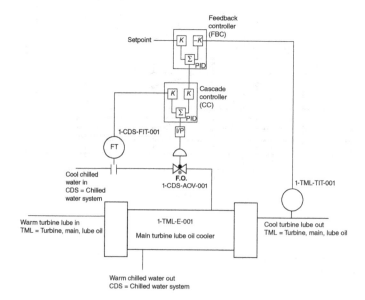

Solution:
Referencing Figure 23.7, we find this to be the feedforward – cooling water signal control system. The answer is B.

23.5 What standard addresses boiler control and burner management system safety requirements?

    A. NEC®

    B. NESC

    C. NFPA 70E

    D. NFPA 85

Solution: From text, the answer is D (NFPA 85).

Chapter 24

24.1 Fill in the blanks: "A current transformer with a rating of 10C400 that has a 5 A-rated secondary, can develop up to _____ V on secondary circuit at 20 times rated secondary current (100 A) and not exceed _____ accuracy."

    A. 120, 10%

    B. 400, 10%

    C. 120, 5%

    D. 400, 5%

Solution:
From the definition of accuracy class for CT used in metering, the answer is B (400, 10%).

24.2 What are at least four physical parameters that are needed to be monitored in a power plant to provide adequate information for process control?

    _____

    _____

    _____

    _____

Solution:
Common answers are below but others also qualify.
    Temperature, Flow, Pressure, Level.

24.3 You need to replace a thermocouple in a plant process and no information is known on the device. However, when locating the device, you see that the conductor color on the positive connection is yellow and the conductor color on the negative connection is red. Knowing that the plant was built to ANSI standards, what is the type thermocouple you need to use to replace the existing device?

    A. Type J

    B. Type K

    C. Type T

    D. Type E

Solution:
Referring to Table 24.1, we find the type of thermocouple with a yellow positive and a red negative to be type K. The answer is B.

24.4 You need to replace a thermocouple in a plant process and no information is known on the device. However, when locating the device, you see that the conductor color on the positive connection is green and the conductor color on the negative connection is white. Knowing that the plant was built to IEC standards, what is the type thermocouple you need to use to replace the existing device?

   A. Type J

   B. Type K

   C. Type T

   D. Type E

Solution:

Referring to Table 24.2, we find the type of thermocouple with a green positive and a white negative to be type K. The answer is B.

24.5 A pressure of 30 psia approximately equals…

   A. 30.0 psig

   B. 15.3 psig

   C. 14.7 psig

   D. 0.0 psig

Solution:

Using Table 24.3, we find the pressure in psig to be 15.3 psig. The answer is B.

   It can also be found by subtracting atmospheric pressure of 14.7 psia from 30 psia.

24.6 With the thermocouple connection shown below, what is the measured temperature (assume cold junction temperature compensation at the measuring end)?

   A. $T_{meas} = T_1 + T_2$

   B. $T_{meas} = T_1 - T_2$

   C. $T_{meas} = T_2 - T_1$

   D. $T_{meas} = -T_1 - T_2$

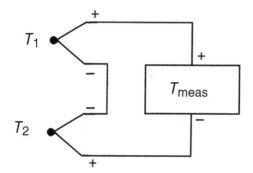

Solution:

Let us start at the positive lead of T_1 and go clockwise around the loop. Doing this, we derive the following equation

$$+T_{meas} + T_2 - T_1 = 0$$

Next, we rearrange the equation to isolate the sensor on one side,

$$T_{meas} = T_1 - T_2$$

This tells us that, with this configuration the sensor, we will detect the difference in temperature between thermocouple T_1 and thermocouple T_2.
The answer is B ($T_{meas} = T_1 - T_2$).

24.7 The typical reference resistance of a Copper RTD is?

    A. 10 ohms

    B. 25 ohms

    C. 100 ohms

    D. 200 ohms

Solution:
From Table 24.3, we find the following.

Copper (10 Ω)

Nickel (25 Ω)

Platinum (100 Ω)

The answer is A (10 ohms).

24.8 Refer to the drawing of a differential pressure manometer (see figure below). A differential pressure manometer is installed across an orifice in a ventilation duct. With the ventilation conditions as shown, the pressure at P_1 is _____ than P_2, and airflow is from _____.

    A. greater; left to right

    B. greater; right to left

    C. less; left to right

    D. less; right to left

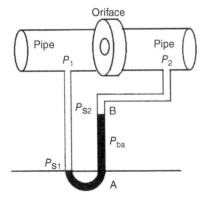

Solution:
A (greater; left to right).

24.9 A water storage tank is vented to atmosphere. The tank is located at sea level and contains 100,000 gal of 80° F water (specific gravity of 1.0). A pressure gauge at the bottom of the tank reads 5.6 psig. What is the approximate water level in the tank?

A. 13 ft

B. 17 ft

C. 21 ft

D. 25 ft

Solution:

Using equation (24.12), we find the following.

$$H(\text{in.}) = [P(\text{psig})/\text{SG}] \times 27.678$$
$$H(\text{in.}) = [5.6 \text{ psig}/1.0] \times 27.678$$
$$H(\text{in.}) = 155 \text{ in. or } 13 \text{ ft}$$

The answer is A (13 ft).

24.10 If a tank contains 30 ft of water at 60°F and top of tank is open to atmosphere and the reference standoff pipe contains 30 ft of water at same temperature and is also open to atmosphere, what is the approximate D/P sensed by the detector?

A. Greater than zero (tank side pressure greater than standoff pipe side)

B. Zero

C. Less than zero (tank-side pressure less than standoff pipe side)

Solution:

Since the height of the fluid is the same on both tank side and standoff pipe side and also since the temperature (and therefore density) of the fluid is the same on both the tank side and the standoff pipe side, the same pressure is sensed on both sides of the differential pressure transmitter and the D/P is zero.
The answer is B (Zero).

24.11 If a tank contains 30 ft of water at 60°F and top of tank is open to atmosphere and the reference standoff pipe contains no water (0 ft) of water and is also open to atmosphere, what is the approximate D/P sensed by the detector?

A. Greater than zero (tank side pressure greater than standoff pipe side)

B. Zero

C. Less than zero (tank-side pressure less than standoff pipe side)

Solution:

Since the height of the fluid in the tank is greater than the height of the fluid in the standoff pipe, the pressure on the tank side is greater than the pressure on the standoff pipe side and the differential pressure transmitter and the D/P is greater than zero.
The answer is A (Greater than zero (tank-side pressure greater than standoff pipe side)).

24.12 If a tank contains 30 ft of water at 600°F and top of tank is open to atmosphere and the reference standoff pipe contains 30 ft of water but at a temperature of 60°F and the standpipe is also open to atmosphere, what is the approximate D/P sensed by the detector?

A. Greater than zero (tank side pressure greater than standoff pipe side)

B. Zero

C. Less than zero (tank side pressure less than standoff pipe side)

Solution:
The height of the fluid is the same on both tank side and standoff pipe side. However, the temperature of the water in the standoff pipe is lower which means the specific density of the water in the standoff pipe is greater, this will lead to a condition where the pressure on the tank side is less than the pressure on the standoff pipe side and the differential pressure transmitter and the D/P is less than zero
The answer is C (Less than zero (tank side pressure less than standoff pipe side)).

24.13 If flow is doubled through an orifice plate, what is the increase in differential pressure across the flow plate?

A. New differential pressure is same as original differential pressure

B. New differential pressure is 2 times original differential pressure

C. New differential pressure is 4 times original differential pressure

D. New differential pressure is 8 times original differential pressure

Solution:
Using equation (24.8), we can determine the relationship between flow and differential pressure. Equation (24.8) states the following.

$$Q = kA[(2ghp)/(GT)]^{1/2}$$

If we only change head and flow, we can simplify this equation to proportionality.

$$Q^2 \propto h$$

Or we can rewrite this as follows.

$$(Q_1/Q_2)^2 = (h_1/h_2)$$

If we let condition 1 be original and condition 2 be new, we are looking for new value of head, h_2 with respect to h_1. We find this by rearranging the above equation to the following.

$$(h_2/h_1) = (Q_2/Q_1)^2$$

We are not given values for flow, but we know flow increased by double. Therefore, the ratio of Q_2 (new flow) to Q_1 (old flow) is as follows.

$$Q_2/Q_1 = 2$$

Substituting this into above equation we finally find,

$$(h_2/h_1) = (2)^2$$
$$(h_2/h_1) = 4$$

The answer is C (New differential pressure is four times original differential pressure).

24.14 A transducer is installed on a tank that is scaled to send a 4–20 mA signal proportional to a tank level of 4–20 ft (such that 4 ft = 4 mA and 20 ft = 20 mA). At five different times, when the tank level was physically measured at 12 ft, the ideal output of the transmitter should have been 12 mA, but actual output was 9 mA all five times sample was tested. The question is, is this transducer accurate, precise, both, or neither?

A. Very accurate but not very precise

B. Not very accurate but very Precise

C. Both accurate and precise

D. Neither accurate nor precise

Solution:

Since all samples agreed with each other, this system is very precise. However, since the overall average deviates from the actual value, this system is not accurate. The answer is B (Not very accurate but very precise).

Chapter 25

25.1 What is the purpose of a check valve?

A. Regulate the rate of flow through the discharge pipe

B. Prevent flow of fluid in reverse direction.

C. Provide overpressure protection for piping system.

D. Prevent clogging of the suction line

Solution:

The check valve is designed to allow flow of fluid in only one direction and not in the reverse direction. Therefore, the answer is B (Prevent flow of fluid in reverse direction).

25.2 What is the purpose of a valve positioner?

A. Provide the driving force to actuate the valve

B. Monitor feedback signal of valve position and drive the valve actuator until the set point and feedback values agree

C. Provide overpressure protection for piping system

D. Prevent flow of fluid in reverse direction

Solution:

The valve actuator drives the valve. The valve positioner takes a feedback signal and if there is an error between the desired valve position and the actual valve position, the positioner drives the valve actuator until the valve desired and actual position agree. Therefore, the answer is B (Monitor feedback signal of valve position and drive the valve actuator until the set point and feedback values agree).

25.3 What is the purpose of a relief valve?

A. Regulate the rate of flow through the discharge pipe

B. Prevent flow of fluid in reverse direction

C. Provide overpressure protection for piping system

D. Prevent clogging of the suction line

Solution:

C (Provide overpressure protection for piping system).

25.4 Given a motor that is running at 10 times faster than the speed of the valve actuator where the motor is providing 10 ft lb torque to the gear reducer, determine the amount of torque that the valve actuator is providing to the valve stem.

Solution:

In the problem above, the ratio of motor speed to valve steam speed (the gear reduction) is given as 10:1. If we define the motor speed as ω_{in} and if we define actuator speed as ω_{out}, then the ratio of motor speed to actuator speed is

$$\omega_{in}/\omega_{out} = 10 : 1$$

Now using equation (25.3), we can find the toque of the actuator compared with the torque of the motor (T_{out} / T_{in}) as follows.

$$T_{out}/T_{in} = \omega_{in}/\omega_{out}$$
$$T_{out}/T_{in} = 10 : 1$$
$$T_{out} = 10 : 1 \times T_{in}$$
$$T_{out} = 10 : 1 \times 10 \text{ ft lb} = 100 \text{ ft lb.}$$

25.5 Given a resistance coefficient of 3.85, a flow through a valve of 5000 gal/min, a valve diameter of 12 in., determine the head loss across this valve and the associated power loss due to losses in this valve given a fluid density of 50 kg/m^3?

Solution:

First, we need to calculate the head loss of the valve. To do this, we can use equation (25.4),

$$HL = (kQ^2)/(D^4 385)$$
$$HL = (3.85(5000 \text{ gpm})^2)/((12'')^4 385)$$
$$HL = 12 \text{ ft}$$

Next, we need to calculate the power loss in the valve, but to do this we need to convert from standard to metric units as shown below.

$$HL = 12 \text{ ft} \times (0.3408 \text{ m}/1 \text{ ft}) = 4.11 \text{ m}$$
$$Q = 5000 \text{ gpm} \times (0.0631 \text{ m}^3/\text{sec}/1000 \text{ gpm}) = 0.3155 \text{ m}^3/\text{sec}$$

Now we can use equation (25.5) to calculate the losses in this valve using the assumed fluid density of 1 kg/m^3.

$$P_v = \rho \text{ g HL } Q$$
$$P_v = (50 \text{ kg/m}^3)(9.80665 \text{ m/ sec}^2)(4.11 \text{ m})(0.3155 \text{ m}^3/ \text{sec})$$
$$P_v = 635.8 \text{ W.}$$

25.6 Design a pneumatic valve control system for an actuator that is connected to a globe valve where the valve is closed when the stem is driven down and the valve is opened when the stem is driven up. For this application, we have defined the failsafe mode for this particular application to be with the valve closed (the failsafe mode is application- and valve-design specific). The valve is to be operated remotely by a solenoid attached to the pilot pneumatic valve manifold to open the valve and, when the solenoid is de-energized, the valve is to automatically close.

Solution:

For this application, we have defined the failsafe mode for this particular application to be with the valve closed (this is application and valve design specific). We also defined the direction of actuation for the main valve such that to close the valve, we drive the stem down and, to open the valve, we drive the stem up. Since the failsafe mode for the valve is with the valve closed, this means that the pneumatic valve actuating cylinder must drive the valve stem down when air is removed from the cylinder. Therefore, we would select a pneumatic valve actuating cylinder that has a spring to force the valve stem down with no air supplied and we would have the actuator drive the valve stem up when air is supplied to the valve actuating cylinder.

Next, we need to select the correct pneumatic pilot valve module. From the design description, the valve is to be operated remotely with the use of a solenoid mounted on the pilot pneumatic valve manifold to open the valve and, when the electric solenoid is released, the valve is to automatically close. Therefore, when we remotely energize the solenoid on the pilot pneumatic valve manifold, this should admit air into the main valve pneumatic actuating cylinder. When we de-energize the solenoid mounted to the pilot pneumatic valve manifold, a spring in the pilot pneumatic valve manifold should return the manifold to a position that exhausts the main valve pneumatic actuating cylinder to atmosphere, allowing the main valve pneumatic actuating cylinder spring to lower the valve shaft closing the valve. The figure below achieves this design.

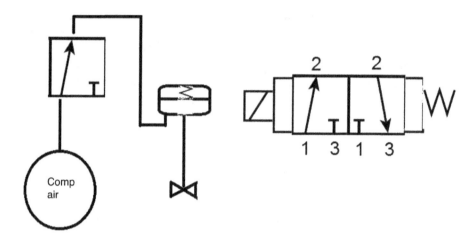

Looking at the figure above, when the pilot pneumatic valve manifold is active by the solenoid, the air supply port 1 is connected to the valve cylinder port 2 and the valve actuator has air pressure applied to the bottom of the diaphragm. In this position, the pneumatic valve manifold port 3 is isolated. This applies air pressure to the bottom of the valve actuator and drives the valve stem up to open the globe valve.

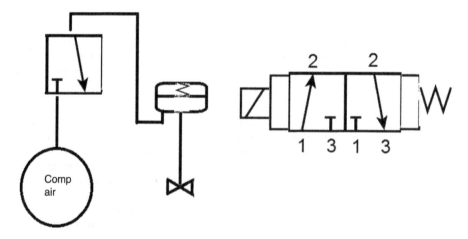

When the solenoid connected to the pilot pneumatic valve manifold is de-energized, the spring in the pneumatic valve manifold will return the pneumatic valve manifold to this normal position where the air supply port 1 is isolated and the cylinder supply port 2 is connected to the exhaust port 3. This removes air pressure from the bottom of the valve actuator by exhausting this chamber to atmosphere and the spring in the valve actuator drives the valve stem down to close the globe valve. This is shown in the figure above. Additionally, the failsafe position of the valve is with the main valve shaft lowered and the main valve closed should air pressure be lost, as required by our system description.

Chapter 26

26.1   Which equation below shows the formation of nitric acid (acid rain)?

   A. $2NO + O_2 \rightarrow 2NO_2$

   B. $3NO_2 + H_2O \rightarrow 2HNO_3 + NO$

   C. $4NO + 4NH_3 + O_2$ –(catalyst) $\rightarrow 4N_2 + 6H_2O +$ heat

   D. $2NO_2 + 4NH_3 + O_2$ –(catalyst) $\rightarrow 3N_2 + 6H_2O +$ heat

   Solution:
   From equation (26.2), answer is B ($3NO_2 + H_2O \rightarrow 2HNO_3 + NO$).

26.2   True or false, Thermal NO_x is generated from nitrogen contained in the fuel that is provided to the combustion process.

   Solution:
   From the Glossary of Terms section, the answer is False.

   Thermal NO_x – is the NO_x generated where the source of nitrogen is the air used in the combustion process.

26.3   For wet scrubbing of exhaust gas to remove SO_2, which equation below shows the sulfation reaction if the reagent used is limestone?

   A. $2CaO + 2SO_2 + O_2 \rightarrow 2CaSO_4$

   B. $2CaO + 2MgO + 2SO_2 + O_2 \rightarrow 2CaSO_4 + 2MgO$

   C. $2CaO + 2SO_2 + O_2 \rightarrow 2CaSO_4$

   D. $2CaO + 2SO_2 + O_2 \rightarrow 2CaSO_4$

   Solution:
   From equation (26.9), the answer is A ($2CaO + S2O_2 + O_2 \rightarrow 2CaSO_4$).

26.4   For wet scrubbing of exhaust gas to remove SO_2, which equation below shows the sulfation reaction if the reagent used is dolomitic limestone?

   A. $2CaO + 2SO_2 + O_2 \rightarrow 2CaSO_4$

   B. $2CaO + 2MgO + 2SO_2 + O_2 \rightarrow 2CaSO_4 + 2MgO$

   C. $2CaO + 2SO_2 + O_2 \rightarrow 2CaSO_4$

   D. $2CaO + 2SO_2 + O_2 \rightarrow 2CaSO_4$

   Solution:
   From equation (26.11), the answer is B ($2CaO + 2MgO + 2SO_2 + O_2 \rightarrow 2CaSO_4 + 2MgO$).

26.5   For wet scrubbing of exhaust gas to remove SO_2, which equation below shows the sulfation reaction if the reagent used is quicklime?

   A. $2CaO + 2SO_2 + O_2 \rightarrow 2CaSO_4$

   B. $2CaO + 2MgO + 2SO_2 + O_2 \rightarrow 2CaSO_4 + 2MgO$

C. $2CaO + 2SO_2 + O_2 \rightarrow 2CaSO_4$

D. $2CaO + 2SO_2 + O_2 \rightarrow 2CaSO_4$

Solution:

From equation (26.12), the answer is C ($2CaO + 2SO_2 + O_2 \rightarrow 2CaSO_4$).

26.6   For wet scrubbing of exhaust gas to remove SO_2, which equation below shows the sulfation reaction if the reagent used is hydrated lime?

A. $2CaO + 2SO_2 + O_2 \rightarrow 2CaSO_4$

B. $2CaO + 2MgO + 2SO_2 + O_2 \rightarrow 2CaSO_4 + 2MgO$

C. $2CaO + 2SO_2 + O_2 \rightarrow 2CaSO_4$

D. $2CaO + 2SO_2 + O_2 \rightarrow 2CaSO_4$

Solution:

From equation (26.14), the answer is D ($2CaO + 2SO_2 + O_2 \rightarrow 2CaSO_4$).

Chapter 27

27.1   The process of adding certain chemicals to the water to allow non-soluble material to coagulate is known as which process?

A.   Clarification

B.   Evaporation

C.   Flocculation

D.   Sedimentation

Solution:

From the Glossary of Terms section, the answer is A (Clarification).

27.2   The process of converting water to steam to separate the pure water from the unwanted ions and soluble materials is known as which process?

A.   Clarification

B.   Evaporation

C.   Flocculation

D.   Sedimentation

Solution:

From the Glossary of Terms section, the answer is B (Evaporation).

27.3   The process where the water is slowly stirred is known as which process?

A.   Clarification

B.   Evaporation

C.   Flocculation

D.   Sedimentation

Solution:

From the Glossary of Terms section, the answer is C (Flocculation).

27.4   The process whereby the water is stored in a certain area where the suspended non-soluble materials will settle out of the water stream is known as which process?

A.   Clarification

B.   Evaporation

C. Flocculation

D. Sedimentation

Solution:

From the Glossary of Terms section, the answer is D (Sedimentation).

27.5 A system averages an influent flow of 10 mgd. What is the average flow rate in gallons per minute that flows into the system?

Solution:

Utilizing equation (27.1), we find the gallons per minute to be

$$GPM = MGD \times 694.4 \text{ (gpm)}$$
$$GPM = 10 \times 694.4 \text{ (gpm)}$$
$$GPM = 6944 \text{ gpm.}$$

27.6 A circular demineralized water storage tank has a diameter of 40 ft and a usable height of 30 ft. What is the usable volume of the tank in cubic feet? What is the usable volume of the tank in gallons?

Solution:

Utilizing equation (27.12), we find the usable volume to be

$$V = 0.7854 \times D^2 \times H \text{ (ft}^3)$$
$$V = 0.7854 \times 40^2 \times 30 \text{ (ft}^3)$$
$$V = 37,700 \text{ ft}^3$$

Utilizing equation (27.10), we can convert from cubic feet to gallons.

$$V \text{ (gal)} = 7.48052 \times V \text{ (ft}^3)$$
$$V \text{ (gal)} = 7.48052 \times 37,700 \text{ (ft}^3)$$
$$V \text{ (gal)} = 282,009 \text{ gal.}$$

27.7 A 200,000 gal tank received 1.5 mgd of flow. What is the detention time of the fluid in the tank?

Solution:

Utilizing equation (27.14), we find that the detention time is

$$\text{Detention time} = 24 \times V \text{ (gal)}/(1,000,000 \times MGD) \text{ (hr)}$$
$$\text{Detention time} = 24 \times 200,000/(1,000,000 \times 1.5) \text{ (hr)}$$
$$\text{Detention time} = 3.2 \text{ hr.}$$

27.8 How many pounds per day of chlorine are needed to provide a dosage of 15.0 mg/L to a water stream that flows at 3.0 MGD?

Solution:

Utilizing equation (27.15), we find that the dosage is

$$\text{Dosage} = \text{Dose} \times MGD \times 8.34 \text{ (lb/day)}$$
$$\text{Dosage} = 15.0 \times 3.0 \times 8.34 \text{ (lb/day)}$$
$$\text{Dosage} = 375.3 \text{ lb/day.}$$

27.9 A clarifier tank has an influent of suspended solids of 500 mg/L and an effluent flow of suspended solids of 100 mg/L. What is the removal efficiency of this process for suspended solids?

Solution:

Utilizing equation (27.16), we find that the removal efficiency of this process is

$$\eta\text{-process} = (\text{Influent} - \text{Effluent})/\text{Influent}$$
$$\eta\text{-process} = (500 \text{ mg/L} - 100 \text{ mg/L})/500 \text{ mg/L}$$
$$\eta\text{-process} = 0.8 \text{ or } 80\%.$$

27.10 Given that a pump supplies a water system with a total dynamic head of 238 ft at a flow rate of 2000 gpm, determine the water horsepower (P_w) that the pump delivers.

Solution:

Utilizing equation (27.17), we find that the pump power is

$$P_w = \text{TDH} \times \text{Flow}/3960$$
$$P_w = 238 \times 2000/3960$$
$$P_w = 120 \text{ HP}.$$

27.11 Given that a pump supplies 120 HP of power to a water system and given that the pump efficiency is 80% and the motor efficiency is 93%, determine the electrical power (kW) required to provide 120 HP water power.

Solution:

Using equation (27.23) and rearranging for P_e, we find.

$$P_e = 0.746 \times P_w/(\eta\text{-pump} \times \eta\text{-motor})$$
$$P_e = 0.746 \times 120/(0.8 \times 0.93)$$
$$P_e = 120.3 \text{ kW}.$$

27.12 If a pump that draws 120.3 kW of electrical power runs for 20 hr per day and the electrical rate for power is $0.15 kWh, how much does it cost to run the pump for 1 year (365 days)?

Solution:

The cost to run the pump is the product of the cost per kWh for the energy, the amount of time the motor is energized, the average electrical power the motor draws, and the time period under consideration.

$$\text{Cost} = P_e \times \text{time energized} \times \text{cost per kWh} \times \text{period of operation}$$
$$\text{Cost} = 120.3 \text{ kW} \times (20/24) \times \$0.15 \text{ kWh} \times 365 \text{ days} \times (24 \text{ hr/day})$$
$$\text{Cost} = \$131,729.$$

Chapter 28

28.1 In a solar cell doped with boron and phosphorous, which layer is the n-layer semiconductor?

A. Boron-doped semiconductor

B. Phosphorous-doped semiconductor

Solution:

From text, the *n*-layer semiconductor material is doped with phosphorous which has 5 outer shell electrons. Therefore, the answer is B.

28.2 Given an air density of 1.2 kg/m³, a blade radius of 15 m, and a wind velocity of 22.3 mph (10 m/sec), determine the theoretical maximum mechanical power that can be transferred by the wind turbine.

Solution:

First, we have to determine the swept area. The radius of the turbine blade was given as 10 m, so the swept area is

$$3.14 \times R^2 \, \text{m}^2$$
$$= 3.14 \times 15^2 \, \text{m}^2$$
$$= 3.14 \times 225 = 707 \, \text{m}^2$$

Utilizing equation (28.1) for maximum mechanical power as a function of wind speed, we find

$$P_{max} = 0.5 \times \rho \times A \times V^3$$
$$P_{max} = 0.5 \times (1.2 \, \text{kg/m}^3) \times (707 \, \text{m}^2) \times (10 \, \text{m/sec})^3$$
$$P_{max} = 424115 \, \text{kg m}^2/\text{sec}^3 = 424 \, \text{kW or } 424 \, \text{kJ/sec}.$$

28.3   Given the same parameters that were used in Problem 28.2, if the turbine blade efficiency is 30%, determine the mechanical power available to the turbine shaft.

Solution:

From Problem 28.2, we found $P_{max} = 424$ kW. Using the definition of turbine blade efficiency, we can find that the mechanical power available by the method shown below.

Utilizing equation (28.2) for the mechanical power available, we find

$$\eta = P_{mech}/P_{max} \qquad (28.2)$$

$$P_{mech} = \eta \times P_{max}$$
$$P_{mech} = 0.3 \times 424 \, \text{kW}$$
$$P_{mech} = 127.2 \, \text{kW}.$$

INDEX

Energy Production Systems Engineering: An Introduction for Electrical Engineers to Electrical Power Generation Facilities, Systems, and Equipment, First Edition. Thomas H. Blair.
© 2017 by The Institute of Electrical and Electronics Engineers, Inc. Published 2017 by John Wiley & Sons, Inc.

IEEE Press Series on Power Engineering

Series Editor: M. E. El-Hawary, Dalhousie University, Halifax, Nova Scotia, Canada

The mission of IEEE Press Series on Power Engineering is to publish leading-edge books that cover the broad spectrum of current and forward-looking technologies in this fast-moving area. The series attracts highly acclaimed authors from industry/academia to provide accessible coverage of current and emerging topics in power engineering and allied fields. Our target audience includes the power engineering professional who is interested in enhancing their knowledge and perspective in their areas of interest.